Selected Papers from SDEWES 2018

Selected Papers from SDEWES 2018 Conferences on Sustainable Development of Energy, Water and Environment Systems

Special Issue Editors

Neven Duić

Mário Costa

Qiuwang Wang

Francesco Calise

Poul Alberg Østergaard

MDPI • Basel • Beijing • Wuhan • Barcelona • Belgrade

Special Issue Editors
Neven Duić
University of Zagreb
Croatia

Qiuwang Wang
Xi'an Jiaotong University
China

Francesco Calise
University of Naples Federico II
Italia

Mário Costa
University of Lisbon
Portugal

Poul Alberg Østergaard
Aalborg University
Denmark

Editorial Office
MDPI
St. Alban-Anlage 66
4052 Basel, Switzerland

This is a reprint of articles from the Special Issue published online in the open access journal *Energies* (ISSN 1996-1073) from 2018 to 2019 (available at: https://www.mdpi.com/journal/energies/special_issues/SDEWES_2018)

For citation purposes, cite each article independently as indicated on the article page online and as indicated below:

LastName, A.A.; LastName, B.B.; LastName, C.C. Article Title. *Journal Name* **Year**, *Article Number*, Page Range.

ISBN 978-3-03921-554-6 (Hbk)
ISBN 978-3-03921-555-3 (PDF)

Contents

About the Special Issue Editors

Neven Duić a full professor of power engineering in the Department of Energy, Power Engineering and Environment, Faculty of Mechanical Engineering and Naval Architecture in the University of Zagreb. He has been a member of the Postgraduate Study Committee since 2008 and is also the Chair for Ph.D. studies in the process-energy field. He was Chair of the Faculty Committee for International Projects from 2010 until 2014. Prof. Duić has published 108 original scientific papers and 7 review papers in scientific journals referenced in SCI and CC publications. Eighty-two of his scientific papers are Q1 papers, 39 of which are in the top 5% of journals in the category (according to ISI's Web of Knowledge database). His papers were cited 2173 times in Scopus and 1779 times in Web of Knowledge. His H-index is 28 in Scopus and 24 in Web of Knowledge. According to CROSBI (the Croatian Scientific Bibliography) Prof. Duić has published the highest number of CC scientific papers in technical science since 2007. He has received a National Award for significant scientific achievement in 2016 in the field of engineering sciences. Prof. Duić has successfully led eight Ph.D. students and is currently mentoring five FMENA Ph.D. students. He has been a member of Ph.D. thesis juries for 11 Ph.D. students, seven of whom have gone on to study at foreign universities (Macquarie University, Aalborg University, University of Limerick, Instituto Superior Tecnico, University of Sarajevo, University of Beograd, University of Santander). He is an editor for *Energy Conversion and Management* (Q1), subject editor for the Q1 journal Energy, and a member of the editorial board of the following journals: *Applied Energy* (Q1), *Clean Technologies and Environmental Policy* (Q2), and *Thermal Science* (Q3). Since 2013, he has been the Editor-in-Chief of the *Journal of Sustainable Development of Energy, Water and Environment Systems* (JSDEWES), which was promptly and successfully indexed in the Scopus database. He has organised a series of conferences on the topic of sustainable development of energy, water and environment systems. He was also a member of organising, scientific, and programming committees of more than 57 research conferences. Prof. Duić is a coordinator of Croatian participation in two projects of the Croatian Foundation for Science and over 40 international scientific research projects. He is a national representative of Horizon 2020 projects for ERC/MSCA/FET. His main areas of interest are energy policy and planning, energy economics, sustainable development policy and resource planning, climate change mitigation, combustion engineering and modelling, research and innovation policy. Full CV: http://powerlab.fsb.hr/nduic/.

Mário Costa is a Full Professor in the area of environment and energy at the Mechanical Engineering Department of Instituto Superior Técnico. He graduated in Chemical Engineering at University of Coimbra in 1984, obtained his Ph.D. in Mechanical Engineering at Imperial College London in 1992, and his Habilitation in Mechanical Engineering at the Technical University of Lisbon in 2009. Currently, he teaches courses on thermodynamics, combustion, renewables energies, and integrated energy systems. He has supervised 8 Postdocs, 18 Ph.Ds., 75 M.Sc. and 35 Diploma students. He has participated in more than 50 national and international projects in the topic area of energy and environment and has (co-)authored a book, over 130 papers in international peer-reviewed journals, over 170 papers in international conferences, and has given more than 30 invited lectures at other universities and international symposia. Currently, he serves as Associate Editor of the Proceedings of the Combustion Institute and belongs to the Editorial Board of the Aerospace, Combustion and Flame, Energy Conversion and Management, Energies and Energy and Fuels. He was the recipient of the Caleb Brett Award of the Institute of Energy in 1991, the Sugden Award of the British Section of the Combustion Institute in 1991, the Prémio Científico Universidade Técnica de Lisboa/Santander Totta in 2010, a Menção Honrosa Universidade de Lisboa/Santander Universidades in 2016, and of the Prémio Científico Universidade de Lisboa/Santander Universidades in 2017.

Qiuwang Wang is a full professor at the School of Energy and Power Engineering, Xi'an Jiaotong University. He received a B.Sc. in Fluid Machinery and a Ph.D. degree in Engineering Thermophysics from Xi'an Jiaotong University in 1991 and 1996, respectively. He was a visiting scholar at City University of Hong Kong (1998-1999), a guest professor at Kyushu University of Japan (2003), and a senior visiting scholar at Rutgers, The State University of New Jersey, USA since 2013. He is currently teaching the course titled "Heat Transfer" for undergraduate students, and the course "Advanced Heat Transfer" for graduate students, respectively. His research interests include heat transfer enhancement and its applications to engineering problems, high-temperature/high-pressure heat transfer and fluid flow, transport phenomena in porous media, numerical simulation, prediction & optimization, etc. He is a recipient of National Funds for Distinguished Young Scientists by the NSF of China (2010) and was granted the title of Changjiang Scholarship Chair Professor by the Ministry of Education of China (2013), leader of Innovation Team in Key Areas of Ministry of Science and Technology (2016), People Plan of Science and Technology Innovation Leading Talents (2017). His research team obtained the 2nd Grade National Award for Technological Invention of China (2015), and the National Science and Technology Progress Innovation Team Award of China (2017). He is now the China Delegate of Assembly for International Heat Transfer Conferences (AIHTC) (2015-), a member of Scientific Council of the International Centre for Heat and Mass Transfer (ICHMT) (2009-), a vice president of Chinese Society of Engineering Thermophysics in Heat and Mass Transfer (2010-), an Associate Editor of Heat Transfer Engineering since 2011, and is an Editorial Board Member for several international journals such as Energy Conversion and Management, Applied Thermal Engineering, Energies, Frontiers in Energy, etc. He is the Initiator and Chairman of International Workshop on Heat Transfer Advances for Energy Conservation and Pollution Control (IWHT) (which has been taking place every two years since 2011). He has also delivered more than 40 Invited/Keynote lectures in international conferences or foreign universities. He has also authored or co-authored 4 books and more than 180 international journal papers, and obtained 25 China Invention Patents and 2 US Patents.

Francesco Calise was born in 1978 and graduated cum laude in mechanical engineering from the University of Naples Federico II, Italy, in 2002. He obtained his Ph.D. degree in mechanical and thermal engineering in 2006. From 2006 to 2014, he was a Researcher and Assistant Professor of applied thermodynamics at the University of Naples Federico II. In 2014, he took the position of Associate Professor at the University of Naples Federico II. His research activity has been mainly focused on the following topics: Fuel cells, advanced optimization techniques, solar thermal systems, concentrating photovoltaic/thermal photovoltaic systems, energy saving in buildings, solar heating and cooling, Organic Rankine Cycles, geothermal energy, dynamic simulations of energy systems, renewable Polygeneration systems, and many others. He was an invited lecturer for some courses/conferences in the UK and Finland. He teaches several courses on energy management and applied thermodynamics at the University of Naples Federico II for B.Sc., M.S. and Ph.D. students. He was a supervisor of several Ph.D. degree theses. He is a reviewer of about

30 international journals. He was involved in several research projects funded by the EU and the Italian Government. He is a member of the Editorial Boards of 10 international journals. He was a Conference Chair and/or member of a scientific committee in several sessions in international conferences, promptly indeed in Scopus. He has organised several of the series of conferences on Sustainable Development of Energy, Water and Environment Systems and was a member of organising, scientific and programming committees of more than 57 other research conferences.

Poul Alberg Østergaard is full professor in energy planning at the Department of Planning at Aalborg University, Denmark. He has worked within the field of energy planning since 1995, with a focus on the simulation of energy systems based on high penetrations of renewable energy sources as well as on the design of renewable energy system scenarios. Poul is engaged in various national and international research projects on 4th generation district heating, renewable energy investment strategies for Denmark in an interconnected European context, and smart renewable energy islands modelling tool development for energy system transition, for example. He has an h-index of 29 (Scopus, August 2019). In addition, he has extensive teaching and supervision experience at all levels and is the programme director of the M.Sc. programme in Sustainable Energy Planning and Management at Aalborg University and editor-in-chief of the International Journal of Sustainable Energy Planning and Management. He is or has also acted as guest/co-editor for a number of other energy-related journals including Energy, Energies, Applied Energy, Renewable Energy, Frontiers in Energy Research, and the Journal of Energy. He is one of the main organisers of the annual conference on 4th Generation District Heating and Smart Energy Systems.

Preface to "Selected Papers from SDEWES 2018 Conferences on Sustainable Development of Energy, Water and Environment Systems"

Several countries have recently realized that the present paradigm of development in environmental and energy fields is unsustainable. As global awareness of environmental issues grows, governments worldwide are promoting policies which aim to limit the harmful effects of human development on the environment. In particular, these novel policies seek to address the rapid increase of the global temperature (especially in the polar regions) as well as to manage human plastic waste accumulating in seas. To resolve these issues will require a number of actions to be implemented; unfortunately, the recent COP 24 Conference was unsuccessful in establishing a global agreement to achieve these actions. In order for the upcoming 2020 COP 26 Conference to be successful in creating such an agreement, the scientific community must support the policymakers' efforts. Within this framework, it is necessary to show the efforts of countries worldwide as they try to negotiate an agreement to increase the energy efficiency and reduce greenhouse gas (GHG) emissions. In addition, the research reports generated can provide quantitative measures of the necessary actions to be implemented in order to address sustainable and efficient energy use. Among other topics, the papers presented in this issue investigate innovations in novel efficient and environmentally friendly technologies mainly based on renewable energy sources. The study also highlights the different sectors involved in this task, such as energy conversion systems, urban areas, mobility, sustainability, water management, social aspects, etc. In this field, the Sustainable Development of Energy, Water and Environment Systems (SDEWES) conference provides the foremost forum for discussion on these topics. The 13th Sustainable Development of Energy, Water and Environment Systems Conference was held in Palermo, Italy in 2018. This Special Issue of Energies, precisely dedicated to 13th SDEWES Conference, features papers in three main topics: Energy policy and energy efficiency in urban areas, energy efficiency in industry, biomass and other miscellaneous energy systems.

Neven Duić, Mário Costa, Qiuwang Wang, Francesco Calise, Poul Alberg Østergaard

Special Issue Editors

Review

Toward an Efficient and Sustainable Use of Energy in Industries and Cities

Francesco Calise [1], Maria Vicidomini [1,*], Mário Costa [2], Qiuwang Wang [3], Poul Alberg Østergaard [4] and Neven Duić [5]

[1] Department of Industrial Engineering, University of Naples Federico II, P.le Tecchio 80, 80125 Naples, Italy
[2] IDMEC, Mechanical Engineering Department, Instituto Superior Técnico, Universidade de Lisboa, 1649-004 Lisboa, Portugal
[3] Key Laboratory of Thermo-Fluid Science and Engineering (Ministry of Education), Xi'an Jiaotong University, Xi'an 710049, China
[4] Department of Planning, Aalborg University, Rendsburggade 14, 9000 Aalborg, Denmark
[5] Faculty of Mechanical Engineering and Naval Architecture, University of Zagreb, Ivana Lučića 5, 10000 Zagreb, Croatia
*   Correspondence: maria.vicidomini@unina.it; Tel.: +39-081-768-5953

Received: 3 July 2019; Accepted: 14 August 2019; Published: 16 August 2019

Abstract: Several countries have recently realized that the present development paradigm is not sustainable from an environmental and energy point of view. The growing awareness of the population regarding environmental issues is pushing governments worldwide more and more to promote policies aiming at limiting harmful effects of human development. In particular, the rapid increase of the global temperature, especially in the polar regions, and the management of human wastes, mainly plastic in seas, are some of the main points to be addressed by these novel policies. Several actions must be implemented in order to limit such issues. Unfortunately, the recent COP 24 Conference was not successful, but hopefully an agreement will be established in 2020 at the COP 26 Conference. The effort performed by policymakers must be mandatorily supported by the scientific community. In this framework, this paper aims at showing that countries worldwide are trying to negotiate an agreement to increase energy efficiency and reduce greenhouse gas (GHG) emissions. In addition, in this paper all the researchers reported can provide quantitative measures of the actions to be implemented in order to address a sustainable and efficient use of energy. Here, innovations in terms of novel efficient and environmentally friendly technologies mainly based on renewable energy sources have been also investigated. The study also highlights different sectors that have been involved for this aim, such as energy conversion systems, urban areas, mobility, sustainability, water management, social aspects, etc. In this framework, specific conferences are periodically organized in order to provide a forum for discussion regarding these topics. In this area the Sustainable Development of Energy, Water and Environment Systems (SDEWES) conference is the most ordinary conference. The 13th Sustainable Development of Energy, Water and Environment Systems Conference was held in Palermo, Italy in 2018. The current Special Issue of Energies, precisely dedicated to the 13th SDEWES Conference, is based on three main topics: energy policy and energy efficiency in urban areas, energy efficiency in industry and biomass and other miscellaneous energy systems.

Keywords: sustainable development; renewable energy; biomass; energy efficiency in industry

---

1. Introduction

On 18 June 2018 negotiators from EU Parliament, Commission and Council reached a new agreement, focused on the development of a climate-friendly, affordable and secure energy system for the EU countries. With respect the previous proposals, which negotiated on the Energy Performance in Buildings Directive and on the revised Renewable Energy Directive, the new agreement is based on considerably strong goals from the energy efficiency and emissions point of view: an increase of 32.5% of the energy efficiency for 2030 and an emissions reduction of 40%. Considering the revision of Energy Building Performance Directive and 32% renewable energy target for the EU for 2030 (see STATEMENT/18/4155), the defined targets will allow EU countries to obtain the goals set by the Paris Agreement and a clean energy transition. Furthermore, these targets could lead to several advantages for EU citizens, such as an enhanced security of the energy production systems, a more efficient energy market, a more healthy and comfortable environment, a considerable reduction of energy bills [1]. To reach such ambitious aims, industry and academia must focus on the analysis and design of novel energy conversion systems and on the coupling of conventional and renewable energy systems [2]. Particularly, in this area numerous academics have been involved [3], analysing the sustainable development initiatives by considering the economic, environmental and energy aspects [4], and developing novel solutions for definite sectors [5–8]. Sustainable development involves extremely numerous different disciplines (renewable, water, energy, electrical and control engineering, etc.). At the beginning of this century, to approach this issue, the succession of the Sustainable Development of Energy, Water and Environment Systems (SDEWES) conferences was established.

In 2018, the 13th SDEWES Conference (SDEWES 2018) was held in Palermo, Italy. The conference brought together 400 scientists, researchers, and experts in the field of sustainable development from around 50 countries. The conference is based on nine special sessions, one special event, three invited lectures and two panel debates with some of the most distinguished specialists of the sector. 80 posters and 330 papers were presented.

The papers in this Special Issue (SI) are based on articles presented at SDEWES 2018 Conference, including different research issues, namely economic, technical, social and environmental studies, including works analysing the sustainability of energy, water, transport, food and environment production systems and their integration and interconnection. From 330 accepted manuscripts, 24 were selected for this SI of *Energies* in a continuation of an ongoing fruitful cooperation between *Energies* and SDEWES. The papers within the present SI can be classified into three main research fields: energy policy and energy efficiency in urban areas (seven papers), energy efficiency in industry (10 papers) and biomass and other miscellaneous energy systems (seven papers).

2. Background

Numerous publications under other and this journal's special sections or volumes dedicated to the SDEWES conference series are examined in this section. The studies included in this section are classified into the following research fields: energy policy and energy efficiency in urban area, energy efficiency in industry, biomass and other miscellaneous energy systems.

2.1. Energy Policy and Energy Efficiency in Urban Areas

During the previous SDEWES conferences, several papers investigated the topic of energy policy and energy efficiency in urban areas. Regarding this topic, numerous studies are available in the previous SDEWES SIs, adopting different approaches. Table 1 lists the topics, methodologies and main outcomes of the papers analysed in this subsection.

Table 1. Topics, methodologies and outcomes of the previous SDEWES papers dealing with energy policy and energy efficiency in urban areas.

Reference	Topic	Methodology	Main Outcomes
[9,10]		Numerical model	Energy Security indicators
[11]	Energy security	EnergyPLAN	Drought impacts on the future Finnish energy system
[12]		Quantitative geo-economic approach	Geo-economic Index of Energy Security
[13]		Literature review	Classic energy security concepts and energy technology changes.
[14]		EnergyPLAN	Renewable energies optimization
[15]	Clean mobility sector	Literature review	Electrification and biofuels
[16]		Life cycle analysis	Decision-makers support
[17–19]	Urban energy sustainability	Comprehensive benchmarking	SDEWES Index
[20]	Metropolitan areas sustainability	Analytic hierarchy process	Multi-criteria decision-making technique
[21]	Local water-energy-food nexus	Quantitative analysis	Relations between the food and water sectors
[22]	Water-energy nexus	Life cycle analysis	Renewable heating and solutions lighting
[23]	District heating systems	Data analysis	Main factors of the consumption
[24]	Nearly zero greenhouse city	Dynamic simulation—EnergyPLAN	Electricity and thermal energy costs
[25]	District analysis	Dynamic simulation	District sustainability indicators and new strategies for retrofitting
[26]	District heating and cooling networks	Dynamic models	Real dynamic controls
[27]		Techno-economic analysis	Energy saving-targeted support mechanisms
[28]	Sustainable and efficient waste management	Numerical model	Optimal selection and location
[29]	Tri-generation: municipal solid waste gasification	Numerical model	Positive economic performance

For example, many studies focused on the problem of energy security becoming a critical issue in several EU countries [30,31]. In this framework, several works investigated the fuel poverty vulnerability of urban neighbourhoods [32], or the energy poverty for vulnerable households or the elderly [33] through the definition of several energy poverty indices, including the evaluation of the security of supply as well as the environmental and social components [34]. The topic of energy sources security in EU countries is also investigated by Matsumoto et al. [9]. In their work, the authors aim at understanding how the EU countries energy security has changed from the energy supply point of view between 1978 and 2014. The highest improvement of the overall energy security is identified in Denmark and Czech Republic due to their diversity of the origins of imports and primary energy sources. Through the calculation of special indicators, three kinds of countries were identified: (I) Countries with significantly high levels of moderate improvements and energy security over time; (II) countries with medium levels of energy security and moderate improvements; and (III) countries with significant improvements and low energy security levels. In particular, countries of the Groups

I and III were interesting because they show exemplar practices while the policies of Groups III, in leading to improvement, represent a guides for other countries.

The energy security on the future Finnish energy system is investigated by Jääskeläinen et al. [11]. By considering that Nordic energy system is particularly dependent on hydropower production, the generation inadequacy in case of a severe drought is analysed by using the EnergyPLAN simulation tool. The indirect impacts of a drought in Finland's neighbouring countries are also considered. During winter peaks an extreme drought shows relatively limited impacts on generation adequacy due to available hydropower storage. Anyway, by considering that Finland's electricity market strongly depends on Sweden and Norway, the Finnish energy system would be affected more strongly by an extreme drought via cross-border electricity trade. A novel geo-economic approach to quantitatively measure the energy security is proposed by Radovanic et al. [12]. By means of this new technique, based on the combination of sovereign credit rating and conventional indicators, authors defined a different Geo-economic Index of Energy Security for the measurement of the political, economic and financial stability. The performed measures of this new index are significantly different from the simple indicators evaluated by the conventional approach. The main conclusions of the authors highlight that the renewable energy production and energy dependence least affect the energy security. Therefore, it is required to examine further the sovereign credit rating and to review the significance and type of the impact of the Energy Dependence indicator as a measure of energy security in general. Other paper regarded the energy security are reported in references [10,13].

Numerous papers focused on a clean transition of the mobility sector [35], in order to reduce emissions and energy demand. In fact, in the EU this sector covers about one third of the total energy consumption, so that important energy measures must be implemented. For example, Dorotić et al. [14] modelled a novel solution to obtain a carbon neutral island, which uses 100% intermittent renewable energy sources and 100% share of smart charge vehicles, by means the integration of the vehicle-to-grid concept, power, heating and cooling sectors and renewable energy sources. The electrical marine transportation is also considered. The EnergyPLAN tool is adopted for this goal, in order to implement an optimization procedure aiming at selecting the power supply capacities of solar and wind technologies for the Croatian Island of Korčula. Results show that the least-cost solution is based on an arrangement including 40 MW of wind and 6 MW of installed solar capacities. The combination with 22 MW of wind capacity and 30 MW of solar capacity shows the lowest quantity of total electricity import and export.

Dominkovic et al. [15] performed a comprehensive literature review aiming at investigating a sustainable clean energy transition in the transportation sector. In this work, the authors state that to reduce the emissions of pollutants four fundamental solutions could be implemented, such as electricity, hydrogen, biofuels and synthetic fuels (electrofuels). The results show that in the EU the most important reduction of both energy demand and emissions can be obtained through the electrification of 72.3% of the transport energy demand. The authors estimated that biofuels could cover the remaining 3069 TWh. In addition, for the replacement of the fossil fuels, by renewable electrofuels, an additional demand of heat and electricity, equal to 925 TWh and 2775 TWh, is estimated. A novel method to support decision-makers in evaluating the uncertainty of the life cycle impacts of different bus technologies is developed by Harris et al. [16]. The advanced Technology Impact Forecasting method, integrating a life cycle model, allows the analysis of eleven scenarios including different combinations of battery technologies—lithium-titanate (LTO)-lithium nickel-ithium-iron phosphate (LFP) and cobalt-manganese (NMC)—charging infrastructure and well-to-tank pathways. Results show that scenarios including electric technologies mitigate the GHG emissions by 58–10% compared to the baseline diesel bus, although their high life cycle costs, ranging from 129–247%. LTO systems are the most effective batteries for the mitigation of GHG emissions. Electric vehicles are investigated also in previous SDEWES SIs: (i) Vialetto et al. [36] performed a thermodynamic study of a shared cogeneration system for Northern Europe climate; (ii) Del Moretto et al. [37] compared diesel and electric tourist trains from environmental, energy and socio-economic points of view in a novel sustainable mobility connection for campsites; (iii) Novosel et al. [38] modelled the energy planning of the Croatian transportation system; (iv) Firak et al. [39] investigated the Croatian future transportation sector based on hydrogen infrastructure and fuel-cell vehicles; (v) Briggs et al. [40] analysed and simulated an inner-city diesel-electric hybrid bus; and (vi) Knez et al. [41] considered policies for the commercialization of low emission electric vehicles.

Numerous papers dealing with energy efficiency in urban areas were involved in SDEWES SIs [42], specifically focusing the trends of European research on district heating and cooling networks [43] and the energy management in municipalities [44]. In this framework, a comprehensive benchmarking of sustainability of urban energy, water and environment systems investigating 120 cities is performed by Kılkış [17] by the definition of the Sustainable Development of Energy, Water and Environment Systems city index (SDEWES Index [18]). The index, considered a useful benchmarking tool, aims at comparing the cities sustainable performance, and includes seven dimensions: (i) renewable energy potential and utilisation, (ii) penetration of energy and CO_2 saving measures, (iii) CO_2 emissions and industrial profile, (iv) water and environmental quality, (v) research and development, innovation and sustainability policy, (vi) city planning and social welfare and (vii) energy consumption and climate. By proposing a scenario where the use of residual energy among near cities, improvements of the index indicators on 60 cities are detected. The benchmarking results can be used by urban planners and local decision-makers to motivate policy-learning opportunities.

A similar work is presented also in [19], where the SDEWES Index is calculated for 26 significantly different cities from around the world, from Europe, Latin America and Africa as well as new cities in Asia. The higher values of the index are detected for cities including effective urban management in environment and water and efficient district energy networks. Copenhagen obtains the highest value of the index. A multi-criteria decision-making technique is implemented by Carli et al. [20] in order to evaluate the sustainability of metropolitan areas by considering their energy, water and environment systems. In this work, the authors aim to demonstrate the novelty and robustness of their approach, based on analytic hierarchy process with respect to the well-established method, such as the above mentioned SDEWES Index. This work covers the research gap concerning the assessment of the sustainability of metropolitan cities and provides their specific improvements, by means of the estimation of qualitative indicators. Their approach, applied to the four metropolitan areas Bitonto, Bari, Molfetta and Mola in South Italy, show the strategic and overriding interventions to obtain sustainable metropolitan cities. The limitation of the proposed approach is due to the vagueness and genuine uncertainty typical of the human decision making.

The role of the water-energy-food (WEF) nexus at local level, as recognised as a factor in implementing the sustainable development goals in small-scale energy projects in developing countries, is investigated by Pfaff et al. [21]. The authors systematically analysed the relations between the food and water sectors, by quantifying and identifying the critical links. The analyses show that the energy needs are indirectly due to the provision of energy to the agricultural sector and directly due to agricultural activities or food. The optimization of the provision of water and energy in two school buildings in order to promote sustainable municipalities is proposed by Gamarra et al. [22]. School buildings are selected due to the high energy consumption, mainly in hot weather conditions, lacking of energy efficiency measures and water scarcity complications. A life cycle assessment is performed in order to evaluate the impacts of the school activities per student. Moreover, different improvement measures, including the integration of the renewable energies or lighting technologies replacement are calculated. Results show that renewable heating solutions and lighting replacement measures can reduce the fossil energy demand by between 64.06% and 78.98% and 12.05% and 9.54%, respectively.

Valuable findings are obtained by Gianniou et al. [23] on the design and optimization of strategies for demand management towards energy-efficiency policies for district heating systems. They analysed the residential heating consumption of 8293 Danish single-family households in Aarhus. Sufficiently constant load profiles with peaks in the evening and in the early morning are obtained and that the most relevant factors of the consumption are the building area, age and family size.

De Luca et al. [24] analysed the potential transition of an Italian city to a zero GHG city by 2030. To this end, several technologies are proposed and modelled in the TRNSYS environment: photovoltaic and thermal solar panels, wind turbines, heat pumps and biogas cogeneration. Subsequently, to carry out their analysis, TRNSYS results are taken as inputs of the EnergyPLAN. The calculated thermal and electric energy prices, equal to 0.12 €/kWh$_t$ and 0.11 €/kWh$_e$, respectively, are very profitable. Sözer and Kükrer [25] evaluated design strategies of retrofitting activities by defining specified district sustainability indicators in order to improve the existing energy condition of a district. The key aim is to recognise the current conditions of the investigated district to develop new strategies for retrofitting. The district consists of three buildings with 20,000 m^2 conditioned area. To assess energy indicators, taking into account GHG emissions, indoor comfort and return on investment, an hourly dynamic simulation software is used. The most relevant result of this work is that by calculating the sustainability indicators with this method, a 5.16 years return on investment and a 69 kg/m^2 year GHG emission reduction are achieved.

Arce et al. [26] developed district heating and cooling networks component models (hot water storage, distribution pipe and network) aiming at representing the real dynamic operation of these components for control purposes. Beccali et al. [27] proposed an analysis aiming at assessing the techno-economic feasibility of the retrofit of existing power plants when the heat recovery is used for district heating and cooling purposes for tertiary and residential buildings. Six different Italian small islands, featured by different weather and demographic conditions, are investigated. Due to the irregular load profiles and low heat density, this solution can be economically profitable only if energy saving–targeted support mechanisms (as white certificates for cogeneration) are considered.

Conversely, the obtained discounted payback time ranging from 17.4 year up to 30 year. The weather conditions do not affect the obtained results. Additional papers were published on this topic in the previous SDEWES SIs, dealing with: the energy-economic analysis by a dynamic simulation model of a district cooling, heating, and domestic hot water plant based on solar and geothermal energy [45], the hourly optimization model of a district heating system based on electric heaters, heat pumps, boilers, solar thermal collectors and thermal energy storage units [46], the evaluation of the economic and thermal efficiency of a coal-fired municipal plant coupled to a district heating system in case of repowering with a gas turbine [47]. The topic of the sustainable and efficient waste management to increase the quality of human life in urban area is also addressed in numerous papers [48]. For example, Santibañez-Aguila et al. [28] developed a mathematical model aiming at planning a sustainable waste management system among multiple different neighbouring cities. The considered waste management system model is general and takes into account different landfills, processing facilities and technologies, processing, utilized raw materials and manufactured products. The model, applied to a suitable case study located in the west region of Mexico, is able to define the optimal location and selection of the waste management system. This model is also used to define the material flows to be sold, stored, processed and transported.

Katsaros et al. [29] designed and modelled a novel tri-generation system consisted of an ammonia-water absorption chiller, municipal solid waste gasification and a solid oxide fuel cell. Such plant is suitable to provide electricity, heating and cooling in hotels and hospitals buildings. They performed a sensitivity analysis aiming at identifying the variation of the air equivalent ratio and the operating temperatures of the gasifier and desorber on the system efficiency. Results show that the developed tri-generation plant could completely cover the cooling and electricity demand and up to 55% of the heat demand. The payback period and net present value were 4.5 years and 5 million euros, respectively. The overall system performance can be enhanced for high gasification temperatures and low air equivalent ratios. Other papers on the efficient waste management regard the definition of a primary energy return index, used for comparing different municipal solid waste management scenarios [49]; the coupling of renewable energy source in a wastewater treatment plant consisted of conventional activated sludge systems [50]; and a novel methodology applied to evaluate urban wastewater treatment plants in terms of energy efficiency [51,52].

This topic was dealt with comprehensively in previous SDEWES SIs focusing on different aspects, namely: the influence of urban form on the performance of road pavement solar collector system [53]; the energy management in a smart municipal energy grid including combined heat and power plants, solar photovoltaic and wind technologies [54]; the analysis of the future energy scenarios on the Danish municipality of Helsingør to obtain a cost-optimal combination between individual heating, district heating and heat savings [55]; a study on the management of dust to ensure that urban environment and industry can coexist in a sustainable and beneficial manner [56]; the adoption, in urban water distribution systems, of energy storage systems to meet the water demand [57]; the evaluation of carbon emissions in highly polluted European cities [58]; the replacement of bituminous roofs as green roofs to make cities more 'future proof' and resilient [59]; the integration of the renewable energy resources to enhance the regional energy efficiency and sustainability [60]; and many other papers concerning the analysis of energy efficiency targets of the member countries of the EU [61–63].

2.2. Energy Efficiency in Industry

Generally, the industry sector covers about one-third of energy and process-related GHG emissions [64]. Therefore, energy measures aiming at increasing the energy efficiency of the involved different processes in this sector represent a pivotal topic. The numerous papers dealing with this topic in previous journal SIs dedicated to the SDEWES conferences are summarised in Table 2 and described as follows.

Table 2. Topics, methodologies and outcomes of the previous SDEWES papers dealing with energy efficiency in industry.

Reference	Topic	Methodology	Main Outcomes
[65]	Heat recovery for the hydrocarbon processing industry	Numerical model	Thermoeconomic and environmental performance
[66]	Energy efficient industrial applications	Simulation model	Prediction of material and energy flows
[67]	Cogeneration in the industrial sector	Energy and economic analysis	Indicator to choose the best cogeneration technology
[68]	Heat production in the pulp and paper industry	Second-law and operating costs analysis	Flexible power production
[69]	Ceramic dust powder as sorbent for heavy metals	Experimental analysis	Sorption capacity of Cu, Pb and Zn
[70]	Marine sediments for the brick industry	Experimental analysis	Suitable additives
[71]	Plybamboo industry	Life cycle assessment	Low environmental impact
[72]	Ruths steam accumulator in industrial processes	Numerical model	Storage improvement by phase change materials
[73]	Energy retrofits in aluminum industry	Life Cycle Assessment and Cost	Energy and environmental indicators
[74]	Optimization of operational aspects of the industry sector	Simulation	Biomass, biogas and renewable gas for fossil fuel reduction.
[75]	Preventing summer overheating in metal processing factory	Dynamic simulation model	Energy efficient measures

Varga and Csaba [65] proposed low-temperature heat recovery to supply an ORC to improve the energy efficiency in the hydrocarbon processing industry. In particular, the ORC was applied as a condenser linked to the main distillation column of a vacuum residue processing unit. The ORC evaporator condenses and cools down vapours from 140 °C to 50 °C. The authors evaluated the thermoeconomic and environmental performance of the proposed system by selecting appropriate working fluids. With isobutane, isopentane and pentane, the obtained ORC power ranges varied between 452 and 678 kW. The highest power was obtained with isobutane and the related avoided CO_2 emissions were 1085 t/y. From an economic point of view, the minimum payback period was 3.4 years, and butane provided the best result regarding this criterion.

Smolek et al. [66] developed an interdisciplinary simulation model applied to energy efficient industrial applications. All aspects of the industry are taken into account, classified into the energy system, building, logistics and production and equipment. The production durations, energy consumption of the overall system by varying operating strategies, production schedules and environmental settings (time of year, climate, etc.) are taken into account. They applied the model to a real production facility producing baked goods, fresh as well as frozen by comparing two different scenarios. In scenario 1, production occurs during summer at temperature between 31 °C and 16 °C. Scenario 2 considers the same production settings during the winter the ambient temperature varying between −14 °C and −6.5 °C. The authors stated that their model is useful to predict material flows and energy demand of any factory.

Gambini and Vellini [67] developed an analysis aiming at selecting and optimally designing cogeneration units in the industrial sector in the light of the new high-efficiency cogeneration regulatory context. A comprehensive characterization of industrial sectors to evaluate the thermal and electric energy consumption, annual production, as well as an analysis of each cogeneration technology is provided. They define a performance indicator suitable to choose the best cogeneration technology suited to an industrial process, and provide the following useful guidelines: (i) internal combustion engines represent an appropriate technology for almost all investigated industrial processes by reaching the best performance for small electric sizes; (ii) combined cycle power plants with condensing turbine are suitable for heat demand and large annual production industrial processes; and (iii) steam power plants attain worse energy, economic and environmental performances, therefore, they are recommend for few industrial processes.

Wolf et al. [68] focused on the performance evaluation of three different heat production processes for supplying industrial processes in the paper and pulp industry. The aim of these processes—a gas turbine equipped by a heat recovery boiler, a high-temperature heat pump recovering waste heat and a gas and steam turbine combined cycle process—is the production of heat as 4 bar (abs) saturated steam. The thermodynamic and economic efficiency of three processes are compared based on costs of heat and exergy flows. The results show that the heat pump (working with a COP of 4) provides a share of between 45% and 76% of the heat and it has higher exergetic efficiency and the best economic technology in case of the natural gas price is 25 €/MWh and the electricity price is lower than 45 €/MWh. The authors state that the quantitative results of this paper can be helpful to industrial plant and electricity grid operators, in order to obtain a flexible power production. In particular, the electricity can be produced during the high electricity prices hours and purchased from the grid to supply heat pumps during low electricity prices, thereby balancing out supply and demand mismatches in the electricity grid.

Keppert et al. [69] dealt with the utilisation of red clay-based ceramic dust powder, generated as a waste product in the production of hollow bricks, as a sorbent for heavy metals (Cu, Pb and Zn). The experiments show a decreased sorption capacity with this sequence: Cu > Pb > Zn. The waste ceramic powder resulted in a very efficient cement substitute although the adsorbed metals, mainly in the case of Cu, significantly reduces the rate of setting and strengthening of concrete.

Baksa et al. [70] tested the marine sediments of the Port of Koper as appropriate raw materials for the brick industry, particularly for the production of clay blocks, roofing and ceramic tiles. They carried out various analyses to determine if the marine sediments are suitable and environmentally friendly for their use in the brick industry. From their tests, aiming at verifying the mechanical properties and the frost-resistance of the materials, the authors showed that without any additives, marine sediments exhibit too much shrinkage in drying and firing (about 12% vs. 3–4% of the usual shrinkage of brick clay), as well as higher water absorption than the normal values of 20%. However, by adding suitable additives, such as clay, marine sediments are suitable for producing clay brick products.

Chang et al. [71] presented a study concerning the environmental benefits/impacts and the carbon storage of the plybamboo industry in Taiwan. They compared several materials, by means a life cycle assessment (LCA), obtaining the following main results: the low environmental impact is reached by plybamboo, but its impact is higher than the concrete one, having the lower environmental impact and higher global warming index.

Dusek and Hofmann [72] modelled the integration of a Ruths steam accumulator with phase change materials and electrical heating elements to increase the efficiency of industrial processes, applying steam as heat transfer medium. In particular, phase change materials surround the Ruths steam accumulator, storing the excess steam in order to consume it subsequently at high charging and discharging rates. After about 50 min the stored energy significantly increases with respect to the Ruths steam accumulator one, 0.34 vs. 0.45 MWh.

Royo et al. [73] proposed different retrofitting scenarios for improving the environmental and energy efficiency in aluminum industry. They focused on the manufacturing of an aluminum billet, namely alloy production, heating, extrusion and finishing, by proponing and comparing a novel technology based on direct current (DC) induction with the reference standard techniques, such as natural gas heating and alternating current (AC) induction heating system. Four typical European electricity country mixes were taken into account (Greece, Spain, Italy and France) to highlight their effect on the environmental impact when DC induction solution is adopted. The novel technology showed a lower impact (up to 23%) in most indicators. Reductions of up to 8% of GHG emissions are obtained in every country. With regard to the conventional global warming indicator, in France (the best-case scenario), the global warming indicator notably decreases about 80% and increases 50% in Greece.

Wiese and Baldini [74] proposed a simulation method to optimize in detail operational features of the industry area. They applied their model to a Danish case study, concentrating on the structure of the energy use by considering the end-use processes (drying, heating/boiling, space heating, lightning), to the geographical mapping of industrial energy consumption, and measures for fossil fuels (electricity, gasoil, kerosene, natural gas) and CO_2 emissions. To cut fossil fuels they recognized the following options: electrification (heat pumps due to their flexibility to combine electricity to provide heat at different temperature levels), energy cascading, biogas, biomass and renewable gas. The potential end-uses to reduce the adoption of fossil fuels are space heating, low temperature and electricity processes. Particularly, excess heat and savings can possibly contribute up to 9 PJ and 38, respectively. Biogas (35–170 PJ), biomass (75–315 PJ) and renewable gas can also represent a great opportunity for fossil fuel reduction. Gourlis and Kovacic [75] developed a dynamic thermal simulation model proposing several passive measures for avoiding summer overheating in an existing single-story metal processing factory in Berndorf, Austria. The examined passive measures were night natural ventilation, the adoption of a water-based elastomeric cool roof coating on the existing roof, with 0.87 infrared emittance and 0.87 solar reflectance, white venetian blinds on the south oriented façade windows, exterior solar shadings and light grey roller shades on the south side of the saddle shaped roof skylights over the main hall, the thermal improvement of the building fabric. Both the exterior shading and cool coating on the roof improved thermal environment with overheating occurring less than 1% of the time.

Other papers presented in the SDEWES SIs concerning this topic were: the recovery of valuable industrial metals from household battery waste [76], the optimal operating strategy of a trigeneration layout for an engine manufacturing facility [77], the exergy-energy analysis of geothermal energy-assisted milk powder production line [78], the waste-heat recovery for low-temperature applications in an electric steelmaking industry [79].

2.3. Biomass and Other Miscellaneous Energy Systems

Biomass is the most promising renewable energy source to obtain a sustainable development as assessed by the several studies concerning this topic included in the previous SDEWES SIs.

In this field, the replacement of the conventional fossil fuels with biofuels (biogas, bioethanol, biodiesel, etc.) represents the most promising solution to reach the decarbonization targets. Consequently, a lots of papers investigating this topic are included in previous SDEWES SIs and summarised in Table 3.

Table 3. Topics, methodologies and outcomes of the previous SDEWES papers dealing with biomass and other miscellaneous energy systems.

Reference	Topic	Methodology	Main Outcomes
[80]	Liquid and gaseous bioethanol	Modelling and experimental analysis	Lower NO_X emissions compared to diesel oil
[81]	Napier grass for bioethanol in heavy metals contaminated soil	Experimental analysis	Sustainable solution
[82]	Biofuel in Finland transportation	Multi-objective dynamic biofuel cycle model	Decrease in fossil fuel consumption for heavier vehicles
[83]	Energy recovery from manure of different livestock farms	Numerical model	Biogas and syngas for cogeneration, waste minimization
[84]	Wood biomass microcogeneration	Experimental analysis	Syngas chromatograph characterization and efficiency evaluation
[85]	Bioethanol production rice paddies and forest sector	Numerical evaluation	Bioenergy supply and carbon sequestration from biomass wastes
[86]	Agroindustrial waste for the biosorption of chromium and lead ions from aqueous solutions	Experimental analysis	Comparative analysis
[87]	Microalge into Fischer–Tropsch liquids, hydrogen, electricity, and thermal energy	Numerical model	Carbon emissions reduction per ton of microalgae
[88]	Microalge into biogas	Experimental analysis	Suggestions for enhancing biogas production
[89]	Spent coffee grounds into bio-oil	Monodimensional model and experiments	Bio-oil production peak at 500 °C
[90]	Poultry wastes into biofuel	Experimental analysis	Pyrolysis processes for enhancing bio-oil production
[91]	Urban rain gardens and herbaceous plants into bioethanol and solid biofuels	Dynamic simulation model	Carbon sequestration evaluation
[92]	Electricity, bioethanol and gasoline vehicles	Life cycle assessment	Carbon emissions reduction

Balog et al. [80] analysed the contaminant emissions of various aqueous bioethanol solutions in gaseous and liquid forms for the combustion in swirl burners. Liquid hydrous ethanol combustion is featured by 56% lower NO_X emissions with respect to the diesel oil combustion. Anyway, diluted alcohols increase production costs.

The Napier grass phytoremediation for bioethanol production is investigated by Chun et al. [81]. In particular, they evaluate the use of soil artificially contaminated with heavy metals (Zn, Cd, and Cr) for biomass production. For biomass containing heavy-metal the fermentation ethanol concentration was higher than the control biomass: 8.69–12.68, 13.03–15.50 and 18.48–19.31 g/L in Zn, Cd, and Cr environments, respectively. Results show that the heavy metals had a positive effect on bacteria fermentation. Thus, for bioethanol production the Napier grass phytoremediation has a positive effect on the sustainability of environmental resources.

Palander et al. [82] investigated the energy performance of road freight transportation in Finland, with the increase of the local biofuel cycles in a 100% carbon-neutral wood procurement. They examined three end-user scenarios of advanced liquid biofuels with a 5%, 15% and 30% decrease in fossil fuel consumption and different heavier vehicles (60, 68 and 76 t) in wood transportation. Due to the fossil fuel decrease, the energy ratio of the total renewable wood energy input divided by the fossil-fuel energy input to drive the system, increased 43% with wood-based transportation.

Milani and Montorsi [83] developed a numerical tool for the efficient exploitation of biomass by applying their model to a suitable reference case. In particular, the energy recovery from manure of different livestock farms in Modena, Italy, is analysed. Different waste to energy technologies, such as anaerobic digestion, gasification and incineration and several vegetable and animal biomasses, are considered, to produce biogas and syngas for a cogeneration producing electric and thermal power. The annual electric production from the animal farming manure proved to exceed by 195.7% the energy requirement of the whole agricultural industry in the province. The system minimizes the amount of waste that has to be disposed to 3–6%.

Villetta et al. [84] present an experimental analysis of a gasifier coupled to a micro internal combustion engine (20 kW$_e$), supplied by wood biomass. The gas chromatograph characterization of the produced syngas revealed that the main system inefficiency occurs in the gasifier. Due to the lower heating value of the produced syngas (equal to 3731 kJ/Nm^3), the combustion engine's electrical efficiency does not exceed 22.5%, and the global electrical efficiency of the plant is about 13.5%.

Chang et al. [85] presented a case study regarding the utilisation of waste biomass derived from an enhanced production from rice paddies and from the forest sector in Taiwan. They aim to replace gasoline by the bioethanol production from biomass. The authors evaluate that biomass wastes from rice paddies and forest sector could generate 31.69 PJ and 222.37 PJ annually, respectively, when bioethanol is used.

Boeykens et al. [86] experimented the adoption agroindustrial waste as low-cost alternative method for the biosorption of lead ions and chromium from aqueous solutions. The evaluated agroindustrial materials were sugarcane bagasse, peanut shell, avocado peel, wheat bran, banana peel and pecan nutshell. Wheat bran obtained the highest percentage of lead removal (89%). For all tested biosorbents, a percentage of chromium generally much lower compared with lead is obtained, the banana peel being the most efficient with a 10% removal.

Graciano et al. [87] developed a model based on the thermochemical conversion of microalgae biomass into Fischer-Tropsch liquids, hydrogen, electricity, and thermal energy via polygeneration. This kind of plant presents better energy performance for the microalgae conversion into liquid transportation fuels compared with conventional biomass-to-liquid-fuels and therefore reduced equivalent CO_2 emissions. In particular, it is obtained microalgae can displace fossil fuels at a rate of 0.23 m^3 of liquid fuels, 16 kg of hydrogen and 430 kWh of electricity per ton of microalgae. The displaced fossil fuels would reduce carbon emissions at a rate of 560 kg of carbon dioxide per ton of microalgae.

Marques et al. [88] developed a method to enhance the biogas production by means the pre-treatment of the microalgal biomass. They detected that the best biogas yields (compared to fresh biomass without hydrolysis) is obtained with the thermochemical hydrolysis during the anaerobic digestion with biomass acidification using CO_2. For fresh biomass and biomass hydrolysed for 60 min and 120 min, the resulted biogas yields were of 97.5, 146.4 and 61.3 mL/L day, respectively. This work proves that the substitution of sulphuric acid with CO_2 as acidification agent and the use of waste energy from flue gas result in an additional potential of CO_2 utilisation.

Codignole et al. [89] developed a monodimensional model for the prediction of spent coffee grounds bio-oil production through the fast pyrolysis in a screw reactor. The obtained oil yields are in the order of 56% with the peak of bio-oil production is obtained experimentally at the intermediate temperature of 500 °C.

Kantarli et al. [90] examined the potential of poultry wastes as feedstock in non-catalytic and catalytic fast pyrolysis processes for their conversion into biofuel. Pyrolysis of poultry meal showed high amounts of bio-oil, while pyrolysis of poultry litter high amounts of solid residue owing to its high ash content. All bio-oil samples from the pyrolysis of poultry wastes contained relatively high amounts of nitrogen (9 wt% in the case of poultry meal and ca. 5–8 wt% in the case of poultry litter) compared with bio-oils from lignocellulosic biomass.

Chan et al. [91] introduced a new dynamic model to simulate the carbon sequestration potential from urban rain gardens with woody and herbaceous plants. They investigated the conversion of carbohydrates into bioethanol and lignin into solid biofuels. The simulation results suggested that the maximum carbon sequestration potential of the studied urban rain garden can increase from 6.7 kg/m^2 to 14.7 kg/m^2 through harvesting and converting the plant-derived biomass into biofuels.

Picirelli et al. [92] performed a comparative environmental life cycle assessment of conventional vehicles with different fuel options (electricity, bioethanol, gasoline), for a sustainable transportation system in Brazil. The replacement of 50% of the internal combustion engine gasoline vehicles fleet by bioethanol vehicles would decrease annually the carbon emissions around 45.8 Mt CO_2-eq. (33% of the total emissions).

Szulczewski et al. [93] investigated a novel method for the estimation of biomass yield of *Miscanthus giganteus* in the course of vegetation. They modelled the biomass growth by simple measurements of the shoot length, mass and diameter. On the basis of experiments data, a correlation of two features of *Miscanthus* shoot volume index and shoot mass was determined. The number of shoots per plant was estimated by shifted Pascal distribution. The accuracy of estimation of is strictly dependent on the number of shoots on which biometric measurements are performed. The authors concluded that the best trade-off is obtained for 10 shoots of *Miscanthus*.

Many papers were presented in the previous SDEWES SIs concerning this topic, namely: hybrid biomass-solar systems [94], marine vehicles supplied by biofuels [95], analysis of active solid catalysts for esterification of tall oil fatty acids with methanol [96], biogas production by an integrated system for sewage sludge drying through solar energy [97], multi-criteria and principal component analyses in soybean biodiesel production [98], hydrothermal conversion for lignocellulosic biomass for the production of lignin, syngas or bio-oil [99], animal waste from tanneries as fuel and for biogas production [100] and valorisation of agroindustrial wastes to produce hydrolytic enzymes by fungal solid-state fermentation [101].

3. Research Topics Represented in This Special Issue

After the review process, 24 papers from 13th SDEWES Conferences were selected for this SI. The main ideas of these papers that are among the best articles presented at the conference are briefly reviewed in the following subsections.

3.1. Energy Policy and Energy Efficiency in Urban Areas

Weiler et al. [102] presented a novel methodology for the design of central energy generation and supply scenarios, including district heating systems. The tool is based on a 3D urban modelling approach. This paper aims at significantly contributing to the development of suitable tools to be used to improve energy and environmental issues in urban areas. In fact, cities are responsible for more than 60% and 70% of the energy demand and of CO_2 emissions, respectively. In particular, according to EU targets, CO_2 emissions must be reduced by 80% by 2050 with respect to the levels of 1990. To achieve this goal, cities' energy demands must be first estimated and then reduced. Several tools are presently available for the simulation of district-scale energy systems. Such systems are considered crucial for the future transition toward a fully renewable energy system. In this framework, heat pumps powered by renewable electricity and cogeneration plants supplied by renewable gas are considered extremely promising to achieve the goals in terms of energy efficiency and reduction of CO_2 emissions.

The present study presents calculation models for district heating systems including both the above-mentioned devices. The model was developed within the simulation engine INSEL 8.2, under development at the University of Applied Sciences Stuttgart. Suitable building simulation models are employed to model the buildings included in the district under investigation. Building heat demand analysis is performed using the German standard DIN 18599. In addition, a special procedure, according to the German standard VDI 4710, is employed to transfer monthly data into hourly ones, required to perform a dynamic simulation of the system under investigation. As for the heat pump, this device is modelled using a data-lookup approach. In particular, a polynomial fit curve is used in order to calculate output data as a function of source and sink temperatures. A similar simplified approach is used to simulate the performance of the combined heat and power (CHP) system. Suitable models are also used to simulate a central storage system and the district heating network.

A case study is developed for a small town in the South of Germany, close to Stuttgart, where the local government aims at achieving climate neutrality by 2050. In particular, the town of Walheim in the district of Ludwigsburg was selected. It includes 3200 inhabitants and it is located next to the River Neckar. The developed 3D CityGML model includes 1610 buildings. Therefore, the size of the CityGML file is relatively small, determining limited calculation times. Authors assumed a future scenario achievable around 2050, when all buildings will be refurbished according to the present German legislation EnEV 2016. Therefore, all the buildings are featured by low-temperature heating devices and they can be supplied by low-temperature district heating systems. As mentioned before, the model first evaluates the monthly heating energy demand and then such data are reported on an hourly basis using VDI 4710. Then, PV systems were suitably designed taking into account roof area availability, their shape, and their orientation. A nominal power of about 6.4 MW was calculated. The CHP unit nominal power was 2.0 MW and it runs for about 4000 h per year.

Results show that the scenario including the heat pump, HP, powered by photovoltaic panels, PV, suffers for the winter operation due to the low solar availability, determining a solar fraction close to 15%. This is due to the fact that PV electricity is mainly available in summer, whereas HP electrical and thermal demand mainly occur in winter. As for the CHP unit, it seems very attractive only when supplied by renewable energy sources (e.g., biogas or gas from P2G). Authors also concluded that a suitable mixture of renewables must be designed (e.g., wind, solar, biomass, etc.) in order to achieve a good match between demand and generation. In this framework, the appropriate selection of different storage management schemes is crucial in order to achieve the above mentioned matching. This paper proved that the development of accurate urban energy simulations models is crucial for building authorities or municipalities in order to calculate user demands and to evaluate possible future scenarios including energy efficiency and renewables.

Dominkovic et al. [103] focused on the selection of the optimal capacity of storage systems in district cooling systems. They started from the same energy policy scenario pointed out in the previous paper. Cities are responsible for the largest amount of total CO_2 emissions. Therefore, all the actions aiming at reducing CO_2 emissions in cities are major drivers for achieving the recent goals established by the Paris Agreement in terms of energy and environment. In addition, cities also suffer for a dramatic problem related to the air pollution which is instrumental in the premature death of about 6.5 million people per year. Once again, both CO_2 and pollutant emission problems may be limited by reducing the amount of fossil fuels converted in different forms of energy. The authors of this paper performed a detailed literature review analysing the recent papers investigating the problem of CO_2 emission reductions in cities. They showed that a number of recent papers analysed different aspects of this problem, such as: energy planning, use of renewables, matching between production and demand, novel and advanced energy storage systems. This last was especially important for the authors who noticed that only a few authors focused on the problem of cold storage in district cooling systems, whereas the problem of heat storage in district heating system is widely investigated. On the basis of the literature review, this paper aims at analysing the optimal selection of storage technology and capacity, while also including socio-economic costs. The calculations are performed using an energy planning model previously developed by the authors.

A suitable case study is presented for the case of Singapore due to its high degree of urbanization. In addition, in Singapore, the space cooling demand is extremely high and district cooling systems are a common selection and the large population density and the high GDP results in a dramatically high energy use per capita. Several different technologies were analysed by the optimization tool: solar thermal collectors, absorption chillers, heat pumps using the waste heat from datacentres as heat sources, thermal energy storage systems, geothermal energy, electrical storage, combined heat and power plants powered by natural gas or renewables, wind turbines, photovoltaic panels, vehicle-to-grid technology, gas and hydrogen storage, fuel cells and solid-oxide electrolysers, syngas, individual chillers and reverse osmosis desalination of seawater. Several optimizations are performed to calculate the optimal configurations. In particular, the authors aimed at calculating the difference the optimal configurations of the energy systems without and with the large-scale energy storages. Similarly, the optimization tool was used in order to calculate the optimal capacities of the components, based on different shares of individual and district cooling. Results show that the optimal district cooling share for the case of Singapore was 30% when PV capacity is selected on the basis of the spatial constraints. A similar result is also obtained when unconstrained PV capacity is considered, assuming nearby space outside of the city borders.

The problem of the limitation of CO_2 and pollutant emission in cities was also addressed by Smajla et al. [104] investigating the use of liquefied natural gas (LNG) in heavy truck traffic. In particular, they focused on the possible benefits achievable by the use of LNG rather than diesel in heavy-duty vehicles for the EU market. They analysed the papers available in the literature investigating this topic, presenting a detailed literature review of the works considering different energy, economic and environmental aspects of the LNG utilisation in vehicles. In this paper the authors also analysed in detail the main characteristics of LNG in terms of physical and chemical properties and they presented the main technique of production and storage along with the main economic data. They focused on the various fields of application of LNG and they also analysed all the financial aspects of the fuel switch. In fact, LNG trucks are about 30–40% more expensive than conventional diesel trucks. However, LNG trucks present a fuel cost of 0.306 USD/km, which is much less than the 0.444 USD/km for diesel trucks.

The authors also analysed the environmental aspects related to the use of LNG in vehicles. They concluded that GHG emissions may be reduced by 67%. From a safety point of view, LNG shows similar features with respect to the other fuels. It is only worth mentioning that LNG is cryogenic and it may be harmful for skin or eyes. Thus, visor, gloves and other forms of protection must be mandatorily used when handling LNG. The authors also performed an investigation of the LNG filling infrastructure in EU and they found that several EU countries—mainly the UK, The Netherlands and Spain—present a reasonable number of LNG filling stations. The technology of those filling stations along with the LNG heavy trucks was also analysed in detail.

A case study was presented for a region in Croatia that has a great geostrategic position. Here, no LNG fuel infrastructure exists and no LNG vehicles are registered in the country. Some prototypical project related to LNG are going to be founded by the Croatian government and a series of legislation acts are going to promote this LNG technology. A specific proposal for the case of Croatia was included in the present work. In summary, the authors show that LNG use in transportation sector is extremely profitable from both environmental and financial points of view. On the other hand, LNG is still scarcely used due to lack of infrastructures. The poor availability of LNG filling stations dramatically limits the use of LNG vehicles, and vice versa. The authors conclude that significant measures must be implemented by EU countries in order to stimulate LNG market.

Testi et al. [105] focused on the optimal operation of microgrids in a cluster of buildings located in a campus combining cogeneration and renewables. In fact, renewable energy sources are often unpredictable whereas cogeneration plants can be dispatched. This technology is also highly supported by EU government, also suggesting combining cogeneration with: decentralized systems based on renewables, district heating and cooling and heat pumps. Several studies are available in the literature analysing this topic, addressing several problems, such as optimal selection of synthesis/design variables, optimal operation and control strategies according to energy, economic and environmental objective functions. In this work, the authors present a novel configuration for smart multi-energy microgrids. These systems include distributed energy units and a centralized cogeneration supplying thermal energy to a micro-district heating network. For this system, the authors investigate the benefits of integrating reversible heat pumps in buildings included in such systems. The heat pumps are crucial in the thermal/electrical balance between production and demand, since they can shift space heating demand from heat to electricity and they can significantly enhance the operative flexibility of the microgrid as well as promote the renewable energy technologies integration.

The authors considered a complex system layout including a plurality of devices, namely: gas-fired boilers, solar thermal collectors, cogeneration unit, photovoltaic panels, wind turbines, electrical grid, thermal storage, electrical chillers and heat pumps. The system supplies space heating and cooling energy, domestic hot water and electrical energy to the users. In order to perform the calculations a suitable simulation model was implemented. A case study was analysed for a campus located in Trieste, Italy. The campus includes: classrooms, offices, dining halls, gyms and dormitories. The optimization was performed aiming at reducing the total annual energy cost. The results proved that the utilisation of heat pumps is beneficial from several points of view. In fact, heat pumps significantly improve the flexibility and cost-effectiveness of the considered energy system. In summary, the proposed system exhibits an 8% total-cost saving, 11% carbon emission reduction and 8% primary energy saving with respect to the centralized reference case. The novel proposed system also allows a 40% reduction in the electricity exchange with the grid.

Kona et al. [106] presented a study related to the analysis of the role of the local authorities and cities as one of the main driver for the energy transition toward a more sustainable system. This paper is a part of the Covenant of Majors initiative and it aims at proposing a new method for indirect accounting of emissions in cities. In fact, in the last few years several local authorities adhered to several different initiatives aiming at mitigating the issues related to the climate change. In particular, the Covenant of Majors initiative aimed at reducing the CO_2 emissions levels by at least 20% by 2020 or at least 40% by 2030. This goal will be achieved through so-called Sustainable Energy Action Plans (SEAP). SEAP has been also combined with climate risk assessment in Sustainable Energy and Climate Action Plans (SECAPs). This paper, analyses in detail climate mitigation action plans for 2020. The CoM is a unique feature of multilevel polycentric governance that goes far beyond transnational city networking [11]. This initiative is supported by the European Commission and the last goal was established at a reduction of 27% of the GHG emissions by 2020, starting from a 23% overall reduction already achieved. A suitable calculation procedure, also based on techniques available in literature, has been implemented in order to calculate indirect emissions. In particular, authors analysed both the location-based method, the market-based method and the efficiency method. Then, authors also presented a detailed overview of the EU energy and climate policies and urban energy and climate government to support sustainable energy and climate actions plans. In this framework, the authors analysed the key measures to be implemented to achieve the above-mentioned goals. They found that those goals could be achieved by a variety of technologies, namely: photovoltaic, solar thermal, wind energy, hydroelectric power, bioenergy, geothermal energy, combined heat and power systems, district heating and cooling, smart grids and waste water management.

Hammad et al. [107] focused on smart cities, presenting an optimization of zoning, land-use and facility location. Once again, the authors start from the analysis of energy consumption in the world, pointing out that the majority of this consumption is due to the urban regions where almost 55% of the overall world population lives. In this framework, this paper focuses on location planning in smart cities. The authors considered three aspects: (i) the guidelines for the construction of a smart city from scratch, considering zoning and land use; (ii) the location of buildings in smart cities; and (iii) the determination of the effects of the selection of such locations. The authors considered appropriate social, environmental and economic cost objective functions. As a consequence, the work provides suggestions regarding the allocation of zones and the assignment of buildings to locations in the region, the expansion decisions related to the road structure of the city and the expansion of the capacity of existing links in the network (if one already exists). In this paper, several mathematical optimisations are implemented in order to model key strategic decisions in smart city design and planning.

The authors first analysed in detail the papers available in literature analysing this topic, pointing out the novelty of their approach with respect to the main techniques previously presented. In fact, the authors aim at providing useful guidelines for the planning and design of city zoning, including the selection of building location and transport networks in smart cities. The model implemented in this work was used for a case study related to the design of the urban structure of a smart city. A lexicographic approach showed that the variations in cost can be up to 52% when the objective function is based on carbon emissions. Considering, the e-constraint method a trade-off cost of up to 471% is obtained when the considered objective functions are simultaneously optimised. In order to examine the computational performance of the proposed approach, a total of 350 instances were solved. The authors also tested computational times, showing that the proposed model was able to solve about 72% of the proposed instances within the 1000 s. The authors also compare GA and PSO algorithms, showing that GA was the fastest, even if its accuracy was about 68% lower with respect to the exact approach.

In the framework of sustainability, Hoehn et al. [108] focused on food loss management strategies. The authors point out that the food supply chain is extremely inefficient and is responsible for a significant amount of pollution. This circumstance is due to the industrial procedures, to the chemical products used in agriculture, to the excessive use of packaging and to the huge transportation costs. All these factors cause significant energy and environmental costs. These costs must be reduced using a different objective functions, also including the environmental one, in order to improve the efficiency and reduce food losses. The approach is based a food waste-to-energy-to-food approach. The authors proposed an empirical index to quantify food losses nutritional energy. They considered different scenarios of utilisation of these food losses: (i) biogas production by landfill; (ii) incineration recovering energy, (iii) anaerobic digestion. The authors found that the production of 1 kJ of nutritional energy requires about 8.7 kJ of primary energy. Such consumption is mainly due to the distribution and agricultural production stages. The authors analysed 11 categories, showing that fish and seafood, vegetables, meat and pulses presented the lowest values of the developed index. In addition, the embodied energy losses are mainly due to the consumption stage, which accounts for more than 66% of the total. The consumption stage is responsible of the highest food energy. The highest primary energy demand is due to the distribution stage, although this stage produces the lowest food energy losses. The authors concluded that the efficiency of the food supply is highly affected by the food category under study. In addition, they also found that anaerobic digestion is the best option for biogas production. Finally, this process maybe further beneficial for the generation of different additional by-products, such as hydrogen or methane recovery.

3.2. Energy Efficiency in Industry

Sakamoto et al. [109] analysed the water reuse in a cooling tower. The water consists of an oil refinery effluent. This study aims at facing the problem of water scarcity by obtaining water using an effluent stream. Several research articles investigate water reuse in different applications, using a variety of different approaches. The novelty of this paper lies in the application of LCA for the evaluation of available technologies for water reuse in a closed-loop process. Thus, the paper calculates the environmental performance of different systems to be used as effluent treatment plant from an oil refinery located in Brazil. The analysis is performed using a 'cradle-to-gate' approach. In particular, the methodology implemented in this paper consists of the following procedures: definition of the quality of the effluent entering the wastewater treatment plants of the refinery; selection of the process which will use the treated water; selection of water recovery strategies; analysis of the operational conditions and technological approach; calculation of emissions and resource consumption; design of mathematical models; application of the LCA technique for each scenario; and critical analysis of the results.

As mentioned before, different arrangements are analysed, including reverse osmosis, evaporation and crystallization. The lowest impacts indices are achieved by the scenario using waste heat to drive the evaporation process. Nevertheless, because the operation of other refinery sectors affects the operation of this arrangement, the alternative is not recommended. Therefore, authors concluded that the scenario including coprecipitation as a pre-treatment technique for reverse osmosis-fed effluent and steam recompression driving the evaporator shows the lowest impact scenario. Finally, the authors noted a certain variability among the observed scenarios. However, the results clearly showed that desalination is an efficient alternative in the reduction of effluent discharge and water consumption. Such results can be further enhanced by using less environmentally aggressive for salts removal in the pre-treatment plant.

Gambini et al. [110] presented a case study for a cogeneration system for a paper industry in Italy. Cogeneration is a very mature and efficient technology which allows one to simultaneously produce thermal, cooling and electrical energy. This technology has been significantly supported by EU governments over the past years since cogeneration allows one to significantly reduce the consumption of fossil fuels thereby simultaneously reducing CO_2 emissions. In particular, the financial support significantly depends on the overall efficiency of the system, in turn depending by the amount of available heat consumed by the user. A case study is developed for the pulp and paper industry, specifically referring to an Italian industry. A suitable simulation model was developed in GateCycle software. The authors analysed different cogeneration systems to be used in the paper mill and pulp industry. The calculations were performed using real data regarding time-dependent energy consumptions. The analysis is performed from thermodynamic, economic, and environmental points of view, comparing different cogeneration technologies, namely: steam power plants with condensing turbine, steam power plants with backpressure turbine, gas turbines, combined cycle power plants and internal combustion engines. High-efficiency cogeneration, defined according to the guidelines issued by the European Commission, has been considered in the economic calculations. Results show that a gas turbine is the best technology for this sector.

Another possible method to save energy is the utilisation of waste heat for different purposes. This possibility was investigated by several authors in this SI. In particular, Xu et al. [111] presented the optimization of waste heat recovery in a sinter vertical tank by a numerical simulation. They focus on the generation of the waste heat in the production process of the steel industry. The authors analyse two different aspects—heat transfer quantity and heat quality—and implement a multi-objective optimization, based on a genetic algorithm. The optimization procedure includes a back-propagation (BP) neural network. This is a feed-forward neural network trained using the error back-propagation algorithm. This method uses the gradient descent method in order to evaluate the minimum of the square of the network error. This technique is applied considering as objective functions exergy destructions caused by heat transfer and heat flows. An appropriate CFD numerical model is implemented to perform the calculations. This model consists in continuity equations, momentum equations, k-equations and ε-equations. In addition, energy equations are also implemented for both gas and solid phases. Once the thermodynamic conditions are obtained, a suitable exergy model allows one to calculate exergy flows and exergy destructions for the simulated systems. The developed model shows good performance in the simulation of the heat recovery of the sinter. The authors also found that the higher the sinter particle diameter, the lower the outlet air temperatures and the higher the outlet sinter temperatures. The exergy analysis shows that with an increase of the air mass flow rate, the exergy destruction due to the fluid flow and the heat transfer both gradually increase. In addition, the higher the sinter flow rate, the lower the exergy destruction due to the fluid flow and heat transfer.

Another paper dealing with heat recovery is presented by Kilkis [112]. Kilkis focused on the possible recovery of waste heat by the flue gases of power plants. This recovery must be performed using forced-draught fans in order to avoid affecting the performance of the power plant. Additional parasitic loads are due to the circulating pumps. Thus, the useful effect is the thermal recovery and the "fuel" is represented by the parasitic electrical loads. This problem was analysed using an exergy analysis. The paper performs a comparative analysis of four technologies for electricity generation: thermoelectric generators; organic-rankine cycle with or without a heat pump. These technologies are further compared with the direct use of the thermal exergy. A new optimization method based on exergy analysis is developed in order to calculate the optimum control strategies. A case study was performed for one of the presented methods, showing that the proposed method provides a number of additional results which cannot be achieved by conventional techniques.

Another numerical study was performed by Wang et al. [113]. They focused on packed beds, which are diffusely used in industrial applications in order to enhance heat transfer. This problem is very difficult from the numerical point of view since the flow structure inside the packed bed is very complex, due to the different distribution paths. This circumstance dramatically depends on the tube-to-particle ratio. Several studies are available in the literature addressing this problem using a variety of different methodologies. Even if several papers focused on this topic, only the analysis of multi-size particle mixing and its distribution are still under debate. Therefore, authors implemented a CFD-DEM method to perform a full numerical simulation is developed for packed beds with low tube-to-particle ratios. The work aims at restraining wall effects, reducing the porosity near the wall, and strengthening the heat transfer in the core area. Entransy dissipation is used as objective function. The authors found that the radial distribution of the particle size dramatically affects the bed velocity and temperature distribution. They also concluded that the heat transfer performance can be enhanced by filling small particles in the near wall region. The increase of distribution thickness can improve the heat transfer reducing the equivalent thermal resistance.

Haflzan et al. [114] analysed a heat exchanger network from both energy and economic points of views. This topic has been widely investigated in literature using many different approaches, including pinch analysis, exergy analysis, thermo-economic optimizations. This paper discusses a technique to manage temperature disturbances in the design of heat exchanger networks, maximizing heat recovery. The paper investigates the impact of the supply temperature variations on heat exchanger sizing, utility consumption and bypass placement. The authors also aim at reducing the consequent fluctuations on downstream heat exchangers. The authors implemented the plus-minus principle for process changes and new heuristics methods for the heat exchanger sizing and bypass placement. The novel methodology proposed by the authors is divided in several steps: stream data extraction with disturbances, maximization of heat recovery at the rated conditions using pinch analysis, construction of the grid diagram, management of temperatures disturbances. This methodology was applied for two case studies and the results show that the configuration of the heat exchanger network is maintained for all the cases. Furthermore, the exchanger areas are designed at the maximum capacity, including a suitable bypass to manage some critical operating conditions. The proposed methodology improved the annualised utility cost by up to 89%.

Kuczyński et al. [115] focused on energy recovery in natural gas regulation stations by using suitable turboexpanders. In fact, natural gas regulation stations feature a significant energy dissipation, due to dissipative pressure reduction. The idea is to use a suitable expander to reduce the pressure, by simultaneously producing electricity. It is worth noting that a certain preheating is required before the expansion in order to limit the possible hydrate formation inside the pipes. Several system layouts were developed and investigated by a suitable model based on energy balances, suitable models for the calculation of fluid properties and an experimental correlation for the performance of the expander. The authors concluded that this technology is extremely profitable, but it suffers for a remarkable seasonal operation. In addition, the expander efficiency significantly decreases at part load conditions which occur for the majority of the operation of the system. The authors also developed some simplified formulas to be used as guidelines for energy and economic feasibility analyses of those systems.

Several ancillary topics were also investigated in this area. Ligus et al. [116] presented an experimental analysis aiming at optimizing the two-phase fluid flow in an airlift pump. These devices are used for the vertical transport of liquids using a gas. The paper focuses on the hydrodynamic effects of this device and the experiments are based on optimal image technique, using image grey-level analysis to identify two-phase flow regimes. The authors aim at evaluating the void fraction and pressure drops, and present a new technique for evaluating the optimum operating regime which can be used to detect stability and efficiency of liquid transport and to evaluate the correlation between the required gas flux and the total lifting efficiency.

Sapinska-Sliwa et al. [117] focused on borehole heat exchangers. They developed a new method based on thermal response tests based on resistivity equations. The study was also supported by experimental analyses to support the robustness of the developed methodology. Martinelli et al. [118] investigated a wave energy converter. This device is used for energy harvesting simultaneously protecting the coast erosion. Lab-scale tests were performed in order to evaluate the efficiency of the developed device. Under short regular waves, such experiments showed a 35% efficiency. On the other hand, the devices show a small effect in terms of coastal protection.

3.3. Biomass and Other Miscellaneous Energy Systems

The topic of the different possible utilisation of biomass for the reduction of primary energy consumption is also diffusely investigated in this SI. Pawlak-Kruczek et al. [119] analyse gasification and torrefaction processes. In particular, they focus on sewage sludge that is a residue of wastewater processing. This stream must be stabilized since it includes a number of pathogens and organic matter and it can be used for energy purposes. In some cases, in order to reduce the costs related to the management of the sewage sludge and to achieve the above-mentioned goals, it is dried by a suitable thermal treatment. An example of these treatments is torrefaction which is performed at high temperature, typically in anaerobic conditions. Oxygen is used only when flue gases are directly used as heat source. The torgas obtained by this process mainly includes hydrocarbons, water and carbon monoxide. As the authors point out in this paper, several works are available in literature presenting different analyses of torrefaction systems. A similar process is gasification which consists in the conversion a solid fuel in a mixture of gases. This process produces a syngas rich in hydrocarbons. However, a large amount of contaminants are produced by the gasification process. In fact, the typical value of tar ranges from 1 to 10 g/m^3. Dozens of papers are available in literature presenting different investigations of gasification processes, including thermodynamic, economic and environmental analyses. The literature review performed by the authors shows that there is a lack of knowledge in terms of the effects of the sewage torrefaction on the successive gasification process. The present study is performed using an experimental approach. In particular, a suitable experimental setup was manufactured in order to analyse the above-mentioned effect. This equipment includes a multistage tape dryer/torrefier, air blowers, pressure regulators, a combustion chamber in ceramic refractory, an allotermal gasifier. The tests showed that torrefaction is a viable solution to decrease the tar content of syngas. However, authors also concluded that a further research must be performed in autothermal gasifiers and that further tests should be performed in a pilot-scale gasifier to confirm the initial results obtained in the present work. Finally, the process must be further optimized in order to enhance its efficiency.

Gliński et al. [120] studied the utilisation of biomass in micro combined heat and power (mCHP) units for power peak shaving. This device will be used in order to mitigate the fluctuations of the electrical system, due to the massive use of unpredictable renewable energy sources. The authors focus on mCHP units fed by biomass-derived fuels such as methanol or externally heated devices (ORC or Stirling) powered by biomass fuels. The paper aims at developing a methodology to design a mCHP control strategy in order to reduce its power consumption. The proposed control strategy was compared with the conventional scenario, the heat and power production of a biomass-fired mCHP unit operating in an exemplary house. The calculation is performed using the data of a boiler powered by wood pellets coupled with a Stirling engine, manufactured by ÖkoFEN. Once again, the authors performed a detailed literature review aiming at showing the novelty of their approach with respect to the methodologies presented in previous papers. A case study was analysed for a 250 m^2 house located near Warsaw showing an energy performance coefficient equal 110 kWh/m^2. For this case study, electrical, heating and domestic hot water demands were experimentally evaluated. The simulations of the system were performed using the energyPRO 4.0 software. Results show that using the new algorithm, the mCHP supplies 71% of the total demand for energy during the morning peak and evening peak. The new control strategy allows one to minimize electricity exchange with the grid. This effect is higher in the mid seasons when mCHP unit was able to provide up to 100% of the total morning peak energy demand and up to 83% of the total evening peak. In the summer, mCHP power production is lower than the demand to the low simultaneous demand of heat.

Hossain et al. [121] investigated the utilisation of nanoadditives on the energy and combustion performance of neat jatropha biodiesel. One of the main drivers to be implemented in order to achieve the decarbonization goals may consist in the utilisation of biomass-based fuels in the transportation sector. In this framework, biodiesel is a very attractive possibility. It can be used in a variety of blends in internal combustion engines. Biodiesel can be either produced by edible or non-edible crops. As pointed out by the authors of the paper, several studies are available in the literature investigating many aspects related to the production, conversion and utilisation of biodiesel. The performance of the biodiesel may be significantly enhanced when suitable nanoparticles are added to the fuel. In particular, in the present work two nanoparticles are considered: cerium oxide and aluminium oxides. The authors produced jatropha biodiesel in the lab in two stages, esterification and transesterification. Nanoadditives-J100 fuel blends are tested in a multi-cylinder engine. The results of the experimental campaign showed that Ce = 2 nanoparticles failed to fully amalgamate with the jatropah biodiesel. In order to achieve such amalgamation, suitable surfactants must be used. The authors also found that the proposed fuel exhibited better performance in terms of CO, UHC and NO_x emissions. In addition, such fuel also shows a better combustion performance by lower smoke opacity values. Finally, the full load and part load thermal performance was comparable with the one achieved for other types of fuel.

Restrepo-Valencia et al. [122] analysed the utilisation of bio-energy with carbon capture. They also included suitable storage systems and they focused on a sugarcane mill in Brazil. This kind of technology is considered by the authors one of the key factors to achieve the recently established goals, in terms of control of global average temperature. In fact, the combined use of biomass-based fuels and the simultaneous utilisation of carbon capture technologies will significantly contribute to achieve the above-mentioned goals. The carbon capture technology includes several subprocesses: conditioning, which separates CO_2 into a pure stream, carbon capture, compression and storage. The authors applied carbon capture and storage technology to a sugarcane mill, presenting both technical and economic analyses. In particular, they focused on a representative Brazilian sugarcane mill, equipped with a cogeneration unit. A non-commercial software is adopted to model the system integration into sugarcane mills. The system was modelled in Suitable models are included for all the components of the system. The analysis showed that both from fermentation and combustion, CO_2 capture is technically feasible. However, in case of combustion, the energy efficiency is limited by the management of the surplus electricity. The authors concluded that other carbon capture technologies must be evaluated in order to limit the amount of electricity sold to the grid. Conversely, carbon capture from the fermentation seems very promising due to the low impact on the mill, the benefits on the ethanol carbon footprint and the moderate capital cost. The authors also show that biomass-based cogeneration units suffer for a lower electrical efficiency when compared to gas-fired engines. Finally, for the produced electricity, the authors suggest to first consider carbon capture technologies and then selling electricity.

Finally, in this research area, another paper is presented by Tanczuk et al. [123]. They analysed the use of chicken manure in the energy industry. They focused on the case of Poland, which is one of the main producer of chicken manure in EU area. The authors aimed at evaluating the technical feasibility of using this material for energy purposes and at evaluating the related energy potential. The goal is to reduce the costs of managing poultry manure as a waste by converting it into energy. The analysis is performed by a simplified approach on the basis of the data available for the case of Poland, in terms of production, composition and other properties of this material. The authors evaluated the heating value of the chicken manure at high moisture as come from the farms. The results of their analysis show that the yearly theoretical energy potential in Poland was about 40.38 PJ. Yearly technical potential of chicken biomass depends on the selected conversion technology and it varies from 27.3 PJ to 9.01 PJ. The highest energy degradation can be obtained when heat and electricity is produced via anaerobic digestion. Finally, the authors also concluded that fluidized bed combustion shows the highest energy efficiency.

In this special session a considerable number of papers also analysed an interesting topic, hydrogen, usually scarcely investigated within SDEWES SIs. Moser et al. [124] analysed the possibility to produce hydrogen, using solar energy and using thermo-chemical cycles. In particular, they focused on concentrated solar power system producing heat to be supplied to a two-step thermochemical cycle. Two different redox materials are used in this cycle: cerium dioxide and nickel ferrite. Once again, this study is developed in the framework of the recent policies aiming at mitigating the effects of climate change and at promoting a full decarbonized energy system. To this end, the production of hydrogen using solar energy is extremely promising since it is a carbon-free fuel, produced by renewable energy sources. Moser et al. present a detailed literature review related to the production of hydrogen, using solar energy, paying special attention to the solar thermolysis, capable to convert water in hydrogen and oxygen. The process consists of two steps. The first one is an endothermic thermal reduction step, the second one is exothermic water splitting. In this paper, they present a simplified model for the comparative analysis of different two-step thermochemical cycles. The model considers both technical and economic aspects. The components are modelled using a simplified 0-D approach, based on energy and mass balances, also including suitable kinetic models for nickel-ferrite and ceria. A case study is presented for a 90 MW solar hydrogen production plant. The authors aimed at minimizing the dumping of solar power. For both cases, the optimum reactors number per module has been calculated. The comparison of the two analysed scenarios showed that nickel-ferrite achieves lower efficiency than ceria (6.4% instead 13.4%). From an economic point of view, the feasibility of the system is limited by the high capital cost of the solar subsystem.

Kuczyński et al. [125] evaluated the possibility to use existing transmission pipelines to transport hydrogen and/or methane/hydrogen-rich gases. In fact, as mentioned in the previous study, hydrogen is a very attractive and promising fuel due to its scarce environmental impact. However, hydrogen infrastructure is still missing and extremely expensive. Therefore, several researchers are investigating the possibility to adopt existing pipelines of natural gas to transport mixtures of hydrogen and methane or hydrogen. The properties of hydrogen, however, are extremely different from those of methane. Therefore, hydrogen pipelines are much more complicated than natural gas pipelines. The authors developed an apposite hydraulic model to evaluate the practicability of transporting hydrogen in natural gas pipelines. The model calculates pressure drops, temperature changes and heat transfer also using a real gas model. The authors assumed an outlet pressure equal to 24 bar (g). The authors also found that the change in temperature depends on the pipeline length; the supposed temperatures were 25 °C and 5 °C. Another important issue is the hydrogen impact on natural gas transmission. The calculations showed that the maximum hydrogen share in natural gas should be lower than 15–20%, in order to preserve natural gas quality.

4. Comments

At the 13th Sustainable Development of Energy, Water, and Environment Systems Conference held in 2018 in Palermo, Italy, interesting topics concerning sustainable development and energy efficiency were discussed and presented. This SI presents some of this work within the following main topics: energy efficiency and energy policy in urban areas, energy efficiency in industry and biomass and hydrogen. The editors of this SI of *Energies*, dedicated to 13th SDEWES Conference, believe that the above reported papers are focused on topics considered pivotal for the *Energies* journal. Such studies can be considered as fundamental and potentially practical tools to promote and disseminate the energy sustainability in different sectors: industries, metropolitan and urban areas, waste heat recovery, waste materials, water-energy nexus, etc.

Information on the future SDEWES Conferences and related activities are available on the website of the International Centre for Sustainable Development of Energy, Water, and Environment Systems (SDEWES Centre).

Author Contributions: F.C. and M.V. prepared an initial draft of the manuscript, which was completed, corrected and reviewed by M.C., N.D., Q.W. and P.A.Ø.

Funding: This research received no external funding.

Acknowledgments: The guest editors thank the authors of the papers that submitted their manuscripts to this special issue for the high quality of their work. We also thank all the reviewers who provided valuable and highly appreciated comments and advice, and the managing editors of *Energies* for their patience and excellent support.

Conflicts of Interest: The authors declare no conflict of interest.

References

1. Berardi, U.; Tronchin, L.; Manfren, M.; Nastasi, B. On the Effects of Variation of Thermal Conductivity in Buildings in the Italian Construction Sector. *Energies* **2018**, *11*, 872. [CrossRef]
2. Urbaniec, K.; Mikulčić, H.; Wang, Y.; Duić, N. System integration is a necessity for sustainable development. *J. Clean. Prod.* **2018**, *195*, 122–132. [CrossRef]
3. Olawumi, T.O.; Chan, D.W.M. A scientometric review of global research on sustainability and sustainable development. *J. Clean. Prod.* **2018**, *183*, 231–250. [CrossRef]
4. Halati, A.; He, Y. Intersection of economic and environmental goals of sustainable development initiatives. *J. Clean. Prod.* **2018**, *189*, 813–829. [CrossRef]
5. Kono, J.; Ostermeyer, Y.; Wallbaum, H. Investigation of regional conditions and sustainability indicators for sustainable product development of building materials. *J. Clean. Prod.* **2018**, *196*, 1356–1364. [CrossRef]
6. Scordato, L.; Klitkou, A.; Tartiu, V.E.; Coenen, L. Policy mixes for the sustainability transition of the pulp and paper industry in Sweden. *J. Clean. Prod.* **2018**, *183*, 1216–1227. [CrossRef]
7. Stoycheva, S.; Marchese, D.; Paul, C.; Padoan, S.; Juhmani, A.-S.; Linkov, I. Multi-criteria decision analysis framework for sustainable manufacturing in automotive industry. *J. Clean. Prod.* **2018**, *187*, 257–272. [CrossRef]
8. Xia, B.; Olanipekun, A.; Chen, Q.; Xie, L.; Liu, Y. Conceptualising the state of the art of corporate social responsibility (CSR) in the construction industry and its nexus to sustainable development. *J. Clean. Prod.* **2018**, *195*, 340–353. [CrossRef]
9. Matsumoto, K.I.; Doumpos, M.; Andriosopoulos, K. Historical energy security performance in EU countries. *Renew. Sustain. Energy Rev.* **2018**, *82*, 1737–1748. [CrossRef]
10. Chung, W.-S.; Kim, S.-S.; Moon, K.-H.; Lim, C.-Y.; Yun, S.-W. A conceptual framework for energy security evaluation of power sources in South Korea. *Energy* **2017**, *137*, 1066–1074. [CrossRef]
11. Jääskeläinen, J.; Veijalainen, N.; Syri, S.; Marttunen, M.; Zakeri, B. Energy security impacts of a severe drought on the future Finnish energy system. *J. Environ. Manag.* **2018**, *217*, 542–554. [CrossRef] [PubMed]
12. Radovanović, M.; Filipović, S.; Golušin, V. Geo-economic approach to energy security measurement—Principal component analysis. *Renew. Sustain. Energy Rev.* **2018**, *82*, 1691–1700. [CrossRef]
13. Proskuryakova, L. Updating energy security and environmental policy: Energy security theories revisited. *J. Environ. Manag.* **2018**, *223*, 203–214. [CrossRef] [PubMed]

14. Dorotić, H.; Doračić, B.; Dobravec, V.; Pukšec, T.; Krajačić, G.; Duić, N. Integration of transport and energy sectors in island communities with 100% intermittent renewable energy sources. *Renew. Sustain. Energy Rev.* **2019**, *99*, 109–124. [CrossRef]

15. Dominković, D.F.; Bačeković, I.; Pedersen, A.S.; Krajačić, G. The future of transportation in sustainable energy systems: Opportunities and barriers in a clean energy transition. *Renew. Sustain. Energy Rev.* **2018**, *82*, 1823–1838. [CrossRef]

16. Harris, A.; Soban, D.; Smyth, B.M.; Best, R. Assessing life cycle impacts and the risk and uncertainty of alternative bus technologies. *Renew. Sustain. Energy Rev.* **2018**, *97*, 569–579. [CrossRef]

17. Kılkış, Ş. Benchmarking the sustainability of urban energy, water and environment systems and envisioning a cross-sectoral scenario for the future. *Renew. Sustain. Energy Rev.* **2019**, *103*, 529–545. [CrossRef]

18. Kılkış, Ş. Sustainable Development of Energy, Water and Environment Systems (SDEWES) Index for policy learning in cities. *Int. J. Innov. Sustain. Dev.* **2018**, *12*, 87–134. [CrossRef]

19. Kılkış, Ş. Application of the Sustainable Development of Energy, Water and Environment Systems Index to World Cities with a Normative Scenario for Rio de Janeiro. *J. Sustain. Dev. Energy Water Environ. Syst.* **2018**, *6*, 559–608. [CrossRef]

20. Carli, R.; Dotoli, M.; Pellegrino, R. Multi-criteria decision-making for sustainable metropolitan cities assessment. *J. Environ. Manag.* **2018**, *226*, 46–61. [CrossRef] [PubMed]

21. Terrapon-Pfaff, J.; Ortiz, W.; Dienst, C.; Gröne, M.-C. Energising the WEF nexus to enhance sustainable development at local level. *J. Environ. Manag.* **2018**, *223*, 409–416. [CrossRef] [PubMed]

22. Gamarra, A.R.; Istrate, I.R.; Herrera, I.; Lago, C.; Lizana, J.; Lechón, Y. Energy and water consumption and carbon footprint of school buildings in hot climate conditions. Results from life cycle assessment. *J. Clean. Prod.* **2018**, *195*, 1326–1337. [CrossRef]

23. Gianniou, P.; Liu, X.; Heller, A.; Nielsen, P.S.; Rode, C. Clustering-based analysis for residential district heating data. *Energy Convers. Manag.* **2018**, *165*, 840–850. [CrossRef]

24. De Luca, G.; Fabozzi, S.; Massarotti, N.; Vanoli, L. A renewable energy system for a nearly zero greenhouse city: Case study of a small city in southern Italy. *Energy* **2018**, *143*, 347–362. [CrossRef]

25. Sözer, H.; Kükrer, E. Evaluation of Sustainable Design Strategies Based on Defined Indexes at a District Level. *J. Sustain. Dev. Energy Water Environ. Syst.* **2018**, *6*, 609–630. [CrossRef]

26. del Hoyo Arce, I.; Herrero López, S.; López Perez, S.; Rämä, M.; Klobut, K.; Febres, J.A. Models for fast modelling of district heating and cooling networks. *Renew. Sustain. Energy Rev.* **2018**, *82*, 1863–1873. [CrossRef]

27. Beccali, M.; Ciulla, G.; Di Pietra, B.; Galatioto, A.; Leone, G.; Piacentino, A. Assessing the feasibility of cogeneration retrofit and district heating/cooling networks in small Italian islands. *Energy* **2017**, *141*, 2572–2586. [CrossRef]

28. Santibañez-Aguilar, J.E.; Flores-Tlacuahuac, A.; Rivera-Toledo, M.; Ponce-Ortega, J.M. Dynamic optimization for the planning of a waste management system involving multiple cities. *J. Clean. Prod.* **2017**, *165*, 190–203. [CrossRef]

29. Katsaros, G.; Nguyen, T.-V.; Rokni, M. Tri-generation system based on municipal waste gasification, fuel cell and an absorption chiller. *J. Sustain. Dev. Energy Water Environ. Syst.* **2018**, *6*, 13–32. [CrossRef]

30. Li, W.; Bao, Z.; Huang, G.H.; Xie, Y.L. An Inexact Credibility Chance-Constrained Integer Programming for Greenhouse Gas Mitigation Management in Regional Electric Power System under Uncertainty. *J. Environ. Inform.* **2018**, *31*. [CrossRef]

31. Guo, Y.; Wang, Q.; Zhang, D.; Yu, D.; Yu, J. A Stochastic-Process-Based Method for Assessing Frequency Regulation Ability of Power Systems with Wind Power Fluctuations. *J. Environ. Inform.* **2018**, *32*. [CrossRef]

32. März, S. Assessing the fuel poverty vulnerability of urban neighbourhoods using a spatial multi-criteria decision analysis for the German city of Oberhausen. *Renew. Sustain. Energy Rev.* **2018**, *82*, 1701–1711. [CrossRef]

33. Okushima, S. Gauging energy poverty: A multidimensional approach. *Energy* **2017**, *137*, 1159–1166. [CrossRef]

34. Radovanović, M.; Filipović, S.; Pavlović, D. Energy security measurement—A sustainable approach. *Renew. Sustain. Energy Rev.* **2017**, *68*, 1020–1032. [CrossRef]

35. Valero-Gil, J.; Allué-Poc, A.; Ortego, A.; Tomasi, F.; Scarpellini, S. What are the preferences in the development process of a sustainable urban mobility plan? New methodology for experts involvement. *Int. J. Innov. Sustain. Dev.* **2018**, *12*, 135–155.

36. Vialetto, G.; Noro, M.; Rokni, M. Thermodynamic Investigation of a Shared Cogeneration System with Electrical Cars for Northern Europe Climate. *J. Sustain. Dev. Energy Water Environ. Syst.* **2017**, *5*, 590–607. [CrossRef]

37. Del Moretto, D.; Colla, V.; Branca, T.A. Sustainable mobility for campsites: The case of Macchia Lucchese. *Renew. Sustain. Energy Rev.* **2017**, *68*, 1063–1075. [CrossRef]

38. Novosel, T.; Perković, L.; Ban, M.; Keko, H.; Pukšec, T.; Krajačić, G.; Duić, N. Agent based modelling and energy planning—Utilization of MATSim for transport energy demand modelling. *Energy* **2015**, *92*, 466–475. [CrossRef]

39. Firak, M.; Đukić, A. Hydrogen transportation fuel in Croatia: Road map strategy. *Int. J. Hydrog. Energy* **2016**, *41*, 13820–13830. [CrossRef]

40. Briggs, I.; Murtagh, M.; Kee, R.; McCulloug, G.; Douglas, R. Sustainable non-automotive vehicles: The simulation challenges. *Renew. Sustain. Energy Rev.* **2017**, *68*, 840–851. [CrossRef]

41. Knez, M.; Obrecht, M. Policies for promotion of electric vehicles and factors influencing consumers' purchasing decisions of low emission vehicles. *J. Sustain. Dev. Energy Water Environ. Syst.* **2017**, *5*, 151–162. [CrossRef]

42. Kılkış, Ş. A Nearly Net-Zero Exergy District as a Model for Smarter Energy Systems in the Context of Urban Metabolism. *J. Sustain. Dev. Energy Water Environ. Syst.* **2017**, *5*, 101–126. [CrossRef]

43. Sayegh, M.A.; Danielewicz, J.; Nannou, T.; Miniewicz, M.; Jadwiszczak, P.; Piekarska, K.; Jouhara, H. Trends of European research and development in district heating technologies. *Renew. Sustain. Energy Rev.* **2017**, *68*, 1183–1192. [CrossRef]

44. Wagner, O.; Berlo, K. Remunicipalisation and Foundation of Municipal Utilities in the German Energy Sector: Details about Newly Established Enterprises. *J. Sustain. Dev. Energy Water Environ. Syst.* **2017**, *5*, 396–407. [CrossRef]

45. Carotenuto, A.; Figaj, R.D.; Vanoli, L. A novel solar-geothermal district heating, cooling and domestic hot water system: Dynamic simulation and energy-economic analysis. *Energy* **2017**, *141*, 2652–2669. [CrossRef]

46. Pavičević, M.; Novosel, T.; Pukšec, T.; Duić, N. Hourly optimization and sizing of district heating systems considering building refurbishment—Case study for the city of Zagreb. *Energy* **2017**, *137*, 1264–1276. [CrossRef]

47. Tańczuk, M.; Skorek, J.; Bargiel, P. Energy and economic optimization of the repowering of coal-fired municipal district heating source by a gas turbine. *Energy Convers. Manag.* **2017**, *149*, 885–895. [CrossRef]

48. Feil, A.; Pretz, T.; Jansen, M.; Thoden van Velzen, E.U. Separate collection of plastic waste, better than technical sorting from municipal solid waste? *Waste Manag. Res.* **2016**, *35*, 172–180. [CrossRef]

49. Tomić, T.; Schneider, D.R. Municipal solid waste system analysis through energy consumption and return approach. *J. Environ. Manag.* **2017**, *203*, 973–987. [CrossRef]

50. Brandoni, C.; Bošnjaković, B. HOMER analysis of the water and renewable energy nexus for water-stressed urban areas in Sub-Saharan Africa. *J. Clean. Prod.* **2017**, *155*, 105–118. [CrossRef]

51. Di Fraia, S.; Massarotti, N.; Vanoli, L.; Costa, M. Thermo-economic analysis of a novel cogeneration system for sewage sludge treatment. *Energy* **2016**, *115*, 1560–1571. [CrossRef]

52. Di Fraia, S.; Massarotti, N.; Vanoli, L. A novel energy assessment of urban wastewater treatment plants. *Energy Convers. Manag.* **2018**, *163*, 304–313. [CrossRef]

53. Nasir, D.S.N.M.; Hughes, B.R.; Calautit, J.K. Influence of urban form on the performance of road pavement solar collector system: Symmetrical and asymmetrical heights. *Energy Convers. Manag.* **2017**, *149*, 904–917. [CrossRef]

54. Batas-Bjelic, I.; Rajakovic, N.; Duic, N. Smart municipal energy grid within electricity market. *Energy* **2017**, *137*, 1277–1285. [CrossRef]

55. Ben Amer-Allam, S.; Münster, M.; Petrović, S. Scenarios for sustainable heat supply and heat savings in municipalities—The case of Helsingør, Denmark. *Energy* **2017**, *137*, 1252–1263. [CrossRef]

56. Koval, S.; Krahenbuhl, G.; Warren, K.; O'Brien, G. Optical microscopy as a new approach for characterising dust particulates in urban environment. *J. Environ. Manag.* **2018**, *223*, 196–202. [CrossRef] [PubMed]

57. Sutherland, S.H.; Smith, B.T. Resilience Implications of Energy Storage in Urban Water Systems. *J. Sustain. Dev. Energy Water Environ. Syst.* **2018**, *6*, 674–693. [CrossRef]

58. Kucbel, M.; Corsaro, A.; Svédová, B.; Raclavská, H.; Raclavský, K.; Juchelková, D. Temporal and seasonal variations of black carbon in a highly polluted European city: Apportionment of potential sources and the effect of meteorological conditions. *J. Environ. Manag.* **2017**, *203*, 1178–1189. [CrossRef]

59. van der Meulen, S.H. Costs and Benefits of Green Roof Types for Cities and Building Owners. *J. Sustain. Dev. Energy Water Environ. Syst.* **2019**, *7*, 57–71. [CrossRef]

60. Liew, P.Y.; Theo, W.L.; Wan Alwi, S.R.; Lim, J.S.; Abdul Manan, Z.; Klemeš, J.J.; Varbanov, P.S. Total Site Heat Integration planning and design for industrial, urban and renewable systems. *Renew. Sustain. Energy Rev.* **2017**, *68*, 964–985. [CrossRef]

61. Knoop, K.; Lechtenböhmer, S. The potential for energy efficiency in the EU Member States—A comparison of studies. *Renew. Sustain. Energy Rev.* **2017**, *68*, 1097–1105. [CrossRef]

62. Moser, S. Overestimation of savings in energy efficiency obligation schemes. *Energy* **2017**, *121*, 599–605. [CrossRef]

63. Pleßmann, G.; Blechinger, P. Outlook on South-East European power system until 2050: Least-cost decarbonization pathway meeting EU mitigation targets. *Energy* **2017**, *137*, 1041–1053. [CrossRef]

64. Norman, J.B. Measuring improvements in industrial energy efficiency: A decomposition analysis applied to the UK. *Energy* **2017**, *137*, 1144–1151. [CrossRef]

65. Varga, Z.; Csaba, T. Techno-economic evaluation of waste heat recovery by organic Rankine cycle using pure light hydrocarbons and their mixtures as working fluid in a crude oil refinery. *Energy Convers. Manag.* **2018**, *174*, 793–801. [CrossRef]

66. Hybrid Building Performance Simulation Models for Industrial Energy Efficiency Applications. *J. Sustain. Dev. Energy Water Environ. Syst.* **2018**, *6*, 381–393. [CrossRef]

67. Gambini, M.; Vellini, M. On Selection and Optimal Design of Cogeneration Units in the Industrial Sector. *J. Sustain. Dev. Energy Water Environ. Syst.* **2019**, *7*, 168–192. [CrossRef]

68. Wolf, M.; Detzlhofer, T.; Proll, T. A Comparative Study of Industrial Heat Supply Based On Second-Law Analysis And Operating Costs. *Therm. Sci.* **2018**, *22*, 2203–2213. [CrossRef]

69. Keppert, M.; Doušová, B.; Reiterman, P.; Koloušek, D.; Záleská, M.; Černý, R. Application of heavy metals sorbent as reactive component in cementitious composites. *J. Clean. Prod.* **2018**, *199*, 565–573. [CrossRef]

70. Baksa, P.; Cepak, F.; Kovačič Lukman, R.; Ducman, V. An Evaluation of Marine Sediments in Terms of their usability in the Brick Industry: Case Study Port of Koper. *J. Sustain. Dev. Energy Water Environ. Syst.* **2018**, *6*, 78–88. [CrossRef]

71. Chang, F.-C.; Chen, K.-S.; Yang, P.-Y.; Ko, C.-H. Environmental benefit of utilizing bamboo material based on life cycle assessment. *J. Clean. Prod.* **2018**, *204*, 60–69. [CrossRef]

72. Dusek, S.; Hofmann, R. A Hybrid Energy Storage Concept for Future Application in Industrial Processes. *Therm. Sci.* **2018**, *22*, 2235–2242. [CrossRef]

73. Royo, P.; Ferreira, V.J.; López-Sabirón, A.M.; García-Armingol, T.; Ferreira, G. Retrofitting strategies for improving the energy and environmental efficiency in industrial furnaces: A case study in the aluminium sector. *Renew. Sustain. Energy Rev.* **2018**, *82*, 1813–1822. [CrossRef]

74. Wiese, F.; Baldini, M. Conceptual model of the industry sector in an energy system model: A case study for Denmark. *J. Clean. Prod.* **2018**, *203*, 427–443. [CrossRef]

75. Gourlis, G.; Kovacic, I. Passive measures for preventing summer overheating in industrial buildings under consideration of varying manufacturing process loads. *Energy* **2017**, *137*, 1175–1185. [CrossRef]

76. Ebin, B.; Petranikova, M.; Steenari, B.-M.; Ekberg, C. Recovery of industrial valuable metals from household battery waste. *Waste Manag. Res.* **2019**, *37*, 168–175. [CrossRef] [PubMed]

77. Calise, F.; Dentice d'Accadia, M.; Libertini, L.; Quiriti, E.; Vanoli, R.; Vicidomini, M. Optimal operating strategies of combined cooling, heating and power systems: A case study for an engine manufacturing facility. *Energy Convers. Manag.* **2017**, *149*, 1066–1084. [CrossRef]

78. Yildirim, N.; Genc, S. Energy and exergy analysis of a milk powder production system. *Energy Convers. Manag.* **2017**, *149*, 698–705. [CrossRef]

79. Chinese, D.; Santin, M.; Saro, O. Water-energy and GHG nexus assessment of alternative heat recovery options in industry: A case study on electric steelmaking in Europe. *Energy* **2017**, *141*, 2670–2687. [CrossRef]

80. Kun-Balog, A.; Sztankó, K.; Józsa, V. Pollutant emission of gaseous and liquid aqueous bioethanol combustion in swirl burners. *Energy Convers. Manag.* **2017**, *149*, 896–903. [CrossRef]

81. Ko, C.-H.; Yu, F.-C.; Chang, F.-C.; Yang, B.-Y.; Chen, W.-H.; Hwang, W.-S.; Tu, T.-C. Bioethanol production from recovered napier grass with heavy metals. *J. Environ. Manag.* **2017**, *203*, 1005–1010. [CrossRef] [PubMed]

82. Palander, T.; Haavikko, H.; Kärhä, K. Towards sustainable wood procurement in forest industry—The energy efficiency of larger and heavier vehicles in Finland. *Renew. Sustain. Energy Rev.* **2018**, *96*, 100–118. [CrossRef]

83. Milani, M.; Montorsi, L. Energy Recovery of the Biomass from Livestock Farms in Italy: The Case of Modena Province. *J. Sustain. Dev. Energy Water Environ. Syst.* **2018**, *6*, 464–480. [CrossRef]

84. La Villetta, M.; Costa, M.; Cirillo, D.; Massarotti, N.; Vanoli, L. Performance analysis of a biomass powered micro-cogeneration system based on gasification and syngas conversion in a reciprocating engine. *Energy Convers. Manag.* **2018**, *175*, 33–48. [CrossRef]

85. Chang, K.-H.; Lou, K.-R.; Ko, C.-H. Potential of bioenergy production from biomass wastes of rice paddies and forest sectors in Taiwan. *J. Clean. Prod.* **2019**, *206*, 460–476. [CrossRef]

86. Boeykens, S.P.; Saralegui, A.; Caracciolo, N.; Piol, M. Agroindustrial Waste for Lead and Chromium Biosorption. *J. Sustain. Dev. Energy Water Environ. Syst.* **2018**, *6*, 341–350. [CrossRef]

87. Graciano, J.E.A.; Chachuat, B.; Alves, R.M.B. Enviro-economic assessment of thermochemical polygeneration from microalgal biomass. *J. Clean. Prod.* **2018**, *203*, 1132–1142. [CrossRef]

88. Marques, A.d.L.; Pinto, F.P.; Araujo, O.Q.d.F.; Cammarota, M.C. Assessment of Methods to Pretreat Microalgal Biomass for Enhanced Biogas Production. *J. Sustain. Dev. Energy Water Environ. Syst.* **2018**, *6*, 394–404. [CrossRef]

89. Codignole Luz, F.; Cordiner, S.; Manni, A.; Mulone, V.; Rocco, V. Biomass fast pyrolysis in screw reactors: Prediction of spent coffee grounds bio-oil production through a monodimensional model. *Energy Convers. Manag.* **2018**, *168*, 98–106. [CrossRef]

90. Kantarli, I.C.; Stefanidis, S.D.; Kalogiannis, K.G.; Lappas, A.A. Utilisation of poultry industry wastes for liquid biofuel production via thermal and catalytic fast pyrolysis. *Waste Manag. Res.* **2019**, *37*, 157–167. [CrossRef] [PubMed]

91. Chan, K.-L.; Dong, C.; Wong, M.S.; Kim, L.-H.; Leu, S.-Y. Plant chemistry associated dynamic modelling to enhance urban vegetation carbon sequestration potential via bioenergy harvesting. *J. Clean. Prod.* **2018**, *197*, 1084–1094. [CrossRef]

92. de Souza, L.L.P.; Lora, E.E.S.; Palacio, J.C.E.; Rocha, M.H.; Renó, M.L.G.; Venturini, O.J. Comparative environmental life cycle assessment of conventional vehicles with different fuel options, plug-in hybrid and electric vehicles for a sustainable transportation system in Brazil. *J. Clean. Prod.* **2018**, *203*, 444–468. [CrossRef]

93. Szulczewski, W.; Żyromski, A.; Jakubowski, W.; Biniak-Pieróg, M. A new method for the estimation of biomass yield of giant miscanthus (Miscanthus giganteus) in the course of vegetation. *Renew. Sustain. Energy Rev.* **2018**, *82*, 1787–1795. [CrossRef]

94. Hussain, C.M.I.; Norton, B.; Duffy, A. Technological assessment of different solar-biomass systems for hybrid power generation in Europe. *Renew. Sustain. Energy Rev.* **2017**, *68*, 1115–1129. [CrossRef]

95. Davis, G.W. Addressing Concerns Related to the Use of Ethanol-Blended Fuels in Marine Vehicles. *J. Sustain. Dev. Energy Water Environ. Syst.* **2017**, *5*, 546–559. [CrossRef]

96. Kocík, J.; Samikannu, A.; Bourajoini, H.; Pham, T.N.; Mikkola, J.-P.; Hájek, M.; Čapek, L. Screening of active solid catalysts for esterification of tall oil fatty acids with methanol. *J. Clean. Prod.* **2017**, *155*, 34–38. [CrossRef]

97. Di Fraia, S.; Figaj, R.D.; Massarotti, N.; Vanoli, L. An integrated system for sewage sludge drying through solar energy and a combined heat and power unit fuelled by biogas. *Energy Convers. Manag.* **2018**, *171*, 587–603. [CrossRef]

98. Interlenghi, S.F.; de Almeida Bruno, P.; Araujo, O.d.Q.F.; de Medeiros, J.L. Social and environmental impacts of replacing transesterification agent in soybean biodiesel production: Multi-criteria and principal component analyses. *J. Clean. Prod.* **2017**, *168*, 149–162. [CrossRef]

99. Özdenkçi, K.; De Blasio, C.; Muddassar, H.R.; Melin, K.; Oinas, P.; Koskinen, J.; Sarwar, G.; Järvinen, M. A novel biorefinery integration concept for lignocellulosic biomass. *Energy Convers. Manag.* **2017**, *149*, 974–987. [CrossRef]

100. Lazaroiu, G.; Pană, C.; Mihaescu, L.; Cernat, A.; Negurescu, N.; Mocanu, R.; Negreanu, G. Solutions for energy recovery of animal waste from leather industry. *Energy Convers. Manag.* **2017**, *149*, 1085–1095. [CrossRef]
101. Marzo, C.; Diaz, A.B.; Blandino Caro, A. Valorization of agro-industrial wastes to produce hydrolytic enzymes by fungal solid-state fermentation. *Waste Manag. Res.* **2019**, *37*, 149–156. [CrossRef] [PubMed]
102. Weiler, V.; Stave, J.; Eicker, U. Renewable Energy Generation Scenarios Using 3D Urban Modeling Tools—Methodology for Heat Pump and Co-Generation Systems with Case Study Application. *Energies* **2019**, *12*, 403. [CrossRef]
103. Dominković, D.F.; Krajačić, G. District Cooling Versus Individual Cooling in Urban Energy Systems: The Impact of District Energy Share in Cities on the Optimal Storage Sizing. *Energies* **2019**, *12*, 407. [CrossRef]
104. Smajla, I.; Karasalihović Sedlar, D.; Drljača, B.; Jukić, L. Fuel Switch to LNG in Heavy Truck Traffic. *Energies* **2019**, *12*, 515. [CrossRef]
105. Testi, D.; Conti, P.; Schito, E.; Urbanucci, L.; D'Ettorre, F. Synthesis and Optimal Operation of Smart Microgrids Serving a Cluster of Buildings on a Campus with Centralized and Distributed Hybrid Renewable Energy Units. *Energies* **2019**, *12*, 745. [CrossRef]
106. Kona, A.; Bertoldi, P.; Kılkış, Ş. Covenant of Mayors: Local Energy Generation, Methodology, Policies and Good Practice Examples. *Energies* **2019**, *12*, 985. [CrossRef]
107. Hammad, A.W.; Akbarnezhad, A.; Haddad, A.; Vazquez, E.G. Sustainable Zoning, Land-Use Allocation and Facility Location Optimisation in Smart Cities. *Energies* **2019**, *12*, 1318. [CrossRef]
108. Hoehn, D.; Margallo, M.; Laso, J.; García-Herrero, I.; Bala, A.; Fullana-i-Palmer, P.; Irabien, A.; Aldaco, R. Energy Embedded in Food Loss Management and in the Production of Uneaten Food: Seeking a Sustainable Pathway. *Energies* **2019**, *12*, 767. [CrossRef]
109. Sakamoto, H.; Ronquim, F.M.; Seckler, M.M.; Kulay, L. Environmental Performance of Effluent Conditioning Systems for Reuse in Oil Refining Plants: A Case Study in Brazil. *Energies* **2019**, *12*, 326. [CrossRef]
110. Gambini, M.; Vellini, M.; Stilo, T.; Manno, M.; Bellocchi, S. High-Efficiency Cogeneration Systems: The Case of the Paper Industry in Italy. *Energies* **2019**, *12*, 335. [CrossRef]
111. Xu, C.; Liu, Z.; Wang, S.; Liu, W. Numerical Simulation and Optimization of Waste Heat Recovery in a Sinter Vertical Tank. *Energies* **2019**, *12*, 385. [CrossRef]
112. Kılkış, B. Development of an Exergy-Rational Method and Optimum Control Algorithm for the Best Utilization of the Flue Gas Heat in Coal-Fired Power Plant Stacks. *Energies* **2019**, *12*, 760. [CrossRef]
113. Wang, S.; Xu, C.; Liu, W.; Liu, Z. Numerical Study on Heat Transfer Performance in Packed Bed. *Energies* **2019**, *12*, 414. [CrossRef]
114. Hafizan, A.M.; Klemeš, J.J.; Wan Alwi, S.R.; Abdul Manan, Z.; Abd Hamid, M.K. Temperature Disturbance Management in a Heat Exchanger Network for Maximum Energy Recovery Considering Economic Analysis. *Energies* **2019**, *12*, 594. [CrossRef]
115. Kuczyński, S.; Łaciak, M.; Olijnyk, A.; Szurlej, A.; Włodek, T. Techno-Economic Assessment of Turboexpander Application at Natural Gas Regulation Stations. *Energies* **2019**, *12*, 755. [CrossRef]
116. Ligus, G.; Zając, D.; Masiukiewicz, M.; Anweiler, S. A New Method of Selecting the Airlift Pump Optimum Efficiency at Low Submergence Ratios with the Use of Image Analysis. *Energies* **2019**, *12*, 735. [CrossRef]
117. Sapińska-Sliwa, A.; Rosen, M.A.; Gonet, A.; Kowalczyk, J.; Sliwa, T. A New Method Based on Thermal Response Tests for Determining Effective Thermal Conductivity and Borehole Resistivity for Borehole Heat Exchangers. *Energies* **2019**, *12*, 1072. [CrossRef]
118. Martinelli, L.; Volpato, M.; Favaretto, C.; Ruol, P. Hydraulic Experiments on a Small-Scale Wave Energy Converter with an Unconventional Dummy Pto. *Energies* **2019**, *12*, 1218. [CrossRef]
119. Pawlak-Kruczek, H.; Wnukowski, M.; Niedzwiecki, L.; Czerep, M.; Kowal, M.; Krochmalny, K.; Zgóra, J.; Ostrycharczyk, M.; Baranowski, M.; Tic, W.J.; et al. Torrefaction as a Valorization Method Used Prior to the Gasification of Sewage Sludge. *Energies* **2019**, *12*, 175. [CrossRef]
120. Gliński, M.; Bojesen, C.; Rybiński, W.; Bykuć, S. Modelling of the Biomass mCHP Unit for Power Peak Shaving in the Local Electrical Grid. *Energies* **2019**, *12*, 458. [CrossRef]
121. Hossain, A.K.; Hussain, A. Impact of Nanoadditives on the Performance and Combustion Characteristics of Neat Jatropha Biodiesel. *Energies* **2019**, *12*, 921. [CrossRef]
122. Restrepo-Valencia, S.; Walter, A. Techno-Economic Assessment of Bio-Energy with Carbon Capture and Storage Systems in a Typical Sugarcane Mill in Brazil. *Energies* **2019**, *12*, 1129. [CrossRef]

123. Tańczuk, M.; Junga, R.; Kolasa-Więcek, A.; Niemiec, P. Assessment of the Energy Potential of Chicken Manure in Poland. *Energies* **2019**, *12*, 1244. [CrossRef]
124. Moser, M.; Pecchi, M.; Fend, T. Techno-Economic Assessment of Solar Hydrogen Production by Means of Thermo-Chemical Cycles. *Energies* **2019**, *12*, 352. [CrossRef]
125. Kuczyński, S.; Łaciak, M.; Olijnyk, A.; Szurlej, A.; Włodek, T. Thermodynamic and Technical Issues of Hydrogen and Methane-Hydrogen Mixtures Pipeline Transmission. *Energies* **2019**, *12*, 569. [CrossRef]

Article

Renewable Energy Generation Scenarios Using 3D Urban Modeling Tools—Methodology for Heat Pump and Co-Generation Systems with Case Study Application [†]

Verena Weiler *,[‡], Jonas Stave [‡] and Ursula Eicker

Centre for Sustainable Energy Technology, University of Applied Sciences Stuttgart, 70174 Stuttgart, Germany; jonas.stave@hft-stuttgart.de (J.S.); ursula.eicker@hft-stuttgart.de (U.E.)
*　Correspondence: verena.weiler@hft-stuttgart.de; Tel.: +49-711-8926-2950
†　This paper is an extended version of our paper published in SDEWES 2018, Palermo, Italy, 30 September–4 October 2018.
‡　These authors contributed equally to this work.

Received: 24 December 2018; Accepted: 24 January 2019; Published: 28 January 2019

Abstract: In the paper, a method was developed to automatically dimensionalize and calculate central energy generation and supply scenarios with a district heating system for cities based on 3D building models in the CityGML format and their simulated heat demand. In addition, the roof geometry of every individual building is used to model photovoltaic energy generation potential. Two types of supply systems, namely a central heat pump (HP) system and a large co-generation (combined heat and power-CHP) system (both with a central storage and district distribution system), are modeled to supply the heat demand of the area under investigation. Both energy generation models are applied to a case study town of 1610 buildings. For the HP scenario, it can be shown that the case study town's heat demand can be covered by a monovalent, low-temperature system with storage, but that the PV only contributes 15% to the HP electricity requirement. For the CHP scenario, only 61% of the heat demand can be covered by the CHP, as it was designed for a minimum of 4000 operating hours. Both the PV and the CHP excess electricity are fully injected into the grid. As a result, the primary energy comparison of both systems strongly depends on the chosen primary energy factors (PEF): with given German regulations the CHP system performs better than the HP system, as the grid-injected electricity has a PEF of 2.8. In the future, with increasingly lower PEFs for electricity, the situation reverses, and HPs perform better, especially if the CHP continues to use natural gas. Even when renewable gas from a power to gas (P2G) process is used for the CHP, the primary energy balance of the HP system is better, because of high conversion losses in the P2G process.

Keywords: urban simulation; photovoltaics; heat pump; co-generation; heat demand analysis; low-temperature district heating

---

1. Introduction

Cites are responsible for more than 60% of the energy demand and about 70% of the CO_2 emissions worldwide [1], more than half of the world's population lives in cities [2]. Therefore, buildings and their energy demand are one of the keys to an efficient reduction of emissions. To reach the ambitious goal of the European Union to reduce 80% of CO_2 emissions by 2050 compared to 1990 [3], a range of measures and scenarios need to be analyzed and evaluated to identify suitable strategies.

In a first step, the current heat demand of the city under investigation is needed. There are different approaches and methods to simulate this demand, as described by Reinhart and Davila [4], Frayssinet et al. [5] or Li et al. [6]. However, all the methods and studies reviewed there focus on the energy demand only, not on the energy generation. Allegrini et al. [7] describe several tools and methods that can simulate district-scale energy systems, some of them also calculate the energy demand. Most of the studies and methods described focus on one or several related energy generation components which are modeled in great detail. Allegrini et al. state that tools are needed which run on simplified yet validated models to achieve faster results to support decision makers in early design stages on an urban level. 3D models of the buildings and cities under investigation can help to deliver more detailed information as the base for demand and supply simulations [8–10].

The World Energy Outlook 2018 by the International Energy Agency [11] states that electrical heating as well as heating systems supplied by gas continue to play a significant role in the future. In general, district heating systems are considered an important part of the transition to renewable energy generation, as they allow to use different (renewable) energy sources as inputs and facilitate storage [12]. Two of the most promising renewable energy systems for urban areas at the moment are co-generation plants operated with renewable gas and HP operated with renewable electricity [13,14]. CHP are often used in district heating networks as they provide heat very efficiently and thus offer a significant contribution to the reduction of CO_2-emissions [15,16]. Large HP in district heating networks have advantages over individual installations in buildings, as the heat can be stored in large storages as well as in the network itself. By this, surplus renewable energy can be used whenever it is available. With the development and implementation of 4th generation low-temperature district heating networks, the integration of HPs becomes a viable alternative. For the efficient use of this kind of heating network, the buildings need to be refurbished and use a heating system that can operate on low supply temperatures. In Sweden, about 10% of the heat needed in district heating networks is supplied by large HP [17,18]. Also, in Denmark, the goal is to be less dependent on fossil fuels by integrating HP in district heating networks [19]. The city of Copenhagen wants to supply their district heating network in a CO_2 neutral way by 2025 [20].

Another important step for reducing emissions is to use the solar irradiance to produce electricity with photovoltaic (PV) systems. There are several studies that show methods on how to analyze the PV potential in urban areas [21,22]. Most of the studies investigating the PV potential are based on the 3D geometry of the buildings and their roofs [23–26] since knowing the exact roof shape and orientation helps to calculate more accurate values for the PV potential energy yield.

In this study, calculation models for district heating systems with either CHP units or large HP are developed. Subsequently, heat demand, PV potential, and both energy generation systems are assessed and compared for a small town in Southern Germany. For this modeling task, the urban modeling tool SimStadt [27] is used to analyze several hundred buildings in parallel.

2. Methodology

For the development of different energy system scenarios, the heat demand of the case study town needs to be calculated first. Additionally, the local PV potential needs to be determined to assess the possible local electricity production. Then, energy generation models need to be developed to dimension and compare systems to meet the heat demand calculated beforehand. In this section, first the simulation environment SimStadt is explained and how it calculates heat demand and PV potential. Then, the development of two models for energy generation systems is shown in detail.

2.1. Simulation Environment

The analyses in this paper are made with the simulation platform SimStadt, using an internal connection to the simulation engine INSEL 8.2 [28]. Both SimStadt and INSEL are under development at the University of Applied Sciences Stuttgart. Figure 1 shows the simulation environment graphically.

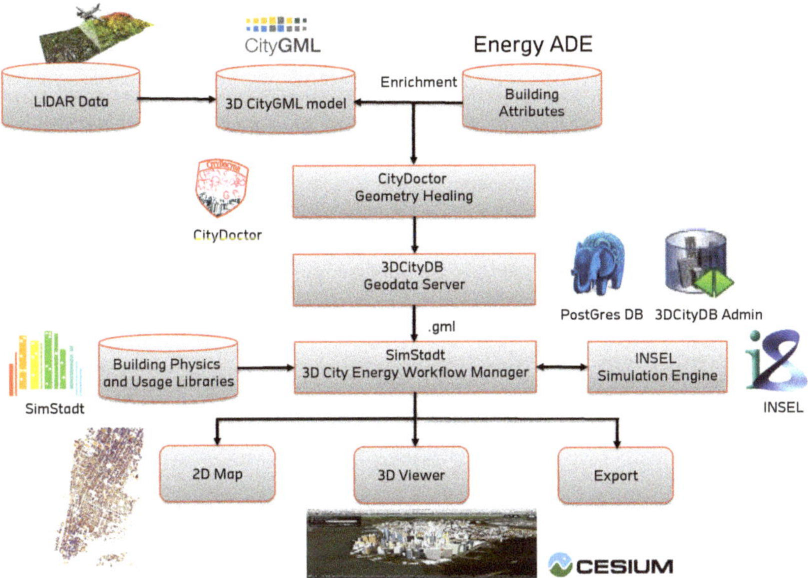

Figure 1. SimStadt simulation environment with data sources.

The upper part of Figure 1 describes the generation of 3D CityGML models from LIDAR data [29] and the possible enrichment of the models with building attributes e.g., from municipal sources or the EnergyADE [30]. The CityGML format can depict existing environments such as buildings, roads, and landscape. Building models can be available in five different level of details (LOD); LOD0 as only a planar shape, LOD1 where the building is represented as a cube with an average building height and a flat roof, LOD2 which has more detailed information about building heights and roof shapes, LOD3 introduces windows and LOD4 has a detailed interior design and information about wall thicknesses (see Figure 2).

Figure 2. Visualization of LOD0 to LOD4 in the CityGML format [31].

The CityGML model is then quality checked by the tool CityDoctor, which is under development at the University of Applied Sciences. This tool checks the 3D CityGML model and can repair possible geometrical errors, e.g., open polygons which prevent the buildings from being recognized properly are closed. The model can then be stored in the 3D CityDB geodata server or directly used for simulation in SimStadt.

SimStadt has a graphical user interface (GUI) and consists of different work flow steps that use information given in the CityGML model, building physics, and usage library and from weather databases. The geometry of every individual building from the CityGML file is used and linked to the different libraries in the SimStadt platform.

The building physics library is based on the typology developed by the German Institute IWU (Institut Wohnen und Umwelt) [32], which classifies buildings according to their type and year of construction. For each building type and period, there is a typical building with its respective wall, roof, and window properties. These properties are applied to the actual building geometry and SimStadt then calculates individual u-values for each building. Additional to the properties of existing buildings, different refurbishment scenarios are available in the building physics library.

The usage library is based on several German norms and standards, focusing on heating set point temperatures, occupancy schedules and internal gains that are different according to the usage (residential, office, retail, etc.) of each building.

INSEL is the simulation engine behind SimStadt and can be purchased from doppelintegral [28]. INSEL is a block diagram simulation system for programming applications for renewable energy technologies. It has some ready-made simulation models included, but it is also possible to design own model extensions or entirely new models.

The results from the SimStadt simulations can be visualized in 2D maps, on a 3D globe [33] or the building-specific results can be exported in a .csv-file.

SimStadt is still in a beta status and therefore not publicly available; however there are several studies and publications where the calculation method of SimStadt is described and the tool was used to calculate the heat demand or PV potential of different case studies [27,34–40].

2.2. Heat Demand Analysis

SimStadt calculates the heat demand of a town or city quarter as a monthly energy balance according to the German standard DIN 18599 [41]. It uses the information in the building physics and usage libraries to determine the heat transmissions through the building envelope for each building according to their building type and age.

For the dimensioning and calculation of the generation systems that are the focus of this paper, however, more time-resolved values are needed. Therefore, the monthly values are transferred into hourly values, depending on the hourly outside temperature and the desired room temperature of 20 °C. This process is described in the German standard VDI 4710 [42]. The heating period in Germany commonly is from October to April; in summer months, the heating system is shut down entirely, so the heat demand (excluding domestic hot water demand) is zero. Additionally, there is a nightly shutdown from 0 to 6 am, where the minimum temperature is 15 °C. Nightly shutdown is a common practice to reduce the heat load and is included in several German norms, e.g., in the DIN V 18599 which also describes the monthly energy balance calculation.

2.3. Photovoltaic Electricity Generation

The PV potential can also be simulated with SimStadt and the same CityGML file from the heat demand analysis is used as input data in combination with the irradiance data of the location. In the first simulation steps, SimStadt is using the 3D CityGML Model to determine the inclination and azimuth for every roof area and calculates the solar irradiance on those surfaces. Different radiation models can be applied, in this case the Hay algorithm is used [43]. It does not take shading of trees or other buildings into account. However, since the building density is quite low and the individual buildings are of similar height, this is an acceptable limitation and trade-off between accuracy and computation time.

The next step is the parametrization for the PV potential analysis. It consists of the building surface suitability and PV system parameters. Suitability settings include for example, the minimum surface area and the minimum annual irradiation on the roof, at which a roof is determined as suitable for a PV system installation (default: $\geq$20 m^2/950 kWh/m^{2}a). The PV system parameters include among others the ratio of the module area to the roof area (default: 30% for flat roofs to reduce shading, 40% for tilted roofs to account for roof windows, chimneys etc.); The module tilt was chosen as 20° for a south-facing installation on a flat roof in Germany, as this represents a good compromise of maximum irradiance, optimal row spacing (shading), self-cleaning and back-ventilation of the PV modules [44].

The default module efficiency and performance ratio are 17% and 80%, respectively. Based on those parameters the yearly energy yield for every suitable roof area is calculated.

With the current SimStadt version, only the annual and monthly cumulated PV energy yield can be simulated. Therefore, with the help of the simulation engine INSEL 8.2, an hourly PV load profile was generated in this study using the same weather data and PV system parameters as in SimStadt. Then, the roof areas of the area under investigation are divided into four roof types (tilted roof orientations (east/south/west) and flat roofs). The average inclination and azimuth of every roof type is calculated, as well as the share of the surface area of every roof type. The PV load profile in INSEL considers the same roof type shares with their average inclination and azimuth. The annual PV yield from SimStadt can then be split per hour via the load profile from INSEL.

3. Extension of the Simulation Environment and Addition of New Functionalities

The simulation environment described above has been developed over several years and different projects at the University of Applied Sciences Stuttgart and was adapted in some ways for the analyses in this study. The focus of this paper is the methodology for the addition of energy generation systems to the simulation environment. Add the end of this section, the current simulation sequence is explained in detail. First, the added functionalities themselves are described.

3.1. Heat Pump Generation

The whole HP system is developed in the INSEL simulation environment. A simplified representation of the model is shown in Figure 3.

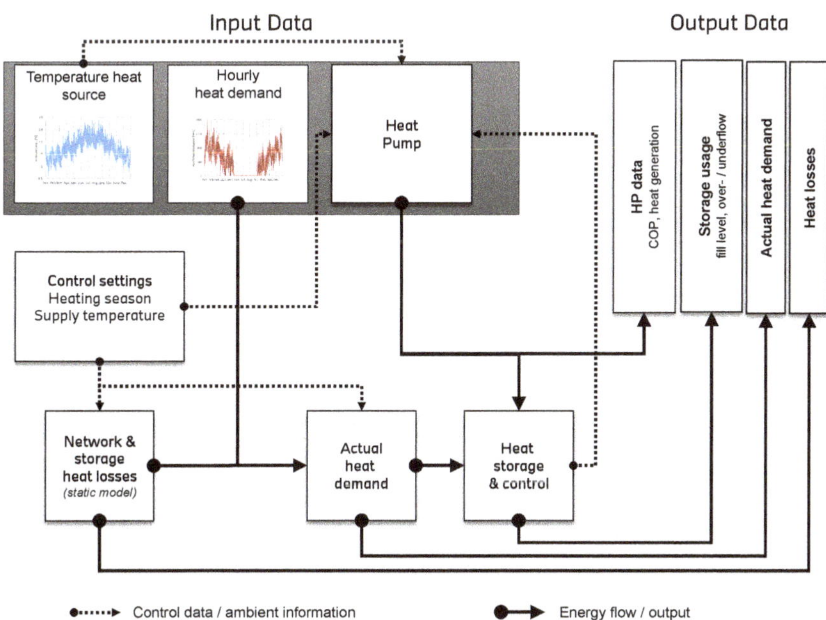

Figure 3. Structure of simplified INSEL HP simulation model.

Depending on the type of HP used, the air, water, or ground temperature of the heat source is needed as an input in an hourly resolution. Also needed is specific HP data such as coefficient of performance (COP) and power characteristics dependent on the heat source temperature and supply temperature of the heating network, which are usually provided by the manufacturer. The HP is modeled by a polynomial fit to this manufacturer data as a function of source and sink temperature. Finally, the monthly heat demand that is calculated in SimStadt and then transferred to hourly values is needed as an input. The months of the heating season can be adjusted in the model to control the operation time of the HP. Also, the needed supply temperature of the heating network can be varied depending on the system configuration.

Different key performance indicators (KPI) are analyzed as outputs of the model. The most important results are the information about the HP performance such as hourly or annual COP and heat generation. Data on the storage usage, hourly fill level, overflow or deficit of storage are also given as outputs. Additionally, values for heat losses (both for the distribution and storage losses) as well as the actual heat demand including those losses are provided for every time step.

3.2. District Heating Network and Central Storage

Additional to the HP, a large central storage is included in the system model to reduce the operating hours of the HP. The storage is sized to supply the average total heat load for three hours. This amount of time is chosen because of the utility blocking time, where the utility company is allowed to stop providing electricity to the HP for two hours in a row for a maximum of three times a day. The additional hour provides extra reserves in case of multiple blocks in a short period of time.

The heat demand is directly met from the storage. If the storage level is lower than 20% of its full capacity, the HP switches on to both meet the current demand and fill up the storage. As soon as the storage level reaches 80% of its full capacity, the HP switches off. With this control, it is ensured that the generation system does not switch on and off all the time, especially in spring and fall months when the demand can be met by the storage alone for a few hours in a row. Storage losses are included and calculated for a steel tank with 10 cm insulation.

Distribution heat losses are dependent on the length of the piping and the supply temperature and are therefore calculated in the model according to these two parameters. The losses are calculated with a simple steady-state model, using double piping with 5 cm insulation. The resulting losses are validated with other sources [45,46].

3.3. Co-Generation

The co-generation system is modeled very similarly to the HP simulation model shown in Figure 3. Most of the parameters remain unchanged (e.g., the storage size and losses, distribution losses, and supply temperature, heating season, and the hourly heat demand). In this model the co-generation unit is operated in a heat-controlled mode and the electricity is seen as the by-product. As input for the CHP unit, the technical information from the manufacturer such as efficiency and nominal power are needed.

For the dimensioning of the co-generation unit, the annual load duration curve of the bulidings under investigation is analyzed. The choice of CHP operating hours is based on current economic feed-in conditions in Germany, where co-generation units should have at least 4000 full load hours to be able to operate cost efficiently. This corresponds to a rated output of the CHP unit to cover 20 to 25% of the maximum heat load [47]. With this typical marked-based design, a coverage of about 50−75% of the annual heat demand of a district heating network is usually achieved [48]. The rest of the demand that cannot be met by the CHP is provided by gas boilers. This mode of operation is very efficient to meet the peak demand [49], since gas boilers can easily and often switch on and off without negative effects on their efficiency.

3.4. Current and Future Calculation Sequence

So far, the new functionalities are not yet added to the SimStadt platform itself and therefore, different tools need to be used to calculate each step. The calculation sequence needed in this study is shown in Figure 4.

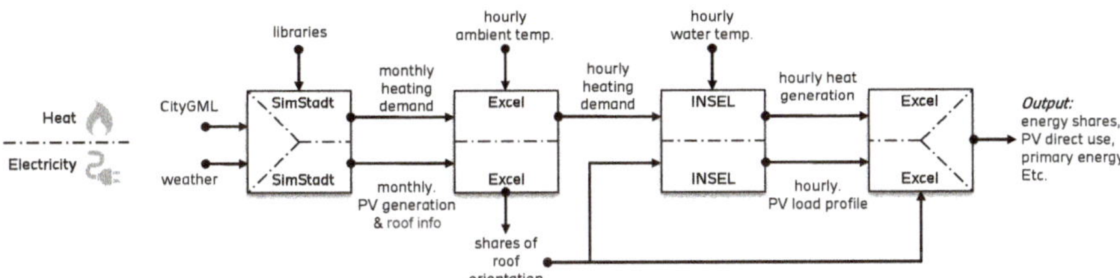

Figure 4. Simulation and calculation sequence (the upper part shows the sequence for calculating the heat demand, the lower part shows the electricity side including the PV potential).

In the first part, different libraries, a weather data set specific to the location and the CityGML file of the area under investigation are needed as inputs in SimStadt. There, the monthly heat demand and PV generation as well as the information about the roof orientation, is given as outputs. This sequence is described in Sections 2.2 and 2.3.

At present Microsoft Excel (or any other spreadsheet tool) is still needed as an interface between the SimStadt simulation results and the INSEL models simulating the hourly energy supply systems to transfer monthly values to hourly values, as described in Section 2.2.

The hourly PV energy yield must be determined from the monthly values and the hourly profile, which is calculated in INSEL with the shares of the roof orientation, as described in Section 2.3.

The results of the INSEL energy generation calculation then need to be exported to Microsoft Excel again for aggregating, analyzing, and visualizing the data output at the end. In the future, this whole process will be integrated as one work flow sequence into SimStadt.

4. Application to Case Study

The proposed methodology is applied to a small town in Southern Germany in the administrative district of Ludwigsburg, which belongs to the metropolitan area of Stuttgart. Since 2015, the district council of Ludwigsburg decided on a climate protection concept to determine the possibilities for reducing CO_2 emissions, to acquire subsidies and financial support based on a catalogue of measures, and decided on the overall goal to make the district climate-neutral by 2050 [50].

The town Walheim in the district of Ludwigsburg that is used as a case study in this analysis has around 3200 inhabitants. Next to Walheim flows the river Neckar, which is the second biggest river in the state of Baden-Württemberg. 1610 buildings are contained in the 3D CityGML model that is used for the simulations (see Figure 5). This leads to a moderate size of the CityGML file and consequently short computation times.

All these factors are very good prerequisites that make Walheim an ideal case study for the application of this methodology.

Most of the buildings are in LOD2, but some of the buildings are represented only in LOD1 (blue in Figure 2), so their actual roof shape is not known. 46% of the buildings are garages or sheds that are not heated. The remaining 872 buildings with a total footprint area of 109,919 m^2 are mainly residential buildings (89%), the rest are either used for industry, offices or retail.

For this study, all buildings are assumed to be refurbished according to the current German Energy Saving Ordinance EnEV 2016 [51]. This assumption is needed to be able to simulate low-temperature district heating for all buildings in the town. Of course, this kind of large-scale refurbishment cannot be expected to take place in just a few years but it fits together nicely with the goals of the district of Ludwigsburg of becoming climate-neutral by 2050. Additionally, there are special programs and incentives from the KfW (Kreditanstalt für Wiederaufbau-German Reconstruction Loan Corporation) to promote building and city quarter refurbishments.

Figure 5. 3D CityGML model of the case study town of Walheim. The CityGML model was created by the State Agency for Spatial Information and Rural Development Baden-Württemberg.

In a first step, the monthly heat demand of every building in the town of Walheim is calculated with SimStadt. Table 1 shows the cumulated heat demand for all buildings of Walheim for every month.

Table 1. Total monthly heat demand of Walheim in MWh.

January	February	March	April	May	June	July	August	September	October	November	December
3115	2280	1337	307	34	1	0	0	17	574	1923	3005

Then, the monthly values are transferred into hourly values as described in Section 2.2. The hourly demand of each building is cumulated for each time step and ordered in a sorted annual heat load curve according to the value of the heat demand of each time step (see Figure 6). The total annual heat demand of Walheim amounts to approximately 12,500 MWh/a with a peak load of 7767 kW in January. The considered heating season is October until April. Domestic hot water is not included in this assessment.

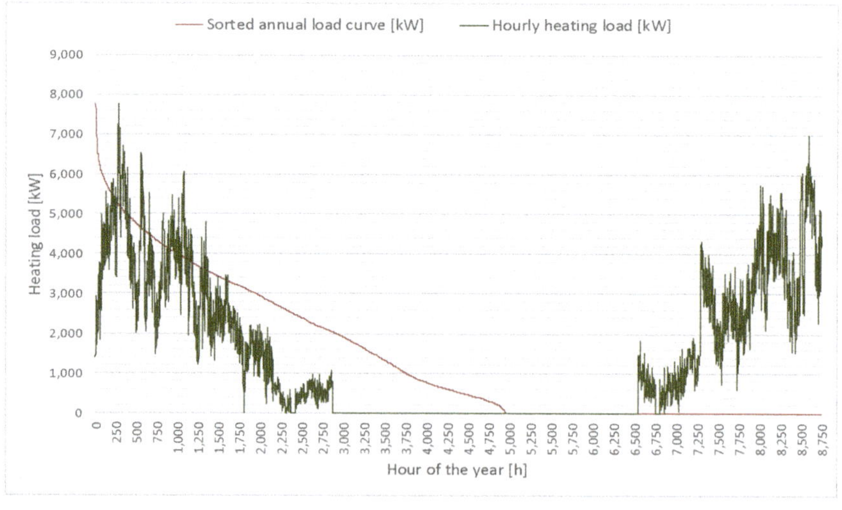

Figure 6. Sorted annual heat load curve (red) and hourly heat load (green) for Walheim (EnEV standard).

In this scenario, as many photovoltaic (PV) modules as possible are put on all suitable roofs of all buildings in Walheim. The German Erneuerbare-Energien Gesetz (EEG, renewable energies law) [52] says that PV systems smaller than 10 kW do not have to pay additional fees when they feed PV electricity to the grid. Therefore, most PV systems on residential buildings are smaller than that, even though there might be space left on the roofs to place additional PV modules. Renowned German scientist Volker Quaschning addresses this issue [53] and demands that all usable roof spaces should be used for PV electricity generation.

The PV potential simulation is carried out as it is described in Section 2.3. The buildings in the CityGML file that are only available in LOD1 are excluded from the PV potential analysis, since their roof shape is not known. Therefore, the orientation cannot be determined.

According to the SimStadt PV simulation this results in 1470 suitable roof areas for Walheim, with a total nominal power of 6417 kWp and an annual energy yield of 6043 MWh. Figure 7 shows the simulation results of a part of Walheim for

- (a) inclination of the roofs in [°]
- (b) annual sum of global irradiation in [kWh/m^{2}a]
- (c) the PV suitability of the roof according to the applied parameters and restrictions and
- (d) the total annual PV energy yield of each roof in [MWh/a].

Figure 8 shows the PV power output and monthly energy yield over the course of one year. The monthly PV energy yield of all suitable roofs in Walheim combined is calculated in SimStadt (blue) and the daily mean PV power output in Walheim is shown in yellow, which is calculated with the specific PV load profile in INSEL.

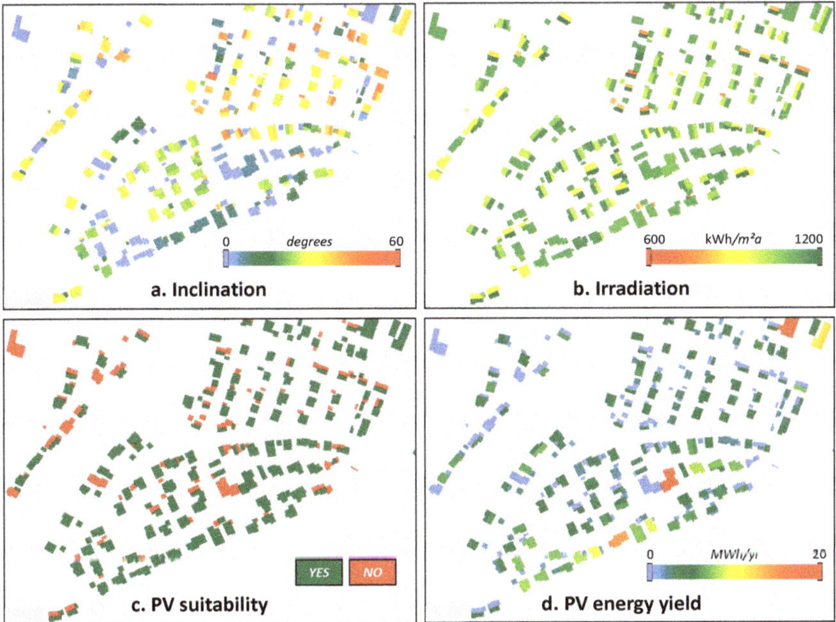

Figure 7. Examples of the simulation results for PV electricity generation in SimStadt for a part of Walheim.

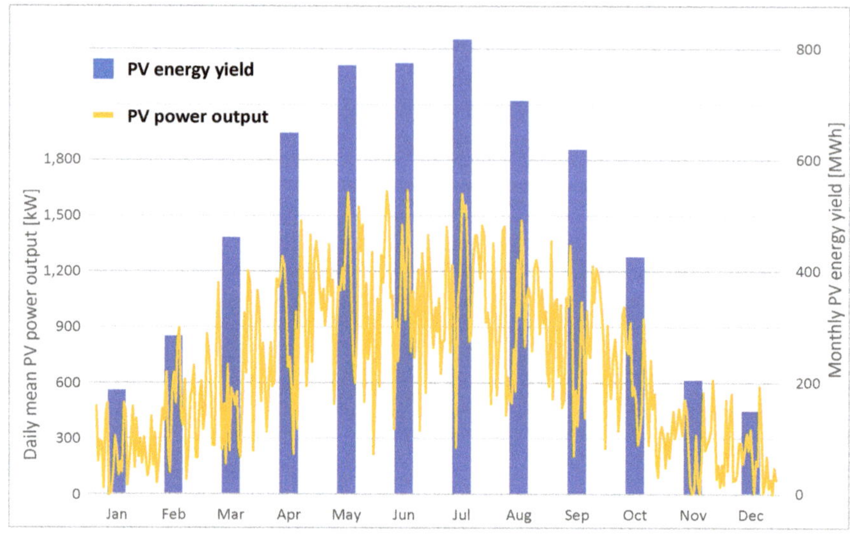

Figure 8. Monthly PV energy yield (blue = SimStadt simulation results) and daily mean PV power output (yellow = via INSEL PV load profile) for Walheim.

The sorted annual heat load from Figure 6 is used to choose the appropriate HP to match the demand of Walheim. Since the river Neckar flows directly next to the town, it serves as the heat source for the HP. The average daily water temperature between 1988 and 2014 at the measuring point in Besigheim, which is only a few hundred meters from the town of Walheim, varies between 4.21 °C in December and 21.71 °C in August (see Figure 9).

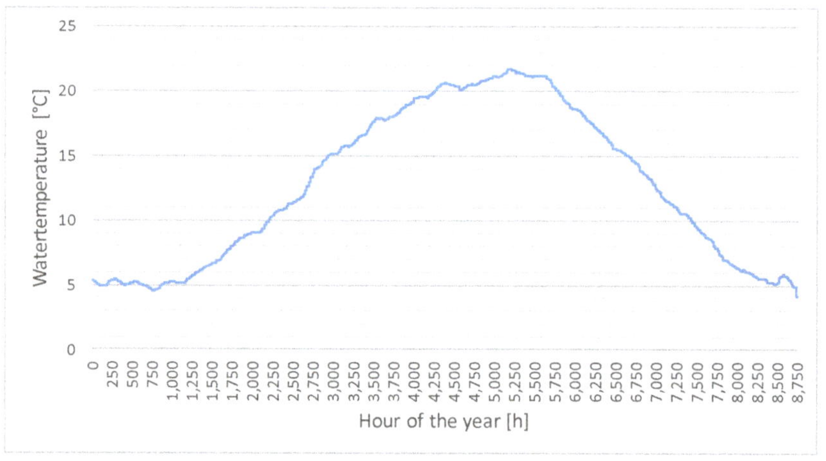

Figure 9. 26-year average water temperature of the river Neckar at Besigheim (data from [54]).

Since the HP are the only source of heating energy for the network and are therefore in a monovalent operation mode, several large industry heat pumps need to be used in this case to supply the whole town of Walheim. After extensive market research, the IWWS 960 ER1a from Ochsner Energie Technik was selected. It is a water-water industry HP and has a nominal heat output of 966 kW. This size is ideal, since eight of the aforementioned HP can meet the peak load of 7767 kW.

Distribution heat losses are dependent on the length of the piping and are therefore calculated according to the total length of the network, which is assumed to be 12 km. Since the supply level of the heating network is only 45 °C, the mean heat losses of 106 kWh/h for the whole network are rather low with 7.5% on average. The temperature of the heating network can be chosen to be this low, because all the buildings connected to the network have been refurbished to a low-energy standard as well as low-temperature heating systems before.

The storage is set to supply the average annual load of the town of 1431 kW for three hours which leads to a size of 15 m in height and a diameter of 4 m of the insulated steel tank. Assuming a discharge temperature difference of 20 K (supply temperature to the network 45 °C, return temperature 25 °C), this results in 4314 kWh storage capacity. Storage losses of 0.851 kWh/h are included in the model.

Based on the dimensioning guidelines described in Section 3.3, a co-generation unit within the range of 2000 kW nominal heat output is selected to supply the heat demand of the town of Walheim. The electricity generated by the CHP is assumed to be fed into the grid.

Since in this study both biogas (from fermentation of biomass or waste water sludge digestion) and gas from power to gas systems (P2G), where surplus electricity is used to produce gas from water with electrolysis are considered as fuel for the CHP, a unit of the manufacturer 2G was selected, which can be operated with both fuels with only a few technical changes (CHP Type: 2G avus 2000c). For natural gas and gas from P2G the nominal heat output of this CHP is 1977 kW with a thermal efficiency of 43.2% and the nominal electrical power output 2000 kW with an efficiency of 43.7%. In the case of biogas, the overall efficiency of the unit is 2.1% lower, which corresponds to 2009 kW/42.5% on the thermal side and 2000 kW/42.3% on the electrical side. The rest of the heat demand that cannot be met by the co-generation unit needs to be supplied by central gas boilers. To ensure the highest possible security of supply, two gas boilers with a nominal power of 4000 kW with an efficiency of 95% each are used. Thus, in the case of maintenance work on the CHP, the gas boilers can cover the entire heat load.

5. Results of Two Scenarios for Energy Generation Simulation

The overall results for the energy generation of both system simulations are compared in Figure 10. The eight HP with a maximum heating power output of 8354 kW can generate 94% of the needed heating energy in only 1890 h. The HP system can store a large amount of heat and use this heat at other times to meet the demand. The heat that can be stored corresponds to the area between the gray and yellow lines in Figure 10.

The CHP unit with a heating power output of constantly 2000 kW runs for 4012 h but only supplies 61% of the heat, the rest needs to be produced by the gas boilers. These shares correspond to 8024 MWh from the CHP unit and 5058 MWh or 39% from the gas boilers. The CHP system cannot store as much heat as the HP system because most of the produced heat is used directly. This is due to the lower nominal heat output of the CHP that produces less surplus heat compared to the eight HPs, even though the same storage size is used. This can be seen in Figure 10 as the area between the blue and yellow lines.

Table 2 shows the KPI for the two systems. The heat demand is the same for both system designs, the heat losses slightly differ because the CHP system can store less heat in the storage and has therefore lower storage losses.

Electricity yield from PV is the same in both scenarios, in the CHP scenario, additional electricity is produced when the CHP is running. All electricity that is not directly used by either the HP or the P2G process is assumed to be fed to the grid. The remaining electricity needed for operation of either system is assumed to come from the grid.

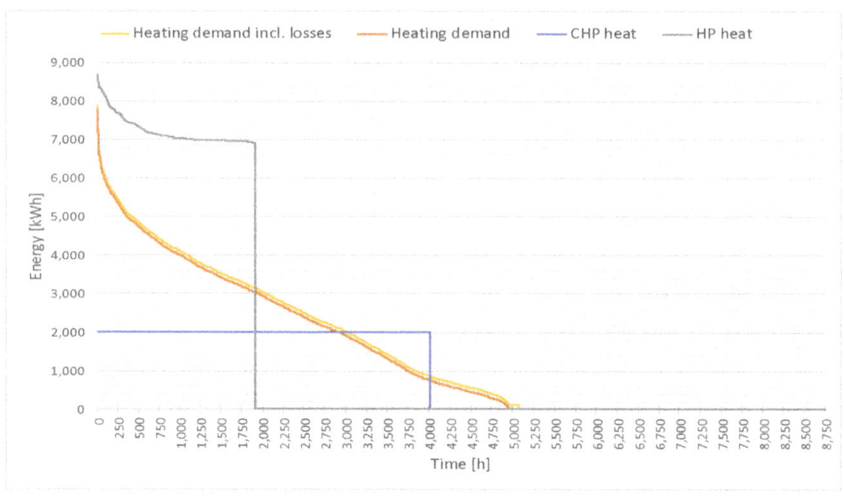

Figure 10. Annual overview of CHP and HP heat generation as well as heat demand.

The CHP unit can either be operated with biogas or with gas from power to gas systems (P2G) fed by an electrolysis process with an efficiency factor of 0.65 kWh/kWh [55]. The electricity needed for the P2G process is assumed to be 100% renewable and all the PV electricity yield in this scenario is used for the P2G process. The overall electricity demand for the simulated CHP system using gas from P2G consequently is about 11 times as high as for the HP system. This balance improves of course if biogas from fermentation is used.

Table 2. Key performance indicators (KPI) for both systems.

KPI	HP	CHP + Gas Boilers (P2G)	CHP + Gas Boilers (Biogas)
Total heat demand [MWh]	12,574	12,574	12,574
Total heat losses [MWh]	546	544	544
Full load hours	1890	4012	4012
Seasonal performance factor	4.1	n.a.	n.a.
Gas demand [MWh]	0	23,898	23,898
Electricity demand [MWh]	3319	36,766	0
Electricity generation [MWh]	6043	6043 (PV) + 8024 (CHP)	6043 (PV) + 8024 (CHP)
Electricity to grid [MWh]	5532	0	14,067
Electricity from grid [MWh]	2808	22,700	0
Electricity balance [MWh]	3724	−22,700	14,067

Figure 11 shows the monthly heat demand of Walheim (orange, in background) and how this demand is met by either the HP or CHP production and the storage (in front). On the left, the HP system plus storage is visualized in green, on the right in blue is the CHP and gas boilers plus storage. The CHP system always meets 100% of the demand, while the HP system has both under- and overproduction for most of the months. Overproduction for the HP occurs, when the storage tank is "full", but the HP is still running for the remainder of the time step of one hour. This amount of energy is usable in reality, but leads to higher storage losses due to the temperature increase. This effect is due to the modeling time step of one hour. The storage (gray) is used in both systems, however the CHP unit can only produce significant surplus heat for storage in March, April, and October, when the monthly heat demand is considerably below 2000 MWh. In the summer months (May–September), the heating network is switched off, therefore only the storage losses remain.

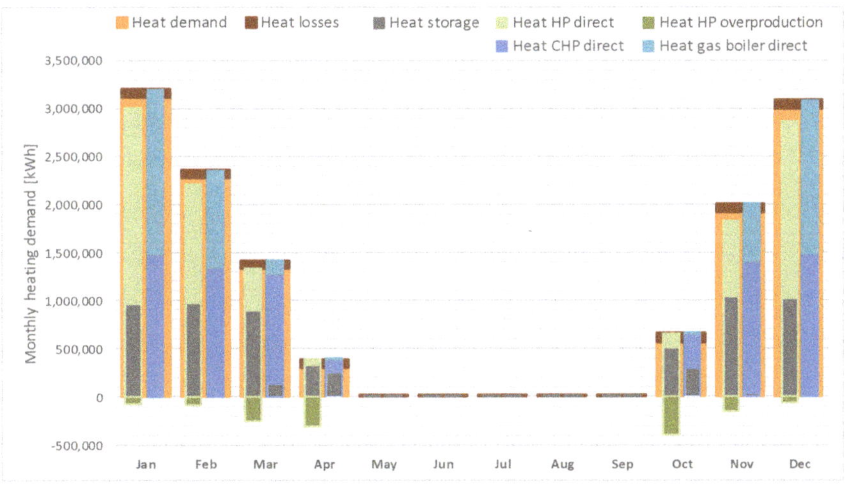

Figure 11. Coverage of heat demand by HP and CHP systems respectively.

Both systems (HP and CHP scenario) generate the same amount of electricity with PV and the CHP system has additional electricity generation on top of the PV yield. The HP uses the PV electricity whenever possible. 15% of the HP electricity demand can thus be met by PV directly. Because of non-simultaneity, the rest of the electricity needed for the operation of the HP needs to come from the grid and 92% of the PV generated electricity is fed to the grid. This excludes the household electricity demand, which is the same in both scenarios and not taken into account here. Own-consumption of the PV electricity for household electricity demand is not assessed in this study.

The monthly comparison between the PV energy yield, the PV energy use by the HP and the remaining electricity needed from the grid for the operation of the HP can be seen in Figure 12. It shows a mismatch of a high PV energy yield (orange) in the summer months and a high HP electricity demand (green plus blue) in the winter months. However, the annual electricity balance of HP demand and PV generation in the town of Walheim is positive (PV electricity yield is 82% higher than the HP electricity demand, see Table 2).

A period of 100 h in January is chosen to highlight the differences in heat generation and other KPIs for both systems (see Figure 13). In this period, the heat demand reaches its peak with a load of 7767 kW. In Figure 13 the modulation of the HP is clearly visible. The loads during hours 281 till 287 cannot be met by the HP directly. In most of those hours however, the demand can be met with heat from the storage, only for three hours the system does not produce enough heat. If the heat demand is lower than what the HP produces (e.g., in the first 10 h of the graph), the surplus heat fills the storage. This stored heat is then used to satisfy the heat demand when the HP turns off. However, when the HP is switched off and the demand is higher than the heat that is available in storage, there is a slight underperformance of the system. The deficit is only during one hour at a time and only once for two consecutive hours. The one-hour deficits are due to the model time step of one hour and do not have practical relevance, as the HP would switch on directly if the storage tank is below the threshold. This fact combined with the low-energy standard of the buildings and their surface heating system leads to the conclusion that this deficit is acceptable.

Figure 13b shows the same time period for the CHP system. The CHP unit runs constantly but can only provide part of the heat demand. This is due to the dimensioning of the CHP unit; the rest of the demand is provided by the gas boilers (not in the figure). The storage is empty since there is no surplus heat from the CHP available.

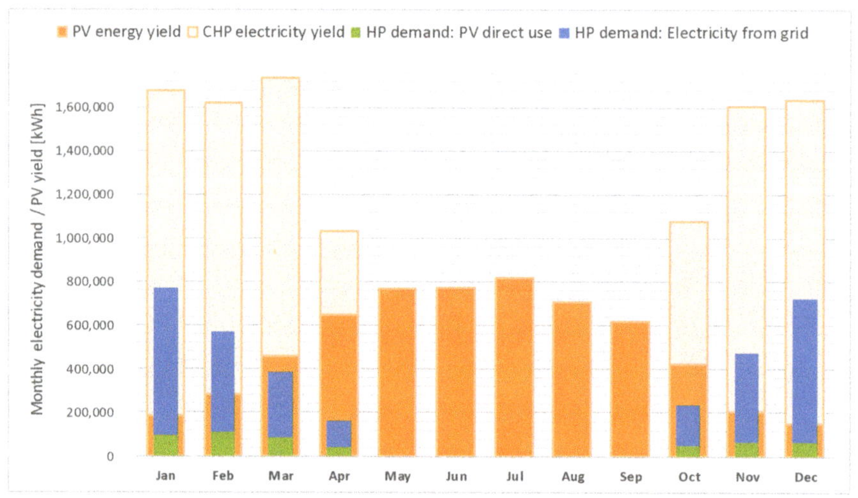

Figure 12. Monthly PV and CHP electricity yield and HP electricity demand.

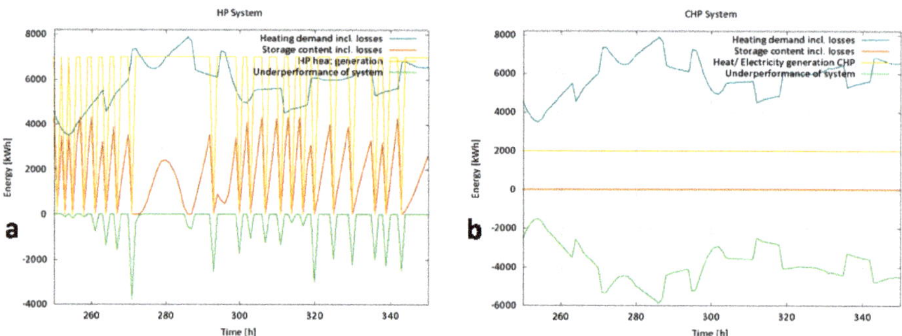

Figure 13. Analysis of the performance of both systems from the hours of the year 250 to 350, which include the peak heat demand ((**a**): HP system, (**b**): CHP system).

In Figure 14, 100 h in mid-October are shown for both systems. In this period, the heat demand is lower than in Figure 13, therefore the systems run differently. Figure 11a shows that the HP system runs only one hour at a time, before switching off for four or more consecutive hours when the heat demand is met from the storage. Since the power of the CHP unit is lower than of the HP, less heat can be loaded into the storage in Figure 14b compared to the HP system. Also, the CHP unit must run for more than one hour at a time and is switched off for only three to four consecutive hours.

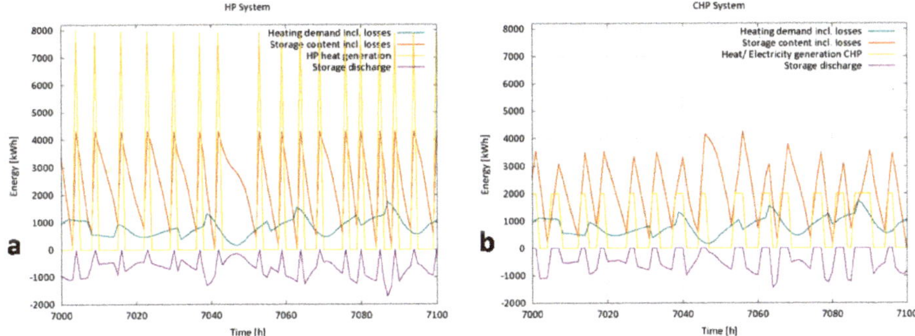

Figure 14. Analysis of the performance of both systems from the hours of the year 7000-7100 ((**a**): HP System, (**b**): CHP System).

6. Assessment of Primary Energy Demand

The primary energy demand is calculated by multiplying the final energy demand with a PEF that takes into account the effort and losses that occur during the whole value chain of the energy carrier.

To demonstrate the influence of the energy carrier used as well as the influence of the current and future grid mix, different scenarios are calculated in Figure 15. The electricity generated by the CHP unit can be assessed in different ways, depending e.g., on how the electricity in the grid is produced and what type of gas is used. If the CHP runs on natural gas and the electricity from the CHP and PV is fed into a grid with predominantly fossil and nuclear electricity generation, it replaces this kind of electricity with renewable or co-generated electricity. The current factor in Germany for this replacement is 2.8 [56], which means that the electricity generated by the CHP and PV that is fed into this grid is valued 2.8 times higher than its actual amount of electricity (Figure 15—status quo). Since the CHP units produce more electricity than the HP, the overall balance of the CHP is better. However, when the electricity mix of the grid changes in the future due to the integration of increasingly renewable electricity (Figure 15—business as usual), this factor will have to change as well. Then, the surplus electricity would be worth less and the HP scenario wins against the CHP. If biogas is used in the CHP, the PEF for gas changes to 0.5. In this scenario (Figure 15—biogas) the CHP has the better primary energy balance. According to a meta-study by the Ökoinstitut and Fraunhofer ISI [57], the PEF for electricity in 2050 in Germany will be 0.3. Assuming an overall efficiency of 0.65 for the P2G process, this results in a PEF of 0.5 for gas from P2G (Figure 15—renewable energy). In this case, the primary energy balance for the HP system is the better one, because the P2G process is subject to high energy losses and the electricity fed into the grid by the CHP is not able to replace any fossil fuels anymore.

In general, the use of PEF to compare different energy systems must be evaluated. With a growing share of renewable electricity that factor loses its validity and might have to be replaced by different indicators [58].

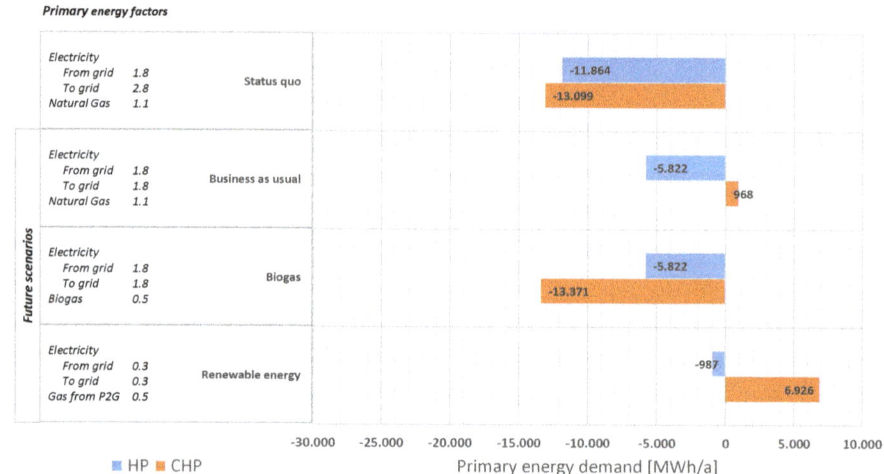

Figure 15. Comparison of different PEFs to assess the HP and CHP systems.

7. Conclusions and Outlook

In this study, two renewable energy generation models are developed and applied to a small German town. The SimStadt simulation environment was extended to include the developed models for a central HP system with a district heating network and central storage as well as a CHP system with a district heating network and central storage. The investigated indicators include the heat demand and heat production as well as the electricity generation and electricity demand of the respective systems.

A HP scenario with PV has the main disadvantage that due to low winter irradiance, the fraction of PV electricity used directly by the HP is only 15%. The main share of PV electricity is produced in summer; the main share of heat demand however occurs in winter.

For the CHP energy generation, it is important to include the source of the gas used. If we assume renewable energy for all energy sources (renewable electricity for the HP system and either biogas or gas from P2G, produced from renewable electricity for the CHP system), a look at the total electricity needed for operation of the systems is essential.

To achieve a good match between demand and generation, it is important that energy generation from other renewable sources, such as wind, is investigated and further models and scenarios are created for the simulation environment. Additionally, different storage management schemes or storage types need to be investigated to maximize the usage of PV electricity by the HP.

In a next step, an economic analysis of the systems in question with their respective energy in- and outputs will be made. With this additional assessment, more KPIs for each system can be assessed and the systems can be compared on a broader level. At present, both HP and CHP system cannot regulate their heating power, so they are either running on full load or are switched off. Models that support partial load operation are under development right now and could provide a more detailed assessment and comparison of the two systems.

In conclusion, urban energy simulations are a very important instrument for analyzing demands and potentials and comparing different scenarios. Based on the simulation results, the overall energy efficiency can be improved, and emissions reduced. The outcome of these simulations can be used to advise city planners, building authorities or municipalities to help design a renewable and sustainable urban future.

Author Contributions: Conceptualization, V.W., J.S. and U.E.; Methodology, V.W., J.S.; Software, V.W., J.S.; Validation, V.W.; Formal analysis, V.W., J.S.; Investigation, V.W., J.S.; Writing-original draft preparation, V.W., J.S.; Writing-review and editing, V.W., J.S. and U.E.; Visualization, V.W., J.S.; Supervision, U.E.

Funding: This research is part of a phd programme called ENRES, supported by LGF and MWK Baden-Württemberg. In addition, the research is part of the IN-SOURCE (funded by the Belmont Forum Joint program initiative on the Food Water Energy Nexus (BMBF grant 01LF1802A)) and SimStadt 2.0 (funded by BMWi, grant number 03ET1459A) projects.

Conflicts of Interest: The authors declare no conflict of interest.

Abbreviations

The following abbreviations are used in this manuscript:

CityGML	City Geography Markup language
CHP	combined heat and power
CO_2	carbon dioxide
EnEV	Energieeinsparverordnung, Energy Saving Ordinance
GUI	graphical user interface
HP	heat pump(s)
KPI	key performance indicator
LOD	level of detail
P2G	power to gas
PEF	primary energy factor
PV	Photovoltaic

References

1. United Nations. *The New Urban Agenda*; United Nations: New York, NY, USA, 2017.
2. United Nations Department of Economic and Social Affairs/Population Division. *World Urbanization Prospects: The 2011 Revision*; United Nations: New York, NY, USA, 2014.
3. European Commission. *Energy-roadmap 2050*; European Commission: Luxembourg, 2012.
4. Reinhart, C.F.; Davila, C.C. Urban building energy modeling—A review of a nascent field. *Build. Environ.* **2016**, *97*, 196–202. [CrossRef]
5. Frayssinet, L.; Merlier, L.; Kuznik, F.; Hubert, J.L.; Milliez, M.; Roux, J.J. Modeling the heating and cooling energy demand of urban buildings at city scale. *Renew. Sustain. Energy Rev.* **2018**, *81*, 2318–2327. [CrossRef]
6. Li, W.; Zhou, Y.; Cetin, K.; Eom, J.; Wang, Y.; Chen, G.; Zhang, X. Modeling urban building energy use: A review of modeling approaches and procedures. *Energy* **2017**, *141*, 2445–2457. [CrossRef]
7. Allegrini, J.; Orehounig, K.; Mavromatidis, G.; Ruesch, F.; Dorer, V.; Evins, R. A review of modelling approaches and tools for the simulation of district-scale energy systems. *Renew. Sustain. Energy Rev.* **2015**, *52*, 1391–1404. [CrossRef]
8. Chen, Y.; Hong, T.; Piette, M.A. Automatic generation and simulation of urban building energy models based on city datasets for city-scale building retrofit analysis. *Appl. Energy* **2017**, *205*, 323–335. [CrossRef]
9. Agugiaro, G. Energy planning tools and CityGML-based 3D virtual city models: Experiences from Trento (Italy). *Appl. Geomat.* **2016**, *8*, 41–56. [CrossRef]
10. Bahu, J.M.; Koch, A.; Kremers, E.; Murshed, S.M. Towards a 3D spatial urban energy modelling approach. *ISPRS Ann. Photogramm. Remote Sens. Spat. Inf. Sci.* **2013**, *II-2/W1*, 33–41. [CrossRef]
11. International Energy Agency. *World Energy Outlook 2018 Executive Summary*; International Energy Agency: Paris, France, 2018.
12. Lund, H.; Werner, S.; Wiltshire, R.; Svendsen, S.; Thorsen, J.E.; Hvelplund, F.; Mathiesen, B.V. 4th Generation District Heating (4GDH). *Energy* **2014**, *68*, 1–11. [CrossRef]
13. ForschungsVerbund Erneuerbare Energien. *Erneuerbare Energien im Wärmesektor-Aufgaben, Empfehlungen und Perspektiven*; Positionspapier des ForschungsVerbunds Erneuerbare Energien; ForschungsVerbund Erneuerbare Energien: Berlin, Germany, 2015.
14. Wärmewende 2030. *Schlüsseltechnologien zur Erreichung der mittel-und langfristigen Klimaschutzziele im Gebäudesektor. Studie im Auftrag von Agora Energiewende*; Wärmewende 2030: Berlin, Germany, 2017.

15. Olsthoorn, D.; Haghighat, F.; Mirzaei, P.A. Integration of storage and renewable energy into district heating systems: A review of modelling and optimization. *Sol. Energy* **2016**, *136*, 49–64. [CrossRef]

16. Lake, A.; Rezaie, B.; Beyerlein, S. Review of district heating and cooling systems for a sustainable future. *Renew. Sustain. Energy Rev.* **2017**, *67*, 417–425. [CrossRef]

17. Eriksson, M.; Vamling, L. Future use of heat pumps in Swedish district heating systems: Short- and long-term impact of policy instruments and planned investments. *Appl. Energy* **2007**, *84*, 1240–1257. [CrossRef]

18. Averfalk, H.; Ingvarsson, P.; Persson, U.; Gong, M.; Werner, S. Large heat pumps in Swedish district heating systems. *Renew. Sustain. Energy Rev.* **2017**, *79*, 1275–1284. [CrossRef]

19. Østergaard, P.A.; Mathiesen, B.V.; Möller, B.; Lund, H. A renewable energy scenario for Aalborg Municipality based on low-temperature geothermal heat, wind power and biomass. *Energy* **2010**, *35*, 4892–4901. [CrossRef]

20. Jensen, J.K.; Ommen, T.; Markussen, W.B.; Elmegaard, B. Design of serially connected district heating heat pumps utilising a geothermal heat source. *Energy* **2017**, *137*, 865–877. [CrossRef]

21. Mainzer, K.; Killinger, S.; McKenna, R.; Fichtner, W. Assessment of rooftop photovoltaic potentials at the urban level using publicly available geodata and image recognition techniques. *Sol. Energy* **2017**, *155*, 561–573. [CrossRef]

22. Vulkan, A.; Kloog, I.; Dorman, M.; Erell, E. Modeling the potential for PV installation in residential buildings in dense urban areas. *Energy Build.* **2018**, *169*, 97–109. [CrossRef]

23. Hofierka, J.; Suri, M. The solar radiation model for Open source GIS: Implementation and applications. In Proceedings of the Open Source GIS—GRASS Users Conference 2002, Trento, Italy, 11–13 September 2002.

24. Gagliano, A.; Patania, F.; Nocera, F.; Capizzi, A.; Galesi, A. GIS-Based Decision Support for Solar Photovoltaic Planning in Urban Environment. In *Sustainability in Energy and Buildings*; Springer: Berlin/Heidelberg, Germany, 2013; pp. 865–874, doi:10.1007/978-3-642-36645-1_77

25. Redweik, P.; Catita, C.; Brito, M. Solar energy potential on roofs and facades in an urban landscape. *Sol. Energy* **2013**, *97*, 332–341. [CrossRef]

26. Lukač, N.; Žlaus, D.; Seme, S.; Žalik, B.; Štumberger, G. Rating of roofs' surfaces regarding their solar potential and suitability for PV systems, based on LiDAR data. *Appl. Energy* **2013**, *102*, 803–812. [CrossRef]

27. Eicker, U.; Nouvel, R.; Schulte, C.; Schumacher, J.; Coors, V. 3D-Stadtmodelle als Grundlage für Wärmebedarfssimulationen. In Proceedings of the BauSIM 2012, Berlin, Germany, 26–28 September 2012; pp. 1–7.

28. Doppelintegral GmbH. *INSEL 8.2*; Doppelintegral GmbH: Stuttgart, Germany, 2018.

29. Bonczak, B.; Kontokosta, C.E. Large-scale parameterization of 3D building morphology in complex urban landscapes using aerial LiDAR and city administrative data. *Comput. Environ. Urban Syst.* **2019**, *73*, 126–142. [CrossRef]

30. Nouvel, R.; Kaden, R.; Bahu, J.M.; Kaempf, J.; Cipriano, P.; Lauster, M.; Benner, J.; Munoz, E.; Tournaire, O.; Casper, E. Genesis of the EnergyADE. In Proceedings of the CISBAT 2015, Lausanne, Switzerland, 9–11 September 2015.

31. Biljecki, F.; Ledoux, H.; Stoter, J. An improved LOD specification for 3D building models. *Comput. Environ. Urban Syst.* **2016**, *59*, 25–37. [CrossRef]

32. Loga, T.; Stein, B.; Diefenbach, N.; Born, R. *Deutsche Wohngebäudetypologie*; IWU—Institut Wohnen und Umwelt: Darmstadt, Germany, 2015.

33. Weiler, V.; Wuerstle, P.; Schmitt, A.; Stave, J.; Braun, R.; Zirak, M.; Coors, V.; Eicker, U. Methoden zur integration von sachdaten in citygml dateien zur verbesserung der energetischen analyse von stadtquartieren und deren visualisierung. In Proceedings of the IBPSA Germany Conference, Karlsruhe, Germany, 26–28 September 2018.

34. Eicker, U.; Nouvel, R.; Duminil, E.; Coors, V. Assessing passive and active solar energy resources in cities using 3D city models. In Proceedings of the 2013 ISES Solar World Congress, Cancun, Mexico, 3–7 November 2013.

35. Nouvel, R.; Brassel, K.H.; Bruse, M.; Duminil, E.; Coors, V.; Eicker, U.; Robinson, D. SimStadt, a workflow-driven urban energy simulation platform for CityGML city models. In Proceedings of the CISBAT 2015 International Conference on Future Buildings and Districts, Lausanne, Switzerland, 9–11 September 2015; EPFL Solar Energy and Building Physics Laboratory: Lausanne, Switzerland, 2015.

36. Nouvel, R.; Zirak, M.; Coors, V.; Eicker, U. The influence of data quality on urban heating demand modeling using 3D city models. *Comput. Environ. Urban Syst.* **2017**, *64*, 6880. [CrossRef]

37. Nouvel, R.; Zirak, M.; Dastageeri, H.; Coors, V.; Eicker, U. Urban energy analysis based on 3D city model for national scale applications. In Proceedings of the IBPSA Germany Conference, Aachen, Germany, 22–24 September 2014.

38. Monien, D.; Strzalka, A.; Koukofikis, A.; Coors, V.; Eicker, U. Comparison of building modelling assumptions and methods for urban scale heat demand forecasting. *Future Cities Environ.* **2017**, *3*. [CrossRef]

39. Harter, H.; Weiler, V.; Eicker, U. Developing a roadmap for the modernisation of city quarters—Comparing the primary energy demand and greenhouse gas emissions. *Build. Environ.* **2017**, *112*, 166–176. [CrossRef]

40. Rodríguez, L.R.; Duminil, E.; Ramos, J.S.; Eicker, U. Assessment of the photovoltaic potential at urban level based on 3D city models: A case study and new methodological approach. *Sol. Energy* **2017**, *146*, 264–275. [CrossRef]

41. *DIN V 18599-10: Energy Efficiency of Buildings-Calculation of the Net, Final and Primary Energy Demand for Heating, Cooling, Ventilation, Domestic Hot Water and Lighting-Part 10: Boundary Conditions of Use, Climatic Data*; DIN Deutsches Institut für Normung e. V.: Berlin, Germany, 2016.

42. *VDI 4710-2: Meteorological Data for Technical Building Sevices Purposes-Degree Days*; VDI-Gesellschaft Technische Gebäudeausrüstung: Düsseldorf, Germany, 2007.

43. Hay, J.E. Calculating solar radiation for inclined surfaces: Practical approaches. *Renew. Energy* **1993**, *3*, 373–380. [CrossRef]

44. Quaschning, V.; Hanitsch, R. Höhere Flächenausbeute durch Optimierung bei aufgeständerten Modulen. In Proceedings of the Symposium Photovoltaische Solarenergie, Staffelstein, Germany, 11–13 March 1998.

45. Doetsch, C.; Taschenberger, J.; Schönberg, I. *Leitfaden Nahwärme*; Fraunhofer IRB Verlag: Stuttgart, Germany, 1995.

46. Rosa, A.D.; Li, H.; Svendsen, S. Method for optimal design of pipes for low-energy district heating, with focus on heat losses. *Energy* **2011**, *36*, 2407–2418. [CrossRef]

47. Nowak, W.; Arthkamp, J. *BHKW Fibel*; ASUE Arbeitsgemeinschaft für sparsamen und umweltfreundlichen Energieverbrauch e.V: Berlin, Germany, 2015.

48. Deutsche Energie-Agentur GmbH (dena). *Leitfaden Biomethan BHKW-Direkt*; Dena: Berlin, Germany, 2013.

49. Fruergaard, T.; Christensen, T.; Astrup, T. Energy recovery from waste incineration: Assessing the importance of district heating networks. *Waste Manag.* **2010**, *30*, 1264–1272. [CrossRef] [PubMed]

50. Landkreis Ludwigsburg. *Klimaschutzkonzept des Landkreises Ludwigsburg*; Landkreis Ludwigsburg: Ludwigsburg, Germany, 2015.

51. *Verordnung über Energiesparenden Wärmeschutz und Energiesparende Anlagentechnik bei Gebäuden (Energieeinsparverordnung-EnEV)*; Federal Ministry of Justice and Consumer Protection: Berlin, Germany, 2015.

52. *Gesetz für den Ausbau Erneuerbarer Energien (Erneuerbare-Energien Gesetz-EEG)*; Federal Ministry of Justice and Consumer Protection: Berlin, Germany, 2017.

53. Quaschning, V.; Weniger, J.; Bergner, J. Vergesst den Eigenverbrauch und macht die Dächer voll! In Proceedings of the PV-Symposium Bad Staffelstein, Bad Staffelstein, Germany, 25–27 April 2018.

54. Landesanstalt für Umwelt Baden-Württtemberg. *Jahreskatalog Fließgewässer*; Landesanstalt für Umwelt Baden-Württtemberg: Karlsruhe, Germany, 2018.

55. Quaschning, V. *Sektorkopplung durch die Energiewende*; University of Applied Sciences Berlin: Berlin, Germany, 2016.

56. *DIN V 4701-10: Energy Efficiency of Heating and Ventilation Systems in Buildings-Part 10: Heating, Domestic Hot Water, Ventilation*; DIN Deutsches Institut für Normung e. V.: Berlin, Germany, 2016.

57. Greiner, B.; Hermann, H. *Sektorale Emissionspfade in Deutschland bis 2050-Stromerzeugung*; Öko-Institut: Freiburg, Germany, 2016.

58. Wuppertal Institut. *Konsistenz und Aussagefähigkeit der Primärenergie-Faktoren für Endenergieträger im Rahmen der EnEV*; Diskussionspapier unter Mitarbeit von Dietmar Schüwer, Thomas Hanke und Hans-Jochen Luhmann: Berlin, Germany, 2015.

Article

District Cooling Versus Individual Cooling in Urban Energy Systems: The Impact of District Energy Share in Cities on the Optimal Storage Sizing

Dominik Franjo Dominković [1,*] and Goran Krajačić [2]

[1] Department of Applied Mathematics and Computer Science, Technical University of Denmark, Matematiktorvet, 2800 Kgs. Lyngby, Denmark

[2] Faculty of Mechanical Engineering and Naval Architecture, University of Zagreb, Ivana Lucica 5, 10000 Zagreb, Croatia; goran.krajacic@fsb.hr

* Correspondence: dodo@dtu.dk; Tel.: +45-93-511-530

Received: 25 December 2018; Accepted: 26 January 2019; Published: 28 January 2019

Abstract: The energy transition of future urban energy systems is still the subject of an ongoing debate. District energy supply can play an important role in reducing the total socio-economic costs of energy systems and primary energy supply. Although lots of research was done on integrated modelling including district heating, there is a lack of research on integrated energy modelling including district cooling. This paper addressed the latter gap using linear continuous optimization model of the whole energy system, using Singapore for a case study. Results showed that optimal district cooling share was 30% of the total cooling energy demand for both developed scenarios, one that took into account spatial constraints for photovoltaics installation and the other one that did not. In the scenario that took into account existing spatial constraints for installations, optimal capacities of methane and thermal energy storage types were much larger than capacities of grid battery storage, battery storage in vehicles and hydrogen storage. Grid battery storage correlated with photovoltaics capacity installed in the energy system. Furthermore, it was shown that successful representation of long-term storage solutions in urban energy models reduced the total socio-economic costs of the energy system for 4.1%.

Keywords: district cooling; energy storage; linear programming; tropical climate; integrated energy modelling; energy system optimization; temporal resolution; energy planning; variable renewable energy sources

---

1. Introduction

The transition to a low-carbon society got an important impetus following the 21st Conference of Parties, held in Paris in 2015, which resulted in the so-called Paris Agreement. The main long-term goal of the agreement is to keep the increase in the average global temperature preferably to 1.5 °C above the pre-industrial levels [1]. A recent Special Report on Global Warming made by the Intergovernmental Panel on Climate Change reached a conclusion that in order to reach the targets of the Paris Agreement, carbon dioxide (CO_2) emissions will need to fall by 45% until 2030 (compared to 2010 levels), and reach carbon neutrality by the year 2050 [2].

Cities are currently responsible for emitting 70% of the total energy-related CO_2 emissions, while they produce 80% of the world's gross domestic product (GDP) [3]. Moreover, more than half of the global population lives in cities currently, and it is expected that this share will rise to 66% by 2050 [4]. Due to larger economic activity, more compact spatial layout with higher population densities, cities play an important role in tackling climate problems. Moreover, cities around the world are included in dedicated programs combat climate change such as the Covenant of Mayors for Climate &

Energy [5], the Cities Alliance—Cities Without Slums [6], the CIVITAS: Cleaner and better transport in cities [7], the Energy Cities—where action & vision meet [5] and the Climate Alliance [8].

Another important problem, which is more pronounced in urban areas than in rural areas is air pollution [9]. According to the International Energy Agency, 6.5 million premature deaths worldwide can be attributed to the air pollution, placing the air pollution to the 4th place of the largest threats to the health of the humans [10]. For the case of Singapore, it was shown that focusing solely on mitigating climate change emissions, can result in increased air pollution, mainly due to increased biomass use [11].

In a traditional framework of energy systems, the demand for energy is fixed and it needs to be met in all times by different energy sources. The latter framework can become increasingly expensive as more variable renewable energy sources, such as photovoltaics (PV) and wind are introduced, as they cannot be dispatched when needed. In order to reduce both air pollution and climate change related emissions, a shift in the traditional framework started to occur, by creating demand when there is available supply [12]. However, in times when there is very low production from variable renewable energy sources, and demand cannot be reduced enough for the available supply or it is more expensive to reduce it, a resulting difference in demand and supply has to be resolved by different storage solutions.

The first problem detected in the literature was that different energy storage types are rarely represented in the same study, thus making it complicated to detect the interdependency of different storage solutions. Energy storage is often discussed in terms of the power system in general. One review paper focused on the storage solutions for electric vehicles [13]. The authors reached a conclusion that there is still insufficient research carried out about the material support, proper disposal, recycling, safety measures and cost [13]. Another study focused on storage in terms of the integration of wind power [14]. The authors distinguished between the positive impact of storage on power applications, such as voltage and frequency control, as well as on energy applications (storing energy itself) [14]. The authors reached the conclusion that the energy storage can be an effective solution to satisfy the stability and reliability requirements of the power system.

In another set of studies, a focus was placed on the ice storage only. One study examined the system consisting of ice storage, an adsorption chiller and a parabolic trough solar collector [15]. The authors showed that the highest coefficient of performance (COP) in the proposed setup was only 0.15 [15]. In order to increase the performance of the system, some authors suggested the use of double effect absorption chillers in a combination with solar heating plant [16]. The simulated COP of absorption chiller was 1.92, while the measured values showed that the real COP was only 1.21 [16]. Furthermore, much higher inlet temperatures are needed for double-effect absorption chiller compared to the single-effect absorption chiller, which can be problematic for heat sources available at lower temperatures [17]. All those papers pointed to the use of ice storage as daily storage (cooling load shift from day to night), and none of them examined the possibility of using the cold storage in a more seasonal manner.

However, energy storage is a wider concept than just the electrical storage or cold storage. One research paper showed the difference between pumped hydro (electrical), thermal, gas, and liquid storage solutions [18]. The authors concluded that liquid fuel, gas and thermal storage solutions are much cheaper than the electrical storage, as well as that those storage types can store energy for longer periods of time with lower losses [18]. Another paper focused on optimal capacities of different storage types when reaching strict CO_2e emission targets [11]. When looking into minimizing the total socio-economic costs of the energy system, the authors showed that thermal energy storage and hydrogen storage have large potential in future urban energy systems with a high share of electricity generated from PV, while the potential of natural gas and grid battery storage will be much lower [11].

The second problem detected from the literature was a lack of the representation of the long-term storage solutions, usually due to poor (coarse) temporal resolution. A problem that arose in connection with the dawn of the intermittent energy sources, such as wind farms and PV, is the decentralized locations of those energy sources in an energy system. In order to successfully represent and optimize the future capacity investments, energy planning models must successfully represent energy systems both spatially and temporally. In recent reports on long-term energy models, the one from the National Renewable Energy Laboratory (NREL) [19] and the other from the International Renewable Energy Agency (IRENA) [20], it was stated that in order for models to capture both temporal and spatial dimensions, as well as to be computationally tractable, the models use different decomposition techniques or so-called time slicing in order to reduce the complexity of the optimization models. Furthermore, the NREL's report has clearly stated that the lost ability to represent large-scale storage solutions should be further addressed.

One paper demonstrated the possibility of using statistics-based methods for selecting the optimal subset of the representative days to capture the variability of solar generation and power load in the distribution system [21]. They significantly reduced the number of days that need to be represented, from 365 to 40. However, although they have captured the solar generation and power demand distributions variability, their approach only allows for modelling of intra-day storages.

In Ref [22], the author concluded that for the economic assessment of the energy system it is especially important to capture the value of flexible resources, such as transmission and storage. However, the author has pointed out that the typical models aggregate intra-annual time segments and capacity blocks [22]. The author's model captured the covariance of resources, operational constraints and regional heterogeneity but energy storage was not able to store seasonal surpluses of renewable energy, according to the author [22].

There are three different papers that tackled the issue of resolution and computational tractability of the energy planning models. The authors in [23] used the National Renewable Energy Laboratory's Resource Planning Model to demonstrate the difference between results on higher and lower temporal distribution. They simulated different dispatch periods for the "peak-day", ranging from one day to four consecutive days. They concluded that the shorter simulation periods resulted in over-scheduling of hydro energy and less installation of combined cycle natural gas plants [23]. However, their temporal resolution was still very small (coarse), they focused only on the power sector and no large-scale storages were represented in the model. The same model (Resource Planning Model) was used in another report [24]. The authors carried out a very detailed spatial representation of the distribution and transmission; however, they were still using only consecutive 4-day periods as a temporal resolution. Moreover, the authors explicitly stated that they did not model new storage investments in the model. They reported that increasing sampled dispatch periods had a dramatic effect on computation time, while the investment decisions did not change when changing model configuration from 96-h sample to 24-h sample [24]. The focus of their research was still solely on the power sector, without proper representation of the long-term storages.

To the best knowledge of the authors of this paper, the most detailed research on temporal and spatial trade-offs so far has been carried out in [25]. The authors used the POWER planning model for scenarios with a high share of variable renewable energy generation. The authors concluded that the most trade-offs yield up to the 15% of cost differences. Concerning the spatial resolution, the authors showed that the uniform buildout case resulted in a 10% reduction in cost compared to the site-by-site buildout case. Focusing on the temporal resolution, the authors showed that the total cost is significantly lower with a coarser temporal resolution. However, two limitations, out of the several ones that the authors detected in their paper, are important for this paper. First, the authors reported that their work has focused only on the power sector and it has not captured the current research trend to include all the energy sectors in order to achieve cross-sectoral synergies and second, due to the temporal disjointedness of most days in subsets, the chronological tracking of storages was limited to the 24 h [25].

This paper tackled the both detected problems, the first regarding the optimal storage portfolio depending on the share of district energy supply, and the second, regarding the difference in optimal technology portfolio and corresponding socio-economic costs in energy systems with and without large-scale storage solutions represented. As many different storage solutions exist, it is important to take a holistic look into the energy system, including all of its sectors, in order to be able to track the interplay between different storage solutions. In the existing literature, there is still not enough research carried out on the optimal storage portfolio dependent on different shares of district energy supply. In order to tackle the latter problem, this paper focused on resulting optimal storage portfolio based on the stepwise increase in the share of district energy supply, in this case on district cooling supply for the case of the city of Singapore.

2. Materials and Methods

There are many different energy planning and/or capacity expansion models. The fundamental and the most important difference is between optimization and simulation models. A comprehensive overview of the existing models can be found in [26]. In a recent paper, a critical review has been stated towards the models that report only one optimal solution of the future energy system [27]. The authors claim that the simulation models are better suited for a decision making about the future investments as they typically report several scenarios, i.e. several alternatives, as opposed to the optimization models that typically represent only one (optimal) solution [27]. Although the author of this paper shares the concern about the reporting of the single (optimal) solution to different stakeholders that do not completely understand the complexity of the energy modelling, the number of assumptions and uncertainties about the data, the problem being researched here is especially suitable to be solved by optimization model due to the two reasons. First, the focus of the research is a representation of the different storage solutions, which can be only handled using the optimization techniques. Even the simulation models, such as EnergyPLAN, handle the storage operation via different optimization methods [28]. Second, the geographical scope of this paper is a city scale, as opposed to the more common national or regional models. A city usually has different peculiarities and specifics that need to be taken into account when modelling its energy system. If the focus had been set to the larger geographical areas such as nations, those specifics of a single city would have been averaged out.

Hence, in this paper, a linear continuous optimization model was used to model the representative city of a hot climate. As a linear optimization model, the developed model has many similarities to the more common Balmorel model [29]. The main difference is that the model developed for this specific issue can be more tailored, allowing for better representation of different storage types, as well as different integration technologies such as electrolysers and fuel cells. The results of the model include the optimal operation of the energy system on the hourly resolution, as well as the optimal capacities of different technologies in the energy system. The developed model is specifically tailored for integrated energy modelling, which includes power, cooling/heating, gas, mobility and water desalination sector, including the interactions among those sectors. The goal of the integrated energy modelling is to capture additional efficiencies through integration of different energy sectors, contrary to the models focusing on a single sector only. Thus, although the focus of this paper was on the cooling sector, all the energy sectors were optimized simultaneously in order to capture interactions between different energy sectors.

The full set of equations can be found in Appendix B of this paper, while the most important equations are presented and discussed in this section. Most of the equations were initially developed in [30], and further updated in [11].

The objective function of the model was to minimize the total yearly socio-economic costs of the energy system (Equation (1)). Contrary to the business-economic calculations, the socio-economic costs usually exclude different taxes as those are considered to be internal redistributions within the society. However, the cost of negative externalities such as air pollution and climate change costs can be internalized when calculating the total socio-economic costs. In this paper, the costs of both CO_2e (CO_2, CH_4, N_2O) emissions and air pollutants (NO_2, SO_2, PM) were taken into account. The CH_4 and N_2O emissions were transferred to the equivalent CO_2 emissions using the global warming potential factors. The objective function included levelized investment costs of different technologies, fixed operating and maintenance costs (O&M) and fuel costs. The costs of infrastructure, such as electrical transmission and gas transmission grids were also modelled:

$$\min Z = \sum_{i=1}^{n} \left(fix\_O\&M_i + lev_{inv_i} \right) x_i + \sum_{j=1}^{m} \left(var\_O\&M_j + \frac{fuel_j}{\eta_j} + CO2e_j \cdot CO2e_{inten_j} + air\_poll_j \cdot air\_poll_{inten,j} \right) x_j + \sum_{k=1}^{p} \left(gas\_imp_k + petr\_imp_k \right) x_k \; [30] \tag{1}$$

Levelized investment costs were calculated using Equation (2). All the infrastructure, including power plants, cogeneration plants, distribution grids and different storage types were annualized and corresponding yearly costs were added to the total yearly socio-economic costs reported in this study:

$$lev\_inv_i = inv_i \cdot \frac{dis\_rate_i}{1 - (1 + dis\_rate_i)^{-lifetime_i}} \tag{2}$$

In total, five different storage solutions were implemented in the model, battery grid storage (Equation (3)), battery storage in vehicles, pit thermal energy storage, hydrogen and natural gas storages:

$$battery\_level_r = battery\_level_{r-1} + x_{j,battery,storage\_ch,r} - \eta_{batt} x_{j,battery,storage_{dis},r} - x_{j,battery,storage\_grid\_dis,r} \tag{3}$$

Storages were modelled using the sets of variables for storage charge, storage discharge and level of energy in the storage. Moreover, storage losses were taken into account. Battery storage of vehicles included another set of variables, the one representing the discharge of electricity to the grid, allowing modelling of the vehicle-to-grid behaviour. All other storage types were modelled in the same manner as the grid battery storage represented by Equation (3).

District cooling balance included district cooling demand that needed to be satisfied, cold generated in absorption chillers, cold from geothermal energy (Equation (4)):

$$abs_{DC} + geothermal_{DC} \geq DC_{demand} \tag{4}$$

while the balance of absorption chillers is given by Equation (5):

$$x_{j,wasteheat,l} + \eta_{th} x_{j,heat,storage\_dis,r} \geq \left(abs_{DC} / COP_{abs} \right) \tag{5}$$

Here it is important to note that the term $x_{j,wasteheat,l}$ represents heat that originates from different sources: gas CHP, waste incineration plant, solar thermal or waste heat from data centres. The waste heat from data centres was assumed to be coupled with heat pumps, which would sufficiently raise the temperature to be fed to single-effect absorption chillers.

A total cooling demand could be satisfied either by district cooling or individual cooling, as presented in Equation (6):

$$DC_{demand} + ind\_cool_{demand} \geq cool_{demand,total} \tag{6}$$

The total individual cooling demand needed to be met by individual chillers is obtained using Equation (7). Individual chillers were assumed to be building level air conditioners:

$$chiller_{individual} \geq ind\_cool_{demand} \tag{7}$$

The remaining part of the model developed in this paper, which was used for representing the other sectors of the energy system, can be found in Appendix B.

The model was built using the Matlab interface and Gurobi optimization solver. Optimizations were carried out on a personal computer, with an i7 processor (1.9 GHz), 16 GB of physical RAM, and

25 GB of virtual RAM, dedicated from 500GB SSD hard disc. A single optimization run took 30 min in the most time-consuming run.

3. Case Study

Singapore was chosen for the case study as it is a 100% urbanized country, at a high level of economic development and suitable for district cooling research as it is located very close to the equator. Being a country—city, a large agglomeration with lots of available data, Singapore represents very well the urban zones of many developed countries, especially those with a similar climate.

Singapore is one of the most densely populated cities in the world, with 7900 people per km^2. It is a very developed region, having the 3rd largest GDP at power purchase parity (PPP), with 93,905 USD per capita according to the World Bank [31]. Large population and high economic activity result in a large energy use per capita. According to the International Energy Agency, Singapore's primary energy demand per capita was 53.8 MWh in 2015 [32]. Singapore is the most developed part of south-east Asia and thus, its energy consumption is already high [33]. Currently, south-east Asia is rapidly developing, both in terms of population and economic activity. Consequently, it can be expected that the neighbouring cities will see a significant increase in primary energy demand to meet the needs of its increasing population. For the case of Malaysia, the IEA forecasts increase in energy demand by a factor of 2 until the year 2040 [33].

Hence, the results implemented for this case study are transferable to other cities located in the warm regions, as it can be expected that most of the cities that are now significantly increasing economic activity will see much higher energy demand for cooling purposes in the future.

Different assumptions were made as a part of this research. CO_2e emissions factors, capacities of energy facilities, energy demand and the description of the transportation sector can be found in great detail in [34] and [35]. The energy supply of Singapore is dominated by gas power only plants, as can be seen in Table 1.

Table 1. The energy supply of Singapore in the reference scenario.

Plant Type	Capacity [MW]	Fuel	Efficiency	Reference
Power plants	2225	Natural gas/LNG	47.5%	[36,37]
Waste-to-energy plants	256.8	Waste	16.7%	[36–38]
PV	33.1	-	12.4% (capacity factor)	[36,38]

The following technologies were predefined in the optimization model: solar heating, absorption chillers, waste heat from data centres coupled with heat pumps to reach sufficient temperatures [39], PTES storage and geothermal energy in district cooling sector; grid battery storage, waste and gas combined heat and power plants, wind turbine, PVs and vehicle-to-grid in the power sector; natural gas transmission grid (import), hydrogen storage, gas storage, solid-oxide electrolysers and fuel cells, syngas to natural gas and syngas to gasoline syntheses and gasoline import in gas and mobility sectors. Finally, other technologies included in the model were individual chillers (individual cooling) and reverse osmosis for desalination of seawater. It is important to notice that all of those technologies were predefined in all the scenarios of this research paper. However, based on the conditions in a specific scenario, different optimal set-ups were possible in different scenarios.

As the focus of this paper was on the optimal portfolio of different storage types depending on different shares of district cooling and individual cooling, it is important to discuss different assumptions regarding the implemented storage types. In this paper, individual cooling included cooling solutions from an apartment level up to the one whole building. To be counted as district cooling, at least two buildings had to be connected to the district cooling grid.

Pit thermal energy storage (PTES) was assumed for district cooling purposes. It is used to store large quantities of warm water before it is fed to single effect absorption chillers. There are two main reasons why pit thermal energy storage was chosen for district cooling sector instead of ice (cold) storage. First, the current literature review shows that no cold storage is currently being used as seasonal storage, while there are plenty of real-world installations of pit thermal energy storage solutions [40]. Second, for ice storage charging, very low temperatures are needed, which cannot be obtained by water-lithium bromide absorption chillers. Instead, the ammonia-water solution should be used. However, the ammonia-water solution was reported to have a lower coefficient of performance than the water-lithium bromide solution [40].

A single-effect water-lithium bromide absorption chiller technology was adopted for this study, in order to make it possible to utilize as much waste heat from different sources as possible. The coefficient of performance (COP) of 0.7 was used. District cooling system was organized as a centralized system, as both PTES and absorption chillers were centrally located. On the other hand, building level air conditioners, designated in this research paper as individual cooling, had a COP of 4.0.

The cost of the district cooling distribution grid with water as a medium flowing through the pipes was adopted from [34], and the price of 37 €/MWh of yearly delivered cold energy was used. The losses of district cooling grid are significantly lower than the district heating grid. One reason is a lower temperature difference between the water flowing through the pipes and the surrounding ground, and the other is a generally larger diameter of distribution pipes in the case of the cooling grid, resulting in a lower area-to-volume ratio and corresponding heat losses. For the tropical region, average losses of 4.8% were reported [34], which is also the number used in this study.

Hydrogen was assumed to be stored in a cavern type of storage at 50 bars. Methane storage was also assumed to be an underground cavern. Grid batteries and vehicle batteries technology were lithium nickel cobalt aluminum oxide (Li-ion) NCA. Assumed investment costs and efficiencies can be seen in Table 2.

Table 2. Investment costs and efficiency of different storage solutions.

	Investment Cost (€/MWh)	Efficiency *	Reference
PTES	540	75%	[41]
Methane storage	99	97%	[41]
Hydrogen storage	11,000	96%	[41]
Grid battery	127,300	91%	[42]
Vehicle batteries	127,300	91%	[42]
** vehicle to grid mode		79% **	[43,44]

* including charging losses, ** including direct current to alternating current transformation

In this paper, a detailed assessment of the potential CO_2 source for natural gas synthesis was out of the scope of the paper. However, the National Climate Change Secretariat of Singapore estimated that large amounts of CO_2 could be extracted from industry and gas power plants in their Carbon Capture and Utilisation roadmap [45]. It is important to keep in mind that CO_2 capture would increase the reported costs of the energy system, which was excluded from this study.

Availability of space for installing large amounts of low-density energy generators, such as PV, is often an issue in highly populated cities in south-east Asia. Singapore has a particularly high density of population, with many competing uses of available space. Regulations for using available space are very tight, and hence, the available area for PV installations has to be constrained. According to the solar photovoltaic roadmap developed for Singapore, the maximum available space for PV installations corresponds to 12,250 MW of peak capacity [46]. The referenced PV capacity already included expected development of the efficiency of PV panels that will consequently require less area for the same capacity than it is the case today. The latter includes also 20% of available seawater where floating PV could be installed. In order to take into account the PV capacity constraint, two different scenarios were developed; the first included the constrained PV capacity (dubbed *PV constrained* scenario) and the second did not include the constraint on PV capacity (dubbed *unconstrained PV* scenario). The latter would mean that Singapore should import one part of the energy generated from PV from nearby regions.

Out of other technologies, solar thermal was constrained to 2000 MW also due to space constraints, while geothermal was constrained to 50 MW_e, in line with the current estimates for geothermal potential [47]. Finally, waste heat from data centres was constrained to 735 MW_{th}, a conservative estimate taken from [35].

4. Results

Several different optimization runs were carried out in order to reach two major conclusions. The first was to detect the difference between the optimal configuration of the energy system with and without the large-scale energy storages, in order to evaluate the potential error when reducing the time resolution in the capacity extension models. This part of the research included spatial constraints influencing maximum PV and solar thermal capacity, as described in the Case study section.

The second gap detected in the literature review was the difference in optimal capacities of different technologies based on different shares of district cooling and individual cooling. In order to reach conclusions on this issue, optimizations were run stepwise (in steps of 10%), starting from 0% of district cooling (100% of individual cooling) to 100% of district cooling (0% of individual cooling). Two scenarios, in this case, *PV constrained* and *unconstrained PV*, were run in order to assess the difference between transition the energy system to an isolated one and non-isolated one.

One can note from Table 3. that total socio-economic costs were 4.1% lower when different storage solutions were included in the urban energy modelling. This difference shows that it is important to be able to represent different storage solutions in energy planning of future urban energy systems. The latter difference would have been even larger if spatial constraints influencing the maximum installed PV and solar thermal capacities had been relaxed. The main difference in costs occurred from capital savings in gas infrastructure, as gas storage replaced one part of gas CHP plant capacity and one part of the gas import capacity.

Table 3. Results of the comparison of energy models with and without large-scale storage representation (*Storages represented* optimization run and *Storages not represented* run had all the predefined technologies the same, except that storage capacities were constrained to zero in the case of *Storages not represented* scenario).

	Storages Represented	*Storages Not Represented*	**Difference**
Total socio-economic costs [mil €]:	8012	7685	−4.1%
Capital costs [mil €]:	3773	3464	−8.2%
Operating costs [mil €]:	3609	3592	−0.5%
CO_2e costs [mil €]:	607	607	−0.1%
CO_2 costs [mil €]:	605	605	−0.1%
Air pollution costs [mil €]:	23	22	−3.9%
NOx emissions [kg]	11,751,113	11,711,620	−0.3%
SOx emissions [kg]	0	0	-
PM emissions [kg]	19,694	19,627	−0.3%
CO_2e emissions [Mt]:	29	29	−0.1%
Primary energy demand (incl. Industry) [GWh]	163,871	163,914	0.0%
Methane consumption [GWh]	73,258	73,021	−0.3%
Oil consumption [GWh]	41	7	−81.8%
Renewable [GWh]	17,662	17,801	0.8%
Waste consumption [GWh]	5989	6166	2.9%
Industry consumption–natural gas [GWh]	66,920	66,920	0.0%

The differences in the optimal configuration of the energy system can be seen in Table 4. One can notice from Table 2 that although different storage solutions resulted in a reduced socio-economic cost in the case of *storages represented* scenario, CO_2 emissions were almost the same. The reason is that the lower share of district cold generation via absorption chillers was replaced by individual electric chillers, which had a high COP value of 4. However, the large share of individual cooling resulted in higher total socio-economic costs.

Table 4. The difference in the optimal energy system configuration with and without representation of large scale storages. Only non-zero capacities are presented (including all the storages). Optimal capacities were the one that resulted in the lowest socio-economic costs, for valid conditions in each of the specific scenarios.

Capacities [MW]	Storages Not Represented	Storages Represented
solar heating	2000	2000
Electric chillers	43	0
Absorption chillers	4780	6074
PTES storage (GWh)	0	417
battery storage (GWh)	0	4
waste CHP	170	170
gas CHP	5929	4785
PVs	12,250	12,250
gas net import	13,950	10,960
Hydrogen storage (GWh)	0	0
gas storage (GWh)	0	112
SOEC	0	0
SOFC	0	0
gasoline import	813	459
electric vehicles battery capacity (GWh)	15	16 *
Reverse osmosis	31,620	34,937
Geothermal DC capacity	50	50
DC capacity	4607	6074
Non DC capacity	9727	9517

* No vehicle-to-grid mode in non-storage optimization run.

By comparing Tables 3 and 4, one can notice that the increased flexibility via the different storage representation in the energy system model resulted in significantly lower total socio-economic costs, which mainly originated from the lower capital costs.

Results of the stepwise increase in the share of district cooling can be looked at in Figures 1–3. The difference in socio-economic costs for different shares of district cooling and individual cooling, for both *PV constrained* and *unconstrained PV* scenarios, can be seen in Figure 1. The lowest socio-economic costs occurred at 30% penetration of district cooling in both cases. For the *PV constrained* scenario, the socio-economic cost at 30% share of district cooling was 7.9 billion EUR, while for *unconstrained PV* scenario the total socio-economic cost was 7.5 billion EUR. Thus, the corresponding difference of allowing larger capacities of PV resulted in a cost reduction of 5.1%, all other assumptions being the same. Increasing the share of district cooling from 50% to 100% resulted in significantly higher total socio-economic costs of the energy system.

Total primary energy demand in *PV constrained* and *unconstrained PV* scenarios can be seen in Figure 2. Primary energy demand started to increase significantly after surpassing the share of district cooling of 50%. At lower shares of district cooling, *unconstrained PV* scenario had generally lower primary energy demand. On the other hand, at higher shares of district cooling, primary energy demand in *PV constrained* scenario was lower than in the *unconstrained PV* scenario. Starting from the district cooling share of 40%, optimal capacities and corresponding costs were the same in both *PV constrained* and *unconstrained PV* scenarios. The reason is that the optimal capacity of PV in all those runs was lower than the PV capacity constraint in *PV constrained* scenario.

The total socio-economic costs were reducing until the share of district cooling of 30% was reached in both scenarios. The main reasons were better utilization of waste heat from gas CHPs and waste incineration plants, which was utilized for a cold generation in absorption chillers, as well as utilization of PTES, a cheaper storage option than the battery storage. Starting from the district cooling share of 40%, socio-economic costs of the energy system started increasing again. The main cause for the latter behaviour was a lack of low-cost waste or excess heat. Without enough heat available, gas CHP needed to consume more fuel in order to generate a sufficient amount of heat needed for absorption chillers. Having more waste heat available could result in a higher optimal share of district cooling than in this case study.

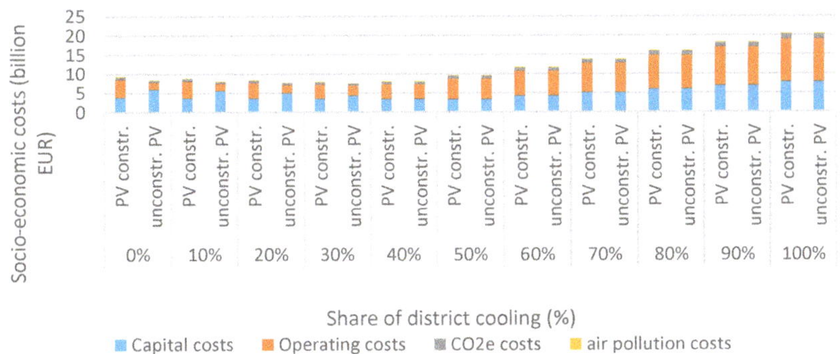

Figure 1. Comparison of the total socio-economic costs in *PV constrained* and *unconstrained PV* scenarios.

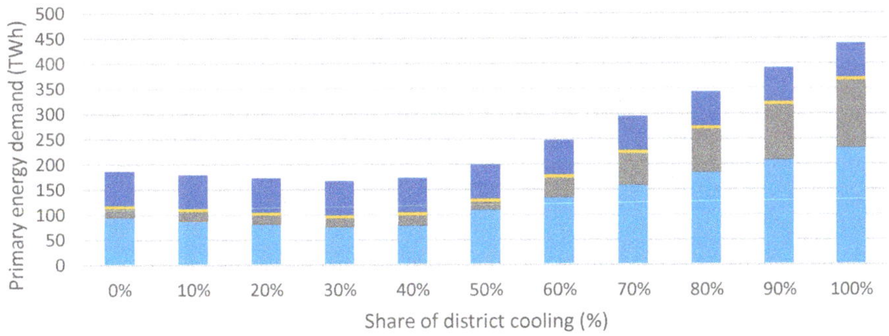

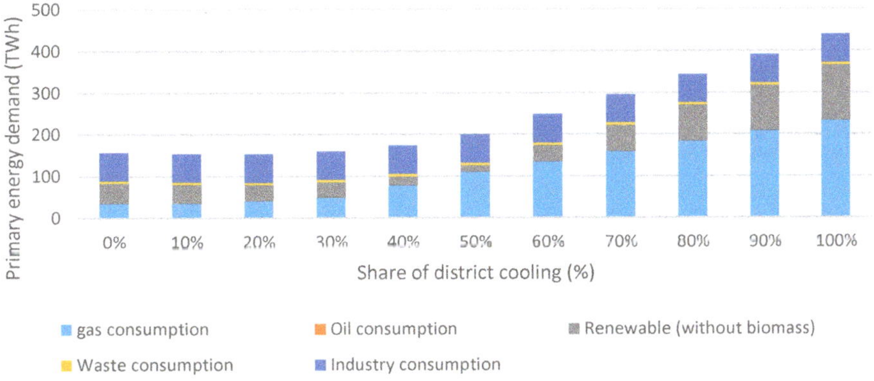

Figure 2. Total primary energy demand in *PV constrained* (**Top**) and *unconstrained PV* scenarios (**Bottom**).

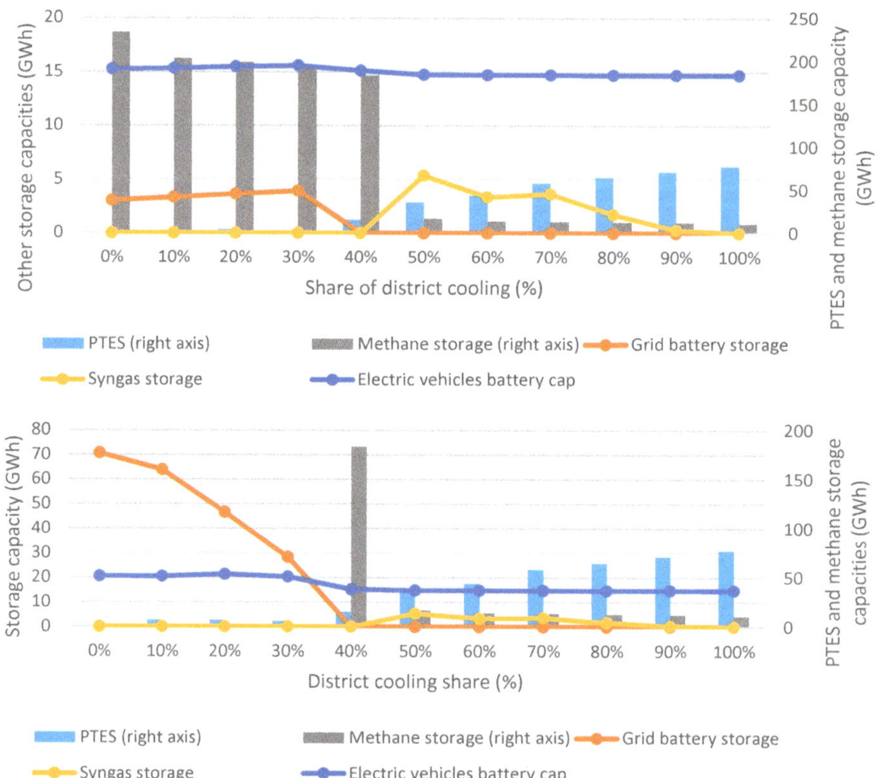

Figure 3. Optimal storage capacities in *PV constrained* (**Top**) and *unconstrained PV* (**Bottom**) scenarios. Note the difference in orders of magnitude between right and left y-axes.

PV constrained scenario had generally a lower share of capital costs and a larger share of operating costs than in the *unconstrained PV* scenario. The main reason is that the *unconstrained PV* scenario had significantly larger capacities of PV and grid batteries, resulting in higher upfront investment and lower running costs.

Both air pollution and CO_2e costs were generally higher in *PV constrained* scenario than in the *unconstrained PV* scenario. The lowest CO_2e emissions and air pollutant emissions were in the *unconstrained PV* scenario, without any district cooling. The reason for the latter is that a significant capacity of PV did not cause any carbon emissions or air pollution, while gas CHP plant that was dominating energy source for larger shares of district cooling contributed to both carbon emissions, as well as air pollution.

Optimal shares of different storage capacities can be seen in Figure 3. The largest capacities were those of methane and PTES storage types. Generally, demand for grid battery storage is low in different cases, as smart charging and vehicle-to-grid technology are already providing lots of flexibility in the energy system. The only exception is the case of *unconstrained PV* scenario, up to the share of district cooling of 30%. In those cases, the grid battery storage capacity was relatively large, as was the capacity of PV. The latter points to the conclusion that grid battery storage can be an economically viable solution if they can be utilized for many day-night charging and discharging cycles, as it was the case when large capacity of PV existed in the energy system. Discharge of electricity for the grid in the vehicle-to-grid mode was maximally utilized in the *unconstrained PV* scenario, at 30% share of district cooling, with the discharge of electricity from vehicle batteries equalling 0.72% of the total electricity generation over the year. The lowest socio-economic costs in both scenarios occurred at 30% share of district cooling.

At 30% of district cooling share in *PV constrained* scenario, optimal storage capacities of PTES, grid battery, methane, hydrogen and electric vehicle batteries were 2.7 GWh, 3.9 GWh, 195.1 GWh, 0 GWh and 15.6 GWh. In the case of *unconstrained PV* scenario, optimal storage capacities of PTES, grid battery, methane, hydrogen and electric vehicle batteries were 5.34 GWh, 28.3 GWh, 0 GWh, 0 GWh and 20.3 GWh. The behaviour of optimal PV capacities and battery storage capacities showed a correlation between those two technologies. Capacities of all the technologies for all the optimization runs can be found in Appendix A.

5. Discussion

Focusing on the first detected issue that was addressed in this paper, looking into Table 3. One can notice significant savings in total socio-economic costs of the energy system that were achieved only by better representing different storage solutions. In that sense, this paper is directly addressing a research gap detected in recent NREL's [19] and IRENA's [20] reports on energy modelling. The main difference for the case of Singapore is more efficient utilization of gas infrastructure. The difference in total socio-economic costs in systems with and without different storage solutions represented would have been even larger if PV capacity had been left unconstrained. A sensitivity analysis was carried out with unconstrained PV capacity in order to check the latter claim. It was found out that the difference in total socio-economic costs of the energy system was 7.9% when different storage solutions were represented. The main difference originated from the optimal capacity of PV, which was 52% larger in the case when storage solutions were successfully represented.

Moreover, one can observe in Table 3 a significant difference in oil consumption in the two cases, however, it should be noted that oil consumption in both cases is very low compared to the current oil consumption of Singapore. The reason for almost no oil consumption is that industrial oil consumption was switched to natural gas consumption and most of the transport sector was electrified, a result of the optimization model used in this research paper.

Focusing on the second issue addressed in this paper, looking into primary energy demand in Figure 2, one can note that at high shares of district cooling, primary energy demand tend to rise significantly. There are two main reasons for significantly higher primary energy demand in cases with a higher share of district cooling. First, at lower shares of district cooling, there is a large capacity of individual air conditioners (building level air conditioners), with a coefficient of performance of 4. Hence, ambient thermal energy that was used in the cooling cycle was not accounted for as primary energy demand. The second reason is that large shares of district cooling consequently had a large demand for natural gas. After all the low-cost heat sources were utilized, further heat generation from gas CHPs was not the optimal procedure in energy efficiency terms. Introduction of SOEC and SOFC in some of the runs with a higher share of district cooling resulted in additional conversion steps, increasing the losses in the energy system.

The optimal share of district cooling in both *PV constrained and unconstrained PV* scenarios was 30%. As the current share of district cooling in Singapore is very low, reaching the share of 30% could be challenging. One option would be to connect very densely populated central parts of the city, which includes mostly commercial buildings, to the district cooling and gradually spread the system towards residential areas. The other option would be first to implement district cooling in areas that are currently being developed, and connect the other buildings to the district cooling grid when major retrofit of buildings is scheduled, the approach proposed in [34].

It is important to discuss here the sensitivity of the results. During the case study representation, it became clear that Singapore is significantly lacking in resources for very large shares of renewable energy in the total energy consumption. Singapore is a city with a large energy demand per capita, and it is a very densely populated city. Moreover, it does not have a good wind potential for either onshore nor for offshore turbines. Due to the lack of available space, PV plants built on the ground are not an option in the current urban development plans of Singapore. All of these reasons contribute to the optimal share of district cooling of 30%. The amount of low priced waste heat is a very sensitive parameter, significantly influencing the results. Cities with similar climate, but higher amounts of waste heat availability, would probably result in a significantly larger optimal share of district cooling.

Larger shares of district cooling caused a higher need for waste heat that could be utilized via absorption chillers. To match that additional demand, optimal capacities of gas CHPs were larger than in the case of higher shares of individual cooling. The latter was also matched with much higher capacities of methane storage compared to other storage types. When maximum PV capacity was unconstrained, battery storage capacities were larger than in the case of *PV constrained* scenario. The reason for the latter is that battery grid storage and vehicle batteries in the form of smart charging were used by the system as day-night storage solutions, while methane, hydrogen and PTES storages solutions were used by the system for much longer periods.

Generally larger capacities of gas and thermal energy storage compared to the grid battery storage confirmed the finding from [18]. Somewhat contrary to the [18], in this paper, optimal grid battery capacities were found to be also rather large, especially in the case of *unconstrained PV* scenario. The latter can be contributed to the specific weather patterns of Singapore, having no clear distinctions between different seasons and only day-night distinctive insolation pattern (without seasonality effect). In the future work, the optimal capacities of grid battery storage should be assessed in a different setup in order to detect its correlation to different variable renewable energy sources; however, with larger optimal capacities of wind energy compared to PV.

Some regions like the Caribbean have a high potential of both wind and solar energy and thus, large penetration of cost-efficient renewable energy technology can be achieved without large district energy supply systems [48]. On the other hand, this case showed that for the case of Singapore, that has large potential only for solar energy and not for wind energy, a 30% share of district cooling was optimal.

Although there are not many research papers dealing with an optimal share of district versus individual cooling taking holistic energy modelling approach, there is a lot of research on the optimal share of district heating. Energy roadmap Europe has found that optimal shares of district heating versus individual heating solutions would be 66% for Italy and Spain, 50% for the Netherlands and Germany, and between 27% and 43% for most other EU countries [49]. The reason for the difference between the larger optimal shares of district heating compared to district cooling can be explained by climate specifics. Warmer regions that have very high cooling density demands usually correlate with high global insolation, having a large potential for solar energy development. Consequently, a combination of solar energy and efficient heat pumps can be more cost-effective than having very large shares of district cooling based on waste heat utilization in absorption chillers. To check the optimal share of district cooling versus individual cooling, future research projects could encompass different cities across the world, using a similar model to the one developed for this study.

6. Conclusions

Two detected research gaps were part of this study. In order to assess the optimal share of district cooling versus individual cooling taking holistic energy modelling into account, several optimization runs were carried out with a stepwise increase in district cooling share. It was shown that optimal district cooling share for the case of Singapore was 30% in both scenarios that were carried out; one with constrained PV capacity that reflected spatial constraints, and the second with unconstrained PV capacity that would include the use of nearby space, outside of the city borders.

In order to improve the research on the second detected research gap, two cases were assessed, one with five different storage types represented, and the second one without long-term storage types represented. The results revealed that an error by not representing storage technologies in energy planning models was 4.1% for the case of Singapore. The latter difference raised to 7.9% when PV capacity constraint was relaxed.

Author Contributions: Conceptualization, D.F.D. and G.K.; Data curation, D.F.D.; Formal analysis, D.F.D.; Funding acquisition, D.F.D.; Investigation, D.F.D. and G.K.; Methodology, D.F.D.; Software, D.F.D.; Supervision, G.K.; Validation, D.F.D.; Writing–original draft, D.F.D.

Funding: This work was financed as a part of the CITIES project No. DSF1305-00027B funded by the Danish Innovationsfonden and KeepWarm project that received funding from the European Union's Horizon 2020 research and innovation programme under grant agreement N°784966. Those contributions are greatly acknowledged. Those projects covered the open access fee for publishing the present paper.

Acknowledgments: This is substantially revised and updated conference paper presented at the 1st Latin American SDEWES conference in Rio de Janeiro, Brazil, 28–31 January 2018.

Conflicts of Interest: The authors declare no conflict of interest.

Nomenclature

abs_{DCc}	Cold production for district cooling (DC) from single phase absorption units, MWh
air_poll_j	Costs of air pollution emissions, €/kg
$air\_poll_{inten,j}$	Air pollution intensity of a certain technology or energy within the system boundaries, kg/MWh
$battery\_level_r$	Level of energy stored in batteries in hour r, MWh
$chiller_{DC}$	Production of cold in DC from centralized electric chillers, MWh
$COP_{DC_{chiller}}$	Coefficient of performance of chillers in DC
$chiller_{individual}$	Production of cold from individual electric chillers, MWh
$cool_{demand,total}$	Total cooling demand, MWh
$COP_{individual}$	Coefficient of performance of individual chillers
COP_{abs}	Coefficient of performance of absorbers
$CO2e\_inten_j$	CO_2 intensity of a certain technology or energy within the system boundaries, ton/MWh
$CO2e_j$	Costs of CO_2 emissions, €/ton
DC_{demand}	DC demand, MWh
dis_rate_i	Discount rate of the technology i, %
el_dem	Electricity demand, MWh
$ele_{transport}$	Electricity demand for electrified part of the transport sector, MWh
$fix\_O\&M_i$	Fixed operating and maintenance costs of energy plants, €/MW
$fuel_j$	Fuel cost of specific energy type, €/MWh$_{fuel}$
gas_dem	Gas demand, MWh
gas_imp_k	Price of import of gas in a specific hour, €/MWh
$gas_{synthesis}$	Synthetic natural gas production from syngas using gas synthesis, MWh
$geothermal_{DC}$	Cold production for DC from geothermal waste heat, MWh
$heat\_level_r$	Heating energy content stored in the energy storage, MWh
i	Energy technology index
ind_cool_{demand}	Individual cooling demand, MWh
inv_i	Total investment in technology i, €
j	Energy technologies that consume fuels and have emissions
lev_inv_i	Levelized cost of investment over the energy plant lifetime, €/MW
$lifetime_i$	Lifetime of the technology i, years
$petr\_dem$	Gasoline demand, MWh
$petr\_imp_k$	Price of import of gasoline in a specific hour, €/MWh
r	Time Index, hour
RO	Fresh water production from sea water desalination using reverse osmosis, m^3
$SOEC$	Hourly production of syngas from solid-oxide electrolysers, MWh
$SOFC$	Hourly production of electricity from solid-oxide fuel cells, MWh
t	hour, h
$var\_O\&M_j$	Variable operating and maintenance costs of energy plants, €/MWh
$water_{demand}$	Demand for fresh water production from sea desalination via RO, m^3
x_i	Capacity variables of energy plants and gas grid, MW

x_j	Generation capacities of energy plants (8,760 variables for each energy plant, representing the generation in each hour during the one year), MWh
$x_{j,EL}$	Hourly generation of technologies which generate electricity, MWh
$x_{j,EL,gas}$	Hourly generation of technologies which generate electricity and are driven by gas, MWh
$x_{j,EL,other}$	Hourly generation of technologies which generate electricity and are driven by other fuel types, or are not fuel-driven (Photovoltaics (PVs) and wind turbines), MWh
$x_{j,battery,storage_{ch}}$	Hourly charge of vehicles battery storage, MWh
$x_{j,battery,storage\_dis}$	Hourly discharge of electricity of vehicles battery storage, MWh
$x_{j,battery,storage\_grid\_dis,r}$	Hourly discharge of electricity of vehicles battery storage to the power grid (vehicle-to-grid (V2G)), MWh
$x_{j,grid_{battery},storage_{ch}}$	Hourly charge of electricity grid battery storage, MWh
$x_{j,grid\_battery,storage\_dis}$	Hourly discharge of electricity grid battery storage, MWh
$x_{j,heat,storage\_ch,t}$	Hourly charge of heat to the heat storage operated in the DH system t, MWh
$x_{j,heat,storage\_dis,t}$	Hourly discharge of heat from the heat storage operated in the DH system t, MWh
$x_{j,wasteheat,l}$	Heat generation needed for absorption chillers; from gas, waste CHPs, solar thermal or waste heat from data centres, MWh
x_k	Import or export across the system boundaries of different types of energy (8760 variables per one type of energy, representing the flow in each hour during the one year), MWh
η_j	Efficiency of technology, $\mathrm{MWh_{energy}/MWh_{fuel}}$
η_{th}	Efficiency of thermal energy storage, $\mathrm{MWh_{heat}/MWh_{heat}}$
η_{batt}	Efficiency of grid battery storage, $\mathrm{MWh_{ele}/MWh_{ele}}$

Appendix A

Table A1. Optimal capacities of technologies in *PV constrained* scenario (MW).

Individual cooling:	100%	90%	80%	70%	60%	50%	40%	30%	20%	10%	0%
District cooling	0%	10%	20%	30%	40%	50%	60%	70%	80%	90%	100%
Solar heating	0	0	0	1042	2000	2000	2000	2000	2000	2000	2000
Absorbers	0	1300	2778	4098	6749	9788	10,949	12,187	13,478	14,759	16,041
Waste heat data centers	0	0	0	0	735	735	735	735	735	735	735
PTES storage m^3	0	24,168	63,006	47,725	258,573	609,594	747,944	994,079	1,107,915	1,217,468	1,328,843
Battery storage	3017	3321	3621	3917	0	0	0	0	0	0	0
Waste CHP	161	161	161	170	170	170	170	170	170	170	170
Gas CHP	6345	5938	5538	5130	5280	6469	7864	9239	10,656	12,074	13,489
Wind turbine	0	0	0	0	0	0	0	0	0	0	0
PVs	12,250	12,250	12,250	12,250	10,167	500	500	500	500	500	500
El grid import	0	0	0	0	0	0	0	0	0	0	0
El grid export	0	0	0	0	0	0	0	0	0	0	0
Gas net import	13,505	12,746	11,963	11,170	11,420	14,994	17,784	20,538	23,372	26,207	29,040
Gas net export	0	0	0	0	0	0	0	0	0	0	0
Hydrogen storage [MWh]	0	0	0	0	0	5347	3350	3637	1741	289	0
Gas storage [MWh]	233,053	202,799	198,216	195,132	182,997	16,518	13,702	13,045	12,542	11,921	10,789
SOEC	0	0	0	0	0	1021	3738	6386	9072	11,783	14,504
SOFC	0	0	0	0	0	613	2271	3873	5464	7073	8703
Syngas to fuel	0	0	0	0	0	0	0	0	0	0	0
Syngas to natural gas	0	0	0	0	0	0	0	0	0	0	0
Gasoline imports	8405	8283	8378	8392	6791	1016	1026	1026	1029	1030	1032
El vehicles battery capacity	15,214	15,307	15,467	15,563	15,099	14,716	14,700	14,700	14,694	14,692	14,690
Reverse osmosis	38,181	38,073	37,873	37,718	34,981	31,499	31,499	31,499	31,497	31,495	31,495
Geothermal DC capacity	0	0	0	50	50	50	50	50	50	50	50
DC capacity	0	1433	2867	4300	5734	7167	8601	10,034	11,452	12,864	14,276
Non-DC capacity	14,334	12,901	11,468	10,034	8601	7167	5734	4300	2883	1471	0

Table A2. Optimal capacities of technologies in *unconstrained PV* scenario (MW).

Individual cooling:	100%	90%	80%	70%	60%	50%	40%	30%	20%	10%	0%
District cooling	0%	10%	20%	30%	40%	50%	60%	70%	80%	90%	100%
Solar heating	0	0	1241	2000	2000	2000	2000	2000	2000	2000	2000
Electric chillers	0	0	0	0	0	0	0	0	0	0	0
Absorbers	0	1796	3125	5673	6749	9788	10,949	12,187	13,478	14,759	16,041
Waste heat data centers	0	0	0	630	735	735	735	735	735	735	735
PTES storage m^3	0	109,431	109,174	91,822	258,573	609,594	747,944	994,079	1,107,915	1,217,468	1,328,843
Battery storage	70,717	63,949	46,685	28,300	0	0	0	0	0	0	0
Waste CHP	161	161	170	170	170	170	170	170	170	170	170
Gas CHP	4993	4685	4372	4070	5280	6469	7864	9239	10,656	12,074	13,489
Wind turbine	0	0	0	0	0	0	0	0	0	0	0
PVs	36,739	34,176	29,295	23,589	10,167	500	500	500	500	500	500
El grid import	0	0	0	0	0	0	0	0	0	0	0
El grid export	0	0	0	0	0	0	0	0	0	0	0
Gas net import	12,080	11,463	10,837	10,233	11,420	14,994	17,784	20,538	23,372	26,207	29,040
Gas net export	0	0	0	0	0	0	0	0	0	0	0
Hydrogen storage [mwh]	0	0	0	0	0	5347	3350	3637	1741	289	0
Gas storage [MWh]	0	0	0	0	182,997	16,518	13,702	13,045	12,542	11,921	10,789
SOEC	0	0	0	0	0	1021	3738	6386	9072	11,783	14,504
SOFC	0	0	0	0	0	613	2271	3873	5464	7073	8703
Syngas to fuel	0	0	0	0	0	0	0	0	0	0	0
Syngas to natural gas	0	0	0	0	0	0	0	0	0	0	0
Gasoline import	2791	2791	9383	9383	6791	1016	1026	1026	1029	1030	1032
El vehicles battery capacity	20,411	20,399	21,245	20,256	15,099	14,716	14,700	14,700	14,694	14,692	14,690
Reverse osmosis	31,606	31,927	32,370	34,382	34,981	31,499	31,499	31,499	31,497	31,495	31,495
Geothermal DC capacity	0	50	50	50	50	50	50	50	50	50	50
DC capacity	0	1433	2867	4300	5734	7167	8601	10,034	11,452	12,864	14,276
Non-DC capacity	14,334	12,901	11,468	10,034	8601	7167	5734	4300	2883	1471	0

B.

Calculation of levelized investment costs:

$$lev\_inv_i = inv_i \cdot \frac{dis\_rate_i}{1 - (1 + dis\_rate_i)^{-lifetime_i}} \tag{8}$$

Electricity balance:

$$
\begin{aligned}
x_{j,EL,gas} + x_{j,EL,other} \quad &+ x_{j,battery,storage\_grid\_dis} + x_{j,grid\_battery,storage\_dis} + SOFC \\
&- x_{j,battery,storage_{ch}} - x_{j,grid_{battery},storage_{ch}} - \frac{chiller_{DC}}{COP_{DC_{chiller}}} - \frac{chiller_{individual}}{COP_{individual}} - 0.1 \\
&\cdot geothermal_{DC} - RO - \frac{SOEC}{\eta_{SOEC}} + flex - flex_{ch} \geq el\_dem
\end{aligned} \tag{9}
$$

Gas balance:

$$
x_{j,an\_dig} + gas\_imp \quad + gas_{synthesis} + x_{j,gas,storage_{dis}} - x_{j,gas,storage\_ch} - \frac{x_{j,EL,gas}}{\eta_j} \geq gas\_dem \tag{10}
$$

Individual cooling balance:

$$chiller_{individual} \geq ind\_cool_{demand} \tag{11}$$

The endogenous choice between shares of the district and individual cooling:

$$DC_{demand} + ind\_cool_{demand} \geq cool_{demand,\ total} - cool_{en,ef} \tag{12}$$

Balance of absorption chillers:

$$x_{j,wasteheat,l} + \eta_{th} x_{j,heat,storage\_dis,r} \geq (abs_{DC} / COP_{abs}) \tag{13}$$

Transport demand balance:

$$petr_{dem} + ele_{transport} \cdot C_1 + methanol \geq transp\_demand \tag{14}$$

Hydrogen balance:

$$SOEC - \frac{SOFC}{\eta_{SOFC}} - \frac{gas_{synthesis}}{\eta_{SNG}} - \frac{methanol}{\eta_m} + x_{j,syngas,storage_{dis}} - x_{j,syngas,storage_{ch}} \geq 0 \tag{15}$$

Gasoline balance:

$$petr_{imp} \geq petr_{dem} \tag{16}$$

Drinkable water production balance:

$$RO \geq water_{demand} \tag{17}$$

Capacity constraints:

$$x_j \leq x_i \cdot t \tag{18}$$

Constraints on transmission capacity (power, gas, district cooling):

$$x_k \leq x_i \cdot t \tag{19}$$

Carbon dioxide emissions constrain:

$$CO2_{inten j} \cdot x_j + CO2_{inten k} \cdot x_k \leq CO2_{cap} \tag{20}$$

Biomass production constrain:

$$\frac{x_{j,EL,biomass}}{\eta_{j,EL}} \leq algae\_prod \tag{21}$$

Heat storage balance equation:

$$heat\_level_r = heat\_level_{r-1} + x_{j,heat,storage\_ch,r} - \eta_{th} x_{j,heat,storage\_dis,r} \tag{22}$$

Starting-end point constraint:

$$heat\_level_1 = heat\_level_{8760} = 0 \tag{23}$$

References

1. United Nations Framework Convention on Climate Change. The Paris Agreement 2015. Available online: https://unfccc.int/process-and-meetings/the-paris-agreement/the-paris-agreement (accessed on 27 January 2019).
2. Intergovernmental Panel on Climate Change. Global Warming of 1.5 °C. 2018. Available online: https://www.ipcc.ch/sr15/ (accessed on 27 January 2019).
3. IEA. *Energy Technology Perspectives 2016: Towards Sustainable Urban Energy Systems*; IEA: Paris, France, 2016; Volume 14. [CrossRef]
4. Dominković, D.F. *Modelling Energy Supply of Future Smart Cities*; Technical University of Denmark: Kongens Lyngby, Denmark, 2018.
5. The European Association of Local Authorities in Energy Transition. Energycities—Where Action & Vision Meet n.d. Available online: http://www.energy-cities.eu/ (accessed on 4 January 2019).
6. Global Partnership Supporting Cities. Cities Alliance—Cities without Slums n.d. Available online: http://www.citiesalliance.org (accessed on 4 January 2019).
7. A Netowork of Cities in Europe and Beyond. CIVITAS—Cleaner and Better Transport Solutions n.d. Available online: https://civitas.eu/ (accessed on 4 January 2019).
8. European Association. Climate Alliance n.d. Available online: https://www.climatealliance.org/ (accessed on 4 January 2019).
9. Hooftman, N.; Oliveira, L.; Messagie, M.; Coosemans, T.; Van Mierlo, J. Environmental analysis of petrol, diesel and electric passenger cars in a Belgian urban setting. *Energies* **2016**, *9*, 84. [CrossRef]
10. IEA. *Energy and Air Pollution*; IEA: Paris, France, 2016.
11. Dominković, D.F.; Dobravec, V.; Jiang, Y.; Nielsen, P.S.; Krajačić, G. Modelling smart energy systems in tropical regions. *Energy* **2018**, *155*, 592–609. [CrossRef]
12. Connolly, D.; Lund, H.; Mathiesen, B.V. Smart Energy Europe: The technical and economic impact of one potential 100% renewable energy scenario for the European Union. *Renew. Sustain. Energy Rev.* **2016**, *60*, 1634–1653. [CrossRef]
13. Hannan, M.A.; Hoque, M.M.; Mohamed, A.; Ayob, A. Review of energy storage systems for electric vehicle applications: Issues and challenges. *Renew. Sustain. Energy Rev.* **2017**, *69*, 771–789. [CrossRef]
14. Zhao, H.; Wu, Q.; Hu, S.; Xu, H.; Rasmussen, C.N. Review of energy storage system for wind power integration support. *Appl. Energy* **2015**, *137*, 545–553. [CrossRef]
15. Li, C.; Wang, R.Z.; Wang, L.W.; Li, T.X.; Chen, Y. Experimental study on an adsorption icemaker driven by parabolic trough solar collector. *Renew. Energy* **2013**, *57*, 223–233. [CrossRef]
16. Riepl, M.; Loistl, F.; Gurtner, R.; Helm, M.; Schweigler, C. Operational performance results of an innovative solar thermal cooling and heating plant. *Energy Procedia* **2012**, *30*, 974–985. [CrossRef]
17. Marugán-Cruz, C.; Sánchez-Delgado, S.; Rodríguez-Sánchez, M.R.; Venegas, M.; Santana, D. District cooling network connected to a solar power tower. *Appl. Therm. Eng.* **2015**, *79*, 174–183. [CrossRef]
18. Lund, H.; Østergaard, P.A.; Connolly, D.; Ridjan, I.; Mathiesen, B.V.; Hvelplund, F.; Thellufsen, J.Z.; Sorknæs, P. Energy Storage and Smart Energy Systems. *Int. J. Sustain. Energy Plan. Manag.* **2016**, *11*, 3–14. [CrossRef]
19. Cole, W.; Frew, B.; Mai, T.; Sun, Y.; Bistline, J.; Blanford, G.; Young, D.; Marcy, C.; Namovicz, C.; Edelman, R.; et al. *Variable Renewable Energy in Long-Term Planning Models: A Multi-Model Perspective Variable Renewable Energy in Long-term Planning Models: A Multi-Model Perspective*; National Renewable Energy Laboratory: Golden, CO, USA, 2017.
20. IRENA. Planning for the Renewable Future. Available online: https://www.irena.org/ (accessed on 30 December 2017).
21. Palmintier, B.; Bugbee, B.; Gotseff, P. Representative Day Selection Using Statistical Bootstrapping for Accelerating Annual Distribution Simulations. In Proceedings of the 2017 IEEE Power & Energy Society Innovative Smart Grid Technologies Conference, Torino, Italy, 26–29 September 2017; pp. 1–5.

22. Bistline, J.E. Economic and technical challenges of flexible operations under large-scale variable renewable deployment. *Energy Econ.* **2017**, *64*, 363–372. [CrossRef]
23. Barrows, C.; Mai, T.; Hale, E.; Lopez, A.; Eurek, K. Considering renewables in capacity expansion models: Capturing flexibility with hourly dispatch. In Proceedings of the 2015 IEEE Power & Energy Society General Meeting, Denver, CO, USA, 26–30 July 2015.
24. Mai, T.; Barrows, C.; Lopez, A.; Hale, E.; Dyson, M.; Eurek, K. *Implications of Model Structure and Detail for Utility Planning: Scenario Case Studies Using the Resource Planning Model Implications of Model Structure and Detail for Utility Planning: Scenario Case Studies Using the Resource Planning Model*; National Renewable Energy Laboratory: Golden, CO, USA, 2015.
25. Frew, B.A.; Jacobson, M.Z. Temporal and spatial tradeoffs in power system modeling with assumptions about storage: An application of the POWER model. *Energy* **2016**, *117*, 198–213. [CrossRef]
26. Connolly, D.; Lund, H.; Mathiesen, B.V.; Leahy, M. A review of computer tools for analysing the integration of renewable energy into various energy systems. *Appl. Energy* **2010**, *87*, 1059–1082. [CrossRef]
27. Lund, H.; Arler, F.; Østergaard, P.A.; Hvelplund, F.; Connolly, D.; Mathiesen, B.V.; Karnøe, P. Simulation versus optimisation: Theoretical positions in energy system modelling. *Energies* **2017**, *10*, 840. [CrossRef]
28. Lund, H. *EnergyPLAN Documentation*; Version 11.4; Aalborg University: Aalborg, Denmark, 2014.
29. Open Source. Balmorel Model n.d. Available online: http://balmorel.com/ (accessed on 14 February 2018).
30. Dominković, D.F.; Bačeković, I.; Sveinbjörnsson, D.; Pedersen, A.S.; Krajačić, G. On the way towards smart energy supply in cities: The impact of interconnecting geographically distributed district heating grids on the energy system. *Energy* **2017**, *20*. [CrossRef]
31. World Bank. PPP, World Development Indicators Database. July 2017. Available online: http://data.worldbank.org/indicator/NY.GDP.PCAP.PP.CD?view=chart (accessed on 2 July 2017).
32. International Energy Agency (IEA). IEA: Singapore, Indicators for 2015. 2015 2018. Available online: https://www.iea.org/classicstats/statisticssearch/report/?country=SINGAPORE&product=indicators&year=2015 (accessed on 21 December 2018).
33. International Energy Agency (IEA). *Southeast Asia Energy Outlook*; IEA: Paris, France, 2015.
34. Dominković, D.F.; Rashid, K.B.; Romagnoli, A.; Pedersen, A.S.; Leong, K.C.; Krajačić, G.; Duić, N. Potential of district cooling in hot and humid climates. *Appl. Energy* **2017**, *208*, 49–61. [CrossRef]
35. Dominković, D.F.; Romagnoli, A.; Fox, T.; Schrøder Pedersen, A. Potential of waste heat and waste cold energy recovery in Singapore for district cooling applications: Impacts on energy system. In Proceedings of the 40th Annual IAEE International Conference, Singapore, 18–21 June 2017.
36. Energy Market Authority. Singapore Energy Statistics 2015. Available online: https://www.ema.gov.sg/cmsmedia/Publications_and_Statistics/Publications/SES2015Chapters/Publication_Singapore_Energy_Statistics_2015.pdf (accessed on 27 January 2019).
37. Wikipedia. Power Stations in Singapore n.d. Available online: https://en.wikipedia.org/wiki/List_of_power_stations_in_Singapore (accessed on 9 February 2017).
38. International Energy Agency (IEA). International Energy Agency: Singapore—Balances for 2014 n.d. Available online: https://www.iea.org/statistics/statisticssearch/report/?country=Singapore&product=balances (accessed on 9 February 2017).
39. Wahlroos, M.; Pärssinen, M.; Manner, J.; Syri, S. Utilizing data center waste heat in district heating—Impacts on energy efficiency and prospects for low-temperature district heating networks. *Energy* **2017**, *140*, 1228–1238. [CrossRef]
40. Gadd, H.; Werner, S. *Thermal Energy Storage Systems for District Heating and Cooling*; Woodhead Publishing Limited: Sawston, Cambridge, UK, 2014.
41. Energinet.dk. Technology Data for Energy Plants. Available online: https://ens.dk/sites/ens.dk/files/Analyser/c_teknologikatalog_for_individuelle_varmeanlaeg_og_energitransport_2012.pdf (accessed on 30 July 2012).
42. IRENA. Electricity Storage and Renewables: Costs and Markets to 2030. Available online: https://www.irena.org/ (accessed on 30 December 2017).
43. Uddin, K.; Jackson, T.; Widanage, W.D.; Chouchelamane, G.; Jennings, P.A.; Marco, J. On the possibility of extending the lifetime of lithium-ion batteries through optimal V2G facilitated by an integrated vehicle and smart- grid system. *Energy* **2017**, *133*, 710–722. [CrossRef]

44. Alonso, M.; Amaris, H.; Germain, J.G.; Galan, J.M. Optimal charging scheduling of electric vehicles in smart grids by heuristic algorithms. *Energies* **2014**, *7*, 2449–2475. [CrossRef]
45. Karimi, I.A.; Shamsuzzaman, F.; Saeys, M.; Chen, W.; Aggrawal, S.; Vasudevan, S.; Chee, S.Y.; Quah, M.; Obbard, J.; Yan, Y.; et al. *Carbon Capture & Storage/Utilisation: Singapore Perspectives*; National University of Singapore: Singapore, 2014.
46. Luther, J.; Reindl, T. *Solar Photovoltaic (PV) Roadmap for Singapore*; Singapore Economic Development Board (EDB) and Energy Market Authority (EMA): Singapore, 2013.
47. Oliver, G.J.H.; Palmer, A.C.; Tjiawi, H.; Zulkefli, F. Engineered geothermal power systems for Singapore. *IES J. Part A Civ. Struct. Eng.* **2011**, *4*, 245–253. [CrossRef]
48. Dominkovic, D.F.; Stark, G.; Hodge, B.M.; Pedersen, A.S. Integrated energy planning with a high share of variable renewable energy sources for a Caribbean Island. *Energies* **2018**, *11*, 2193. [CrossRef]
49. Connolly, D.; Lund, H.; Mathiesen, B.V.; Werner, S.; Möller, B.; Persson, U.; Boermans, T.; Trier, D.; Østergaard, P.A.; Nielsen, S. Heat roadmap Europe: Combining district heating with heat savings to decarbonise the EU energy system. *Energy Policy* **2014**, *65*, 475–489. [CrossRef]

Article

Fuel Switch to LNG in Heavy Truck Traffic

Ivan Smajla, Daria Karasalihović Sedlar *, Branko Drljača and Lucija Jukić

Faculty of Mining, Geology and Petroleum Engineering, University of Zagreb, Pierottijeva 6, 10 000 Zagreb, Croatia; ivan.smajla@rgn.hr (I.S.); branko27493@gmail.com (B.D.); lucija.jukic@rgn.hr (L.J.)
* Correspondence: daria.karasalihovic-sedlar@rgn.hr; Tel.: +385-1-5535-829

Received: 27 December 2018; Accepted: 2 February 2019; Published: 6 February 2019

Abstract: Liquefied natural gas (LNG) use as a fuel in road and maritime traffic has increased rapidly, and it is slowly entering railroad traffic as well. The trend was pushed by the state administrations of mainly EU countries and international organizations seeing LNG as a cost-effective and environmentally friendly alternative to diesel. Different infrastructural projects for the widespread use of LNG in transport have been launched around the world. The main goal of this paper was to analyze use of LNG as a fuel for heavy trucks. Different aspects of LNG chain were analyzed along with economical and ecological benefits of LNG application. Filling stations network for LNG were described for the purpose of comparative analysis of diesel and LNG heavy trucks. Conclusion has shown that using LNG as propellant fuel has numerous advantages over the use of conventional fuels. The higher initial investment of the LNG road vehicles could be amortized in their lifetime use, and in the long-term they are more affordable than the classic diesel vehicles. In addition to cost-effectiveness, LNG road vehicles reduce CO_2 emissions. Therefore, the environmental goals in transport, not only of the member states but worldwide, could not be met without LNG in heavy truck traffic.

Keywords: heavy truck traffic; road transport; liquefied natural gas (LNG); alternative fuels

---

1. Introduction

The last two decades the global natural gas market has undergone major changes triggered by increased production of unconventional gas in the United States (US) [1]. The unconventional gas has influenced the price of natural gas globally, especially on the US market. The price affects the balance of the supply and demand and the market share of liquefied natural gas (LNG) on the global market is mainly driven by the price, but more and more it becomes the strategic and geopolitical issue [2,3]. Global gas demand is expected to grow 1.6% per annum over the next four years, reaching a level of nearly 4 billion cubic meters by 2022, compared to 3.63×10^{12} m^3 in demand in 2016 [4]. Industrial sector demand is becoming the main driver of gas consumption growth, replacing the electricity production sector, where renewable sources and coal compete the natural gas. Predictions show that the US, as the world's largest consumer and natural gas producer, will account for 40% of the world's additional gas production by 2022 thanks to the remarkable growth of domestic production. Not lately than 2022, US production would amount to 890×10^9 m^3, or more than a one fifth of global gas production. Therefore, the large amounts of produced gas could be turned into LNG and sent worldwide for exports [4]. Large quantities of export LNG in the US could result in lower prices for natural gas in Europe, especially lower prices for LNG shipments coming to Europe [5]. Even without lower prices the number of LNG import terminals is increasing in Europe and with lower prices the expansion of the LNG market could be even bigger and LNG could be very favorable for use as a fuel for heavy trucks consequently [6,7]. So far, many studies have shown the economic and environmental benefits of using LNG as a fuel for heavy trucks. Arteconi et al. in their study from 2009 have analyzed the life-cycle greenhouse gas of LNG as a heavy vehicle fuel in Europe. They have compared the life cycle in terms of greenhouse gasses (GHG) emissions of diesel and LNG used as fuels for heavy-duty vehicles at the European market (then EU 15). They study revealed that numerous studies have reported different results depending on different assumptions. In their study, two possible LNG procurement strategies were considered in the case of LNG being traded at the regasification terminal or in the case of small-scale LNG. They determined that the use of terminal enables a 10% reduction in GHG emissions compared to diesel, while the emission from small-scale LNG are comparable to diesel ones [8].

Ou and Zhang in their study from 2013 have examined the primary fossil energy consumption and GHG emissions for natural gas based alternative vehicle fuels in China. The results of their study have shown that compressed natural gas (CNG) and LNG fueled vehicles have similar well-to-wheels energy use compared to conventional gasoline and diesel vehicles, but difference is emerging along with mileage. They also concluded that CNG and LNG vehicles emit 10–20% less GHG emissions than gasoline and 5–10% less than diesel vehicles. Due to lower content of carbon in natural gas than in petroleum, gas to liquid vehicles emit approximately 30% more GHG emissions than conventional vehicles but the carbon emission intensity of the LNG energy chain is highly sensitive to efficiency of natural gas liquefaction and process used [9]. The environmental impacts and benefits of LNG use in heavy trucks will be furthermore investigated in following sections.

Other studies during literature research reveal economic aspects of LNG use in transport sector. Different research that were related to use of natural gas in transport resulted in fact that LNG use is sensitive to the fuel price differentials. An increase in natural gas price will reduce the economic appeal of using LNG in transport. In the Energata study, the important factors that affect the economics of LNG in transport focusing on LNG vehicles versus diesel have been highlighted along with opinion on the effects of price increase on the future of natural gas vehicles. The study results showed that using LNG for transport would continue to make economic sense even in the case of price increase but resulting on payback time for a LNG vehicle owner. Overall, it was concluded that increasing environmental awareness, more government support like subsidies and stricter environmental regulations would influence on increase of natural gas in transport sector [10].

Madden et al. have conducted exhaustive research for CEDIGAZ, the International Association for Natural Gas, predicting that LNG as a fuel will capture a significant market share in the transport sector by 2035. They have observed the greatest potential in road transport, were annual demand was projected to 96 million tons per year in the base scenario, while demand in the marine transport of LNG was estimated to 77 million tons per annum (mtpa). The rail sector could add another 6 mtpa to global demand according to their predictions. They also stated that the development of LNG as a transport fuel faces a number of challenges including going hand in hand with the development of fueling infrastructure [11]. The fact that filling stations network have not been developed at all in South East Europe will be further more analyzed in following sections.

The Pfoser and al. in their latest research from 2018 discussed acceptance of LNG as an alternative fuel and have analyzed determinants and policy issues. Their study implied that transport sector contributed a quarter of the total greenhouse gas emissions in the EU-28 in 2015. Road transport was responsible for almost 73% of the total greenhouse gas emissions from transport [12]. Authors stated that several alternative fuel technologies have emerged recently. The main issue concerning the most of them are considerable restrictions in case of heavy trucks traffic or long distance transport. Electric vehicles offer short ranges and long recharging time limiting their application to urban transport and short distances. The high potential of hydrogen in reducing GHG emissions is on the other hand decreased with production costs of hydrogen that are still very high limiting wider application. The use of biofuels such as bioethanol and biodiesel is problematic due to land availability for their production [12]. Furthermore they also analyzed environmental benefits of introducing LNG as alternative fuel including the clean combustion of LNG causing nearly 99% less particle (PM) and sulfur oxide (SO_x) emissions, around 80% less nitrogen oxides (NO_x) and around 20% less CO_2 compared to diesel fuel. They stated that emissions reduction could be even further enhanced by mixing liquefied methane from renewable resources into the fuel. Since transport sector represents the severe obstacles to compliance with European policy environmental targets, and road transport in particular, it is clear that any greenhouse gasses (GHG) reduction represents the step forward, which was the main reason for conducting proposed research concerning possibility of use of LNG in transport sector in South East European region. Since, measures for increasing share of LNG in transport use, not only road but also marine, river and railroad could bring environmental benefits and consequently increase competitive advantages of LNG it was necessary to investigate specific data of LNG road transport use and possibility of introducing LNG into transport sector of South East Europe [12,13]. In the Central and South Europe and also in the Republic of Croatia as a part of this region, there is currently no infrastructure for LNG use, but according to European regulations, the Government of the Republic of Croatia has adopted the law for alternative fuels and the LNG infrastructure development plan [14,15]. The similar situation is in other countries in region. Besides Spain, France, and West European countries, the LNG infrastructure network for LNG use in EU in transport is least-developed or it is not developed at all. Due to the facts that the increased volumes of LNG are available for export worldwide, that the flexibility of LNG carriers has increased lately, that the share of long-term contracts among LNG suppliers decreased, the world order of the global LNG market changes which make LNG more and more accessible and affordable and therefore suitable for widespread use.

2. LNG as a Fuel

Liquefaction of natural gas reduces its volume by approximately 600 times making it economical for transport. It is colorless and odorless, it is not toxic and does not corrode. LNG is produced from natural gas in the liquefaction plants where the liquefaction process begins with the separation of water, acid gases, and heavy hydrocarbons that would freeze in the process and become solid or damage the plant otherwise. LNG typically contains between 85 to 95% of methane, but may also contain other components, such as ethane, propane, butane, and nitrogen. Purified gas is then cooled into liquid state at $-162\,^{\circ}C$, to reduce the volume and ease the transport. Liquefaction of one kilogram of natural gas or biomethane requires around 0.65–0.9 kWh of energy. LNG is then transported using ships specially designed for LNG transport. Upon reaching its destination most of LNG is turned back to gaseous state in regasification terminal, where LNG is heated and further transported by pipelines for industry consumers, households and energy transformation needs [16–18]. A smaller amount of LNG is used as a fuel in maritime and road transport [19]. Although the prices of such vehicles are higher, LNG as a fuel has many advantages over conventional diesel and compressed natural gas (CNG), which will be discussed later in this paper.

LNG's highest heating value is estimated at 24 MJ/l while the lowest higher heating value of LNG is estimated at 21 MJ/l. LNG energy density is 2.4 times higher than compressed natural gas (CNG) and it is comparable to the energy intensity of propane, but it is only 60% of the diesel energy density and 70% of the gasoline energy density [20,21].

2.1. Field of Application

Both CNG and LNG are lighter than gasoline or diesel, but have a lower energy content, which means that a larger tank would be required for the same traveling distance. One liter of diesel fuel has the same energy value as 1.7 liters of LNG. Most significant difference between CNG and LNG is in density (LNG has a density of 466 kg/m^3 at $-162\,^{\circ}C$ and atmospheric pressure while CNG has a density of 215 kg/m^3 at pressure of 250 bar and room temperature). It takes 2.4 liters of CNG to produce the same amount of energy as one liter of LNG. This makes the LNG better fuel when it comes to larger vehicles and vehicles that travel long distances. LNG has several advantages in transport, especially for high power engines [11,22]. The liquid state of the LNG allows a flow similar to the flow of diesel fuel thus resulting in faster filling up a tank comparing to CNG. So far, natural gas has been used as a fuel for heavy vehicles only in vehicles for local use, up to 450 kilometers from filling to filling, such as public transport and garbage removal. The reason is that 1 liter of diesel is equivalent to 5 liters of natural gas compressed to 250 bar. For this reason, the compressed gas fuel tank volume should be about five times larger than the diesel tank, which reduces the convenience for larger distances. This highlights the benefits of LNG, with 1.7 liters being enough to replace a liter of diesel. Also high power engines require high fuel injection pressure in the cylinder, and pressurizing LNG is far more efficient than the equivalent process for CNG [20,23].

The application of LNG in transport has its disadvantages and one of them that has to be taken into consideration is that its application has distance limitation. The 950- to 1200-kilometer fuel cycle between fuel filling is distance for the future filling stations networks planning. There are over 4.6 million heavy trucks and buses running for 340 days a year on average, which is an ideal market for LNG [24]. In addition, LNG has already been applied in maritime traffic, and in Norway it is also considered to be used in railway transport. Most likely, the CNG will be developed further in urban transport systems and delivery trucks, where the CNG is considered an attractive option. Nonetheless, there is a type of filling stations that can quickly and economically profitable generate large volumes of CNG from LNG (LNG-CNG filling station), which means that LNG could enter the CNG market as raw material. The LNG-CNG filling station is economically far more cost-effective than the CNG filling station as far as capital investment and maintenance is concerned and can fill the vehicle's cargo tank much faster because the LNG is a liquid and requires less pressure drops for the same mass flow compared to gas. It should also be noted that the price of LNG is slightly higher than the CNG because it is technologically more demanding to produce [20]. Overall number of natural gas powered vehicles is still relatively small, and it is estimated to be below 5% of all vehicles [25].

2.2. Financial Aspects

Financial aspects of fuel use are the main reason for fuel switch and it is obvious that wider acceptance and consequently application of LNG in transport use should be analyzed firstly from the economical point of view. The LNG truck was about 30–40% more expensive than the diesel truck in 2013 according to energy analysts Enerdata. In the analysis a comparison of the fuel price per kilometer for an LNG truck and a diesel truck was made [10]. The data of the Enerdata analysis showed that the LNG truck is spending 0.138 USD per kilometer less than the diesel truck (Table 1).

Table 1. Comparison of fuel costs per km for diesel and liquefied natural gas (LNG) truck.

Fuel Type	Fuel Consumption Per km	Fuel Price	Fuel Cost Per km
LNG	0.46 m^3/km	0.666 USD/m^3	0.306 USD/km
Diesel	0.39 l/km	1.125 USD/l	0.444 USD/km

Differential fuel costs are reduced to 28% (from the current level of 31%) if half of the price increase (rounded to 0.05 USD/km) is passed on to the end user. If a possible rise in fuel prices is calculated, the difference in fuel costs per kilometer would be reduced to 25%. Although the money-repayment period would be prolonged in the latter case, analysts believe that it is economically viable switching from diesel to LNG trucks [10]. In first example, if a new truck is taken into account (2017 Kenworth T370) that costs 158,943.00 $ assuming that the price of an equivalent LNG truck was 40% higher (222,520.00), the LNG truck would become more cost-effective than a conventional truck after 4 years and 2.5 months. After a decade the total savings would be around 87,000 USD. In the budget analysis, the annual distance travelled is 109,620.50 km [24].

In the second example, it was assumed that the new diesel truck would cost 80,000 euros, so its LNG alternative would cost 108,000 euros and that LNG truck would be 20% more expensive for maintenance. In this scenario, it was taken into account that the truck consumes 30 liters of diesel fuel per 100 kilometers or 25 kg of LNG per 100 kilometers. The price of diesel was 1.15 € per liter while the price of LNG is calculated as follows: at the Spanish hub, the LNG average price for the first quarter of 2017 was about 0.28 €/kg. According to the Competition Agency data, between 30–35% of the retail fuel price is at a wholesale price. If it is assumed that the wholesale price is 33% of the retail price, the price of 0.84 €/kg is obtained. In both cases, the trucks exceeded 110,000 km per year in the period of 15 years. Annual fuel and maintenance costs were calculated and shown in Table 2. Over 15 years, LNG truck would have saved around 138,000 euros and would become more cost effective after 2.16 years.

Table 2. Annual fuel and maintenance costs comparison for diesel and LNG truck.

Fuel Type	Fuel Cost	Maintenance Cost	Total
LNG	23 100 €	11 484 €	34 584 €
Diesel	37 950 €	9570 €	47 520 €

National Laboratory Argonne tested 18 LNG trucks over a 15-month period, and they concluded that LNG as a fuel allows significant reduction of costs and harmful emissions. Similar fuel consumption on the basis of equivalent energy during the study period was achieved in LNG trucks. On average, the fuel costs for LNG truck was around 48% less than for diesel trucks. With such cost estimates, the payback period is less than three years [26].

2.3. Ecological Aspects

The main switch in primary energy use or fuel use as it was previously mentioned was driven by economical aspects of it use. The main driver for fuel price in the future will not be supply and demand mechanism but more and more ecological aspects of fuel use. EU energy policy and environmental targets in energy sector represent the base for considering wider application of LNG in energy consumption in transitional period. The transport sector, as it was previously mentioned, represents one of the problematic issue since electrification of heavy-duty trucks in road traffic or marine transport is not an alternative in near future. Therefore, use of LNG represents alternative in transitional period towards sustainable energy society. The transport sector in the EU countries in 2009 was responsible for 20% of all emissions. One of the most obvious advantages of LNG as a fuel is visible if environmental considerations are taken into account [8,12,27–33]. The European Union encourages the development of alternative fuels to reduce carbon footprint [34]. Greenhouse gas emissions studies have been conducted during the lifecycle of diesel and LNG used in heavy road vehicles in Europe. The lifetime of the fuel is divided into three phases. The first phase is production, which also includes collecting and transportation, as well as diesel refining and LPG liquefaction, second phase is distribution, and the last phase is combustion. LNG liquefied at a large terminal produces about 10% less emissions than diesel fuel and LNG liquefied in a small plant produces around 3% less emissions than diesel fuel [8]. Apart from reducing CO_2 emissions, SO_X, NO_X and particle emissions are also significantly reduced [35,36]. According to Kofod and Stephenson well to wheel GHG emissions from LNG trucks are around 19% lower than the ones from diesel truck and by using bio-LNG it's possible to reduce GHG emissions up to 67% [37,38]. In China, the emissions were calculated based on the Tsinghua Fuel Lifecycle Analysis Model, a specifically designed fuel analysis instrument in China where the conclusion was that LNG powered vehicles have 5 to 10% less emissions than gasoline or diesel powered vehicles [9,27]. According to the research done by Song et al. usage of LNG trucks resulted in slight GHG reductions compared to a diesel trucks, estimated at approximately 8.0% based on the full life cycle [39]. Hao et al. said that one of the most effective strategies toward sustainable transportation is the promotion of natural gas vehicles with LNG as a very promising alternative. Compared to diesel vehicles, LNG vehicles tailpipe emissions can be more successfully controlled [40]. It is also important to notice that LNG powered heavy duty vehicles emit significantly less fine particles, which have a negative effect on human health, than conventional trucks [41]. Even though reductions of said greenhouse gasses are significant, climate benefits may be reduced or even delayed for decades due to loss of methane from gas supply chain. Methane impact on climate change is more than 25 times greater than same mass of CO_2 over a 100-year period [42,43]. Thus, switch to LNG trucks may have negative impact on climate for next 200 years, even though usage of natural gas in other sectors may be beneficial at all time frames [44]. When switching from diesel to LNG heavy duty trucks, reduction in methane losses to the atmosphere may be needed to ensure climate benefits at all time frames. Until better data is available on magnitude of methane loss, impact of fuel switch is still uncertain [31]. Misra et al. report five tested technologies and results show that LNG trucks with the three-way catalyst have lower brake-specific NOx emissions. On the other hand, minimization of natural gas leakage into atmosphere during fueling and storing is crucial [45].

The main switch in primary energy use or fuel use as it was previously mentioned was driven by economical aspects of it use. The main driver for fuel price in the future will not be supply and demand mechanism but more and more ecological aspects of fuel use. EU energy policy and environmental targets in energy sector represent the base for considering wider application of LNG in energy consumption in transitional period. The transport sector, as it was previously mentioned, represents one of the problematic issue since electrification of heavy-duty trucks in road traffic or marine transport is not an alternative in near future. Therefore, use of LNG represents alternative in transitional period towards sustainable energy society. The transport sector in the EU countries in 2009 was responsible for 20% of all emissions. One of the most obvious advantages of LNG as a fuel is visible if environmental considerations are taken into account [8,12,27–33]. The European Union encourages the development of alternative fuels to reduce carbon footprint [34]. Greenhouse gas emissions studies have been conducted during the lifecycle of diesel and LNG used in heavy road vehicles in Europe. The lifetime of the fuel is divided into three phases. The first phase is production, which also includes collecting and transportation, as well as diesel refining and LPG liquefaction, second phase is distribution, and the last phase is combustion. LNG liquefied at a large terminal produces about 10% less emissions than diesel fuel and LNG liquefied in a small plant produces around 3% less emissions than diesel fuel [8]. Apart from reducing CO_2 emissions, SO_X, NO_X and particle emissions are also significantly reduced [35,36]. According to Kofod and Stephenson well to wheel GHG emissions from LNG trucks are around 19% lower than the ones from diesel truck and by using bio-LNG it's possible to reduce GHG emissions up to 67% [37,38]. In China, the emissions were calculated based on the Tsinghua Fuel Lifecycle Analysis Model, a specifically designed fuel analysis instrument in China where the conclusion was that LNG powered vehicles have 5 to 10% less emissions than gasoline or diesel powered vehicles [9,27]. According to the research done by Song et al. usage of LNG trucks resulted in slight GHG reductions compared to a diesel trucks, estimated at approximately 8.0% based on the full life cycle [39]. Hao et al. said that one of the most effective strategies toward sustainable transportation is the promotion of natural gas vehicles with LNG as a very promising alternative. Compared to diesel vehicles, LNG vehicles tailpipe emissions can be more successfully controlled [40]. It is also important to notice that LNG powered heavy duty vehicles emit significantly less fine particles, which have a negative effect on human health, than conventional trucks [41]. Even though reductions of said greenhouse gasses are significant, climate benefits may be reduced or even delayed for decades due to loss of methane from gas supply chain. Methane impact on climate change is more than 25 times greater than same mass of CO_2 over a 100-year period [42,43]. Thus, switch to LNG trucks may have negative impact on climate for next 200 years, even though usage of natural gas in other sectors may be beneficial at all time frames [44]. When switching from diesel to LNG heavy duty trucks, reduction in methane losses to the atmosphere may be needed to ensure climate benefits at all time frames. Until better data is available on magnitude of methane loss, impact of fuel switch is still uncertain [31]. Misra et al. report five tested technologies and results show that LNG trucks with the three-way catalyst have lower brake-specific NOx emissions. On the other hand, minimization of natural gas leakage into atmosphere during fueling and storing is crucial [45].

2.4. Safety Aspects

During history, LNG was misperceived as unsafe fuel since its liquefaction process in the first half of the 20th century was characterized by fatal accidents. But development of LNG industry after the 1950s and all scientific record show that LNG has been used for years without a major incident, so it can be considered as a relatively safe fuel [18,23]. As mentioned before, LNG has no color, is non-toxic or carcinogenic, and may explode only indoors at a concentration of 5% to 15% [46]. However, some special security measures are needed. LNG is cryogenic and dangerous if it comes into contact with skin or eyes, which is why it is imperative to use visor, gloves and other forms of protection when working with LNG. Compared to other fuels, LNG is very safe when all the regulations are followed including best business practices [47].

2.5. Comparison to Other Alternatives

At this moment, there is no alternative powertrain or fuel that is economically attractive for long haul application, except for LNG trucks. Diesel fuel economy in traversed distance per equivalent of liter of diesel of LNG trucks with compression ignition is similar to diesel trucks, while LNG trucks with spark ignition have somewhat lower fuel economy (22% to 28% lower than diesel trucks). Hybridization of both LNG and diesel trucks improve fuel economy, but not significantly for long haul (3%–6%).

When it comes to CO_2 emissions, LNG trucks can compete with diesel hybrid trucks and LNG trucks with compression ignition fare better than even fuel cell and battery electric trucks.

When it comes to breakeven price, as mentioned before, LNG trucks fare better than conventional diesel trucks and hybrids are not attractive option, except for diesel vehicles with a day cycles [48]. CNG and CNG hybrid trucks are also economically attractive options, but only for medium duty and urban applications [49].

3. LNG Application in Road Traffic

LNG vehicles are increasingly being considered and used, in road, maritime, railway traffic. In the use of such vehicles, the most significant is China, which in 2017 had more than 230,000 LNG trucks and around 3000 filling stations (CNG and LNG). There was 540 percent increase of sales of LNG heavy trucks in first seven months of the 2017 which accounts around 39,000 new trucks. Also, China plans to increase a number of filling stations to around 12,000 stations by 2020. Locations and types of this filling stations are carefully chosen using premade studies [50]. That way LNG trucks would spend 40% of the total Chinese consumption of LNG [51–54]. Vehicle use of natural gas is considered to be a high priority by policy enacted in a year 2007 by National Development and Reform Commission of China [55]. US currently has 144 LNG filling stations across the country (76 public) and plans to build 38 new ones [56]. In Europe there are around 1500 LNG trucks along with around 100 LNG filling stations. Filling infrastructure is most developed in Northwest Europe (UK, Netherlands and Norway) [57]. In 2016, 81% of LNG-CNG filling stations in Europe were built in the Netherlands, the United Kingdom, and Spain. In the rest of the Europe, the building of LNG and LNG-CNG stations begun in 2014 [58–62]. Current number of filling stations can be seen on Figure 1 (yellow are filling stations that are planned or in construction) [63].

Figure 1. Current and planned LNG filling stations in Europe (EU) [63].

Within this paper, an analysis of the impact of LNG trucks on ecological and financial aspects for the European Union has been conducted. According to Kofod and Stephenson, well to wheel GHG emissions from LNG trucks are on average 19% less than the diesel ones. They claim that LNG trucks for 1 MJ of energy used produce 211.7 grams of CO_2 equivalent (CO_{2eq}), while a diesel trucks produce 262 grams of CO_2 equivalent for 1 MJ of energy used [37]. These numbers were used in the calculation of annual GHG emission for one truck. The annual GHG emissions were obtained by multiplying annual mileage, average fuel consumption, and the heating value of one liter of diesel or one kilogram of LNG and said CO_2 emission equivalent for 1 MJ of energy used. The Table 3 shows the selected parameters for the calculation of annual GHG emissions.

Table 3. Parameters for calculation of annual greenhouse gasses (GHG) emissions.

Annual mileage (km)	70,000–130,000
Average diesel fuel consumption (l/100 km)	30
Average LNG consumption (kg/100 km)	25
Diesel lower heating value (MJ/l)	36
LNG lover heating value (MJ/kg)	48.6
Diesel GHG emissions (gCO_{2eq}/MJ)	262
LNG GHG emissions (gCO_{2eq}/MJ)	211.7

Annual GHG emissions per truck and emission reduction per truck depending on annual mileage expressed in tonnes of CO_{2eq} are shown in the Table 4.

Table 4. Annual liquefied natural gas (LNG) truck GHG emission reduction.

Annual Mileage (km)	70,000	80,000	90,000	100,000	110,000	120,000	130,000
Diesel missions (tonne of CO_{2eq})	198.07	226.37	254.66	282.96	311.26	339.55	367.85
LNG emissions (tonne of CO_{2eq})	180.05	205.77	231.49	257.22	282.94	308.66	334.38
Emission reduction per truck (tonne of CO_{2eq})	18.02	20.60	23.17	25.74	28.32	30.89	33.47

Due to the lack of data on the average annual mileage at EU level, several values ranging from 70,000 to 130,000 km are assumed for annual mileage. The value of 70,000 km was chosen by the author's experience on the national market, while the value of 130,000 km was selected due to average annual mileage for Dutch distribution services [64]. Average fuel consumption and heating values were chosen based on author's experience.

Over the past two years, around 300,000 new trucks are registered annually in the EU, considering that number, it was possible to calculate annual emission reduction presuming that some of the new diesel trucks would be replaced by LNG trucks [65]. An analysis was made based on several different LNG truck shares in the total number of new trucks in the EU with the assumption that all other trucks are diesel trucks. Proposed shares of LNG trucks were 5%, 10%, 15%, 20%, 25%, 30%, and 50%. GHG emissions reduction depending on the annual mileage of the vehicle and LNG truck share expressed in 1000 tonnes of CO_{2eq} is shown in Table 5.

Table 5. GHG emission reduction for different LNG truck shares on EU level.

		New Trucks Emission Reduction (1000 tonnes of CO_{2eq})						
Annual mileage (km)		70,000	80,000	90,000	100,000	110,000	120,000	130,000
	5%	270.32	308.93	347.55	386.17	424.78	463.40	502.02
	10%	540.63	617.87	695.10	772.33	849.57	926.80	1004.04
Share of LNG	15%	810.95	926.80	1042.65	1158.50	1274.35	1390.20	1506.05
trucks in new	20%	1081.27	1235.74	1390.20	1544.67	1699.14	1853.60	2008.07
trucks	25%	1351.59	1544.67	1737.75	1930.84	2123.92	2317.01	2510.09
	30%	1621.90	1853.60	2085.30	2317.01	2548.71	2780.41	3012.11
	50%	2703.17	3089.34	3475.51	3861.67	4247.84	4634.01	5020.18

Emission reduction was obtained by multiplying the emission reduction per truck, total number of new trucks in the EU and LNG truck share.

As noted above, apart from the ecological, the financial impact of LNG trucks in the EU was also analyzed in this paper. Within this analysis, annual fuel costs per truck for five different diesel prices and five different LNG prices were calculated depending on the annual mileage. The annual fuel costs per truck cost was obtained by multiplying the annual mileage, fuel price and fuel consumption shown in Table 3. These annual costs depending on annual mileage and fuel prices are shown in Table 6.

Table 6. Annual fuel costs for diesel and LNG truck.

		Annual Fuel Costs Per Truck (€)						
Annual mileage (km)		70,000	80,000	90,000	100,000	110,000	120,000	130,000
	0.90	18,900	21,600	24,300	27,000	29,700	32,400	35,100
	1.10	22,050	25,200	28,350	31,500	34,650	37,800	40,950
Diesel price (€/l)	1.30	27,300	31,200	35,100	39,000	42,900	46,800	50,700
	1.50	30,450	34,800	39,150	43,500	47,850	52,200	56,550
	1.70	33,600	38,400	43,200	48,000	52,800	57,600	62,400
	0.80	14,000	16,000	18,000	20,000	22,000	24,000	26,000
	0.98	16,333	18,667	21,000	23,333	25,667	28,000	30,333
LNG price (€/kg)	1.16	20,222	23,111	26,000	28,889	31,778	34,667	37,556
	1.33	22,556	25,778	29,000	32,222	35,444	38,667	41,889
	1.51	24,889	28,444	32,000	35,556	39,111	42,667	46,222

After calculating fuel costs per truck for several different fuel prices, five different price scenarios were created. In these scenarios, it is assumed that all 300,000 trucks in EU are LNG trucks. From the Figure 2 it can be seen that scenario "Diesel = 1.70 €/l; LNG = 1.16 €/kg" has the greatest savings because of the biggest diesel–LNG price ratio. Using this scenario with the assumption that all the trucks in the EU are LNG trucks with the annual mileage of 100,000 km, annual fuel cost savings on the EU level would be over 6.5 billion Euros. On the other side, using scenario "Diesel = 1.10 €/l; LNG = 1.51 €/kg" with the same assumptions as for the previous scenario, annual fuel cost savings would be negative, i.e. losses would be around 1.43 billion Euros on the EU level.

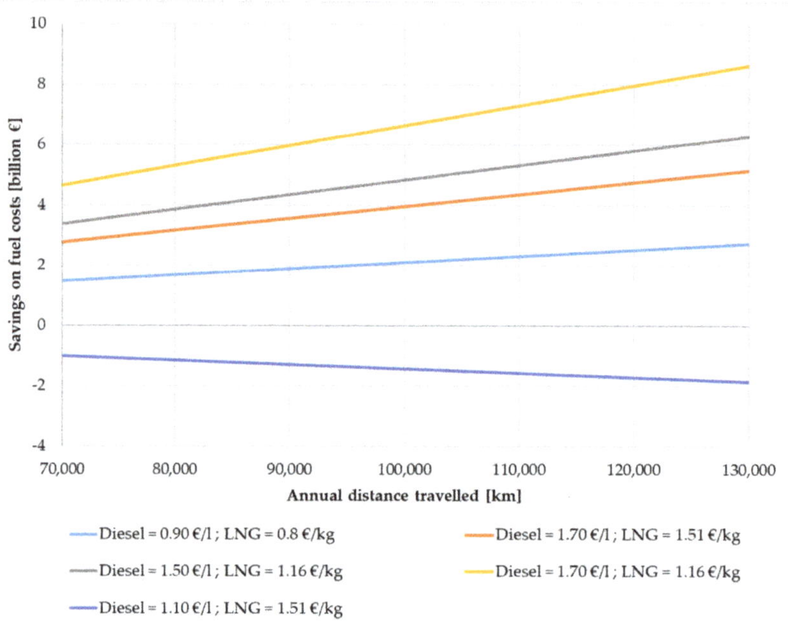

Figure 2. Saving on fuel costs for five different price scenarios on EU level.

Since it is not practical to assume that all trucks in the EU will be LNG trucks, therefore by using the above stated calculations it is possible to calculate fuel savings for any LNG truck share and price scenarios. For example: A diesel truck that exceeds 124,000 km per year has an annual fuel cost of 55,800 € if the price of diesel is 1.5 €/l. The LNG truck for same mileage as the diesel one has annual fuel costs around 41,300 € if the price of LNG is 1.33 €/kg. It is clear that LNG truck annual fuel cost savings are around 14,500 €. For quantifying this savings for a larger area or region, calculated savings per truck needs to be multiplied with the number of LNG trucks in that area or region.

3.1. Filling Stations

Type of LNG stations can be divided in 3 groups: those that can fill LNG as liquid, as gas (LNG-CNG) or as both (LNG and LNG-CNG) [54,66,67]. Each of these filling stations can be made in a permanent, mobile, or semi-mobile configuration. Each configuration has its own advantages and disadvantages [68,69]. Permanent stations are usually the most technologically advanced and can support the provision of more vehicles but they require larger initial investment. The ideal type of station depends on the current and anticipated demand for LNG and the available technology. Countries that have larger fleets of LNG trucks and have some form of LNG charging infrastructure in already established routes, have simpler selection of location for a permanent station [68].

Typical LNG station consists of cryogenic tank for LNG, in the size of 60–70 m^3, cryogenic submerged centrifugal pumps, dispenser with the proper certificates, connected to the system for charging, cooling system that keeps the LNG cold, gas discharge detection equipment, the control system that supports remote access, evaporator, odorizing unit, and dispenser for CNG [68]. LNG stations are structurally similar to conventional diesel and gasoline stations, since they both deliver liquid fuel [70] (Figure 3).

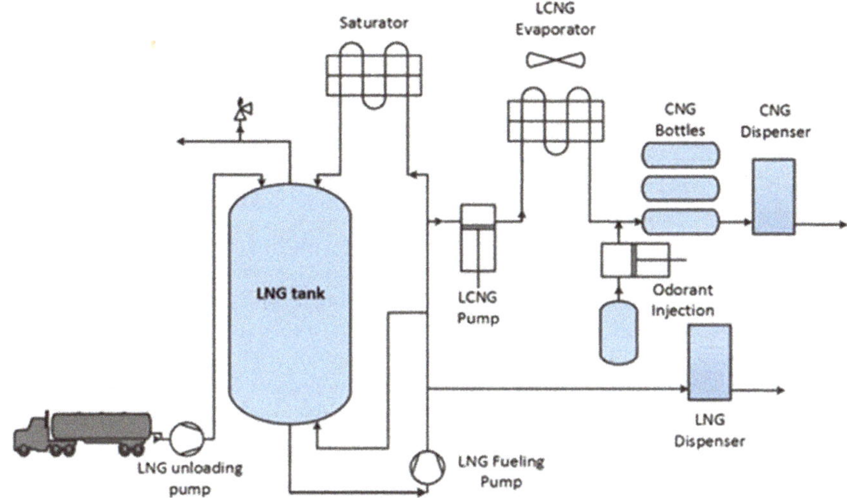

Figure 3. Schematics of typical LNG/CNG (compressed natural gas) filling station [71].

3.2. LNG Heavy Trucks

Most popular LNG heavy trucks on the European market that are currently available are Iveco Stralis Natural Power, Volvo FM Metan Diesel, Hardstaff Mercedes Benz Actros and Scania G 340 LNG. Most of these trucks have several parallel LNG fuel tanks set up similarly to the classic diesel side rail. Containers are set to operate in parallel and the operator does not have to fill them separately [72–74]. The system has one connection for fuel delivery and one connection for ventilation to the filling station. Other parts of the system include interlaced pipe and pipe systems, heat exchanger, and several fuel control devices such as automatic fuel shut-off valve, pressure regulator, tank fuel level indicator, and lights that will light up in the cabin in a case of low temperature or low pressure in the tank. Some systems use pumps, and some, such as the Agility Fuel Solutions system, use the pressure of the tank itself to successfully supply the engine with the required fuel [75].

Large number of LNG systems in trucks don't use pumps. When the engine is running, the natural gas under pressure goes out of the tank to the engine. Cold pressurized fuel passes through a heat exchanger [76] that uses heat from the engine coolant to evaporate the liquid and turn it into gas. When it comes out of the heat exchanger, the fuel is hot, under pressure of the tank, ready to burn in the engine [77]. An additional big advantage of LNG is that the cold energy of the fuel can be used for air conditioning [78]. It is important to note that most of LNG engine designs are not from "clean sheet", but are converted versions of existing conventional fuel engine design. For that reason, there were many issues with commercial LNG engines compared to diesel engines, like poor torque at low speed and bad performance during acceleration [79]. One issue that causes this is low volumetric efficiency due to gasification process as well as bad boosting pressure at low speed and turbo-lag during acceleration. Certain changes were suggested to improve overall performance of LNG engines, especially during accelerating, or driving uphill. These problems can be mended by introducing intake air supply device coupled to a LNG engine intake system which could significantly improve engine torque and transient response. Performance could be further improved by adjusting the fuel-injection quantity, as well as optimizing spark timing [80]. Also, the LNG engine performance depends on the proportion of methane in the LNG mixture [81].

4. LNG Application in Croatia

Central and East Europe (East from Austria) only have few LNG filling stations. To describe the situation in the region Croatia was used since it has a great geostrategic position with the location on the East–West road map with the connection sea ports connection. In the Republic of Croatia, LNG fuel infrastructure does not exist at the moment and there are not any registered vehicles powered by LNG. In last few years Croatian government founded company that is currently working on building the LNG regasification terminal on island Krk, Croatia. The purpose of that terminal will be to increase diversification and security of natural gas supply [82]. On the other side, in 2016 there were 208 registered cars, 84 trucks, 10 mopeds, 6 motorcycles, 108 buses, and 11 CNG powered tractors, along with the infrastructure of two filling stations, in Zagreb and Rijeka [14]. Also, in 2016, a total of 57,911 vehicles were used in the Republic of Croatia that are using liquefied petroleum gas as a fuel, out of which 56,914 cars, 875 trucks, 8 mopeds and motorcycles, 16 buses, and 98 tractors and non-road mobile machinery [15]. From this, it is evident that there is a willingness of consumers to use alternative fuels. In 2017 there were 28,672 heavy trucks registered in Croatia and the number of new heavy trucks doubled in last five years [83]. Currently there are around 2000 new heavy trucks every year. Considering this, it is evident that there is a great potential for using LNG as a fuel for heavy trucks in road traffic. One Croatian company is currently working on the construction of the first LNG filling station in Croatia. Depending on the interest of transport companies Zagreb was chosen as a first location. The European Union encourages the construction of the said station with a 50% co-financing from the Connecting Europe Facility (CEF) program. The same company plans to build additional 11 filling stations throughout Croatia by 2025. The problem in Croatia, unlike in Germany, France and some other EU member states, is a shortage of incentives for purchasing LNG vehicles that would encourage transport companies to buy these vehicles.

In order to align with the EU Directive 2014/94/EU Croatian Parliament has passed the Law on Establishment of Alternative Fuels Infrastructure (09/12/2018). The third article of the law defines the terms to which the law applies, and LNG and CNG are defined as alternative fuels, along with hydrogen, liquefied petroleum gas, biomethane, and others [14]. The fourth article states that the National Policy Framework is to be established within six months of its entry into force. The National Policy Framework was adopted at a government session on the 6 April 2017. One part of the National Policy Framework related to the establishment of the LNG fuel infrastructure for heavy trucks requires following: "With the aim of facilitating the traffic of heavy trucks on main roads in the Republic of Croatia the infrastructure for the supply of LNG for heavy trucks must be available at the outskirts of the cities of Zagreb and Rijeka by 31 December 2025 and by 31 December 2030 in cities Split, Ploče, Slavonski Brod, Zadar, and Osijek (Figure 4), unless it is shown by 2020 for Zagreb and Rijeka and by 2025 for the other listed cities, that there is complete lack of demand. In the case of sufficient demand, it is possible to foresee the setting up of mobile units for the supply of heavy trucks by LNG on motorways at the edge of the cities near the motorway" [15]. The main problem of the road transport not only in Croatia but in all other Eastern Countries is that it has been redirected to marine transport via Kopar and Trieste port (Slovenia and Italy) replacing the great share of heavy road transport. The stricter environmental protection legislation for marine transport dating from 2020 will certainly increase the price of marine transport. On the other hand, some other infrastructure projects like Pelješki most in Croatia (that is on EU PCI—Project of Common Interest) list will result in finishing highway infrastructure in Croatia. Furthermore, Montenegro, and Albania highway projects will result in an Ionian–Adriatic road direction that will compete not only with marine transport but also other road routes like Italian route and therefore will have direct result in a new road direction development requiring filling stations infrastructure. Higher density of road transport will consequently lead to development of alternative fuels infrastructure development.

Figure 4. Infrastructure for the supply of LNG for heavy trucks from the National Policy Framework [6].

The use of LNG as a fuel for heavy trucks would result in a reduction of emissions, thus supporting the aims of several strategic documents of the European Union member states but also nonmember states such as: Europe 2020, White paper 2011—Roadmap to a Single European Transport Area, Clean Energy for All Europeans and National Energy Strategies of the Countries [84–87].

As for the EU, an analysis of the impact of LNG trucks on ecological and financial aspects for the Republic of Croatia has been developed in this chapter. The analysis methodology is the same as in the previous chapter and all parameters except annual mileage from Table 3 were also used in this analysis. For annual mileage, data was taken from the Center for Vehicles of Croatia database stating the average annual mileage for trucks in the Republic of Croatia is around 52,000 km [83]. Due to the accurate data on the annual mileage, a smaller range of annual mileage was taken as opposed to the EU example. Annual GHG emissions per truck and emission reduction per truck depending on annual mileage expressed in tonnes of CO_{2eq} for the Republic of Croatia are shown in the Table 7.

Table 7. Annual LNG truck GHG emission reduction.

Annual Mileage (km)	50,000	52,000	54,000	56,000
Diesel emissions (tonne of CO_{2eq})	141.48	147.14	152.80	158.46
LNG emissions (tonne of CO_{2eq})	128.61	133.75	138.90	144.04
Emission reduction (tonne of CO_{2eq})	12.87	13.39	13.90	14.42

In the past few years, about 2000 new trucks are registered annually in the Republic of Croatia and concerning this data, it was possible to calculate annual GHG emission reduction for the same LNG trucks shares like in the EU example. GHG emissions reduction depending on the annual mileage of the vehicle and LNG truck share expressed in 1000 tonnes of CO2eq is shown in Table 8.

Table 8. GHG emission reduction for different LNG truck shares on Republic of Croatia level.

New Trucks Emission Reduction (1000 tonnes of CO_{2eq})					
Annual Mileage (km)		50,000	52,000	54,000	56,000
	5%	1.29	1.34	1.39	1.44
	10%	2.57	2.68	2.78	2.88
	15%	3.86	4.02	4.17	4.33
Share of LNG trucks in new trucks	20%	5.15	5.35	5.56	5.77
	25%	6.44	6.69	6.95	7.21
	30%	7.72	8.03	8.34	8.65
	50%	12.87	13.39	13.90	14.42

For financial aspects, the same methodology, prices and parameters were used as in the EU example except for annual mileage, which was taken from the Center for Vehicles of Croatia. The annual fuel costs per truck depending on annual mileage and fuel prices are shown in Table 9.

Table 9. Annual fuel costs for diesel and LNG truck.

Annual Fuel Costs Per Truck (€)					
Annual Mileage (km)		50,000	52,000	54,000	56,000
	0.90	13,500	14,040	14,580	15,120
	1.10	15,750	16,380	17,010	17,640
Diesel price (€/l)	1.30	19,500	20,280	21,060	21,840
	1.50	21,750	22,620	23,490	24,360
	1.70	24,000	24,960	25,920	26,880
	0.80	10,000	10,400	10,800	11,200
	0.98	11,667	12,133	12,600	13,067
LNG price (€/kg)	1.16	14,444	15,022	15,600	16,178
	1.33	16,111	16,756	17,400	18,044
	1.51	17,778	18,489	19,200	19,911

After calculating average truck fuel costs for several different fuel prices, five different price scenarios were created (Figure 5). These scenarios are a bit different from the EU ones, with the same assumption that all trucks in Croatia were LNG trucks. As previously, some scenarios predict a substantial saving on fuel costs, while some predict losses. In the author's opinion, the most realistic price scenario for Croatia is "Diesel = 1.30 €/l; LNG = 1.16 €/kg". This scenario predicts annual fuel costs savings for LNG truck around 5250 €. For quantifying costs savings, the same methodology as before can be used.

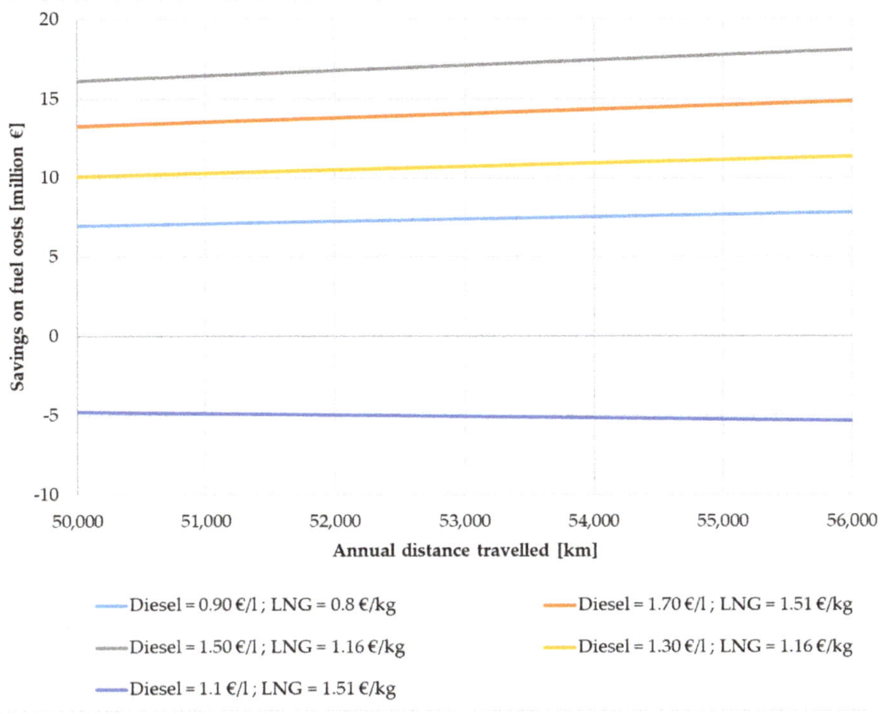

Figure 5. Saving on fuel costs for five different price scenarios on Republic of Croatia level.

5. Conclusions

The use of LNG in heavy trucks traffic certainly represents the contribution of the gas industry in CO_2 emissions reduction in the transportation sector. The environmental advantages over conventional fuels for heavy trucks, despite higher initial investment price for the vehicles, give an advantage to LNG over diesel fuels, especially in terms of legislation banning diesel in EU countries. As the reduction of exhaust emissions and decarbonization of the energy sector are among the strategic goals of the European Union, LNG and other alternative fuels have been promoted through different projects and legal incentives. As was analyzed in the paper, the transport sector represents the severe obstacles to compliance with European policy environmental targets, especially in road transport, therefore calculated greenhouse gases' reduction represents contribution to EU emission reduction goals. The results of conducted research show that increased share of LNG use in road transport could bring environmental and financial benefits and consequently increase competitive advantages of LNG. However, the expansion of LNG as a fuel in the European Union has not been widely spread and not introduced at all in of Southeast Europe, especially due to lack of infrastructure. The undeveloped fuel filling infrastructure and the lack of LNG use as a fuel in road traffic in this region represent Circulus Vitiosus. The LNG vehicles have not been purchased due to lack of infrastructure, and there is no infrastructure because there are not enough vehicles to provide this infrastructure. The undefined provisions of the legal regulations certainly do not help and therefore the harmonization of the regulations along with wider LNG promoting actions with incentives implemented will be needed in the future. Numerous projects implemented across the European Union can be a good example of measures that can be undertaken to allow the use of LNG as a fuel. Southeastern Europe, truck fleet operators, vessels, barges, and a wider public could have increased environmental and economical benefits from encouraging LNG as an alternative fuel, especially considering the possibility of building

LNG terminal in Croatia that will possibly supply all filling stations in the Southeastern European region. Development of an LNG filling station network could connect the whole Europe in LNG road chain terms and therefore would result in increasing role of natural gas as a transitional fuel in transportation sector. This must be also accompanied by broader LNG acceptance, which is an important requirement for developing adequate and effective strategy measures for the introduction of LNG into a specific market in this case Southeast Europe. Measures should include stimulation of LNG demand, increasing the availability of LNG by introducing new terminals and filling stations and by diversification of LNG supply routes and finally improving ecological effects of LNG use as alternative fuel, as was also suggested in the research study conducted by Pfoser et al in 2018. Methodology developed in the paper including results of this research could help in developing national and regional energy scenarios and stating energy strategy goals concerning alternative fuels at specific market.

Author Contributions: Conceptualization, D.K.S. and B.D.; Methodology, I.S.; Software, I.S.; Validation, I.S., D.K.S. and B.D.; Formal Analysis, I.S. and L.J.; Investigation, I.S., D.K.S., B.D. and L.J.; Resources, D.K.S. and L.J.; Data Curation, I.S. and B.D.; Writing-Original Draft Preparation, I.S. and B.D.; Writing-Review & Editing, I.S., D.K.S., B.D. and L.J.; Visualization, I.S. and L.J.; Supervision, D.K.S.; Project Administration, I.S. and L.J.; Funding Acquisition, D.K.S.

Funding: This research was conducted within ESCOM Project. Project is funded by the Environmental Protection and Energy Efficiency Fund with the support of the Croatian Science Foundation grant number PKP-2016-06-6917. Publication process is supported by the Development Fund of the Faculty of Mining, Geology and Petroleum Engineering, University of Zagreb.

Conflicts of Interest: The authors declare no conflict of interest. The funders had no role in the design of the study; in the collection, analyses, or interpretation of data; in the writing of the manuscript, or in the decision to publish the results

References

1. Karasalihović Sedlar, D.; Hrnčević, L.; Brkić, V. Impact of unconventional gas production on LNG supply and demand. *Rudarsko-geološko-naftni zbornik* **2010**, *22*, 37–45.
2. Nosić, A.; Karasalihović Sedlar, D.; Jukić, L. Oil and Gas Futures and Options Market. *Rudarsko-geološko-naftni zbornik* **2017**, *32*, 38.
3. Ledenko, M.; Velić, J.; Karasalihović Sedlar, D. Analysis of oil reserves, production and oil price trends in 1995, 2005, 2015. *Rudarsko-geološko-naftni zbornik* **2018**, *33*, 42. [CrossRef]
4. IEA. IEA Sees Global Gas Demand Rising to 2022 as Us Drives Market Transformation. 2017. Available online: http://www.iea.org/newsroom/news/2017/july/iea-sees-global-gas-demand-rising-to-2022-as-us-drives-market-transformation.html (accessed on 28 August 2018).
5. Asche, F.; Oglend, A.; Osmundsen, P. Gas versus oil prices the impact of shale gas. *Energy Policy* **2012**, *47*, 117–124. [CrossRef]
6. Smajla, I.; Karasalihović Sedlar, D.; Drljača, B.; Jukić, L. Application of liquefied natural gas in road traffic in the Republic of Croatia. In Proceedings of the 13th SDEWES conference, Palermo, Italy, 30 September–4 October 2018.
7. Calderon, M.; Illing, D.; Veiga, J. Facilities for bunkering of liquefied natural gas in ports. *Transp. Res. Procedia* **2016**, *14*, 2431–2440. [CrossRef]
8. Arteconi, A.; Brandoni, C.; Evangelista, D.; Polonara, F. Life—Cycle greenhouse gas analysis of LNG as a heavy vehicle fuel in Europe. *Appl. Energy* **2010**, *87*, 2005–2013. [CrossRef]
9. Ou, X.; Zhang, X. Life-cycle Analyses of Energy Consumption and GHG Emissions of Natural Gas-Based Alternative Vehicle Fuels in China. *J. Energy* **2013**, *2013*, 268263. [CrossRef]
10. Enerdata. Effect of Price Reforms on the Demand of LNG in Transport in China. 2014. Available online: https://www.enerdata.net/publications/executive-briefing/china-lng-price-reforms-effets.html (accessed on 28 August 2018).
11. Madden, M.; White, N.; Le Fevre, C. *LNG in Transportation*; INIS-FR—15-0528; CEDIGAZ: Paris, France, 2014; Volume 46.
12. Pfoser, S.; Schauer, O.; Costa, Y. Acceptance of LNG as an alternative fuel: Determinants and policy implications. *Energy Policy* **2018**, *120*, 259–267. [CrossRef]

13. Ozbilen, A.; Dincer, I.; Hosseini, M. Chapter 4.1—Comparative Life Cycle Environmental Impact Assessment of Natural Gas and Conventional Vehicles. *Exergetic Energetic Environ. Dimens.* **2018**, 913–934.

14. Novine, N. *Zakon o Uspostavi Infrastrukture za Alternativna Goriva*; Narodne Novine: Zagreb, Croatia, 2016.

15. Novine, N. *Odluka o Donošenju Nacionalnog Okvira Politike za Uspostavu Infrastrukture i Razvoj Tržišta Alternativnih Goriva u Prometu*; Narodne Novine: Zagreb, Croatia, 2017.

16. Andersson, H.; Christiansen, M.; Fagerholt, K. Transportation Planning and Inventory Management in the LNG Supply Chain. *Energy Nat. Resour. Environ. Econ.* **2010**, 427–439. [CrossRef]

17. GIIGNL. *Basic Properties of LNG*; LNG Information Paper No. 1; GIIGNL: Paris, France, 2009.

18. Foss, M.M. *An Overview on Liquefied Natural Gas (LNG), Its Properties, the LNG Industry, and Safety Considerations*; Center for Energy Economics: Austin, TX, USA, 2012.

19. Wan, C.; Yan, X.; Zhang, D.; Yang, Z. A novel policy making aid model for the development of LNG fuelled ships. *Transp. Res. Part A Policy Pract.* **2019**, *119*, 29–44. [CrossRef]

20. Bloomers, P.; Oulette, P. *LNG as a Fuel for Demanding High Horsepower Engine Applications: Technology and Approaches*; Westport Innovations Inc.: Vancouver, BC, Canada, 2013.

21. Coetzer, J. A new high energy density battery system. *J. Power Sources* **1986**, *18*, 377–380. [CrossRef]

22. Cheenkachorn, K.; Chedthawut, P.; Choi, G.H. Performance and emissions of a heavy-duty diesel engine fuelled with diesel and LNG (liquid natural gas). *Energy* **2013**, *53*, 52–57. [CrossRef]

23. Siu, N.; Herring, J.S.; Cadwallader, L.; Recce, W.; Byers, J. *Qualitative Risk Assessment for an LNG Refueling Station and Review of Relevant Safety Issues*; No. INEEL/EXT-97-00827-Rev. 2; Idaho National Engineering Lab: Idaho Falls, ID, USA, 1998.

24. AFDC. Average Annual Vehicle Miles Traveled of Major Vehicle Categories. 2017. Available online: https://www.afdc.energy.gov/data/10309 (accessed on 28 August 2018).

25. Chala, G.T.; Abd Aziz, A.R.; Hagos, F.Y. Natural Gas Engine Technologies: Challenges and Energy Sustainability Issue. *Energies* **2018**, *11*, 2934. [CrossRef]

26. Argonne National Laboratory. *Case Study—Liquefied Natural Gas, Argonne*; Argonne National Laboratory: Lemont, IL, USA, 2013.

27. Yan, X.; Crookes, R.J. Life cycle analysis of energy use and greenhouse gas emissions for road transportation fuels in China. *Renew. Sustain. Energy Rev.* **2009**, *13*, 2505–2514. [CrossRef]

28. Burel, F.; Taccani, R.; Zuliani, N. Improving sustainability of maritime transport through utilization of Liquefied Natural Gas (LNG) for propulsion. *Energy* **2013**, *57*, 412–420. [CrossRef]

29. Banawan, A.A.; Morsy El Gohary, M.; Sadek, I.S. Environmental and economical benefits of changing from marine diesel oil to natural-gas fuel for short-voyage high-power passenger ships. *Proc. Inst. Mech. Eng. Part M J. Eng. Marit. Environ.* **2010**, *224*, 103–113. [CrossRef]

30. Pfoser, S.; Aschauer, G.; Simmer, L.; Schauer, O. Facilitating the implementation of LNG as an alternative fuel technology in landlocked Europe: A study from Austria. *Res. Transp. Bus. Manag.* **2016**, *18*, 77–84. [CrossRef]

31. Camuzeaux, J.R.; Alvarez, R.A.; Brooks, S.A.; Browne, J.B.; Sterner, T. Influence of methane emissions and vehicle efficiency on the climate implications of heavy-duty natural gas trucks. *Environ. Sci. Technol.* **2015**, *49*, 6402–6410. [CrossRef]

32. Wegrzyn, J.; Gurevich, M. *Liquefied Natural Gas for Trucks and Buses*; No. 2000-01-2210; SAE International: Warrendale, PA, USA, 2000.

33. Osorio-Tejada, J.L.; Llera-Sastresa, E.; Scarpellini, S. Liquefied natural gas: Could it be a reliable option for road freight transport in the EU? *Renew. Sustain. Energy Rev.* **2017**, *71*, 785–795. [CrossRef]

34. European Commission. *Clean Transport—Support to the Member States for the Implementation of the Directive on the Deployment of Alternative Fuels Infrastructure—Good Practice Examples*; European Commission: Brussels, Belgium, 2016.

35. Baltic Transport Journal 2011. The Best Solution Is LNG. Available online: http://cleantech.cnss.no/wp-content/uploads/2011/11/LNG-articles-Baltic-Transport-Journal-October-2011.pdf (accessed on 28 August 2018).

36. Sharafian, A.; Talebian, H.; Blomerus, P.; Herrera, O.; Merida, W. A review of liquefied natural gas refueling station designs. *Renew. Sustain. Energy Rev.* **2017**, *69*, 503–513. [CrossRef]

37. Kofod, M.; Stephenson, T. Well-to Wheel Greenhouse Gas Emissions of LNG Used as a Fuel for Long Haul Trucks in a European Scenario. *Conf. Pap.* **2013**. [CrossRef]

38. Alamia, A.; Magnusson, I.; Johnsson, F.; Thunman, H. Well-to-wheel analysis of bio-methane via gasification, in heavy duty engines within the transport sector of the European Union. *Appl. Energy* **2016**, *170*, 445–454. [CrossRef]

39. Song, H.; Ou, X.; Yuan, J.; Yu, M.; Wang, C. Energy consumption and greenhouse gas emissions of diesel/LNG heavy-duty vehicle fleets in China based on a bottom-up model analysis. *Energy* **2017**, *140*, 966–978. [CrossRef]

40. Hao, H.; Liu, Z.; Zhao, F.; Li, W. Natural gas as vehicle fuel in China: A review. *Renew. Sustain. Energy Rev.* **2016**, *62*, 521–533. [CrossRef]

41. Giechaskiel, B. Solid Particle Number Emission Factors of Euro VI Heavy-Duty Vehicles on the Road and in the Laboratory. *Int. J. Environ. Res. Public Health* **2018**, *15*, 304. [CrossRef] [PubMed]

42. Anderson, B.; Bartlett, K.B.; Frolking, S.; Hayhoe, K.; Jenkins, J.C.; Salas, W.A. *Methane and Nitrous Oxide Emissions from Natural Sources*; United States Environmental Protection Agency: Washington, DC, USA, 2010.

43. Clark, N.N.; McKain, D.L.; Johnson, D.R.; Wayne, W.S.; Li, H.; Akkerman, V.; Sandoval, C.; Covington, A.N.; Mongold, R.A.; Hailer, J.T.; et al. Pump-to-Wheels Methane Emissions from the Heavy-Duty Transportation Sector. *Environ. Sci. Technol.* **2017**, *51*, 968–976. [CrossRef]

44. U.S. Environmental Protection Agency. *Inventory of U.S. Greenhouse Gas Emissions and Sinks: 1990−2010, EPA Publication 430-R-12-001*; U.S. Environmental Protection Agency: Washington, DC, USA, 2012.

45. Misra, C.; Ruehl, C.; Collins, J.F.; Chernich, D.; Herner, J. In-use NOx emissions from diesel and liquefied natural gas refuse trucks equipped with SCR and TWC respectively. *Environ. Sci. Technol.* **2017**, *12*, 6981–6989. [CrossRef]

46. Kumar, S.; Kwon, H.T.; Choi, K.H.; Lim, W.; Cho, J.H.; Tak, K.; Moon, I. LNG: An eco-friendly cryogenic fuel for sustainable development. *Appl. Energy* **2011**, *88*, 4264–4273. [CrossRef]

47. Luo, D.-X. Safety Guarantee System for Natural Gas Supply Based on Multi-functional LNG Station. *Gas Heat* **2008**, *3*, 024.

48. Zhao, H.; Burke, A.; Zhu, L. *Analysis of Class 8 Hybrid-Electric Truck Technologies Using Diesel, LNG, Electricity, and Hydrogen, as the Fuel for Various Applications; Research Report—UCD-ITS-RR-13-25*; Institute of Transportation Studies, World Electric Vehicle Symposium and Exhibition (EVS27): Barcelona, Spain, 2013.

49. Burke, A.; Zhu, L. *Analysis of Medium Duty Hybrid-Electric Truck Technologies Using Electricity, Diesel, and CNG/LNG as the Fuel for Port and Delivery Applications*; University of California-Davis, Institute of Transportation Studies: Davis, CA, USA, 2014.

50. Li, Y.; Li, W.; Yu, Y.; Bao, L. Planning of LNG Filling Stations for Road Freight: A Case Study of Shenzhen. *Transp. Res. Procedia* **2017**, *25*, 4580–4588. [CrossRef]

51. NGV Global. China Heavy Duty Natural Gas Truck Demand Surges. 2017. Available online: https://www.ngvglobal.com/blog/china-heavy-duty-natural-gas-truck-market-surges-0930 (accessed on 29 August 2018).

52. Xing, Y.; Liu, M.-E. Status quo and prospect analysis on LNG industry in China. *Nat. Gas Ind.* **2009**, *1*, 043.

53. He, T.; Huang, H.; Lin, X.; Li, L. Application analysis and marketing proposals of the LNG vehicles and gas filling stations in China. *Nat. Gas Ind.* **2010**, *9*, 026.

54. Chen, S.P.; Xie, G.F.; Li, Q.Y.; Chang, K. Comparison among LNG, CNG and L-CNG Filling Stations. *Gas Heat* **2007**, *7*, 006.

55. Ma, L.; Geng, J.; Li, W.; Liu, P.; Li, Z. The development of natural gas as an automotive fuel in China. *Energy Policy* **2013**, *62*, 531–539. [CrossRef]

56. AFDC. Alternative Fueling Station Locator. 2018. Available online: https://www.afdc.energy.gov/stations/#/find/nearest (accessed on 14 December 2018).

57. GIE. LNG Map. 2017. Available online: https://www.gie.eu/index.php/gie-publications/maps-data/lng-map (accessed on 14 December 2018).

58. Le Fevre, C.N. *The Prospects for Natural Gas as a Transportation Fuel in Europe*; Oxford Institute for Energy Studies: Oxford, UK, 2014.

59. Giamouridis, A.; Paleoyannis, S. *Security of Gas Supply in South Eastern Europe: Potential Contribution of Planned Pipelines, LNG and Storage*; Oxford Institute for Energy Studies: Oxford, UK, 2011.

60. Simmer, L.; Aschauer, G.; Schauer, O.; Pfoser, S. LNG as an alternative fuel: The steps towards European implementation. *Wit Trans. Ecol. Environ.* **2014**, *186*, 887–898.

61. Vehrs, T. LNG as future transport fuel in Europe. In Proceedings of the 4th Annual Baltic Energy Summit, Tallinn, Estonia, 14–15 November 2012.
62. Prontera, A. *European Energy Security and the Politics of LNG Development in Southern Europe and the Mediterranean*; Conference Paper; UACES: London, UK, 2016.
63. LNG Blue Corridors. LNG Stations in Europe. Available online: http://lngbc.eu/ (accessed on 17 December 2018).
64. Vermeulen, R.; Verbeek, R.; van Goethem, S.; Smokers, R. Emissions Testing of Two Euro VI LNG heavy-Duty Vehicles in the Netherlands: Tank-to-Wheel Emissions. TNO report 2017. TNO 2017 R11336. Available online: https://repository.tudelft.nl/view/tno/uuid:4b1cb389-868b-4a0c-91e2-81599242db5c/ (accessed on 23 January 2019).
65. ACEA. Commercial Vehicle Statistical Press Releases. Available online: https://www.acea.be/statistics/tag/category/commercial-vehicles-registrations (accessed on 23 January 2019).
66. He, H.; Lin, W.; Gu, A.Z. Preliminary Technical Study on L/CNG Stations. *Nat. Gas Ind.* **2007**, *27*, 126.
67. Luo, D.-X. Technical Study of Multifunction LNG Station. *Gas Heat* **2008**, *6*, 034.
68. Gallego, J.; Lebrarto, J.; Leclercq, N. *LNG Stations Regulations. State of the Art*; LNG-BC D4; European Commission: Brussels, Belgium, 2013.
69. Qian, W.B.; Zhang, Y.; Li, D.; Yuan, G. Techno-economic Comparison between L-CNG Filling Station and CNG Filling Station. *Gas Heat* **2012**, *4*, 027.
70. Boyer, M.L. CNG/LNG Filling Station. U.S. Patent No. 9,551,461, 2017. Available online: https://patents.google.com/patent/US20150013831 (accessed on 7 December 2018).
71. SafBon. CNG/LNG (Compressed Natural Gas/Liquefied Natural Gas) Filling Station. Available online: http://www.safbon.com/Index/lists/catid/34.html (accessed on 16 December 2018).
72. Kalet, G.; Gustafson, K. LNG Delivery System. U.S. Patent No. 5,373,702, 20 December 1994.
73. Bassi, A. *Liquefied Natural Gas (LNG) as Fuel for Road Heavy Duty Vehicles Technologies and Standardization*; No. 2011-24-0122; SAE International: Warrendale, PA, USA, 2011.
74. Chandler, K.; Norton, P.; Clark, N. *Alternative Fuel Truck Evaluation Project-Design and Preliminary Results*; No. 981392; SAE International: Warrendale, PA, USA, 1998.
75. Agility Fuel Solutions, LNG Fuel Systems. 2018. Available online: https://agilityfuelsolutions.com/ (accessed on 30 August 2018).
76. Neeser, T.A.; Hedegard, K.W. LNG Delivery System for Gas Powered Vehicles, Minnesota Valley Engineering, Inc. U.S. Patent 5,127,230, 18 February 1992.
77. Chart Industries. LNG Vehicle Fuel Tank System Operations Manual. 2013. Available online: http://files.chartindustries.com/3835849_LNG-Operations-Manual-Final-Draft_010515_web.pdf (accessed on 30 August 2018).
78. Yan, F.; Yi, W.; Xu, B.; Luo, Y. Study on the Characteristics of Heat Exchanger for Cold Energy Recovery in LNG Vehicles. *Energy Procedia* **2016**, *104*, 487–491. [CrossRef]
79. Nwafor, O.M.I. Effect of advanced injection timing on the performance of natural gas in diesel engine. *Sadhana* **2000**, *25*, 11–20. [CrossRef]
80. Tang, Q.; Fu, J.; Liu, J.; Zhou, F.; Yuan, Z.; Xu, Z. Performance improvement of liquefied natural gas (LNG) engine through intake air supply. *Appl. Ther. Eng.* **2016**, *103*, 1351–1361. [CrossRef]
81. Chen, Z.; Zhang, F.; Xu, B.; Zhang, Q.; Liu, J. Influence of methane content on a LNG heavy-duty engine with high compression ratio. *Energy* **2017**, *128*, 329–336. [CrossRef]
82. Veselić, M.; Karasalihović Sedlar, D.; Hrnčević, L. LNG regasification terminals access capacity analysis for security of European natural gas supply. *Rudarsko-geološko-naftni zbornik* **2011**, *23*, 25–38.
83. Centar za vozila Hrvatske. Statistika. Available online: https://www.cvh.hr/tehnicki-pregled/statistika/ (accessed on 24 January 2019).
84. European Commission. *EUROPA 2020*; European Commission: Brussels, Belgium, 2010.
85. European Commission. *White Paper 2011—Roadmap to a Single European Transport Area*; European Commission: Brussels, Belgium, 2011.
86. European Commission. *Clean Energy for All Europeans*; European Commission: Brussels, Belgium, 2016.
87. Hrvatski Sabor. *Strategija Energetskoj Razvoja Republike Hrvatske*; Hrvatski Sabor: Zagreb, Croatia, 2009.

 *energies*

Article

Synthesis and Optimal Operation of Smart Microgrids Serving a Cluster of Buildings on a Campus with Centralized and Distributed Hybrid Renewable Energy Units

Daniele Testi *, Paolo Conti, Eva Schito, Luca Urbanucci and Francesco D'Ettorre

DESTEC (Department of Energy Engineering), University of Pisa, 56122 Pisa, Italy; paolo.conti@unipi.it (P.C.); eva.schito@for.unipi.it (E.S.); luca.urbanucci@ing.unipi.it (L.U.); f.dettorre@studenti.unipi.it (F.D.)
* Correspondence: daniele.testi@unipi.it; Tel.: +39-050-2217109

Received: 30 December 2018; Accepted: 20 February 2019; Published: 23 February 2019

Abstract: Micro-district heating networks based on cogeneration plants and renewable energy technologies are considered efficient, viable and environmentally-friendly solutions to realizing smart multi-energy microgrids. Nonetheless, the energy production from renewable sources is intermittent and stochastic, and cogeneration units are characterized by fixed power-to-heat ratios, which are incompatible with fluctuating thermal and electric demands. These drawbacks can be partially overcome by smart operational controls that are capable of maximizing the energy system performance. Moreover, electrically driven heat pumps may add flexibility to the system, by shifting thermal loads into electric loads. In this paper, a novel configuration for smart multi-energy microgrids, which combines centralized and distributed energy units is proposed. A centralized cogeneration system, consisting of an internal combustion engine is connected to a micro-district heating network. Distributed electric heat pumps assist the thermal production at the building level, giving operational flexibility to the system and supporting the integration of renewable energy technologies, i.e., wind turbines, photovoltaic panels, and solar thermal collectors. The proposed configuration was tested in a hypothetical case study, namely, a University Campus located in Trieste, Italy. The system operation is based on a cost-optimal control strategy and the effect of the size of the cogeneration unit and heat pumps was investigated. A comparison with a conventional configuration, without distributed heat pumps, was also performed. The results show that the proposed configuration outperformed the conventional one, leading to a total-cost saving of around 8%, a carbon emission reduction of 11%, and a primary energy saving of 8%.

Keywords: energy microgrids; energy system integration; energy system optimization; smart building clusters; hybrid renewable systems; heat pumps; district heating; cogeneration

---

1. Introduction

Cogeneration of useful heat and electrical power for an urban district or a cluster of buildings is a technically mature, environmentally-friendly and cost-effective solution, supported by the European Union Directive [1] on energy efficiency, together with the use of renewable energy sources (RES). Indeed, the Directive [2] on energy performance of buildings indicates four high-efficiency technologies, whose feasibility should be evaluated prior to construction of any new building: (a) decentralized energy supply systems based on RES; (b) cogeneration; (c) district or block heating or cooling; and (d) heat pumps.

In view of this, several different configurations of distributed energy systems (DES) have been investigated in recent years, mostly focusing on cogeneration units and RES technologies. Pagliarini and Ranieri [3] studied the effectiveness of thermal storage coupled with a cogeneration engine to satisfy the energy requirements of a university campus, and stressed the importance of the sizing of the storage. Bracco et al. [4] dealt with the topic of distributed generation, by presenting the University of Genoa polygeneration microgrid, which is based on RES and cogeneration units. A tool for the optimal integrated design and operation of a trigeneration system serving a cluster of buildings was proposed by Piacentino et al. [5,6]. The optimal design and operation of a hybrid renewable energy system based on an internal combustion engine and photovoltaic panels was investigated by Destro et al. [7].

In addition to the synthesis and design problems, the optimal operational strategy of energy microgrids has also received considerable attention. Indeed, the adoption of smart control techniques can significantly improve the economic and environmental performances of those systems [8]. For example, Roldán-Blay et al. [9] developed an algorithm for the optimal management of a RES-based electric microgrid. Similarly, Phan et al. [10] investigated schedule strategies to minimize the operating cost of a building energy system with photovoltaic panels and a wind micro-turbine. Asaleye et al. [11] proposed a decision-support tool that identifies the optimal operation of renewable energy microgrids by considering forecast of climate variables.

All the above-mentioned works show the energy, environmental and economic effectiveness of cogeneration systems and renewable energy technologies in smart energy grids. Nevertheless, some unresolved issues remain. Indeed, the intermittent and stochastic nature of RES limits their use, and cogeneration units are characterized by a fixed power-to-heat ratio, thus, they fail to match both fluctuating thermal and electric demands. In this context, electrically driven heat pumps may represent an interesting solution due to their ability to shift thermal loads into electric loads. Moreover, heat pumps are a mature and efficient technology, and they are especially suited to the implementation of smart control strategies [12].

For those reasons, the present work discusses a novel configuration for smart multi-energy microgrids, which consists of distributed energy units and a centralized cogeneration unit feeding a micro-district heating network. Specifically, we investigate the benefits of integrating reversible heat pumps for heating and cooling purposes at the building level. The heat pumps represent an interconnection between the electricity and heating networks, therefore, they can be used to increase the operational flexibility of the microgrid and support the integration of renewable energy technologies, i.e., wind turbine, photovoltaic panels, and solar thermal collectors.

The paper is structured as follows. Section 2 presents the design and modeling methodology of the smart multi-energy microgrid. Sections 3 and 4 present the case study and the optimization problem and methodology, which are used in Section 5 to compare the proposed configuration using distributed heat pumps, with a more conventional solution that employs a centralized CHP (Combined Heat and Power) system and natural-gas boilers. Finally, Section 6 presents the concluding remarks.

2. Energy System Overview and Modeling

In this work, we refer to multi-energy microgrids of medium dimensions with different buildings and loads, using an integrated thermal and electrical energy production system fed by traditional and renewable sources to concurrently satisfy various services (heating, cooling, electrical energy and domestic hot water). The considered generators are: (i) a CHP consisting in an internal combustion engine (ICE); (ii) natural gas boilers; (iii) heat pumps and chillers; (iv) solar thermal collectors; (v) wind turbines; and (vi) photovoltaic modules. Thermal storage is also considered. Figure 1 shows a simplified classification scheme of the reference energy system.

As is well-known, the traditional design approach based on a separate analysis of each component represents a suboptimal design method for multi-energy systems [13]. The so-called simulation-based optimization methods are the most recognized procedures to investigate the best synthesis, sizing and control of integrated system through the simulation of the operative performances. Therefore, in the following sub-section we present the operative dynamic model of each block listed in Figure 1.

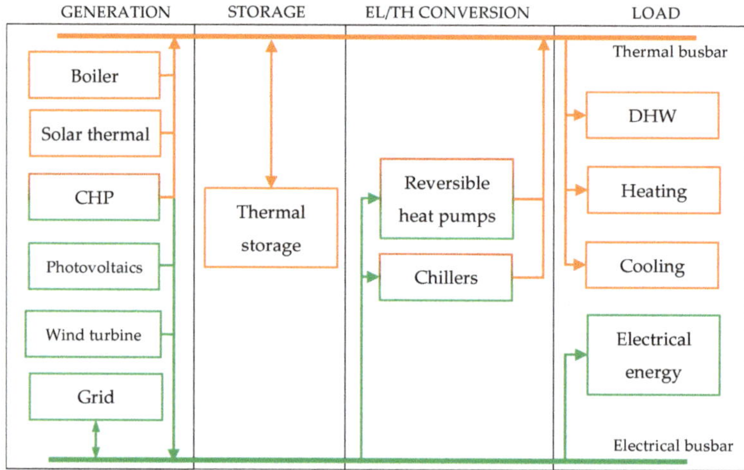

Figure 1. Schematic of the energy system.

Modeling of the System Components

The components models must be based on a proper trade-off between the accuracy of the results and computational effort. The latter feature is essential to allow their employment within the optimization procedure to identify the most efficient design and operation of the smart multi-energy microgrid.

The ICE is modeled through performance curves taken from [14], which provide the thermal and electric efficiency, η_{el} and η_{th}, respectively, as a function of the engine load factor L_{ICE}. The electric (E_{ICE}) and thermal (Q_{ICE}) output are evaluated based on the load factor, L_{ICE}, and the ICE nominal electric power capacity (E_{ICE}^{nom}), as follows:

$$\begin{cases} E_{ICE} = E_{ICE}^{nom} \cdot L_{ICE} \\ Q_{ICE} = [\eta_{th}(L_{ICE})/\eta_{el}(L_{ICE})] \cdot E_{ICE}^{nom} \cdot L_{ICE} \end{cases} \tag{1}$$

The boilers can be modeled with a constant efficiency (η_B) over their whole operating range. Photovoltaic panels are simulated through the model provided by [15], which considers the PV performance as a function of the solar irradiance, PV characteristics and cell array temperature:

$$E_{PV} = n_{PV} S_{PV} \eta_{PV} \eta_{inv} I_{sol,PV} \tag{2}$$

$$\eta_{PV} = \eta_{PV,ref} \left[1 - \beta_{T,PV} \left(T_{PV} - T_{PV,ref} \right) \right] \tag{3}$$

$$T_{PV} = T_{ext} + (219 + 819 K_t) \frac{NOCT - 20}{200} \tag{4}$$

The thermal performances of the ST are evaluated through the classical model illustrated in [16], based on the characteristics of the panel in terms of transmittance and absorptance factors for normal irradiance ($< \tau\alpha >_n$), removal factor (F_r) and frontal losses (U_l). The equations read:

$$\eta_{ST} = F_r(\tau\alpha)_n \left[1 - b_0 \left(\frac{1}{cos\theta} - 1 \right) \right] - \frac{F_r U_L (T_{ST,in} - T_{ext})}{I_{sol,ST}} \tag{5}$$

$$Q_{ST} = n_{ST} S_{ST} \eta_{ST} I_{sol,ST} \tag{6}$$

The electrical power generated by the wind turbine varies as the cube of the wind speed, between a cut-in speed and a nominal speed; the latter corresponds to the nominal electrical power generated by the wind turbine. The nominal electrical power is generated between the nominal wind speed and a cut-out speed. The equations, in accordance with [17], read:

$$E_{WT} = \begin{cases} 0, & w \langle w_{cut-in} \text{ or } w \rangle w_{cut-out} \\ kw^3, & w_{cut-in} \leq w \leq w_{nom} \\ kw_{nom}^3, & w_{nom} \leq w \leq w_{cut-out} \end{cases} \tag{7}$$

where k is a coefficient that depends on the characteristic curve of the generator.

The internal energy variation in the thermal storage is calculated considering the thermal fluxes provided to the water volume from all the connected generators and the heat delivered to the load:

$$V_{TS}\rho_W c_W \Delta T_{TS} = \sum_i Q_{TS,in,i} - \sum_j Q_{TS,out,j} - Q_{TS,ls} \tag{8}$$

where the heat losses of the storage tank are evaluated as

$$Q_{TS,ls} = UA_{TS}(T_{TS} - T_{ext,TS}) \tag{9}$$

Reversible heat pumps and chiller performance are evaluated by means of the so-called second-law efficiency [18]. The method reads:

$$COP = \eta_H^{II} \cdot COP_{id} = \eta_H^{II} \frac{T_{cond}}{T_{cond} - T_{eva}} \tag{10}$$

$$EER = \eta_C^{II} \cdot EER_{id} = \eta_H^{II} \frac{T_{eva}}{T_{cond} - T_{eva}} \tag{11}$$

where COP_{id} and EER_{id} are the coefficients of performance of a reversed Carnot cycle operating between the source and sink temperatures. According to manufacturers, both η_H^{II} and η_C^{II} can be assumed as constant.

The generators are connected to the thermal storages and the buildings through a district heating network (DHN), whose heat losses can be modeled as follows:

$$Q_{DHN,ls} = U_{DHN} L_{DHN} \left(T_{avg,DHN} - T_{ground} \right) \tag{12}$$

Finally, the heating/cooling loads of the buildings can be evaluated through a model that correlates the sol-air temperature [19] with the energy load of the building, based on the standard EN 15306 [20]. This model is further improved by considering the effect of the building thermo-physical properties in shifting the influence of the external climate on the heating/cooling load.

$$Q_{th,H/C} = P_{H/C} \left(1 - \frac{\overline{T_{ext}^*} - T_{des,H/C}}{T_{off,H/C} - T_{des,H/C}^*} \right) \tag{13}$$

$$\overline{T_{ext}^*}(t) = \frac{1}{\overline{\phi}} \sum_{i=0}^{\overline{\phi}} T_{ext}^*(t - \overline{\phi} + 1) \tag{14}$$

$$\overline{\phi} = \sum_i \frac{(UA)_i \phi_i}{[\Sigma_i (UA)_i + H_{ve}]} \tag{15}$$

$$T_{ext}^*(t) = T_{ext}(t) + \frac{\alpha_S}{h_e} I_{sol}(t) \tag{16}$$

Further details on Equations (13)–(16) can be found in [21]. This model represents a good trade-off between simplified models (e.g., the energy signature method [22]), which simply correlates external temperature and heating/cooling load, and dynamic models (e.g., TRNSYS or EnergyPlus), which include the building inertia characteristics, solar radiation, and internal loads, providing more accurate results, but requiring a detailed knowledge of the building envelope and heat gain profiles.

3. Case Study

In this work, we refer to an integrated energy system serving a hypothetical campus, located in Trieste, Italy. This city has a favorable climate, where RES (solar thermal, photovoltaic modules, wind turbines) can provide a significant amount of energy. The Italian Thermotechnical Committee (CTI) provides hourly profiles of external temperature, global solar irradiance on the horizontal plane and wind speed [23]. The monthly-average values of the external temperature and irradiance on horizontal plane are reported in Figure 2.

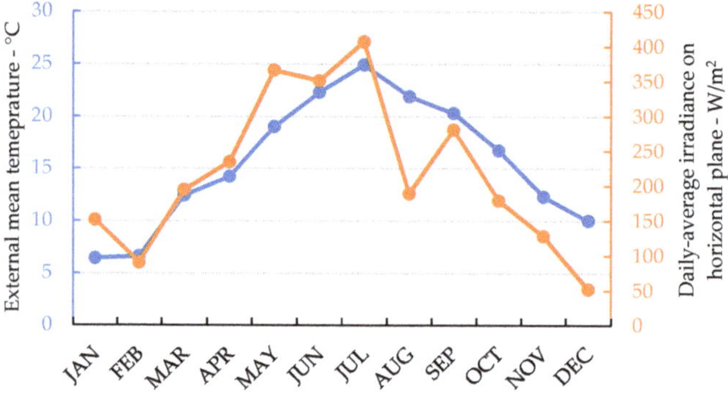

Figure 2. Average monthly temperature and daily irradiance on horizontal plane.

The campus is located far enough from the city to avoid airflow obstructions and shading, which would reduce the renewable energy share. As a whole, the 1000-student campus occupies a surface of 1×0.5 km^2 and includes five dormitories, a dining hall, a gym, a students' center with classrooms and administrative offices. A schematic representation of the campus is shown in Figure 3.

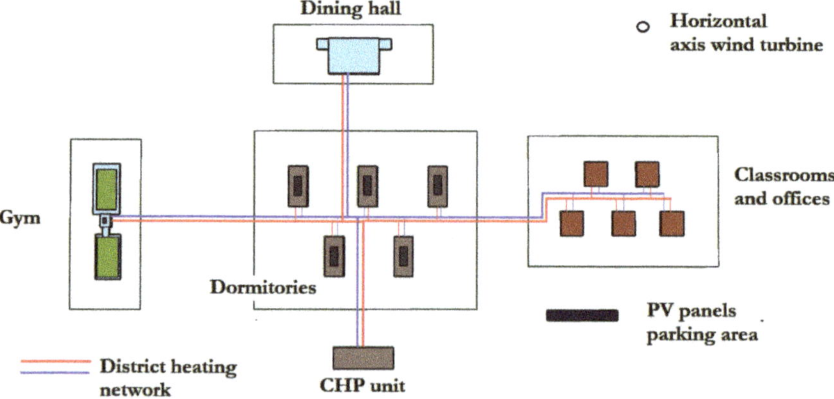

Figure 3. Scale representation of the campus: buildings, district heating, and generation systems.

All the buildings have a similar structure in terms of thermal transmittance of the walls, roofs, floors, and windows (0.29 W/m^2·K, 0.19 W/m^2·K, 0.26 W/m^2·K, 1.80 W/m^2·K, respectively). Specific profiles of internal gains and electrical energy requirements were chosen for each building type, according to their use, periods of presence, and appliances. The internal gains and electrical energy requirements have lower values during weekends and holidays. The buildings have different terminal units that require two supply temperature levels (i.e., low and medium). The power peak and total energy for the heating, cooling, domestic hot water (DHW) and electrical energy services are reported in Table 1.

Table 1. Peak values and energy needs for the four services of the campus and for the five types of buildings.

Tag	Dormitories	Gym	Offices	Dining Hall	Classrooms
Total floor surface, m^2	10,800	260	3168	3740	2112
Terminal units	Radiant panels (Low-temp)	Fancoils (Mid-temp)	Fancoils (Mid-temp)	Fancoils (Mid-temp)	Fancoils (Mid-temp)
Heating load, kW	140	6	20	54	25
Heating demand, MWh	214	6	22	72	29
Cooling load, kW	71	12	32	65	114
Cooling demand, MWh	19	4	24	33	108
DHW load, kW	81	54	0	0	0
DHW demand, MWh	109	75	0	0	0
Power load, kW	50	10	20	20	20
Power demand, MWh	223	45	90	90	90

In this work, we compare a "centralized" configuration (see Figure 4a), in which the DHN is fed by a cogeneration unit and a centralized gas boiler, with a "distributed" one (see Figure 4b), in which reversible heat pumps are installed in the buildings. In both configurations, the sizing of the solar and wind generators remains the same, while an optimization analysis is performed for the CHP and heat pumps, together with the optimal control strategy (see Section 4).

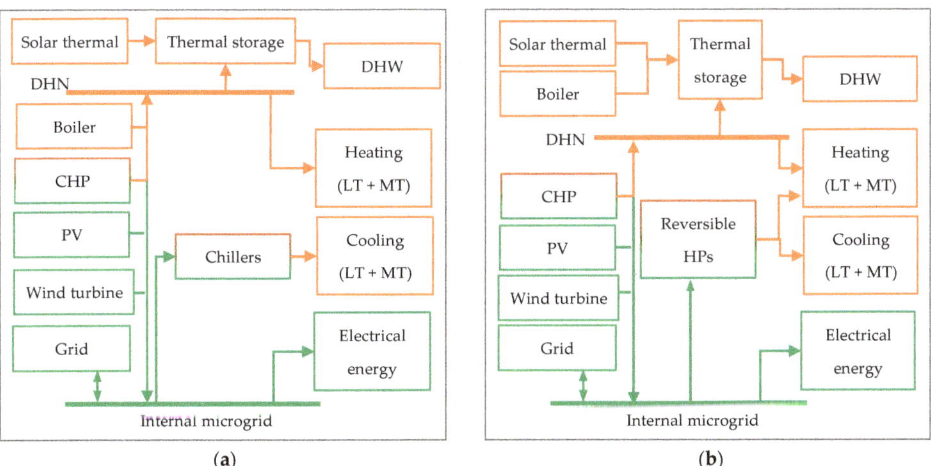

Figure 4. Analyzed configuration for the reference multi-energy microgrid: (**a**) centralized, (**b**) distributed.

The generators concur to satisfy the heating/cooling/DHW/electrical energy requirements of the campus with the following strategy:

- The electrical demand consists of the energy input of the electrical appliances, DHN circulation pumps, and heat pumps or chillers. Electrical energy is purchased from the grid (E_P) when the power generated by the photovoltaic modules (E_{PV}), the wind turbine (E_{WT}) and ICE (E_{ICE}) is not sufficient to satisfy the requirements. On the contrary, if the electrical production is higher than loads, the overproduction is sold to the grid (E_S).
- The DHW service ($Q_{HT,D}$) is required only for dormitories and the gym. The thermal storages are heated by the solar thermal panels (Q_{ST}), and by the DHN or back-up boilers (Q_B) when the temperature drops below the setpoint value (i.e., 55 °C). In the centralized configuration, the DHN represents the back-up generator, while in the distributed configuration both DHN or local boilers can be used (see Figure 4). The thermal storages are only used for the DHW service.
- The heating service is required at two different levels of temperature (low and medium, $Q_{LT,D}$ and $Q_{MT,D}$). In the centralized configuration, it is delivered by the DHN, in the distributed layout it is provided by the DHN or by the local heat pumps (Q_{HP}).
- Also, the cooling service is required at two different levels of temperature ($C_{LT,D}$ cooling requirement through radiant panels, $C_{MT,D}$ cooling requirement through fan-coils). In the centralized configuration, air-to-water electrically-driven chillers are used whereas in the distributed configuration, reversible heat pumps are used.
- In the distributed configuration, the reversible heat pumps are in the buildings and provide both heating and cooling service at all temperature levels.

Table 2 reports all the parameters of the analyzed smart multi-energy microgrid.

Table 2. Parameters used in the analysis.

Parameter	Value	Parameter	Value	Parameter	Value
PV collectors Number	800	ST panels Number	30	Building Time shift, $\bar{\phi}$	3 h
PV Single surface, S_{PV}	1.5 m^2	ST Single surface, S_{ST}	3 m^2	Supply heating temperatures	MT: 45 °C LT: 35 °C
PV coefficient, $\beta_{T,PV}$	0.507%/K	ST Removal factor, F_R	0.8	Supply cooling temperatures	MT: 15 °C LT: 7 °C
PV Reference operational temperature, $T_{ref,PV}$	25 °C	ST Frontal losses, U_L	5 W/(m^2·K)	-	
Overall efficiency of the PV electronic converter, η_{inv}	0.90	ST $(\tau\alpha)_n$	0.7		
PV Nominal operation cell temperature $NOCT_{PVT}$	45 °C	ST angle modifier, b_0	0.1		
Wind turbine Capacity, $P_{WT,nom}$	150 kW	Reversible HPs Second-law efficiency, $\eta^{II}_{H/C}$	0.4/0.35	DHN Length, L_{DHN}	1750 m
WT Nominal speed, w_{nom}	12 m/s	Thermal storage Volume, V_{TS}	20 m^3	DHN Loss coefficient, U_{DHN}	0.15 W/(m·K)
WT Cut-in speed, $w_{cut\text{-}in}$	3.5 m/s	TS Loss coefficient, UA_{TS}	0.02 W/K	Ground temp, T_g	15.6 °C
WT Cut-out speed, $w_{cut\text{-}out}$	20 m/s	TS Set point	55 °C	-	-
Boiler efficiency, η_b	0.90	TS Maximum temperature	90 °C	-	-

4. Optimization Problem and Methodology

In this work, we aim to compare the centralized and the distributed configurations defined in Section 3, from an economical point of view. The selected performance index is the annual total cost, TC, defined as:

$$TC = \frac{INV}{t_{life}} + O\&M + OC \tag{17}$$

where INV/t_{life}, $O\&M$, and OC are the yearly capital, maintenance, and energy operational costs, respectively, and t_{life} is the considered lifetime of the microgrid (i.e., 20 years). Since we are making a comparative analysis, TC only includes the costs that differ in the two configurations, i.e., the purchased equipment cost (PEC) for the ICE unit, reversible HPs, chillers, and boilers, the associated operations and maintenance costs (O&M), and the net cost of the energy purchased from the gas and power grids.

We do not consider the sizing of the RES technologies, which are assumed to have the same design and energy production in both the distributed and centralized configurations; therefore, they are not included in the economic analysis and optimization process. The terms in Equation (17) read:

$$INV = PEC_{ICE} + PEC_{HP} + PEC_C + PEC_B \tag{18}$$

$$O\&M = O\&M_{ICE} + O\&M_{HP} + O\&M_C + O\&M_B \tag{19}$$

$$OC = \sum_i^{8760} OC^i = \sum_{i=1}^{8760} \left(c_F^i F_B^i + c_F^i F_{ICE}^i + c_{el,P}^i E_P^i - c_{el,S}^i E_S^i \right) \tag{20}$$

The timestep length adopted for the energy system simulation is one hour. The investment and maintenance cost functions are presented in Table 3. The costs of HPs, chillers and boilers were obtained through a linear regression of actual manufacturers' data.

Table 3. Cost functions.

Generator	PEC (€)	O&M (€)	References
ICE	$7789\left(E_{ICE}^{nom}\right)^{0.6}$	$0.015 \cdot \sum_i^{8760}\left(E_{ICE}^i\right)$	[24–26]
HPs	$206Q_{HP}^{nom} + 10{,}000$	$0.02 \cdot PEC_{HP}$	[27,28]
Chillers	$206Q_C^{nom} + 3824$	$0.02 \cdot PEC_C$	Manufacturers data and [27]
Boilers	$56Q_B^{nom} + 2222$	$0.02 \cdot PEC_B$	Manufacturers data and [27,28]

According to the operating strategy described in Section 3, both the sizing and control optimization can be written as a function of the ICE nominal electrical capacity, E_{ICE}^{nom}, and load factor profile, L_{ICE}^i.

1. We assumed a set of 201 possible values of E_{ICE}^{nom} in the range between 0 and 1000 kW$_{el}$, namely, $E_{ICE,n}^{nom} \in \{0,\ 5,\ 10\ldots995,\ 1000\} \qquad n = 1,2,3\ldots201$
2. For each *n-th* ICE size, an exhaustive search is performed to find the ICE load factor, $L_{ICE,j}^i$, which minimizes the operational energy cost at any *i-th* timestep, OC^i, (i.e., control optimization). Further details on this point are provided in Section 4.1.
3. The optimal $L_{ICE,n}^i$ sequence determines the thermal and electrical output profiles of the ICE. Subsequently, also the other generators output can be easily found as the residual load. The maximum output times a precautionary factor of 1.1 gives the nominal capacity of each generator.
4. Capital and annual maintenance costs are evaluated for any *n-th* sizing and corresponding optimal control sequence. Then, the total annual cost is evaluated.
5. Finally, we selected the best sizing and corresponding control strategy as the one with the minimum total annual cost, evaluated at point 4.

4.1. Operational Optimization Problem

The operational optimization problem consists in identifying the scheduling of the generators that meets the energy demands at minimum cost (i.e., cost for purchasing electricity from the grid, income for selling electricity to the grid, cost of natural gas). The operational optimization problem is therefore defined as the minimization of the total annual energy cost.

$$min\{TC\} = min\left\{ \sum_{i=1}^{8760} c_F^i F_B^i + \sum_{i=1}^{8760} c_F^i F_{ICE}^i + \sum_{i=1}^{8760} +c_{el,P}^i E_P^i - \sum_{i=1}^{8760} c_{el,S}^i E_S^i \right\} \tag{21}$$

where $i = 1, \ldots, 8760$ timesteps, c_F is the fuel price (0.04 €/kWh), and $c_{el,P}$ and $c_{el,S}$ are the prices of purchased and sold electricity (0.18 and 0.04 €/kWh), respectively. Consequently, the following decision variables are considered:

$$E^i_{ICE},\ Q^i_{ICE},\ Q^i_{B,j},\ Q^i_{HP,LT,k},\ Q^i_{HP,MT,k},\ C^i_{C,LT,k},\ C^i_{C,MT,k},\ E^i_S,\ E^i_P. \tag{22}$$

and demand constraints and balance equations and inequalities are defined as follows

$$Q^i_{DHN,HT,j} + Q^i_{B,j} - Q^i_{HT,netD,j} = 0 \tag{23}$$

$$Q^i_{DHN,MT,k} + Q^i_{HP,MT,k} - Q^i_{MT,D,k} = 0 \tag{24}$$

$$Q^i_{DHN,LT,k} + Q^i_{HP,LT,k} - Q^i_{LT,D,k} = 0 \tag{25}$$

$$Q^i_{ICE} - Q^i_{DHN,ls} - \sum_j Q^i_{DHN,HT,j} - \sum_k Q^i_{DHN,MT,k} - \sum_k Q^i_{DHN,LT,k} \geq 0 \tag{26}$$

$$C^i_{C,LT,k} - C^i_{LT,D,k} = 0 \tag{27}$$

$$C^i_{C,MT,k} - C^i_{MT,D,k} = 0 \tag{28}$$

$$E^i_P - E^i_S - \sum_k E^i_{HP,k} - \sum_k E^i_{C,k} + E^i_{ICE} + E^i_{PV} + E^i_{WIND} - E^i_D - E^i_{aux} = 0 \tag{29}$$

To solve the optimal operation problem, an ad-hoc dispatch strategy algorithm has been developed, based on the following considerations:

- the problem can be considered "static";
- three orders of priority must be considered for the DHN dispatch.

Indeed, in the energy system under investigation, the overall optimum coincides with the sum of optimums of every single timestep, since the behavior of the TSs (Thermal Storages) linked to the solar thermal is independent from the operational control. For this reason, as already shown in [29], the overall operational problem can be split into 8760 subproblems, one for each timestep, and the problem can be considered "static". Moreover, the DHN must always satisfy with higher priority the medium-temperature heat demand $Q^i_{MT,D}$, as opposed to the low-temperature heat demand $Q^i_{LT,D}$, since the HPs operate with higher COP at lower supply temperatures. Therefore, only the three following combinations must be evaluated:

(1) HT–MT–LT
(2) MT–HT–LT
(3) MT–LT–HT

For each possible dispatch priority, once the L_{ICE} is set and the amount of electricity and heat produced by the ICE is defined, the thermal losses and the net amount of heat available at the DHN are known from Equations (23) and (24). Then, Equations (25)–(27) state that the boiler and the HP production must meet the remaining heat demand, if any. Furthermore, Equations (27) and (28) require that the electrical chillers or reversible heat pumps meet the chilled water demand, and Equation (29) defines the electrical energy exchange with the grid. Therefore, as mentioned, the nine decision variables are bound to each other and the problem is conveniently reduced to finding the dispatch priority order and the optimal ICE load factor L_{ICE} that minimize the cost of energy at each timestep. Eleven discrete values of L_{ICE} have been considered, and an exhaustive search algorithm was adopted to identify the optimal solution, among all the possible combinations. This allows the development of a low computational-cost algorithm, compatible with the need for a quick response for real-time implementation [30] and further advanced analyses (e.g., optimal design and uncertainty analysis, as in [29]).

5. Results and Discussion

Figure 5 shows the total annual costs depending on the nominal electrical capacity of the CHP unit. We note that the optimal CHP size is practically the same for both of the configurations (75 kW$_{el}$), with a total cost reduction of about 8% for the distributed solution. Economic details are presented in Table 4. The size of all generators are shown in Table 5.

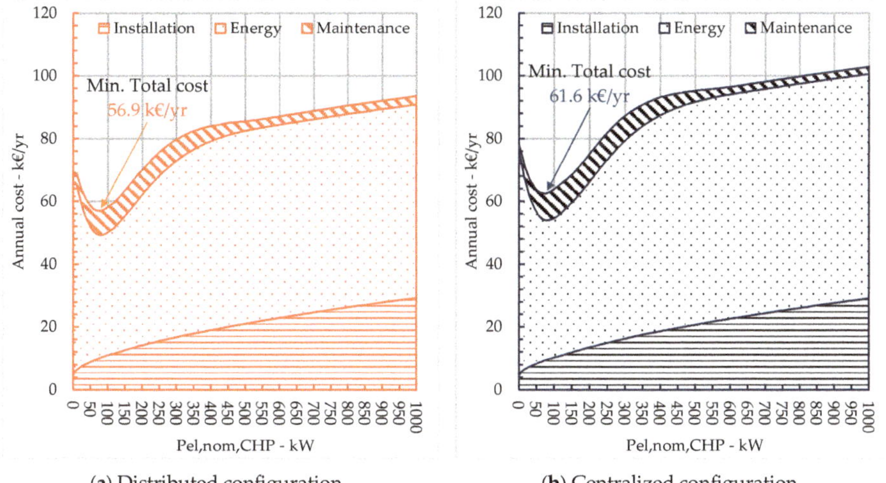

(**a**) Distributed configuration.   (**b**) Centralized configuration.

Figure 5. Installation, energy, and maintenance annual costs using the optimal control strategy.

Table 4. Annual costs of the optimal distributed and centralized configurations, k€/yr.

Tag	Distributed Configuration	Centralized Configuration
PEC_{ICE}	5.2	5.2
PEC_{HP}	4.2	-
PEC_C	-	3.5
PEC_B	0.5	0.6
Total	*9.9*	*9.3*
OC_{ICE}	35.8	45.1
OC_B	2.1	5.7
$OC_{el,P}$	7.1	4.5
$OC_{el,S}$	-5.6	-11.4
Total	*39.4*	*43.8*
O&M	*7.6*	*8.5*
Total Cost	**56.9**	**61.6**

Table 5. Optimal sizes for the distributed and centralized configurations, kW.

Units	Tag	Distributed Configuration	Centralized Configuration
	$P_{el,nom}$	75	75
CHP	$P_{th,nom}$	107	107
	$P_{in,nom}$	200	200
HPs	$P_{th,H,nom}$	370	-
	$P_{th,C,nom}$	-	323
Chillers	$P_{th,C,nom}$	-	323
Boilers	$P_{th,nom}$	150	180

Figure 6 shows some examples of the heat and power profiles resulting from the optimization procedure in three weeks of the year for the distributed configuration. In addition to the areas and lines explained in the chart legend, we specify that the white area under the blue curve in the thermal plots represents the thermal overproduction by the CHP unit; the white area under the green, orange, or yellow lines in the electricity plots quantifies the electrical energy sold to the grid.

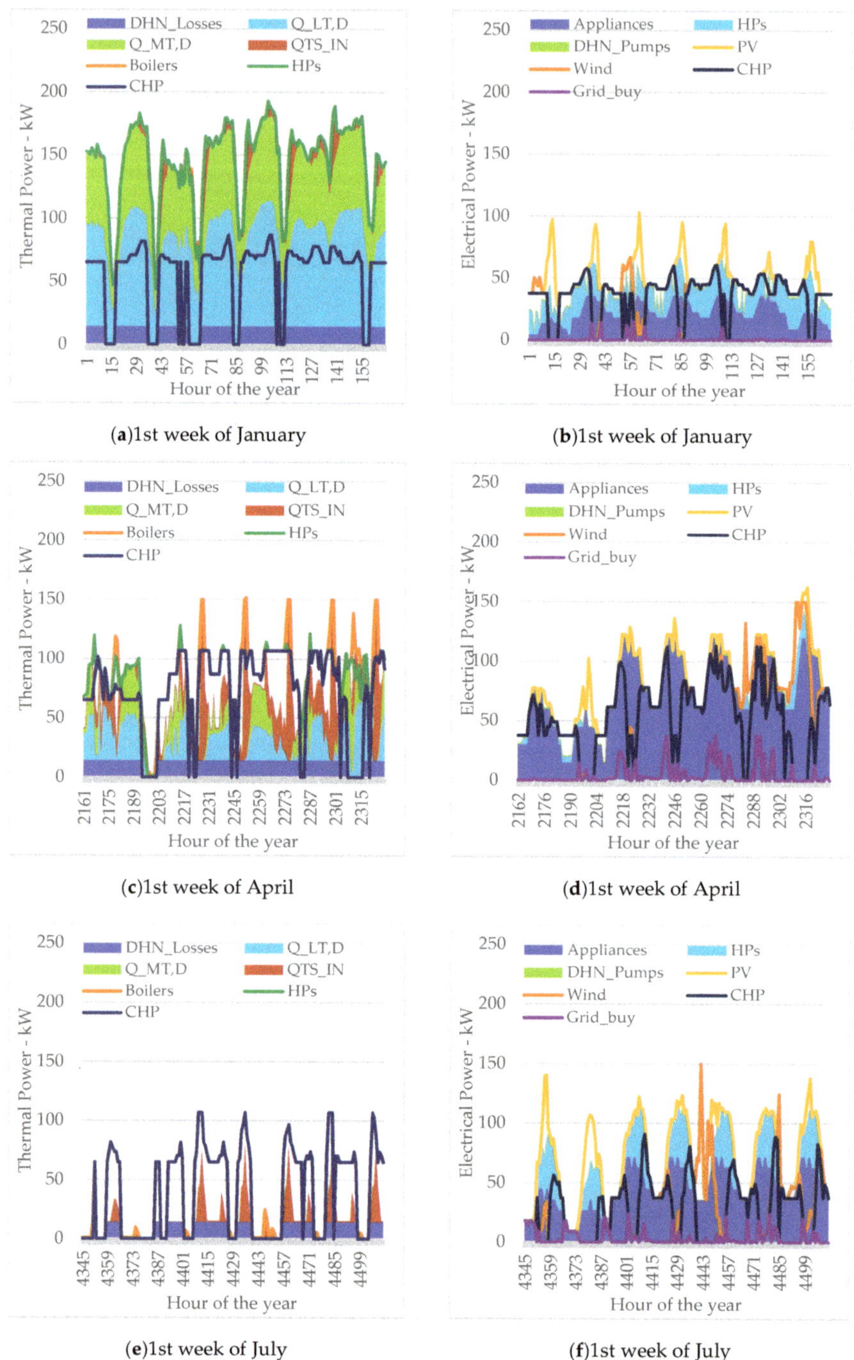

Figure 6. Heat and power profiles in 3 weeks of the year (areas are for consumption, lines for production).

The CHP delivers the base load during the winter season and the heat pumps meet the heating demand. The boilers are rarely used as the electricity produced by the renewables sources and CHP is enough to run the heat pumps. During the mid-seasons, the CHP mainly meets the high electricity demand of the appliances. The HPs are rarely used as it is more convenient to use the boilers to meet the few thermal peaks with respect to purchasing more electrical energy from the grid. In summer, the CHP still operates at high load factors to meet the residual electrical load from the variable solar and wind production. The thermal overproduction mainly occurs during the mid and summer seasons because it is more convenient to maximize the electrical production of the CHP unit, with respect to purchasing electricity from the grid.

The annual values reported in Figures 7 and 8 highlight the energy and economic advantages of the distributed configuration. The distributed configuration reduces both the DHN thermal losses and the CHP overproduction (−27% and −9%, respectively), thus, reducing the overall thermal energy production (−6%). The heat pumps deliver about 24 % of the heating load and the boilers reduce their output to about 68%. In the centralized configuration, the CHP unit delivers 77% of the thermal load, being more cost-effective than the gas boiler. However, the corresponding total electricity production (CHP, PV, and wind generator) is significantly greater than the electrical load, resulting in an unprofitable overproduction (about 28% of the total electricity production is sold to the grid at an uneconomical price). On the contrary, the introduction of the heat pumps shifts part of the thermal load to the electrical one, increasing the self-consumption of electricity and reducing both the CHP energy production (−20% of thermal and −21% of electrical energy, respectively) and the power sold to the grid (−51%). In total, the presence of the HPs reduces the power exchange with the electrical grid from 259 to 159 MWh (−40%) considering both sold and purchased quantities.

Although the objective function of the optimization process refers to an economic index, the energy and the environmental benefits of the distributed configuration are shown in Table 6. The net no-RES primary energy consumption was reduced by about 8%, and the equivalent CO_2 emissions were reduced by about 11%. These values were evaluated considering the primary energy factors and the specific CO_2 emissions of the Italian energy systems [31,32]. According to a grid perspective, the energy system can be thought of as an electricity generator with a specific emission of 232 g/kWh in the distributed configuration and 238 g/kWh in the centralized one. We note that the average value of specific CO_2 emissions for Italian power production is about 313 g/kWh.

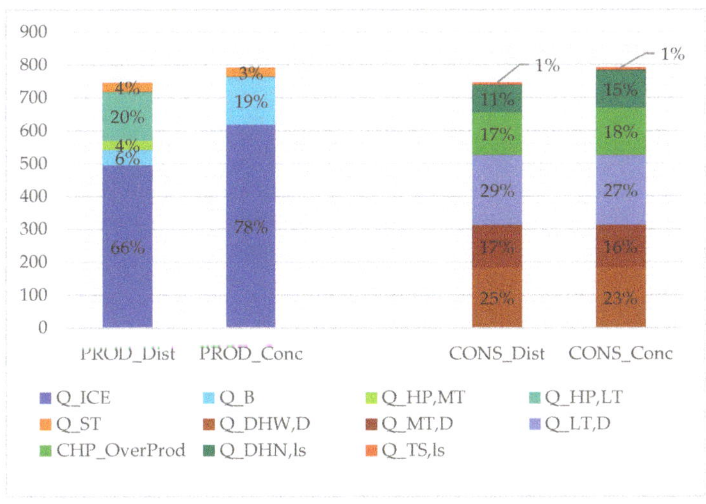

Figure 7. Thermal energy balance (MWh/yr).

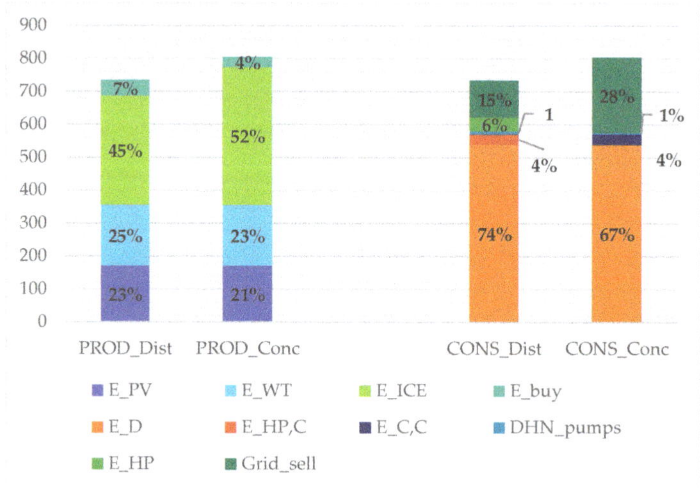

Figure 8. Electrical energy balance (MWh/yr).

Table 6. Energy and environmental performance indexes.

MWh/yr	Value	Value	ton/yr	Value	Value
$EP_{no\text{-}RES,IN,CHP}$	940.3	1182.6	$CO_{2,CHP}$	180.86	227.47
$EP_{no\text{-}RES,IN,,Boilers}$	54.8	148.6	$CO_{2,Boilers}$	10.55	28.58
$EP_{no\text{-}RES,IN,Grid}$	91.8	58.8	$CO_{2,Grid}$	15.31	9.81
$EP_{no\text{-}RES,OUT,Grid}$	−218.6	−445.7	$CO_{2,Grid}$	−36.45	−74.33
TOT $EP_{no\text{-}RES,IN}$	**868.3**	**944.4**	**TOT CO_2**	**170.26**	**191.53**

Overall, the results show that the proposed hybrid centralized-distributed configuration outperforms the more conventional centralized configuration from an economic, environmental, and efficiency perspectives. Indeed, the introduction of heat pumps at the building level enhances the operational flexibility of the system by enabling the interconnection between the thermal and electric networks. In this way, RES-based energy production can be used mainly on-site–instead of being sold to the regional grid–and the use of inefficient technologies, such as natural-gas boilers, can be drastically reduced.

6. Conclusions

An innovative configuration for smart multi-energy microgrids serving clusters of buildings has been presented. The energy system combines both centralized and distributed generation units, optimally integrating cogeneration-based micro-district heating, RES technologies, and reversible heat pumps.

The proposed system was tested in a hypothetical case study, namely, a University Campus located in Trieste (Italy). A detailed modeling of the building load demands, district heating network, and all energy units has been provided in order to simulate the energy system in a reference-year scenario. Moreover, an operational optimization algorithm was specifically developed to identify the generator scheduling that meets the energy demand while minimizing the operational cost. The optimal size of the cogeneration unit and reversible heat pumps has also been found.

The proposed configuration was compared to a more conventional layout based completely on centralized heat production. The results show how the introduction of distributed heat pumps to assist the thermal production at the building level enhances the flexibility and cost-effectiveness of the energy system. Indeed, an 8% total-cost saving, 11% carbon emission reduction, and 8% primary energy saving were achieved compared to the centralized reference case. Moreover, the proposed configuration significantly reduced the electric energy exchange with the regional grid (around 40% less).

Future work will address the current limitations of the work: the optimal sizing of the whole system will be investigated, the effect of uncertainty in weather conditions and economic parameters will be analyzed, and the effectiveness of energy storage managed by predictive control will be evaluated.

Author Contributions: Conceptualization, D.T.; Data curation, D.T., P.C., E.S., L.U. and F.D.; Methodology, D.T., P.C., E.S., L.U. and F.D.; Software, D.T., P.C., E.S., L.U. and F.D. Supervision, D.T.; Writing–original draft, D.T., P.C., E.S., L.U. and F.D.; Writing–review & editing, D.T., P.C., E.S., L.U. and F.D.

Funding: This research was funded by the University of Pisa (PRA 2017–18, Grant n. 33).

Acknowledgments: We would like to thank our colleagues at the University of Pisa involved in energy systems integration activities, who contributed to the discussion and development of the mathematical and engineering models, with a special acknowledgement to Professors Marco Raugi, Davide Poli, Davide Aloini, and Antonio Frangioni.

Conflicts of Interest: The authors declare no conflict of interest.

Nomenclature

Acronyms

CHP	Combined Heat and Power
DHN	District Heating Network
DHW	Domestic Hot Water
ICE	Internal Combustion Engine
INV	Investment
O&M	Operations and Maintenance
PEC	Purchased Equipment Cost
RES	Renewable Energy Sources

Parameters

c	Cost, €/kWh
COP	Coefficient of performance, dimensionless
OC	Annual operational cost, €
TC	Total annual energy cost, €
UA	Overall heat transfer coefficient, kW/K
η	Efficiency, dimensionless

Continuous variables

C	Cooling energy, kWh
E	Electric energy, kWh
F	Energy content of the consumed fuel, kWh
L	Load factor, dimensionless
P	Power, kW
Q	Thermal energy, kWh
S	Surface, m^2
T	Temperature, °C
U	Global heat transfer coefficient, kW/(m^2·K)
V	Volume, m^3
w	Wind speed, m/s
λ	Thermal Conductivity, kW/(m·K)
ρ	Density, kg/m^3

Subscripts

avg	Average
aux	Auxiliary
B	Boiler
ICE	Internal combustion engine
C	Chiller
D	Demand
DHN	District Heating Network
el	Electric
EC	Electric Chiller
ext	External
F	Fuel
HP	Heat Pump
HT	High-temperature level
i	i-th hourly timestep
j	j-th building with DHW requirements
k	k-th building with heating/cooling requirements
Ls	Losses
LT	Low-temperature level
MT	Medium-temperature level
P	Purchased
PEG	Electricity purchased by the grid
PV	Photovoltaic
ref	Reference
S	Sold
ST	Solar Thermal
SEG	Electricity sold to the grid
th	Thermal
TS	Thermal Storage
w	Water
WT	Wind Turbine

Superscripts

Nom	Nominal
II	Second-law

References

1. European Parliament and Council. *Directive 2012/27/EU on Energy Efficiency*; European Parliament and Council: Brussels, Belgium, 2012.
2. European Parliament and Council. *Directive 2010/31/EU on the Energy Performance of Buildings*; European Parliament and Council: Brussels, Belgium, 2010.
3. Pagliarini, G.; Rainieri, S. Modeling of a thermal energy storage system coupled with combined heat and power generation for the heating requirements of a University Campus. *Appl Therm. Eng.* **2010**, *30*, 1255–1261. [CrossRef]
4. Bracco, S.; Delfino, F.; Pampararo, F.; Robba, M.; Rossi, M. The University of Genoa smart polygeneration microgrid test-bed facility: The overall system, the technologies and the research challenges. *Renew. Sustain. Energy Rev.* **2013**, *18*, 442–459. [CrossRef]
5. Piacentino, A.; Barbaro, C.; Cardona, F.; Gallea, R.; Cardona, E. A comprehensive tool for efficient design and operation of polygeneration-based energy µgrids serving a cluster of buildings. Part I: Description of the method. *Appl. Energy* **2013**, *111*, 1104–1121.
6. Piacentino, A.; Barbaro, C. A comprehensive tool for efficient design and operation of polygeneration-based energy µgrids serving a cluster of buildings. Part II: Analysis of the applicative potential. *Appl. Energy* **2013**, *111*, 1122–1138.

7. Destro, N.; Benato, A.; Stoppato, A.; Mirandola, A. Components design and daily operation optimization of a hybrid system with energy storages. *Energy* **2016**, *117*, 569–577. [CrossRef]

8. Murphy, M.D.; O'Mahony, M.J.; Upton, J. Comparison of control systems for the optimisation of ice storage in a dynamic real time electricity pricing environment. *Appl. Energy* **2015**, *149*, 392–403. [CrossRef]

9. Roldán-Blay, C.; Escrivá-Escrivá, G.; Roldán-Porta, C.; Álvarez-Bel, C. An optimisation algorithm for distributed energy resources management in micro-scale energy hubs. *Energy* **2017**, *132*, 126–135. [CrossRef]

10. Phan, Q.A.; Scully, T.; Breen, M.; Murphy, M.D. Determination of optimal battery utilization to minimize operating costs for a grid-connected building with renewable energy sources. *Energy Convers. Manag.* **2018**, *174*, 157–174. [CrossRef]

11. Asleye, D.A.; Breen, M.; Murphy, M.D. A decision support tool for building integrated renewable energy microgrids connected to a smart grid. *Energies* **2017**, *10*, 1765. [CrossRef]

12. Fischer, D.; Madani, H. On heat pumps in smart grids: A review. *Renew. Sustain. Energy Rev.* **2017**, *70*, 342–357. [CrossRef]

13. Grassi, W.; Conti, P.; Schito, E.; Testi, D. Solutions to improve energy efficiency in HVAC for renovated buildings. In *Handbook of Energy Efficiency in Buildings*; Asdrubali, F., Desideri, U., Eds.; Elsevier: Amsterdam, The Netherlands, 2019; Chapter 9.3.

14. Ebrahimi, M.; Keshavarz, A. *Combined Cooling Heating and Power: Decision-Making, Design and Optimization*; Elsevier: Amsterdam, The Netherlands, 2015.

15. Evans, D.L. Simplified method for predicting photovoltaic array output. *Sol. Energy* **1981**, *27*, 555–560. [CrossRef]

16. Duffie, J.A.; Beckman, W.A. *Solar Engineering of Thermal Processes*, 4th ed.; John Wiley & Sons, Inc.: Hoboken, NJ, USA, 2013.

17. Burton, T.; Sharpe, D.; Jenkins, N.; Bossanyi, E. *Wind Energy Handbook*; John Wiley & Sons: Hoboken, NJ, USA, 2002.

18. CEN. *Heating Systems in Buildings–Method for Calculation of System Energy Requirements and System Efficiencies–Part 4-2: Space Heating Generation Systems, Heat Pump Systems*; EN 15316-4-2; CEN: Brussels, Belgium, 2008.

19. O'Callaghan, P.W.; Probert, S.D. Sol-air temperature. *Appl. Energy* **1977**, *3*, 307–311. [CrossRef]

20. CEN. Energy performance of buildings. In *Overall Energy Use and Definition of Energy Ratings*; EN 15306; CEN: Brussels, Belgium, 2008.

21. Testi, D.; Schito, E.; Conti, P. Cost-optimal sizing of solar thermal and photovoltaic systems for the heating and cooling needs of a nearly Zero-Energy Building: Design methodology and model description. *Energy Procedia* **2016**, *91*, 517–527. [CrossRef]

22. Testi, D.; Rocca, M.; Menchetti, E.; Comelato, S. Criticalities in the NZEB retrofit of scholastic buildings: Analysis of a secondary school in Centre Italy. *Energy Procedia* **2017**, *140*, 252–264. [CrossRef]

23. CTI (Italian Thermotechnical Committee). *National Typical Meteorological Years*; CTI: Milan, Italy, 2016.

24. Bejan, A.; Tsatsaronis, G.; Moran, M. *Thermal Design and Optimization*; Wiley: Hoboken, NJ, USA, 1995.

25. Paradigma Catalogue. 2018. Available online: http://www.paradigmaitalia.it/sites/default/files/2018-03/lp_041_rev.00_listino_compact_power_2018_3.pdf (accessed on 18 February 2019).

26. Energy Efficiency Report, Politecnico di Milano. 2013. Available online: http://www.energystrategy.it/report/efficiency-report.html (accessed on 18 February 2019).

27. Renewable Energy Report, Politecnico di Milano. 2018. Available online: http://www.energystrategy.it/report/renewable-energy-report.html (accessed on 18 February 2019).

28. CEN. *Energy Performance of Buildings–Overall Energy Use and Definition of Energy Ratings*; EN 15459-1; Annex, D., Ed.; CEN: Brussels, Belgium, 2017.

29. Urbanucci, L.; Testi, D. Optimal integrated sizing and operation of a CHP system with Monte Carlo risk analysis for long-term uncertainty in energy demands. *Energy Convers. Manag.* **2018**, *157*, 307–316. [CrossRef]

30. Urbanucci, L.; Testi, D. An operational optimization method for a complex polygeneration plant based on real-time measurements. *Energy Convers. Manag.* **2018**, *170*, 50–61. [CrossRef]

31. D.M. 26-06-2015-Applicazione delle metodologie di calcolo delle prestazioni energetiche e definizione delle prescrizioni e dei requisiti minimi degli edifici; Adeguamento del decreto del Ministro dello sviluppo economico, 26 giugno 2009-Linee guida nazionali per la certificazione energetica degli edifici (In Italian). Available online: https://www.mise.gov.it/index.php/it/normativa/decreti-interministeriali/2032966-decreto-interministeriale-26-giugno-2015-applicazione-delle-metodologie-di-calcolo-delle-prestazioni-energetiche-e-definizione-delle-prescrizioni-e-dei-requisiti-minimi-degli-edifici (accessed on 22 February 2019).
32. ISPRA. *Fattori di Emissione Atmosferica di Gas a Effetto Serra e Altri Gas nel Settore Elettrico*; ISPRA: Roma, Italy, 2018. (In Italian)

Article

Covenant of Mayors: Local Energy Generation, Methodology, Policies and Good Practice Examples

Albana Kona [1,*], Paolo Bertoldi [1] and Şiir Kılkış [2]

[1] European Commission, Joint Research Centre (JRC), 21027 Ispra, Italy; paolo.bertoldi@ec.europa.eu
[2] The Scientific and Technological Research Council of Turkey (TÜBİTAK), Ankara 06100, Turkey; siir.kilkis@tubitak.gov.tr
* Correspondence: albana.kona@ec.europa.eu

Received: 30 December 2018; Accepted: 3 March 2019; Published: 13 March 2019

Abstract: Local authorities and cities are at the forefront of driving the energy transition, which plays a crucial role in mitigating the effects of climate change. The greenhouse gas emissions in cities, due to energy consumption, are placed into two categories: direct emissions generated from the combustion of fossil fuels mainly in buildings and transport sectors, and indirect emissions from grid-supplied energy, such as electricity and district heating and/or cooling. While there is extensive literature focused on direct greenhouse gas emissions accounting in cities' inventories, research has focused to a lesser extent on allocation methods of indirect emissions from grid-supplied energy. The present paper provides an updated definition for the concept of local energy generation within the Covenant of Mayors initiative and proposes a new methodology for indirect emission accounting in cities' greenhouse gas emission inventories. In addition, a broader policy framework in which local action is taken is discussed based on the European Union energy and climate policies, and over 80 exemplary Covenant of Mayors good practices are identified across the technology areas of local energy generation and four modes of urban climate governance. The contributions of the paper demonstrate that local authorities have the capacity to support and mobilize action for local energy generation investments through the multiple modes of urban climate governance to update and strengthen climate action

Keywords: Covenant of Mayors; indirect emission allocation; local energy generation

---

1. Introduction

An increasing number of cities and local governments adhere to transnational initiatives that are active on climate change mitigation. Cities that adhere to transnational networks on climate change by making emission inventories and climate action plans publicly available, although in the absence of obligation, render themselves accountable both globally as well as locally [1]. Their performance and identity are increasingly scrutinized in terms of global impact and exploited in the scientific literature [2]. This includes factors influencing the cities' participation in the networks and multilevel governance models that include observed [3–12] drivers influencing the emissions and target setting [13–16], tools and strategies for the redaction of the climate action plans [17–22] and benchmarking methods [23–26]. The factors further involve the assessments of the global contribution of local climate mitigation actions [27–34].

In the European Union (EU), local authorities (LA) and cities have a crucial role in building public support for the EU's energy and climate goals, while deploying more decentralized and integrated energy systems. The Covenant of Mayors (CoM) is an EU based initiative, which started in 2008. The CoM has been a disrupting phenomenon in the arena of transnational initiatives, which have expanded tremendously over the past 10 years [35], covering more than 7,850 local authorities and 252 million inhabitants as of December 2018. The CoM at the time of writing the paper is part of the

Global Covenant of Mayors initiative, while this paper specifically addresses the experience gathered in the CoM in Europe between 2008 and 2018.

When LAs join the initiative, the CoM signatories (all of whom participate voluntarily) commit to reduce the levels of carbon dioxide (CO_2) emissions in their territories by at least 20% by 2020 or at least 40% by 2030 through the implementation of a Sustainable Energy Action Plan (SEAP). Recently, actions on adaptation (climate risk assessment) have also been included in addition to those on mitigation. The combined plans are called Sustainable Energy and Climate Action Plans (SECAPs). In this paper, the focus is on climate mitigation action plans with commitment targets for 2020, i.e., the SEAPs. The CoM is a unique feature of multilevel polycentric governance that goes far beyond transnational city networking [11]. The initiative is supported by the European Commission and managed jointly by the Covenant of Mayors' Office (CoMO), a consortium of cities' networks and the Commission's Joint Research Centre (JRC), which provides scientific and methodological support. The latest overall assessment of the initiative by JRC [36] shows that the signatories' overall commitment to reducing greenhouse gas (GHG) emissions is 27% by 2020, i.e., 7 percentage points above the minimum requested target of 20%. Based on data from 315 implementation reports accompanied by a monitoring emissions inventory (MEI) (covering 25.5 million inhabitants and mainly for the period 2012–2014), a 23% overall reduction in emissions is observed to be already achieved.

The CoM reporting framework requires a three-step approach to the signatories: i) submission of emission inventories according to their standards, ii) setting a mitigation target as well as drawing a climate action plan, and lastly, iii) monitoring the progress towards the targets. The minimum requirement includes all the direct emissions that are produced within geographical boundaries (buildings and urban transport sectors), as well as indirect emissions associated with the final consumption of grid electricity and of heating/cooling networks. The local generation of energy and associated direct emissions are not part of the activity sectors that are included in the emission inventory but are considered in the calculation of the local emission factors to be applied to the consumption of grid supplied energy [37].

While accounting of direct emissions sources generally follows a coherent approach to those of the Intergovernmental Panel on Climate Change (IPCC), indirect emissions accounting is more complex and challenging as it requires methods on assessing the average emission factor of grid supplied energy. Grid supplied energy can take the form of electricity and district heating/cooling carriers. The identification of the emission sources of heat/cold production could be straightforward due to the limited number of generation units fueling the networks and certainly because heat can be transported effectively only for short distances from generation units to the users, compared to electricity. Therefore, in defining the criteria for the classification of local energy generation units, the focus herein is only on electricity, and the CoM methodology regarding accounting for the indirect emissions associated with grid supplied energy consumptions in cities is introduced.

Further within the method, a broader policy framework in which local action is taken is discussed based on the EU energy and climate policies. Over 80 exemplary good practices are overviewed across the technology areas of local energy generation ranging from photovoltaics, solar thermal, wind energy, hydroelectric power, bioenergy, geothermal energy, combined heat and power (CHP), district heating and/or cooling (DH/C) and smart grids, as well as energy generation from waste and wastewater based on the CoM Signatories' good practices. These good practices are associated with the urban climate governance options that have been put into action by the CoM signatories. Overall, the paper addresses multiple gaps in the literature by updating the approach for indirect emissions accounting and the linkage of good practices in the CoM Signatories' good practices database to the four modes of urban climate governance.

2. Materials and Methods

The methods are presented in the subsequent sections based on the definition of local energy generation, followed by the accounting of the indirect emissions in the CoM framework. The technical metrics that are important for an accurate and realistic accounting of indirect GHG emissions are further supported with the identification of good practices that are necessary to increase local energy generation. In this context, the method of the research work is summarized in Figure 1 with the aim of allowing authorities at the local level the opportunity to update and strengthen climate action. The broader policy framework in which local action is taken is also discussed based on EU energy and climate policies, the effective interaction of which is necessary for policy alignment.

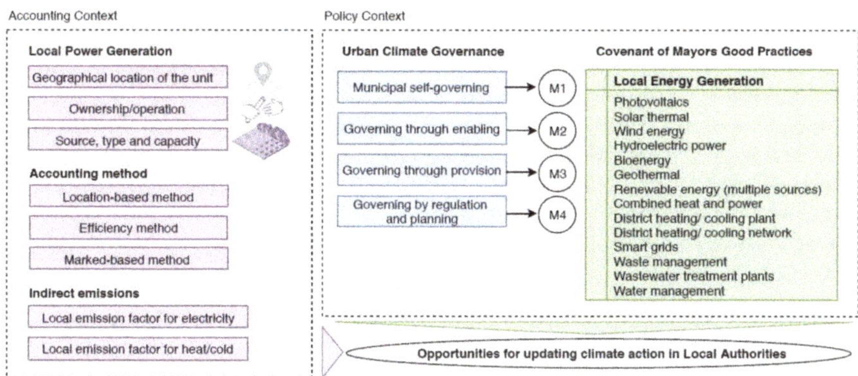

Figure 1. Overview of the accounting context and the policy context for local energy generation. Source: own elaboration.

2.1. Covenant of Mayors Definition of Local Power Generation

The grid-supplied electricity is produced by a plurality of generation sources, ranging from distributed to centralized power plants, which can be located inside or outside the geographical boundaries of the cities' emission inventory.

Distributed generation is generally defined in the literature as electric power generation connected to the electrical distribution networks or on the customer side of the network, while usually, the connection of generation units to the transmission network is typical of centralised generation. The central idea of distributed generation, however, is to locate generation close to the load, hence on the distribution network or on the customer side of the meter. The distributed power facilities may differ according to "the purpose, the power scale, the power delivery, the technology, the environmental impact, the mode of operation, and the penetration of distributed generation" [38–41].

Based on the above definitions, a number of questions arise: what is local generation and how can it be differentiated from the national ones? Can it be assimilated into the definition of distributed generation? Generally, the term "distributed generation" refers to renewable energy technologies and it is often interchangeable with the term "local generation".

In relation to GHG emissions accounting, the present paper deals with the concept of local energy generation, which differs mainly from distributed generation. Within the CoM initiative that deals mainly with climate change actions at the local level, the environmental impact of power generation is of high importance, as well as the capability of the local governments to properly address it in their climate action plans. The regional context also affects the criteria, taking into account the national framework of energy and climate governance, characterized by different jurisdictions between state and non-state actors. Hence, there is a need to clearly define the criteria on which a power generation facility can be classified as local generation, and therefore to be accounted in the cities' emission inventories within the CoM framework.

The following main aspects were identified to be discussed in defining local production of electricity (LPE) within the scope of the CoM framework, and more precisely the location, the ownership, the source, the type of the technology and the capacity or power scale of the local generation facilities.

- Geographical location of the unit: The location of the energy unit in the local territory is the first criterion. The geographical boundaries of the "local territory" are the administrative ones of the entity (municipality, region) governed by the local authority, which is a signatory to the CoM. Hence, electricity that is produced by installations/plants located inside the local territory has to be included in assessment of the LPE.
- Ownership/operation: All plants/installations under the direct control of the LA (operated and/or at least partly owned by the municipality) should be accounted for in the calculation of the LPE. Therefore, electricity produced by installations/plants located outside the local territory can also be optionally included if they are under the direct control of the LA. The amount of the electricity production can be assessed according to the responsibility of the LA and the share of ownership of all partners (municipalities or commercial partners), which avoids double counting inside and outside the territory. The motivation for taking into account generation facilities under municipality control is that the LA should lead by example on climate actions starting from their own/operated facilities.
- Source, type and capacity of the local electricity production unit: The method described in this paper recommends including all the individual electricity generation units in the local territory and also any plant outside the local territory that is owned and/or operated by LA, classified by the type and capacity as below:

  - Local electricity production from renewable sources and combustible renewables are classified regardless of the technology and capacity, with the exclusion of the electricity sold to third parties that are located outside the local administrative boundaries and are identified through disclosed attributes. The rationale behind this is similar to the concept of the "residual mix" used by member states (MS) in the EU for assessing the grid emission factor. When determining the residual mix, MS often exclude the cancellation of electricity attributes (purchased via a Guarantee of Origin (GO) certificate in Europe) from the grid emission average [42]. The GOs are tracking instruments, introduced in 2009 by the Renewable Energy Directive (RED 2009/28/EC), that provide a means of demonstrating the origin of renewable electricity to consumers. The GOs system is a virtual one where the renewable attribute of energy trades separately from the physical energy. The usage is limited within 12 months of production of the corresponding energy unit, and is cancelled once it has been used.
  - Local electricity production from non-renewable sources, classified by types and capacity:

    - All combined heat and power plants, without capacity limit: the CHP system can be defined as local generation, as the second product (thermal energy) is consumed locally. Combined cycle gas turbines, internal combustion engines, combustion turbines, biomass gasification, geothermal, and Stirling engines, as well as fuel cells, are suitable for CHP processes. The heat demand usually drives the operation process, unless a back-up system for the heat production is in place.
    - Electricity-only with a capacity limit of 20 MW of thermal input: According to the principles that are laid out in the CoM, the inventory is not meant to be an exhaustive inventory of all emission sources in the territory but focuses on the energy consumption side and on the sectors and activities (buildings and transport) upon which the local authority has a potential influence. Large industrial power plants, covered by cap and trade schemes, such as the European Union Emission Trading Scheme (EU ETS), are not under LAs competence, but regulated by the ETS directive (2003/87/EC).

The amount of electricity to be reported in emission inventories as local electricity production will have a direct influence on the value of the local emission factor for electricity, and consequently, on the emissions that are associated with the consumption of electricity.

2.2. Covenant of Mayors Methodology on Accounting the Indirect Emissions

Two methods exist in the literature to allocate the emissions generated via electricity production to the final energy consumer of a given grid: location-based and market-based [43]. The location-based method is grounded on average generation emission factors for defined territories, considering local, subnational, or national boundaries, therefore reflecting the average emission intensity of the grid. In the marked-based method, the emission factors are derived from electricity purchases bundled with contractual instruments. The market-based method presents a higher degree of accuracy in the assessment of the emission factor. However, data collection in the context of market-based methods remains the main challenge when compared to location-based approaches, where local governments need to focus on assessing the indirect emissions of a multitude of consumers in their territories.

In order to calculate the indirect CO_2 emissions that are to be attributed to the local consumption of electricity, JRC developed a specific methodology as described in this paper by estimating the local emission factor for electricity (EFE), taking into account both location- and market-based methods, but also an efficiency method for emission allocations in the case of CHPs.

2.2.1. The Location-Based Method

The location-based method is used for assessing the amount of electricity and the associated GHG emissions excluding the electricity attributes (i.e., purchased via GOs certificates in Europe). This amount of electricity is associated with the physical electricity that is consumed in the local territory coming from the national energy mix and the local generation units and associated emission factors as below:

- National or European emission factor for electricity consumption (NEEFE): These emission factors can be derived either from international databases, such as the International Energy Agency (IEA) [44], the IPCC emissions factor database [45], or provided by national agencies of MS in the EU (e.g., Italy [46]). The European Commission Joint Research Centre (EC JRC), in the framework of the CoM initiative, also provides regular updates of the NEEFE [47].
- The local energy generation emission factor can be derived directly or is assessable from the local (private or public) electricity provider, costumer and/or unit operator. In case of CHPs, a method has been developed to allocate the emissions based on the energy inputs required to produce separately (not in cogeneration) the same amount of outputs of heat and electricity (as in the CHP power plant output), which is named an efficiency method in the case of CHPs.

2.2.2. The Efficiency Method in Case of CHPs

Within the CoM reporting framework, the "efficiency" method for allocating CO_2 emissions is the recommended method for LAs to be used in case of CHP power plants located within the geographical boundaries of the territory. The rationale behind this choice is to have a consistent method for emissions accounting with the method recommended in the European Energy Efficiency Directive (EED) (2012/27/EU) for determining the efficiency of the cogeneration process and primary energy savings. This method, called the "efficiency method", uses a reference system to allocate the output.

In a first step, the ratio of primary energy savings (PES) in comparison to a reference system is calculated. The methodology used to calculate the PES corresponds to the method defined in Annex II of the EED directive. According to Annex II of the EED directive, PES is defined as follows (Equation (1)):

$$PES = \left(1 - \frac{1}{\frac{\eta_{CHP,heat}}{\eta_{REF,heat}} + \frac{\eta_{CHP,el}}{\eta_{REF,el}}}\right) \qquad (1)$$

where:

- $\eta_{REF,el}$ is the efficiency reference value for separate electricity production at national level;
- $\eta_{REF,heat}$ is the efficiency reference value for separate heat production;
- $\eta_{CHP,el}$ is the efficiency of electricity production with the CHP power plant, measured as a ratio between the annual amount of electricity produced in output ($P_{CHP,el}$) and the total annual amount of primary energy in input to the CHP power plant ($P_{CHP,TOT}$): $\eta_{CHP,el} = \frac{P_{CHP,el}}{P_{CHP,TOT}}$;
- $\eta_{CHP,heat}$ is the efficiency of heat production with the CHP power plant, measured as a ratio between the annual amount of heat produced in output ($P_{CHP,heat}$) and the total annual amount of primary energy in input to the CHP power plant ($P_{CHP,TOT}$): $\eta_{CHP,heat} = \frac{P_{CHP,heat}}{P_{CHP,TOT}}$.

In a second step, the share of primary energy attributed to each of the two outputs electricity and heat can be calculated as follows. Regarding the share of allocation to heat, this amounts to a re-arrangement of Equation (1) based on the term $(1 - PES)$ multiplied by the ratio of primary energy used for heat production in a reference scenario ($\frac{P_{CHP,heat}}{\eta_{REF,heat}}$) on the primary energy used in the cogeneration scenario to produce the same amount of heat ($P_{CHP,TOT}$), as ($\frac{P_{CHP,heat}}{\eta_{REF,heat} \times P_{CHP,TOT}}$). Therefore, the ratio of primary energy allocated to heat is obtained using Equation (2):

$$\text{ratio allocated to heat} = (1 - PES) \times \left(\frac{P_{CHP,heat}}{\eta_{REF,heat} \times P_{CHP,TOT}} \right) = \frac{\frac{P_{CHP,heat}}{\eta_{REF,heat}}}{\left(\frac{P_{CHP,heat}}{\eta_{REF,heat}} + \frac{P_{CHP,el}}{\eta_{REF,el}} \right)} \tag{2}$$

This method allocates the emissions based on the energy inputs required to produce separately (not in cogeneration) the same amount of outputs of heat and electricity (as in the CHP power plant output) as follows:

$$CO_2 \text{ emissions allocated to heat} = CO_{2CHP,heat} = \frac{\frac{P_{CHP,heat}}{\eta_{REF,heat}}}{\left(\frac{P_{CHP,heat}}{\eta_{REF,heat}} + \frac{P_{CHP,el}}{\eta_{REF,el}} \right)} \times CO_{2CHP,TOT} \tag{3}$$

$$CO_2 \text{ emissions allocated to electricity} = CO_{2CHP,el} = CO_{2CHP,TOT} - CO_{2CHP,heat} \tag{4}$$

where:

- $CO_{2CHP,TOT}$ is the total amount of CO_2 emissions in the CHP power plant (tCO_2);
- $CO_{2CHP,heat}$ is the total amount of CO_2 emissions allocated to heat production (tCO_2);
- $CO_{2CHP,el}$ is the total amount of CO_2 emissions allocated to electricity production (tCO_2);
- The recommended value in the CoM of the typical efficiency of separate electricity production ($\eta_{REF,el}$) to be used is set in the national efficiency factor for electricity generation and/or the average of EU regularly published by Eurostat (46%) in Reference [48];
- The recommended value of the typical efficiency of separate heat production ($\eta_{REF,heat}$) to be used in the CoM framework is 90%.

2.2.3. The Market-Based Method

The market-based method is used for assessing the amount of electricity and GHG emissions that are associated, purchased and sold through energy attributes as certified electricity (CE) [49]. Certified electricity is the electricity that meets the criteria for GO of electricity produced from renewable energy sources as set in Article 15 of RED Directive.

- Instead of purchasing the "mixed" electricity from the grid, the local authority/other local actors can decide to purchase certified electricity. The LA will report the amount of purchased electricity ($\sum CE_{purchased}$), which is not already reported under LPE.

- The amount of renewable energy produced by facilities that are located inside the local territory for which the GO of electricity produced from renewable sources is sold to third parties outside the administrative boundaries should not be accounted for as local energy production ($\sum CE_{sold}$).
- $\sum CE$ is the certified electricity accounted for in the inventory as given by Equation (5):

$$\sum CE = \sum CE_{purchased} - \sum CE_{sold} \tag{5}$$

2.2.4. Indirect Emissions due to Local Electricity Consumption

Based on the considerations and assumptions presented, the EFE at the local level should be calculated as follows:

- In the case where the local authority would not be a net exporter of electricity (TCE $\geq$ LPE + CE), the average emission factor will be equal to the total amount of emissions from electricity consumption assessed with (location-based + market-based instruments) over the total amount of electricity consumption as provided in Equation (6):

$$EFE = \frac{[(TCE - \sum LPE - \sum CE) * NEEFE + \sum CO_{2LPE} + \sum CO_{2CE}}{TCE} \tag{6}$$

- In the case where the local authority would be a net exporter of electricity (TCE < LPE + CE), Equation (7) will apply:

$$EFE = \frac{\sum CO_{2LPE} + \sum CO_{2CE}}{\sum LPE + \sum CE} \tag{7}$$

where:

- EFE is the local emission factor for electricity consumption $\left(\frac{tCO_2}{MWh}\right)$
- TCE = total electricity consumption (MWh) in the local territory
- $\sum LPE$ = local electricity production from RES and non-RES facilities (MWh)
- $\sum CE$ = certified electricity accounted in the inventory
- $NEEFE$ = national or European emission factor for electricity consumption $\left(\frac{tCO_2}{MWh}\right)$
- $\sum CO_{2LPE}$ = CO_2 emissions due to local energy production (tCO$_2$)
- $\sum CO_{2CE}$ = CO_2 emissions (tCO$_2$) due the purchase/sold of CE certified electricity

2.2.5. Indirect Emissions from Local District Heating and Cooling Consumption

Indirect emissions from the consumption of heat/cold are estimated based on the emissions that are occurring due to the production of locally consumed heat/cold. If a part of the heat/cold that is produced in the local territory is exported, then the corresponding share of CO_2 emissions should be deducted when calculating the emission factor for heat/cold (EFH). In a similar manner, if heat/cold is imported to the local territory from a plant that is situated outside the local territory, then the share of CO_2 emissions from this plant that corresponds to the heat/cold consumed in the local territory should be accounted for when calculating the emission factor for heat/cold.

In principle, the total amount of heat/cold produced is different than the quantity of heat/cold that is consumed locally. Differences may occur due to auto-consumption of heat/cold by the utility producing it and or due to the transport and distribution losses of heat/cold.

The following formula (Equation (8)) should be applied to calculate the CO_2 emission factor for heat/cold (EFH), taking the above-mentioned issues into consideration:

$$EFE = \frac{\sum CO_{2LPH} + \sum CO_{2IH} - \sum CO_{2EH}}{\sum LHC} \tag{8}$$

where:

- EFH is emission factor for heat/cold (tCO_2/MWh or tCO_2-eq/MWh)
- $\sum CO_{2\ LPH}$ is total CO_2 emissions (tCO_2 or tCO_2-eq) due to the local production of heat/cold
- $\sum CO_{21H}$ is CO_2 emissions related to any imported heat/cold from outside the local territory (tCO_2 or tCO_2-eq)
- $\sum CO_{2EH}$ is CO_2 emissions related to any heat/cold that is exported outside of the local territory (tCO_2 or tCO_2-eq)
- ΣLHC is local heat/cold consumption (MWh)

In the case of CHP plants, it is first required to distinguish between the emissions due to heat and electricity production. District cooling, i.e., purchased chilled water, is in principle a similar product as purchased district heating. However, the process to produce district cooling is different from the process to produce district heating, and there is a larger variety of production methods. If local production of district cooling occurs, or if district cooling is consumed as a commodity by end-users, then the local authority is recommended to contact the district cooling provider for information on the use of fuels or electricity to provide cooling. Accordingly, the emission factors for fuels and electricity presented in this paper can be applied. The approach as put forth in this paper is essential for enabling LAs to account for indirect emissions accurately and realistically.

3. EU Energy and Climate Policies

EU energy and climate policies have large influences on the transition to sustainable local energy systems. In the following, a brief description of the main updated EU energy and climate policies as a total governance system, which strongly influences the municipal policies for climate actions, is reported.

A first major milestone of climate change in the EU policy was the launch of the EU emission trading scheme (ETS) (2003/87/EC) in 2005, establishing a scheme for greenhouse gas emission allowance trading within the EU. According to the ETS directive and subsequent amendments, all combustion installations above 20 MW of thermal input should be part of the scheme with the exception of installations exclusively using biomass, installations for the incineration of hazardous or municipal waste and installations used for research, development and testing of new products. Some installations are also temporary or conditionally excluded from ETS, such as hospitals and some installations below 35 MW. Having an EU established scheme for regulating the emissions of such large combustion installations means the jurisdiction of the LAs is limited. Therefore, the emissions from the ETS installations are not recommended to be accounted in the LAs emission inventories, unless the local government is somehow involved through ownership or operation, or the installation is a CHP power plant.

A second milestone relates to the "A Clean Energy for all Europeans" package, launched in November 2016. By the end of December 2018, a political agreement for the adoption of this package has been reached, formulating three key energy and climate targets by 2030: (1) to reduce greenhouse gas emissions further by at least 40% by 2030 as compared with 1990, (2) a binding renewable target of 32% and (3) an energy efficiency target of 32.5%. The important role of the Covenant is mentioned and acknowledged in the package as a collaborative platform that allows local authorities to learn from one another, as also highlighted in literature [12,23–26]. The package is introduced and is updating a set of legislations including the new Governance Regulation of the Energy Union, the revised Renewable Energy Directive, the revised Energy Efficiency Directive, the revised Energy Performance in Buildings Directive and the revised Electricity Directive and Regulation. The main aspects of these updated EU policies to supporting local energy policies are analyzed in the following:

- The new regulation on governance of the energy union and climate action is based on integrated national energy and climate plans (NECPs), covering 2021–2030 and all the five dimensions of the Energy Union (i.e., decarbonisation, energy efficiency, energy security, internal energy market and research, innovation and competitiveness). The Governance Regulation establishes a clear and transparent regulatory framework, ensuring that the objectives of the EU's 2030 energy and climate targets are achieved. Similarly, in the CoM, the governance of integrated energy and climate action plans has been in place for a decade through transparent and robust framework of reporting and monitoring of the so-called sustainable energy and climate action plans in municipalities. Although in the absence of obligation, cities adhere voluntarily to the initiative and they render themselves accountable both globally as well as locally [1]. The largest difference perhaps between national and transnational systems, such as CoM, on the governance is related to possibilities for sanctioning non-compliance with soft measures such as removal of support or suspension from the initiative [35].

- The RED II directive establishes a clear and stable framework for citizens and communities as stakeholders in the energy system, acknowledging their involvement individually, or through renewable energy communities (RECs). RECs are legal entities based on open and voluntary participation, effectively owned and controlled by the members (including citizens, small and medium-sized enterprises (SMEs), local authorities or municipalities), with the primary purpose of providing environmental, economic or social community benefits for its members or for the local areas where it operates, rather than financial profits. Moreover, the revision requires MS to put in place enabling frameworks, and simplify administration procedures to support citizens and communities investing in renewables. The LAs, through their capacity as self-governing and facilitator may participate as members or operators through their own utilities in renewable energy communities, building legitimacy and political endorsements for their climate action plans. The LA, through its capacity as a regulator and enabler, may further facilitate and accelerate the permit-granting procedure for renewable energy projects and/or facilitate the uptake of RECS in new urban development or in large renovation of multi apartment blocks.

- The Energy Efficiency Directive (EED) and its revision (2018/2002/EC), has two articles of the directive that are of particular interest to local authorities:

(I) Article 14 with the overall objective to encourage the identification of cost-effective potential for delivering energy efficiency, principally through the use of cogeneration, efficient district heating and cooling and the recovery of industrial waste heat or, when these are not cost-effective, through other efficient heating and cooling supply options, and the delivery of this potential. An LA, through its capacity as a regulator and enabler, may facilitate and participate in the process of planning cost-effective heating and cooling networks, identifying possible recovery of waste heat in facilities located within the territory.

(II) Article 7 on energy savings obligations: MS, in designing the policy measures to fulfill the obligations to achieving energy savings [50], should take into account the need to alleviate energy poverty through giving priority to financing the implementation of measures among vulnerable households, and where appropriate in social housing. The LAs are key player in this process of identifying and giving priority of implementing energy savings among vulnerable households.

- The Energy Efficiency in Buildings Directive (EPBD) and its revision (2018/844/EC): First and foremost, all new buildings must be nearly zero-energy buildings by 31 December 2020 (public buildings by 31 December 2018) and the low amount of energy that these buildings require have to be supplied mostly from renewable sources. Meanwhile, the 2018 EPBD revision requires MS to establish a long term renovation strategy to fully decarbonize the national building stocks by 2050. As highlighted in this directive, the local authority itself assumes an exemplary role in the implementation of these actions. Committing to highly efficient buildings and adopting renewable sources in their own facilities are ways local authorities can reduce emissions. By developing energy efficiency and renewable projects and in their buildings, local authorities set an example to the local community, inspiring citizens to adopt sustainable and low-carbon practice. Moreover, local authorities empowered with the jurisdiction to build upon national policies in the building sector can implement codes and regulation with more stringent requirements than national ones. Through these regulations, integrated actions to improve energy efficiency in the buildings are provided and the use of renewable sources for space heating and cooling is fostered.

In conclusion, the revised EU policy governance on energy and climate and its implementation at the national level calls for an enhanced communication process for boosting the multilevel and polycentric governance of urban climate action plans.

4. Urban Energy and Climate Governance to Support Sustainable Energy and Climate Action Plans

The accurate and realistic accounting of indirect emissions by LAs is important within the policy cycle at the local level from planning to taking necessary action. At the same time, policy support from a synthesis of urban climate governance options [51] is required to transform the local energy structure. Governance relates to mechanisms directed toward the coordination of multiple forms of state and non-state action across scales (from local/municipal authorities to national governments), as well as through networks and partnerships that operate within and between cities [12]. The CoM is a unique feature of multilevel polycentric governance that goes far beyond transnational city networking [11]. Such governance has a crucial role in demonstrating, guiding and influencing key measures for achieving emissions reductions through efficient electricity and local heat/cold production. The transition towards a more sustainable urban environment at the local level begins with a common understanding that there is significant potential to curb the city's CO_2 emissions. This understanding provides a basis upon which political leadership instigates a process of exploring possibilities and discussing different options with a wide range of stakeholders towards selecting, detailing, implementing and monitoring local action. In this process, LAs have the capacity to support and mobilize action for local energy generation investments through several modes of urban climate governance

In the following, four modes of urban energy and climate governance are investigated that summarize the scope of each mode, along with the main tools and exemplary actions to support local energy sustainability. The modes of urban energy and climate governance, which are based on definitions from Reference [52], can be mainly summarized as:

- municipal self-governing (M1)
- municipal enabling (governing through enabling) (M2)
- governing through provision (M3)
- governing by regulation and planning (authority) (M4).

Overall, the barriers that can be addressed with each main tool under these modes of governance are different. For this reason, it is often necessary to combine multiple modes of governance to reinforce and align incentives for particular objectives. This must be supported by an analysis of the legal, physical, social and economic barriers hindering local energy generation prior to considering corrective actions and measures. In the following, a collection of good practices on local energy generation based on the CoM Signatories' Benchmarks of Excellence and the literature [53] is provided, especially those that are connected to best practices in CoM signatories.

4.1. Municipal Self-Governing

Municipal self-governing involves aspects that are related to the management of the LAs estate to increase local energy generation, renewable energy demonstration projects in public facilities and public procurement. Prior to the good practices, these opportunities are put forth below:

- Management of the LAs estate to increase local energy generation: The LA requires renewable energy generation to provide for a high share of the building energy needs in the design of new public buildings and the retrofit of existing public buildings. Town halls with innovative solar energy façades and schools that are powered by photovoltaic panels may be given as initial examples. When possible, the LA can also require DH/C grids in public buildings areas, including through contract to connect municipal buildings to the district heating and/or cooling network.

-   Renewable energy demonstration projects in public facilities: Public buildings provide important opportunities as demonstration sites for renewable energy technologies. LAs can publicly test and show the success of renewable energy measures that are implemented in public buildings. Moreover, priority may also be given to less wide-spread technologies, such as low-power absorption chillers and micro-cogeneration to share the results with the stakeholders. The pilot projects can also attract the interest of private stakeholders upon which similar projects can be replicated across the city. Additional opportunities include the use of biogas from wastewater treatment facilities in a CHP or in public vehicles fleet driven by biogas/natural gas.

-   Public procurement: Public procurement can be used to prescribe a share of renewable supply in the case of service contracts to public facilities. One of the prerequisites for the use of public procurement as a strategic tool to increase local energy generation in municipality-owned assets is the identification of appropriate public buildings and facilities. Inefficient heating plants and boilers at these sites can be replaced by cogeneration, trigeneration, renewable energy installations or a combination thereof according to the energy demand profile for heating and cooling. Actions at these sites can also have a high replication potential across private sectors, such as the food industry and hotels, among others.

4.2. Municipal Enabling

Municipal enabling represents opportunities that provide additional policy support for mobilising actors, such as public–private partnerships as well as awareness-raising and training activities.

-   Public–private partnerships: Co-operation between the municipality, local investors and local citizens are deemed to be vital factors of success for realising the transition to 100% renewable energy systems [54]. The leadership of local governments can play an important role in forging partnerships and pooling resources across the public and private sectors. Examples include public–private partnerships for anaerobic digestion of biowaste for CHP-based district heating and the co-financing of public energy upgrading between local and regional authorities and private investors. Especially in the case of the bioenergy sector, the supply of urban biowaste can depend on citizen awareness and motivation to put aside organic waste for separate collection. For this reason, it is also important to motivate citizens in partaking in waste management strategies to enable the use of organic waste to produce biogas.

-   Awareness raising and training activities: Training material, such as those of European projects' [55], include promotional campaigns for solar energy campaigns. Similar campaigns can be effectively combined with supporting tools, including the provision of a solar atlas and solar land registry (Berlin Solar Atlas (Germany), Paris Solar Land Registry (France), Vlaams–Brabant Climate Map (Belgium), etc.)

-   Community cooperatives for local energy projects: Community cooperatives for local energy projects can enable citizens to have collective ownership and management of projects, including those based on renewable energy generation. At the same time, community cooperatives require a certain level of citizen engagement and empowerment. For this reason, awareness building activities can be used to mitigate perceptions of risk of renewable energy co-operatives that can hinder the profitability of such initiatives as well as their contribution to energy transition objectives. In this case, awareness building in support of local energy generation can also be used to satisfy needs to empower community participation and buy-in within such initiatives [56].

4.3. Governing through Provision

Governing through provision encompasses the process of making available direct energy infrastructure investments, as well as incentives and grants for local energy generation.

- Direct energy infrastructure investments: The utilisation of renewable energy sources in the urban built environment may be limited due to insufficiencies in the energy infrastructure. In this case, direct investments at the local level can include those for modernising and expanding DH/C networks. In combination with regulatory means of governance, LAs can further require that connections to the DH/C network are compulsory for buildings located in related zones. In the case of municipally owned utilities, minimum quotas for renewable energy sources or co-generation can be set. LAs can also involve utility companies in new projects for local energy generation to take advantage of their experience, facilitate greater access to the grid and reach a larger share of individual consumers. Public housing further provides a venue to promote the integrated use of renewable energy sources, including solar thermal, solar photovoltaic (PV), biomass and micro-cogeneration.

- Incentives and grants for local energy generation: LAs can issue municipal green bonds and create funds for renewable energy deployment. The provision of financial opportunities such as these can overcome market failures and address economic barriers for the widespread deployment of related technologies. In addition, LAs can provide subsidies for connections to the district heating network as well as those that may involve the use of any low temperature heating systems that can reduce the operating temperature of the network to increase its efficiency [57].

- Subsidies for local electricity and thermal energy production based on CHP plants as well as the provision of financing for demonstration projects on smart grids can also accelerate progress towards CO_2 mitigation targets based on local energy generation.

4.4. Governing by Regulation and Planning

Regarding ordinances on the mandatory use of renewable energy, Solar Thermal Ordinances (STOs) represent one of the most prevalent forms of mandatory regulations for renewable energy. A growing number of European municipalities, regions and countries has adopted such obligations [58]. In addition, municipalities can require mandatory installations of photovoltaic systems among other renewable energy technologies. Regulatory measures can further require households and private companies to purchase green electricity through obligations on local energy suppliers. Other tools for policy action at the local level also include revision of urban planning regulation to consider the necessary infrastructures required for the development of the DH/C.

In addition to setting regulations, strategic energy planning tools and decisions provide a means for local authorities to evaluate and enforce decisions to promote local energy generation. The following steps exemplify instances in which strategic energy planning would be necessary to promote the generation and utilisation of local energy resources, including those of residual heat from the industry, data centres and wastewater treatment plants. Local maps with information on heat demand densities and the locations and magnitudes of residual heat from industry and power generation can largely facilitate this process.

- Evaluation of geothermal energy potential considering legal and technical barriers of ground perforation and the environmental effect on the underground water layer.
- With regard to the use of biomass, making a technical and economical evaluation of the potential of the biomass harvested in public spaces, companies and citizens' properties, the potential impacts of biomass combustion on air quality and health should also be evaluated.
- Considering the integration of residual heat into the district and cooling network, including sources of residual heat from the industry, data centres, wastewater treatment plants and waste incinerators.

In addition, land use planning should be considered for large-scale solar plants and wind turbines. These aspects call for integrated urban planning processes to support local energy generation decisions as the basis for additional action, such as:

- Establishing an integrated urban planning process to promote renewable energy generation deployment and identifying possible sites to install local energy generation installations, such as those for solar, wind, small hydro and biogas, will ensure the availability and compatibility of public and private space to achieve projects. Some European local authorities offer rooftops of public buildings to private companies for rent to produce energy by means of photovoltaic collectors [59]. Establishing integrated urban planning processes, include those to promote DH/C networks and cogeneration plants, should be supported with mapping tools of thermal energy demand from buildings based on reliable data from utilities.

5. Key Measures for Transition to Sustainable Local Energy Systems

The above modes of urban climate governance, which are classified as M1 to M4, need to be used in combination to provide effective policy support. A holistic understanding of the key measures and technological options that are available at the local level is required to support the design and implementation of policies to promote local energy generation. For this reason, key measures are described with the aim of providing guidance towards potential application areas. Insight from signatories that have already undertaken the technological options is summarised to underline the rapid transition that is taking place at the local level based on the promotion of local energy generation with renewable energy.

The EU has the ambition to be the world number one in renewable energy [60]. To fulfil this objective, the next generation of renewable technologies must be developed and the energy that is produced from renewable sources must be integrated into the energy system in an efficient and cost-effective manner. In this context, there is increasing interest in the decentralisation of the energy supply with more local ownership. Local energy supply options can take the form of district energy systems, local power generation utilities and energy services companies (ESCo). LAs can be whole or partial owners of these utilities and promote community partnership. The relevant modes of urban climate governance that are involved showcase the integrated approach that is needed for supporting particular renewable energy solutions, which is also summarized in Figure 2. Examples for the policy measures are based on compilations from the Covenant of Mayors Signatories' Good Practices database [61].

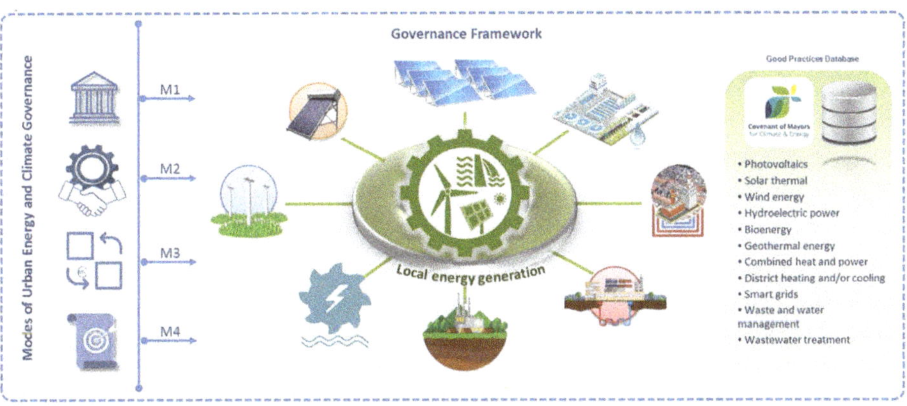

Figure 2. Necessity for a coherent policy mix for local energy generation. Source: own elaboration.

In particular, Tables A1–A3 in the Appendix A are formed as a representation of the 1059 key actions of CoM signatories in the good practices database [61] as "benchmarks of excellence" for local electricity and local heat/cold production. The good practices were exhaustively scanned and categorised into the renewable energy source and technology areas in Figure 2. The key actions were then combined into 82 representative statements with examples from CoM signatories alongside an identification of the modes of governance M1 to M4 that the key action involves. Based on this overview, it is possible to observe the diversity of governance modes that support renewable energy technologies. Such an overview also extends discussions on modes of governance from the perspective of good practices from CoM signatories with a specific focus on local energy generation.

5.1. Photovoltaic

As a widespread measure, CoM signatories are financing and owning photovoltaic (PV) pilot plants on public buildings and facilities based on rooftop PV and building-integrated PV systems (M1). Equally valid measures include installations on the roofs of bus sheds (e.g., 968 kW in Mantova, Italy, as an action that provides 400 tonnes of CO_2 reductions and 990 MWh of local electricity generation [62]), parking lots, and other available areas. Other CoM signatories have constructed a PV park on the ground of a municipal property at a former landfill site, such as in Torrile, Italy, or Évora, Portugal, with about 4000 MWh of local electricity generation based on 3,388 € per tonne CO_2 reduced at the time of implementation [63]. The active use of municipal areas for PV technologies also extends to approaches that combine both municipal self-governing (M1) and governing by provision (M3). CoM signatories are giving a concession of surface rights and renting of rooftop areas in public buildings for PV installations and/or promoting PV installations in public buildings based on collaboration with the ESCo and third-party financing for PV systems in school buildings.

As enablers and providers (M2 and M3), CoM signatories are involved in public–private partnerships for photovoltaic solar parks (e.g., 24.2 MW in Coruche, Portugal), as well as city supported photovoltaic campaigns. In Hannover, Germany, the target of the photovoltaic campaign is to reach 1 million square metres of solar modules by 2020 [64]. In the capacity of regulators (M4), LAs are using their authority to put forth energy supplier obligations for PV systems, such as mandates that PV system installations should be equal to a given share of the total installed power in the municipality. In this respect, it is observed that the complete spectrum of governance modes from M1 to M4 is utilised to various extents in ways that promote PV technologies in cities.

Other measures for PV that involve various governance modes as marked in Table A1 include a municipality bonus for PV installation on citizens' roofs, interest-free loans for associations or schools that install PV panel installations (e.g., Bree, Belgium [65]), the provision of PV systems in a civic center that also supplies electric vehicle charging stations (135 kW in Poole, United Kingdom, with a cost of about 3,587 € per tonne CO_2 reduction at the time of implementation [66]), and real-time data sharing on electricity generation based on the PV systems of the City Council for purposes of awareness-building (e.g., Málaga, Spain, with 609 MWh of local electricity generation [67]).

In other aspects, awareness-building and planning supporting tools for solar energy are actively promoted based on a solar land registry for roof-top PV (or solar thermal) installations in Paris, France [68], while an online solar chart for identifying preferable areas for solar energy technologies is put into use for the benefit of the local actors in Lisbon, Portugal, with an implementation cost of about 10,000 € [69]. Numerous other CoM signatories are providing online solar roof cadasters for every building in the municipality, e.g., Bremen [70], Fürstenfeldbruck [71] and Hannover [64] in Germany, Barcelona in Spain [72], and others. Important complementary measures include public awareness-building to reach annual increase targets for PV in the private buildings and land use planning for utility-scale PV plants in the municipality within broader aspects of urban planning.

5.2. Solar Thermal

Municipal buildings and facilities can provide a point of acceleration for upscaling solar energy technologies at the urban level, which is also valid for solar thermal technologies. In this capacity (M1), CoM signatories are actively increasing the use of solar collectors on the rooftops of municipal buildings, swimming pool facilities, sport buildings and schools, including both flat-plate and parabolic solar collectors. Multiple CoM signatories are also taking the opportunity to replace electrical heaters and boilers in public buildings based on solar collectors. As an enabler (M2), CoM signatories are also mobilising purchasing groups to allow the widespread diffusion of solar thermal technology.

As authoritative measures, CoM signatories are utilizing their capacities as regulators (M4) to put forth an ordinance for installing solar collectors. Zagreb, Croatia requires the use of solar collectors in all buildings in the health care sector as an action that provides a CO_2 reduction of 2077 tonnes and local heat production of 9344 MWh [73]. Loures, Portugal, requires solar thermal systems in 100% of schools that include south-facing facades and terraces [74]. These and related measures are supporting LAs to reach targets for increasing the use of solar thermal technologies.

5.3. Wind Energy

Local ownership of local energy generation is relevant for any energy source. Wind energy represents one of the energy sources in which it is implemented based on such examples as the promotion of locally owned wind turbines (e.g., Ringkøbing-Skjern, Denmark [75]). In addition, in Nijmegen, the Netherlands, a wind and solar farm was established with citizen cooperation (M2) as a measure with a CO_2 reduction of 23,488 tonnes at 1,426 € per tonne and local energy generation of 58,300 MWh [76]. The role of LAs in triggering local energy generation investments was also observed in the case of public procurement of municipality owned wind turbines (M1). In Eskilstuna, Sweden, the public procurement of 4×3.3 MW wind turbines that are owned by the municipality enabled about 40% of the municipal electricity load to be satisfied from wind energy [77]. In Lund, Sweden, the business model for local ownership is realised based on a co-ownership of wind-power plants [78]. As enablers (M2), CoM signatories are also pursuing the attraction of companies that want to generate electricity from wind energy based on such measures as prioritised case handling and licensing of wind turbines. Land use planning is also valid for wind turbines in which CoM signatories are involved in aspects of the use of regulatory and planning capacities (M4).

5.4. Hydroelectric Power

Alongside other renewable energy sources, the use of municipally owned facilities for local energy generation (M1) extends to hydroelectric power. For example, in Ronchi Valsugana, Italy, mini-hydroelectric plants are constructed on municipal waterworks [79]. Other CoM signatories in which mini-hydroelectric power plants are constructed include the Italian cities of Mazzin at 2908 € per tonne CO_2 reduced [80], Rosà with 2×20 kW [81] and others in which investment is attracted to realize an in-stream 10 MW tidal hydro power plant. Other cases in the database includes the CoM signatory of Roman in Romania in which an equal amount of electricity that is needed for public building and public lighting loads is generated from a run-of-river hydroelectric plant [82].

5.5. Bioenergy

The utilisation of bioenergy for local energy generation can involve the waste and wastewater sectors, as well as the sectors of agriculture and forestry in the vicinity. The diversity of bioenergy sources is similarly represented among the CoM signatories based on local energy generation from bioenergy. Multiple signatories are utilising the opportunity to construct new anaerobic digestion plants in publicly owned waste recovery and treatment companies, which can represent a municipal asset (M1). At the same time, public–private partnerships between the municipality and waste management utilities are being established for the anaerobic digestion of biowaste for CHP-based district heating in such signatories as Este, Italy [83]. Similarly, Annicco, Italy, has established a biogas cogeneration plant for electricity and thermal energy provision based on anaerobic digestion that produces 3819 MWh [84]. There are numerous other CoM signatories that have established biogas cogeneration based on zootechnical wastewater and silage cereals and/or a bioenergy (biogas or biomass) driven district heating networks. An example can be given from Banja Luka, Bosnia and Herzegovina, that involves a 6 MW biomass based district heating network [85].

In Bagnolo San Vito, Italy, a consortium for a cogeneration plant based on waste that is produced locally from local consortium companies exemplifies the role of the LA as an enabler (M2). Through the local consortium, 28,350 MWh is generated from sources of local farm sewage and industry at about 1058 € per tonne CO_2 reduced [86]. In signatory municipalities, such as Liepāja, Lithuania, the installation of wood chip boilers in the CHP plant that produces 22,550 MWh is providing carbon neutral district heating at about 1934 € per tonne CO_2 reduced [87]. In Málaga, Spain, 120 wells for degassing biogas capture and network piping recovers methane gas from landfills to produce electricity with a cost of about 423 € per tonne CO_2 reduced [67]. The practice of collecting and recycling used cooking oil for biodiesel production and the use of biomethane in waste collection trucks are other examples that link various sources of waste with an energy service in the transport sector. As can be observed from these examples, CoM signatories are directly involved in promoting the use of bioenergy for local energy generation as well as alternative fuel for vehicles.

5.6. Geothermal Energy

The availability of high-, medium- or low-grade geothermal energy potential is shaping the local response by CoM signatories in the scope of geothermal energy that is or is yet to be utilised. Some signatories are directly constructing geothermal power plants for local electricity generation or taking the opportunity to use low enthalpy geothermal energy resources for the heating of residential buildings. Other signatories provide City Council grants and subsidies to renewable energy technologies, including PV, solar thermal, biomass and ground source heat pumps.

5.7. Multiple Renewable Energy Sources

The orchestration of multiple renewable energy sources is an asset for reaching such targets as net-zero or positive energy targets at the local level in the path of decarbonising the energy sector. In turn, it is important that all modes of urban energy and climate governance are also orchestrated in this direction. In aspects of municipal self-governing (M1), public buildings that are self-sufficient based on on-site renewable energy include a self-sufficient town hall based on bioenergy and PV in Baradili, Italy [88]. A widespread measure among CoM signatories is the application of bioclimatic design principles and renewable energy utilization in public buildings and/or public social housing complexes. Buildings are renovated and equipped with solar thermal collectors and/or biomass in Karlovac, Croatia, with 50% city co-financing [89], and Kozani, Greece, has attained a daycare center that utilises solar and geothermal energy alongside bioclimatic and healthy building design [90].

One of the most prevalent measures among CoM signatories is the purchasing of certified renewable power for public buildings and public lighting. A joint framework agreement for purchasing additional 100% green electricity is also implemented among multiple signatories in the Province of Limburg in the Netherlands. As enablers (M2), LAs are actively involved in awareness building activities, including experimental sessions on renewable energy for students and training campaigns organised by the local energy utilities and agencies. As providers (M3), LAs can be involved in the provision of grants for solar collector and heat pump installations, e.g., Alken, Belgium [91], and subsidies to renewable heat sources in residential buildings, e.g., 25% in Gdynia, Poland [92]. Numerous other CoM signatories established clean technology funds for renewables at the local level or co-financing schemes between local and regional authorities for public energy upgrading. These include a co-financing scheme in Castelnuovo Rangone, Italy, for solar thermal systems [93].

Complementary aspects of regulation and planning (M4) include measures for promoting distributed energy generation based on Urban Building Regulations and authorization procedures.

Demonstrations of net or nearly zero energy buildings is another means of stimulating the local ecosystem for contributions to local energy generation, which include net zero-energy public schools (M1) in Göteborg, Sweden [94], and the Viikki Environment House as a nearly zero-energy office in Helsinki, Finland, with multiple renewable energy sources and district heating connection [95], among others. A pilot public school built according to the Nearly Zero Energy Standard in the Winkelomheide parish of Geel, Belgium [96], and a co-financing of a near zero-energy school building (Scuola Pascoli) with local and national funds represent the interaction of policy tools.

Brownfield urban developments with renewable energy technologies for sustainable districts are rapidly closing the gap between local energy generation and urban planning. In the CoM signatory of Ravenna, Italy, a former port and industrial area is transformed into a new sustainable district [97]. Other examples include onshore power supply based on the renewable energy mix to docking ships in the port to displace fossil fuel usage [94] while in Stockholm, Sweden, the Sustainable Järva project involves 10,000 m^2 of solar cells [98]. Such measures combine multiple governance modes at the local level, including M2 based on an enabling role for urban foresight and planning as M4.

5.8. Combined Heat and Power

The simultaneous production of heat and power in cogeneration plants is being implemented in municipal buildings as a form of municipal self-governing (M1). Biomass-based combined heat and power plants are also contributing to local energy generation in CoM signatories, including 340 GWh$_t$ and 130 GWh$_e$ in Jönköping, Sweden [99]. Other cogeneration plants are being modernised, including fuel flexibility to run on waste and bioenergy, such as in Västerås, Sweden [100]. Low energy houses will also be connected to a low-temperature district heating network. As the provider of energy services, utility companies are also investing in new cogeneration plants with both district heating and cooling infrastructure. In Fürstenfeldbruck, Denmark, the roof of a near CO_2-neutral public CHP plant is also co-located with PV panels for extra electricity supply [71]. In some CoM signatories, subsidies for CHP electricity production take place as a relevant policy measure (M3).

5.9. District Heating and/or Cooling

The presence of a district heating and/or cooling plants and networks either connected to such plants or CHP are an asset to support local energy generation in CoM signatories. One of the policy measures that LAs have in their authority is to establish a contract to connect municipal buildings and schools to the district heating network (M1 and M4). In Milan, Italy, such a contract is also combined with a commitment to invest 10% of the contract sum to energy retrofitting and maintenance [101]. In other cases, LAs have cooperated with the local energy utility to establish a district heating network (M2). Other CoM signatories provide initiatives to increase the purchased volume of energy from the district heating network, including subsidies and obligations for connection to district heating (M3 and M4). The interconnection of district heating networks and extension of distribution piping are other measures that require urban planning and regulation.

Good practices in the database include the Marstal District Heating in Aeroe, Denmark, that contains large-scale solar thermal solutions in district heating systems [102] and Kristianstad, Sweden, in which the connection of buildings to the district heating network is increased [103]. Public buildings in Vittorio Veneto, Italy, will have integrated heating systems [104] (M1) and the share of renewable energy sources in the district heating network will increase from 40% to 95% in Ringsted, Denmark, based on local planning [105] (M4). In addition to district heating works, the connection of buildings and industries to the district cooling network includes an energy efficient data center with PV on the server hall roof in Växjö, Sweden [106]. District networks also enable access to sources of waste heat. Among CoM signatories, residual heat from urban wastewater is utilised in Aachen, Germany [107], waste heat from the local steel industry is recovered in Finspång, Sweden [108], and the use of natural gas is substituted based on the connection of buildings to a district heating network that utilises the available waste heat from a pulp mill in Judenburg, Austria [109].

In addition to establishing and extending district energy networks, their modernisation and rehabilitation is another aspect that requires action based on urban governance. In Bielsko-Biala, Poland, the remote monitoring of pipelines and insulation has reduced heat losses from 30% to 12% [110]. In Rijeka, Croatia, thermal energy distributors and thermostatic radiator valves were installed in the district heating network based on city co-financing with an estimated impact of saving 3140 tonnes of CO_2 emissions [111] (M3). In Riga, Latvia, flue-gas heat recovery is applied to increase the efficiency of heat production [112]. As a relatively unique case, Lerum, Sweden, has cooperated with local actors as an enabler (M2) to establish noise barriers for road and rail traffic that is equipped with solar energy collectors to support local energy generation for the district heating system [113].

5.10. Smart Electricity Grids

In addition to thermal grids, the policy measures of LAs have involved the stimulation of smart electricity grids, including cooperation with the district network operator for demand side management. In Glasgow, the United Kingdom, monitoring and response to peak load in public buildings are implemented towards a future smart grid (M1). Local, regional, national and EU funds are also being utilised to finance pilot projects on smart grids and demonstration sites.

5.11. Waste and Water Management

The efficient planning of available waste flows can allow LAs to oversee that higher recycling and bioenergy stocks are obtained. Numerous CoM signatories are upholding the need for separate waste collection to increase the recycling of municipal solid waste and the use of organic waste for biogas production (M4), which is also supported by awareness-building (M2). In Lakatamia, Cyprus, green waste is used for the production of compost and pellets [114]. In other signatories, organic waste is chosen to be utilised for composting rather than incineration in waste-to-energy plants.

In cross-cutting aspects of water management and local energy generation, renewable sources are integrated for supplying power to pumping tapwater. In addition, electricity usage for pumping is reduced based on reductions in water losses in the drinking-water distribution network in Seixal [115] and Bilbao [116], Portugal. In this way, LAs are able to provide more energy-efficient water services (M3). In order to support related aspects of public awareness, Voznesensk in Ukraine established an information system for energy and water use in the public sector (M1) [117]. Wastewater provides a valuable source of bioenergy upon which Neumarkt in der Oberpfalz in Germany established a self-sufficient wastewater facility based on methane driven CHP plant [118].

6. Conclusions

As the closest level of government to citizens, local authorities play a crucial role in building public support for the European Union's energy and climate goals while deploying more decentralised and integrated energy systems. The European Union Covenant of Mayors initiative has been one of the disrupting phenomena in the arena of transnational initiatives, which have expanded tremendously over the past 10 years. The first version of CoM methodology published in 2010 [119] has been an important guiding document for cities in Europe on the elaboration of SEAPs. After 8 years from the first publication of the CoM guidebook, JRC is in the process of revising the CoM methodology. Gaining the opportunity from this unique momentum, the present work summarises the main methodological updates that are proposed in the new version of the CoM guidebook published in 2018 [120–122].

It is important for local authorities to have a clear framework and pathway for the redaction of the climate action plans; therefore, this updated methodology and the rationale behind it could be helpful in supporting them. The main aspects developed in this paper relate to:

- An updated definition of local energy generation, favouring the inclusion of the new developments of distributed generation, especially from renewable energy sources and cogeneration technologies.
- An updated methodology for indirect emission accounting, taking into account the increasing participation of local citizens along with local authorities in the energy transition.
- An updated overview of the EU energy and climate policies influencing local action in energy generation.
- An updated guidance to local authorities on the modes of energy and climate governance. In this process, local authorities have the capacity to support and mobilise action for local energy generation investments. The four modes of urban energy and climate governance are

investigated and a policy matrix that summarises the scope of each mode along with the main tools, and exemplary actions to support local energy sustainability are provided.

- The exemplary actions to support local sustainable energy generation can be used to further promote city-to-city policy learning based on the Benchmarks of Excellence. The cities will observe that there are a multitude of approaches that can strengthen urban energy and climate governance in a coherent way, including the application of multiple modes simultaneously for combined impact.

Within the Covenant of Mayors framework, it is highly advisable that the CoM signatories utilise the opportunity of revision based on the new version of the CoM guidebook to update and strengthen their climate action. The leadership of cities is crucial to the success of climate mitigation action to address the urgency of global climate change. Cities have shown a significant potential for greater leadership through accelerated action to support local energy generation in the pathway of realising the sustainable energy transition.

Author Contributions: Conceptualization and methodology: A.K., P.B., Ş.K.; formal analysis, investigation and writing—original draft preparation: A.K. and Ş.K.; supervision and project administration: P.B.

Funding: This research received no external funding.

Acknowledgments: We are grateful to the local authorities who make public their engagement in climate action planning through their participation in the Covenant of Mayors initiative. The authors would like to thank European Commission Directorate General for Energy, the CoM Office and JRC colleagues working in the CoM initiative for their support in giving visibility and effectiveness to the effort of cities and local governments in the climate change action. A special acknowledgment to our colleagues Jean François Dallemand and Christian Thiel for reviewing of the CoM methodology and to Gianluca Fulli for discussion on methodological challenges.

Appendix A

Tables A1–A3 represent policy measures regarding renewable energy source and technology, as well as the relevant modes of urban climate governance where M1 is municipal self-governing, M2 is governing through enabling, M3 is governing by provision and M4 is governing by regulation and planning.

Table A1. Renewable energy sources.

Area of Intervention	Policy Measure	M1	M2	M3	M4
Local electricity generation: Photovoltaic	Municipal financing and ownership of PV pilot plants on public buildings (rooftop PV and building-integrated PV systems)	√			
	PV installations on the roofs of bus sheds (968 kW in Mantova, Italy) or parking lots	√			
	Construction of a PV park on ground of municipal property at a former landfill site (994 kW in Torrile, Italy; Évora, Portugal)	√			
	Concession of surface rights and renting of rooftop areas in public buildings for PV	√		√	
	PV installations in public buildings based on collaboration with the ESCo and third-party financing for PV systems in school buildings	√		√	
	Public-private partnership for Photovoltaic Solar Park (24.2 MW in Coruche, Portugal)		√	√	
	City supported photovoltaic campaign • One million square metres of solar modules by 2020 (Hannover, Germany)		√	√	
	Energy supplier obligations for PV systems • Mandate for PV system installations equal to a given share of the total installed power in the municipality				√
	Municipality bonus for photovoltaic and solar thermal installation on citizen's roof			√	
	Interest-free loans for associations or schools for PV panel installations (Bree, Belgium)			√	
	PV systems that supply electric vehicle charging stations (135 kW in Poole, U.K.)	√		√	
	Real time electricity generation data on PV systems of the City Council (Málaga, Spain) and visual consoles on CO_2 reductions	√	√		
	Awareness building and planning supporting tools for solar energy • Solar land registry for roof-top photovoltaic or solar thermal installations (Paris, France) • Solar chart for identifying preferable areas for solar energy technologies (Lisbon, Portugal) • Solar roof cadasters (Bremen, Germany; Fürstenfeldbruck, Germany; Hannover, Germany; Barcelona, Spain, and others)		√	√	√
	Public awareness to reach annual increase targets for PV in the private buildings		√		√
	Land use planning for utility-scale PV plants in the municipality				√
Local heat generation: Solar thermal	Solar collectors on rooftops of municipal buildings, swimming pool facilities, sport buildings and schools (including flat-plate and parabolic solar collector installations) • Replacement of electrical heaters and boilers in public buildings	√			
	Purchasing groups to allow widespread diffusion of solar thermal technology		√		
	Ordinance for installing solar collectors • Solar collectors in all buildings in the health care sector (Zagreb, Croatia) • Solar thermal systems in 100% of schools that include south-facing facades and terraces (Loures, Portugal)				√
	Targets to increase the area of solar thermal in the municipality				√

Table A1. *Cont.*

Area of Intervention	Policy Measure	M1	M2	M3	M4
Local electricity generation: Wind energy	Public procurement of municipality owned wind turbines (4 × 3.3 MW in Eskilstuna, Sweden at 40% of the municipal electricity load)	√			
	Installation of wind power farms • Co-ownership of wind-power plants (municipal company in Lund, Sweden) • Promotion of locally owned wind turbines (Ringkøbing-Skjern, Denmark)	√	√		
	Wind and solar farm with citizen cooperation (Nijmegen, Netherlands)		√		
	Attraction of companies that want to generate electricity from wind energy • Prioritised case handling and licensing of wind turbines		√		
	Land use planning for wind turbines				√
Local electricity generation: Hydroelectric power	Mini-hydro plants on municipal waterworks • Ronchi Valsugana, Italy	√			
	Hydroelectric power plant construction (Manerbio, Italy; Mazzin, Italy; Rosà, Italy)	√	√		
	Attraction of investment to realize an in-stream tidal hydro power plant (10 MW)		√		
	Run-of-river hydroelectric plants • Produces the amount of electricity needed for public building and public lighting loads (Roman, Romania)	√			
Bioenergy	New anaerobic digestion plant in public waste recovery and treatment company	√			
	Public–private partnership between the municipality and waste management utility for anaerobic digestion of biowaste for CHP-based district heating (Este, Italy)		√		
	Biogas cogeneration plant for electricity and thermal energy provision based on anaerobic digestion (Annicco, Italy)			√	
	Biogas cogeneration based on zootechnical wastewater and silage cereals			√	
	Biogas-driven district heating network			√	
	Biomass-based district heating network and/or biomass boilers for replacing diesel boilers (local wood chips < 60 km from sustainable management of forests) • 6 MW in Banja Luka, Bosnia Herzegovina			√	
	Recovery of methane gas from landfills to produce electricity based on gas engines • 120 wells degassing biogas capture and network piping (Málaga, Spain)	√		√	
	Consortium for a cogeneration plant based on biomass certified as sustainable (waste produced locally or from local consortium companies (Bagnolo San Vito, Italy)			√	√
	Installation of wood chip boilers in the CHP plant for carbon neutral district heating • Fuel switching in Liepāja, Lithuania, and other signatory municipalities			√	
	Collection and recycling of used cooking oil for biodiesel production (Loures, Portugal)				√
Geothermal energy	Construction of a geothermal power plant	√			
	Low enthalpy geothermal heating for municipal residential building	√			

Table A1. *Cont.*

Area of Intervention	Policy Measure	M1	M2	M3	M4
	Public buildings that are self-sufficient based on on-site renewable energy • Self-sufficient town hall based on bioenergy and PV (Baradili, Italy)	√			
	Public buildings with bioclimatic design principles and renewable energy utilisation • Public social building complex • Daycare center with solar and geothermal energy (Kozani, Greece)	√			
	Purchasing of certified renewable power for public buildings and public lighting • Joint framework agreement for purchasing 100% green electricity (Province of Limburg, Netherlands)	√			
	Energy renovation of buildings including solar thermal collectors and/or biomass with 50% city co-financing (Karlovac, Croatia)			√	
	Awareness building actions • Experimental sessions on renewable energy for students • Training campaigns organised by the local energy utility/agency		√		
Renewable energy (multiple sources)	City council grants and subsidies for renewable energy (PV, solar thermal, biomass, ground source heat pumps) • Grants for solar collector and heat pump installations (Alken, Belgium) • Subsidy to renewable heat sources in residential buildings (Gdynia, Poland) • Subsidy per square meter of solar thermal collector area (Bonn, Germany)			√	
	Clean technology funds for renewables			√	
	Co-financing between local and regional authorities for public energy upgrading • Co-financing of solar thermal systems on public buildings (Castelnuovo Rangone, Italy)		√	√	
	Promotion of distributed energy generation based on Urban Building Regulations and simplified building authorization procedures		√		√
	Demonstrations of net or nearly zero-energy building with renewable energy • Net zero-energy schools (Göteborg, Sweden) • Viikki Environment House as nearly zero-energy office (Helsinki, Finland) • Pilot public school built according to the Nearly Zero Energy (NZE) Standard (Winkelomheide, Belgium) • Co-financing of a near zero energy school building (Scuola Pascoli, Italy) with local and national funds (Codigoro, Italy)	√	√		
	Brownfield urban development with renewables and sustainable districts • Transformation of former port and industrial area into a new sustainable district (Ravenna, Italy) • Sustainable Järva with 10,000 m^2 of solar cells (Stockholm, Sweden)	√	√		√
	Onshore power supply with high-voltage for docking ships in the port (Göteborg, Sweden)			√	

Table A2. Combined heat and power, district heating and cooling and smart grids.

Area of Intervention	Policy Measure	M1	M2	M3	M4
Combined heat and power	Cogeneration plant for municipal buildings	√			
	Biomass-based combined heat and power plant to support the district heating system (Jönköping, Sweden, with 340 GWh$_t$ and 130 GWh$_e$)			√	
	Modernization of the cogeneration plant with fuel flexibility for waste and bioenergy (Västerås, Sweden)			√	
	Investment of the public utility company in a new cogeneration plant with both district heating and cooling infrastructure • Bioenergy based plant with co-location of PV panels on the roof of the plant (Fürstenfeldbruck, Denmark)	√		√	
	Subsidies for CHP electricity production			√	
District heating/cooling plant	Large-scale solar thermal solutions in district heating systems (Marstal District Heating in Aeroe, Denmark)			√	
	Flue-gas heat recovery to increase efficiency of heat production (Riga, Latvia)			√	
District heating/cooling network	Contract to connect municipal buildings and schools to the district heating network • Commitment to invest 10% of the contract sum to energy retrofitting and maintenance (Milan, Italy)	√			√
	Integrated heating systems between public buildings (Vittorio Veneto, Italy)	√			
	Initiative to increase the purchased volume of energy from the district heating network • Subsidies and/or obligations for connection to district heating			√	√
	Modernization and rehabilitation of district heating and/or cooling networks • Remote monitoring of pipelines and insulation to reduce heat losses (30% to 12% in Bielsko-Biala, Poland)			√	
	Co-financing of thermal energy distributors and thermostatic radiator valves in the district heating network (Rijeka, Croatia)			√	
	Connection of low energy houses to a low-temperature district heating network (Västerås, Sweden)			√	
	Connection of buildings and industries to the district cooling network • Energy efficient data center with PV on the server hall roof (Växjö, Sweden)			√	
	Utilisation of residual heat from urban wastewater (e.g., Aachen, Denmark, and others)			√	
	Utilisation of industrial waste heat • Recovery of waste heat from the local steel industry (Finspång, Sweden) • Substitution of the use of natural gas based on the use of waste heat from a pulp mill (Judenburg, Austria)			√	
	Co-operation to establish noise barriers for road and rail traffic equipped with solar energy collectors to support the local district heating system (Lerum, Sweden)		√		
	Cooperation with the local energy utility to establish a district heating network		√		
	Interconnection of district heating networks and extension of distribution piping			√	√
	Urban planning to increase the connection of buildings to the municipally-owned district heating network (Kristianstad, Sweden)			√	√
	Increase in the share of renewable energy sources in the district heating network • From 40% to 95% in Ringsted, Denmark				√
Smart electricity grids	Cooperation with the district network operator for demand side management • Monitoring and response to peak load in public buildings towards a future smart grid (Glasgow, U.K.)	√	√		
	Financing of pilot projects on smart grids and allocation of local demonstration sites (local, regional, national and EU funds)			√	

Table A3. Waste and water management (including wastewater treatment).

Area of Intervention	Policy Measure	M1	M2	M3	M4
Waste management	Separate waste collection to increase the recycling of municipal solid waste and the use of organic waste for biogas production			√	√
	Use of green waste for the production of compost and pellets (Lakatamia, Cyprus)			√	√
	Utilisation of organic waste for composting rather than waste-to-energy incineration			√	√
Wastewater treatment plants	Self-sufficient wastewater facility based on methane driven combined heat and power plant (Neumarkt in der Oberpfalz, Germany)			√	√
Water management	Integration of renewable sources for supplying power to pumping tapwater			√	√
	Reduction in electricity usage for pumping based on reductions in water losses in the drinking-water distribution network(Seixal, Portugal; Bilbao, Portugal)			√	√
	Information system for energy and water use in the public sector (Voznesensk, Ukraine)	√	√		

References

1. Gordon, J.D. The Politics of Accountability in Networked Urban Climate Governance. *Glob. Environ. Politics* **2016**, *16*, 82–100. [CrossRef]
2. Byrne, J.; Taminiau, J.; Seo, J.; Lee, J.; Shin, S. Are solar cities feasible? A review of current research. *Int. J. Urban Sci.* **2017**, *21*, 239–256. [CrossRef]
3. Bansard, J.S.; Pattberg, P.H.; Widerberg, O. Cities to the rescue? Assessing the performance of transnational municipal networks in global climate governance. *Int. Environ. Agreem. Politics Law Econ.* **2017**, *17*, 229–246. [CrossRef]
4. Christoforidis, G.C.; Chatzisavvas, K.C.; Lazarou, S.; Parisses, C. Covenant of Mayors initiative—Public perception issues and barriers in Greece. *Energy Policy* **2013**, *60*, 643–655. [CrossRef]
5. Pablo-Romero, M.; Sánchez-Braza, A.; González-Limón, J.M. Covenant of Mayors: Reasons for Being an Environmentally and Energy Friendly Municipality. *Rev. Policy Res.* **2015**, *32*, 576–599. [CrossRef]
6. Lombardi, M.; Pazienza, P.; Rana, R. The EU environmental-energy policy for urban areas: The Covenant of Mayors, the ELENA program and the role of ESCos. *Energy Policy* **2016**, *93*, 33–40. [CrossRef]
7. Fecondo, P.; Moca, G. The ELENA Programme in the Province of Chieti—A Public Private Partnership Best Practice Improving Energy Efficiency of Buildings and Public Lighting Systems. *J. Sustain. Dev. Energy Water Environ. Syst.* **2015**, *3*, 230–244. [CrossRef]
8. Melica, G.; Bertoldi, P.; Kona, A.; Iancu, A.; Rivas, S.; Zancanella, P. Multilevel governance of sustainable energy policies: The role of regions and provinces to support the participation of small local authorities in the Covenant of Mayors. *Sustain. Cities Soc.* **2018**, *39*, 729–739. [CrossRef]
9. Rashidi, K.; Patt, A. Subsistence over symbolism: The role of transnational municipal networks on cities' climate policy innovation and adoption. *Mitig. Adapt. Strateg. Glob. Chang.* **2017**, *23*, 507–523. [CrossRef]
10. Bulkeley, H.; Andonova, L.; Bäckstrand, K.; Betsill, M.; Compagnon, D.; Duffy, R.; Kolk, A.; Hoffmann, M.; Levy, D.; Newell, P.; et al. Governing climate change transnationally: Assessing the evidence from a database of sixty initiatives. *Environ. Plan. C Gov. Policy* **2012**, *30*, 591–612. [CrossRef]
11. Kern, K. Cities as leaders in EU multilevel climate governance: Embedded upscaling of local experiments in Europe. *Environ. Politics* **2019**, *28*, 125–145. [CrossRef]
12. Broto, V.C. Urban Governance and the Politics of Climate change. *World Dev.* **2017**, *93*, 1–15. [CrossRef]
13. Pablo-Romero, M.; Pozo-Barajas, R.; Sánchez-Braza, A. Understanding local CO_2 emissions reduction targets. *Renew. Sustain. Energy Rev.* **2015**, *48*, 347–355. [CrossRef]
14. Khan, F.; Sovacool, B.K. Testing the efficacy of voluntary urban greenhouse gas emissions inventories. *Clim. Chang.* **2016**, *139*, 141–154. [CrossRef]
15. Pablo-Romero, M.; Pozo-Barajas, R.; Sánchez-Braza, A. Analyzing the effects of Energy Action Plans on electricity consumption in Covenant of Mayors signatory municipalities in Andalusia. *Energy Policy* **2016**, *99*, 12–26. [CrossRef]
16. Croci, E.; Lucchitta, B.; Janssens-Maenhout, G.; Martelli, S.; Molteni, T. Urban CO_2 mitigation strategies under the Covenant of Mayors: An assessment of 124 European cities. *J. Clean. Prod.* **2016**, *169*, 161–177. [CrossRef]

17. Yu, W.; Pagani, R.; Huang, L. CO_2 emission inventories for Chinese cities in highly urbanized areas compared with European cities. *Energy Policy* **2012**, *47*, 298–308. [CrossRef]

18. Delponte, I.; Pittaluga, I.; Schenone, C. Monitoring and evaluation of Sustainable Energy Action Plan: Practice and perspective. *Energy Policy* **2017**, *100*, 9–17. [CrossRef]

19. Leal, V.M.S.; Azevedo, I. Setting targets for local energy planning: Critical assessment and a new approach. *Sustain. Cities Soc.* **2016**, *26*, 421–428. [CrossRef]

20. Dall'O', G.; Norese, M.; Galante, A.; Novello, C. A Multi-Criteria Methodology to Support Public Administration Decision Making Concerning Sustainable Energy Action Plans. *Energies* **2013**, *6*, 4308–4330. [CrossRef]

21. Di Leo, S.; Salvia, M. Local strategies and action plans towards resource efficiency in South East Europe. *Renew. Sustain. Energy Rev.* **2017**, *68*, 286–305. [CrossRef]

22. Gagliano, A.; Nocera, F.; D'Amico, A.; Spataru, C. Geographical information system as support tool for sustainable Energy Action Plan. *Energy Procedia* **2015**, *83*, 310–319. [CrossRef]

23. Kilkiş, Ş. Sustainable development of energy, water and environment systems index for Southeast European cities. *J. Clean. Prod.* **2016**, *130*, 222–234. [CrossRef]

24. Kılkış, Ş. Composite index for benchmarking local energy systems of Mediterranean port cities. *Energy* **2015**, *92*, 622–638. [CrossRef]

25. Doukas, H.; Papadopoulou, A.; Savvakis, N.; Tsoutsos, T.; Psarras, J. Assessing energy sustainability of rural communities using Principal Component Analysis. *Renew. Sustain. Energy Rev.* **2012**, *16*, 1949–1957. [CrossRef]

26. Messori, G.; Morello, E.; Caserini, S. Indice di Implementazione dei Piani d'Azione per l'Energia Sostenibile: Metodologia e Applicazione alla Città Metropolitana di Milano. *Ing. dell'Ambiente* **2018**, *5*, 302–320.

27. Seto, K.C.; Dhakal, S.; Bigio, A.; Blanco, H.; Delgado, G.C.; Dewar, D.; Huang, L.; Inaba, A.; Kansal, A.; Lwasa, S.; et al. Chapter 12. Human Settlements, Infrastructure, and Spatial Planning. In *Edenhofer, Climate Change 2014: Mitigation of Climate Change. Contribution of Working Group III to the Fifth Assessment Report of the Intergovernmental Panel on Climate Change*; Cambridge University Press: Cambridge, UK, 2014.

28. Arup and C40 Cities Climate Leadership Group. *Global Aggregation of City Climate Commitments—Methodological Review*; ARUP: London, UK, 2014.

29. Reckien, D.; Flacke, J.; Dawson, R.J.; Heidrich, O.; Olazabal, M.; Foley, A.; Hamann, J.J.; Orru, H.; Salvia, M.; Hurtado, S.D.G.; et al. Climate change response in Europe: what's the reality? Analysis of adaptation and mitigation plans from 200 urban areas in 11 countries. *Clim. Chang.* **2014**, *122*, 331–340. [CrossRef]

30. Erickson, P.; Tempest, K. *Advancing Climate Ambition: How City-Scale Actions Can Contribute to Global Climate Goals. No. 2014–6*; Stockholm Environment Institute: Stockholm, Sweden, 2014.

31. International Energy Agency (IEA). *Energy Technology Perspectives 2014*; International Energy Agency: Paris, France, 2014.

32. Ürge-Vorsatz, D.; Petrichenko, K.; Antal, M.; Staniec, M.; Labelle, M.; Ozden, E.; Labzina, E. *Best Practice Policies for Low Energy and Carbon Buildings: A Scenario Analysis*; Research report prepared by the Center for Climate Change and Sustainable Policy (3CSEP) for the Global Buildings Performance Network May 2012; Central European University Press: Budapest, Hungary, 2012.

33. Replogle, M.A.; Fulton, L.M. *A Global High Shift Scenario: Impacts and Potential for More Public Transport, Walking, and Cycling with Lower Car Use*; The National Academies of Sciences, Engineering, and Medicine: Washington, DC, USA, 2014; p. 35.

34. Kona, A.; Bertoldi, P.; Monforti-Ferrario, F.; Rivas, S.; Dallemand, J.F. Covenant of mayors signatories leading the way towards 1.5 degree global warming pathway. *Sustain. Cities Soc.* **2018**, *41*, 568–575. [CrossRef]

35. Widerberg, O.; Pattberg, P. Accountability Challenges in the Transnational Regime Complex for Climate Change. *Rev. Policy Res.* **2017**, *34*, 68–87. [CrossRef]

36. Kona, A.; Melica, G.; Bertoldi, P.; Calvete, S.R.; Koffi, B.; Iancu, A.; Zancanella, P.; Janssens-Maenhout, G.; Dallemand, J.F. *Covenant of Mayors in Figures: 8-Year Assessment*; Publications Office of the European Union: Luxembourg, 2017.

37. Bertoldi, P.; Kona, A.; Rivas, S.; Dallemand, J.F. Towards a Global Comprehensive and Transparent Framework for Cities and Local Governments enabling an Effective Contribution to the Paris Climate Agreement. *Curr. Opin. Environ. Sustain.* **2018**, *30*, 67–74. [CrossRef]

38. Ackermann, T. Distributed generation: A definition. *Electr. Power Syst. Res.* **2001**, *57*, 195–204. [CrossRef]

39. Von Wirth, T.; Gislason, L.; Seidl, R. Distributed energy systems on a neighborhood scale: Reviewing drivers of and barriers to social acceptance. *Renew. Sustain. Energy Rev.* **2018**, *82*, 2618–2628. [CrossRef]
40. Singh, B.; Sharma, J. A review on distributed generation planning. *Renew. Sustain. Energy Rev.* **2017**, *76*, 529–544. [CrossRef]
41. L'Abbate, A.; Fulli, G.; Starr, F.; Peteves, S. *Distributed Power Generation in Europe: Technical Issues for Further Integration*; JRC European Commission Scientific and Technical Report; European Commission: Brussels, Belgium, 2007.
42. Klimscheffskij, M.; van Craenenbroeck, T.; Lehtovaara, M.; Lescot, D.; Tschernutter, A.; Raimundo, C.; Seebach, D.; Timpe, C. Residual Mix Calculation at the Heart of Reliable Electricity Disclosure in Europe—A Case Study on the Effect of the RE-DISS Project. *Energies* **2015**, *8*, 4667–4696. [CrossRef]
43. WRI. *GHG Protocol Scope 2 Guidance*; WRI: Washington, DC, USA, 2015; p. 151.
44. International Energy Agency (IEA). *Emission Factors*; International Energy Agency (IEA): Paris, France, 2018.
45. IPCC. *Emission Factor Database*; IPCC: Geneva, Switzerland, 2018.
46. ISPRA. *Fattori di Emissione Atmosferica di gas ad Effetto Serra e Altri gas nel Settore Elettrico*; ISPRA: Rome, Italy, 2018.
47. Covenantofmayors.eu. Available online: https://www.covenantofmayors.eu/support/library.html (accessed on 1 December 2018).
48. Eurostat/Web/Energy/Data/Shares. Available online: http://ec.europa.eu/eurostat/web/energy/data/shares (accessed on 1 December 2018).
49. Bertoldi, P.; Huld, T. Tradable certificates for renewable electricity and energy savings. *Energy Policy* **2006**, *34*, 212–222. [CrossRef]
50. Fawcett, T.; Rosenow, J.; Bertoldi, P. Energy efficiency obligation schemes: Their future in the EU. *Energy Effic.* **2019**, *12*, 57–71. [CrossRef]
51. Kern, K.; Bulkeley, H. Cities, Europeanization and multi-level governance: Governing climate change through transnational municipal networks. *J. Common Mark. Stud.* **2009**, *47*, 309–332. [CrossRef]
52. Kern, K.; Alber, G. Governing Climate Change in Cititⁱes: Modes of Urban Climate Governance in Multi-level Systems. In Proceedings of the Competitive Cities and Climate Change, OECD Conference Proceedings, Milan, Italy, 9–10 October 2008; pp. 171–196.
53. Covenant of Mayors for Climate and Energy Signatories' Good Practices. Available online: https://www.covenantofmayors.eu/plans-and-actions/good-practices.html (accessed on 1 December 2018).
54. Young, J.; Brans, M. Analysis of factors affecting a shift in a local energy system towards 100% renewable energy community. *J. Clean. Prod.* **2017**, *169*, 117–124. [CrossRef]
55. Energy Efficiency Training of Trainers (EETT). Available online: https://ec.europa.eu/energy/intelligent/projects/en/projects/eett. (accessed on 1 December 2018).
56. Herbes, C.; Brummer, V.; Rognli, J.; Blazejewski, S.; Gerick, N. Responding to policy change: New business models for renewable energy cooperatives—Barriers perceived by cooperatives' members. *Energy Policy* **2017**, *109*, 82–95. [CrossRef]
57. Verda, V.; Caccin, M.; Kona, A. Thermoeconomic cost assessment in future district heating networks. *Energy* **2016**, *117*, 485–491. [CrossRef]
58. Solar Thermal Ordinances = Making A Commitment to Local Sustainable Energy. Available online: http://www.estif.org/fileadmin/estif/content/projects/prosto/downloads/prostobrochure.pdf (accessed on 1 December 2018).
59. CEMR, Climate Alliance, Energie-Cités. *Guide for Local and Regional Governments 'Save the Energy, Save the Climate, Save Money'*; CEMR: Brussels, Belgium, 2008.
60. Energy Union Research, Innovation and Competitiveness Priority, No1 in Renewable Energy. Available online: https://setis.ec.europa.eu/low-carbon-technologies/no1-renewables (accessed on 1 December 2018).
61. Covenant of Mayors Good Practices Database. Available online: https://www.covenantofmayors.eu/plans-and-actions/good-practices.html (accessed on 1 December 2018).
62. Mantova Key Actions. Available online: https://www.covenantofmayors.eu/about/covenant-community/signatories/key-actions.html?scity_id=6703 (accessed on 1 December 2018).
63. Évora Key Actions. Available online: https://www.covenantofmayors.eu/about/covenant-community/signatories/key-actions.html?scity_id=3822 (accessed on 1 December 2018).

64. Hannover Key Actions. Available online: https://www.covenantofmayors.eu/about/covenant-community/signatories/key-actions.html?scity_id=1610 (accessed on 1 December 2018).
65. Bree Key Actions. Available online: https://www.covenantofmayors.eu/about/covenant-community/signatories/key-actions.html?scity_id=5231 (accessed on 1 December 2018).
66. Poole Key Actions. Available online: https://www.covenantofmayors.eu/about/covenant-community/signatories/key-actions.html?scity_id=6026 (accessed on 1 December 2018).
67. Málaga Key Actions. Available online: https://www.covenantofmayors.eu/about/covenant-community/signatories/key-actions.html?scity_id=1699 (accessed on 1 December 2018).
68. Paris Key Actions. Available online: https://www.covenantofmayors.eu/about/covenant-community/signatories/key-actions.html?scity_id=1796 (accessed on 1 December 2018).
69. Lisboa Key Actions. Available online: https://www.covenantofmayors.eu/about/covenant-community/signatories/key-actions.html?scity_id=1872 (accessed on 1 December 2018).
70. Bremen Key Actions. Available online: https://www.covenantofmayors.eu/about/covenant-community/signatories/key-actions.html?scity_id=1603 (accessed on 1 December 2018).
71. Fürstenfeldbruck Key Actions. Available online: https://www.covenantofmayors.eu/about/covenant-community/signatories/key-actions.html?scity_id=2167 (accessed on 1 December 2018).
72. Barcelona Key Actions. Available online: https://www.covenantofmayors.eu/about/covenant-community/signatories/key-actions.html?scity_id=1950 (accessed on 1 December 2018).
73. Zagreb Key Actions. Available online: https://www.covenantofmayors.eu/about/covenant-community/signatories/key-actions.html?scity_id=2019 (accessed on 1 December 2018).
74. Loures. Available online: https://www.covenantofmayors.eu/about/covenant-community/signatories/action-plan.html?scity_id=3118 (accessed on 1 December 2018).
75. Ringkøbing-Skjern Key Actions. Available online: https://www.covenantofmayors.eu/about/covenant-community/signatories/key-actions.html?scity_id=4302 (accessed on 1 December 2018).
76. Nijmegen Key Actions. Available online: https://www.covenantofmayors.eu/about/covenant-community/signatories/key-actions.html?scity_id=2120 (accessed on 1 December 2018).
77. Eskilstuna. Available online: https://www.covenantofmayors.eu/about/covenant-community/signatories/action-plan.html?scity_id=2725 (accessed on 1 December 2018).
78. Lund. Available online: https://www.covenantofmayors.eu/about/covenant-community/signatories/key-actions.html?scity_id=2456 (accessed on 1 December 2018).
79. Ronchi Valsugana Key Actions. Available online: https://www.covenantofmayors.eu/about/covenant-community/signatories/key-actions.html?scity_id=6050 (accessed on 1 December 2018).
80. Mazzin Key Actions. Available online: https://www.covenantofmayors.eu/about/covenant-community/signatories/key-actions.html?scity_id=6001 (accessed on 1 December 2018).
81. Rosà Key Actions. Available online: https://www.covenantofmayors.eu/about/covenant-community/signatories/key-actions.html?scity_id=5533 (accessed on 1 December 2018).
82. Roman. Available online: https://www.covenantofmayors.eu/about/covenant-community/signatories/action-plan.html?scity_id=8732 (accessed on 1 December 2018).
83. Este Key Actions. Available online: https://www.covenantofmayors.eu/about/covenant-community/signatories/key-actions.html?scity_id=4901 (accessed on 1 December 2018).
84. Annicco Key Actions. Available online: https://www.covenantofmayors.eu/about/covenant-community/signatories/key-actions.html?scity_id=5596 (accessed on 1 December 2018).
85. Banja Luka Key Actions. Available online: https://www.covenantofmayors.eu/about/covenant-community/signatories/key-actions.html?scity_id=2018 (accessed on 1 December 2018).
86. Bagnolo San Vito Key Actions. Available online: https://www.covenantofmayors.eu/about/covenant-community/signatories/key-actions.html?scity_id=3121 (accessed on 1 December 2018).
87. Liepāja Key Actions. Available online: https://www.covenantofmayors.eu/about/covenant-community/signatories/key-actions.html?scity_id=6214 (accessed on 1 December 2018).
88. Baradili Key Actions. Available online: https://www.covenantofmayors.eu/about/covenant-community/signatories/key-actions.html?scity_id=5534 (accessed on 1 December 2018).
89. City of Karlovac Key Actions. Available online: https://www.covenantofmayors.eu/about/covenant-community/signatories/key-actions.html?scity_id=2943 (accessed on 1 December 2018).

90. Kozani Key Actions. Available online: https://www.covenantofmayors.eu/about/covenant-community/signatories/key-actions.html?scity_id=5540 (accessed on 1 December 2018).
91. Alken Key Actions. Available online: https://www.covenantofmayors.eu/about/covenant-community/signatories/key-actions.html?scity_id=5225 (accessed on 1 December 2018).
92. Gdynia Key Actions. Available online: https://www.covenantofmayors.eu/about/covenant-community/signatories/key-actions.html?scity_id=3949 (accessed on 1 December 2018).
93. Castelnuovo Rangone Key Actions. Available online: https://www.covenantofmayors.eu/about/covenant-community/signatories/key-actions.html?scity_id=1828 (accessed on 1 December 2018).
94. Göteborg Key Actions. Available online: https://www.covenantofmayors.eu/about/covenant-community/signatories/key-actions.html?scity_id=1884 (accessed on 1 December 2018).
95. Helsinki Key Actions. Available online: https://www.covenantofmayors.eu/about/covenant-community/signatories/key-actions.html?scity_id=1765 (accessed on 1 December 2018).
96. Winkelomheide Key Actions. Available online: https://www.covenantofmayors.eu/about/covenant-community/signatories/key-actions.html?scity_id=7909 (accessed on 1 December 2018).
97. Ravenna Key Actions. Available online: https://www.covenantofmayors.eu/about/covenant-community/signatories/key-actions.html?scity_id=1839 (accessed on 1 December 2018).
98. Stockholm. Available online: https://www.covenantofmayors.eu/about/covenant-community/signatories/key-actions.html?scity_id=1888 (accessed on 1 December 2018).
99. Jönköping. Available online: https://www.covenantofmayors.eu/about/covenant-community/signatories/action-plan.html?scity_id=2447 (accessed on 1 December 2018).
100. Västerås Key Actions. Available online: https://www.covenantofmayors.eu/about/covenant-community/signatories/key-actions.html?scity_id=6427 (accessed on 1 December 2018).
101. Milan Key Actions. Available online: https://www.covenantofmayors.eu/about/covenant-community/signatories/key-actions.html?scity_id=1834 (accessed on 1 December 2018).
102. Aeroe. Available online: https://www.covenantofmayors.eu/about/covenant-community/signatories/action-plan.html?scity_id=6640 (accessed on 1 December 2018).
103. Kristianstad Key Actions. Available online: https://www.covenantofmayors.eu/about/covenant-community/signatories/key-actions.html?scity_id=2495 (accessed on 1 December 2018).
104. Vittorio Veneto Key Actions. Available online: https://www.covenantofmayors.eu/about/covenant-community/signatories/key-actions.html?scity_id=6534 (accessed on 1 December 2018).
105. Ringsted Key Actions. Available online: https://www.covenantofmayors.eu/about/covenant-community/signatories/key-actions.html?scity_id=2594 (accessed on 1 December 2018).
106. Växjö Key Actions. Available online: https://www.covenantofmayors.eu/about/covenant-community/signatories/key-actions.html?scity_id=1889 (accessed on 1 December 2018).
107. Aachen Key Actions. Available online: https://www.covenantofmayors.eu/about/covenant-community/signatories/key-actions.html?scity_id=1601 (accessed on 1 December 2018).
108. Finspång Key Actions. Available online: https://www.covenantofmayors.eu/about/covenant-community/signatories/key-actions.html?scity_id=2548 (accessed on 1 December 2018).
109. Judenburg Key Actions. Available online: https://www.covenantofmayors.eu/about/covenant-community/signatories/key-actions.html?scity_id=4681 (accessed on 1 December 2018).
110. Bielsko-Biala Key Actions. Available online: https://www.covenantofmayors.eu/about/covenant-community/signatories/key-actions.html?scity_id=1865 (accessed on 1 December 2018).
111. Rijeka Key Actions. Available online: https://www.covenantofmayors.eu/about/covenant-community/signatories/key-actions.html?scity_id=1815 (accessed on 1 December 2018).
112. Riga. Available online: https://www.covenantofmayors.eu/about/covenant-community/signatories/action-plan.html?scity_id=1851 (accessed on 1 December 2018).
113. Lerum. Available online: https://www.covenantofmayors.eu/about/covenant-community/signatories/action-plan.html?scity_id=7303 (accessed on 1 December 2018).
114. Lakatamia Key Actions. Available online: https://www.covenantofmayors.eu/about/covenant-community/signatories/key-actions.html?scity_id=6195 (accessed on 1 December 2018).
115. Seixal Key Actions. Available online: https://www.covenantofmayors.eu/about/covenant-community/signatories/key-actions.html?scity_id=3801 (accessed on 1 December 2018).

116. Bilbao Key Actions. Available online: https://www.covenantofmayors.eu/about/covenant-community/signatories/key-actions.html?scity_id=2909 (accessed on 1 December 2018).
117. Voznesensk Key Actions. Available online: https://www.covenantofmayors.eu/about/covenant-community/signatories/key-actions.html?scity_id=1900 (accessed on 1 December 2018).
118. Neumarkt in der Oberpfalz Key Actions. Available online: https://www.covenantofmayors.eu/about/covenant-community/signatories/key-actions.html?scity_id=1617 (accessed on 1 December 2018).
119. Bertoldi, P.; Cayuela, D.B.; Monni, S.; de Raveschoot, R.P. *How to Develop A Sustainable Energy Action Plan (SEAP)—Guidebook*; Publication Office of the European Union: Luxembourg, 2010; p. 120.
120. Bertoldi, P.; Rivas, S.; Melica, G.; Palermo, V.; Dallemand, J.F. *Guidebook "How to Develop A Sustainable Energy and Climate Action Plan (SECAP)": Part 1—The SECAP Process, Step by Step Towards Low Carbon and Climate Resilient Cities by 2030*; EUR 29412; Publications Office of the European Union: Luxembourg, 2018.
121. Bertoldi, P.; Iancu, A.; Kona, A.; Suvi, M.; Muntean, M.; Lah, O.; Rivas, S. *Guidebook "How to Develop A Sustainable Energy and Climate Action Plan (SECAP)": PART 2—Baseline Emission Inventory (BEI) and Risk and Vulnerability Assessment (RVA)*; EUR 29412; Publications Office of the European Union: Luxembourg, 2018.
122. Bertoldi, P.; Kona, A.; Palermo, V.; Zangheri, P.; Serrenho, T.; Rivas, S.; Labanca, N.; Kilkis, S.; Lah, O.; Glancy, R.; et al. *Guidebook "How to Develop A Sustainable Energy and Climate Action Plan (SECAP)": PART 3—Policies, Key Actions, Good Practices for Mitigation and Adaptation to Climate Change and Financing SECAP(s)*; EUR 29412; Publications Office of the European Union: Luxembourg, 2018.

Article

Sustainable Zoning, Land-Use Allocation and Facility Location Optimisation in Smart Cities

Ahmed WA Hammad [1],*, Ali Akbarnezhad [2], Assed Haddad [3] and Elaine Garrido Vazquez [3]

[1] Faculty of Built Environment, UNSW Sydney, Sydney 2052, Australia
[2] School of Civil Engineering, the University of Sydney, Sydney 2006, Australia; ali.nezhad@sydney.edu.au
[3] Departamento de Construção Civil, Escola Politécnica da Universidade Federal do Rio de Janeiro, Athos da Silveira Ramos 149, Ilha do Fundão, RJ, Brazil; assed@poli.ufrj.br (A.H.); elainevazquez@poli.ufrj.br (E.G.V.)
* Correspondence: a.hammad@unsw.edu.au

Received: 2 January 2019; Accepted: 15 March 2019; Published: 5 April 2019

Abstract: Many cities around the world are facing immense pressure due to the expediting growth rates in urban population levels. The notion of 'smart cities' has been proposed as a solution to enhance the sustainability of cities through effective urban management of governance, energy and transportation. The research presented herein examines the applicability of a mathematical framework to enhance the sustainability of decisions involved in zoning, land-use allocation and facility location within smart cities. In particular, a mathematical optimisation framework is proposed, which links through with other platforms in city settings, for optimising the zoning, land-use allocation, location of new buildings and the investment decisions made regarding infrastructure works in smart cities. Multiple objective functions are formulated to optimise social, economic and environmental considerations in the urban space. The impact on underlying traffic of location choices made for the newly introduced buildings is accounted for through optimised assignment of traffic to the underlying network. A case example on urban planning and infrastructure development within a smart city is used to demonstrate the applicability of the proposed method.

Keywords: sustainable smart city; mathematical optimisation; urban design; bilevel modelling; location theory; traffic assignment; infrastructure expansion; building location

---

1. Introduction

Existing statistics state that almost 55% of the population of the world currently resides in urban regions [1,2], with estimates that this rate will increase to 70% by 2050 [3]. As this progression towards greater urban centres continues to increase, a need has emerged to find ways for supporting this growth in a sustainable manner. Furthermore, there is the challenge of dealing with the pollution levels that result from exacerbated activities in cities [4]. Along with the surging rates of urbanisation and pollution, the world has also experienced a breakthrough in the use of technologies, specifically those related to information and communication technologies (ICT) [5]. Updates in connectivity between various electronic platforms has led to the development of the Internet of Things (IoT), which is based on networks formed between physical devices and appliances to allow data transfer and exchange for enhanced operations [6].

The integration of ICT and IoT is thought to lead to an enhanced system for the management of cities. As a result, the notion of smart cities has arisen in response to the need for sustainable cities that can accommodate the growing population numbers, hence enhancing cities' liveability and the wellness and living standards of citizens [7]. Even though universal agreement on a specific definition of a 'smart city' is still lacking, its main domain lies in the use of information and ICT in sectors such as infrastructure, buildings and energy [8]. In particular, concepts from ICT and IoT are being increasingly reflected in the operations of existing cities, resulting in an interrelated platform between large numbers of citizens, transport networks, services and urban assets and facilities [9].

Decision-making in planning and operations of smart cities needs to be structured around two main considerations, namely strategic and tactical decisions [10]. Yet, current emphasis in the literature is on the technical interfaces making up the various data-exchange enabling platforms placed within the cities, with little emphasis on the strategic and tactical urban planning aspects of smart cities. Within the area of strategic decision-making in city planning, zoning of the city and the location of its operating facilities, including schools, hospitals and so forth, are both of significant relevance [11]. An appropriate selection of zone clusters and the subsequent selection of locations to place buildings in form the main components that are involved in city design. In cities, the positioning of new buildings leads to the generation of traffic demand in the existing network structure [12]. This causes additional traffic loadings on the existing network, and if not well planned for, can result in major transportation delays to network users. Traffic congestion is thought to result in over \$121 billion in losses [13] and can increase the amount of carbon emissions from traffic by more than 53% [14]. Another important strategic consideration is the development of the underlying travel network of the region. Specifically, the development of the transport network will be based on the capacities required to handle the initiated travel on the transport networks, while the operations of the network will in turn be associated with the established capacity of the network, along with the traffic loading patterns on the links (roads) of the network. Decisions related to the transport network are determined based on population rates and estimated travel via the various transport modes utilised in the region [15]. As a result, the zoning of the city will have a direct impact on the locations available for the facilities required to service the underlying population, which in turn would also impact the traffic and operation of the transportation network [16]. It is vital to thus integrate the decision-making that is involved in the zoning, facility location and transport network capacity design of the underlying smart city.

The attention in this article is directed towards the concept of location planning in smart cities. The proposed approach can be divided into three main areas: (i) the need for establishing a framework for the construction of a smart city from scratch, where zoning and land use need to be specified; (ii) the location of buildings in smart cities and the investment decisions made regarding the expansion of the existing road network structure and capacity, which involves the consideration of attributes that influence the decision of positioning buildings such as schools, hospitals and offices; and (iii) the determination of the resulting impacts caused by such location decisions based on the triple bottom-line of sustainability, via formulation of appropriate social, environmental and economic cost objective functions. As a result, three main decisions, which form the essence of urban planning and design in smart cities are targeted: namely, the decisions made regarding the allocation of zones and the assignment of buildings to locations in the region, the expansion decisions related to the road structure of the city and the expansion of the capacity of existing links in the network (if one already exists).

In this paper, a range of mathematical optimisation problems are integrated, including the clustering problem [17], the assignment problem [18], the facility location problem [19] and the urban traffic design problem [12], in order to model key strategic decisions in smart city design and planning. The work proposed herein is expected to form an integral component of the urban design of smart cities. In particular, such a framework will find applicability in prospective smart city planning designs, to ensure a sustainable and integrated city structure where buildings and road networks are strategically planned for. The framework can be used both for new smart city development and for decisions to be made within an existing smart city and which impact the urban design morphology of the existing city.

Remainder of this article is divided as follows: the next section provides a review of the literature on smart city planning in terms of zoning, facility location and transport network design. The proposed mathematical optimisation framework for strategic planning of smart cities is then presented. Following that, the algorithm and formulations developed are outlined. A numerical example of a case project is used to validate the proposed framework. Concluding remarks are provided at the end.

2. Literature Review

A significant number of research studies examine the concept of smart city and its use for sustainable urban planning. The literature reviewed herein falls under, but is not limited to, one of the following fields that are imperative building blocks in smart cities: energy management, city structure zoning and location, intelligent transport systems and smart infrastructure.

Key aspects regarding IoT and its use in smart cities were reviewed in [20]. A bibliometric analysis in [21] examines frequent city categorisation, including smart cities, used for sustainable urbanisation. A comprehensive review was conducted in [22] for energy management and sustainable planning in smart cities [23]. Spatiotemporal forecasting methods, which exploit time series data from various locations within the context of smart cities, and their applications for smart city transport and building management were evaluated in [24].

Several studies focused on energy management in smart cities [25]. Wojnicki and Kotulski [26] proposed an outdoor lighting control system for smart cities. An activity-aware system to automate building systems in smart cities was developed in [27]. A framework to optimise energy management on smart campuses was proposed in [28]. A heating and cooling modelling system was proposed for minimising electricity consumption in smart cities [29]. A smart city architecture was developed in [30] for addressing challenges in smart grid distribution. A comparative assessment of smart energy systems to ensure sustainable, clean and reliable energy in smart cities is found in [31].

In terms of location optimisation, the authors of [32] proposed an optimisation approach that relies on integrating geographic information systems (GIS) with a fuzzy-analytical hierarchy process (FAHP) for choosing suitable wind farm sites. Impacts of location choices made on wind turbine operations were discussed in [33]. In the context of infrastructure planning for smart cities, the authors of [34] developed a city navigation approach for electric vehicles, where locations of charging stations were assumed to be unknown. A business model tool for smart cities to facilitate undertaking strategic decision-making promptly was proposed in [35]. In [36], a localisation-based key management system for meter data encryption in smart cities was proposed. The utilisation of smart parking lots in smart cities was investigated in [37].

In terms of urban planning and structuring, an integrated model which considers land use, transportation and energy systems for future smart cities was presented in [38]. A framework for a smart city in China, which focuses on use of big data for infrastructure city planning and management, was developed by the authors of [39]. Use of smart technology for enhancing the sustainability of the construction sector through targeting demolition waste was discussed in [40]. Even though an investigation on algorithms deployed for sustainable transport policy in cities has been carried out, as detailed in [41], there is no link that has been developed to account for location decisions of new buildings and their impacts on the smart city urban design structure. New ways to achieve systematic-based solutions that augment the process adopted in urban design and planning, and which can lead to future viability and prosperity in metropolitan regions, need to be developed.

As is apparent, there is little focus on developing mathematical optimisation models that integrate location and transportation decisions for use in urban design of smart cities. In particular, there is an apparent lack in studies that focus on addressing operational research problems that are relevant to the strategic planning of urban areas. This is especially imperative in urban design of smart cities, where emphasis on the interconnectivity of key features within the city environment, including transportation systems and adjoining city layout, is highly regarded. As a result, in this work, the aspect within smart city design which will be examined refers to an automated and systematic urban planning procedure that relies on the use of mathematical optimisation frameworks. Such mathematical frameworks can be utilised for a range of applications in smart cities, including intelligent transport planning systems, intelligent energy monitoring and delivery and smart urban design approaches. The main contributions of this study are as follows: (i) the development of a mathematical method for dividing the proposed smart city region into zonal clusters; (ii) formulation of an assignment problem for land-use selection in zonal clusters; and (iii) development of a mathematical framework for enhancing the sustainable planning of location decisions made regarding building placement, and for measuring their impact on the routing of traffic within the smart city, through automated infrastructure investment decisions. The research presented herein thus integrates strategic aspects of location planning and traffic assignment involved in the urban design of smart cities. The next section outlines the main components making up the developed framework.

3. Smart City Zoning, Facility Location and Transport Network Planning Framework

The main motivation behind the framework proposed in this study is to ensure the efficient planning and design of city zoning, building location and transport network of smart cities, while accounting for environmental, social and economic considerations. As was previously discussed, the layout and location planning decision in smart cities considers two main planning aspects, namely strategic and operational planning; this is displayed in Figure 1. As can be seen, in terms of the strategic planning decisions, the main variables that need to be modelled in the proposed framework are the zoning of the region, the link expansion variables on the existing network, network extension through addition of new links, and the positioning of buildings. Apart from future population growth, which creates a slight increase in the demand induced in the regions, the focus in the developed framework is mostly on traffic demand generated by the placement of buildings.

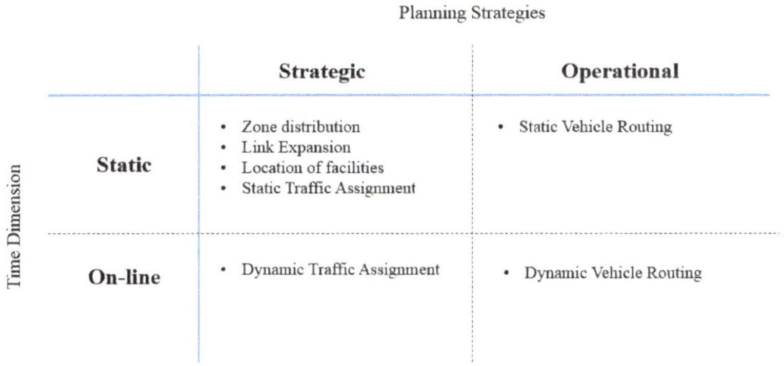

Planning Strategies

Time Dimension		**Strategic**	**Operational**
	Static	• Zone distribution • Link Expansion • Location of facilities • Static Traffic Assignment	• Static Vehicle Routing
	On-line	• Dynamic Traffic Assignment	• Dynamic Vehicle Routing

Figure 1. Some of the operational research problems that can be considered when planning for urban design.

An outline of the proposed framework in this paper is shown in Figure 2. Three main decisions, which form the essence of urban planning and design in smart cities are targeted: namely, the decisions made regarding allocation of zones and the assignment of buildings to locations in the region, the expansion decisions related to the road structure of the city and the expansion of the capacity of existing links in the network (if one already exists). The main concept introduced via the developed framework is the vital integration of all three decisions into a single model that simultaneously optimises the decision-making process involved.

The strategic decision for smart cities starts with the division of the region into zones, assuming the subject is a new city that requires zoning. This step also involves determining the land-use patterns in the region. In the case that an already existing smart city is considered, then the zoning procedure can be neglected, and the existing zonal configuration can be adopted instead. The second step involves the positioning of the buildings within the zones defined and in accordance with the allocated land-use patterns. Examples of buildings that need to be located include offices, retail shops, hospitals and schools. The third step is that related to infrastructure development and expansion. Such decisions will be influenced by the previous two steps and so it is necessary to assimilate both decisions together in a fashion that permits the translations of the impacts that the zoning and location decisions have on the infrastructure investment decisions made by the planners.

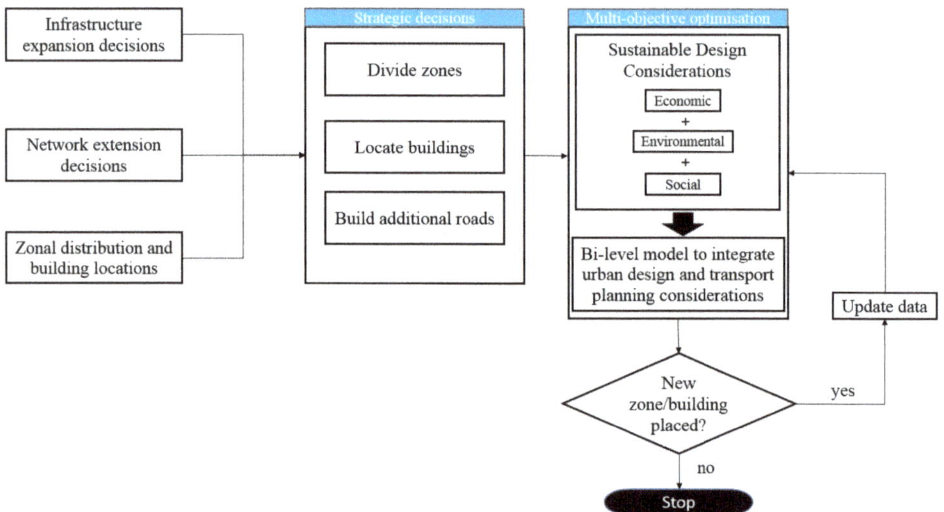

Figure 2. Proposed framework for planning in smart cities.

Given that a chief consideration in smart cities is ensuring effective mobility and robust decision-making for improving the transport of goods and people, developing an approach that can incorporate a forward-looking method for assignment of traffic based on newly introduced buildings is thus imperative. To enable this to happen, a multi-objective optimisation model is developed, based on a bilevel structure [42]. The bilevel structure is needed in order to model the decision spaces of the two main decision-makers in the model: namely, the urban planners and the transportation network users. The importance of generating sustainable solutions that target the triple bottom-line of sustainability is also accounted for through considering objective functions in the optimisation model developed, with focus on environmental, social and economic impacts of the locations and infrastructure decisions made.

The model can then be adapted to continue to be utilised for the strategic decisions to be made within the smart city, whenever a change in the structure of the city is induced. The change that is emphasised in the framework is related to the introduction of a new zone or building within the region. As shown in Figure 2, the procedure loops back to the optimisation model, whose associated parameters are updated in response to the induced changes, and a new solution is generated. Otherwise, if no change is induced, the algorithm ceases.

The proposed framework can also be linked to other automated systems that rely on the use of ICT in daily management of the city, such as online estimation of origin–destination (OD) matrices within the city for enhanced traffic assignment and real-time traffic state estimation and updates [43]. The work in this article is specifically targeted towards enhancing the intelligence of the transport systems through considering the impacts of newly positioned buildings on the underlying network.

4. Mathematical Optimisation Models

In this section, the mathematical formulations that are integrated in the proposed framework are outlined. The mathematical optimisation models can be divided into three main types: the first is associated with the zone division and clustering in the region, the second relates to the assignment of land-use patterns to the zone clusters and the third is related to the location of buildings and traffic assignment to the underlying network, based on infrastructure decisions made in the region. Notations adopted in the proposed mathematical models are given in Nomenclature.

4.1. Zoning Regions

Assuming an input of free land dispersed in the region is provided by the decision-maker (urban planner), the first step in zoning a smart city involves clustering the land into discrete zones. This step will involve the use of a clustering algorithm in order to yield a layout of area nodes clustered into zones. The second step involves assigning a land-use zone to each cluster created via the clustering algorithm. Each of these steps will now be explained.

4.1.1. Area Clustering

Each available area is clustered into a set of zones via the use of a clustering algorithm. In this study, principles from the k-means clustering approach are adopted [44]. The algorithm is summarised in Figure 3.

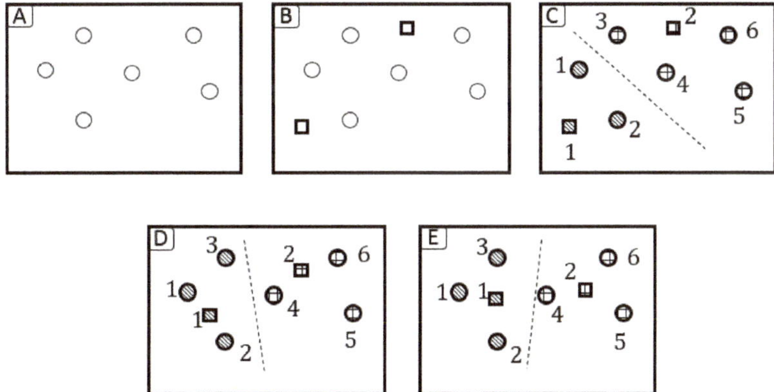

Figure 3. Steps involved in the k-means clustering approach adopted for zoning of the region. (**A**) Nodes in the region for positioning new buildings are identified a priori (i.e., circles); (**B**) centroids of zones to be created are placed randomly in the region (i.e., squares); (**C**) each node in the region is assigned to a centroid based on distance proximity (each node is numbered and assigned to a numbered centroid; matching displayed through the shading in the figure); (**D**) the location of the centroid is recalculated and new nodes are assigned/removed in response; (**E**) the final clusters are created.

Let A be the coordinate of the area nodes and let Q denote the centroid of the zones to be created in the region. The underlying urban space should already contain the available spaces for region development, referred to as the nodes (see Figure 3A). The algorithm starts by placing an input number of zone centroids in the urban space at random, as demonstrated in Figure 3B. The aim is to then group the nodes into clusters, forming the discrete zones of the region. The algorithm iterates through all nodes present in the region, finding the nearest centroid to each node, according to Equation (1).

$$Q_j = \arg \min_j D(A_i, Q_j) \tag{1}$$

where $D(A_i, Q_j)$ is a distance function.

Once all nodes have been iterated through, a centroid is allocated to one or more nodes, as demonstrated via the hatching displayed in Figure 3C. The algorithm then recalculates the position of each centroid based on the following equation, Equation (2):

$$Q_j = \frac{1}{n_j} \sum_{i \in V_j} A_i \qquad \forall j \in \Pi \tag{2}$$

where set V_j is the set of all area nodes allocated to the centroid $j \in \Pi$ in the previous step.

Equation (2) considers the average sum of all area node coordinates clustered around a specific zone centroid as the determining attribute of the updated centroid position.

4.1.2. Assignment Model

Once the zone clusters are formed, the next step involves developing an assignment model that allocates each zone to a specific land use in accordance with a given set of criteria. As an example, the criteria can be based on distance to existing roads, soil surface type of each zone cluster, distance to cities/towns nearby, etc. In this study, the assignment model developed is represented via a binary integer programming formulation, where an objective function based on a defined set of criteria is utilised. One of the common criteria used in zoning is the travel distance between zones. As a result, the objective function in this study is formulated to minimise the travel between the different land-use patterns, based on predications of travel of people within the region. Let Π and Λ denote the set of land-use and zonal clusters available, let a_{kl} denote the people that are expected to travel between land use k and l (e.g., between commercial and residential zones) and let b_{vo} denote the distance of travel between zone cluster v and o. The integer variable w_{kv} is defined to equal to 1 if land use $k\epsilon\Pi$ is assigned to zone cluster $v\epsilon\Lambda$, and 0 otherwise. The objective function to be minimised (total distance of travel between land uses allocated to zones), is given as Equation (3):

$$\sum_{v,o \in \Pi} \sum_{k,l \in \Lambda} a_{kl} b_{vo} w_{kv} w_{lo} + \sum_{v \in \Pi} \sum_{k \in \Lambda} g_{kv} w_{kv} \tag{3}$$

Essential constraints that are defined are assignment constraints, which specify that each land use that planners intend to position in a given region are assigned to a particular land zone, as given by Equation (4):

$$\sum_{V \in \Pi} w_{kv} = 1 \qquad \forall k \in \Lambda \tag{4}$$

The domain of the binary variables is defined in Equation (5):

$$w_{kv} \in \{0,1\} \qquad \forall k \in \Lambda, \forall v \in \Pi \tag{5}$$

Additional sets of constraints that specify other requirements, such as distances that are required between land uses and required connections to existing roads, etc., can also be formulated. It is also important to note that other objective functions that target other criteria can be formulated for assignment of land-use areas to zonal clusters identified. The distance criterion was adopted in this study due to its high relevance in zone planning in urban regions.

4.2. Building Location and Infrastructure Model

Once the zonal configuration and land use specified for each zone is obtained, the next step involves formulating a model to (i) locate buildings in the region and (ii) determine the investment expansions required for existing infrastructure in response. A bilevel model [42] is proposed which accounts for the decision of two key decision-makers in a smart city setting, namely the urban planners and the transportation network users. The decision space of urban planners is modelled through optimising decisions related to the sustainability of the locations chosen for the buildings within the urban region, along with optimising the infrastructure investment decisions. The decision space of the network users is modelled through optimising their choice of links within the network in response to congestion created and demand generated when an urban planner places new buildings. Since the transport users respond to changes induced by decisions made by urban planners, the model proposed is developed based on a two-level hierarchical system, where the upper level represents the optimisation of the urban planners' decision, while the lower level models the behaviour of users in response to decisions made by urban planners.

The upper level of the proposed model is described next.

4.2.1. Upper Level

The main decision variables in the upper level are: (i) the location decision, represented by the binary variable z_{fs} and which equals 1 if building f is placed in location s, and 0 otherwise; (ii) the binary variable y_{ij}, which specifies whether link (i, j) is constructed or not; and (iii) the continuous variable ϕ_{ij}, which indicates whether an existing link of the network is expanded or not.

Upper-Level Objective Functions

The upper-level model involves the formulation of three objective functions; each function targets one specific measure of sustainability. The first equation modelled is a proxy for the social pillar of sustainability (Equation (6)); it minimises the total noise pollution experienced in each zone of the smart city. Noise is generated by the buildings to be positioned in the region, as measured by the parameter M_{rs}.

$$\min_z \sum_{t \in T} \sum_{f \in F_t} \sum_{s \in P_t} \sum_{r \in P} z_{fs} M_{rs} \tag{6}$$

Equation (7) targets the economic aspect of sustainability, where the cost of constructing buildings in the zones of the urban region, $\overline{C}_s$, is minimised.

$$\min_z \sum_{t \in T} \sum_{f \in F_t} \sum_{s \in P_t} z_{fs} \overline{C}_s \tag{7}$$

The final objective function, Equation (8), considers the minimisation of the total carbon emissions from users on the traffic network. Equation (8) accounts for the emissions from different transportation modes, ε_m, which is multiplied by (i) the distance of the links of the network, d_{ij}; (ii) the flow on the links, x_{ij}; and (iii) the time of travel on the links of the network, which considers the congestion impacts on the roads, as given by Equations (9) and (10).

$$\min_x \sum_{(i,j) \in L^R} \sum_{m \in \Gamma} \varepsilon_m d_{ij} x_{ij} t_{ij}(x_{ij}) \tag{8}$$

$$t_{ij}(x_{ij}) = T^0_{ij} \left(1 + 0.15 \left(\frac{x_{ij}}{k^0_{ij} + \phi_{ij}} \right)^4 \right) + (1 - y_{ij}) M \qquad \forall (i,j) \in L^N \tag{9}$$

$$t_{ij}(x_{ij}) = T^0_{ij} \left(1 + 0.15 \left(\frac{x_{ij}}{k^0_{ij} + \phi_{ij}} \right)^4 \right) \qquad \forall (i,j) \in L^R \backslash L^N \tag{10}$$

where T^0_{ij} denotes the free flow travel time, while k^0_{ij} and ϕ_{ij} denote the existing capacity and the upgraded capacity of link (i, j), respectively.

In particular, Equations (9) and (10) represent the BPR link cost function developed by the Bureau of Public Roads (BPR) [45], which accounts for congestion. Equations (9) and (10) encompass the decisions related to the expansion of the network in order to determine the impacts on congestion levels in the network. Specifically, Equation (9) considers the link addition decisions, y_{ij}, while Equation (10) applies for all other link types that fall into L^R, apart from the new links L^N.

4.2.1.0. Upper-Level Constraints

A number of constraints are defined in the upper-level model to delineate part of the decision space of the urban planners. In particular, Equation (11) specifies that each building is to be positioned in a node within the region.

$$\sum_{p \in P_t} z_{fp} = 1 \qquad \forall t \in T, \forall f \in F_t \tag{11}$$

Equation (12) indicates that each node should host at most a single building:

$$\sum_{f \in F_t} \sum_{t \in T} z_{fp} \leq 1 \qquad \forall p \in P \tag{12}$$

Equation (13) is a budget constraint to ensure that investment decisions related to network expansion are kept under control.

$$\sum_{(i,j) \in L^E} c_{ij}\phi_{ij} + \sum_{(i,j) \in L^N} c_{ij}y_{ij} \leq B \tag{13}$$

The domain of the upper-level variables is defined by Equations (14)–(17).

$$z_{fp} \in \{0,1\} \qquad \forall p \in P, \forall f \in F \tag{14}$$

$$0 \leq \phi_{ij} \leq k_{ij}^0 \qquad \forall (i,j) \in L^E \tag{15}$$

$$\phi_{ij} = 0 \qquad \forall (i,j) \in L^N \tag{16}$$

$$y_{ij} \in \{0,1\} \qquad \forall (i,j) \in L^N \tag{17}$$

4.2.2. Lower Level

People within the smart city will attempt to reduce their individual travel times when travelling on the transportation network. These decisions will highly depend on the changes induced by decisions made by urban planners, in terms of both the location of new buildings and network expansion decisions related to infrastructure investment.

Lower-Level Objective Functions

Since the transport network users will attempt to minimise their individual travel times, this sort of selfish behaviour of users can be modelled via a user equilibrium (UE) traffic assignment model [46], such as Equation (18).

$$\min_x \quad \sum_{(i,j) \in L^R} \int_0^{x_{ij}} t_{ij}(\omega)d\omega \tag{18}$$

Lower-Level Constraints

The lower-level constraints focus on the flow variable, to assign traffic to different links within the network. To ensure flow conservation at each node within the network, Equation (19) is defined.

$$\sum_{\substack{j \in D \cup P: \\ (i,j) \in L^R \cup L^V}} x_{ij}^u - \sum_{\substack{j \in D \cup P: \\ (i,j) \in L^R \cup L^V}} x_{ji}^u = q_{iu} \qquad \forall i \in D \cup P, \forall u \in U, i \neq b, (i,u) \in W \tag{19}$$

In order to link the decision variable of the upper level with the decisions made by the lower level, Equations (20) and (21) are defined. In particular, Equation (20) states that flow from the location of the building to the sink node (which accumulates total travel on the network to the particular building type, i.e., single destination) is only possible if that specific building is located on the node from which the link emanates. Equation (21) states that total flow to all destinations on a proposed link within the network, x_{ij}, is not made possible unless the link is assigned to be constructed.

$$x_{pu}^{u} \leq z_{fp}M \quad \forall t \in T, \forall p \in P^{t}, \forall f \in F^{t}, \forall u \in D^{t} \tag{20}$$

$$x_{ij} \leq y_{ij}M' \quad \forall(i,j) \in L^{N} \tag{21}$$

Equation (22) is a definitional constraint which specifies that the total flow on a given link is the sum of all flows heading towards all destinations, u, on that respective link:

$$x_{ij} = \sum_{u \in U} x_{ij}^{u} \quad \forall(i,j) \in L^{R} \tag{22}$$

The domain of the lower-level decision variable is defined via Equation (23).

$$x_{ij}^{u} \geq 0 \quad \forall(i,j) \in L^{R} \cup L^{V}, \forall u \in U \tag{23}$$

5. Solution Approach

The lower-level constraints focus on the flow variable in order to assign traffic to different links within the network. To ensure flow conservation at each node within the network, Equation (19) is defined.

In order to solve the proposed bilevel model above, a procedure which relies on converting the bilevel formulation into a single level model is adopted. A flow chart depicting the major steps undertaken is presented in Figure 4. In particular, the Karush–Kuhn–Tucker (KKT) equivalent conditions are used to reformulate the lower-level model, resulting in a single-level representation. The resulting single level is a mixed integer nonlinear programming (MINLP) model, which is then linearised through implementing a scheme that is based on piecewise approximation of the convex BPR function. Given that multiple objectives are considered at the upper level, a multi-objective optimisation solving approach is required. Lexicographic optimisation [47], which assumes a particular preference order over the criteria included, is adopted. The next section outlines the use of Karush-Kuhn-Tucker (KKT) conditions to reformulate the bilevel model.

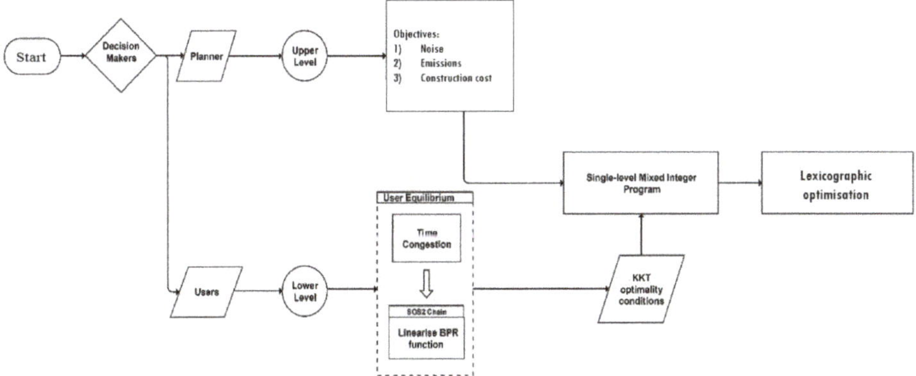

Figure 4. Solution approach adopted for the bilevel airport location model.

5.1. Equivalent Lower-Level Model

The UE conditions of the lower-level program can be represented by a set of first-order equivalent constraints, namely the KKT conditions [42]. A dual variable μ_{iu} is defined for Equation (19). Complementary slackness conditions of KKT, which are equivalent to the UE of the lower level and which require either $(t_{ij} - \mu_{iu} + \mu_{ju}) = 0$ or $x_{ij}^u = 0$, are enforced by Equations (24) and (25).

$$t_{ij} - \mu_{iu} + \mu_{ju} \geq 0 \qquad \forall(i,j) \in L^R, u \in U \tag{24}$$

$$(t_{ij} - \mu_{iu} + \mu_{ju})x_{ij}^u = 0 \qquad \forall(i,j) \in L^R, u \in U \tag{25}$$

Since Equation (25) involves the multiplication of two variables and is hence nonlinear, the single-level model cannot be solved using a linear solver. To overcome this, an appropriate linearisation scheme to reformulate Equation (25) needs to be applied, as demonstrated in the next section.

5.2. Linearisinng the KKT Conditions

Let ω_{iju} be an auxiliary binary integer variable, which equals 1 if $t_{ij} - \mu_{iu} + \mu_{ju} = 0$, and 0 otherwise. The complementary slackness condition, Equation (25), is replaced with the following set of constraints, Equations (26)–(28), resulting in the linearisation of KKT conditions:

$$x_{ij}^u \leq \omega_{iju}O \qquad \forall(i,j) \in L^R, u \in U \tag{26}$$

$$t_{ij} - \mu_{iu} + \mu_{ju} \leq (1 - \omega_{iju})O' \qquad \forall(i,j) \in L^R, u \in U \tag{27}$$

$$\omega_{iju} \in \{0,1\} \qquad \forall(i,j) \in L^R, u \in U \tag{28}$$

where O and O' are large positive constants.

5.3. Linearising the BPR Function

A chain of linked special ordered sets (SOS) conditions is implemented for linearising the BPR functions in Equations (9) and (10) [48]. The principle idea behind the SOS linearisation scheme adopted is shown in Figure 5. The domains of x_{ij} and ϕ_{ij} (i.e., the two continuous variables in the BPR function) are partitioned into $h \in H$ and $e \in E$ regions, respectively, based on a grid point structure. With each grid point, a continuous variable, namely ψ_{ijhe}, is associated. A necessary condition is imposed on ψ_{ijhe}, which states that no more than four adjacent grid points can be nonzero. The flow and link capacity variables can then be represented by Equations (29) and (30), respectively:

$$x_{ij} = \sum_{e \in E}\sum_{h \in H} \overline{x}_{ijh}\psi_{ijhe} \qquad \forall(i,j) \in L^R \tag{29}$$

$$\phi_{ij} = \sum_{e \in E}\sum_{h \in H} \overline{\phi}_{ije}\psi_{ijhe} \qquad \forall(i,j) \in L^R \tag{30}$$

where $\overline{x}_{ijh}$ and $\overline{\phi}_{ije}$ are predefined, fixed values of flow and capacity, respectively, used in the piecewise linearisation of the BPR function.

The BPR function is then approximated by the linear formulation through Equations (31) and (32):

$$t_{ij} = \sum_{e \in E}\sum_{h \in H} T_0\left(1 + 0.15\left(\frac{\overline{x}_{ijh}}{k_{ij}^0 + \overline{\phi}_{ije}}\right)^4\right)\psi_{ijhe} + (1 - y_{ij})M \qquad \forall(i,j) \in L^N \tag{31}$$

$$t_{ij} = \sum_{e \in E}\sum_{h \in H} T_0\left(1 + 0.15\left(\frac{\overline{x}_{ijh}}{k_{ij}^0 + \overline{\phi}_{ije}}\right)^4\right)\psi_{ijhe} + (1 - y_{ij})M \qquad \forall(i,j) \in L^R\backslash L^N \tag{32}$$

The conditions imposed on ψ_{ijhe} are given by Equations (33)–(36):

$$\sum_{e \in E} \sum_{h \in H} \psi_{ijhe} = 1 \qquad \forall (i,j) \in L^R \tag{33}$$

$$\xi_{ijh} = \sum_{e \in E} \psi_{ijhe} \qquad \forall (i,j) \in L^R, \forall h \in H \tag{34}$$

$$\eta_{ije} = \sum_{h \in h} \psi_{ijhe} \qquad \forall (i,j) \in L^R, \forall e \in E \tag{35}$$

$$\xi_{ijh}, \eta_{ije} \in SOS2 \qquad \forall (i,j) \in L^R, \forall h \in H,, \forall e \in E \tag{36}$$

Equation (33) is the usual convex combination requirement in piecewise linear approximation. Two auxiliary continuous variables, ξ_{ijh} and η_{ije}, are defined and these are embedded within Equations (34)–(36), so that at most, four adjacent variables of ψ_{ijhe} can be nonzero. The combination of Equations (34) and (36) specifies that two adjacent ψ_{ijhe} at most in the $h \in H$ direction can be nonzero, while Equations (35) and (36) state that at most two adjacent ψ_{ijhe} at most in the $\forall e \in E$ can be nonzero. In particular, Equation (36) states that the variables, ξ_{ijh}, η_{ije}, are of a special ordered set (SOS) of Type 2 (i.e., SOS2), where a maximum of two of the latter variables that are adjacent can be nonzero. This becomes obvious from Figure 5, since for the grid structure shown and in accordance with the latter equations enforced, not more than two adjacent variables of ξ_{ijh} and η_{ije} can be nonzero in the x and y directions, respectively. The SOS2 conditions are specified as follows in Equations (37)–(40):

$$\xi_{ijh} \leq \xi_{ijh-1} + \xi_{ijh:h \in \overline{H}} \qquad \forall (i,j) \in L^R, \forall h \in H \tag{37}$$

$$\sum_{h \in \overline{H}} \xi_{ijh} = 1 \qquad \forall (i,j) \in L^R \tag{38}$$

$$\eta_{ije} \leq \gamma_{ije-1} + \gamma_{ije:e \in \overline{E}} \qquad \forall (i,j) \in L^R, \forall e \in E, \tag{39}$$

$$\sum_{e \in \overline{E}} \gamma_{ije} = 1 \qquad \forall (i,j) \in L^R \tag{40}$$

The domain of the variables used to mimic SOS2 is given as follows by Equations (41)–(43):

$$\xi_{ijh} \in \{0,1\} \qquad \forall (i,j) \in L^R, \forall h \in H \tag{41}$$

$$\gamma_{ije} \in \{0,1\} \qquad \forall (i,j) \in L^R, \forall e \in E \tag{42}$$

$$\psi_{ijhe} \qquad \forall (i,j) \in L^R, \forall h \in H, \forall e \in E \tag{43}$$

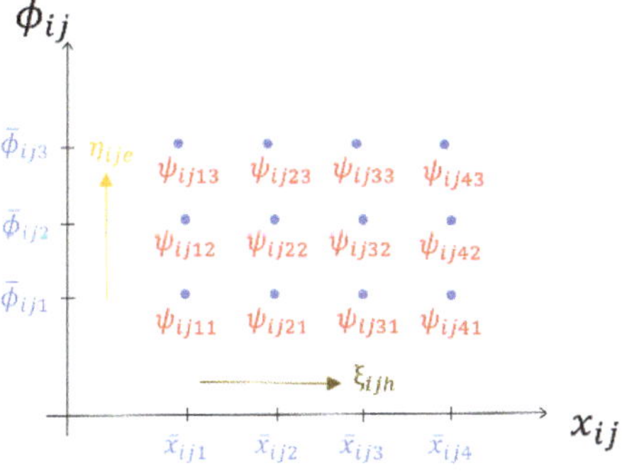

Figure 5. Grids defined for piecewise linearisation of the BPR function.

5.4. Linearing Carbon Emissions Objection Function

To linearise the carbon emissions objective function, Equation (8) is replaced by the equivalent Equation (44):

$$\min_x \sum_{i \in P} \sum_{u \in U} \sum_{m \in \Gamma} \varepsilon_m d_{iu} \pi_{iu} q_{iu} \tag{44}$$

where π_{iu} highlights the shortest travel time between origin i and destination u.

5.5. Lexicographic Optimisation

Given that the bilevel model proposed for the urban design of smart cities contains multiple objectives that need to be satisfied, no single solution will optimise all criteria at once. As a result, the concept of optimality adopted in single-objective optimisation is replaced with the concept of Pareto optimality.

A solution z^* of a multi-objective optimisation problem is said to be Pareto optimal if there is no other feasible solution $\bar{z}$ such that $f_\theta(\bar{z}) \leq f_\theta(z^*) \quad \forall \theta \in \Theta$ and $f_\rho(\bar{z}) \leq f_\rho(z^*)$ for at least one index $\rho \in \Theta, \theta \neq \rho$, where Θ is the set of objective functions solved in the multi-objective problem.

Lexicographic optimisation involves assigning a preference order over all objective functions considered and solving the problem over a number of stages [49]. In this paper, the lexicographic optimisation approach is adopted, given that it is likely that urban designers have a preference order defined over certain criteria when structuring a region. The algorithm developed is displayed in Algorithm 1.

Algorithm 1 Lexicographic Optimisation

```
Input: Criteria preference
```
$f_1^* = min_{x \in X} f_1(x)$
```
for  θ = 2, ... , Θ
```
$\qquad f_\theta^* = min_{x \in X} \{f_\theta(x) : f_\rho(x) \leq f_\rho^* \; \forall \rho = 1, ..., \theta - 1\}$
```
end for
Lexicographic minimiser:
```
$\qquad x^* \epsilon \{x \epsilon X : f_\rho(x) \leq f_\rho^* \; \forall \rho = 1, ..., \theta\}$

The notation adopted for the lexicographic optimisation process is given by the term $lex \min[B_v, B_w]$, which indicates that the model is first solved by minimising the highest ranked objective, B_v. Once an optimal solution is yielded, the model is re-solved by adopting B_w as the objective function and by including the constraint $B_v \leq B_v^*$ in the model, where B_v^* is the optimal solution of B_v obtained at the initial stage. The final solution is that attained once all $|\Theta| - 1$ objective functions have been included as constraints, where Θ is the set of all objective functions involved in the model.

6. Computational Results

In this section, the computational experiments utilised to demonstrate the applicability of the proposed optimisation framework are explained. In the first set of experiments, labelled Scenario 1, the framework is tested on a realistic example of a region being developed into a smart city. The structure of the city is displayed in Figure 6A. Figure 6B displays the available locations for positioning different buildings in the region. The type of buildings considered in the example include schools, hospitals, residential dwellings, offices, bus stops and factories. In the second set of experiments, multiple instances of network structures are generated in order to examine the performance of the proposed model. For both sets of experiments, the proposed mathematical optimisation models are coded in AMPL [50]; Python [51] is used as the programming language to generate instances in Scenario 2. The model is run on a personal computer with Intel Core i7, 2.2 GHz CPU and 8 GB RAM. CPLEX 12.7 is adopted as the linear solver [52].

(A)   (B)

Figure 6. (**A**) Region examined in the case example; (**B**) available locations for placing buildings in the region.

6.1. Scenario 1

The first step of the framework involves partitioning the region into a number of zonal clusters. The k-means clustering algorithm of Figure 3 is utilised; the resulting zonal cluster is displayed in Figure 6A. A total of eight zonal clusters within the region have been identified. The next step involves assigning a land use to each zonal cluster. This enables the allocation of zones for positioning different building types in. This is achieved via the assignment model represented by Equations (3)–(5). It is desired to place four land-use zone types, namely three residential zones, two mixed-use zones, two commercial zones and one industrial zone. The resulting zones assigned are displayed in Figure 7B and Table 1. The building types desired to be placed, along with the associated maximum noise generated and noise sensitivity thresholds for each building, are displayed in Table 2.

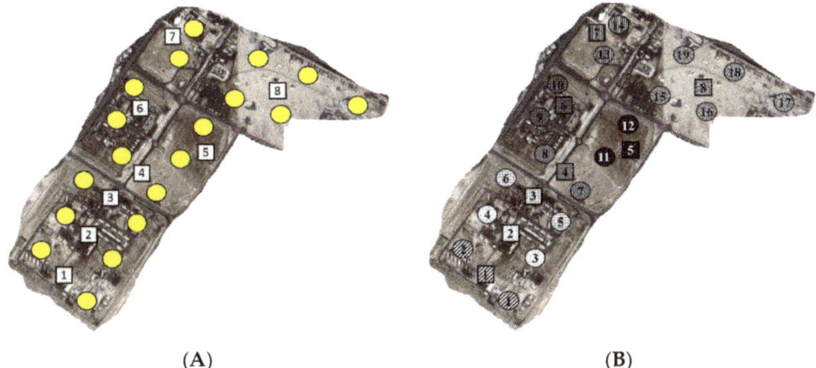

(A)   (B)

Figure 7. (**A**) Resulting zonal cluster where location nodes (in yellow) are grouped to the zone clusters (square); (**B**) Pattern matching between zones and allocated locations.

Table 1. Land-use distribution.

Land Use	Zone Cluster	Location Nodes
Residential	1	1, 2
Residential	2	3, 4
Residential	3	5, 6
Mixed use	4	7, 8
Commercial	5	11, 12
Mixed use	6	9, 10
Commercial	7	13, 14
Industrial	8	15, 16, 17, 18, 19

Table 2. Building type and associated noise characteristics.

Building Type	Maximum Noise Generated (dB(A))	Noise Sensitivity Threshold (dB(A))
Office	70	70
School	80	55
Residential	75	60
Hospital	65	40
Bus Stop	85	75
Factory	85	75

The construction cost associated with each node is as follows: for nodes 1, 5, 7, 10, 17 and 18, construction cost is given as \$503,124; for nodes 2, 3, 6, 13, 14, 15 and 19, construction cost is \$209,353; and for nodes 4, 8, 9, 11, 12 and 16, the construction cost is given as \$100,111.

The third stage involves an implementation of the strategic decision-support model for allocating buildings and determining the traffic assignment and any investments required in the connecting infrastructure of the region. Since the region is new, no existing network is present. A sample network shape utilised to outline the expected linking structure between the eight zones identified is depicted in Figure 8.

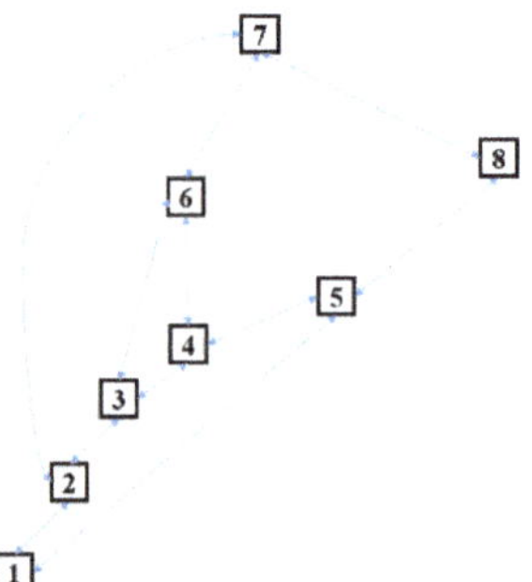

Figure 8. Transportation network of case example, where the numbered squares are the zones, and the arrows indicated the travel networks between the zones.

Lexicographic Optimisation Results

Let:

$$B_1 = \min_z \sum_{t \in T} \sum_{f \in F_t} \sum_{s \in P_t} z_{fs} \overline{C_s} \tag{45}$$

$$B_2 = \min_z \sum_{t \in T} \sum_{f \in F_t} \sum_{s \in P_t} \sum_{r \in P} z_{fs} M_{rs} \tag{46}$$

$$B_3 = \min_x \sum_{(i,j) \in L^R} \sum_{m \in \Gamma} \varepsilon_m d_{ij} x_{ij} t_{ij}(x_{ij}) \tag{47}$$

The preference assumed over the objective functions is given as follows: $B_3 \succ B_2 \succ B_1$, where the relationship $a \succ b$ highlights the preferential ranking of a in comparison to b. In the first stage, the carbon emissions on the transport network, B_3, are minimised (via minimising the total system travel time of the network). The emission factors associated with each transportation mode analysed in the smart city are given in Table 3, as obtained from [53]. Preference in the second stage is given to minimising the total sum of noise pollution within the region: B_2. In the final stage of the lexicographic approach, the construction cost involved with constructing buildings at each location is minimised. Through applications of the lexicographic algorithm, the lexicographic Pareto optimal solutions obtained are displayed in Table 4. As is displayed, in the initial run, carbon emissions are minimised, while the rest of the objectives are evaluated (without being optimised yet). In the second lexicographic run, the carbon emissions remain at their minimum level, while noise pollution decreases by 45% and construction costs increase by 21%, in comparison to the first stage of the lexicographic run. In the third lexicographic run, both carbon emissions and noise pollution stay at their minimum levels (due to the constrained optimisation), while construction cost cannot be minimised further without violating the constraints placed on the carbon emissions and noise pollution functions.

Table 3. Emission factors for each transportation mode analysed.

Transportation Mode	Emission Factor (kg CO_2-eq pkm^{-1})
Car	0.183
Bus	0.056
Intercity Train	0.041

Table 4. Lexicographic optimisation with the order $B_3 \succ B_2 \succ B_1$.

	B_3 (Carbon-Equivalent kg CO_2-eq)	B_2 (dB(A))	B_1 (AUD \$)	Constructed Links
$lex\,\min[B_3]$	19,786	679	5,287,456	All links
$lex\,\min[B_3, B_2]$	19,786	375	6,427,212	All links
$lex\,\min[B_3, B_2, B_1]$	19,786	375	6,287,456	All links

6.2. Scenario 2

In the second set of experiments, the case example of Figure 5 is slightly modified to allow for an extensive computational analysis of the model developed. A total of 350 random instances are generated for examining the behaviour of the bilevel model. The size of the networks considered starts at 10 nodes and is incremented by 5 nodes until 40 nodes are reached. Travel distances are assumed to be proportional to the Euclidian distance between the zones, while buildings to be placed are assumed to be 40–90% of the number of nodes considered in the instance generated, in order to generate an encompassing set of scenarios. Figure 9 displays the average computational time required to yield an optimum solution, along with the percentage of instances solved to optimality within a 1000 s time limit. As can be seen, beyond 20 nodes, as the instance size increases, the average computation time increases and the percentage of instances solved to optimality decreases.

6.3. Comparison with Other Metaheuristics

In this section, common optimisation algorithm approaches adopted in the literature, including genetic algorithms (GA) [54] and particle swarm optimisation (PSO) [55], are contrasted with the proposed exact approach. Based on rigorous tests, the population size, mutation rate and crossover rate of 250, 0.05 and 0.7, respectively, were adopted in the GA, whereas for PSO, population size was set as 250, while acceleration constant and weight parameters were set as 3 and 0.4, respectively. As can be seen from Figure 10A, the solving time of the GA is better than both PSO and the proposed exact approach, though solution accuracy is better in the proposed exact approach. The results of Figure 10B highlight that even though the metaheuristic approaches can be faster in producing a solution, the solution quality of the exact approach will always be better. In addition, for the case considered herein, the fastest approach to generate a solution is obtained using a GA, though PSO can be slightly more accurate in terms of solution quality in contrast to the GA.

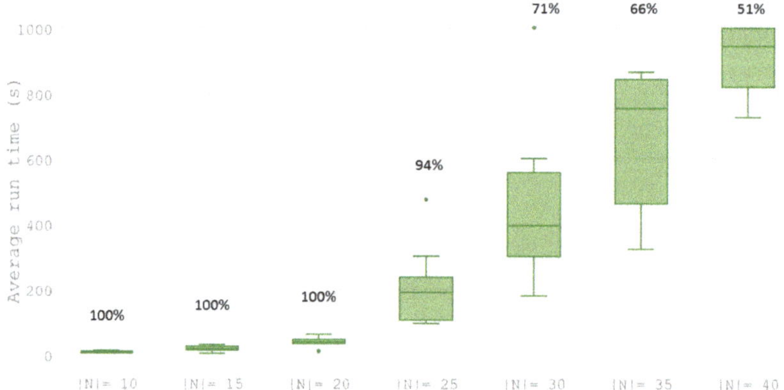

Figure 9. Box plot highlighting performance of the model proposed as instance size increases.

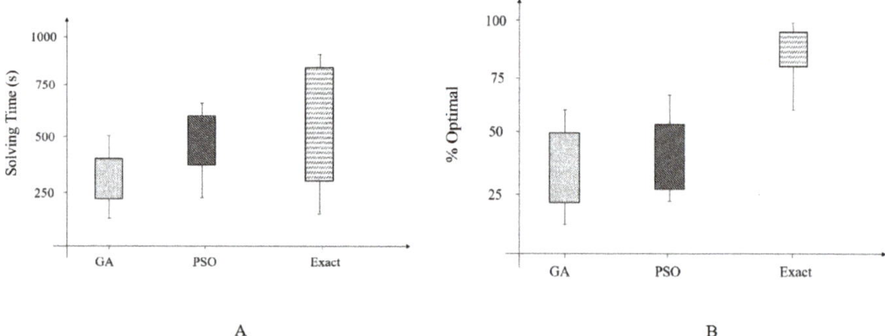

Figure 10. (**A**) Solving time and (**B**) percentage (%) optimality of solution quality comparison between the genetic algorithm (GA), particle swarm optimisation (PSO) and the exact approach.

6.4. Multi-objective Variant

A pure Pareto-based formulation is applied, which relies on optimising all the objective functions simultaneously. The solution strategy adopted herein is referred to as the ϵ-constraint method, which relies on optimising a single objective function while accounting for the remaining objectives using constraints. A parametric variation of the right-hand side (RHS) of the constrained functions then ensues to generate the efficient Pareto points on the frontier. The reader is referred to [56] for additional information on the implementation of the ϵ-constraint method adopted.

The results obtained on application of the ϵ-constraint method to the case study presented above are displayed in Figure 11A–C. In addition, Table 5 presents the optimised values of extreme points used to plot the tradeoff curve. In Figure 11A, a clear tradeoff exists between minimising noise and minimising carbon-equivalent emissions. In Figure 11B, a tradeoff is shown between minimising layout cost and minimising noise, while Figure 11C displays the tradeoff between minimising layout cost and minimising carbon-equivalent emissions.

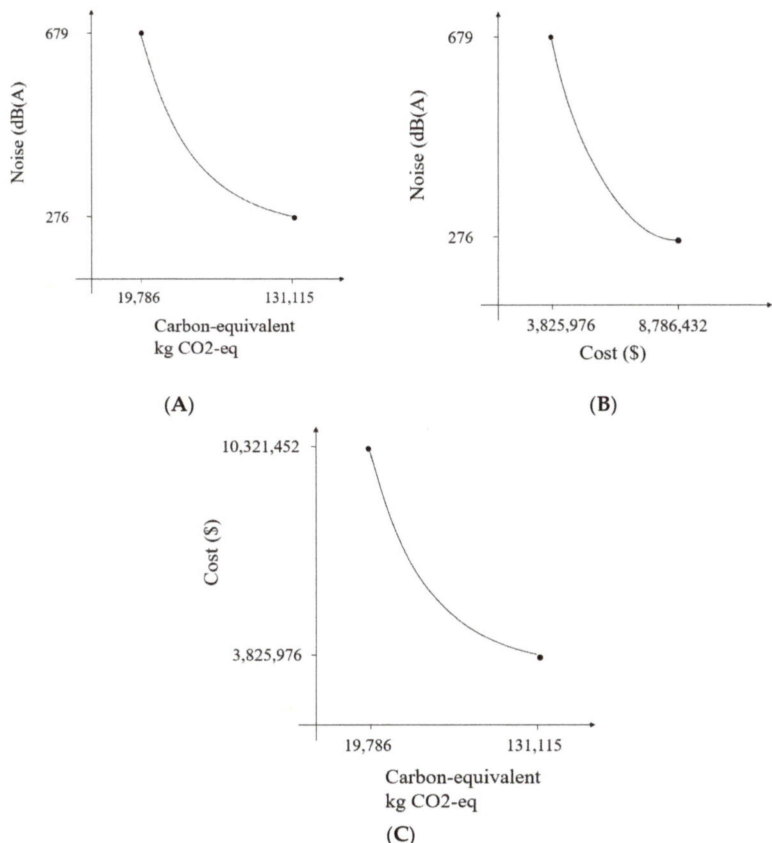

Figure 11. (A) Pareto curve highlighting the trade-off between noise minimisation and carbon-equivalent emission minimisation; (B) Pareto curve highlighting the trade-off between noise minimisation and layout cost minimisation; (C) Pareto curve highlighting the trade-off between layout cost minimisation and carbon-equivalent emission minimisation.

Table 5. Payoff table between all objective functions considered.

Objective Functions	B1 = Layout Cost ($)	B2 = Noise (dB(A))	B3 = Carbon-Equivalent (CO_2-eq)
Min B1	3,825,976	679	131,115
Min B2	8,786,432	276	131,115
Min B3	10,321,452	679	19,786

The importance of the Pareto curves produced lies in the fact that the decision-maker is now able to visualise the magnitude of the impact associated with each efficient solution produced.

6.5. Discussion and Insight

In comparison to some of the approaches in the literature, the proposed framework targets key strategic operational research problems that are encountered in the urban design of smart cities. For instance, in [57], the authors consider using integer optimisation to maximise floor area while accounting for sunlight in urban design. A mixed integer program was developed in [58] for designing building interiors. Integer programming was utilised in [59] for urban street network design. As can be noticed, there is a lack of focus on optimisation problems encountered in the strategic design of urban areas, where traffic and building layout are both integrated. The proposed approach in this article tackles this gap by proposing a mathematical optimisation model which integrates the latter. In addition, multiple objectives are rarely adopted in urban design optimisation [12]. As a result, through the proposed framework herein using multi-objective optimisation, the decision-maker can visualise the different tradeoffs that result when incorporating all objective functions considered.

7. Conclusions

Globally, we have witnessed a shift towards developing smart cities to deal with the challenges of rising population and urbanisation rates, along with the necessity of ensuring sustainable development. It seems therefore necessary to incorporate intelligent and robust urban planning frameworks that can simultaneously target transport and land-use considerations in smart cities.

In this paper, a framework, based on mathematical optimisation for the strategic planning of zoning, facility location and transport network design in smart cities was proposed. The framework combines some strategic operational research problems, including the clustering problem, the assignment problem, the facility location problem and the network design problem, in a systematic fashion. An algorithm based on k-means clustering is applied to divide a given region into zones, and an assignment problem is then solved to determine the land-use types within the region. The final stage of the framework involves solving a bilevel model that accounts for the hierarchical decision making of urban planners and travel network users. The proposed bilevel model considers the location decisions of buildings within a smart city setting, along with the investment decisions related to expansion of the underling transportation network. Multiple objective functions were formulated to target the triple bottom-line of sustainability in order to ensure a sustainable urban layout of the smart city. In particular, as a social factor, noise pollution in the region was minimised; as an environmental factor, carbon-equivalent emissions on the transportation network were minimised; and as an economic factor, the construction cost of buildings was minimised. A solution approach based on converting the bilevel model into a single-level model was outlined, along with a linearization procedure. A lexicographic optimisation approach was utilised to handle the multi-objective nature of the developed model. In addition, the ϵ-constraint method, which generates the Pareto front when considering all objective functions involved, was also adopted.

The proposed model was applied on a realistic case example for the design of the urban structure of a smart city. A lexicographic approach highlighted variations in cost of up to 52% when carbon emissions are given first preference by decision-makers. The ϵ-constraint method highlighted that a trade-off cost of up to 471% can result when simultaneously optimising the objective functions involved. In order to examine the computational performance of the proposed approach, a total of 350 instances were solved. Results demonstrate that solving time increased rapidly once the transportation network size of the instance generated exceeded 30 nodes. On average, the proposed model was able to solve 72% of the proposed instances within the imposed 1000 s time limit.

A comparison was also conducted between the proposed exact approach and metaheuristic solving algorithms including a GA and PSO. Results indicated that the GA was the fastest in terms of solution time, although solution accuracy was on average reduced by 68% compared to the results obtained when utilising the exact approach.

The proposed framework integrates several key operational research problems that are representative of certain aspects of the urban design problem involved in smart cities. However, several limitations are associated with the proposed approach. First, a deeper investigation into all facets of urban design that are associated with smart cities is lacking, since only several operational research problems are tackled in the framework developed. Second, the incorporation of tactic decision-making problems, such as vehicle routing, was missing from the proposed framework. Future works will thus look at these two areas.

Author Contributions: Conceptualization, A.WA.H. and E.G.V.; Data curation, A.WA.H.; Formal analysis, A.WA.H; Funding acquisition, E.G.V. and A.H.; Investigation, A.WA.H.; Methodology, A.WA.H, A.A., and A.H.; Supervision, E.G.V. and A.H.; Validation, A.H.; Writing—original draft, A.WA.H and A.A.

Funding: Authors would like to acknowledge the Brazilian Government for their support by the CNPq (National Council for Scientific and Technological Development)

Conflicts of Interest: The authors declare no conflict of interest.

Nomenclature

Notation	Meaning
$t \in T$	Type of building to be positioned in smart city
$v \in \Pi$	Set of zone centroids to position in a region
$k \in \Lambda$	Set of land uses to define within the region
$m \in \Gamma$	Set of transport modes on the transport network
$D = \cup_{t \in T} D_t$	Set of dummy nodes, acting as sink nodes for each building type
F_t	Set of all building types
L^C	Existing links that cannot be expanded
L^E	Existing links that can be expanded
L^N	New links that can be formed
$L^R \cup L^C \cup L^E \cup L^N$	All link types that form the travel network
L^V	Virtual links that connect to the sink node
P_t	Set of potential locations for all building types
$P : \cup_{t \in T} P_t$	Set of all nodes in the region
U	Set of all destination nodes in the region
H	Set of segments for linearising x_{ij}
E	Set of segments for linearising ϕ_{ij}
$\overline{H} = \{h \in H : ord(h) < |H|\}$	Set of segments for linearising x_{ij}, less 1
$\overline{E} = \{e \in E : ord(e) < |E|\}$	Set of segments for linearising ϕ_{ij}, less 1
A_n	Set of area node coordinates identified in the region, where $n \in P_t$
Q_v	Set of zonal coordinates of the region, where $v \in \Pi$
V_v	Set of all area nodes allocated to the centroid of a zone
$u \in U$	Set of all destinations
W	Origin–destination (OD) matrix
$\overline{C}_s$	Cost of constructing a building at node $s \in P$
M_{rs}	Noise pollution assessed at node r in the region, resulting due to a building placed at node $s \in P$
ε_m	Emission factors of transport mode $m \in \psi$ in kg CO2 $-$ eq pkm^{-1}
d_{ij}	Distance between nodes
k^0_{ij}	Capacity of link $(i,j) \in L^R$
B	Available budget for expansion of network
$\overline{x}_{ijh}$	A fixed parameter for the flow on link $(i,j) \in L^R$, used in the piecewise Linearisation of the BPR function
$\overline{\phi}_{ije}$	A fixed parameter for the expanded capacity on link$(i,j) \in L^E$, used in the piecewise linearisation of the BPR function
M, M', O, O'	Large positive constants
w_{kv}	Binary variable, which equals 1 if land use $k \in \Lambda$ is assigned to zone cluster $v \subset \Pi$, and 0 otherwise
z_{fs}	Binary variable, which equals 1 if type t building, $f \in F_t$, is located at node $s \in P$, and 0 otherwise
y_{ij}	Binary variable, which equals 1 if link is added to the network, and zero otherwise
x_{ij}	Flow on link $(i,j) \epsilon L^R$
ϕ_{ij}	Amount of additional capacity added to link $(i,j) \epsilon L^E$

Notation	Meaning
ω_{iju}	Auxiliary binary variable for linearising the KKT complementary slackness conditions
ψ_{ijhe}	Auxiliary binary variable for linearising the BPR function
ξ_{ijh}	Auxiliary continuous variables for linearising the BPR function, for link $(i,j)\epsilon L^R$ and segment
π_{iu}	Continuous variable which highlights the shortest travel time between origin i and destination u
η_{ije}	Auxiliary continuous variables for linearising the BPR function $(i,j)\epsilon L^R$
ζ_{ijh}	Binary variable for mimicking SOS2 behaviour, defined over segments $h \in H$
γ_{ije}	Binary variable for mimicking SOS2 behaviour, defined over segments $e \in E$

Bureau of public roads (BPR); Karush-Kuhn-Tucker (KKT); Special Ordered Set 2 (SOS2)

References

1. Li, B.; Yao, R. Urbanisation and its impact on building energy consumption and efficiency in China. *Renew. Energy* **2009**, *34*, 1994–1998. [CrossRef]
2. United Nations. *World Urbanisation Prospects: The 2018 Revision*; UN DESA: New York city, NY, USA, 2018.
3. UNICEF. Unicef Urban Population Map. 2017. Available online: https://www.unicef.org/sowc2012/urbanmap/# (accessed on 26 May 2018).
4. Raimbault, M.; Dubois, D. Urban soundscapes: Experiences and knowledge. *Cities* **2005**, *22*, 339–350. [CrossRef]
5. Vanolo, A. Is there anybody out there? The place and role of citizens in tomorrow's smart cities. *Futures* **2016**, *82*, 26–36. [CrossRef]
6. Slaughter, R.A. The IT revolution reassessed part three: Framing solutions. *Futures* **2018**, *100*, 1–19. [CrossRef]
7. Ferraris, A.; Santoro, G.; Papa, A. The cities of the future: Hybrid alliances for open innovation projects. *Futures* **2018**, *103*, 51–60. [CrossRef]
8. Kramers, A.; Höjer, M.; Lövehagen, N.; Wangel, J. Smart sustainable cities—Exploring ICT solutions for reduced energy use in cities. *Environ. Model. Softw.* **2014**, *56*, 52–62. [CrossRef]
9. Park, E.; del Pobil, A.P.; Kwon, S.J. The Role of Internet of Things (IoT) in Smart Cities: Technology Roadmap-oriented Approaches. *Sustainability* **2018**, *10*, 1388. [CrossRef]
10. Bekta, T.; Crainic, T.G.; van Woensel, T. From Managing Urban Freight to Smart City Logistics Networks. In *Network Design and Optimization for Smart Cities*; World Scientific: Quebec City, QC, Canada, 2016; Volume 8, pp. 143–188.
11. Hammad, A.W.A.; Rey, D.; Akbarnezhad, A. A Bi-level Mixed Integer Programming Model to Solve the Multi-Servicing Facility. In *Data and Decision Sciences in Action*; Springer: Cham, Switzerland, 2018; pp. 381–395.
12. Hammad, A.W.A.; Akbarnezhad, A.; Rey, D. Sustainable urban facility location: Minimising noise pollution and network congestion. *Transp. Res. Part E Logist. Transp. Rev.* **2017**, *107*, 38–59. [CrossRef]
13. Djahel, S.; Jabeur, N.; Barrett, R.; Murphy, J. Toward V2I communication technology-based solution for reducing road traffic congestion in smart cities. In Proceedings of the 2015 International Symposium on Networks, Computers and Communications (ISNCC), Hammamet, Tunisia, 13–15 May 2015; pp. 1–6.
14. Bharadwaj, S.; Ballare, S.; Rohit; Chandel, M.K. Impact of congestion on greenhouse gas emissions for road transport in Mumbai metropolitan region. *Transp. Res. Procedia* **2017**, *25*, 3538–3551. [CrossRef]
15. Giles-Corti, B.; Vernez-Moudon, A.; Reis, R.; Turrell, G.; Dannenberg, A.L.; Badland, H.; Foster, S.; Lowe, M.; Sallis, J.F.; Stevenson, M.; et al. City planning and population health: A global challenge. *Lancet* **2016**, *388*, 2912–2924. [CrossRef]
16. Cui, S.; Zhao, H.; Zhang, C. Locating Charging Stations of Various Sizes with Different Numbers of Chargers for Battery Electric Vehicles. *Energies* **2018**, *11*, 3056. [CrossRef]
17. Brimberg, J.; Mladenović, N.; Todosijević, R.; Urošević, D. Local and Variable Neighborhood Searches for Solving the Capacitated Clustering Problem. In *Optimization Methods and Applications*; Springer: Cham, Switzerland, 2017; pp. 33–55.
18. Patriksson, M. *The Traffic Assignment Problem: Models and Methods*; Courier Dover Publications: Utrecht, Netherlands, 2015.

19. Karatas, M.; Yakıcı, E. An iterative solution approach to a multi-objective facility location problem. *Appl. Soft Comput.* **2018**, *62*, 272–287. [CrossRef]

20. Talari, S.; Shafie-khah, M.; Siano, P.; Loia, V.; Tommasetti, A.; Catalão, J.P.S. A Review of Smart Cities Based on the Internet of Things Concept. *Energies* **2017**, *10*, 421. [CrossRef]

21. De Jong, M.; Joss, S.; Schraven, D.; Zhan, C.; Weijnen, M. Sustainable–smart–resilient–low carbon–eco–knowledge cities; making sense of a multitude of concepts promoting sustainable urbanization. *J. Clean. Prod.* **2015**, *109*, 25–38. [CrossRef]

22. Calvillo, C.F.; Sánchez-Miralles, A.; Villar, J. Energy management and planning in smart cities. *Renew. Sustain. Energy Rev.* **2016**, *55*, 273–287. [CrossRef]

23. Deakin, M.; Reid, A. Smart cities: Under-gridding the sustainability of city-districts as energy efficient-low carbon zones. *J. Clean. Prod.* **2018**, *173*, 39–48. [CrossRef]

24. Tascikaraoglu, A. Evaluation of spatio-temporal forecasting methods in various smart city applications. *Renew. Sustain. Energy Rev.* **2018**, *82*, 424–435. [CrossRef]

25. Maier, S. Smart energy systems for smart city districts: Case study Reininghaus District. *Energy Sustain. Soc.* **2016**, *6*, 23. [CrossRef]

26. Wojnicki, I.; Kotulski, L. Improving Control Efficiency of Dynamic Street Lighting by Utilizing the Dual Graph Grammar Concept. *Energies* **2018**, *11*, 402. [CrossRef]

27. Thomas, B.L.; Cook, D.J. Activity-Aware Energy-Efficient Automation of Smart Buildings. *Energies* **2016**, *9*, 624. [CrossRef]

28. Barbato, A.; Bolchini, C.; Geronazzo, A.; Quintarelli, E.; Palamarciuc, A.; Piti, A.; Rottondi, C.; Verticale, G. Energy Optimization and Management of Demand Response Interactions in a Smart Campus. *Energies* **2016**, *9*, 398. [CrossRef]

29. Sharma, S.; Dua, A.; Singh, M.; Kumar, N.; Prakash, S. Fuzzy rough set based energy management system for self-sustainable smart city. *Renew. Sustain. Energy Rev.* **2018**, *82*, 3633–3644. [CrossRef]

30. Ruiz-Romero, S.; Colmenar-Santos, A.; Mur-Pérez, F.; López-Rey, Á. Integration of distributed generation in the power distribution network: The need for smart grid control systems, communication and equipment for a smart city—Use cases. *Renew. Sustain. Energy Rev.* **2014**, *38*, 223–234. [CrossRef]

31. Dincer, I.; Acar, C. Smart energy systems for a sustainable future. *Appl. Energy* **2017**, *194*, 225–235. [CrossRef]

32. Ali, S.; Lee, S.-M.; Jang, C.-M. Determination of the Most Optimal On-Shore Wind Farm Site Location Using a GIS-MCDM Methodology: Evaluating the Case of South Korea. *Energies* **2017**, *10*, 2072. [CrossRef]

33. Ali, S.; Lee, S.-M.; Jang, C.-M. Techno-Economic Assessment of Wind Energy Potential at Three Locations in South Korea Using Long-Term Measured Wind Data. *Energies* **2017**, *10*, 1442. [CrossRef]

34. García, R.M.; Prieto-Castrillo, F.; González, G.V.; Tejedor, J.P.; Corchado, J.M. Stochastic Navigation in Smart Cities. *Energies* **2017**, *10*, 929. [CrossRef]

35. Díaz-Díaz, R.; Muñoz, L.; Pérez-González, D. The Business Model Evaluation Tool for Smart Cities: Application to SmartSantander Use Cases. *Energies* **2017**, *10*, 262. [CrossRef]

36. Parvez, I.; Sarwat, A.I.; Wei, L.; Sundararajan, A. Securing Metering Infrastructure of Smart Grid: A Machine Learning and Localization Based Key Management Approach. *Energies* **2016**, *9*, 691. [CrossRef]

37. Lanza, J.; Sánchez, L.; Gutiérrez, V.; Galache, J.A.; Santana, J.R.; Sotres, P.; Muñoz, L. Smart City Services over a Future Internet Platform Based on Internet of Things and Cloud: The Smart Parking Case. *Energies* **2016**, *9*, 719. [CrossRef]

38. Yamagata, Y.; Seya, H. Simulating a future smart city: An integrated land use-energy model. *Appl. Energy* **2013**, *112*, 1466–1474. [CrossRef]

39. Wu, Y.; Zhang, W.; Shen, J.; Mo, Z.; Peng, Y. Smart city with Chinese characteristics against the background of big data: Idea, action and risk. *J. Clean. Prod.* **2018**, *173*, 60–66. [CrossRef]

40. Iacovidou, E.; Purnell, P.; Lim, M.K. The use of smart technologies in enabling construction components reuse: A viable method or a problem creating solution? *J. Environ. Manag.* **2018**, *216*, 214–223. [CrossRef] [PubMed]

41. De Paula, L.B.; Marins, F.A.S. Algorithms applied in decision-making for sustainable transport. *J. Clean. Prod.* **2018**, *176*, 1133–1143. [CrossRef]

42. Bard, J.F. *Practical Bilevel Optimization: Algorithms and Applications*; Springer Science & Business Media: Berlin, Germany, 2013.

43. Nigro, M.; Cipriani, E.; del Giudice, A. Exploiting floating car data for time-dependent Origin–Destination matrices estimation. *J. Intell. Transp. Syst.* **2018**, *22*, 159–174. [CrossRef]

44. Hartigan, J.A.; Wong, M.A. Algorithm AS 136: A k-means clustering algorithm. *J. R. Stat. Soc. Ser. C Appl. Stat.* **1979**, *28*, 100–108. [CrossRef]

45. Transportation Research Board, National Research Council. *Highway Capacity Manual*; Transportation Research Board, National Research Council: Washington, DC, USA, 2000.

46. Wardrop, J.G.; Whitehead, J.I. Some theoretical aspects of road traffic research. *Proc. Inst. Civ. Eng.* **1952**, *1*, 767–768. [CrossRef]

47. Lam, T.B. *Multi-Objective Optimization in Computational Intelligence: Theory and Practice: Theory and Practice*; IGI Global: Pennsylvania, PA, USA, 2008.

48. Beale, E.M.L. Integer Programming. In *Computational Mathematical Programming*; Schittkowski, K., Ed.; Springer: Berlin/Heidelberg, Germany, 1985; pp. 1–24.

49. Ehrgott, M. *Multicriteria Optimization*; Springer Science & Business Media: Berlin, Germany, 2013.

50. Fourer, R.; Gay, D.; Kernighan, B. *Ampl: A Modeling Language for Mathematical Programming*; DUXBURY, THOMSON Boyd & Fraser: Illinois, IL, USA, 1993; Volume 119.

51. Van Rossum, T.G. Python tutorial, technical report. Technical report, Centrum voor Wiskunde en Informatica (CWI), Amsterdam, CS-r9526. May 1995.

52. IBM, "CPLEX Optimizer," 25 March 2019. Available online: https://www.ibm.com/analytics/cplex-optimizer (accessed on 5 January 2019).

53. Spielmann, M.; Bauer, C.; Dones, R.; Tuchschmid, M. Transport Services. Swiss Centre for Life Cycle Inventories Dübendorf, Ecoinvent report 14. 2007.

54. Konak, A.; Coit, D.W.; Smith, A.E. Multi-objective optimization using genetic algorithms: A tutorial. *Reliab. Eng. Syst. Saf.* **2006**, *91*, 992–1007. [CrossRef]

55. Kennedy, J.; Eberhart, R.C. Particle swarm optimization. In Proceedings of the International Conference on Neural Networks, Perth, WA, Australia, 27 November–1 December 1995;–1948; Volume 1942–1948.

56. Mavrotas, G. Effective implementation of the ε-constraint method in Multi-Objective Mathematical Programming problems. *Appl. Math. Comput.* **2009**, *213*, 455–465. [CrossRef]

57. Hua, H.; Hovestadt, L.; Tang, P.; Li, B. Integer programming for urban design. *Eur. J. Oper. Res.* **2019**, *274*, 1125–1137. [CrossRef]

58. Wu, W.; Fan, L.; Liu, L.; Wonka, P. MIQP-based Layout Design for Building Interiors. *Comput. Graph. Forum* **2018**, *37*, 511–521. [CrossRef]

59. Peng, C.-H.; Yang, Y.-L.; Bao, F.; Fink, D.; Yan, D.-M.; Wonka, P.; Mitra, N.J. Computational network design from functional specifications. *ACM Trans. Graph. TOG* **2016**, *35*, 131. [CrossRef]

Article

Energy Embedded in Food Loss Management and in the Production of Uneaten Food: Seeking a Sustainable Pathway

Daniel Hoehn [1,*], María Margallo [1], Jara Laso [1], Isabel García-Herrero [1], Alba Bala [2], Pere Fullana-i-Palmer [2], Angel Irabien [1] and Rubén Aldaco [1]

[1] Department of Chemical and Biomolecular Engineering, University of Cantabria, Avda. De los Castros s/n, 39005 Santander, Spain; maria.margallo@unican.es (M.M.); jara.laso@unican.es (J.L.) isabel.garciaherrero@unican.es (I.G.H.); irabienj@unican.es (A.I.); aldacor@unican.es (R.A.)

[2] UNESCO Chair in Life Cycle and Climate Change ESCI-UPF, Universitat Pompeu Fabra, Pg. Pujades 1, 08003 Barcelona, Spain; alba.bala@esci.upf.edu (A.B.); pere.fullana@esci.upf.edu (P.F.P.)

* Correspondence: daniel.hoehn@unican.es; Tel.: +34-942-846531; Fax: +34-942-201591

Received: 29 December 2018; Accepted: 19 February 2019; Published: 25 February 2019

Abstract: Recently, important efforts have been made to define food loss management strategies. Most strategies have mainly been focused on mass and energy recovery through mixed food loss in centralised recovery models. This work aims to highlight the need to address a decentralised food loss management, in order to manage the different fractions and on each of the different stages of the food supply chain. For this purpose, an energy flow analysis is made, through the calculation of the primary energy demand of four stages and 11 food categories of the Spanish food supply chain in 2015. The energy efficiency assessment is conducted under a resource use perspective, using the energy return on investment (EROI) ratio, and a circular economy perspective, developing an Energy return on investment – Circular economy index ($EROI_{ce}$), based on a food waste-to-energy-to-food approach. Results suggest that the embodied energy loss consist of 17% of the total primary energy demand, and related to the food categories, the vegetarian diet appears to be the most efficient, followed by the pescetarian diet. Comparing food energy loss values with the estimated energy provided for one consumer, it is highlighted the fact that the food energy loss generated by two to three persons amounts to one person's total daily intake. Moreover, cereals is the category responsible for the highest percentage on the total food energy loss (44%); following by meat, fish and seafood and vegetables. When the results of food energy loss and embodied energy loss are related, it is observed that categories such as meat and fish and seafood have a very high primary energy demand to produce less food, besides that the parts of the food supply chain with more energy recovery potential are the beginning and the end. Finally, the $EROI_{ce}$ analysis shows that in the categories of meat, fish and seafood and cereals, anaerobic digestion and composting is the best option for energy recovery. From the results, it is discussed the possibility to developed local digesters at the beginning and end of the food supply chain, as well as to developed double digesters installations for hydrogen recovery from cereals loss, and methane recovery from mixed food loss.

Keywords: anaerobic digestion and composting; circular economy; energy return on investment; hydrogen bioenergy; food waste hierarchy

---

1. Introduction

The food supply chain is one of the most polluting daily activities when impacts along product life cycles are considered [1]. This is mainly due to several factors, such as the high degree of mechanization, the use of agrochemical products in agriculture, the long distances in distribution routes, the overpacking of products, and the growth of consumption of processed food, especially the so-called fourth and fifth range products (formed by products that are both ready to be consumed and sold refrigerated). These factors have entailed an increase in the energy consumption throughout the entire supply chain, transforming it from a net producer of energy, to a net consumer of energy [2]. However, this is not a new phenomenon; in fact, in the energy crisis of the 70s, Pimentel et al. [3] found that the energy efficiency of modern food production was declining. Over time, the energy inputs began to be higher than the energy outputs [4], and according to Cuellar and Webber [5], Lin et al. [6] and Vittuari et al. [7], nowadays the food supply chain requires 10–15 kJ of fossil fuel to produce 1 kJ of food. From the whole supply chain, the high-energy intensity of agriculture has meant an enormous increase in the consumption of fossil fuels; however, this is common in all phases of the food supply chain and it varies depending on the type of product and processing level. Therefore, the energy intensity of modern food systems represents a major issue in a current framework of decreasing limited resources, and growing population [8].

Due to the growing awareness regarding these problems, social pressure has been increasing in order to overcome these problems through the development of energy technologies that lead to sustainable development [9]. In this sense, although energy and food have a well-known connection from the perspective of chemical energy contained in food products, the energy resources embedded for food production is less explored, and the available related information is scarce. Moreover, estimations are often limited to the first stages of production, without taking into account the fact that the food supply chain consists of several successive steps, and each one of them needs energy for its specific processes. It is estimated that around 30% of the world´s total energy consumption is due to the food system [10]. According to the European Commission [11], industrial activities related to food systems require approximately 26% of the European Union´s final energy consumption. Thus, it is essential to focus on the reduction of energy used in food production systems by improving their efficiency, as it could be one of the most important drivers for development of sustainable food production systems. Searching for that efficiency, food losses (FLs) have central consequences on the energy balance on the food supply chain, additionally leading to a significant environmental impact in terms of inefficient use of natural resources, biodiversity and habitat loss, soil and water degradation, and greenhouse gas emissions [7]. According to the Organization for Economic Cooperation and Development [12], more than a third of the food produced is wasted, involving around 38% of the energy embedded in its production. Specifically, Spain has the seventh highest level of food wastage in the European Union, with 7.7 million tones. According to Garcia-Herrero et al. [13], each Spanish citizen is estimated to throw away in household consumption 88 kg of food per year, being this step the one that more FL generates. Besides, FL is directly related to food security and presents nutritional and ethical issues, as 795 million people suffer from undernourishment [14], and it is projected that by 2050 the world population will reach 9.8 billion persons [15]. Kummu et al. [16] estimated that the nutritional energy lost in the food supply chain would be enough to feed around 1.9 billion people, and approximately half of those losses could be prevented. Thus, FL supposes a missed opportunity to feed the world's growing population [17].

Related to this, it is necessary to consider the fact that with the FL, two types of energy are also lost: food energy loss (FEL), which is the nutritional energy of the FL, and embodied energy loss (EEL), which is the primary energy invested in producing FL. In addition, it is necessary to take into account the energy used in the management of FL after it has been disposed. Regarding to the latter, the efficiency in energy recovery through different management strategies, can vary considerably depending on the strategy and the FL composition. In summary, precise accountings of energy use for the production of consumed (and non-consumed) food are extremely challenging for developing strategies to mitigate energy losses [7].

In this overall framework, this paper aims to develop a novel model in the field of study, to analyse alternative FL management strategies under a circular economy concept based on a food waste-to-energy-to-food approach (Figure 1). While most studies in the literature are focused on the efficiency assessment of the food supply chain, either from a mass [18], an energy [19], or more than one point of view [20]; through the model proposed in this work, it is intended to go further and contribute to the development of integrated waste management strategies for energy-smart food systems. Thereby, the Food and Agricultural Organisation proposal [10] is followed, which focuses on the diversification of renewable energy sources through integrated food production systems, to ensure the access to energy and food security. Moreover, it is projected to follow two of the Sustainability Development Goals for 2030 established by the United Nations Member States [21]: (i) the seventh goal, whose objective is to reach at least a 27% share of renewable energy consumption by 2030; and (ii) the twelfth goal, which aims at halving FL at the retail and consumer level as well as reducing the FL along food production systems. On the other hand, the circular economy package adopted by the European Commission in 2015 is guided by the European Union waste hierarchy, which ranks waste management options according to their sustainability, and gives top priority to preventing and recycling of waste, placing the anaerobic digestion as an always-preferable option to incineration [22]. This ranking aims to identify the options most likely to deliver the best overall environmental outcome, and has been adopted worldwide as the principal waste management framework [23]. However, the waste hierarchy proposal considers FL as a set without taking into account the different specific fractions or at which points along the food supply chain are they produced. Thus, this paper aims also to develop the debate about the statement that the waste hierarchy is a too general proposal. This is in the same line as the thesis of Cristobal et al. [24], who highlighted the fact, that when more criteria are considered along with the environmental one, other tools are needed for making the decision of which FL management strategy is the most optimal.

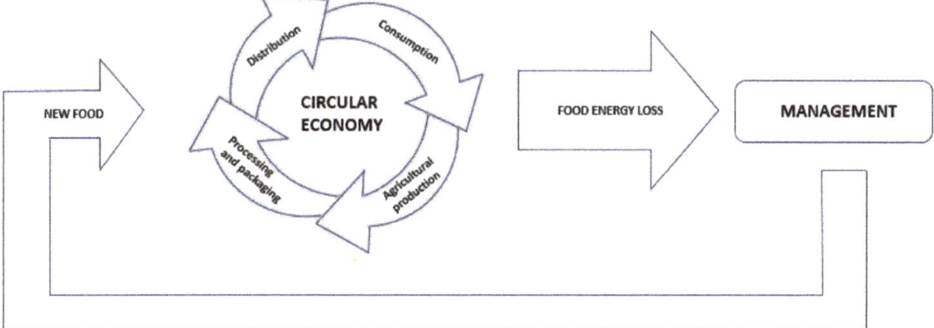

Figure 1. Conceptual diagram of the proposal of this work, recovering energy from food loss with a circular economy approach.

The paper is structured in two main parts. Firstly, Section 2 describes a detailed description of the methodology taking into account a Life Cycle Assessment approach, as well as the material flow analysis enabling to perform the energy flow analysis, conducting to the energy assessment along the food supply chain. Secondly, Sections 3 and 4 introduce the main results and discussion of the study. In particular, the first part includes a full discussion of the main parameters and the influence on the energy efficiency of the food supply chain. While last part reviews the FL management options and the expected improvement measures. The paper ends with a deployment of main conclusions and the future research.

2. Methodology

2.1. Goal and Scope

The main goal of the study is to develop a novel model to define alternative FL management strategies under a circular economy concept based on a food waste-to-energy-to-food approach.

For this objective, an empirical index so-called $EROI_{ce}$, is proposed, which quantifies the amount of nutritional energy that is recovered from the FL of each category of food under study, based on its treatment in three different scenarios: (i) landfill with biogas recovery (L), (ii) incineration with energy recovery (I) and (iii) anaerobic digestion and composting (AD&C). The results are expected to provide an interesting field for discussion about the best energy recovery strategy for the different fractions of FL, trying to develop the path to less generic energy recovery proposals. In view of the results, it is expected to open a debate around a new framework of decentralised FL collection strategies, instead, or as a complement to current centralised strategies.

2.2. Function, Functional Unit and System Boundaries

This work is conducted following the international standards 14,040 [25] and 14,044 [26] from the International Organisation for Standardization. The main function of the study is to determine what type of management strategy from the three different scenarios under study, is most appropriate for the FL management of the categories analysed, through the development of the $EROI_{ce}$ index. The functional unit is defined as the daily intake of an 11,493 kJ per capita and per day diet, by a Spanish citizen for 2015, which is obtained through an energy flow analysis (Table 1).

Table 1. Primary energy demand, nutritional energy provided to consumer and energy return on investment. Values expressed in kilojoules per capita per day and percentage.

Food Category	PED (kJ/cap/day)	Energy Provided to Consumer (kJ/cap/day)	EROI (%)
Eggs	5426	574	10.6
Meat	28,002	1901	6.8
Fish and seafood	16,243	209	1.3
Dairy	7230	938	13.0
Cereals	13,922	3827	27.5
Sweets	799	490	61.3
Pulses	2511	226	9.0
Vegetable oils	3674	2202	60.0
Vegetables	16,894	268	1.6
Fruits	3535	540	15.3
Roots	1691	318	18.8
Total	99,926	11,493	11.5

The system boundaries of this study includes the steps of agricultural production, processing and packaging, distribution and consumption, being therefore realised from "cradle to consumer" (Figure S1). As this study relies heavily on the loss percentages reported by the Food and Agriculture Organisation [27], the definition of FL is based on their latest definition provided in 2014 [28]: "food loss refers to any substance, whether processed, semi-processed or raw, which was initially intended for human consumption but was discarded or lost at any stage of the supply chain. It concerns to every non-food use, including discarded food that was originally produced for human consumption and then recycled into animal feed." Therefore, this work uses the terminology "food losses" to encompass both FL and food waste occurring at every stage, as done by Garcia-Herrero et al. [13].

2.3. Allocations

The scenarios under study are multi-outputs processes in which the management of FL is the main function of the system and the production of electricity and compost are additional functions. The environmental burdens must be allocated among the different functions. To handle this problem the International Organisation for Standardization [25] establishes a specific allocation procedure in which system expansion is the first option. Regarding the landfill scenario, since electricity generation depends on the methane concentration in the landfill biogas, electricity recovered from FL was allocated to the amount of total carbon available in the disposed organic residue.

The incineration process was modelled based on Margallo et al. [29], and in this sense, energy produced is calculated from the high heating value of each FL fraction and the amount that is incinerated.

In the anaerobic digestion scenario, methane is assumed to be combusted with a 25% efficiency of the low heating value of the biogas to generate electricity [30]. The delivering residue of the anaerobic digestion, i.e., digestate, is transferred to a composting plant for the production of biocompost. The compost is assumed to replace mineral fertilizer, with a substitution ratio of 20 kg N equivalent per ton of compost [31]. Energy intensity for fertilizer production as total N is obtained from Thinkstep's Database [32].

2.4. Life Cycle Inventory

For developing the energy flow analysis, data from different sources has been reviewed: the Department of Agriculture and Fishery, Food and Environment [33], the Spanish Institute for the Diversification and Saving of Energy [34], the Spanish Association of Plastics Industry [35], the Spanish Association of Pulp, Paper and Cardboard Manufacturers [36], a magazine specialised in informing about the life cycle of packaging [37], and the Foreign Trade Database [38]. Data for 48 representative commodities were sourced from the consumption database of the Spanish Department of Agriculture and Fishery, Food and Environment [33]. Items were grouped into 11 food categories (eggs, meat, fish and seafood, dairy, cereals, sweets, pulses, vegetable oils, vegetables, fruits and roots), which can be consulted in more detail in Table S1. It has been used several mass-to-energy conversion factors from different sources (Table S2). All the results of the PED, EEL and FEL by each food category under study, and on each food supply chain stage, can be consulted in Tables S3 and S4. Nutritional data for the EROI and the $EROI_{ce}$ estimation, were obtained from the Bedca Database [39] and can be consulted in Table S5. Food products or ingredients not available in that database were sourced from the National Nutrient Database for Standard Reference of the United States Department of Agriculture [40]. In practice, it has been assumed that the nutritional energy do not vary across the supply chain owing to the lack of data. The allocation, conversion and FL factors used (Tables S6 and S7), are based on Gustavsson et al. [27]. The exception were some products, such as apples and bananas, for which specific FL factors were available in Vinyes et al. [41] and Roibás et al. [42].

2.5. Assessment of Food Loss Management Scenarios

Based on Laso et al. [43], the electricity recovered in all the scenarios is assumed to be 100% sent to the grid, displacing electricity from the average electricity mix in Spain, and used for producing new food (Figure 2). This value could be lower if energy losses and its use for other purposes are considered. The analysis of these aspects would correspond to a consequential LCA, which could be analysed in future works.

Scenario 1: landfill with biogas recovery (L). This scenario describes landfilling of FL including biogas recovery. The landfill is composed of biogas and leachate treatment and deposition. The sealing materials (clay, mineral coating, and PE film) and diesel for the compactor is included. Leachate treatment includes active carbon and flocculation/precipitation processing. This scenario has been modelled based on the averages of municipal household FL on landfill process from Thinkstep's Database [32] for Spain, Portugal and Greece. According to the model, a 17% of the biogas naturally released from landfill is assumed to be collected, treated and burnt in order to produce electricity. The remaining biogas is flared (21%) and released to the atmosphere (62%). A rate of 50% transpiration/run off and a 100 years lifetime for the landfill are considered. Additionally, a net electricity generation of 0.0942 MJ per ton of municipal solid FL is assumed [32].

Scenario 2: incineration with energy recovery (I). The considered incineration plant, based on Margallo et al. [29], is composed of one incineration line with a capacity of 12.0 t/h. The combustion is conducted in a roller grate system reaching 1,025°C. Flue gases are treated by means of a selective non-catalytic reduction system (for NOx), bag filter (dust, dioxins, etc.) and semidry scrubbers (acid gases). The main solid residues are fly and bottom ashes. The latter is subjected to magnetic separation to recover the ferrous materials. The inert materials are assumed to be landfilled close to the incineration plant. Fly ashes, classified as hazardous material, are stabilised and sent to an inert landfill. Energy produced in combustion is transferred to flue gases for energy generation. Energy produced is calculated from the high heating value of each FL fraction and the incinerated amount. High heating values are obtained from the Thinkstep's Database [32]. For example, average values of 4832, 14,758 and 4179 kJ/kg have been obtained for fish and seafood, cereals and vegetables.

Scenario 3: anaerobic digestion and composting (AD&C). This scenario considers the combination of AD&C of the solid fraction of digested matter, and is modelled using the life cycle inventory reported by Righi et al. [31]. The anaerobic digestion plant consists of a continuous two-steps process, where the first stage is a high-solid plug-flow reactor operating at thermophilic temperature and the second a completely stirred tank reactor at mesophilic temperature. The total retention time of substrates is about 100 days. The main product of anaerobic digestion is biogas, with an assumed 60% methane content. After it, methane is combusted in an engine to produce electricity. The delivering FL of the anaerobic digestion, i.e., digestate, is transferred to a composting plant for the production of biocompost. The potential production of methane for each food category is calculated using the procedure reported by Eriksson et al. [44], according to which the theoretical methane production is estimated as described in Equation (1):

$$Nm^3_{CH_4,i} = DS_i \cdot VS_i \cdot F_i \tag{1}$$

where $Nm^3_{CH_4,i}$ is the theoretical methane production of food category i; DS_i is the dry matter content; VS_i is the percentage of volatile solids in food category i expressed in dry matter terms; F_i is an specific production factor of methane expressed in $Nm^3_{CH_4,i}$ per ton of volatile solids. These values are sourced from Carlsson and Uldal [45].

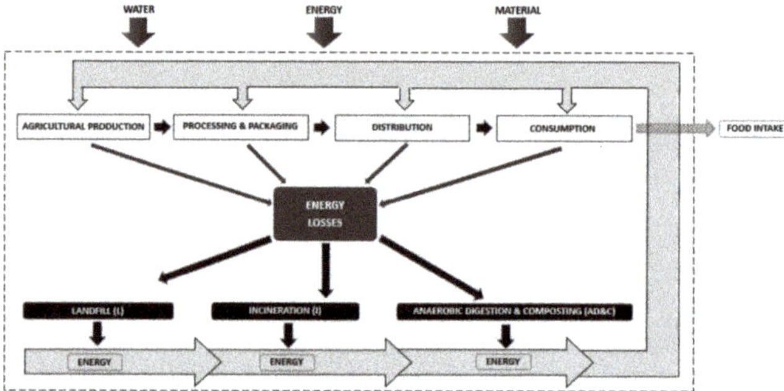

Figure 2. System boundaries, including the outline of the different considered scenarios.

2.6. Material and Energy Flow Analysis

A material flow analysis quantifies the mass/resources flow, loss in a system, and facilitates in data reconciliation in a well-defined space and time [46]. As seen in Equation (2), the material flow analysis consider the food losses occurring along the supply chain as follows:

$$FL_{i,j} = \frac{F_{i,1} \cdot \alpha_{i,j}}{\prod_{j=1}^{j} 1 - \alpha_{i,j}} \tag{2}$$

where $F_{i,j}$ is the food available for human consumption of category i leaving the supply chain sector j ($j = 1$ agricultural production, $j = 2$ processing and packaging, $j = 3$ distribution, $j = 4$ consumption). $\alpha_{i,j}$, is the percentage of food losses generated on each stage j for food category i. $F_{i,1}$ describes the daily intake of food category i for a 11,493 kJ per capita per day diet (Table 1). For this study, the material flow analysis made by Garcia-Herrero et al. [13], has been used as a reference. The energy flow analysis is developed through the combination of the material flow analysis and the calculated primary energy demand (PED) for each food category along the supply chain.

2.7. Energy Impact Assessment

In this work, it has been introduced as energy impact assessment the $EROI_{ce}$ index in order to quantify the amount of nutritional energy that is recovered from the FL of each category of food under study. The $EROI_{ce}$ index is based on a food waste-to-energy-to-food approach, assuming that the energy that is recovered from FL is reintroduced into the food supply chain in form of food (Figure 2). For its development, the proposed methodology (Figure 3) firstly develops an energy flow analysis through determining the PED of each of the four stages in which the food supply chain is divided (agricultural production, processing and packaging, distribution and consumption), as seen in Equation (3):

$$PED_{i,j} = W_{i,j} \cdot AP_{i,j} \tag{3}$$

where $W_{i,j}$ is the weighted average of energy intensity by mass of each category i, on each supply chain stage under study j (j = 1 agricultural production, j = 2 processing and packaging, j = 3 distribution, j = 4 consumption), in kJ/kg. $AP_{i,j}$ is the annual production of each category i, on each stage under study j, in kg.

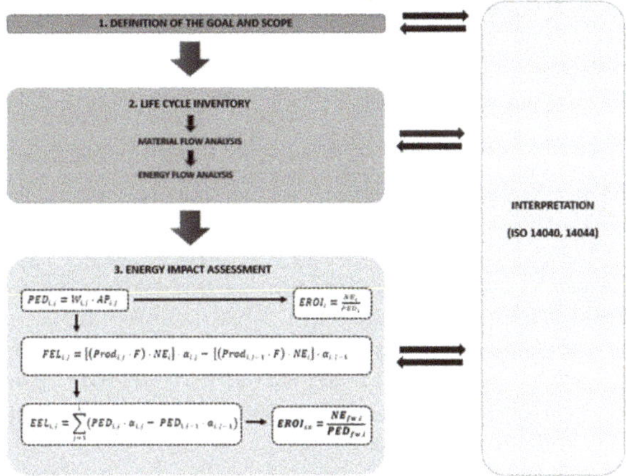

Figure 3. Methodology of the study.

Secondly, the EEL is computed, which means, the primary energy that was used to produce the food that is loss. EEL is calculated as stated in Equation (4):

$$EEL_{i,j} = \sum_{j=1}^{i} \left(PED_{i,j} \cdot \alpha_{i,j} - PED_{i,j-1} \cdot \alpha_{i,j-1} \right) \tag{4}$$

To calculate it, the sum of the $PED_{i,j}$ multiplied by their respective percentages of loss $\alpha_{i,j}$ is performed. From the second stage, these results are subtracted from the previous stage multiplied by their respective previous loss percentages $\alpha_{i,j-1}$.

Once these data have been obtained, the FEL of each food category i under study is calculated. Following the Food and Agriculture Organization concept for FL [28], FEL can be defined, as the amount of chemical energy contained in food and initially addressed to human consumption that, for any reason is not destined to its main purpose. It has been estimated according to Equation (5):

$$FEL_{i,j} = \left[\left[Prod_{i,j} \cdot F \right] \cdot NE_i \right] \cdot \alpha_{i,j} - \left[\left[Prod_{i,j-1} \cdot F \right] \cdot NE_i \right] \cdot \alpha_{i,j-1} \tag{5}$$

where $Prod_{i,j}$ is the production of each category of food, which is multiplied by F, which are the factors of allocation and conversion proposed by Gustavsson et al. [27] to represent the amount of food that is used for human consumption and that is considered edible. These values are firstly multiplied by the nutritional energy, and next by the percentages of losses considered in the literature $\alpha_{i,j}$. From the second stage, the previously lost amount is subtracted, multiplied by the conversion factor of the previous stage $\alpha_{i,j-1}$. Then, it has been calculated the EROI of each food category under study i, and each step j. EROI is the estimation of the quantity of energy delivered by a production technology relative to the quantity of energy invested [47]. Although it was initially devised to the assessment of energy systems, the concept has been adapted (Equation (6)) to quantify ratios of food energy output relative to food production energy inputs. This ratio can be estimated as follows:

$$EROI_i = \frac{NE_i}{PED_i} \tag{6}$$

where $NE_{i,}$ is the nutritional energy contained in each food category i, and PED_i is the primary energy demand for the production of each category i. Finally, the $EROI_{ce}$ is calculated. For it, the electricity recovered from the management of FL is transformed into its equivalent amount of primary energy, and assumed to be redirected to the production of food. As shown in Equation (7), this index consist in the division between the nutritional energy $NE_{fw,i}$, obtained from the transformation into nutritional energy of the primary energy that is recovered through each FL management strategy, and each FL fraction of a specific food category; between the primary energy demand $PED_{fw,i}$ that was used in the management of FL:

$$EROI_{ce} = \frac{NE_{fw,i}}{PED_{fw,i}} \tag{7}$$

3. Results

3.1. Energy Flow Analysis

Results from the energy flow analysis are shown in the Sankey diagram of Figure 4. The diagram represents the inputs and outputs of primary energy along the entire chain, using the reference unit (kJ/cap/day). By calculating the primary energy balance until the end of the chain (99,926 kJ) which is need to produce the 11,493 kJ/cap/day of nutritional energy provided to consumer on average by each Spanish citizen; it is suggest that in the Spanish food supply chain, 8.7 kJ of primary energy is needed to produce 1 kJ of nutritional energy. In the agricultural production stage, the allocated flow to FL is distinguished from the resulting flow assigned to non-food uses. The net domestic supply after considering agricultural production, imports, exports and stock variation is 24,476 kJ/cap/day. From this, 4970 kJ/cap/day (20%) are invested in producing animal feed, seed and other non-food uses such as oil and wheat for bio-energy. The other 19,506 kJ/cap/day of the primary energy (80%) are used for food for human consumption. In this diagram, it is highlighted the fact that the stages with a higher PED are distribution (which in addition to distribution places, also includes national and international import transportation, as well as consumer transport to go to the markets) and agricultural production, followed by the stage of processing and packaging. These results could reinforce the thesis that the more local, seasonal and unprocessed the consumption, the lower expenditure of energy in transport and distribution. It is, however, important to note that a lower energy expenditure in transport and distribution does not necessarily mean a lower total energy expenditure in food production. There are a number of other factors that should be analysed in future works in this field, as for example, the use of agrochemicals or tillage machinery.

When analysing the food categories studied, it is observed that the ones requiring the highest PED for their production are meat, vegetables, fish and seafood and cereals, respectively (Table 1). Of the four categories, meat is the one with the highest PED (28,002 kJ/cap/day), doubling the value of the other three, and representing alone the 28% of the PED for all categories. These results could reinforce the thesis of the need to reduce the consumption of meat due to the energy costs that its production requires, as stated by Popkin [48] and Laso et al. [43]. In addition, if the values for fish and seafood, eggs and dairy categories are added to meat, more than half of the total PED comes from the production of food of animal origin (56,901 kJ/cap/day). In contrast, some categories, especially sweets and roots, have very low values.

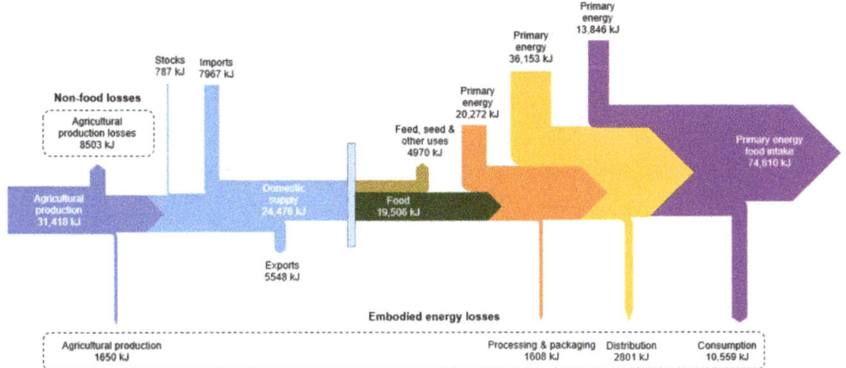

Figure 4. Sankey diagram for primary energy demand of the different food categories throughout the food supply chain. Values expressed in kilojoules per capita per day.

Regarding the values of EROI, sweets (61.3%) and vegetable oils (60.0%) are the food categories with the largest EROI, which indicates that these categories are the most efficient, although not necessarily the healthiest. It must be remarked that this work only assesses nutritional content in terms of energy; other nutritional features are not studied. They are followed by cereals and roots, with 27.5% and 18.8% EROI ratios, respectively. On the opposite side, fish and seafood, vegetables, meat and pulses have the lowest EROI, which indicates a very low energy efficiency in its production process. This agrees with results in the literature [3], which state that animal and animal derived food products consume large amounts of energy resources. Likewise, they reinforce the thesis of Popkin [48] and Laso et al. [43] on the environmental benefits of eating less meat and fish, since there is a huge potential for PED reduction.

3.2. Energy Food Losses Quantification

The energy flow analysis reveals that in terms of EEL, which means the primary energy invested in producing FL, meat, cereals, vegetables and fish and seafood are, respectively, the categories with the highest EEL values. Accordingly, they are the food categories most affected by the energetic inefficiencies in the food supply chain. Their EEL were estimated at 4027, 3259, 3143 and 2650 kJ/cap/day, respectively, which together accounts for almost 84% of the total Spanish EEL (Table 2).

In addition, once again, if the four categories of products of animal origin are added, it is highlighted the fact that around 50% of the total EEL is due to these products. In contrast, the categories with the lowest EEL values are sweets and vegetable oils, which represents values 20 times lower than the category with a higher value (meat). If the EEL is analysed in the different stages, it can be clearly perceived that the stage of consumption is the one in which the highest EEL is produced, representing more than 66% of the total in the whole food supply chain (Figure 5). The total sum of the EEL values obtained, were around 17% of the total PED in the entire food supply chain.

Table 2. Food energy loss and embodied energy loss.

Food Category	FEL		EEL	
	(kJ/cap/day)	(%)	(kJ/cap/day)	(%)
Eggs	113	2	521	3
Meat	553	11	4027	26
Fish and seafood	80	2	2650	17
Dairy	126	3	510	4
Cereals	2386	46	3259	21
Sweets	398	8	159	1
Pulses	96	2	421	3
Vegetable oils	687	13	233	2
Vegetables	176	3	3143	20
Fruits	381	7	661	4
Roots	155	3	331	2
Total	5151	100	15,915	100

In terms of the FEL, the categories of cereals, vegetable oils and meat, represent the highest values (Table 2). As this sequence coincides with the results of the energy provided to consumer (Table 1), these high values could be due to the high percentage of the European diet, which is based on cereals, vegetable oils and meat. On the other side, the categories with the lowest FEL are fish and seafood, pulses and eggs. This sequence agrees again with the results of the energy provided to consumer (Table 1), with the exception of eggs. Thus, the low values of FEL could be also related to the European diet, although other factors not analysed in this work could influence them. Regarding the different stages of the food supply chain, the results show that the stage of consumption is the one with the highest values (Figure 5). Moreover, agricultural production plus processing and packaging together would be the part of the food supply chain with the highest FEL. The distribution stage, despite being the one that requires the most PED, is at the same time the one that clearly generates less FEL (7.4%). When it comes to recover energy from FL, the qualitative and quantitative composition of FL is essential [13], and in this sense, from a quantitative point of view, these results suggest that the largest amount of FEL from which to recover energy occurs at the beginning and end of the food supply chain, being 1130 and 1290 kJ/cap/day the FEL in the stages of agricultural production and processing and packaging, and 2349 kJ/cap/day in the stage of consumption. The total results of the FEL highlighted that approximately 5154 kJ/cap/day are thrown away, which means that from a FEL point of view, for the consumption of two to three persons in Spain, one more person could eat.

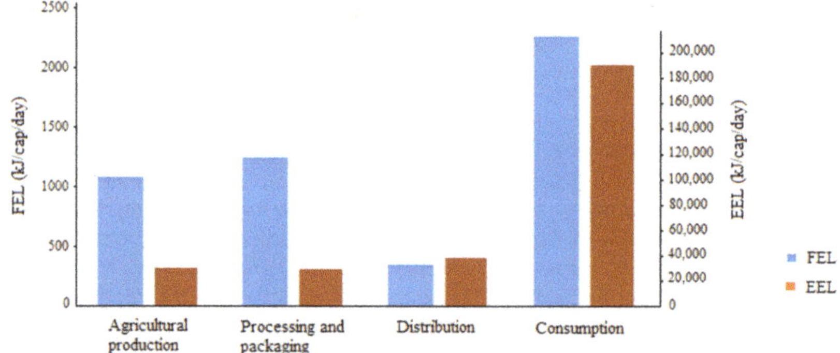

Figure 5. Food energy loss (FEL) and embodied energy loss (EEL) by stage of the food supply chain. Values expressed in kilojoules per capita per day (left and right ordinate axis).

3.3. Nutritional Assessment of the Energy Food Loss

The food categories under study have been classified according to four different diets: vegetarian, pescetarian, Mediterranean and omnivorous. A vegetarian diet includes cereals, roots and tubers, sweets, vegetable oils, vegetables, fruits, pulses, dairy and eggs. A pescetarian diet is a vegetarian diet that includes fish and seafood. A Mediterranean diet is similar to the pescetarian, but includes moderate amounts of meat. Omnivorous diets consider all food groups.

Figures 6 and 7 represent the values obtained from FEL (kJ/cap/day) and EEL (kJ/cap/day), respectively, for the different food categories (abscissa axis) and the different stages (different colors in each column), being the numerical values signified on the ordinate axis.

If the FEL values for each category and stage of the food supply chain are related, it is clear that the category of cereals is the most wasteful one. From a quantitative point of view, it suggests that cereals should be the main category for placing the focus when developing FL management strategies. Moreover, regarding the results, the change of the diet would not imply a significant change in terms of FEL, as can be seen in Figure 6:

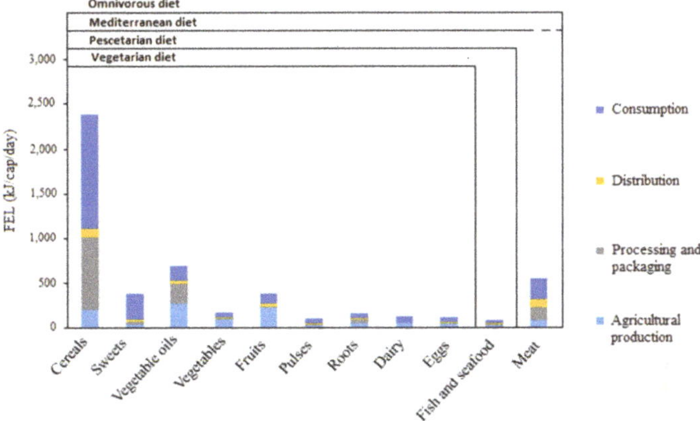

Figure 6. Food energy loss (FEL) of the different food categories throughout the supply chain. Values expressed in kilojoules per capita per day.

On the other hand, Figure 7 displays the EEL values for each category and stage of the food supply chain. From figure, it is observed that the type of diet does have a clear influence. The meat category presents the largest EEL values, followed closely by cereals, vegetables and fish and seafood, respectively. In terms of EEL, the vegetarian diet appears to be the one which the highest amount of primary energy saves, followed by the pescetarian diet. The consumption of meat in the Mediterranean and omnivorous diets supposes a significant increase of EEL.

Taking into account an overall results overview, it is suggested that due to the more mass losses of cereals, their value stands out against the others. However, in case of meat and fish and seafood, when analysing the energy used in its production, those categories have a very high PED to produce low levels of food.

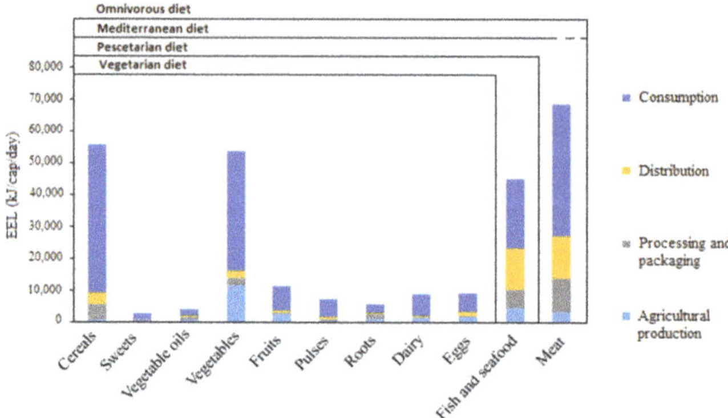

Figure 7. Embodied energy loss (EEL) of the different food categories throughout the supply chain. Values expressed in kilojoules per capita per day.

3.4. Energy Return on Investment–Circular Economy Index

Figure 8 shows a general trend for decreasing PED demand with higher priority levels in the food waste hierarchy. Negative values of $EROI_{ce}$ indicate that the energy recovered from the management of FL is larger than the energy requirements for its management. As shown, landfilling with biogas recovery (Scenario 1: L) do not recover enough energy to compensate the energy expenses of the treatment. Anaerobic digestions and composting (Scenario 3: AD&C) seems to be the best option for the food categories assessed. An exception is suggested for vegetables FL, for which a larger PED is observed for Scenario 2 (I), involving higher energy recovery from the incineration treatment. This may be due to the fact that the fermentation period is longer than the rest of the categories and therefore requires a higher energy consumption.

Afterwards, $EROI_{ce}$ scores have been assessed. Results from Figure 9 suggest that AD&C is the best FL management strategy. On the other hand, it is highlighted that cereals is the category with the highest potential for energy recovery, with values between 20 and 28 times higher than the rest of the categories, regardless of the scenario. This is undoubtedly influenced by the fact that it has the highest FEL value, representing 44% of the total. Finally, it is observed that vegetables appear again as the less energy efficient category, owing to the low energy recovered from its FL management, which could be due to a low carbon content. The numerical results can be consulted in Table S8.

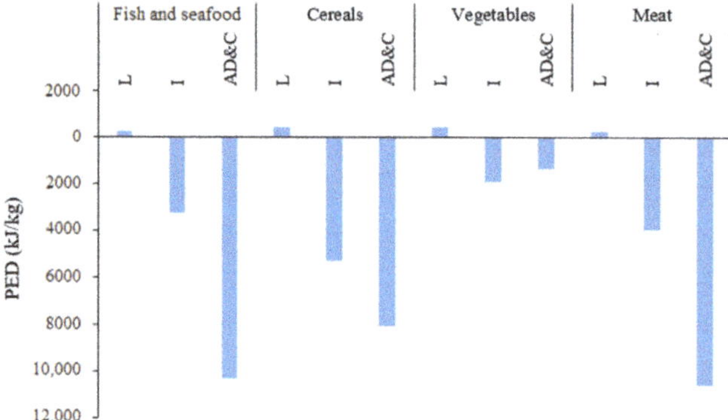

Figure 8. Primary energy demand values for the considered scenarios expressed in kilojoules per kilogram.

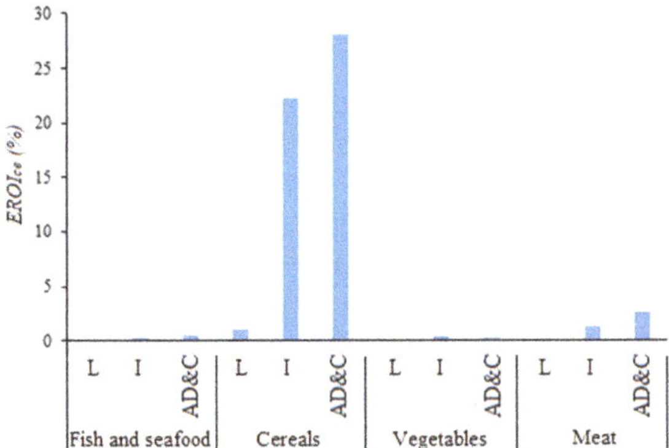

Figure 9. Energy return on investment–Circular economy index for the considered scenarios.

4. Discussion

The results of the energy flow analysis determined a total EEL value of 17% in relation to the total PED along the entire supply chain, showing the consumption stage as the most inefficient one. This is in accordance with Vittuari et al. [7], who assume that embodied energy builds up along the chain, so the latter the FL occurs, the greater the energy loss. The EEL results indicate that in the final part of the food supply chain, which means the sum of the distribution stage plus the consumption stage, the highest amount of EEL is concentrated. The FEL results point out that the stage of consumption is the one with the highest values. Moreover, if the FEL values for agricultural production and processing and packaging are added, it is suggested that the first part of the food supply chain accumulate the highest FEL. These results highlight the option of decentralise the energy recovery strategies, which could improve the efficiency in the FL management systems, by installing energy recovery plants at the beginning and at the end of the FSC.

Regarding the nutritional assessment in terms of EEL, vegetarian and pescetarian diets appear to be the most efficient ones. In this sense, several studies have supported similar thesis taking into account different approaches such as the greenhouse gas emissions [49] and the economic value of FL [13].

From FEL results, the high loss value generated by the cereals category (44%) is remarkable. After assessing the $EROI_{ce}$ scores, results also suggest that cereals is the category with higher potential for energy recovery. In addition, in three of the four categories analysed, results show a general trend for decreasing PED with higher priority levels in the food waste hierarchy [44], standing out the AD&C as the most appropriate for FL management. This reinforces the thesis that FL is an attractive substrate for AD&C because of its low total solids and high content of soluble organics, as stated by David et al. [50]. In this sense, the development of decentralised energy recovery strategies through AD&C could be proposed, as opposed to centralised strategies, which are large scale for the treatment of FL [51].

Following the previous context, new strategies for the different fractions of FL and its compositions could be introduced in order to meet the transition towards a more circular economy [52]. In this case, the cereal fraction stands out in terms of the amount of FEL and the amount of food that can be reintroduced into the food supply chain. In this sense, until now, AD&C has usually been focused on the recovery of biogas in form of methane mainly. In view of the high energy recovery potential of cereals and their high level of hydrocarbons in their chemical composition; it is proposed their separately management, based on the works of Kibbler et al. [53] and Bernstad and La Cour [54]. Due to its composition, it is considered that they have a high potential for the recovery of bioenergy in form of hydrogen. Therefore, this proposal of decentralisation would include the development of two types of AD&C digesters: one for the cereal fraction with hydrogen recovery, and another for the rest of FL, with methane recovery, as can be seen in Figure 10.

Decentralised AD&C plants of biogas production from organic waste and FL, could have clear advantages in concrete contexts like rural regions, and other local economies which are far away from power sources [55]. This has already been tested in many rural contexts around the world, existing good and diverse examples, as the works developed by Raha et al. [56] in India, and Kelebe and Olorunnisola [57] in Ethiopia. Another argument in favour of this decentralisation option is the fact that valorisation in form of biogas is, generally, more applicable when there is homogeneity of the waste [58], and homogeneous FL streams are most likely generated before being mixed with the rest of the FL [59]. In this sense, there are several technological challenges that require future research in order to deploy this technology for small and medium applications.

One of the main barrier for those strategies is the wide variation of feedstock and environmental conditions (e.g., temperature) over space and time, which are more difficult to control through small-decentralised digesters. Additionally, it is important to know that from an energetic point of view, small scale AD&C hardly can perform a strong separation between biodegradable and non-biodegradable fraction. If a stronger pre-treatment is demanded, local anaerobic digestion can become impracticable from both an energy and economic point of view [51].

On the other hand, the decentralised management option could also be applied to the consumption stage, as it is a very simple system [60]. It could be an especially interesting alternative in buildings where a large number of people are living, receiving a high and constant source of power to produce energy, for self-consumption in the first instance, and to sell to the electricity grid if consumption is less than production. As a practical example, a recent study in this field, carried out by Walker et al. [61], analysed systems of micro-scale anaerobic digesters in London, showing that this technology could provide a useful means of processing FL in urban areas.

The proposed change of strategies poses the debate of the 'sustainable de-growth' sustained by Amate [2] and Latouche [62], which emerged as a strategy that aims to generate new social values and new policies capable of satisfying human requirements whilst reducing the consumption of resources. It is also intended to support the European Union action plan for the transition to a more circular economy [63], and the Bioeconomy Strategy [64], contributing to meet the objectives of bioenergy and the sustainable use of renewable sources, through the replacement of fossil fuel by renewable raw materials and the replacement of chemical processes by biological ones.

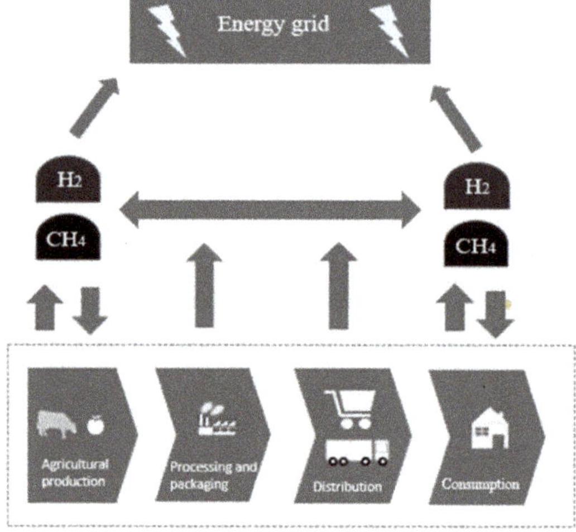

Figure 10. Outline of the proposed energy recovery strategies.

5. Conclusions

The energy flow analysis developed in this work suggest that to produce 1 kJ of nutritional energy, 8.7 kJ of primary energy is required, being the distribution and agricultural production stages the ones that require the most primary energy, respectively. From the 11 categories studied, the ones with the lowest EROI are fish and seafood, vegetables, meat and pulses. In terms of EEL, consumption is the stage with the highest values, representing more than 66% of the total in the whole food supply chain. The total sum of the obtained EEL results was 17% of the total PED. Meat, cereals, vegetables and fish and seafood have the highest values, which together accounts for almost 84% of the total Spanish EEL. If the four categories of products of animal origin are added, it is highlighted the fact that around 50% of the total EEL is due to these products. In terms of FEL, cereals, vegetable oils, meat and sweets, represent the highest values. The stage of consumption is clearly the one with the highest FEL value, although the beginning of the food supply chain would represent a higher FEL if agricultural production and processing and packaging values are added. The distribution stage, despite being the one that requires the most PED, is at the same time the one that clearly generates less FEL (7.4%).

The study suggests that the efficiency of energy of the agri-food supply depends heavily on the food category under study. Meat and fish and seafood have a very high PED to produce less food. Also, according to the $EROI_{ce}$ it is highlighted that cereals is the category with the highest potential for energy recovery from FL, with values between 20 and 28 times higher than the rest of the categories.

Related to the results, it is suggested that energy recovered from FL can contribute considerably to the national energy grid, as well as to energy self-consumption throughout the food supply chain. This could contribute to reduce the environmental costs, the demand of other types of non-clean energies such as coal- and nuclear- energy, and to produce new food from the recovered energy.

Although up to now the collection of FL is usually done in a centralised way, the use of AD&C for decentralised biogas production is, according to this work, one of the most potential technologies of bioenergy generation. It offers a good option of local FL management, which reduces the environmental impact due to transport, and encourages self-consumption, as well as benefiting the economy of local actors. Moreover, the recovery of energy in form of biogas can occur through the generation of different products. In this sense, a proposal of possible treatment strategies for residues of cereals with hydrogen recovery and mixed FL with methane recovery, is made. It is considered that the diversification and decentralisation in FL energy recovery strategies could facilitate the transition to a more circular economy. The efficiency of the proposed strategies could be further improved by intensifying research and optimisation studies. Thus, basic research is critical in order to advance the development of those technologies.

Results from the study allows to facilitate the decision-making process for the proper FL management, developing a general awareness on the need of energy-smart strategies or policies, which are decentralised and adapted to each stage of the food supply chain and the different fractions of food. This claim is in contrast to the waste hierarchy of the European Union, which is considered as a too generic proposal. Specifically, this work aims to highlight the need to address a decentralised and diverse FL management, in order to manage more efficiently the different fractions, and at each of the different stages of the food supply chain. Future works should: (i) simulate different scenarios of decentralised management, (ii) put into practice the cases of pilot studies already carried out, and (iii) optimize systems on a larger scale through the intervention of small-scale systems throughout the food supply chain, for which it is fundamental to establish regional strategies that support the already established global ones. Thus, the general objective of this research field is to follow strategies that act locally to achieve global development.

Supplementary Materials: The following are available online at http://www.mdpi.com/1996-1073/12/4/767/s1, Figure S1: Outline of the assumed division in stages of the food supply chain, Table S1: Food commodities included in the study, based on Garcia-Herrero et al., Table S2: Mass-to-energy conversion factors and life cycle inventory sources, Table S3: Results in petajoules per year in Spain of the primary energy demand by each food category under study, and on each food supply chain stage. The values are related to the percentages assumed, based on

Laso et al., Table S4: Results in MJ/cap/day of the embodied energy loss and in kJ/cap/day of the food energy loss by each food category under study, on each stage, Table S5: Proteins, carbohydrates and energetic content for the food categories under study. Table S6: Allocation and conversion factors used for calculating the edible part of food production which is used for human consumption, Table S7. Food losses percentages for each food category as a percentage of what enters on each supply chain stage. Unless stated otherwise, percentages are obtained from Garcia-Herrero et al. and Gustavsson et al. for Europe region, Table S8: Results of the Energy return on investment–Circular economy index ($EROI_{ce}$) on fish and seafood, cereals, vegetables and meat, on each of the considered scenarios.

Author Contributions: Conceptualisation: R.A.; Investigation: D.H., I.G.H., and J.L.; Methodology and formal and technical analysis: P.F.P., A.B., A.I., and R.A.; Supervision: R.A. and M.M.; Writing and editing of manuscript: D.H., I.G.H. and J.L.

Funding: This work has been made under the financial support of the Project Ceres-Procom: Food production and consumption strategies for climate change mitigation (CTM2016-76176-C2-1-R) (AEI/FEDER, UE) financed by the Ministry of Economy and Competitiveness of the Government of Spain.

Acknowledgments: Daniel Hoehn thanks the Ministry of Economy and Competitiveness of Spanish Government for their financial support via the research fellowship BES-2017-080296. The authors also thank to the UNESCO Chair in Life Cycle and Climate Change.

Conflicts of Interest: The authors declare no conflicts of interest.

Abbreviations

AD&C	Anaerobic Digestion and Composting
EEL	Embodied Energy Loss
EROI	Energy Return On Investment
EROIce	Energy Return On Investment—Circular Economy Index
FEL	Food Energy Loss
FL	Food Loss
I	Incineration with energy recovery
L	Landfill with biogas recovery
PED	Primary Energy Demand

References

1. Carlsson-Kanyama, A.; Ekström, M.; Shanahan, H. Food and life cycle energy inputs: Consequences of diet and ways to increase efficiency. *Ecol. Econ.* **2003**, *44*, 293–307. [CrossRef]
2. Infante-Amate, J.; González de Molina, M. "Sustainable de-growth" in agriculture and food: An agro-ecological perspective on Spain´s agri-food system (year 2000). *J. Clean. Prod.* **2013**, *38*, 27–35. [CrossRef]
3. Pimentel, D.; Pimentel, M.H. *Food, Energy and Society*, 3rd ed.; CRC Press, Taylor and Francis Group: Boca Raton, FL, USA, 2008.
4. Martinez-Alier, J. The EROI of agriculture and its use by the Via Campesina. *J. Peasant Stud.* **2011**, *38*, 145–160. [CrossRef]
5. Cuellar, D.; Webber, E. Wasted Food, Wasted Energy: The Embedded Energy in Food Waste in the United States. *Environ. Sci. Technol.* **2010**, *44*, 6464–6469. [CrossRef] [PubMed]
6. Lin, B.; Chappell, M.; Vandermeer, J.; Smith, G.; Quintero, E.; Bezner-Kerr, R.; Griffith, D.; Ketcham, S.; Latta, S.; McMichae, P.; et al. Effects of industrial agriculture on global warming and the potential of small-scale agroecological techniques to reverse those effects. *CAB Rev. Perspect. Agric. Vet. Sci. Nutr. Nat. Resour.* **2011**, *6*, 020.
7. Vittuari, M.; De Menna, F.; Pagani, M. The Hidden Burden of Food Waste: The Double Energy Waste in Italy. *Energies* **2016**, *9*, 660. [CrossRef]
8. Markussen, M.V.; Østergård, H. Energy Analysis of the Danish Food Production System: Food-EROI and Fossil Fuel Dependency. *Energies* **2013**, *6*, 4170. [CrossRef]
9. Tanczuk, M.; Skorek, J.; Bargiel, P. Energy and economic optimization of the repowering of coal-fired municipal district heating source by a gas turbine. *Energy Convers. Manag.* **2017**, *149*, 885–895.
10. Food and Agriculture Organization of the United Nations (FAO). *FAO Climate-Smart*; FAO: Rome, Italy, 2011.

11. European Commission. *Commission Staff Working Document. European Research and Innovation for Food and Nutrition Security*; European Commission: Brussels, Belgium, 2016.

12. OECD. *Improving Energy Efficiency in the Agro-Food Chain, OECD Green Growth Studies*; OECD Publishing: Paris, France, 2017.

13. Garcia-Herrero, I.; Hoehn, D.; Margallo, M.; Laso, J.; Bala, A.; Batlle-Bayer, L.; Fullana, P.; Vazquez-Rowe, I.; Gonzalez, M.J.; Durá, M.J.; et al. On the estimation of potential food waste reduction to support sustainable production and consumption policies. *Food Policy* **2018**, *80*, 24–38. [CrossRef]

14. FAO. *The State of Food Insecurity in the World. Meeting the 2015 International Hunger Targets: Taking Stock of Uneven Progress*; FAO: Rome, Italy, 2015.

15. Department of Economic and Social Affairs, Population Division, United Nations. *Probabilistic Population Projections based on the World Population Prospects: The 2017 Revision*; United Nations: New York, NY, USA, 2017.

16. Kummu, M.; De Moel, H.; Porkka, M.; Siebert, S.; Varis, O.; Ward, P.J. Lost food, wasted resources: Global food supply chain losses and their impacts on freshwater, cropland, and fertiliser use. *Sci. Total Environ.* **2012**, *4438*, 477–489. [CrossRef] [PubMed]

17. MAGRAMA. *Spanish Strategy "More Food, Less Waste"*; Program to Reduce Food Loss and Waste and Maximize the Value of Discarded Food; MAGRAMA: Madrid, Spain, 2013.

18. Corrado, S.; Sala, S. Food waste accounting along global and European food supply chains: State of the art and outlook. *Waste Manag.* **2018**, *79*, 120–131. [CrossRef] [PubMed]

19. Infante-Amate, J.; Aguilera, E.; González de Molina, M. *La gran Transformación del Sector Agroalimentario Español, Un Análisis Desde la Perspectiva Energética (1960–2010)*; Working Papers Sociedad Española de Historia Agraria: Santiago de Compostela, Spain, 2014.

20. Canning, P.; Charles, A.; Huang, S.; Polenske, K.R. *Water and Energy Use in the U.S. Food System*; U.S. Department of Agriculture, Economic Research Service: Washington, D.C., USA, 2010.

21. United Nations. *The Sustainable Development Goals Report*; United Nations: New York, NY, USA, 2018.

22. European Commission. *Communication from the Commission to the European Parliament, the Council, the European Economic and Social Committee and the Committee of the Regions. The Role of Waste-to-Energy in the Circular Economy*; European Commission: Brussels, Belgium, 2017.

23. Papargyropoulou, E.; Lozano, R.; Steinberger, J.K.; Wright, N.; Bin Ujang, Z. The food waste hierarchy as a framework for the management of food surplus and foodwaste. *J. Clean. Prod.* **2014**, *76*, 106–115. [CrossRef]

24. Cristobal, J.; Castellan, V.; Manfredi, S.; Sala, S. Prioritizing and optimizing sustainable measures for food waste prevention and management. *Waste Manag.* **2018**, *72*, 3–16. [CrossRef] [PubMed]

25. *ISO 14040 Environmental Management—Life Cycle Assessment—Principles and Framework*; ISO: Geneva, Switzerland, 2006.

26. *ISO 14044 Environmental Management—Life Cycle Assessment—Requirements and Guidelines*; ISO: Geneva, Switzerland, 2006.

27. Gustavsson, J.; Cederberg, C.; Sonesson, U.; Emanuelsson, A. *The Methodology of the FAO Study: "Global Food Losses and Food Waste–Extent, Causes and Prevention"—FAO, 2011*; The Swedish Institute for Food and Biotechnology (SIK): Göteborg, Sweden, 2013.

28. FAO. *Save Food: Global Initiative on Food Loss and Waste Reduction. Definitional Framework of Food Loss*; Working Paper; FAO: Rome, Italy, 2014.

29. Margallo, M.; Aldaco, R.; Irabien, A.; Carrillo, V.; Fischer, M.; Bala, A.; Fullana, P. Life cycle assessment modelling of waste-to-energy incineration in Spain and Portugal. *Waste Manag. Res.* **2014**, *32*, 492–499. [CrossRef] [PubMed]

30. Manfredi, S.; Cristobal, J. Towards more sustainable management of European food waste: Methodological approach and numerical application. *Waste Manag. Res.* **2016**, *34*, 957–968. [CrossRef] [PubMed]

31. Righi, S.; Oliviero, L.; Pedrini, M.; Buscaroli, A.; Della-Casa, C. Life Cycle Assessment of management systems for sewage sludge and food waste: Centralized and decentralized approaches. *J. Clean. Prod.* **2013**, *44*, 8–17. [CrossRef]

32. Thinkstep. *Gabi 6 Software and Database on Life Cycle Assessment*; Thinkstep: Leinfelden-Echterdingen, Germany, 2017.

33. Ministry of Agriculture, Fishery, Food and Enviroment, MAPAMA. *Informes de Consumo de Alimentación en España*; MAGRAMA: Madrid, Spain, 2015.

34. Instituto de Diversificación y Ahorro de Energía (IDAE). *Memoria Annual*; IDAE: Sevilla, Spain, 2015.

35. Asociación Española de Industriales de Plásticos (ANAIP). *La Plasticultura en España*; ANAIP: Madrid, Spain, 2015.

36. Spanish Association of Pulp, Paper and Cardboard Manufacturers. Available online: http://www.aspapel.es/ (accessed on 5 July 2018).

37. INFOPACK. Packaging and Industrial Labelling Magazine. Available online: http://www.infopack.es/es (accessed on 4 July 2018).

38. DataComex. Spanish Statistics on International Trade. Available online: http://datacomex.comercio.es/ (accessed on 5 July 2018).

39. Bedca Database. Spanish Food Composition Database. Available online: http://www.bedca.net/ (accessed on 15 June 2018).

40. *USDA Food Composition Databasesp*; USDA: Washington DC, USA, 2018.

41. Vinyes, E.; Asin, L.; Alegre, S.; Muñoz, P.; Boschmonart, J.; Gasol, C.M. Life Cycle Assessment of apple and peach production, distribution and consumption in Mediterranean fruit sector. *J. Clean. Prod.* **2017**, *149*, 313–320. [CrossRef]

42. Roibás, L.; Elbehri, A.; Hospido, A. Carbon footprint along the Ecuadorian banana supply chain: Methodological improvements and calculation tool. *J. Clean. Prod.* **2016**, *112*, 2441–2451. [CrossRef]

43. Laso, J.; Hoehn, D.; Margallo, M.; García-Herrero, I.; Batlle-Bayer, L.; Bala, A.; Fullana-i-Palmer, P.; Vázquez-Rowe, I.; Irabien, A.; Aldaco, R. Assessing Energy and Environmental Efficiency of the Spanish Agri-Food System Using the LCA/DEA Methodology. *Energies* **2018**, *11*, 3395. [CrossRef]

44. Eriksson, M.; Strid, I.; Hansson, P. Carbon footprint of food waste management options in the waste hierarchy—A Swedish case study. *J. Clean. Prod.* **2015**, *93*, 115–125. [CrossRef]

45. Carlsson, M.; Uldal, M. *Substrathandbok för Biogasproduktion [Substrate Handbook for Biogas Production]*; Rapport SGC 200; Svenskt Gastekniskt Center: Malmö, Swedish, 2009.

46. Padeyanda, Y.; Jang, Y.-C.; Ko, Y.; Yi, S. Evaluation of environmental impacts of food waste management by material flow analysis (MFA) and life cycle assessment (LCA). *Mater. Cycles Waste Manag.* **2016**, *18*, 493–508. [CrossRef]

47. Pelletier, N.; Audsley, E.; Brodt, S.; Garnett, T.; Henriksson, P.; Kendall, A.; Krammer, K.J.; Murphy, D.; Nemecek, T.; Troell, M. Energy Intensity of Agriculture and Food Systems. *Annu. Rev. Environ. Resour.* **2011**, *36*, 223–246. [CrossRef]

48. Popkin, B.M. Reducing Meat Consumption Has Multiple Benefits for the World's Health. *Arch. Intern. Med.* **2009**, *169*, 543. [CrossRef] [PubMed]

49. Berners-Lee, M.; Hoolohan, C.; Cammack, H.; Hewitt, C.N. The relative greenhouse gas impacts of realistic dietary choices. *Energy Policy* **2012**, *43*, 184–190. [CrossRef]

50. David, A.; Govil, T.; Kumar, T.A.; McGeary, J.; Farrar, K.; Kumar, S.R. Thermophilic Anaerobic Digestion: Enhanced and Sustainable Methane Production from Co-Digestion of Food and Lignocellulosic Wastes. *Energies* **2018**, *11*, 2058. [CrossRef]

51. Wang, J. Decentralized biogas technology of anaerobic digestion and farm ecosystem: Opportunities and challenges. *Front. Energy Res.* **2014**, *2*, 10. [CrossRef]

52. Arushanyan, Y.; Björklund, A.; Eriksson, O.; Finnveden, O.; Söderman, M.L.; Sundqvist, J.O.; Stenmarck, A. Environmental Assessment of Possible Future Waste Management Scenarios. *Energies* **2017**, *10*, 247. [CrossRef]

53. Kibbler, K.; Reinhart, D.; Hawkins, C.; Motlagh, A.; Wright, J. Food waste and the food-energy-water nexus: A review of food waste management alternatives. *Waste Manag.* **2018**, *74*, 52–62. [CrossRef] [PubMed]

54. Bernstad, A.; La Cour, J. Review of comparative LCAs of food waste management systems—Current status and potential improvements. *Waste Manag.* **2012**, *32*, 2439–2455. [CrossRef] [PubMed]

55. De Souza, G.C.; Rodrigues da Silva, M.D.; Gonçalves, S.E. Construction of Biodigesters to Optimize the Production of Biogas from Anaerobic Co-Digestion of Food Waste and Sewage. *Energies* **2018**, *11*, 870.

56. Raha, D.; Mahanta, P.; Clarke, M.L. The implementation of decentralised biogas plants in Assam, NE India: The impact and effectiveness of the Nationa Biogas and Manure Management Programme. *Energy Policy* **2014**, *68*, 80–91. [CrossRef]

57. Kelebe, H.E.; Olorunnisola, A. Biogas as an alternative energy source and a waste management strategy in Northern Ethiopia. *Biofuels* **2016**, *7*, 479–487. [CrossRef]

58. Girotto, F.; Peng, W.; Rafieenia, R.; Cossu, R. Effect of Aeration Applied During Different Phases of Anaerobic Digestion. *Waste Biomass Valoriz.* **2016**, *9*, 161–174. [CrossRef]

59. De Laurentiis, V.; Corrado, S.; Sala, S. Quantifying household waste of fresh fruit and vegetables in the EU. *Waste Manag.* **2018**, *77*, 238–251. [CrossRef] [PubMed]
60. Lundie, S.; Peters, G.M. Life cycle assessment of food waste management options. *J. Clean. Prod.* **2005**, *13*, 275–286. [CrossRef]
61. Walker, M.; Theaker, H.; Yaman, R.; Poggio, D.; Nimmo, W.; Bywater, A.; Blanch, G.; Pourkashanian, M. Assessment of micro-scale anaerobic digestion for management of urban organic waste: A case study in London, UK. *Waste Manag.* **2017**, *61*, 258–268. [CrossRef] [PubMed]
62. Latouche, S. *Le pari de la Décroissance*; Fayard: Paris, France, 2006.
63. European Commission. *Communication from the Commission to the European Parliament, the Council, the European Economic and Social Committee and the Committee of the Regions. Closing the Loop—An EU Action Plan for the Circular Economy*; COM (2015) 614 Final; European Commission: Brussels, Belgium, 2015.
64. European Commission. *Communication from the Commission to the European Parliament, the Council, the European Economic and Social Committee, and the Committee of the Regions. Innovating for Sustainable Growth: A Bioeconomy for Europe*; COM (2012) 60 Final; European Commission: Brussels, Belgium, 2012.

Article

Environmental Performance of Effluent Conditioning Systems for Reuse in Oil Refining Plants: A Case Study in Brazil

Hugo Sakamoto, Flávia M. Ronquim, Marcelo Martins Seckler and Luiz Kulay *

Chemical Engineering Department, Polytechnic School of the University of São Paulo, Avenida Professor Lineu Prestes, 580, Bloco 18—Conjunto das Químicas, São Paulo 05508-000 SP, Brazil; hugo.sakamoto@usp.br (H.S.); flavia.mronquim@yahoo.com.br or flaviamr@usp.br (F.M.R.); marcelo.seckler@usp.br (M.M.S.)
* Correspondence: luiz.kulay@usp.br; Tel.: +55-11-3091-2233

Received: 12 December 2018; Accepted: 14 January 2019; Published: 21 January 2019

Abstract: This study aims to evaluate the environmental and energy effects of the reuse of 1.0 m^3 of water in a cooling tower obtained from an oil refinery effluent. An arrangement comprising reverse osmosis (RO), evaporation (EV), and crystallization (CR) was created for water desalination. Six process routes were evaluated; for this purpose, each of them was converted into an specific scenario of analysis: S1: pre-treatment with Ethylenediaminetetraacetic acid (EDTA) + RO + EV (multi-effect distillation) + CR; S2: S1 with pre-treatment by $BaSO_4$; S3: with $Ca(OH)_2/CaCO_3/HCl$; S4: S3 with waste heat to supply the thermal demand of EV; S5: S3 with steam recompression in EV; and, S6: S3 with HNO_3 in place of HCl. The analysis was carried out by attributional LCA for primary energy demand (PED) and global warming (GW) impacts. The comparison was carried out for a reference flow (RF) of: add 1.0 m^3 of reused water to a cooling tower with quality to proper functioning of this equipment. S4 presented the best performance among the analyzed possibilities (PED: 11.9 MJ/RF; and GW: 720 $gCO_{2,eq}$/RF). However, dependence on other refinery sectors makes it inadvisable as a regular treatment option. Thus, S5 appears as the lowest impact scenario in the series (PED: 17.2 MJ/RF; and GW: 1.24 $kgCO_{2,eq}$/RF), given the pre-treatment technique of RO-fed effluent, and the exclusive use of steam recompression to meet total EV energy demands. Finally, an intrinsic correlation was identified between RO water recovery efficiency and the accumulated PED and GW impacts on the arrangements that operate with heat and electricity.

Keywords: water recycling; wastewater treatment; environmental and energy performance; life cycle assessment (LCA); crude oil refining

---

1. Introduction

The oil sector continues to be the main primary energy source, contributing with about 35% of global fuel consumption in 2017 [1]. Meeting the demand for oil products, mainly in the industry and fuel sectors, requires large amounts of water, mostly for thermal exchange. This scenario is mostly observed in the countries with the largest volume of oil manufactured goods production, such as Brazil, which occupies the 9th position in the world rank [1]. However, due to Brazil's local reality—of irregular water density and, therefore, scarcity of this natural resource in large urban centers at certain times of the year, several regulatory agencies have increased their restrictions on water collection and effluent disposal, in order to mitigate the depletion of the national water network. This scenario has stimulated recent research aimed at increasing the efficiency of effluent treatments and, consequently, decrease damage to the extraction and disposal sites of this resource.

Water reuse is an approach that has been gaining ground in this area, consisting in treating part, or even, when possible, the whole effluent and subsequently destining it for the supply of industrial or domestic needs. This practice can equate water demand problems. Nevertheless, water reuse can also lead to environmental impacts as energy resource consumption (fluid heating and pumping), from the synthesis of process inputs and waste generation due to the water reconditioning.

Current literature reports varied contributions on water reuse practices. When the investigations addressed the environmental domain, approaches were once again diversified. Chang et al. [2] evaluated energy use and greenhouse gas (GHG) emissions for urban water reuse systems in Korea, while Hendrickson et al. [3] evaluated the same parameters for an operating Living Machine (LM) wetland treatment system, which recycles wastewater in an office building. The study also assessed the performance of the local utility's centralized wastewater treatment plant, which was found to be significantly more efficient than the LM. The life-cycle approach was adopted for cases in which the specification of the environmental variable required a systemic scope [4–11]. Morera et al. [4] applied the Water Footprint to assess the consumption of water resources in wastewater treatment plants (WWTP). Their findings indicated that the choice for WWTP systems leads to a significant decrease in grey resources but, on the other hand, also generates a small blue water footprint in comparison to no-treatment scenarios. Cornejo et al. [5] carried out a comprehensive literature review of carbon footprint (CF) reports from water reuse and desalination systems. The study recognized general CF trends associated with the technologies and recommended improvements to this method based on limitations, challenges and knowledge gaps identified throughout the analysis.

The life cycle assessment (LCA) itself was applied to identify impacts resulting from control strategies in wastewater treatment plants [6], verify the environmental performance of large [7] and small scale [8] WWTP, and compare decentralized wastewater treatment alternatives for non-potable urban reuse [9]. LCA was also used to subsidize environmental and economic assessments performed to determine optimal water reuse in a residential complex [10] and develop an eco-efficiency analysis (EEA) framework for the evaluation of treatment systems proposed for greywater recycling in domestic buildings for non-potable uses [11].

Baresel et al. [12] quantified the environmental effects of using the energy grids from different countries (USA, Spain, and Sweden), as well as the option of applying sludge resulting from effluent treatments as fertilizer within water reuse systems for agriculture and industry. Their findings indicate that altering external treatment aspects can impact process performance, compared to seeking out improvements for a given technology.

O'Connor et al. [13] evaluated 14 different process arrangements for the treatment of a pulp and paper mill effluent by the commutative use of six operations: flotation, clarification, use of activated sludge, an upflow anaerobic sludge blanket (UASB) reactor, ultrafiltration, and reverse osmosis (RO). The study assessed climate change (CC), freshwater ecotoxicity (FEC), eutrophication (Eut) and water recovery impacts. The results demonstrate the significant influence of the solid waste originating from the treatment and the use of electricity in CC impacts, as well as the higher RO efficiency for the quality of the reuse water compared to the ultrafiltration technology.

On the other hand, no optimal scenario for all evaluated categories has been determined. Pintilie et al. [14] tested the feasibility of reuse practices in the industrial sector regarding effluents recovered by a treatment plant in Spain, concluding that electricity consumption displays the greatest weight concerning the environmental impacts and that water reuse can be an adequate alternative for non-potable uses, such as applications in the industrial sector.

To the best of our knowledge, however, no studies in the literature are available regarding LCA application in the evaluation of technologies applied for water reuse in the closed looping process itself. This study contributes to the theme, verifying the environmental performance of different scenarios conceived to conditioning the treated effluent from an oil refinery located in Brazil, so that it can be reused in processes within the facility itself. To pursue this aim, attributional LCA was applied according to a 'cradle-to-gate' approach.

Apart from identifying which reuse treatment bottlenecks should be analyzed in further detail in order to reduce environmental impacts, such an analysis can subsidize information to water resources managements in situations in which water collection is prohibited by legal regulations.

Environmental performance has the potential to become a management criterion under reuse practices. In addition to its close correlation with economic aspects, the adoption of this approach can support optimization actions in procedural arrangements, and for technology selection.

2. Materials and Methods

The method applied in this study encompasses six steps: (i) specification of the effluent quality upstream from the industrial wastewater treatment plants (IWTP) in the refinery, and the choice of a process use for which recovered water is intended for; (ii) definition of water recovery strategies and setting the analysis scenarios; (iii) description of the recovery systems in terms of their technological approach and operational conditions, as well as resource consumption and emissions; (iv) designing of mathematical models to represent each system, from the data and information obtained in the previous step; (v) application of the LCA technique to establish an environmental diagnosis for each scenario concerning primary energy demand and global warming; and (vi) perform a critical review of the obtained results.

2.1. Effluent Specification and Destination of Recovered Water

The refinery effluent is submitted to a conventional treatment at the IWTP. The primary step consisting of an API oil–water separator and a floating filter removes suspended solids, oils, and greases. The API separator is an equipment used to separate gross amounts of oil and/or suspended solids from the effluents of oil refineries. Its design is based on the specific gravity difference between the oil and the wastewater, which is smaller than that one between suspended solids and water. Thus, the suspended solids will settle to the bottom of the separator, the oil will rise to top, and the wastewater will occupy the middle layer, being recovered separately from the other components [15]. Dissolved solids that assign organic load to the effluent are treated during the secondary step, comprising aeration ponds (complete and facultative) and biodisks.

Finally, in the tertiary step, contaminants that result in color and conductivity, as well as nutrients, metals, non-biodegradable compounds, and volatile suspended solids, are removed from the liquid stream by passing through a clarifier and an activated carbon filter. Table 1 shows the characteristics of a typical effluent from an IWTP. The IWTP output effluent characteristics can vary significantly depending on the type of oil processed, the refinery configuration, and the operating procedures used in the treatment [16]. On the other hand, such information is not easily accessible. Because of this, Table 1 shows concentration ranges of contaminants present in that stream. The limits of each interval were established from data of Gripp [17], Moreira [18], and Pantoja [19] for refineries of the same technological concept but at different periods.

The effluent presents salt characteristics due to the earlier IWTP treatment, where suspended solid, greases, oils and organic loads have been removed. As the scope of this study begins precisely at this point, the environmental IWTP performance was disregarded.

The process adopted to determine the purpose of the reclaimed water took into account two criteria: (i) volume consumed by productive sector or activity; and (ii) water quality restrictions for each type of use. The worldwide reference for specific water consumption in refining processes ranges from 0.7–1.2 m^3 water/ m^3 processed crude oil [20,21]. Regarding Brazilian plants, this demand is distributed as follows: 46% replace losses occurring in cooling towers; 26% serve the steam production in boilers; 9.0% are allocated to fire-fighting systems; and the remaining (~19%) is absorbed in chemical preparation (dilution), cleaning, and for human consumption [22].

The boiler feedwater requires the most restrictive quality standards in terms of salt, organic matter, and dissolved gas concentrations. The cooling water makeup predisposes intermediate quality levels; in such cases limits are established in order to regulate (or prevent) scale development, corrosion, and slime and algae formation. In general, water used for firefighting does not require any treatment [23]. Considering these arguments, in addition to water recovery system performance regarding the quality of the final product and the effectiveness with which such thresholds can be met, the feed from one of the refinery's cooling towers was established for the final use of the desalinated water.

2.2. Setting the Analysis Scenarios and Description of the Recovery Systems

The effluent quality at the outlet of the tertiary IWTP step (Table 1) obligates a complementary removal of metals (Ca^{2+}, Ba^{2+}, and Na^+), chlorides ($Cl^{\mp}$), and carbonates (CO_3^{2-}) to reconditioning this flow as makeup water in the cooling tower. In this regard, six water recovery arrangements were defined. RO, evaporation (EV) and crystallization (CR) operations are common to all strategies. In contrast, these vary in how the effluent is pre-treated before entering the RO and due to the technology adopted in the EV process.

Table 1. Characteristic values of indicators in effluent. IWTP: industrial wastewater treatment plants.

Analytes	IWTP Effluent [17–19]	IWTP Effluent (Defined Value)
	(Ranges of Concentration, ppm)	(ppm)
Ba^{2+}	0.20–0.50	0.30
Ca^{2+}	38.0–63.0	53.1
Al^{3+}	0.00–0.01	<0.01
Sr^{2+}	1.10–1.83	1.32
SiO_2	0.80–19.4	7.33
Fe (total)	0.00–0.01	<0.01
Mg^{2+}	4.00–7.52	6.56
Na^+	179–283	255
$Cl^{\mp}$	311–425	385
HCO_3^-	55.9–308	216
K^+	6.00–11.5	7.76
NH_4^+	0.50–7.60	4.56
PO_4^{3-}	0.00–3.74	2.12
NO_3^-	22.4–207	101
SO_4^{2-}	85.6–163	110
F^-	0.10–0.69	0.20
TDS (as, NaCl) [1]	359–1103	762
pH	6.74–8.81	6.74

[1] TDS: total dissolved solids.

Based on precipitation principles, pre-treatment aims to remove ions that may reduce, or even compromise, RO performance. The definition of the chemical agents applied in this stage took into account their removal potential, through mechanisms that affect ion solubility in aqueous solutions. Regarding the EV technology, two alternatives were investigated: (i) multi-effect distillation, and (ii) steam recompression. Each recovery strategy resulted in an analysis scenario, coded as S1 to S6. Table 2 presents some specificities for each scenario regarding technological conditions and the origin of the energy supply.

S1 was defined as the baseline scenario and, therefore, no pretreatment technique was charged for this situation. Due to the high ion concentrations in the effluent, a risk for encrustations in the osmosis membrane is noted, making it necessary to add anti-fouling reagents to the liquid effluent at the entrance of the RO module. In S1, this situation was addressed by dosing EDTA, often applied in such cases. However, EDTA excesses can cause side effects, i.e., biofouling, harmful in extreme circuit closure situations, such as the Zero Liquid Discharge regime [24].

In S2, pre-treatment occurred by desupersaturation through the addition of $BaSO_4$ seeds (BaDs). This strategy aims to remove barium sulphate itself through its accumulation in the presence of the crystals formed by this salt [25]. $BaSO_4$ is commonly found at supersaturation concentrations in industrial effluents (particularly in oil refineries), and is one of the main fouling agents in RO plants [26].

Table 2. Technical characteristics and energy sources for each scenario.

Parameter	Technological Conditions	S1	S2	S3	S4	S5	S6
Technological approach	Reverse osmosis (RO)	+	+	+	−	+	+
	Evaporative crystallization (EV + CR)	+	+	+	+	+	+
Pre-treatment method	Barium desupersaturation (BaDs)	−	+	−	−	−	−
	Coprecipitation (CPT)	−	−	+	−	+	+
Energy source	Electricity	BR grid [1]	BR grid	BR grid	BR grid + WH	BR grid	BR grid
	Heat	NG [2]	NG	NG	WH [3]	−	NG
Yield (η_{Si})	Water recovery at RO (%)	84.6	87.0	95.8	−	95.8	95.8

[1] BR grid: Brazilian electricity matrix; [2] NG: natural gas; [3] WH: waste heat.

The coprecipitation method (CPT) by alkalization to pH = 11 applying $Ca(OH)_2$ and $CaCO_3$ seed addition [24] was adopted in S3 for pre-treatment. This allows for the precipitation of not only a large part of the calcium carbonate in solution, but SiO_2 as well, which is also a limiting component in RO, due to its decreasing solubility with increasing acidity levels (pH). CPT may also lead to the removal of metal ions—Ba^{2+}, Sr^{2+}, Fe^{2+}, Mn^{2+}, and Cd^{2+}—by incorporation into the precipitating $CaCO_3$ via isomorphic adsorption, absorption or substitution [27]. The supernatant arising from the CPT stage should be acidified to pH = 8.0 to prevent scales mainly in the RO system. In S3 and S5, this occurs via the addition of HCl (1.0 M) and in S6, with HNO_3 (1.0 M).

S4 assessed the situation in which the thermal energy consumed by EV and CR results from waste heat. In this case, part of the heat released to the atmosphere from the boiler chimneys, or even by the cooling tower, is reused by the desalination unit. This option is economically attractive as it reduces heat generation costs and waives the use of RO—and, therefore, of also any pre-treatment method—by the system [28].

S5 examined a substitutive arrangement of S3, in which the energy demand of EV is supplied exclusively by electricity. This scenario assumes a technological change in the evaporation stage, from which the multi-effect distillation, adopted regularly by scenarios S1–S4 and S6, is replaced by steam recompression [29]. This technology applies steam produced in the evaporator itself in order to lead the evaporation phenomenon at that equipment. For this purpose, a compressor driven by an electric engine is used to raise the functions of state (temperature and pressure) of the steam. In addition to evaporating some of the water existing in the saline solution, the vapor stream is also used to increase the temperature of effluent fed in the evaporator [30].

Since electrolysis, a technology conventionally adopted to obtain HCl, requires high electricity consumption [31], a variant of S3—coded S6—was also investigated, in which the acidification of the supernatant leaving CPT was carried out by the addition of HNO_3, as described previously. The use of nitric acid becomes viable because the ion NO_3^- is, with a few exceptions, quite soluble in water. Therefore, its presence in the effluent does not revert to the risk of fouling in the subsequent treatment stages.

Finally, it should be noted that effluent pretreatment by BaDs increases the water recovery rate by 2.8% at the RO stage (η_{S2}) in comparison to that observed in the baseline scenario. In cases where the action occurs via CPT (S3, S5 and S6) the efficiency gain is even more expressive, surpassing η_{S1} in about 13%.

2.3. Mathematical Model Design

The Hydranautics® 'IMS Design' computational tool was used to model the RO unit. The design of this stage considered an average limit flow of $\overline{Q}$ = 20 Lmh, maximum limiting polarization concentration (βmax = 1.18), and a standard high saline rejection membrane CPA7 MAX. It was also assumed that RO pumps achieved average yields $\overline{\eta}$ = 87%. Except for S5, all the assessed scenarios adopted the multi-effect distillation technology for the EV stage. In such cases, a heat consumption of 230 MJ/m^3 of effluent to be treated was required to place the steam under process conditions, and 9.00 MJ/m^3 of electricity were employed in pumping operations [31]. In S5, evaporation occurs by steam recompression, therefore expending only electricity. A unit that operates with this technology achieves a consumption of 115 MJ/m^3 [28]. Regarding the CR stage, the effluent flow fed to the desalination system (250 m^3/h) also justifies the use of steam recompression [32]. The energy demand related to the crystallization technology is of 238 MJ/m^3 [29]; this performance was evenly considered for modeling all scenarios.

The waste heat that meets the thermal demand of S4 was modeled as an elementary flow. In methodological terms, this decision represents the elimination of all the environmental burdens associated with this flow (and the impacts arising from it) generated by anthropic interventions prior to its use by the desalination system. This is supported by the hypothesis that heat, one of the crude oil refining residues, is revalued in S4, when it becomes a process fluid. For EV, the energy decrease assumed in other scenarios (9.00 MJ/m^3) was also adopted for S4. In contrast, the modeling of CR stage admitted that fluid pumping would consume about 576 MJ/t of obtained salt. This value was estimated by Jongema [33] from a thermodynamic approach based on an exergy analysis.

The precipitation estimations were performed using an OLI Systems Inc. simulator®, which provides the approximate compositions of the formed salts. For S1, this calculation included an additional rate of 4.0 ppm EDTA. In S2, it was assumed that the exchange of BaSO$_4$ seed is performed once a year (C ~ 10 g/L). Moreover, a consumption of 79.2 kJ/m^3 for agitation in the precipitation stage and a monthly cleaning of the equipment with HCl solution (pH = 3.0) were also considered.

S3 includes an electricity consumption of 317 kJ/m^3 from shaking in the coprecipitator, the insertion of 12.5 g/L CaCO$_3$ seeds (annual renewal) and the acidification of the supernatant obtained from the precipitation stage (addition of HCl, 1.0 M) to achieve pH = 8.0, in order to prevent alkaline incrustations.

The transports considered in the models were the piping that leads the recovered water from the desalination plant to the cooling tower (L ~ 200 m) and the salt from the precipitation and crystallization stages to a landfill located 20 km from the refinery. Finally, the Brazilian energy matrix (BR grid) was the exclusive source of electricity supply from the desalination system in all analyzed scenarios.

2.4. Life-Cycle Modeling

2.4.1. Scope Definition

The environmental diagnoses were performed by attributional LCA, under a 'cradle-to-gate' approach and in line with the guidelines provided by ISO 14044 standard [34]. A reference flow (RF) of 'add 1.0 m^3 of reused water to a cooling tower within quality requirements that allow the proper functioning of this equipment' was defined to carry out the analyses.

Figure 1 displays an overview of the set of elements, highlighting all the interconnections among the refining process, the IWTP, and the water recovery system. The product system that represents scenarios S1–S4 and S6 is contained in the dark gray rectangle. Figure 2 details the same arrangement for S5. Modeling of the Product System was based on primary and secondary data, that express local conditions. No multifunctionality was found.

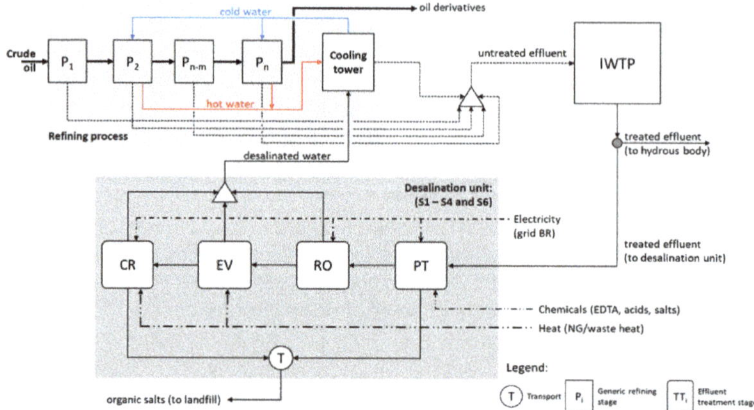

Figure 1. General view of effluent recovery systems (S1–S4 and S6) for reuse in a cooling tower.

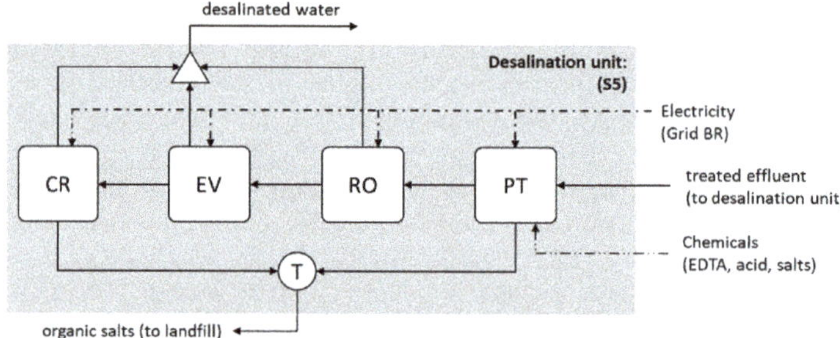

Figure 2. Detail of the desalination system for scenario S5.

The life cycle impact assessment was carried out in two levels. First, the energy consumption of the process in terms of primary energy demand (PED) was quantified by the cumulative energy demand (CED) method—v1.10 [35], which addresses the contributions of the different energy sources, both renewable (biomass: RB; wind: RWD; water: RWA) and non-renewable (fossil: NRF; nuclear: NRN; biomass: NRB). In the second level, the environmental effect of global warming (GW) was calculated by the ReCiPe 2016 Midpoint (H) method—v1.13 [36].

2.4.2. Life Cycle Inventory (LCI)

The effluent concentrations at the input of the process train (Table 1) are primary data obtained from a refinery. In contrast, electric energy and natural gas (NG) supplies were modeled from secondary data. The electricity generation (BR grid) was edited for Brazilian 2015 conditions [37]. Hydropower still remains the most expressive source of energy supply (64% of the BR grid) and biomass (8.0%) was expressed by sugarcane bagasse. For coal (4.0%), a mixed supply model (national and Colombian) was created considering mining operation and the distances of the Brazilian mines to the main thermoelectric plants [38]. For the natural gas from the BR grid (13%) and for heat supply, a model obtained from the Ecoinvent database was tailored to Brazilian conditions considering local offshore extraction, activities involved in refining raw natural gas and transport of final product.

In Brazil, inland transport occurs, mainly, by road. The diesel consumed in these actions was customized from the LCI 'Crude oil, in refinery/US' available at the USLCI database, [39] once again considering the procedural and technological requirements practiced in the country. The inputs and auxiliary materials, such as EDTA, barite and calcium hydroxide were edited from the Ecoinvent datasets entitled 'EDTA, ethylenediaminetetraacetic acid, at plant/RER U' [40], 'barite, at plant/RER' [41] and 'lime, hydrate, loose weight at plant/RER' [42] to more closely reproduce the national reality.

The same approach was applied to HCl production (adjusted from the Ecoinvent: 'Hydrochloric acid, 30% in H_2O, at plant/RER U') [31], HNO_3 (whose LCI 'nitric acid, 50% in H_2O, at plant/RER U' existing in Ecoinvent was adapted) [31], and water (converted from Ecoinvent: 'Tap water, at user/U') [31] used for dilution of concentrated acids to 1.0 M. As in the case of diesel, the original energy inputs—electricity, NG, the diesel itself and other petroleum derivatives—were replaced by inventories created specifically for this study.

3. Results and Discussion

Table 3 describes the environmental PED and GW impacts of the systems included in the S1–S6 scenarios, generated for the supply of 1.0 m^3 of recovered water for a cooling tower at the refinery. S4 displayed the lowest impacts of the entire series for both categories, mainly due to the use of waste heat as a thermal energy source for the system. On the other hand, S1 (baseline scenario) displayed the worst performance indices in the same analyzed dimensions, due to the absence of effluent pretreatment admitted for RO. This condition overloads the osmosis system, which then presents reduced water recovery rates (Table 1), and causes all the elements of the arrangement to require greater heat and electricity amounts in order to generate water within the quality standards required by the cooling tower. The adoption of pre-treatment methods improved the PED performance of all scenarios. The use of BaDs (S2) led to a 13% reduction in impact with respect to S1, whereas the gains were even more expressive with CPT, reaching 55%, 66%, and 54% attenuation, respectively for S3, S5 and S6.

Table 3. Impacts related to primary energy demand (PED) (by subcategory and total impact) and global warming (GW) for scenarios S1–S6.

	Impact Category	S1	S2	S3	S4	S5	S6
	Non-renewable, fossil	44.4	37.6	15.7	3.49	5.97	16.8
	Non-renewable, nuclear	0.51	0.51	0.57	0.57	0.88	0.57
PED	Renewable, biomass	0.78	0.78	0.78	0.99	1.32	0.69
(MJ/RF)	Renewable, wind	0.32	0.32	0.32	0.42	0.50	0.32
	Renewable, water	4.79	4.79	5.33	6.43	8.53	4.62
	Total	50.8	44.0	22.7	11.9	17.2	23.0
	GW ($kgCO_{2,eq}$)	2.93	2.56	1.52	0.72	1.24	1.93

The technological change applied to EV also brought benefits to the category, since the accumulated PED for S5 was 24% lower than for S3. On the other hand, the use of HNO_3 (S6) instead of HCl (S3) did not generate significant energy behavior variations.

This exchange in positions is justified because S1–S3 and S6 present finite demands for thermal energy in EV (Table 4), which are supplied by natural gas, while S4 uses more heat during the EV process than the other scenarios (230 MJ/RF). However, due to waste heat, this utility does not contribute to PED impacts. S5 does not make use of thermal flows during this stage of the process. On the other hand, S4 and S5 present the highest electric consumption rates among the evaluated scenarios (13.1 and 10.4 MJ/RF). The BR grid, predominantly hydroelectric energy, meets these demands [37], with RW as the impact agent.

Table 4. Total energy consumption per stage of the system for each scenario (in MJ/RF). RO: reverse osmosis; EV: evaporation; CR: crystallization.

Stage	Energy Source	S1	S2	S3	S4	S5	S6
RO	Electricity	1.12	1.33	2.12	-	2.12	2.12
EV	Heat	35.9	30.3	9.74	-	-	9.76
	Electricity	1.40	1.19	0.40	9.00	4.86	0.40
	Waste Heat	-	-	-	230	-	-
CR	Electricity	4.07	4.07	3.38	0.11	3.38	3.53
	Waste Heat	-	-	-	3.96	-	-

As depicted in Table 4, natural gas consumption for heat generation in S1/EV constitutes the main NRF contribution source, and thus, PED for this scenario. The effluent pretreatment raises electrical S2/RO consumption in 19% in relation to S1. In contrast, the use of BaDs makes $\eta_{S2} > \eta_{S1}$ (Table 2), leading to a 16% decrease in thermal (heating) and electric (pumping) EV demands, relative to evaporation in S1. The combination of these effects is favorable for S2, since, as it uses lower amounts of natural gas, displays NRF contributions about 15% lower than those observed for S1. Since the contributions to the other subcategories are equivalent for both scenarios (Table 3), PED_{S2} overcame PED_{S1} in a little over 13%.

S3 follows similar trends as those observed for S2. In this case, however, the fact that η_{S3} is higher than η_{S1} due to CPT, leads to higher electric consumption in RO compared to that expended by S1, while also reducing EV energy demands even more intensely. Thus, natural gas consumption was dampened to sufficient levels so that the total NRF_{S3} was lower than NRF_{S1} in 64%, which immediately resulted in a 55% decrease in PED_{S3} compared to the baseline scenario. However, a closer examination of S2 and S3 RW results or indicates an 11% increase in this source of impact due to the replacement of BaDs by CPT. This disparity can be explained by the environmental loads associated with the HCl manufacture.

Electrolysis is an energy intensive technology and the most widely used methods for synthesis of Cl_2. The manufacture of Cl_2 from sea water consumes from 414 to 500 kJ/kg Cl_2 of electricity [31]. Therefore, the manufacture cycle of HCl displays a contribution of 757 kJ/RF, corresponding to 14% of the entire RW value for S3.

The waste heat option reduced PED_{S4} concerning impacts generated on the account of the electrical demand of the system by the BR grid. Therefore, it was already expected that RW impacts would represent the largest share (54%) of the total. Regarding the process, the individual electricity expenditure to pump the effluent to EV (6.23 MJ/RF) is highlighted.

The change in evaporation technology implemented in S5—from multiple-effect to steam recompression—dismissed the use of natural gas at this stage of the system, but increased electricity consumption. Although 22% of the BR grid is composed of non-renewable sources [37], the effect of the discontinuation of the use of natural gas in EV on NRF_{S5} performance prevailed over the inputs that the subcategory received due to the increased electricity use. Moreover, steam recompression caused RW to become the dominant PED_{S5} precursor, with 49% contribution.

Substitution of HCl by HNO_3 introduced in S6 had a positive effect. By inserting electrolysis in the HCl production process, the system's accumulated electricity consumption was reduced $RW_{S6} < RW_{S3}$. On the other hand, HNO_3 synthesis has as its essential raw material ammonia obtained through natural gas steam reforming (0.63 m^3 NG/kg NH_3). In addition, as both the NH_3 and HNO_3 processes are endothermic, another portion of natural gas will be added to the system to meet those needs. The reconciliation between these elements explains the fact that NRF_{S6} supplanted NRF_{S3} at 1.08 MJ/RF, in spite of the decreased global electric consumption observed in S6.

Finally, a collective analysis of the findings described above indicates an intrinsic correlation between actions carried out for the process and its performance in terms of PED. With respect to scenarios S1–S3 and S6, which differ only in the applied pretreatment technology, the increased η_{Si}

(Table 2) is reverted to mitigation of the overall PED impacts (Table 3). EDTA (S1) substitution by BaDs (S2), and of BaDs (S2) by CPT (S3 and S6), led to an accentuated water recovery efficiency in RO. This trend is based on the (high) efficacy achieved by CPT in the effluent-dissolved salt removal compared to other possibilities.

The increased η_{Si} led to improve power consumption in RO (Table 4), due to the additional resistance that the solute formed by the pre-treatment imposes to the solvent flow (water in use by the cooling tower) through the membrane. However, the concentrate flows transferred to EV were reduced. This effect led to higher PED benefits than the deleterious effects caused by increases in the boiling temperature of the solutions, resulting in decreased heat consumption of the multi-effect distillation (Table 4). A similar phenomenon was noted for the concentrate streams routed to CR, which, due to the low flows, provided decreased electrical demands for this stage.

The heat production by natural gas burning resulted in more significant PED contributions than the electricity generation from the BR grid. Therefore, the gain in η_{Si}, provided from the greater effectiveness of the pretreatment is reverted to contribution retention for the category in all the arrangements that depend on such utilities to operate. S5's success can be explained by the same reasoning. Although it displays higher electrical consumption compared to the other scenarios (4.90 MJ/RF), the synergy established between S5 water recovery rate (η_{S5} = 98.5%), and the exemption of the use of heat in EV, compensates this disadvantage.

The situation described by S4 is quite specific. Due to waste heat, the desalination process is limited to the EV-CR set, and even so, this scenario achieved the best PED result among the other options. It should be noted, however, that, in addition to relying on other refinery sectors, the operation of this arrangement requires special care to serve the purposes for which it is intended. This condition makes it not recommended for regular use. S4 can, therefore, be characterized as the lowest PED impact level to be achieved by the system concerning its base technologies.

S4 also presented the lowest GW performance index among the assessed options, again due to the use of waste heat. Under these circumstances, electricity generation from the BR grid becomes the main source of impact for the category, accounting for 98% of the contributions for this scenario.

S3 prevails over the other scenarios that make use of natural gas for heat generation, followed by S6 and S2, with S1 being the most impactful alternative of the set. In addition, none of these options surpassed S5 performance, which applied electricity only during the steam recompression operation. This finding corroborates Pintilie et al. [14], who state that GW impacts are more sensitive to the intensity oscillation of thermal demands than to the electrical requirements of the treatment systems.

In scenarios with lower water recovery rates in RO (S1 and S2), CO_2 emissions of fossil origin ($CO_{2,f}$) derived from the combustion of natural gas for heat production in EV were the main source of impact. According to Cornejo et al. [5], this is not only due to the use of fossil sources for the thermal energy supply of the stage, but also due to the fact that EV is naturally more energy intensive than RO. In S1 and S2, the $CO_{2,f}$ losses to air represented 78% and 75% of their total impacts as GW. The disparity between the results for these scenarios is due to the thermal demand fluctuations caused by the use of BaDs, which decreased from 1.03 m^3/RF in S1 to 0.87 m^3/RF in S2.

The reduction of natural gas consumption caused by the CPT led to significant benefits for S3 concerning GW. However, even if dampened in comparison to S1 and S2 performances, $CO_{2,f}$ emissions originating from the heat generation still accounted for 41% of the GW_{S3}. Another relevant focus of this scenario's impact is on GHG emissions from electricity generation. Responsible for 38% of the GW_{S3}, these contributions originate (once again) from $CO_{2,f}$, summed to CH_4 losses, which occur in the life cycle of the natural gas that feeds thermoelectric plants. Impact precursors comprise the releases of dinitrogen oxide (N_2O) and CO_2 from land transformation ($CO_{2,LT}$) from bioelectricity generation, a source that represents 8.0% of the BR grid [37]. In addition to intervening in S3 to meet RO, EV, and CR energy requirements, the BR grid acts indirectly on the system by participating in the manufacturing stages of the osmosis membrane (147 kJ/m^2) and Ca(OH)$_2$ (22.9 kJ/kg), as well as, mainly, in the HCl production chain. As mentioned previously, this action focuses on the electrolysis of sodium chloride, which, due to its energy intensive character contributes to about 11% of the total impact of the category. Finally, the use of $Ca(OH)_2$ in CPT also brought a significant contribution to GW_{S3}, because of the regular technology applied to obtain quicklime (CaO)—thermal decomposition of limestone, a material that contains calcium carbonate ($CaCO_3$)—in a lime kiln. The calcination of $CaCO_3$ releases 909 gCO_2/kg CaO [42], corresponding to an impact of 146 gCO_{2eq}/RF.

The option of the use of steam compression for water evaporation was promising for GW as effects of increases in electric EV demands, also in this category, were compensated by the suppression of natural gas burning and, therefore, of the GHG emissions that this operation would entail. Thus, the impact observed by S5 was about 18% lower than that achieved by S3. In any case, the increased electricity consumption during the evaporation stage transformed the BR grid into the main source of GW_{S5}, accumulating 74% of the total impact.

Unlike observed for PED, the exchange of HCl for HNO_3 led to worse GW performance in S6 compared to S3. This is due to the N_2O emissions generated during acid synthesis (the Ostwald process), in which anhydrous NH_3 is oxidized to HNO_3 by metal catalysts and stringent temperature and pressure conditions. The dinitrogen oxide ends up being formed because a small portion of the NH_3 is partially oxidized [31]. The high impact factor of N_2O for GW (298 $kgCO_{2,eq}/kg$, [43]), led to these losses (8.40 g/kg HNO_3) representing 22% of the GW_{S6} impact.

Due to the similarities between the primary energy demand and global warming precursors, the correspondence diagnosed for PED between type of pretreatment, η_{Si}, and the cumulative performance of each system also remains valid concerning GW impacts.

4. Conclusions

This study evaluated the reuse of a saline effluent from an oil refinery to supply a cooling tower inside the chemical plant, in a closing loop movement. An arrangement consisting of reverse osmosis (RO), evaporation (EV), and crystallization (CR) was defined for water recovery. Six scenarios were assessed, mainly observing pre-treatment options (desupersaturation or coprecipitation), and different approaches for energy supplying during EV. These are: (i) the use of waste heat to replace the natural gas burning fumes to meet the thermal demand of the system, and (ii) the application of steam recompression in substitution of multi-effect distillation.

The estimation of the environmental effects of the treatments was carried out by attributional LCA, according to a scope from 'cradle-to-gate', for the primary energy demand (PED) and global warming (GW) impact categories.

The scenario that makes use of waste heat as a source of thermal energy for EV (S4) presented the lowest impacts indices among the analyzed possibilities (PED: 11.9 MJ/RF; and GW: 720 $gCO_{2,eq}/RF$). However, the vulnerability of this arrangement, because its operation is subordinate to the operation of other refinery sectors, makes the alternative not recommended. Thus, S5, that applies coprecipitation as a pretreatment technique for RO-fed effluent and adopts steam recompression to meet EV energy demands appears as the lowest impact scenario of the series (PED: 17.2 MJ/RF; and GW: 1.24 $kgCO_{2,eq}/RF$). This occurs because the BR grid provides lower contributions in terms of PED and GW than the natural gas life cycle for heat-generating purposes.

A comprehensive analysis of the research findings identified an intrinsic correlation between water recovery rates in RO and overall PED and GW impacts in the arrangements that use heat and electricity for their operation.

Despite the performance variability observed among the scenarios, the benefits identified by the analysis indicate that water desalination is an environmentally efficient alternative in the reduction of water consumption and effluent discharge. These results can still be improved by the application of less aggressive compounds in the environment to remove salts during the pre-treatment phase and the reuse of residual energy sources.

Author Contributions: F.M.R. designed the evaporator and crystallizer models; M.M.S. provided technical information for crystallizer modeling; H.S. designed the RO simulations and the LCA assessment; L.K. and H.S. analyzed results, carried out the discussion, wrote the paper and reviewed the text.

Funding: This study was financed in part by the Coordenação de Aperfeiçoamento de Pessoal de Nível Superior—Brazil (CAPES) Finance Code 001.

Conflicts of Interest: The authors declare no conflict of interest.

References

1. International Energy Agency (IEA). *Key World Energy Statistics 2018*; IEA: Paris, France, 2018; 51p.
2. Chang, J.; Lee, W.; Yoon, S. Energy consumptions and associated greenhouse gas emissions in operation phases of urban water reuse systems in Korea. *J. Clean. Prod.* **2017**, *141*, 728–736. [CrossRef]
3. Hendrickson, T.P.; Nguyen, M.T.; Sukardi, M.; Miot, A.; Horvath, A.; Nelson, K.L. Life-Cycle Energy Use and Greenhouse Gas Emissions of a Building-Scale Wastewater Treatment and Nonportable Reuse System. *Environ. Sci. Technol.* **2015**, *49*, 10303–10311. [CrossRef] [PubMed]
4. Morera, S.; Corominas, L.; Poch, M.; Aldaya, M.M.; Comas, J. Water footprint assessment in wastewater treatment plants. *J. Clean. Prod.* **2016**, *112*, 4741–4748. [CrossRef]
5. Cornejo, P.K.; Santana, M.V.E.; Hokanson, D.R.; Mihelcic, J.R.; Zhang, Q. Carbon footprint of water reuse and desalination: A review of greenhouse gas emissions and estimation tools. *J. Water Reuse Desal.* **2014**, *4*, 238–252. [CrossRef]
6. Meneses, M.; Concepción, H.; Vrecko, D.; Vilanova, R. Life Cycle Assessment as an environmental evaluation tool for control strategies in wastewater treatment plants. *J. Clean. Prod.* **2015**, *107*, 653–661. [CrossRef]
7. McNamara, G.; Horrigan, M.; Phelan, T.; Fitzsimons, L.; Delaure, Y.; Corcoran, B.; Doherty, E.; Clifford, E. Life Cycle Assessment of Wastewater Treatment Plants in Ireland. *J. Sustain. Dev. Energy Water Environ. Syst.* **2016**, *4*, 216–233. [CrossRef]
8. Garfí, M.; Flores, L.; Ferrer, I. Life Cycle Assessment of wastewater treatment systems for small communities: Activated sludge, constructed wetlands and high rate algal ponds. *J. Clean. Prod.* **2017**, *161*, 211–219. [CrossRef]
9. Opher, T.; Friedler, E. Comparative LCA of decentralized wastewater treatment alternatives for non-potable urban reuse. *J. Environ. Manag.* **2016**, *182*, 464–476. [CrossRef]
10. García-Montoya, M.; Sengupta, D.; Nápoles-Rivera, F.; Ponce-Ortega, J.M.; El-Halwagi, M.M. Environmental and economic analysis for the optimal reuse of water in a residential complex. *J. Clean. Prod.* **2016**, *130*, 82–91. [CrossRef]
11. Lam, C.-M.; Leng, L.; Chen, P.-C.; Lee, P.-H.; Hsu, S.-C. Eco-efficiency analysis of non-potable water systems in domestic buildings. *Appl. Energy* **2017**, *202*, 293–307. [CrossRef]
12. Baresel, C.; Dalgren, L.; Almemark, M.; Lazic, A. Environmental performance of wastewater reuse systems: Impact of system boundaries and external conditions. *Water Sci. Technol.* **2016**, *73*, 1387–1394. [CrossRef] [PubMed]
13. O'Connor, M.; Garnier, G.; Batchelor, W. Life cycle assessment comparison of industrial effluent management strategies. *J. Clean. Prod.* **2014**, *79*, 168–181. [CrossRef]
14. Pintilie, L.; Torres, C.M.; Teodosiu, C.; Castells, F. Urban wastewater reclamation for industrial reuse: An LCA case study. *J. Clean. Prod.* **2016**, *139*, 1–14. [CrossRef]
15. American Petroleum Institute (API). *Design and Operation of Oil-Water Separators*; Monographs on Refinery Environmental Control—Management of Water Discharges: Washington, DC, USA, 1990; 54p.
16. Diya'uddeen, B.H.; Daud, W.M.A.W.; Aziz, A.R.A. Treatment technologies for petroleum refinery effluents: A review. *Process Saf. Environ. Prot.* **2011**, *89*, 95–105. [CrossRef]
17. Gripp, V.S. Environmental, Energetic and Economic Analysis of a Process Design for Water Reuse in Petroleum Refinery. Master's Thesis, University of São Paulo, São Paulo, Brazil, 2013.
18. Moreira, R.H. Desenvolvimento de um Processo de Reúso do Efluente de Refinaria Baseado em um Sistema de Osmose Reversa Combinado com Precipitação. Master's Thesis, University of São Paulo, São Paulo, Brazil, 2017.
19. Pantoja, C.E. Cristalização Assistida por Destilação por Membranas Aplicada ao Reúso de água: Comparação com Outros Métodos de Reuso, Análise do Processo e Projeto Hierárquico de Processo. Ph.D. Thesis, University of São Paulo, São Paulo, Brazil, 2015.

20. Diepolder, P. Is zero discharge realistic? *Hydrocarb. Process.* **1992**, *71*, 129–160.

21. Anze, M.; Alves, R.M.B.; Nascimento, C.A.O. Optimization of Water Use in Oil Refinery. In Proceedings of the 2010 Annual AIChE Meeting, Salt Lake City, UT, USA, 7–12 November 2010.

22. Pombo, F.R. Management of Water Demand in the Oil Refining Industry: Challenges and Opportunities for Rationalization. Ph.D. Thesis, Federal University of Rio de Janeiro, Rio de Janeiro, Brazil, 2011.

23. International Petroleum Industry Environmental Conservation Association (IPIECA). *Petroleum Refining Water/Wastewater Use and Management*; IPIECA Operations Best Practice Series: London, UK, 2010; 60p.

24. Rahardianto, A.; Gao, J.; Gabelich, C.J.; Williams, M.D.; Cohen, Y. High recovery membrane desalting of low-salinity brackish water: Integration of accelerated precipitation softening with membrane RO. *J. Memb. Sci.* **2007**, *289*, 123–137. [CrossRef]

25. Bremere, I.; Kennedy, M.D.; Johnson, A.; van Emmerick, R.; Witkamp, G.-J.; Schippers, J.C. Increasing conversion in membrane filtration systems using a desupersaturation unit to prevent scaling. *Desalination* **1998**, *119*, 199–204. [CrossRef]

26. Bremere, I.; Kennedy, M.; Michel, P.; van Emmerick, R.; Witkamp, G.J.; Schippers, J. Controlling scaling in membrane filtration systems using a desupersaturation unit. *Desalination* **1999**, *124*, 51–62. [CrossRef]

27. Astilleros, J.M.; Pina, C.M.; Fernández-Díaz, L.; Prieto, M.; Putnis, A. Nanoscale phenomena during the growth of solid solutions on calcite {101$^-$4} surfaces. *Chem. Geol.* **2006**, *225*, 322–335. [CrossRef]

28. Heins, W.; Schooley, K. Achieving Zero Liquid Discharge in SAGD heavy oil recovery. *J. Can. Petrol. Technol.* **2004**, *43*, 37–42. [CrossRef]

29. Mickley, M. *Survey of High-Recovery and Zero Liquid Discharge Technologies for Water Utilities*; WateReuse Foundation: Alexandria, VA, USA, 2008; 180p.

30. Al-Karaghouli, A.; Kazmerski, L. Energy consumption and water production cost of conventional and renewable-energy-powered desalination processes. *Renew. Sustain. Energy Rev.* **2013**, *24*, 343–356. [CrossRef]

31. Althaus, H.-J.; Chudacoff, M.; Hischier, R.; Jungbluth, N.; Osses, M.; Primas, A. *Life Cycle Inventories of Chemicals*; Ecoinvent Report No. 8, v2.0; EMPA Dübendorf, Swiss Centre for Life Cycle Inventories: Dübendorf, Switzerland, 2007; 957p.

32. Veza, J.M. Mechanical vapour compression desalination plants—A case study. *Desalination* **1995**, *101*, 1–10. [CrossRef]

33. Jongema, P. Optimization of the Fuel Consumption of an Evaporation Salt Plat with the Aid of the Exergy Concept. In Proceedings of the Sixth International Symposium on Salt, Toronto, ON, Canada, 24–28 May 1983; Salt Institute: Alexandria, VA, USA, 1983; pp. 463–469.

34. International Organization for Standardization. *ISO 14044, Environmental Management—Life Cycle Assessment—Requirements and Guidelines*, 1st ed.; International Organization for Standardization: Genève, Switzerland, 2006; 46p.

35. Hischier, R.; Weidema, B.; Althaus, H.-J.; Bauer, C.; Doka, G.; Dones, R.; Frischknecht, R.; Hellweg, S.; Humbert, S.; Jungbluth, N.; et al. *Implementation of Life Cycle Assessment Methods*; Ecoinvent report No. 3, v. 2.2; Swiss Centre for Life Cycle Inventories: Dübendorf, Switzerland, 2010; 176p.

36. Huijbregts, M.A.J.; Steinmann, Z.J.N.; Elshout, P.M.F.; Stam, G.; Verones, F.; Vieira, M.D.M.; Hollander, A.; Zijp, M.; van Zelm, R. *ReCiPe 2016: A Harmonized Life Cycle Impact Assessment Method at Midpoint and Endpoint Level Report I: Characterization*; Department of Environmental Science, Radbound University: Nijmegen, The Netherlands, 2016; 194p.

37. Empresa de Pesquisa Energética. *Brazilian Energy Balance 2016—Year 2015*; Empresa de Pesquisa Energética: Rio de Janeiro, Brazil, 2016; 294p. (In Portuguese)

38. Empresa de Pesquisa Energética. *Energia Termelétrica: Gás Natural, Biomassa, Carvão, Nuclear*; Empresa de Pesquisa Energética: Rio de Janeiro, Brazil, 2016; 417p. (In Portuguese)

39. Franklin Associates. *Data Details for Petroleum Refining*; Franklin Associates; Eastern Research Group, Inc.: Prairie Village, Kansas, 2003.

40. Hischier, R. *Life Cycle Inventories of Packaging and Graphical Papers*; Ecoinvent report No. 11, v2.0; Swiss Centre for Life Cycle Inventories: Dübendorf, Switzerland, 2007; 17p.

41. Jungbluth, N. Erdöl. In *Sachbilanzen von Energiesystemen: Grundlagen für den ergleich von Energiesystemen und den Einbezug von Energiesystemen in Ökobilanzen für die Schweiz*; Ecoinvent Report No. 6-IV, v 2.0; Dones, R., Ed.; Swiss Centre for Life Cycle Inventories: Dübendorf, Switzerland, 2007; 327p. (In German)

42. Kellenberger, D.; Althaus, H.-J.; Jungbluth, N.; Künniger, T.; Lehmann, M.; Thalmann, P. *Life Cycle Inventories of Building Products*; Final Report Ecoinvent No. 7, v2.0; EMPA Dübendorf, Swiss Centre for Life Cycle Inventories: Dübendorf, Switzerland, 2007; 914p.
43. Stocker, T.F.; Qin, D.; Plattner, G.-K.; Tignor, M.; Allen, S.K.; Boschung, J.; Nauels, A.; Xia, Y.; Bex, V.; Midgley, P.M. (Eds.) *Intergovernmental Panel on Climate Change. Climate Change 2013: The Physical Science Basis. Contribution of Working Group I to the Fifth Assessment Report of the Intergovernmental Panel on Climate Change*; Cambridge University Press: Cambridge, UK; New York, NY, USA, 2013; 1535p.

Article

High-Efficiency Cogeneration Systems: The Case of the Paper Industry in Italy

Marco Gambini, Michela Vellini, Tommaso Stilo *, Michele Manno and Sara Bellocchi

Department of Industrial Engineering, University of Rome Tor Vergata, 00133 Rome, Italy;
gambini@ing.uniroma2.it (M.G.); vellini@ing.uniroma2.it (M.V.); michele.manno@uniroma2.it (M.M.);
sara.bellocchi@uniroma2.it (S.B.)
* Correspondence: tommaso.stilo@alumni.uniroma2.eu; Tel.: +39-06-72597200

Received: 20 December 2018; Accepted: 18 January 2019; Published: 22 January 2019

Abstract: In January 2011, the introduction of high-efficiency cogeneration in Europe radically modified the incentive scheme for combined heat and power (CHP) plants. Since then, the techno-economic feasibility of new cogeneration plants in different areas of application (industry, service, residential, etc.), along with the definition of their optimal operation, have inevitably undergone a radical change. In particular, with reference to the Italian case and according to the most recent ministerial guidelines following the new EU regulation, in the event that cogeneration power plants do not reach an established value in terms of overall efficiency, their operation has to be split into a CHP and a non-CHP portion with incentives proportional to the energy quantities pertaining to the CHP portion only. In the framework of high-efficiency cogeneration, the present study compares different CHP solutions to be coupled with the paper industry that, among all the industrial processes, appears to be the best suited for cogeneration applications. With reference to this particular industrial reality, energy, environmental, and economic performance parameters have been defined, analysed, and compared with the help of GateCycle software. Among the proposed CHP alternatives, results show that gas turbines are the most appropriate technology for paper industry processes.

Keywords: high-efficiency cogeneration; primary energy saving; electricity from cogeneration; paper industry; GateCycle; CO_2 emissions avoided

---

1. Introduction

The term "cogeneration" refers to the simultaneous production of process heat and electricity from a single energy input. All types of power plants can be employed (e.g., steam power plants, gas turbines, combined cycle power plants, internal combustion engines); the extracted heat largely varies in entity and temperature depending on technology, size, and operation parameters.

Such a system has valuable properties to enhance the efficiency of fuel use: the combined production of electricity and heat turns out to be more efficient than the separate production of these two forms of energy. These qualities are a fact, but difficulty in evaluating the efficiency of combined heat and power CHP has been observed since the breakout of this technology. Havelský [1] pointed out the problem back in 1999; a strict regulation of such plants had thus to be created. In the European Union, a well-structured incentive scheme made a contribution to the growth of cogeneration. In 2004, the EC Directive 2004/8 [2] introduced the concept of high-efficiency cogeneration (HEC); the Annexes II and III of the Directive provided it with a quantitative definition establishing that a cogeneration production characterized by primary energy savings (PES) of at least 10% compared to the reference values for separate production (or >0% for small scale cogeneration units, with installed capacity below 1 MWe, and for micro-cogeneration units) is highly efficient and has the right of access to incentives. After the publication of the Directive, the Commission Decision of 19 November 2008 [3] stated that a cogeneration unit operates in full cogeneration mode when the maximum technically possible heat recovery is attained and all produced electricity is thus considered CHP electricity. When a plant does not operate in full cogeneration mode, it is required to distinguish the amount of electricity and heat not produced under cogeneration mode from CHP electricity and heat, which are the only quantities able to receive incentives. The Directive 2017/27/EU [4] repealed Directive 2004/8, confirming all of its points in the Annexes I and II.

Italy implemented the Directive by enacting three decrees [5–7], followed by the publication of ministerial guidelines [8]. The authors studied the transposition of the Directive 2004/8 EC in the Italian context, referring to the methodology indicated in the abovementioned guidelines for dealing with cogeneration units in Reference [9]. That paper was an opportunity to better clarify the methodology to evaluate plants in a high-efficiency cogeneration framework and do the groundwork for estimating their primary energy savings, in a similar fashion as the studies carried out by Kanoglu and Dincer [10]. Starting from this, in Reference [11], the authors assessed the performance of Italian cogeneration plants in terms of the effects of the power loss coefficient on CHP electricity and power-to-heat ratio. Plants with a non-zero power loss coefficient display a lower CHP electricity production, in an equivalent total electric output.

The calculation of the power loss coefficient and the definition of the reference efficiency are critical aspects of the implementation of the Directive, as verified by Gvozdenac et al. [12]; they studied an actual 150 MW capacity CCGT plant of independent power producers (IPPs) in Thailand, and the investigation showed that the reference value of high-efficiency cogeneration has a large effect on the percentage of CHP electricity output. The impact of the power loss coefficient is not as large, but it influences CHP electricity and CHP fuel energy. These factors are crucial in determining a plant's profitability, so Gvozdenac et al. [12] proposed a modified procedure for the assessment of a CHP plant's efficiency in line with the work of Urošević et al. [13].

Coupling an industrial facility with a combined heat and power system in order to increase the overall efficiency of the plant is now very common practice. For instance, Wang et al. [14] optimized the performance of different cogeneration plants to be implemented in the cement industry in order to recover the available waste heat. The authors reviewed several industry sectors, identifying the most appropriate CHP technology for each of them according to the facility size. The pulp and paper industry is characterized by large electric and heat consumptions, and this means that it lends itself to the implementation of cogeneration.

The Pulp and Paper Industry in Europe: Production and Energy Consumptions

Pulp is a cellulose-based material produced by separating lignin and cellulose fibers from wood and it is the basic constituent for the preparation of paper. The process of separation is chemical when pulp is obtained by cooking chipped wood in an autoclave, with the addition of sodium sulphate (Kraft or sulphate pulping process) or magnesium bisulphite (sulphite pulping process); the process is mechanical when fibers are separated from the trunk by means of an abrasive rotary grindstone. The Joint Research Centre (JRC) of the European Commission released the BAT Reference Document for the Production of Pulp, Paper, and Board [15] to make each of the above processes more efficient and CHP systems, based on different thermal power plants, are one of the main available techniques in this respect.

According to the data illustrated in this document, Europe is one of the main actors in the pulp and paper industry. The annual production of pulp in Europe was about 41.0 million t/y, constituting 22% of world's total wood pulp production; this made Europe the second largest pulp producer. Finland and Sweden together covered 57% of European total wood pulp production. Paper and board production totaled 390.9 million t/y worldwide; 25.3% of this amount was produced in Europe. Italy was one of the leading paper and paperboard producers (9.5 million t/y). The number of paper mills in Europe was 887, located mainly in Italy, Germany, France, Spain, and the UK [15].

In order to clarify how much energy this industry sector consumes, it is possible to refer to the IEA's "Energy Technology Perspectives" 2017 report [16], which states that Organisation for

Economic Co-operation and Development (OECD) Europe's final energy use by pulp, paper, and printing, a quantity adding up all energy supplied to the final consumer, was 1.36 EJ in the year 2014, representing 1.7% of the world's total industry energy consumption. The electric and thermal energy demand by the pulp and paper industry in Europe is bound to increase. Szabó et al. [17] elaborated a world model to reproduce technology and market developments of the pulp and paper industry (PULPSIM) starting from current trends. According to their results, paper demand in Europe is expected to grow up to 120 million t/y by 2030. At the national level, the most recent data regarding the Italian pulp and paper industry come from Assocarta's "Rapporto ambientale dell'industria cartaria" 2016 report [18] and are dated for 2014. According to this document, 154 paper facilities are present on Italian territory; data concerning consumption and impact are reported in Table 1.

Table 1. Energy and environmental impact of the Italian paper industry.

Energy Impact			Environmental Impact	
Consumed Energy (TWh)	Generated Energy (TWh)	Employed Steam (TWh)	Direct Emission (t_{CO2})	Indirect Emission (t_{CO2})
7.01	5.54	11.94	4.88	0.71

The pulp and paper industry is obviously responsible for the emission of greenhouse gases in the atmosphere; CO_2 emissions from energy production are considered direct, those coming from energy purchase are considered indirect.

The need for an environmental optimization of paper mills is a long-felt concern, since Thompson's [19] works on paper mills. Monte et al. [20] analyzed different paper industry production processes, proposing, for each of them, beneficial waste management approaches. Furthermore, considering the issue of emissions from a paper industry facility, Bhander and Jozewicz [21] developed a model to estimate changes in emission when switching fuel, installing air pollution equipment, and implementing energy efficiency measures. Combined heat and power systems make a contribution when it comes to complying with the Kyoto Protocol as well: as observed by Chen et al. [22], the pulp and paper industry is characterized by high fossil energy consumption which is strictly linked to high emission of greenhouse gases. In the wake of this, Boharb et al. [23] outlined a methodology to perform energy audits in the pulp and paper industry.

The perception of the need for improved efficiency is now well established in the management of production processes; so, the harsh international competition to which the pulp and paper industry is subjected made energy management a necessary practice for preserving competitiveness. Kong et al. [24] reviewed different energy-efficient technologies to be implemented in the paper industry, creating a data collection useful to assess the most suited ones to each process: CHP systems are perfectly placed in this context, having the ability to improve the global efficiency of the plant. Cogeneration systems, with their improved overall efficiency, can also be a solution to the problems raised by Posch et al. [25].

Returning to the European case, the Joint Research Centre estimates, in its BAT Reference Document for the paper industry, that CHP systems enable paper mills to save about 30% of the energy consumption of a separate production conventional technology and to reduce greenhouse gas emissions. Therefore, the present paper aims at proposing and comparing the most efficient CHP setups which can be employed in the pulp and paper industry, taking into consideration a particular industrial reality. Results can be compared to those achieved by Shabbir et al. [26], taking into account the different production processes considered.

2. Methods

2.1. Quantitative Definition of the Analysed Process

Table 2 displays electric ($C_{s,e}$) and thermal ($C_{s,t}$) specific consumptions for a fully-integrated bark-fired pulp and paper mill, according to what is reported in the above-mentioned JRC BAT Document [15].

Table 2. Electric and heat specific consumptions of a typical Kraft fully-integrated bark-fired pulp and paper mill.

Process	Heat Consumption (kWh/t)	Power Consumption (kWh/t)
Pulp mill process	5800–31,000	320–640
Paper mill process	4200–21,000	380–760
Total	10,000–52,000	700–1400

While electricity is employed to operate machines, heat, in the form of steam, is deployed in both pulp production and drying. With this regard, this paper proposes a methodology to analyze different CHP technologies potentially suitable for paper mills and evaluated in an HEC framework.

According to Reference [18], the paper industry accounts for 154 facilities in Italy with an overall paper production of 56,000 t/y. Supposing a total of 7200 operating hours (equivalent to 300 days a year), the average hourly production of a single Italian paper mill can be set around 8 t/h ($\dot{m}_{paper}$).

The related electric and thermal power requirements were derived from a year-long measurement campaign conducted for a particular Italian industrial reality. Thus, the whole facility operation refers to this particular year. This facility makes use of electricity provided by a power plant, steam from the power plant and natural gas to satisfy the heat demand; therefore, the days for which all three values were able to be measured were the only ones accounted for. About 33% of the thermal demand is satisfied by the thermal power plant steam; the remainder is fulfilled by natural gas burnt in a boiler. It is known that the conditions of the steam provided to the paper mill are a pressure of 8 bar and a temperature of 170 °C. The considered Italian mill consists only in a continuous papermaking machine, which has the purpose of forming sheets starting from pulp produced in batches; this machine requires steam at a specific temperature and pressure condition, which implies a single heat extraction. It is also known that the enthalpy of the returned condensate is 417.5 kJ/kg. The sum of these values allows for the calculation of the total amount of heat required by the paper production process. The daily electric load diagram was derived assuming that the electric demand is constant over a period of 1 h. Operating days were divided into "stationary" days (when the average hourly production was around the average value of produced paper in this facility) and "abnormal" days (when the average hourly production deviated from that value, for example because of failures or maintenance). The above measurements allowed for the construction of the typical day consumption diagram, reported in Figure 1, where only stationary days were taken into consideration and a paper production of 8 t/h was assumed. The electric and thermal power demand (then used for power plant design) were calculated by averaging over the abovementioned load diagram: P_e = 9.6 MWe and P_t = 32 MWt. Starting from these values it is possible to estimate the specific electric and thermal power consumptions of the plant:

$$C_{s,t} = \frac{P_t}{\dot{m}_{paper}} = 40300 \text{ kWh/t}$$

$$C_{s,e} = \frac{P_e}{\dot{m}_{paper}} = 1200 \text{ kWh/t}$$

as power is the product of mass flow rate by the specific consumption of the paper mill.

Falling in the ranges reported in Table 2, the obtained values can be considered validated. Table 3 sums up the input data common to all plant types.

The following subsections illustrate the procedures to carry out a design, an HEC and economic analysis of the CHP technologies taken into consideration.

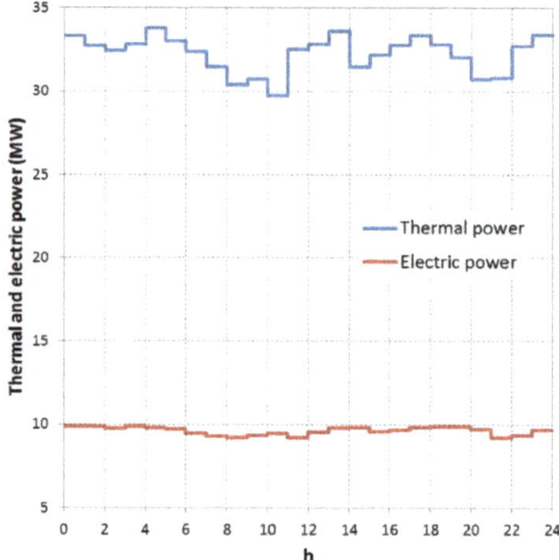

Figure 1. Typical day load diagram. It shows the electric and thermal power demand with values assumed to be constant over a period of 1 h.

Table 3. Input data common to all plant types.

Input Data Common to All Plants	
Operating hours (h/y)	7200
Daily paper production (t/d)	192
Steam pressure required by user (bar)	8
Steam temperature required by user (°C)	170
Returned condensate enthalpy (kJ/kg)	417.5
High-efficiency cogeneration reference enthalpy (1 bar, 15 °C) (kJ/kg)	63
Specific thermal consumption (kWh/t)	4000
Specific electric consumption (kWh/t)	1200
Annual heat demand (MWh/y)	280,320
Annual self-consumed electric energy (MWh/y)	69,120

2.2. Operation Mode and Simulation of the CHP Units

When designing a CHP plant, it is essential to set a mode of operation: in fact, a CHP plant can be operated satisfying electric demands or thermal demands.

A plant satisfying electric demands can cover the user electric demand at full load. In this configuration, however, the heat demand might not be fully covered, and an auxiliary boiler is needed to fulfill the remaining load. This mode of operation is bound to make the plant self-sufficient in terms of electric loads, avoiding energy exchanges with the grid.

Depending on power values, the components of the system are selected from specialized catalogues; starting from the relevant operational parameters indicated, components are modelled and, as a consequence, plant operation is simulated by means of GateCycle software. GateCycle is a PC-based software consisting of a block diagram environment that performs detailed steady-state design and off-design analyses of thermal power systems. Simulations run by means of this software allow the calculation of all the different pressure, temperature, and enthalpy conditions with respect to the working fluid in the system, and as a consequence, a quantitative evaluation of the plant's performance in terms of net power, overall efficiency, primary fuel flow rate, and supplementary fuel flow rate. The variation of the system performance parameters in off-design conditions is evaluated as well, by modelling it in different seasonal situations with GateCycle software.

2.3. Preliminary Design of CHP Technologies for the Analysed Process

First of all, the ratio P_t/P_e of the specific process was calculated: the result was $P_t/P_e = 3.33$.

Average values of the useful-heat-to-electricity ratio ($R_{H/E}$) for Italian plants, similar to the ones calculated by the authors in Reference [27], are available from Reference [28]. They represent the current national state of the art of CHP plants in Italy. Such ratios will be recalculated and employed in the following.

The CHP technologies selected for the present study are internal combustion engine (ICE), gas turbine (GT), steam power plant with back-pressure turbine (SPP-BPT), steam power plant with condensing turbine (SPP-CT), and combined cycle power plants (CCPP). Renewable energy sources were not taken into consideration. In fact, in Italy, renewable energy sources (RES) are not employed to satisfy the electric demand of industrial facilities, which require also heat. In the industry sector, systems capable of providing both process heat and electricity are installed; the size of such systems is determined with the aim of achieving the values for energy parameters imposed by current regulations to gain access to remuneration. Italian regulations provide for dispatching priority for both RES-based plants and cogeneration units: this means that, for these two plant types, electricity is fed into the grid before the amount of energy produced by fossil-fueled thermal power plants bound to the production of electricity only. However, the electricity produced by cogenerators is remunerated with a price which is not as competitive as that set for RES.

If the plant is operated satisfying the electric load, after setting the rated electric power equal to P_e, the rated thermal power is calculated through the value of $R_{H/E}$ corresponding to the chosen technology. If the calculated value is lower than P_t, a supplementary steam flow rate must be estimated.

If the plant is operated satisfying the thermal load, after setting the rated thermal power equal to P_t, the rated electric power is calculated through the value of $R_{H/E}$ corresponding to the chosen technology. If the calculated value is lower than P_e, the resulting electric power deficiency must be compensated for by purchasing electricity from the grid; if the calculated value is higher than P_e, the resulting electric power surplus must be released to the grid.

An optimal design of plants is carried out bearing in mind two hypotheses:

- the useful heat produced by the CHP unit has to be exploited at its most, granting the cogenerator the achievement of the maximum H_{CHP}/E ratio, with this condition being essential to access the best HEC performances;
- electricity has not to be fed into the national grid; this means that the plant size cannot exceed an electric power equal to the electric demand of the process.

In light of this, the calculated value of the process ratio P_t/P_e must be compared to the ratio $R_{H/E}$ related to each CHP technology.

As a consequence of these two hypotheses, if:

$$\frac{P_t}{P_e} > R_{HIE}$$

the considered cogeneration technology is able to satisfy the electric demand of the production process, but not the thermal demand; this implies that the CHP unit has to be designed following the electric loads and auxiliary boilers shall be provided;

Otherwise, if:

$$\frac{P_t}{P_e} < R_{HIE}$$

the considered cogeneration technology is able to satisfy the thermal demand of the production process, but not the electric demand; this implies that the CHP unit has to be designed following the thermal loads and additional electricity shall be drawn from the national grid.

An alternative design of CHP technologies (that is, the plants sized according to the electric load will be sized according to the thermal load and vice versa) can be performed as well, in order to corroborate the validity of the design methodology subject of the present paper and to provide a more complete overview of cogeneration units.

2.4. HEC Energy Performance Parameters

The HEC analysis is conducted by following the guidelines reported in References [7,9,29]. In short, the overall efficiency η_g must be calculated as follows:

$$\eta_g = \frac{E + H_{CHP}}{F - F_{non\text{-}CHP,H}} \tag{1}$$

where E is the total electrical/mechanical energy, H_{CHP} is the CHP useful heat energy (total useful heat energy H reduced by the amount of non-combined useful heat energy $H_{non\text{-}CHP}$), and $F - F_{non\text{-}CHP,H}$ is the amount of fuel energy used in cogeneration. If η_g achieves or exceeds the values of 0.80 for combined cycle gas turbines with heat recovery and steam condensing extraction turbine-based plants or 0.75 for other types of cogeneration units, the plant does not produce any non-CHP electric energy $E_{non\text{-}CHP}$.

For lower values of η_g, the amount of $E_{non\text{-}CHP}$ and the related fuel energy ($F_{non\text{-}CHP,E}$) can be assessed through the evaluation of the power loss coefficient β representing the electricity loss caused by steam extraction for heat production. This parameter allows for the calculation of both the efficiency of non-combined electric energy generation $\eta_{non\text{-}CHP,E}$ and the power-to-heat ratio C_{eff} (C_{actual} in the European legislation). C_{eff} is required to calculate CHP electric energy E_{CHP} starting from H_{CHP}. Being known $\eta_{non\text{-}CHP,E}$, fuel energy for non-CHP electric energy generation $F_{non\text{-}CHP,E}$ can be evaluated; fuel energy for CHP electric energy generation F_{CHP} is then determined by residual. Primary energy saving (PES) can be calculated as follows:

$$PES = \left(1 - \frac{1}{\frac{CHP\ H_\eta}{Ref\ H_\eta} + \frac{CHP\ E_\eta}{Ref\ E_\eta}} \right) \times 100 \tag{2}$$

where $CHP\ E_\eta$ is the electric efficiency of the CHP portion, $CHP\ H_\eta$ is the thermal efficiency of the CHP portion, and $Ref\ H_\eta$ and $Ref\ E_\eta$ are the reference efficiencies for heat and electric production, respectively, provided by the guidelines. This parameter represents electric and thermal energy savings in cogeneration arrangement with respect to the case in which electric and thermal energy are produced separately. It is fundamental to prove that a cogeneration unit is really profitable; in fact, Italian regulations establish that only plants for which $PES \geq 0.1$ ($PES > 0$ for small and micro-CHP units) are eligible to access incentives.

Figure 2 shows a schematized CHP unit with all energy flows named according to the Directive.

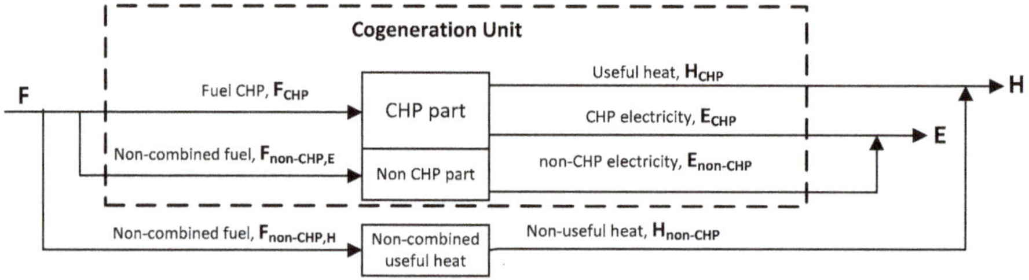

Figure 2. Scheme of a combined heat and power (CHP) unit with energy flows. The guidelines describe the procedure to split the cogeneration unit in a CHP and a non-CHP part for efficiencies below the minimum required threshold.

2.5. Economic Performance Parameters

A preliminary step in evaluating the economic yield of a CHP unit consists in the calculation of the RISP parameter, an expression of the savings fundamental to determine the number of white certificates (WCs) (the securities attesting the savings, in toe, subjected to the incentive scheme):

$$RISP = \frac{E_{CHP}}{\eta_{E,ref}} + \frac{H_{CHP}}{\eta_{H,ref}} - F_{CHP} \tag{3}$$

$$WC = RISP \times 0.086 \times K \tag{4}$$

where K is the harmonization factor, depending on the size of the plant, and $\eta_{E,ref}$ and $\eta_{H,ref}$ are the actual reference efficiencies.

The economic analysis is performed through the assessment of net present value (NPV) and pay-back period (PBP) for the considered plants, taking into account all positive and negative cash flows. However, when considering a comparison between different technologies, the ratio of the net present value to the total investment (NPV/I) may be better suited to assess the overall plant profitability: it is obvious that a plant for which the profit, after a set period of time, carries more weight than the initial outlay is a solution to be preferred.

2.6. Environmental Performance Parameters

The environmental performance of a CHP unit is evaluated in terms of the avoided amount of CO_2 emissions, $CO_{2,a}$. This parameter is calculated by subtracting CO_2 emissions related to a cogeneration plant from those produced by a plant satisfying the same loads in separate production arrangement. The current emission factor attributable to the mix of power plants regularly generating electricity is 330 g_{CO2}/kWh [30] in Italy.

2.7. Total Key Performance Indicator

The most interesting contribution of Reference [27] that has been reapplied in this paper is the definition of the total key performance indicator (TKPI) that may be conveniently used to select, among a variety of possible CHP options, the technology best suited to a specific application:

$$TKPI_i = \frac{PES_i}{PES_{max}} + \frac{PBP_{min}}{PBP_i} + \frac{(NPV/I)_i}{(NPV/I)_{max}} + \frac{CO_{2,a,i}}{CO_{2,a,max}} \tag{5}$$

where $CO_{2,a}$ represents the avoided CO_2 emissions. With reference to the i-th CHP technology, TKPI adds together four key performance indicators referring to energy savings (PES_i/PES_{max}), economic parameters (PBP_{min}/PBP_i and $(NPV/I)_i/(NPV/I)_{max}$) and environmental impact ($CO_{2a,i}/CO_{2,a,max}$). Quantities with subscript *max* (*min*) refer to the maximum (minimum) value of the considered parameter that can be attained in the whole pool of CHP technologies. Each ratio (key performance parameter) can be equal to 1 at most; this means that total key performance parameter value cannot exceed four. It is a comparative measure of a plant's performance and particularly useful to make comparisons as it combines all the relevant parameters that must be evaluated to select a technology when an investment has to be made.

3. Results

The outcome of the design of a CHP plant for supplying energy to a paper mill, contemplating all plant technologies, will be shown in the following, along with related performance analyses.

3.1. Detailed Optimal Design of Cogeneration Units for a Paper Mill

3.1.1. ICE-Based Cogeneration Unit

The cogeneration unit was designed selecting an internal combustion engine model from a Wärtsilä catalogue [31]. The chosen model was W 9L46DF, which is a 4-stroke, non-reversible turbocharged and intercooled engine with direct fuel injection, characterized by a rated power of 10.305 MW. According to the methodology described in Section 2.3, such a plant must be designed depending on the electric load; its scheme is shown in Figure 3. This system was modelled by means of GateCycle software starting from the parameters reported in Section 2.1. As the plant was sized according to the electric load, the heat production is insufficient and must be integrated with an auxiliary boiler in order to fully satisfy the demand ("non-cogeneration heat production").

The simulated plant displays an electric efficiency of 47.5%, a thermal efficiency of 20.5%, and an overall efficiency of 68%; the latter value is lower than the threshold efficiency, this meaning that an HEC portion must be extracted from the CHP unit ($\beta = 0$). The resulting energy balance is illustrated in Table 4. The CHP portion shows an electric efficiency of 47.5% and thermal efficiency of 27.5%. On the basis of the energy balance, knowing from the guidelines that $Ref\,H_\eta = 90\%$ and $Ref\,E_\eta = 52.5\%$, the calculated PES is equal to 17.37%. In Table 4, H_{CHP} and $H_{non\text{-}CHP}$, according to the guidelines, are calculated with a reference enthalpy value for the returned condensate lower than the actual one of 417.5 kJ/kg.

Regarding economic and environmental analysis, the considered input data are reported in Tables 5 and 6. Table 5 shows parameters common to all considered technologies.

The capital cost of the installed unit and the average O&M cost are derived from Reference [32]. Fuel was assumed to be natural gas in all cases and its cost was taken from Reference [33]; fuel cost in CHP arrangement is lower because cogeneration is encouraged by lower excise duties that make fuel, basically, discounted. Fuel cost in separate production arrangements is used to assess the annual cost attributable to the quantity of fuel employed in the production of non-CHP energy amounts; it is also used to evaluate the total annual cost in the reference case of separate production against which the savings produced by cogeneration are estimated. Its variation was evaluated, but the results here presented are calculated for the values of 0.0243 €/kWh (separate production) and 0.0193 €/kWh (CHP arrangement). The cost of the electricity purchased from the grid was derived from Reference [33] as well; calculations were performed for a cost of 0.156 €/kWh. The result was a savings of 8.84 M€ per year, which produces a PBP of 1.12 years and an NPV/I ratio equal to 3.69.

Concerning environmental analysis, the ICE-based cogeneration plant avoids the emission of 14,099 t_{CO2}/y with respect to the case of separate production.

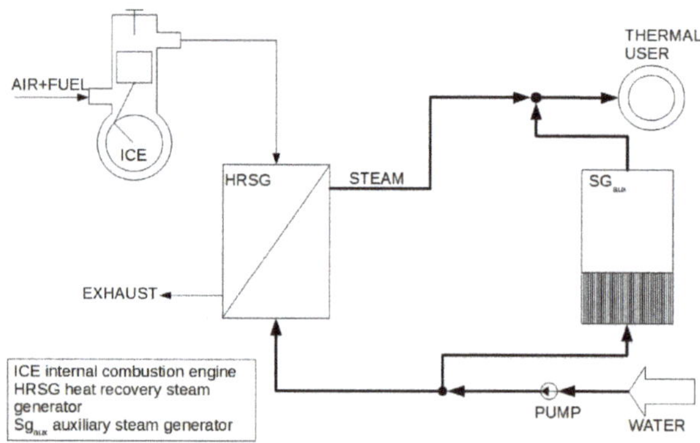

Figure 3. Scheme of the internal combustion engine cogeneration unit taken into consideration.

3.1.2. GT-Based Cogeneration Unit

The cogeneration unit was designed selecting a gas turbine model from a Siemens catalogue [34]. The chosen model was SGT-400, which is a twin-shaft engine characterized by a rated power of 9.7 MW. According to the methodology described in Section 2.3, such a plant must be designed depending on the electric load; its scheme is shown in Figure 4. Results of the simulations are shown in Table 4.

Table 4. Properties and energy balances for the considered cogeneration units (optimal design).

Property\Technology	Internal Combustion Engine	Gas Turbine	Steam Power Plant with Back-Pressure Turbine	Steam Power Plant with Condensing Turbine
design	electric load	electric load	thermal load	electric load
$P_{e,nom}$ (MW)	10.3	9.7	7.2	10.1
SG_{aux}	yes	yes	no	yes
Electricity from grid	no	no	yes	no
η_E	47.5%	34.2%	16.5%	23.1%
η_T	27.5%	76.3%	84.7%	63.4%
η_g	75.0%	>75.0%	>75.0%	86.5%
F (GWh)	372.9	309.0	312.1	368.5
F_{unit} (GWh)	145.5	201.9	312.1	298.6
F_{CHP} (GWh)	108.3	201.9	312.1	298.6
$F_{non-CHP,E}$ (GWh)	37.2	0	0	0
$F_{non-CHP,H}$ (GWh)	227.4	107.1	0	69.9
$E_{process}$ (GWh)	69.1	69.1	69.1	69.1
E_{CHP} (GWh)	51.4	69.1	51.6	69.1
$E_{non-CHP}$ (GWh)	17.7	0	0	0
E_{aux} (GWh)	0	0	17.5	0
H (GWh)	264.5	264.5	264.5	264.5
H_{CHP} (GWh)	29.8	154.0	264.5	189.3
$H_{non-CHP}$ (GWh)	234.7	110.5	0	75.2
Primary energy saving (PES)	17.4%	33.3%	20.4%	12.7%
saving (M€)	8.84	10.52	8.71	9.73
Ppay-back period (PBP) (years)	1.12	1.54	1.41	2.00
NPV/I	3.69	7.03	9.44	6.37
$CO_{2,a}$ (t$_{CO2}$/y)	14,099	26,751	18,605	14,972

Table 5. Input data for economic and environmental analysis common to all technologies.

Input Data for Economic and Environmental Analysis	
Fuel cost in CHP arrangement (€/kWh)	0.0150–0.0315
Fuel cost in separate production arrangement (€/kWh)	0.020–0.032
Grid electricity cost (€/kWh)	0.085–0.170
Discount rate (%)	6
Emission factor (t$_{CO2}$/TJ)	55

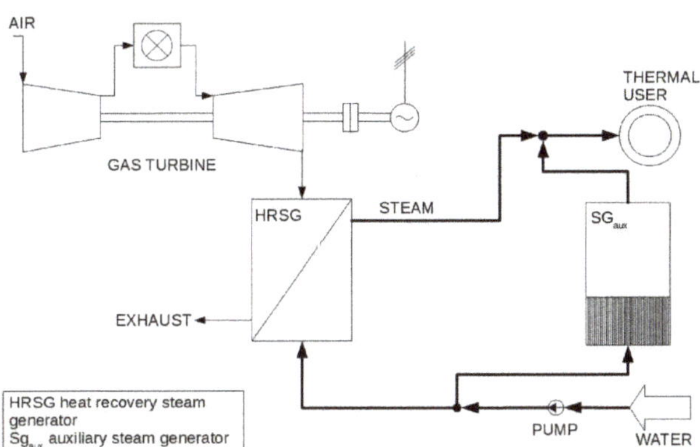

Figure 4. Scheme of the gas turbine cogeneration unit taken into consideration.

Table 6. Input data for economic analysis (different technologies).

Cost\Technology	ICE	GT	SPP-BPT	SPP-CT	CCPP
Unit capital cost (€/Kw)	900	1165	1600	1800	1000
Average operation & maintenance cost as fraction of the installed capital cost (%)	6	6	4	4	4

3.1.3. SPP-BPT-Based Cogeneration Unit

A steam power plant with backpressure turbine cogeneration unit was set-up with a rated power of 7.17 MW. According to the methodology described in Section 2.3, such a plant must be designed depending on the thermal load; its scheme is shown in Figure 5. The electricity production is insufficient and must be integrated with an auxiliary amount of energy purchased from the grid (E_{aux}) to fully satisfy the demand. Results of the simulations are shown in Table 4.

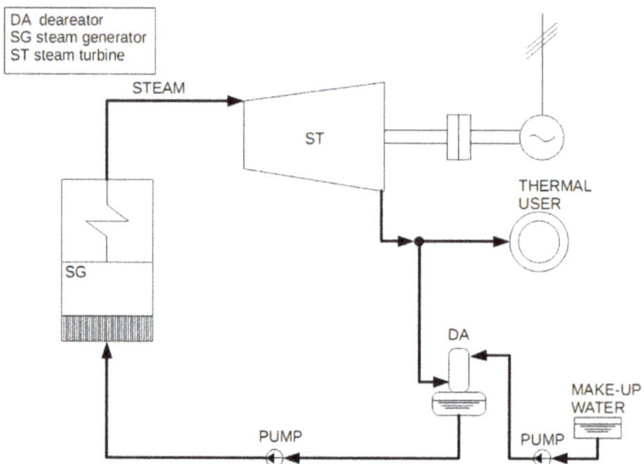

Figure 5. Scheme of the steam power plant with back-pressure turbine cogeneration unit taken into consideration.

3.1.4. SPP-CT-Based Cogeneration Unit

A steam power plant with condensing turbine cogeneration unit was set-up with a rated power of 10.1 MW. According to the methodology described in Section 2.3, such a plant must be designed depending on the electric load; its scheme is shown in Figure 6. Results of the simulations are in Table 4.

3.1.5. CCPP-Based Cogeneration Unit

On the basis of the methodology described in Section 2.3, a combined cycle power plant must be designed depending on the electric load; however, the rated electric power, resulting from the typical heat-to-electricity ratio of this technology, would be much lower than the minimum value for such a plant to be viable (that is 20 MW, according to Reference [35]). Therefore, a CC-based cogeneration unit will not be taken into consideration for the application here discussed.

3.2. Detailed Alternative Design of Cogeneration Units for a Paper Mill

The aim of this section is to study the alternative design of the previously analyzed CHP technologies in order to corroborate the validity of the design methodology subject of the present paper and to provide a more complete overview of cogeneration units.

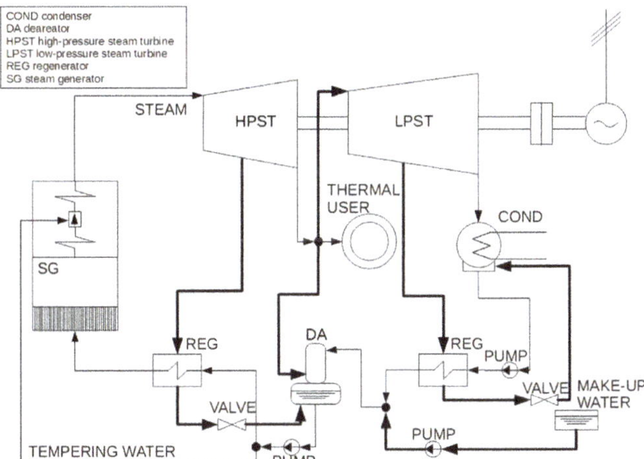

Figure 6. Scheme of the steam power plant with condensing turbine cogeneration unit taken into consideration.

3.2.1. ICE-Based Cogeneration Unit

The size of an ICE cogeneration unit designed depending on the thermal load would be too large for this technology, given its typical heat-to-electricity ratio. The cogeneration unit would have to be composed of eight engines run in parallel; all engines would operate at maximum electric load (in order to accomplish heat requirements) and a great part of electric power would have to be delivered to the grid. Therefore, since this situation is not profitable in an HEC context, an ICE-based cogeneration unit shall not be considered for the application here discussed.

3.2.2. GT-Based Cogeneration Unit

A GT-based cogeneration unit designed supplying the thermal load was modelled by means of GateCycle software starting from the assumptions reported previously, being known at a rated power of 12.9 MW: the outcome of the simulation is the energy balance illustrated in Table 7.

With reference to the abovementioned assumptions, this plant displays an overall efficiency of 92.9% and a PES equal to 23.4%. This setup shows a negative E_{aux}: this means that it produces a surplus amount of electric energy that must be released to the grid. The economic analysis was carried out by resorting to the reference price for purchased electricity in Italy (national unique price, [36]), that is 0.042 €/kWh, to price the electricity fed into the grid (this trade constitutes an incoming cash flow); however, simulations in the cost range 0.042–0.077 €/kWh were also performed. Taking into account the values reported in Tables 5 and 6, and in Section 3.1.1, the result was a savings of 11.94 M€ per year (with respect to the case of separate production), which produces a PBP of 1.35 years and a NPV/I ratio equal to 8.11.

Concerning environmental analysis, it is possible to assume that the release of electricity to the grid causes a decrease in the CO_2 emissions related to the mix of power plants regularly generating electricity. As a consequence, a GT-based cogeneration plant avoids the emission of 60,647 t_{CO2}/y, with respect to the case of separate production.

Table 7. Properties and energy balances for the considered cogeneration units (alternative design).

Property\Technology	Gas Turbine	Steam Power Plant with Back-Pressure Turbine	Combined Cycle Power Plant
design	thermal load	thermal load	thermal load
$P_{e,nom}$ (MW)	12.9	13.33	59.5
SG_{aux}	no	no	no
electricity from grid	no	no	no
η_E	34.5%	23.1%	43.6%
η_T	58.5%	63.8%	36.4%
η_g	92.9%	87.0%	80.0%
F (GWh)	452.3	414.4	934.7
F_{unit} (GWh)	452.3	414.4	934.7
F_{CHP} (GWh)	452.3	414.4	727.5
$F_{non\text{-}CHP,E}$ (GWh)]	0	0	207.2
$F_{non\text{-}CHP,H}$ (GWh)	0	0	0
$E_{process}$ (GWh)	69.1	69.1	69.1
E_{CHP} (GWh)	155.9	95.9	317.5
$E_{non\text{-}CHP}$ (GWh)	0	0	104.1
E_{aux} (GWh)	-86.8	-26.8	-352.5
H (GWh)	264.5	264.5	264.5
H_{CHP} (GWh)	264.5	264.5	264.5
$H_{non\text{-}CHP}$ (GWh)	0	0	0
PES	23.4%	13.1%	19.1%
saving (M€)	11.94	10.09	11.28
PBP (years)	1.35	2.64	6.53
NPV/I	8.11	4.79	0.84
$CO_{2,a}$ (t$_{CO2}$/y)	60,647	25,122	155,743

3.2.3. SPP-BPT-Based Cogeneration Unit

In Section 3.1.3, it was noted that a SPP-BPT-based cogeneration unit requires the purchase of electricity from the grid if designed supplying the thermal load: this means that a design of this kind of plant based on the electric demand would produce a huge waste of exhaust heat, which is pointless in a CHP context. Therefore, a SPP-BPT-based cogeneration unit shall not be considered for the application here discussed.

3.2.4. SPP-CT-Based Cogeneration Unit

A SPP-CT-based cogeneration unit designed supplying the thermal load was modelled by means of GateCycle software starting from the assumptions reported previously: the outcome of the simulation is the energy balance illustrated in Table 7.

3.2.5. CCPP-Based Cogeneration Unit

A CCPP-based cogeneration unit designed supplying the thermal load was modelled by means of GateCycle software starting from the assumptions reported previously: its scheme is shown in Figure 7 (rated electric power equal to 59.5 MW).

The simulated plant displays an overall efficiency of 73.4%: this value is lower than the threshold efficiency, therefore an HEC portion must be extracted from the CHP unit (β = 0.1813). The resulting energy balance and other results of the simulations are illustrated in Table 7. On the basis of the energy balance, the calculated PES is equal to 19.05%.

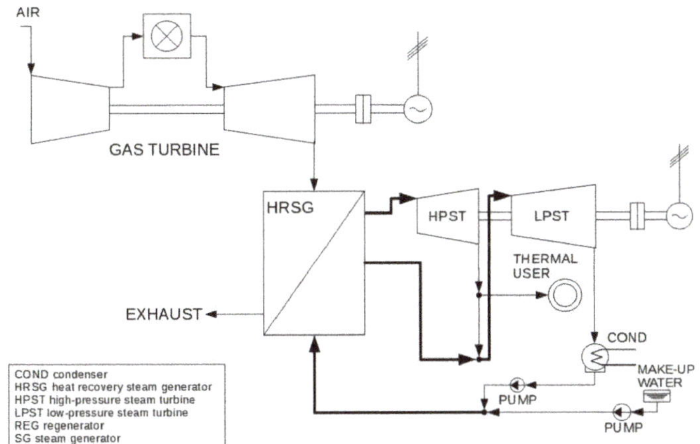

Figure 7. Scheme of the combined cycle cogeneration unit taken into consideration.

4. Discussion

In this section, TKPI is evaluated for each of the examined technologies from the performance parameters assessed in the previous section. Results presented in this section are obtained assuming the following values for costs: 0.0243 €/kWh (fuel cost in separate production arrangement), 0.0193 €/kWh (fuel cost in CHP arrangement), 0.156 €/kWh (cost of purchased electricity), 0.042 €/kWh (cost of electricity fed into the grid). The maximum (minimum) reference values for the indicators appearing in TKPI formula are selected considering the two sizing methodologies (optimal and alternative) separately.

With respect to the optimal design option, Figure 8 shows the resulting total key performance indicator values, broken down into their different contributions.

With a TKPI equal to 3.47, gas turbine turns out to be the most performing technology, followed by steam power plants with backpressure turbine (TKPI equal to 3.10). Internal combustion engines are disadvantaged by the impossibility of exploiting all heat produced; in fact, the amount of heat delivered by cooling water jacket and oil system cannot be harnessed being at much lower temperature levels than what is required by the paper mill's processes. Steam power plants with condensing turbines are the least profitable units for the application here discussed, showing a TKPI of 2.16.

Regarding the alternative design, the breakdown of the resulting total key performance indicators is displayed in Figure 9. As for optimal design methodology, gas turbines and steam power plants with condensing turbine rank respectively first and last among the proposed solutions. Combined cycles are extremely favored by their excellent environmental performance, improved by the huge quantity of electricity fed into the grid, which contributes to the reduction of CO_2 emissions in the model utilized in the present analysis; the environmental indicator plays a major role on the total key performance indicator for this technology that would be otherwise dramatically affected by its poor economic performance, consequence of the very high initial investment (CC size is 4–5 times larger than those of the other technologies). Considering also that the optimal design is not feasible for CC-based cogeneration (see Section 3.1), it can be inferred that this kind of plant is not viable for being coupled with paper mills.

However, where the two design methodologies can be compared, it is clear that the optimal design must be preferred, for the same technology, to higher TKPI values (TKPIs in Figure 9 would be smaller if calculated by using reference values of TKPIs in Figure 8). This statement reaffirms the validity of the sizing methodology proposed by the authors.

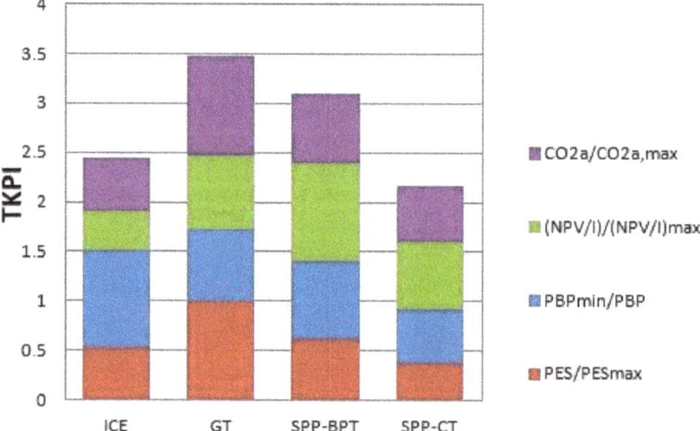

Figure 8. Total key performance indicator (TKPI) values and breakdown for the considered cogeneration technologies (optimal design).

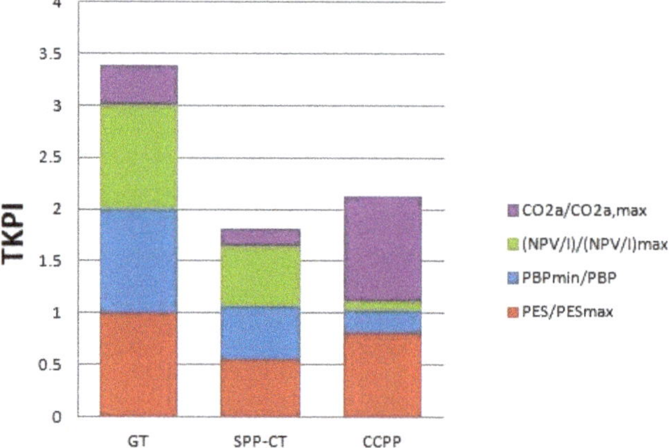

Figure 9. TKPI values and breakdown for the considered cogeneration technologies (alternative design).

The choice of considering two different design criteria responds to the intent to set two separate possible scenarios.

If the optimal methodology is deployed, the plant size is defined so as not to produce a surplus amount of electricity with respect to process requirements, possibly employing auxiliary boilers or purchasing electricity from the grid if cogeneration thermal or electric power are lower than the required ones.

The alternative approach is based, instead, on the admissibility of feeding electricity into the grid. Previous studies developed by the authors, as well as the present one, reveal that cogeneration plants not releasing electricity are today more profitable (in contrast with the situation established by the previous Italian legislation). In fact, results of the present analysis confirm that plants shifted towards to production of heat (satisfaction of the thermal demand) make more sense in an HEC context—it is evident that the optimal design indeed respects the condition of avoided release of electric energy; it was also possible to analyze plants excluded from the optimal design case and to theoretically study contexts remunerating the release of electricity to the grid. The validity of the optimal design criterion is confirmed by the diagram in Figure 10.

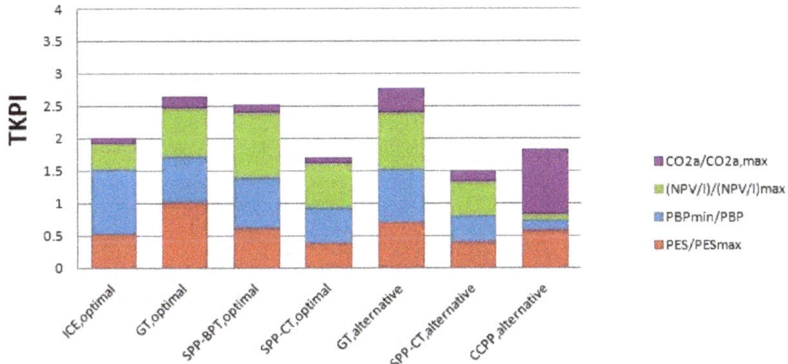

Figure 10. TKPI values and breakdown for the considered cogeneration technologies with white certificates taken into account (combining optimal design and alternative design).

This diagram was realized by considering the maximum (minimum) values of PES, PBP, NPV/I, and $CO_{2,a}$ in a set composed of all seven designed units (four optimal and three alternative), thus not separating the optimally designed technologies from the alternatively designed ones. Gas turbines designed according to the alternative criterion show the highest TKPI; however, given the strong assumptions under which the environmental contribution was evaluated (plants releasing electricity to the grid much benefit from it), such a parameter can be reasonably disregarded and the optimally designed gas turbine is thus confirmed to be the winning technical solution (TKPI = 2.469 against TKPI = 2.390, excluding the environmental contribution).

In the analysis conducted thus far, white certificates are not taken into account in the economic analysis. White certificates are marketable securities which attest primary energy savings achieved in energy final uses through efficiency improvement measures (including cogeneration). Such securities are dispensed by Gestore dei Servizi Energetici (GSE, Energy Service Manager) and are a real income for plants. This means that they indirectly affect the results of the economic calculations here presented, representing a positive value to be added to cash flows. According to the GSE annual report on white certificates [37], in Italy, the average value of incentive securities is equal to 267 €/toe for the year 2017; on the basis of this value, it is possible to price energy savings for each technology (RISP parameter), generating a further positive cash flow. The TKPI parameters can be therefore recalculated: the results are reported in Figures 11 and 12.

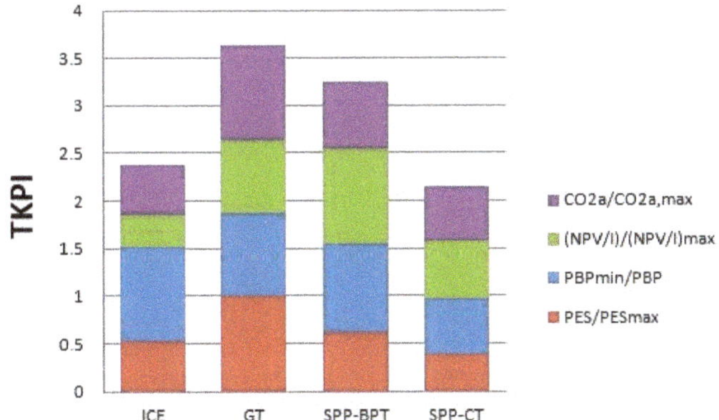

Figure 11. TKPI values and breakdown for the considered cogeneration technologies with white certificates taken into account (optimal design).

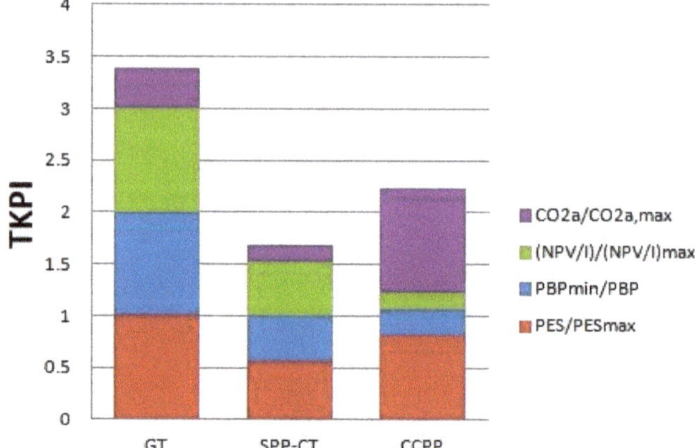

Figure 12. TKPI values and breakdown for the considered cogeneration technologies with white certificates taken into account (alternative design).

The difference from the previous case is slight: since all technologies benefit from incentives, the rankings established in the previous case are maintained. Gas turbines, whose TKPI increases to 3.63, are confirmed to be the most suitable technology; combined cycle economic performance is enhanced but not enough to make it a winning solution.

Besides, it must be noted that white certificates are subject to certain variability in the individual national context and that incentives vary in different legislative realities: this is an element that can be taken into consideration by carrying out a parametric analysis as shown in Figures 13 and 14. It is clear that the change in terms of performance of the individual technology is overall negligible, precisely because performance slightly improves for GT and SPP-BPT, and stays constant for SPP-CT and even somewhat worsens for ICE: this is caused by the growth of the maximum NPV/I ratio, which is faster than that of the NPV/I ratio for the individual technology, with increasing incentives. In any case, gas turbines are hereby confirmed to be the winning technical solution for a CHP unit coupled with a pulp and paper mill and this statement is in line with the outcome of the research conducted by Shabbir et al. [26].

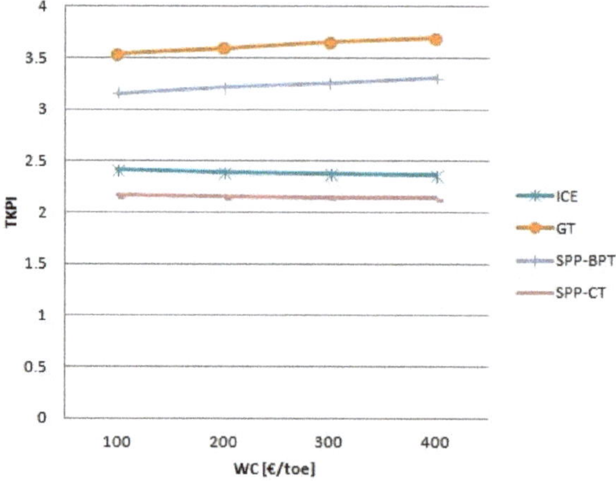

Figure 13. TKPI sensitivity analysis for varying incentive values (optimal design).

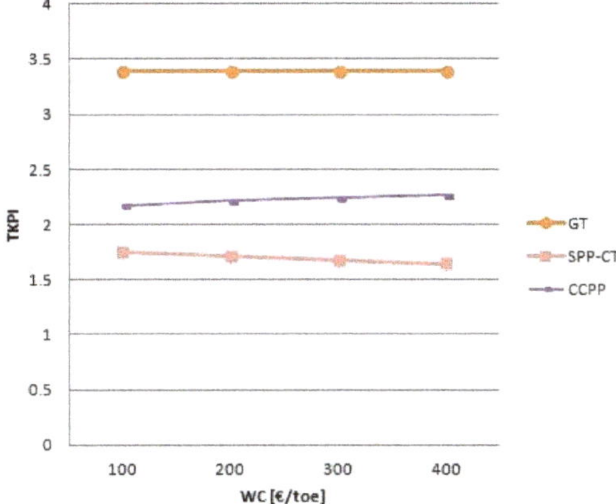

Figure 14. TKPI sensitivity analysis for varying incentive values (alternative design).

Moreover, the entire range of variation for the costs of fuel (cogeneration and non-cogeneration arrangements) and electricity (purchased and sold to the grid), as indicated in Table 5 and in Section 3.2.2, has been explored leading to a variety of scenarios; in all of them gas turbine proved to be the best suited technology. In the different analyses conducted by the authors, technology hierarchy results were confirmed and so was the role of gas turbines as the preferable technology among the possible CHP options to be deployed in the pulp and paper industry.

5. Conclusions

The present paper examines a variety of possible cogeneration technologies to be coupled with the pulp and paper mill industry in the context of high-efficiency cogeneration. The analysis was based on the actual production amounts and load diagrams for a particular industrial reality in Italy and was carried out from a thermodynamic, economic, and environmental perspective, overviewing internal combustion engines, gas turbines, steam power plants with backpressure turbine, steam power plants with condensing turbine and combined cycle power as possible CHP solutions. The plant was sized and modelled according to the available data. Thermodynamic simulations were performed by means of GateCycle software and the energy balances of the considered CHP plants were derived accordingly. The economic analysis was conducted in accordance with the principles of high-efficiency cogeneration (HEC), thus following the guidelines issued by the European Commission and making use of the previously obtained energy balances. As a result, energy savings related to CHP plants were properly estimated and economically valued with respect to separate energy production. Results of the different analysis were quantified through energy, economic, and environmental parameters, added together in a total key performance indicator (TKPI) thus allowing an objective comparison among the different cogeneration technologies taken into account. The calculations indicate gas turbines as the winning technical solution, with a TKPI of 3.47, that can be increased up to 3.63 when white certificates are included in the analysis.

The main achievement of the present study is to provide a solid methodology for a simple and robust evaluation of CHP performances allowing different types of power plants to be compared. The approach illustrated in this paper applies specifically to paper mills; however, being based on general evaluation criteria, the research can be potentially extended and successfully applied to other industry sectors as well.

Author Contributions: Conceptualization, M.G. and M.V.; Methodology M.G. and M.V.; Software, M.V. and T.S.; Validation, M.M. and S.B.; Investigation, M.G., M.V. and T.S.; Writing—Original Draft Preparation, T.S.; Writing—Review & Editing, T.S., M.V. and S.B.; Supervision, M.G.

Funding: This research received no external funding.

Conflicts of Interest: The authors declare no conflict of interest.

Nomenclature

CC	combined cycle
$CCPP$	combined cycle power plants with condensing turbine
C_{eff}	power-to-heat ratio from the Italian legislation
CHP	combined heat and power
CHP	E_η electric efficiency of the CHP portion
CHP	H_η thermal efficiency of the CHP portion
$C_{O\&M}$	operation and maintenance costs
$CO_{2,a}$	CO_2 emissions avoided
$C_{s,e}$	specific electric demand
$C_{s,t}$	specific thermal demand
E	electricity
E_{aux}	auxiliary electricity
E_{CHP}	electricity from cogeneration
$E_{non\text{-}CHP}$	electricity not from cogeneration
$E_{process}$	electricity required by the process
F	fuel input in a CHP system
F_{CHP}	fuel input to produce useful heat and electricity from cogeneration
$F_{non\text{-}CHP,E}$	fuel input to produce electricity not from cogeneration
$F_{non\text{-}CHP,H}$	fuel input to produce heat not from cogeneration
GT	gas turbine
H	heat
h	specific enthalpy
H_{CHP}	useful heat from cogeneration
HEC	high-efficiency cogeneration
$H_{non\text{-}CHP}$	useful heat not from cogeneration
I	total investment
ICE	internal combustion engine
K	harmonization factor
$\dot{m}_{paper}$	produced paper mass flow rate
NPV	net present value
PBP	pay-back period
P_e	electric power
$P_{e,nom}$	nominal electric power
PES	primary energy saving
P_t	thermal power
$Ref\,E_\eta$	reference efficiency for electric production according to the guidelines
$Ref\,H_\eta$	reference efficiency for heat production according to the guidelines
$R_{H/E}$	useful-heat-to-electricity ratio
$RISP$	parameter expressing energy savings in the Italian legislation
SG_{aux}	auxiliary steam generator
$SPP\text{-}BPT$	steam power plants with backpressure turbine
$SPP\text{-}CT$	steam power plants with condensing turbine
$TKPI$	total key performance indicator
WC	white certificates

Greek Letters

β	power loss factor by a heat extraction at a steam turbine
η_{CHP}	cogeneration efficiency
η_E	electric efficiency
$\eta_{E,EQ}$	electric equivalent efficiency
$\eta_{E,ref}$	actual reference electric efficiency
η_g	overall efficiency
$\eta_{H,ref}$	actual reference thermal efficiency
$\eta_{non\text{-}CHP,E}$	efficiency of non-combined electrical/mechanical energy generation
η_T	thermal efficiency
$\eta_{threshold}$	threshold efficiency

References and Notes

1. Havelský, V. Energetic efficiency of cogeneration systems for combined heat, cold and power production. *Int. J. Refrig.* **1999**, *22*, 479–485. [CrossRef]
2. DIRECTIVE 2004/8/EC of the European Parliament and of the Council of 11 February 2004.
3. COMMISSION DECISION 2008/952/EC of 19 November 2008.
4. DIRECTIVE 2012/27/EU of the European Parliament and of the Council of 25 October 2012.
5. Legislative Decree n. 20-8 February 2007-Transposition of Directive 2004/8/EC. Available online: http://www.bosettiegatti.eu/info/norme/statali/2007_0020_cogenerazione.htm (accessed on 30 April 2018).
6. Ministerial Decree of Environment Ministry-4 August 2011. Available online: https://www.mise.gov.it/images/stories/normativa/DM-4-AGOSTO-2011-2.pdf (accessed on 30 April 2018).
7. Ministerial Decree of Ministry of Economic Development-5 September 2011. Available online: https://www.mise.gov.it/images/stories/normativa/DM-5-SETTEMBRE2011.pdf (accessed on 30 April 2018).
8. *High-Efficiency Cogeneration—Guidelines*; Ministry of Economic Development: Roma, Italy, February 2012.
9. Gambini, M.; Vellini, M. High-efficiency cogeneration: Performance assessment of industrial cogeneration power plants. *Energy Procedia* **2014**, *45*, 1255–1264. [CrossRef]
10. Kanoglu, M.; Dincer, I. Performance assessment of cogeneration plants. *Energy Convers. Manag.* **2009**, *50*, 76–81. [CrossRef]
11. Gambini, M.; Vellini, M. High-efficiency cogeneration: Electricity from cogeneration in CHP plants. *Energy Procedia* **2015**, *81*, 430–439. [CrossRef]
12. Gvozdenac, D.; Urošević Gvozdenac, B.; Menke, C.; Urošević, D.; Bangviwat, A. High-efficiency cogeneration: CHP and non-CHP energy. *Energy* **2017**, *135*, 269–278. [CrossRef]
13. Urošević, D.; Gvozdenac, D.; Grković, V. Calculation of the power loss coefficient of steam turbine as a part of the cogeneration plant. *Energy* **2013**, *59*, 642–651. [CrossRef]
14. Wang, J.; Dai, Y.; Gao, L. Exergy analyses and parametric optimizations for different cogeneration power plants in cement industry. *Appl. Energy* **2009**, *86*, 941–948. [CrossRef]
15. Suhr, M.; Klein, G.; Kourti, I.; Gonzalo, M.R.; Giner Santonja, G.; Roudier, S.; Delgado Sancho, L. *Best Available Techniques (BAT) Reference Document for the Production of Pulp, Paper and Board*; JRC Science and Policy Reports; Publications Office of the EU: Luxembourg, 2015.
16. International Energy Agency. *Tracking Clean Energy Progress 2017, Energy Technology Perspectives 2017 Excerpt*; International Energy Agency: Paris, France, 2017.
17. Szabó, L.; Soria, A.; Forsström, J.; Keränen, J.T.; Hytönen, E. A world model of the pulp and paper industry: Demand, energy consumption and emission scenarios to 2030. *Environ. Sci. Policy* **2009**, *12*, 257–269. [CrossRef]
18. Assocarta. *Rapporto ambientale dell'industria cartaria italiana—Dati 2013–2014*; Assocarta: Milano, Italy, 2016.
19. Thompson, G.; Swain, J.; Kay, M.; Forster, C.F. The treatment of pulp and paper mill effluent: A review. *Bioresour. Technol.* **2001**, *77*, 275–286. [CrossRef]
20. Monte, M.C.; Fuente, E.; Blanco, A.; Negro, C. Waste management from pulp and paper production in the European Union. *Waste Manag.* **2009**, *29*, 293–308. [CrossRef] [PubMed]
21. Bandher, G.; Jozewicz, W. Analysis of emission reduction strategies for power boilers in the US pulp and paper industry. *Energy Emiss. Control Technol.* **2017**, *5*, 27–37. [CrossRef]

22. Chen, H.W.; Hsu, C.H.; Hong, G.B. The case study of energy flow analysis and strategy in pulp and paper industry. *Energy Policy* **2012**, *43*, 448–455. [CrossRef]

23. Boharb, A.; Allouhi, A.; Saidur, R.; Kousksou, T.; Jamil, A. Energy conservation potential of an energy audit within the pulp and paper industry in Morocco. *J. Clean. Prod.* **2017**, *149*, 569–581. [CrossRef]

24. Kong, L.; Hasanbeigi, A.; Price, L. Assessment of emerging energy-efficiency technologies for the pulp and paper industry: A technical review. *J. Clean. Prod.* **2016**, *122*, 5–28. [CrossRef]

25. Posch, A.; Brudermann, T.; Braschel, N.; Gabriel, M. Strategic energy management in energy-intensive enterprises: A quantitative analysis of relevant factors in the Austrian paper and pulp industry. *J. Clean. Prod.* **2015**, *90*, 291–299. [CrossRef]

26. Shabbir, I.; Mirzaeian, M. Feasibility analysis of different cogeneration systems for a paper mill to improve its energy efficiency. *Int. J. Hydrogen Energy* **2016**, *41*, 16535–16548. [CrossRef]

27. Gambini, M.; Vellini, M. On Selection and Optimal Design of Cogeneration Units in the Industrial Sector. *J. Sustain. Dev. Energy Water Environ. Syst.* **2019**, *7*, 168–192. [CrossRef]

28. Ministry of Economic Development. Relazione Annuale Sulla Cogenerazione in Italia. 2014. Available online: http://enerweb.casaccia.enea.it/enearegioni/UserFiles/Rel-2016-Cogenerazione.pdf (accessed on 30 April 2018).

29. Gestore Servizi Energetici. *Guida alla Cogenerazione ad Alto Rendimento (CAR)*; Gestore Servizi Energetici: Roma, Italy, 2018.

30. International Energy Agency. World Energy Outlook 2017. Available online: https://www.iea.org/weo2017/ (accessed on 30 April 2018).

31. Wärtsilä. *Wärtsilä 46 DF Product Guide*; Wärtsilä: Helsinki, Finland, 2016.

32. Frangopoulos, C.A. *Cogeneration—Technologies, Optimisation and Implementation*; IET Energy Engineering: London, UK, 2017; Volume 87.

33. Autorità per l'energia elettrica, il gas e il sistema idrico, Relazione annuale sullo stato dei servizi e sull'attività svolta. 2017. Available online: https://www.autorita.energia.it/allegati/relaz_ann/17/RAVolumeI_2017.pdf (accessed on 30 April 2018).

34. Siemens. *Siemens Gas Turbine Portfolio*; Siemens: Berlin, Germany, 2018.

35. Lozza, G. *Turbine a Gas e Cicli Combinati*; Società Editrice Esculapio: Bologna, Italy, 2016.

36. Gestore Mercati Energetici. PUN Index. Available online: http://www.mercatoelettrico.org/It/default.aspx (accessed on 30 April 2018).

37. Gestore Servizi Energetici. Rapporto Annuale Certificati Bianchi. 2017. Available online: https://www.gse.it/documenti_site/Documenti%20GSE/Rapporti%20Certificati%20Bianchi/Rapporto_annuale_CB_2017.pdf (accessed on 30 April 2018).

Article

Numerical Simulation and Optimization of Waste Heat Recovery in a Sinter Vertical Tank [†]

Chenyi Xu, Zhichun Liu *, Shicheng Wang and Wei Liu *

School of energy and power engineering, Huazhong university of science and technology, Wuhan 430074, China; cy_xu@hust.edu.cn (C.X.); shichengwang@hust.edu.cn (S.W.)
* Correspondence: zcliu@hust.edu.cn (Z.L.); w_liu@hust.edu.cn (W.L.);
   Tel.: +86-27-8754-2618 (Z.L.); +86-27-8754-1998 (W.L.)

† This article belongs to the Special Issue selected papers from SDEWES 2018: 13th Conference on Sustainable Development of Energy Water and Environmental Systems. Based on the conference paper: Numerical simulation and optimization of waste heat recovery in sinter vertical tank. (SDEWES2018.00348).

Received: 26 December 2018; Accepted: 21 January 2019; Published: 25 January 2019

Abstract: In this paper, a two-dimensional steady model is established to investigate the gas-solid heat transfer in a sinter vertical tank based on the porous media theory and the local thermal non-equilibrium model. The influences of the air flow rate, sinter flow rate, and sinter particle diameter on the gas-solid heat transfer process are investigated numerically. In addition, exergy destruction minimization is used as a new principle for heat transfer enhancement. Furthermore, a multi-objective genetic algorithm based on a Back Propagation (BP) neural network is applied to obtain a combination of each parameter for a more comprehensive performance, with the exergy destruction caused by heat transfer and the one caused by fluid flow as the two objectives. The results show that the heat dissipation and power consumption both gradually increase with an increase of the air mass flow rate. Additionally, the increase of the sinter flow rate results in a decrease of the heat dissipation and an increase of the power consumption. In addition, both heat dissipation and power consumption gradually decrease with an increase of the sinter particle diameter. For the given structure of the vertical tank, the optimal operating parameters are 2.99 kg/s, 0.61 kg/s, and 32.8 mm for the air flow rate, sinter flow rate, and sinter diameter, respectively.

Keywords: sinter; porous media; local thermal non-equilibrium; exergy destruction minimization; BP neural network; genetic algorithm

---

1. Introduction

In the production process of the steel industry, a large amount of waste heat resources, including sinter sensible heat, is generated. The recycling of sinter waste heat resources is one of the effective ways to reduce energy consumption in the sintering process [1,2].

The research on recycling and utilization of sinter waste heat resources has been carried out for a long time, mainly focusing on sinter ring coolers and sinter belt coolers, which have been applied to engineering projects [3,4]. Shi et al. [5] established a one-dimensional unsteady mathematical model to investigate the gas-solid heat transfer process in sinter ring coolers, and studied the influence of the cold air flow rate and trolley movement rate. Zhang et al. [6] developed a three-dimensional model to investigate the effects of several parameters on the heat transfer process of sinter ring coolers based on porous media theory and the local thermal non-equilibrium model. However, the way of recycling sinter waste heat resources in existing ring cooling machines has some inevitable defects, like the high air leakage rate in the cooling system, low air outlet quality, and low waste heat recovery quantity, resulting in a low efficiency. Therefore, a sinter vertical tank has been put forward as a new way for efficient sinter waste heat recovery [7–10].

Different from the unsteady heat transfer process in sinter ring cooling, the gas-solid heat transfer process in a vertical tank is steady. Due to the complexity and instability of gas flow in the sinter bed layer, the analysis of the gas-solid heat transfer process in the sinter vertical tank is still in the theoretical and experimental research stage. Leong et al. [4] studied the effect of sinter layer porosity distribution on flow and temperature fields in a sinter cooler. Liu et al. [11] experimentally studied gas flow characteristics in a vertical tank for sinter waste heat recovery. Feng et al. [12–14] investigated gas flow characteristics and the modification of Ergun's correlation in a vertical tank for sinter waste heat recovery experimentally. The influence of various structural parameters and operating parameters on heat transfer and evaluation of the process is lacking in systemic research and analysis. Kong et al. [15] numerically investigated the heat transfer and flow process in dry quenching furnace by building a one-dimensional mathematical model. The variation of outlet temperatures of circulating gas and coke under different working conditions and different gas-to-material ratios were obtained. To evaluate the recovery performance of waste heat in a system, the quality of waste heat was considered in a recent study. Feng et al. [16] established a steady gas-solid heat transfer model to numerically analyze the effects of different operating parameters on the cooling air outlet exergy in a sinter vertical tank and optimized parameters by the mixed orthogonal experimental method. In a subsequent study, Gao et al. [17] focused on investigating the resistance characteristics of the gas stream passing through the waste heat recovery tank bed layer by building a homemade experimental bench, then applied the weighted comprehensive scoring method to optimize parameters by comprehensively evaluating cooling air outlet exergy and sinter bed layer resistance loss.

Liu et al. [18–24] proposed the concept of a local exergy destruction rate based on convective heat transfer, obtaining the expression of the local exergy destruction rate, which can be used to represent the irreversible loss in the convective heat transfer process, then applied the exergy destruction minimization as the optimization criterion in the exchanged heat transfer tube.

In this paper, to comprehensively evaluate the waste heat recovery process in the sinter vertical tank from two aspects of heat transfer quantity and heat quality, the theory of exergy destruction minimum is applied based on the two-dimensional steady local thermal non-equilibrium two-equation model and porous medium theory.

To optimize parameters and comprehensively evaluate the flow and heat transfer performance, a multi-objective genetic algorithm [25–29] based on the Back Propagation (BP) neural network is applied. The BP neural network is a feedforward neural network trained according to the error back propagation algorithm. Structurally, it has an input layer, a hidden layer, and an output layer. In essence, it adopts the gradient descent method to calculate the minimum value of the square of the network error. In this paper, the BP neural network is used to train the obtained data for the fitting function. In addition, the traditional method for multi-objective optimization, like the weighted comprehensive scoring method, has some defects, because the allocation of each objective weighted value is subjective and there is no standard for it. Therefore, the multi-objective genetic algorithm is applied to the neural network, with the exergy destruction caused by heat transfer and the exergy destruction caused by fluid flow as the two objectives. Then, the most suitable combination of the three parameters is obtained after the iteration and evolution of the population.

2. Numerical Methodology

2.1. Physical Model

Different from the unsteady process of a gas-solid cross flow fixed bed in a sinter ring cooler, the gas-solid heat transfer process in a sinter vertical tank is stable from the gas-solid countercurrent moving bed essentially. In this paper, we concentrate on how to recycle waste heat efficiently, so its focus is the heat transfer enhancement between gas and solids. Considering that this work mainly focuses on the heat transfer mechanism in the vertical tank, a relatively miniaturized physical model is established. Figure 1 shows the simplified schematic of the computational domain for waste heat recovery in a vertical tank. The inner diameter and height of the cooling section are 1 m and 1.8 m, respectively. The sinter particles with high temperatures fall into the vertical tank from the upper entrance slowly, and then exchange heat with the cooling air from the bottom of the tank. In this way, the high-grade air can be utilized later, such as power generation.

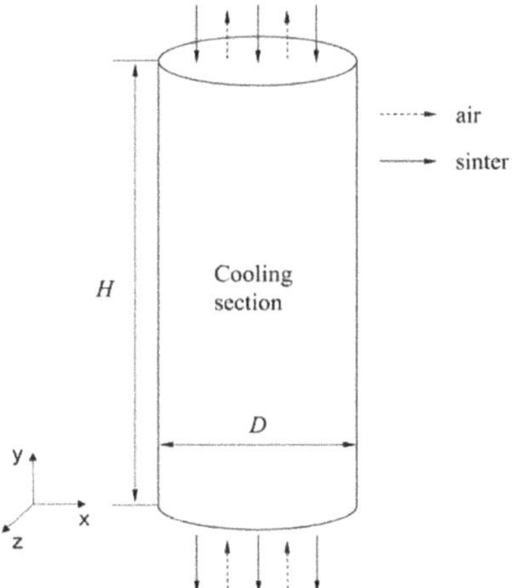

Figure 1. Simplified physical model of the sinter vertical tank.

2.2. Mathematical Model and Exergy Destruction Minimization

2.2.1. Mathematical Model

Due to the wide range of sizes and irregular shapes of sinter particles, it is difficult to mathematically describe and numerically investigate the gas solid heat transfer process precisely. In this paper, sinter particles are considered to be spherical particles of equal size and the cooling section of the vertical tank is assumed to be the porous zone. Simultaneously, the radiant heat transfer between/in the sinter particles and the gas is ignored and the wall of the vertical tank is assumed to be insulated. Considering the turbulent flow of the cooling air in the bed layer, the standard turbulent model is adopted. The following equations are employed to describe the gas-solid heat transfer:

Continuity equation:

$$\frac{\partial}{\partial x_j}(\rho_g u_i) = 0 \tag{1}$$

Momentum equation:

$$\frac{\partial}{\partial x_j}(\rho_g u_i u_j) = -\frac{\partial}{\partial x_i}(p + \frac{2}{3}\rho k) + \frac{\partial}{\partial x_j}[(\mu + \mu_t)(\frac{\partial u_i}{\partial x_j} + \frac{\partial u_j}{\partial x_i} - \frac{2}{3}\delta_{ij}\frac{\partial u_l}{\partial x_l})] + S_i \tag{2}$$

k equation:

$$\rho_g u_j \frac{\partial k}{\partial x_j} = \frac{\partial}{\partial x_j}\left[\left(\mu + \frac{\mu_t}{\sigma_k}\right)\frac{\partial}{\partial x_j}\right] + \mu_t \frac{\partial u_i}{\partial x_j}\left(\frac{\partial u_j}{\partial x_i} + \frac{\partial u_i}{\partial x_j}\right) - \rho_g \varepsilon \tag{3}$$

ε equation:

$$\rho_g u_j \frac{\partial \varepsilon}{\partial x_j} = \frac{\partial}{\partial x_j}[(\mu + \frac{\mu_t}{\sigma_\varepsilon})\frac{\partial}{\partial x_j}] + \frac{c_1 \varepsilon}{k}\mu_t \frac{\partial u_i}{\partial x_j}(\frac{\partial u_i}{\partial x_j} + \frac{\partial u_j}{\partial x_i}) - c_2 \rho_g \frac{\varepsilon^2}{k} \tag{4}$$

The momentum equation of the porous media has additional momentum source terms, which are the momentum loss essentially consisting of two parts, the viscous loss term and the inertial loss term:

$$S_i = -(\frac{\mu}{\alpha}u_i + \frac{1}{2}C_2\rho_g|u|u_i) \tag{5}$$

The viscous resistance coefficient and inertial resistance coefficient are calculated by the following two equations, respectively [6]:

$$\frac{1}{a} = \frac{85.4(1-\varepsilon^2)}{\varepsilon^3 d_p{}^2} \tag{6a}$$

$$C_2 = \frac{0.632(1-\varepsilon)}{\varepsilon^3 d_p} \tag{6b}$$

Considering the temperatures between the air and the sinter in the vertical tank are different after the heat exchange and this tends to be stable, the local thermal non-equilibrium model is adopted to calculate the heat transfer process. The cooling air as the gas phase and the sinter as the solid phase have independent energy equations. The two energy equations are shown below [16]:

Gas phase:

$$\frac{\partial}{\partial x_i}(\rho_g c_g u_g T_g) = \varepsilon \frac{\partial}{\partial x_i}(\lambda_g \frac{\partial T_g}{\partial x_i}) + h_v(T_s - T_g) \tag{7}$$

Solid phase:

$$\frac{\partial}{\partial x_i}(\rho_s c_s u_s T_s) = (1-\varepsilon)\frac{\partial}{\partial x_i}(\lambda_s \frac{\partial T_s}{\partial x_i}) - h_v(T_s - T_g) \tag{8}$$

where ρ, c, and λ are the density, specific heat, and thermal conductivity, respectively. Their special values are listed in Table 1. Additionally, h_v is the volume heat transfer coefficient, which is calculated with the correlation below:

$$h_v = \frac{6h(1-\varepsilon)}{d_p} \tag{9}$$

The Nusselt number (Nu) is calculated as follows by referring to [16]:

$$Nu = \frac{h d_p}{\lambda_g} = 0.2\varepsilon^{0.055}\mathrm{Re}_p{}^{0.657}\mathrm{Pr}^{1/3} \tag{10}$$

Then, according to the solid phase energy equation, the velocity of the solid is added to the left side of the equation as a source term, which can achieve the slow falling process of sinter particles in the vertical tank and obtain the goal of steady-state gas-solid heat transfer. The source terms in the momentum equation and energy equation are applied into the solution equations through the user-defined functions, as well as the variations of the physical parameters of the air and sinter with temperature.

2.2.2. Exergy Destruction Minimization

As we know, available potential is a state variable, characterizing the ability to do the work of a fluid. The convective heat transfer process can be comprehensively studied from two aspects of heat transfer quantity and heat quality when the available potential is used for the physical analysis of fluid particles. Additionally, exergy is a process variable, representing the change of the available potential and the maximum ability to do the work of a process. It is described as follows [19]:

$$ex = (h - T_0 s) - (h_0 - T_0 s_0) = e - e_0 \tag{11}$$

where h_0 and s_0 are the enthalpy and entropy in the environmental state, respectively, and e_0 is the available potential of the fluid in the environmental state.

The change of available potential comes from two parts, the heat flow exergy from the outside and the loss generated during the heat transfer. The local destruction rate represents the loss of input exergy during the heat transfer process, thus it is called exergy destruction. The local exergy destruction rate can be expressed as follows [22]:

$$e_{xd} = T_0 \frac{\lambda(\nabla T)^2}{T^2} + U \cdot (\rho U \cdot \nabla U - \mu \nabla^2 U) \tag{12}$$

where the first item reacts with the irreversible heat loss. The irreversible source is the local temperature gradient. Additionally, it will never be negative, reflecting the irreversibility of the temperature difference heat transfer process. The next item is shown as Equation (13b). In steady-state flow, it is expressed as the product of the velocity and pressure gradient. Liu et al. [22] considered its physical meaning as the pumping work in the flow process, which includes the change of kinetic energy and viscous loss; that is, mechanical work consumed during the flow. This also meets the understanding in thermodynamics that the mechanical work lost during the flow is all mechanical exergy.

$$e_{xd,\Delta T} = T_0 \frac{\lambda(\nabla T)^2}{T^2} \tag{13a}$$

$$e_{xd,\Delta p} = U \cdot (\rho U \cdot \nabla U - \mu \nabla^2 U) \tag{13b}$$

The proposed local exergy destruction rate provides the possibility to quantitatively analyse the irreversibility of each point during the process. The total exergy destruction caused by heat transfer and fluid flow can be synthesized:

$$E_{xd} = \iiint_\Omega e_{xd} dV = \iiint_\Omega \left[T_0 \frac{\lambda(\nabla T)^2}{T^2} + U \cdot (\rho U \cdot \nabla U - \mu \nabla^2 U) \right] dV \tag{14}$$

According to Equation (14), the total exergy destruction consists of the thermal dissipation owing to a temperature difference and the power consumption owing to a pressure drop, which are adopted to respectively represent the irreversible heat loss and the irreversible pressure loss of the gas-solid heat transfer process.

$$E_{xd,\Delta T} = \iiint_\Omega T_0 \frac{\lambda(\nabla T)^2}{T^2} dV \tag{15a}$$

$$E_{xd,\Delta p} = \iiint_\Omega U \cdot (\rho U \cdot \nabla U - \mu \nabla^2 U) dV \tag{15b}$$

3. Results and Discussion

3.1. Grid Independence Analysis and Model Validation

3.1.1. Grid independence Analysis

A two-dimensional structured symmetrical grid with hexahedral elements was conducted in CFD software ICEM. The multi-layer encrypted grid is used for the strong heat transfer section of the wall boundary layer. To avoid the effect of the quality of the grid and to improve the accuracy of calculation results, the grid independence should be analyzed. The main parameters and their values are shown in Table 1. The boundary conditions are the mass-flow-inlet for the gas inlet, the pressure-out for the gas outlet, and the adiabatic-wall for the walls of the computational domain. Combined with the additive source term in the solid energy equation, the solid inlet is set to be the isothermal-wall boundary condition, which can achieve the slow falling process of sinter particles.

Table 1. Main parameters of one condition.

Parameters	Value
Inlet air temperature (K)	293
Inlet sinter temperature (K)	893
Air mass flow rate (kg/s)	1.25
Sinter mass flow rate (kg/s)	1.40
Sinter particle diameter (mm)	25
Sinter density (kg/m^3)	3400
Sinter specific heat (J/(kg·K))	$337.03(T_s - 273)^{0.152}$
Sinter thermal conductivity (W/(m·K))	2.87

According to the results of the grid independence shown in Table 2, the number of the grid elements varies from 12,000 to 40,000. The calculation using the mesh system with 12,000 grid elements is different to converge unless the relaxation factor of energy is adjusted to a very low level, which will greatly increase the time for calculation convergence. Therefore, the model with 12,000 grid elements was been adopted. The relative errors between the results of 14,400, 18,000, and 28,000 grid elements and the result of 40,000 grid elements are all less than 1%, small enough to achieve the requirement for accuracy. Furthermore, because the calculation time required for grids with 14,400 and 18,000 elements is similar, so the model with 18,000 grid elements is selected for the consecutive numerical calculation.

Table 2. Grid independence verification.

Grid Number	$E_{ex,\Delta T}$(W)	Error (%)	$E_{ex,\Delta p}$ (W)	Error (%)
12000	/	/	/	/
14400	492.9	0.42	944.2	0.2
18000	494.7	0.06	946.0	0.01
28000	494.9	0.02	946.1	0
40000	495.0	/	946.1	/

3.1.2. Model Validation

To validate the accuracy and reliability of the numerical model in this study, some results in [13,14] are applied. In [14], gas-solid heat transfer characteristics were experimentally studied in the sinter vertical tank through a homemade experimental setup, and an experimental correlation for describing the gas-solid heat transfer characteristics was obtained. In [13], the Ergun's correlation in a vertical tank was experimentally modified. The comparison of the Nusselt number (Nu) between the numerical study in this paper and the experimental results in [14] under different Re_p is shown in Figure 2a. Additionally, Figure 2b shows the comparison of the fraction factor (f) obtained in the present work and the experimental results in [13]. Excluding the air mass flow rate, the other main parameters and their values of the numerical conditions are shown in Table 1. As can be seen from Figure 2, the deviations of Nu between the numerical results and the experimental results are all less than 5% and the maximum is 4.18%, and the maximum error off is 10.69%. Due to the inhomogeneity and irregularity of sinter particles, the uncertainty in the experimental measurement and the idealization of numerical simulation, the deviations are within the acceptable range and the computational model in the present work is certified to be reasonably reliable.

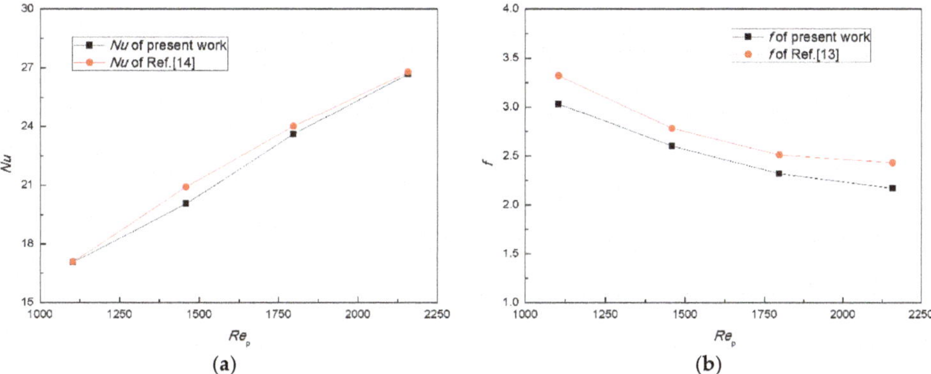

Figure 2. Comparisons of (**a**) *Nu* and (**b**) *f* between the numerical and experimental results.

3.2. Analysis of Influencing Parameters

As is known from the previous numerical and experimental studies, the inlet gaseous velocity and the porosity are the main influencing factors on the solid cooling process. Since the sinter particles are assumed to be equal-diameter spherical particles, the porosity of the sinter bed layer is mainly determined by the particle diameter. Therefore, under the specified structure of the sinter vertical tank in this work, the air mass flow rate, sinter mass flow rate, and sinter particle diameter are the three operating parameters to be studied numerically to obtain the impacts on the waste heat recovery efficiency of the gas-solid heat transfer process. Figure 3 shows the air temperature distribution, sinter temperature distribution, and the pressure distribution in the cooling section when the gas-solid heat exchange is stable. Since the numerical simulation in the present work is based on a two-dimensional axisymmetric model, Figure 3 only shows half of the computational domain and the y axis is the direction of the gas flow. The settings of the partial parameters are listed in Table 1. As we can see in Figure 3, the air temperature in the top of cooling section is obviously higher than in the bottom, which is contrary to the result of the sinter, and the pressure from the bottom to the top is significantly reduced. Figure 4 shows the average cross-sectional temperatures of the air and sinter along the height of the sinter bed layer. Additionally, the temperatures of the air and sinter on the same radial cross section are different. However, since the gas-solid heat transfer process in the cooling section is steady, the air outlet temperature only rises to 760 K lower than 893 K and the sinter outlet temperature drops to 330 K higher than 293 K. In addition, the gas-solid temperature differences of different sections along the height of the bed are approximately equal due to the steady heat transfer process.

The influences of a single parameter on the flow and heat transfer performances evaluated by the thermal dissipation and the power consumption were investigated in the model described above by changing one parameter to several different levels while keeping the other parameters fixed. The values of the parameters at the four levels are shown in Table 3. The ranges of these parameters are set by reference to the homemade experimental setup in [13,14] and the considerations of the quality and quantity of waste heat recovery. The results and analysis are as follows.

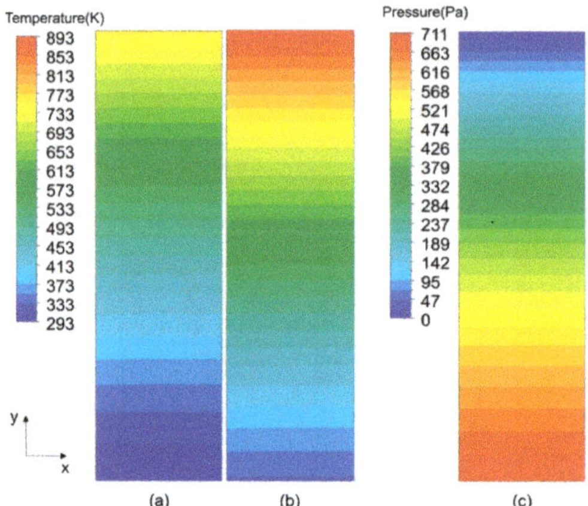

Figure 3. (**a**) Air temperature contour; (**b**) sinter temperature contour; (**c**) pressure contour.

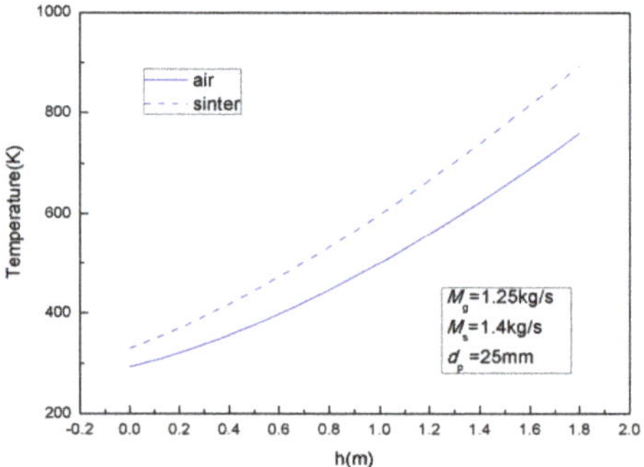

Figure 4. The temperature distributions of air and sinter along the height of the sinter bed layer.

Table 3. Parameters of different levels.

Parameters	1	2	3	4
Air mass flow rate (kg/s)	1.0	1.25	1.5	1.75
Sinter mass flow rate (kg/s)	1.0	1.2	1.4	1.6
Sinter particle diameter (mm)	15	25	35	45

3.2.1. Influence of Air Flow Rate

Figure 5 shows the trends of the outlet air temperature and outlet sinter temperature with the increase of air mass flow rate, when the sinter mass flow rate is 1.4 kg/s and sinter particle diameter is 25 mm. As seen from the figure, the outlet air temperature and outlet sinter temperature both decrease while the air flow rate is increasing. According to the energy conservation law, for the given inlet sinter temperature and given sinter falling speed, the outlet air temperature decreases with the increase of the air flow rate. Then, the increase of the air flow rate results in an increase of the inlet air velocity, which is in favor of improving the convective heat transfer coefficient between the air and sinter, and causes an increase of the sinter cooling rate. Therefore, the outlet sinter temperature also decreases gradually. In addition, Figure 6a shows the average cross-sectional temperature distribution of the air and sinter along the height of the sinter bed layer for different air mass flow rates. Figure 6b shows the temperature differences between the sinter and air along the height of the sinter bed layer. For a given air mass flow rate, although the gas-solid temperature difference is not the same at the different heights of the bed, the temperature trends of the gas and solid are the same. This is an embodiment of the gas-solid steady heat transfer process based on the local non-equilibrium model.

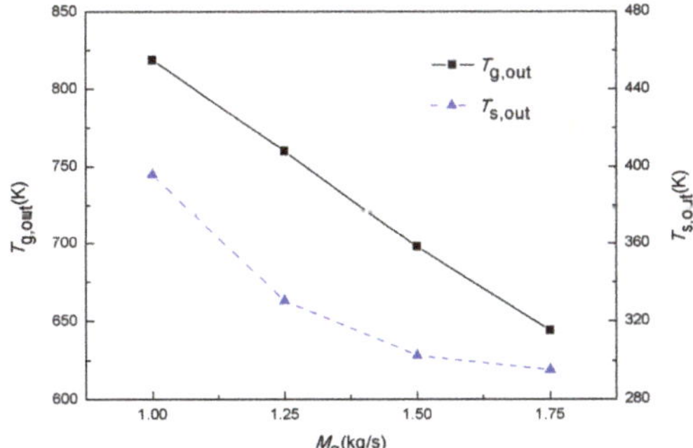

Figure 5. Influence of the air flow rate on the outlet air temperature and outlet sinter temperature.

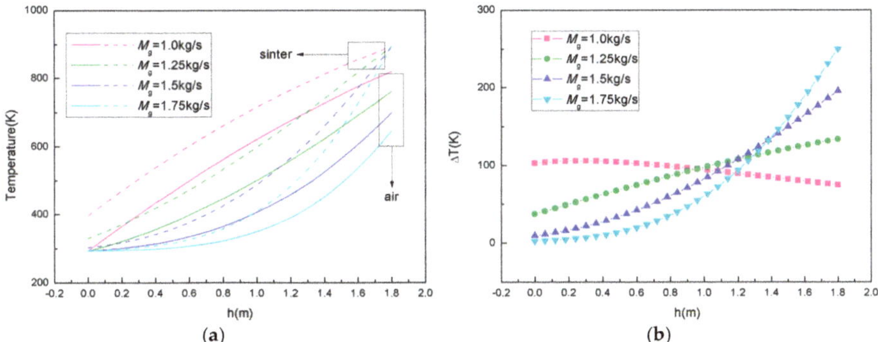

(a)

(b)

Figure 6. (a) The average cross-sectional temperature distribution of the air and sinter; (b) the temperature differences between the sinter and air along the height of the sinter bed layer for different air flow rates.

Figure 7 shows the influence of the air flow rate on the pressure drop of the sinter bed layer. As we can see, the pressure drop increases gradually with an increase of the air flow rate. This can be explained as the increased air flow rate leads to increased resistance and then results in an increase of the pressure drop in the vertical tank.

To comprehensively analyze the heat transfer quantity and energy quality of the convective heat transfer process, the influences of the air flow rate on the irreversible heat loss and the irreversible pressure loss were investigated. Figure 8 shows the distributions of the local heat transfer exergy destruction rate of the solid domain when the air flow rate is 1.0 kg/s, 1.25 kg/s, 1.5 kg/s, and 1.75 kg/s, respectively. According to Equation (13a), the local heat transfer exergy destruction rate is mainly caused by a temperature difference and it is related to the thermal conductivity. Since the thermal conductivity of sinter is much higher than that of air, for the convenience of the analysis, only the local heat transfer exergy destruction rate of sinter is given here. By analyzing and comparing Figures 6a and 8, it can be found that the temperature gradient trends of sinter correspond to the distributions of the local heat transfer exergy distribution rate. Figure 9 shows the exergy destruction caused by heat transfer and by fluid flow changing with the increase of the air mass flow rate, respectively. The local exergy destruction rate gradually increases when the air flow rate rises from 1.0 kg/s to 1.75 kg/s, as shown in Figure 8, which corresponds to the result of the exergy destruction caused by heat transfer in the Figure 9. As for the exergy destruction caused by fluid flow, it can be explained as the increase of the air flow rate leads to an increase of resistance during the flow, and then results in an increase of power consumption. Therefore, the exergy destruction by fluid flow gradually increases with the increase of the air flow rate.

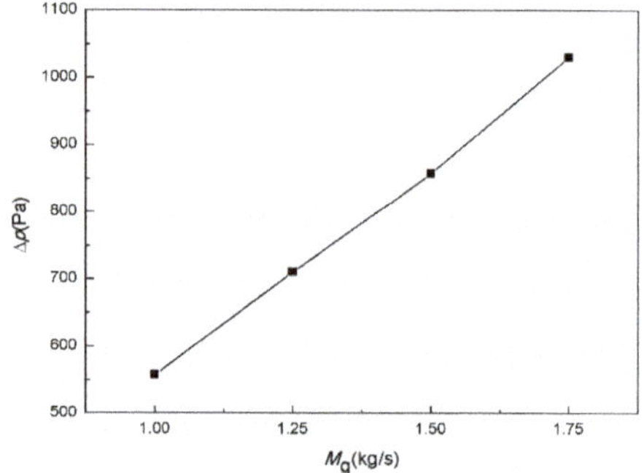

Figure 7. Influence of the air flow rate on the pressure drop.

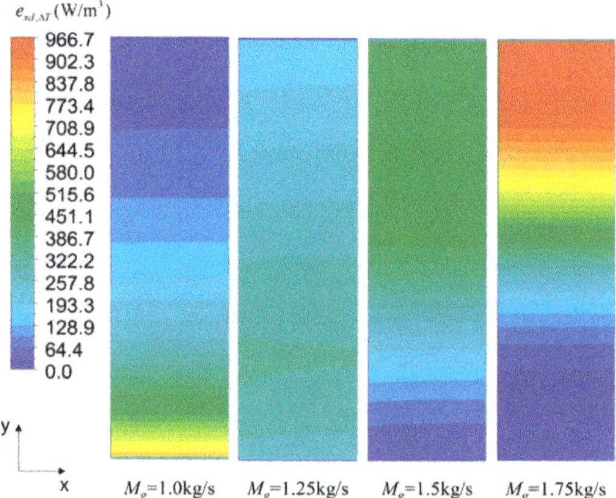

Figure 8. The distributions of the local heat transfer exergy destruction rate of sinter for different air mass flow rates.

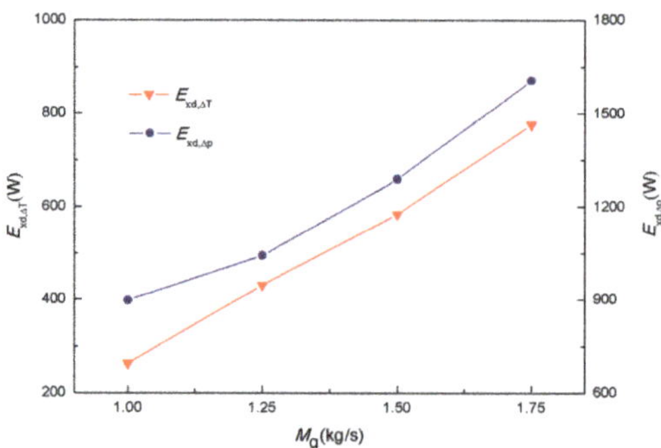

Figure 9. Influence of the air flow rate on exergy destruction caused by heat transfer and caused by fluid flow.

3.2.2. Influence of the Sinter Flow Rate

Figure 10 shows the variations of the outlet air temperature and outlet sinter temperature with the sinter mass flow rate, when the air mass flow rate is 1.25 kg/s and the sinter particle diameter is 25 mm. As we can see, with the increasing of the sinter flow rate, the outlet air temperature and outlet sinter temperature all increase. It can be explained as the increase of the sinter flow rate leads to a decrease of the resident time of sinter particles in the vertical tank. The sinter particles have not yet fully exchanged heat with the air and quickly fall to the bottom of the vertical tank. Therefore, the outlet sinter temperature increases with an increase of the sinter flow rate. Then, for a given air flow rate and sinter particle diameter, the outlet air temperature will increase because of the increase of the sinter average temperature. Furthermore, Figure 11a shows the average cross-sectional temperature distribution of air and sinter along the height of the sinter bed layer for different sinter mass flow rates. Figure 11b shows the temperature differences between sinter and air along the height of the sinter bed layer.

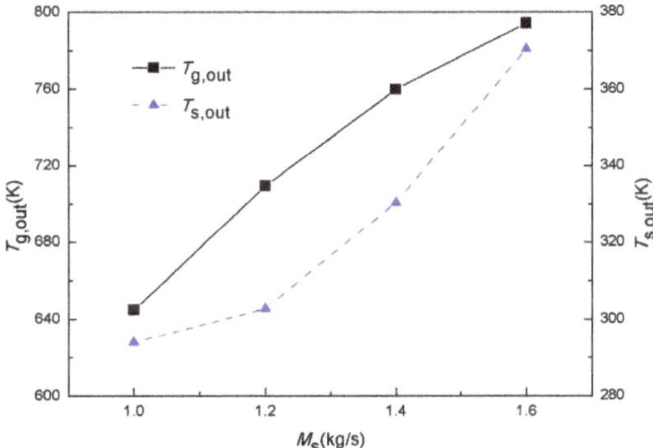

Figure 10. Influence of the sinter flow rate on the outlet air temperature and outlet sinter temperature.

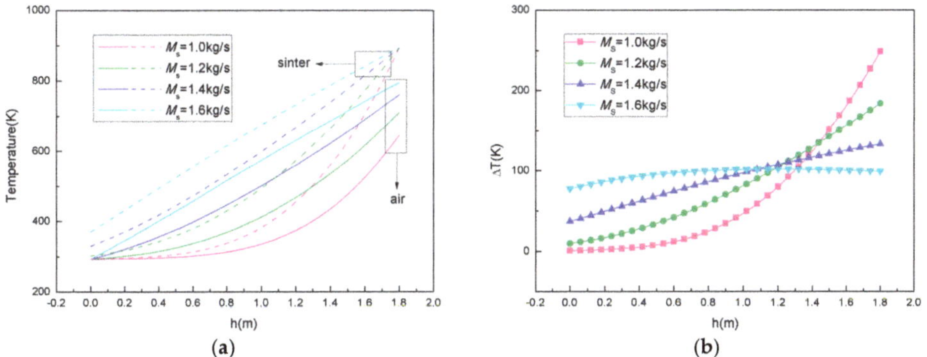

Figure 11. (a) The average cross-sectional temperature distribution of air and sinter; (b) the temperature differences between sinter and air along the height of the sinter bed layer of different sinter flow rates.

Figure 12 shows the influence of the sinter flow rate on the pressure drop from the bottom of the sinter bed layer to the top. Similar to the result caused by increasing the air flow rate, the pressure drop gradually increases with the increase of the sinter flow rate due to the increased resistance.

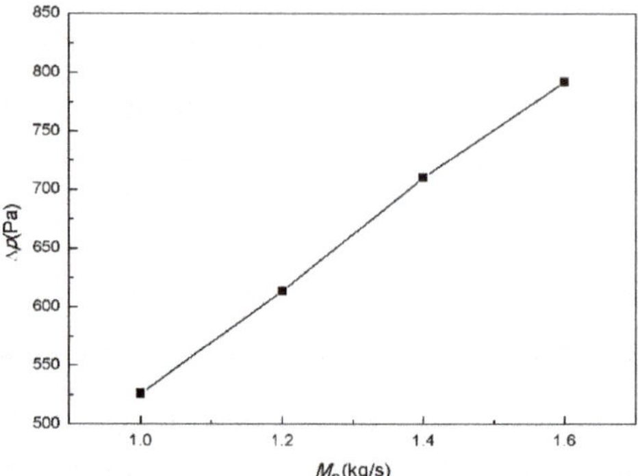

Figure 12. Influence of the sinter flow rate on the pressure drop.

Similarly, the influences of the sinter flow rate on thermal consumption and power consumption were investigated in this work. Figure 13 shows the distributions of the local heat transfer exergy destruction rate of the sinter for different sinter mass flow rates. When the sinter mass flow rate is 1.0 kg/s and 1.2 kg/s, the local heat transfer exergy destruction rate of the sinter increases along the height of the sinter bed layer. This result can be verified from the trend of sinter temperature variation in Figure 11a. Then, the local exergy destruction rate gradually decreases when the sinter flow rate rises from 1.0 kg/s to 1.6 kg/s, corresponding to the result of the exergy destruction caused by heat transfer in Figure 14. It indicates that the irreversible heat loss of the cooling section in the sinter tank decreases with the increase of the sinter mass flow rate. Figure 14 also shows the variety of exergy destruction caused by fluid flow with different sinter flow rates. Similar to the effect of the air flow rate, the increase of the sinter flow rate leads to an increase of the resistance during the flow, and then results in an increased power consumption and exergy destruction caused by the fluid flow increasing.

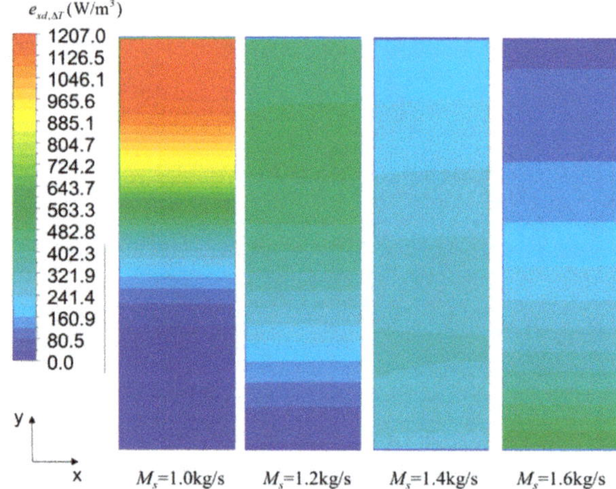

Figure 13. The distributions of the local heat transfer exergy destruction rate for different sinter mass flow rates.

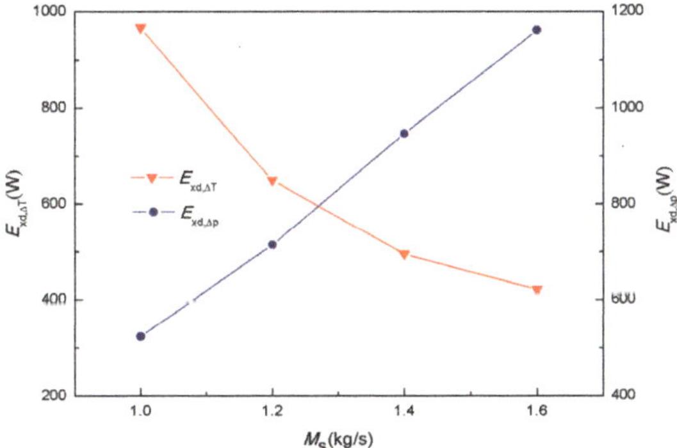

Figure 14. Influence of the sinter flow rate on exergy destruction caused by heat transfer and caused by fluid flow.

3.2.3. Influence of Sinter Particle Diameter

Figure 15 shows the variations of the outlet air temperature and outlet sinter temperature with the sinter particle diameter when air mass flow rate is 1.25 kg/s and sinter mass flow rate is 1.4 kg/s. With the increasing of the sinter particle diameter, the outlet air temperature decreases and the outlet sinter temperature increases. This is because the conduction heat transfer in the sinter particles declines due to the increase of particle thermal resistance caused by the increase of the sinter particle diameter. At the same time, the gas-solid heat transfer area decreases while the sinter particle diameter increases. Therefore, the sinter cooling rate decreases, and results in a decrease of the outlet air temperature and an increase of the outlet sinter temperature. In addition, Figure 16a shows the average cross-sectional temperature distribution of air and sinter along the height of the sinter bed layer for different sinter particle diameters. Figure 16b shows the temperature differences between sinter and air along the height of the sinter bed layer. Different from the condition with a sinter particle diameter of 15 mm, when the sinter particle diameter is 25 mm, 35 mm, and 45 mm, the temperature variation in the vertical tank is similar; the temperature difference between the sinter and air along the sinter bed layer height rises more gently.

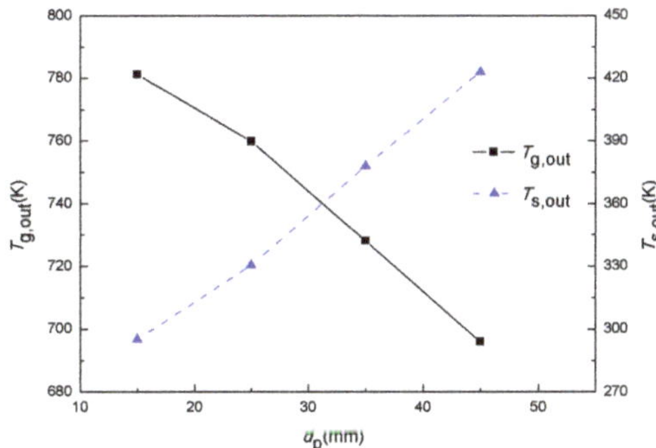

Figure 15. Influence of the sinter particle diameter on the outlet air temperature and outlet sinter temperature.

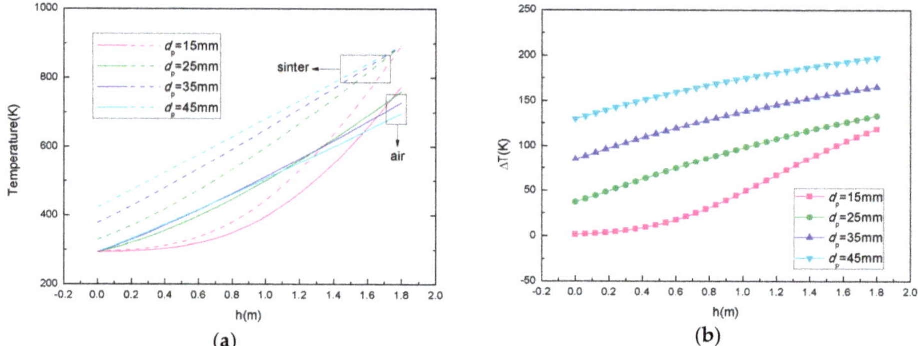

Figure 16. (**a**) The average cross-sectional temperature distribution of air and sinter; (**b**) the temperature differences between sinter and air along the height of the sinter bed layer for different sinter particle diameters.

Then, Figure 17 shows the influence of sinter particle diameter on the pressure drop from the bottom of the sinter bed layer to the top. The increase of the sinter particle diameter leads to an increase of the porosity of the sinter bed layer, which reduces the resistance and pressure drop of the air flow.

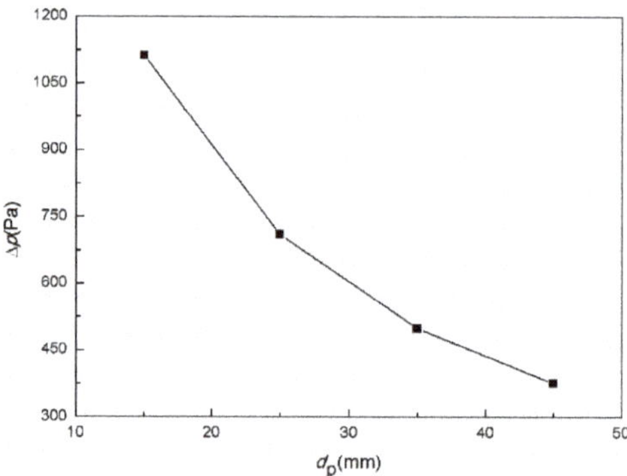

Figure 17. Influence of the sinter particle diameter on pressure drop.

We also performed exergy destruction analysis. Figure 18 shows the distributions of the local heat transfer exergy destruction rate of sinter when the sinter particle diameter is 15 mm, 25 mm, 35 mm, and 45 mm respectively. Additionally, the influences of sinter particle diameter on exergy destruction caused by heat transfer and caused by fluid flow can be seen in Figure 19. As we can see, the local exergy destruction rate gradually decreases when the sinter particle diameter rises from 15 mm to 45 mm, corresponding to the result of the exergy destruction caused by heat transfer in Figure 19. Figure 19 also shows that the exergy destruction caused by the pressure drop gradually decreases with the sinter diameter increasing due to the increased porosity of the sinter bed layer. It indicates that the smaller the particle diameter, the greater the heat dissipation and power consumption.

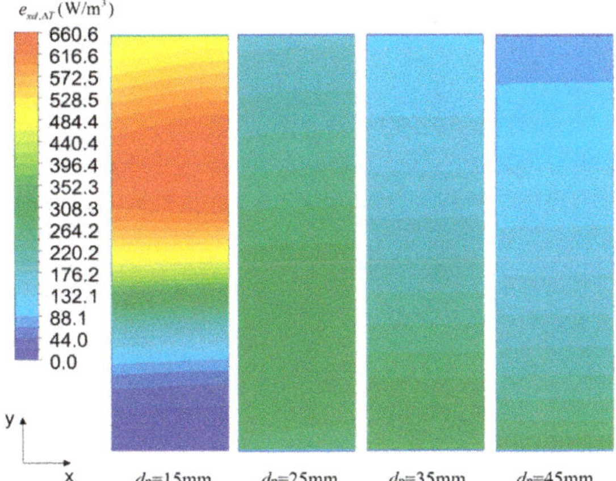

Figure 18. The distributions of the local heat transfer exergy destruction rate of sinter for different particle diameters.

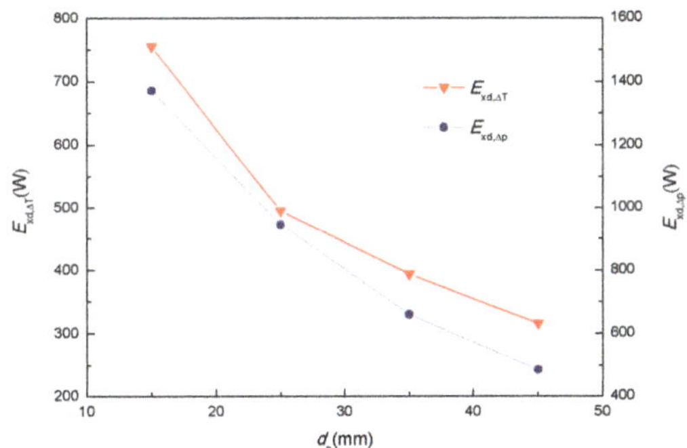

Figure 19. Influence of the sinter particle diameter on exergy destruction caused by heat transfer and fluid flow.

4. Parameter Optimization

The results of a single parameter's influences on the flow and heat transfer performance, respectively representing the irreversible heat loss and irreversible pressure loss of the waste heat recovery process, were obtained. Considering the two objectives are not independent, optimizing one of the goals must be at the expense of the other one during the parameter optimization. To obtain the suitable combination of the air mass flow rate, the sinter mass flow rate, and the sinter particle diameter, a multi-objective genetic algorithm adept in searching globally combined with a BP neural network adept in searching locally was employed in the parameter optimization, which improves the convergence accuracy and rate and reduces the workload greatly.

The structure of the BP neural network in the parameter optimization of the gas-solid heat transfer is shown in Figure 20. Sixty-four sets of samples were selected, of which 51 were used as training samples and 13 were used as test samples in the two neural networks with exergy destruction caused by heat transfer and by fluid flow, respectively. Each set of sample consists of four data, of which the air flow rate, the sinter flow rate, and the sinter particle diameter were used as the input layer neurons and the exergy destruction was used as output layer neurons.

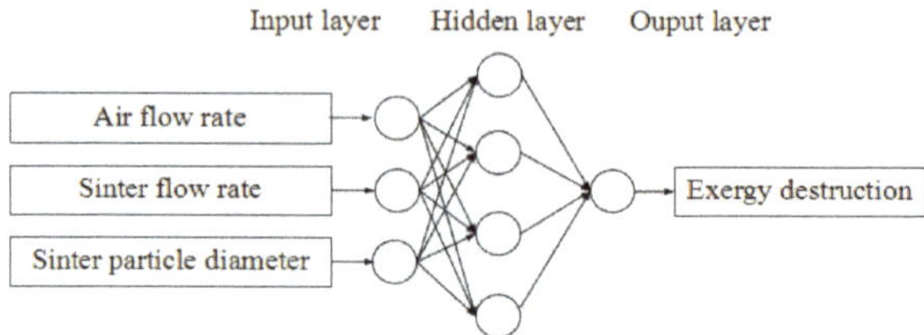

Figure 20. Structure of the back propagation (BP) neural network.

The multi-objective genetic algorithm with the exergy destruction caused by heat transfer and by fluid flow as the objectives sets the population size to 70, the generation to 300, and the convergence criterion to 10^{-4}. The initial range settings are 0.3 kg/s to 3 kg/s, 0.3 kg/s to 3 kg/s, and 3 mm to 50 mm for the air flow rate, sinter flow rate, and sinter particle diameter, respectively. The Pareto frontier derived from the multi-objective optimization algorithm is shown in Figure 21. The results show that $E_{xd,\Delta T}$ of Pareto solutions varies from 248.41 W to 278.67 W, while $E_{xd,\Delta p}$ varies from 230.16 W to 238.48 W, respectively. However, the best one cannot be chosen from Pareto front based only on the size of the two function values. Then, a decision maker technique, Technique for Order Preference by Similarity to an Ideal Solution (TOPSIS), was applied determine the best solution. The basic principle of TOPSIS is that the chosen one should have the shortest distance from the positive ideal solution and the longest distance from the negative ideal solution [30]. The results can be seen in Figure 21. The exergy destruction caused by heat transfer and by fluid flow of the optimal solution are 251.16 W and 232.53 W, respectively, and the corresponding value of the three parameters are 2.99 kg/s, 0.61 kg/s, and 32.8 mm, respectively, with the structural parameters fixed. The optimal solution minimizes the exergy destruction and minimizes the irreversible loss.

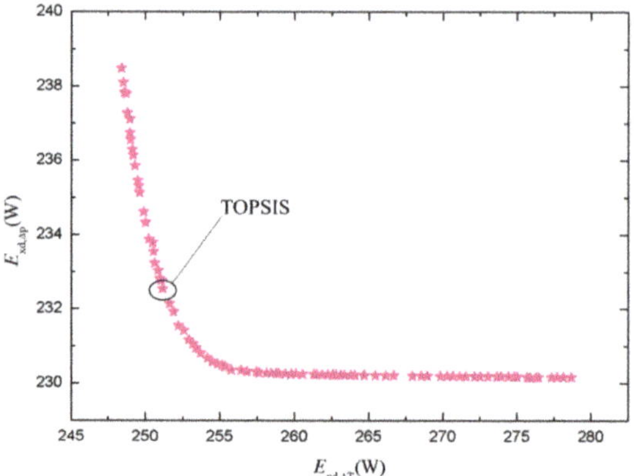

Figure 21. Pareto front.

5. Conclusions

The goal of this study was to investigate numerically how several operating parameters affect the gas-solid heat transfer process in a vertical tank for sinter waste heat recovery, and then to obtain an optimal combination of these operating parameters of a vertical tank with the given structural parameters through suitable evaluation and an optimization method. The following conclusions were acquired.

(1) The two-dimensional gas-solid heat transfer mathematical model, which was established based on the porous media model, standard turbulence model, and local thermal non-equilibrium model, can be used to investigate the heat recovery of the sinter.

(2) The numerical results show that the outlet air and sinter temperatures increase while decreasing the air flow rate or increasing the sinter flow rate. By increasing the sinter particle diameter, the outlet air temperatures decrease and the outlet sinter temperatures increase. The pressure drops decrease while increasing the air flow rate or the sinter particle diameter and decreasing the sinter flow rate or the sinter initial temperature.

(3) Using the exergy destruction minimization as the evaluation method, the exergy destruction caused by heat transfer and by fluid flow both gradually increase with an increase of the air mass flow rate. Additionally, the increase of the sinter flow rate results in a decrease of the exergy destruction caused by heat transfer and an increase of the exergy destruction caused by fluid flow. In addition, the exergy destruction caused by heat transfer and by fluid flow both gradually decrease with an increase of the sinter particle diameter.

(4) The optimal solution for exergy destruction minimization was obtained through the multi-objective genetic algorithm with a BP neural network. For the given structure of a vertical tank, the optimal operating parameters are 2.99 kg/s, 0.61 kg/s, and 32.8 mm for the air flow rate, sinter flow rate, and sinter diameter, respectively.

Adopting the principle of convective heat transfer enhancement based on exergy destruction minimization could provide a guide to develop new heat transfer techniques in waste heat recovery. For further engineering applications, the optimization methods mentioned in this paper can be used to optimize operating parameters when the structure of a vertical tank is given and can also be applied to obtain the most suitable structural parameters, such as the inner diameter and height of the vertical tank when the output of the sinter or other particulates is fixed. Therefore, the applications of exergy destruction minimization and the genetic algorithm would be helpful for further theoretical research on the enhancement of gas solid heat transfer and engineering applications of particulate waste heat recovery.

Author Contributions: Conceptualization, Z.L.; software, Z.L. and W.L.; methodology, formal analysis, Z.L., C.X. and S.W.; data curation, S.W.; investigation, C.X.; writing—original draft preparation, C.X.; writing—review and editing, C.X.; supervision, Z.L. and W.L.

Funding: This research was funded by the National Key R&D Program of China (No.2017YFB0603501).

Conflicts of Interest: The authors declare no conflict of interest.

Nomenclature

H	height of cooling section (m)
D	inner diameter of cooling section (m)
h	area heat transfer coefficient (W/(m^2·K))
h_V	volume heat transfer coefficient (W/(m^3·K))
$1/a$	viscous resistance coefficient (–)
C_2	inertial resistance coefficient (–)
c	specific heat (J/(kg·K))
ε	bed layer voidage (–)
d_p	sinter particle diameter (m)
ρ	Density (kg/m^3)
μ	dynamic viscosity (kg/(m·s))
λ	thermal conductivity (W/(m·K))
e	the available potential (J)
e_{xd}	the local exergy destruction rate (W/m^3)
$E_{xd,\Delta T}$	exergy destruction caused by heat transfer (W)
$E_{xd,\Delta p}$	exergy destruction caused by pressure drop (W)
u	superficial velocity (m/s)

Subscripts

g	air
s	sinter
in	inlet
out	outlet

References

1. Cai, J.J.; Wang, J.J.; Chen, C.X.; Lu, Z.W. Waste heat recovery and utilization in iron and steel industry. *Iron Steel* **2007**, *42*, 1–7. (In Chinese)
2. Ammar, Y.; Joyce, S.; Norman, R.; Wang, Y.D.; Roskilly, A.P. Low grade thermal energy sources and uses from the process industry in the UK. *Appl. Energy* **2012**, *9*, 3–20. [CrossRef]
3. Liu, Y.; Yang, J.; Wang, J.; Cheng, Z.L.; Wang, Q.W. Energy and exergy analysis for waste heat cascade utilization in sinter cooling bed. *Energy* **2014**, *67*, 370–380. [CrossRef]
4. Leong, J.C.; Jin, K.W.; Shiau, J.S.; Jeng, T.M.; Tai, C.H. Effect of sinter layer porosity distribution on flow and temperature fields in a sinter cooler. *Int. J. Min. Metall. Mater.* **2009**, *16*, 265–272. [CrossRef]
5. Shi, H.Z.; Wen, Z.; Zhang, X.; Lou, G.F. Numerical simulation and parameter analysis of high temperature sinter gas-solid heat transfer process. *Ironmak. Steelmak.* **2011**, *33*, 339–345.
6. Zhang, X.H.; Chen, Z.; Zhang, J.Y.; Ding, P.X.; Zhou, J.M. Simulation and optimization of waste heat recovery in sinter cooling process. *Appl. Therm. Eng.* **2013**, *54*, 7–15. [CrossRef]
7. Pan, L.S.; Wei, X.L.; Peng, Y.; Ma, Y.J.; Li, B. Theoretical study on the cooling procedure for vertical flow sinters. *Appl. Therm. Eng.* **2017**, *127*, 592–601. [CrossRef]
8. Feng, J.S.; Dong, H.; Gao, J.Y.; Liu, J.Y.; Liang, K. Exergy transfer characteristics of gas-solid heat transfer through sinter bed layer in vertical tank. *Energy* **2016**, *111*, 154–164. [CrossRef]
9. Caputo, A.C.; Cardarelli, G.; Pelagagge, P.M. Analysis of heat recovery in gas solid moving beds using a simulation approach. *Appl. Therm. Eng* **1996**, *16*, 89–99. [CrossRef]
10. Wen, Z.; Shi, H.Z.; Zhang, X.; Lou, G.F.; Liu, X.L.; Dou, R.F.; Su, F.Y. Numerical simulation and parameters optimization on gas-solid heat transfer process of high temperature sinter. *Ironmak. Steelmak.* **2011**, *38*, 525–529. [CrossRef]
11. Liu, Y.; Wang, J.Y.; Cheng, J.L.; Yang, J.; Wang, Q.W. Experimental investigation of fluid flow and heat transfer in a randomly packed bed of sinter particles. *Int. J. Heat Mass Transf.* **2016**, *99*, 589–598. [CrossRef]
12. Feng, J.S.; Dong, H.; Liu, J.Y.; Liang, K.; Gao, J.Y. Experimental study of gas flow characteristics in vertical tank for sinter waste heat recovery. *Appl. Therm. Eng.* **2015**, *91*, 73–79. [CrossRef]
13. Feng, J.S.; Dong, H.; Dong, H.D. Modification of Ergun's correlation in vertical tank for sinter waste heat recovery. *Powder Technol.* **2015**, *280*, 89–93. [CrossRef]
14. Feng, J.S.; Dong, H.; Gao, J.Y.; Liu, J.Y.; Liang, K. Experimental study of gas–solid overall heat transfer coefficient in vertical tank for sinter waste heat recovery. *Appl. Therm. Eng.* **2016**, *95*, 136–142. [CrossRef]
15. Kong, N.; Wen, Z.; Feng, J.X. Study on Mathematical Model of Flow and Heat Transfer Process in Dry Quenching Furnace. *Jin Autom.* **2004**, *3*, 27–30. (In Chinese)
16. Feng, J.S.; Dong, H.; Gao, J.Y.; Li, H.Z.; Liu, J.Y. Numerical investigation of gas-solid heat transfer process in vertical tank for sinter waste heat recovery. *Appl. Therm. Eng.* **2016**, *107*, 135–143. [CrossRef]
17. Gao, J.Y.; Feng, J.S.; Dong, H. Study of evaluation index for thermal parameters optimization of vertical tank for sinter waste heat recovery. *Metall. Energy* **2017**, *36*, 46–49. (In Chinese)
18. Liu, W.; Liu, Z.; Jia, H.; Fan, A.; Nakayama, A. Entransy expression of the second law of thermodynamics and its application to optimization in heat transfer process. *Int. J. Heat Mass Transf.* **2011**, *54*, 3049–3059. [CrossRef]
19. Jia, H.; Liu, W.; Liu, Z.C. Enhancing convective heat transfer based on minimum power consumption principle. *Chem. Eng. Sci.* **2012**, *69*, 225–230. [CrossRef]
20. Wang, J.; Liu, Z.; Yuan, F.; Liu, W.; Chen, G. Convective heat transfer optimization in a circular tube based on local exergy destruction minimization. *Int. J. Heat Mass Transf.* **2015**, *90*, 49–57. [CrossRef]
21. Wang, J.B.; Liu, W.; Liu, Z.C. The application of exergy destruction minimization in convective heat transfer optimization. *Appl. Therm. Eng.* **2015**, *88*, 384–390. [CrossRef]

22. Wang, J.B.; Liu, Z.C.; Liu, W. Evaluation of convective heat transfer in a tube based on local exergy destruction rate. *Sci. China Technol. Sci.* **2016**, *59*, 1494–1506. [CrossRef]
23. Liu, W.; Liu, P.; Wang, J.B.; Zheng, N.B.; Liu, Z.C. Exergy destruction minimization: A principle to convective heat transfer enhancement. *Int. J. Heat Mass Transf.* **2018**, *122*, 11–21. [CrossRef]
24. Liu, W.; Liu, P.; Dong, Z.M.; Yang, K.; Liu, Z.C. A study on the multi-field synergy principle of convective heat and mass transfer enhancement. *Int. J. Heat Mass Transf.* **2019**. [CrossRef]
25. Ge, Y.; Liu, Z.C.; Liu, W. Multi-objective genetic optimization of the heat transfer for tube inserted with porous media. *Int. J. Heat Mass Transf.* **2016**, *101*, 981–987. [CrossRef]
26. Ge, Y.; Shan, F.; Liu, Z.; Liu, W. Optimal Structural Design of a Heat Sink With Laminar Single-Phase Flow Using Computational Fluid Dynamics-Based Multi-Objective Genetic Algorithm. *J. Heat Transf.* **2018**, *140*, 022803. [CrossRef]
27. Ge, Y.; Liu, Z.C.; Sun, H.; Liu, W. Optimal design of a segmented thermoelectric generator based on three-dimensional numerical simulation and multi-objective genetic algorithm. *Energy* **2018**, *147*, 1060–1069. [CrossRef]
28. Ge, Y.; Wang, S.C.; Liu, Z.C.; Liu, W. Optimal shape design of a minichannel heat sink applying multi-objective optimization algorithm and three-dimensional numerical method. *Appl. Therm. Eng.* **2019**, *148*, 120–128. [CrossRef]
29. Sun, H.N.; Ge, Y.; Liu, W.; Liu, Z.C. Geometric optimization of two-stage thermoelectric generator using genetic algorithms and thermodynamic analysis. *Energy* **2019**, *171*, 37–48. [CrossRef]
30. Yoon, K.P.; Kim, W.K. The behavioral TOPSIS. *Expert Syst. Appl.* **2017**, *89*, 266–272. [CrossRef]

Article

Development of an Exergy-Rational Method and Optimum Control Algorithm for the Best Utilization of the Flue Gas Heat in Coal-Fired Power Plant Stacks †

Birol Kılkış

Turkish Society of HVAC&R Engineers, Ankara 06680, Turkey; baskan@ttmd.org.tr

† Submitted to the SDEWES Special Issue of Energies based on the conference paper "An Exergy-Based Optimum Control Algorithm for Rational Utilization of Waste Heat from the Flue Gas of Coal-Fired Power Plants" in the Proceedings of the 13th SDEWES Conference as underlined in the Acknowledgements.

Received: 8 January 2019; Accepted: 12 February 2019; Published: 25 February 2019

Abstract: Waste heat that is available in the flue gas of power plant stacks is a potential source of useful thermal power. In reclaiming and utilizing this waste heat without compromising plant efficiency, stacks usually need to be equipped with forced-draught fans in order to compensate for the decrease in natural draught while stack gas is cooled. In addition, pumps are used to circulate the heat transfer fluid. All of these parasitic operations require electrical power. Electrical power has unit exergy of almost 1 W/W. On the contrary, the thermal power exergy that is claimed from the low-enthalpy flue gas has much lower unit exergy. Therefore, from an exergetic point of view, the additional electrical exergy that is required to drive pumps and fans must not exceed the thermal exergy claimed. Based on the First-Law of Thermodynamics, the net energy that is saved may be positive with an apparently high coefficient of performance; however, the same generally does not hold true for the Second-Law. This is a matter of determining the optimum amount of heat to be claimed and the most rational method of utilizing this heat for maximum net exergy gain from the process, under variable outdoor conditions and the plant operations. The four main methods were compared. These are (a) electricity generation by thermoelectric generators, electricity generation with an Organic-Rankine Cycle with (b) or without (c) a heat pump, and (d) the direct use of the thermal exergy that is gained in a district energy system. The comparison of these methods shows that exergy-rationality is the best for method (b). A new analytical optimization algorithm and the exergy-based optimum control strategy were developed, which determine the optimum pump flow rate of the heat recovery system and then calculate how much forced-draft fan power is required in the stack at dynamic operating conditions. Robust design metrics were established to maximize the net exergy gain, including an exergy-based coefficient of performance. Parametric studies indicate that the exergetic approach provides a better insight by showing that the amount of heat that can be optimally recovered is much different than the values given by classical economic and energy efficiency considerations. A case study was performed for method (d), which shows that, without any exergy rationality-based control algorithm and design method, the flue gas heat recovery may not be feasible in district energy systems or any other methods of utilization of the heat recovered. The study has implications in the field, since most of the waste heat recovery units in industrial applications, which are designed based on the First-Law of Thermodynamics, result in exergy loss instead of exergy gain, and are therefore responsible for more carbon dioxide emissions. These applications must be retrofitted with new exergy-based controllers for variable speed pumps and fans with optimally selected capacities.

Keywords: flue gas heat recovery; exergy; coefficient of performance; thermoelectric generator; organic rankine cycle; district energy systems

---

1. Introduction

About two-thirds of the energy of the input fuel in conventional thermal power plants is wasted in stacks and cooling towers. Although, from environmental and health points of view, thermal power plants should be located relatively far from cities, in many developing countries, especially old coal-fired thermal power plants with quite low thermal efficiencies (even less than 30%) are located in close proximity to metropolitan cities, such as in New Delhi. Air pollution is a significant concern in these areas, including the Badarpur coal power plant [1]. For example, the air-quality index rose to 1010 on 8 November 2017 [2]. While the only reason for pollution is not the coal-fired power plants, the waste heat from these plants may be utilized in district energy systems to partly offset air pollution by substituting the need for thermal energy, like the heating of buildings, if the reclaimed heat from the power plants that can be delivered to the built environment is not small [3,4]. In typical coal-fired power plants, the condenser is run through an open-loop water circuit while using river water. Although river water is returned to the river without much loss, the water temperature is substantially increased. Flue combustion gases are rejected through a stack. Other industries that have the potential for waste heat recovery in their stacks, in addition to thermal power plants, include the textile industry [5]. In other studies, flue gas heat recovery is found to be beneficial, especially in high-moisture coals that are based on the First-Law of Thermodynamics [6]. The same type of approach is applicable to studies regarding pressurized pulverized coal combustion [7].

Findings that lead to an indication of environmental and economic benefits are mostly based on energy savings. Researchers have also looked into exergy analyses, but almost all of the research was limited to the component basis in order to determine the major exergy destruction points and the overall exergy efficiency [8]. Kaushik et al. [9] determined that the exergy analyses were used to determine the components with the greatest exergy destruction, especially the boiler in coal-based thermal power plants. Heat recovery steam generators and flue gas exhausts to the stack were the focus of other analyses, e.g., [10]. In a typical coal-fired power plant, almost two-thirds of the energy of the coal that is consumed is wasted in the form of heat [11]. In Figure 1, the combined-cycle thermal power plant with a bottoming cycle, which is shown in inset (a) generates only power and all of the waste heat is rejected by some means to the environment, mostly by cooling towers to the atmosphere, which also wastes water. This conflicts with the environment, energy, and water nexus needs of today's world [12]. In inset (b) of Figure 1, however, a part of the heat in the flue gas is recovered in the stack. Condenser heat may also be reclaimed in this process, which reduces the need and the size of the cooling towers. However, if the flue gas is cooled too much in the heat recovery process, then the natural draught in the stack decreases to a point that a draught fan might be needed, which consumes electrical energy. The difference between the unit exergies of electric power and reclaimed thermal power is an important performance metric, which is often ignored. The reclaimed heat then may be utilized in a district heating loop, in addition to the power that is delivered to the customers.

According to EU/2004/8/EC [13], because heat is delivered to customers that are located outside in a metered and useful manner, such a plant qualifies as a cogeneration plant (CHP). According to the same Directive, this plant may qualify for a high-efficiency plant if such a plant results in at least 10% fuel savings in terms of Primary Energy Savings Ratio (*PES*). However, the district heating system needs a pumping station (PS) in the district in order to circulate the heating water in the district loop and a heat exchanger (HE) at the interface of the two closed thermal loops, namely D and P. To recover the heat from the flue gas (in loop P) a pumping system is required. Exergy is gained in the form of heat, but at the same time, exergy is required in the pumps, which is mostly driven by electric power. If the exergy gain in the heat recovery process is less than the exergy demand of the draught fan and circulating pumps, then this system will not contribute from an exergy rationality point of view. The amount of heat that may be recovered has an optimum value, which depends on the stack height (*H*). The stack height may be chosen to be higher during the design stage of the power plant instead of installing a fan to the stack. Yet, this requires embedded exergy of additional material and additional construction work for the stack. This requires a careful optimization for the stack height. Stack height, H, is crucial both in the performance of the power plant and waste heat recovery. In thermal power plants, especially if coal is used, then H may be quite high in order to keep the pollutants elevated enough in the atmosphere. Therefore, H in meters is also a function of the pollutant (Sulfur) content. Today, in large thermal power plants, the stack height may reach up to 150 m. The minimum stack height is calculated by the following equation in the literature:

$$H \geq 4.33 \cdot (F \cdot S)^{0.3} \tag{1}$$

In Equation (1), 4.33 is a factor of the rule of thumb in order to assure sufficient natural draught, P_D (Also see Equation (12) in Section 2.6.2), F is the coal consumption rate in kg per hour, and S is the Sulphur content in the percentages of the coal used. If, for example, a thermal power plant consumes 40 ton (40,000 kg) per hour of coal with 0.4% Sulphur content, then the necessary stack height will be 79 m.

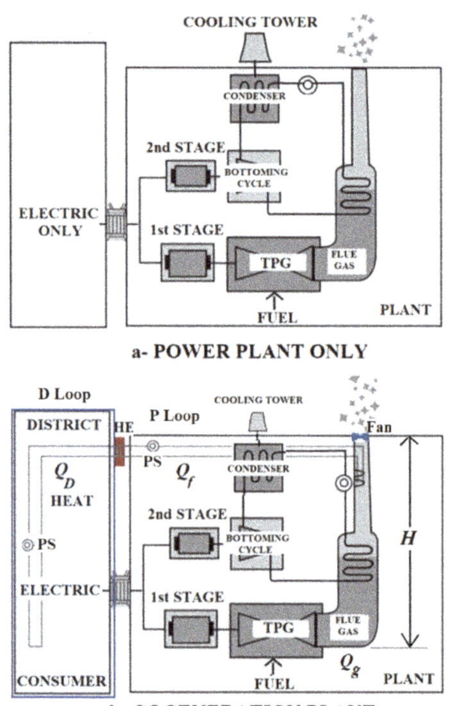

Figure 1. Adaptation of a coal plant to a cogeneration system.

There are four main methods that are available in the literature for utilizing the thermal power exergy claimed along the stack height, once it is reclaimed from the flue gas. These methods, which appear in the literature, may be grouped in the following manner, as will be compared later in the manuscript:

(a) electricity generation by Thermo-electric generator (TEG) arrays [14,15],

(b) electricity generation by an Organic-Rankine Cycle (ORC) without a heat pump [16,17],

(c) electricity generation by ORC turbines with a heat pump [18,19], and

(d) direct utilization of the thermal exergy in the stack in a district energy system [20,21].

In all of the above methods (a)–(d), the forced-draught fan in the stack and fluid circulation pumps are necessary for claiming the thermal power exergy, in addition to the power conversion equipment.

If, for all models, the thermal energy claimed is fixed (100 kW in following methods) and the flue gas inlet temperature is the same for all models, then the forced-draught fan power demand is the same, but the circulation pump power demand changes, depending upon the equipment that is used in each model.

Aims of the Research Work

The main objective of this research is to first close the gap in the literature by a new model concerning the missing investigation of exergy rationality regarding the industrial waste heat utilization with parasitic losses. In so-doing, the study compares the different specific methods of utilizing the heat that is available in the flue gas of coal-fired thermal power plant stacks by developing exergy-based new metrics and a control algorithm in order to maximize the benefits of utilizing such waste heat in the industry and the built environment. Each method was analyzed in terms of the First and Second-Law efficiency, the coefficient of performance (*COP*), and exergy-based *COP* (*COPEX*) prior to the analysis of a case study for the selected methods.

2. Materials and Methods

Waste heat recovery from hot flue gas flowing across the stacks of power plants have a wide potential, but such systems must be exergetically investigated, designed, and controlled in order to achieve a net positive exergy gain and net-negative carbon dioxide (CO_2) emissions responsibility in practice. The comparison of the methods (a) to (d) is provided in the following.

2.1. Method (a): Power and Heat Generation with Thermoelectric Generators

Oswaldo et al. [14] presented the design and development of a new solid-state TEG using thermoelectric modules for the experimental analysis of the technical viability for the uses of waste heat in industrial processes, like forging, hot rolling, industrial refrigeration systems, boilers, and ceramic kilns for a maximum temperature gradient of 1073 K. Their prototype test results that were based on the First-Law of Thermodynamics showed the feasibility of TEGs in reclaiming the industrial waste heat. However, the study did not directly consider the Second-Law and the parasitic exergy demand due to the use of a cooling circuit with electrically driven pump. Memon and Khawaja [15] have also developed a direct heat harvesting method while using a hot plate with heat-stove TEG for electrical performance testing. They acknowledged the presence of electric power demand for cooling the TEG units by forced-air circulation. Yet, their analysis did not address the exergy difference between the thermal exergy that was reclaimed and the electrical power exergy.

In this context, the use of an array of TEG modules facing the flue gas in the stack that are externally cooled by a hydronic circuit with a pump is depicted in Figure 2 for further analysis in this study. The heat that is recovered by the cooling circuit and the electric power that are generated by the TEG array are to be connected to external loads. The arrangement in Figure 2 comprises method (a).

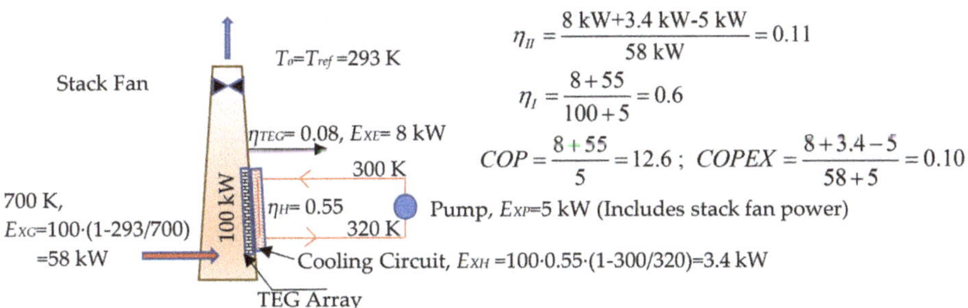

Figure 2. Exergy claim with thermo-electric generator (TEG) arrays cooled with an external hydronic loop.

According to Figure 2, 8 kW of electrical power is generated by the TEG array that has a First-Law efficiency of 0.08. This amount is, at the same time, the amount of electrical power exergy because the unit exergy of electricity is virtually 1 kW/1 kW. Heat from the back side of TEG array is reclaimed by a hydronic circuit and heat exchangers, with an efficiency of 0.55 by using pump(s), which demand 5 kW of electric power (including stack fan power demand), while 100 kW of thermal power at 700 K drives the system. An additional draught fan compensates for the natural draft loss due to the cooling of the flue gas. The Second-Law efficiency is 0.11, while the First-Law efficiency is 0.60 and the *COP* is 12.6. Such a relatively high *COP* leads to the impression that such a system is beneficial and techno-economically attractive. Yet, if *COP* is re-defined in terms of exergy (Equation (2)), namely *COPEX*, then the exergo-environmental properties of the model changes substantially.

$$COPEX = \frac{\text{Net Exergy Gain}}{\text{Total Exergy Input}} = \frac{8\ \text{kW} + 3.4\ \text{kW} - 5\ \text{kW}}{58\ \text{kW} + 5\ \text{kW}} = 0.10 \tag{2}$$

2.2. Method (b): Power Generation with Organic Rankine Cycle Turbines

Another method that is investigated in the literature is driving an ORC turbine after reclaiming the heat of the flue gas and transporting it to the external of the stack. A low-temperature external load through a hydronic circuit that is driven by a pump cools the ORC. The electric power that is generated by ORC turbines depends upon the source temperature while the First-Law efficiency is around 10%. Lecompte et al. [16] applied ORC technology to an electric arc furnace with a First-Law efficiency of about 13%. This efficiency is higher when compared to other lower source temperature applications. Another study investigated the technical aspects of modular ORC systems over a broad range of heat source temperatures with different working fluids [17]. A payback period that was slightly less than five years was predicted at a source temperature of around 700 K based on the First-Law and economic benefits of ORC systems. The principle of this method (b), as modeled in this study, is extended in Figure 3.

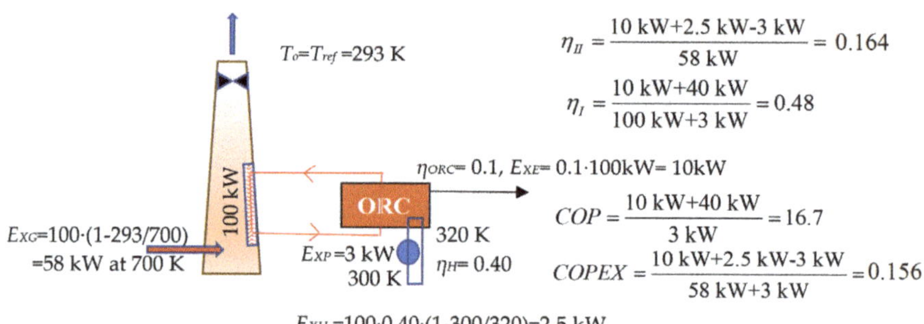

Figure 3. Exergy claim by Organic-Rankine cycle (ORC) and its waste heat.

2.3. Method (c): Power Generation with Organic Rankine Cycle and then Converting to Heat by a Heat Pump

By converting electric power to thermal power with a COP greater than 1, a heat pump may be tailored to the ORC, depending upon the dominant load on the demand side. This method (c) is shown in Figure 4. Re-converting claimed power exergy to thermal power exergy even with a COP, of 4.0 in this case, is not rational. For a break-even condition, the COP of the heat pump needs to be $1/(1 - 300\ \text{K}/320\ \text{K}) = 16$, which is practically not possible in the field with today's technology. This indicates that several heat recovery systems may not be exergetically rational and practical while they are responsible for more CO_2 emissions than saved. This gap is also evident in the study of He et al. [18], in which the feasibility of reclaiming additional electrical power from the waste heat of a fuel cell by an ORC system coupled with a heat pump was investigated, thus establishing a bottoming cycle. This study claimed that, at an optimum operating temperature, the results are favorable according to the First-Law. Among the few studies that involved the exergy concept in the evaluation of medium-temperature heat recovery from industrial gases with ORC technology and CO_2 transcritical cycles, Ayachi et al. [19] have presented their pinch analysis for source temperatures between 438 K and 420 K. However, the case was only taken in a pinch problem domain.

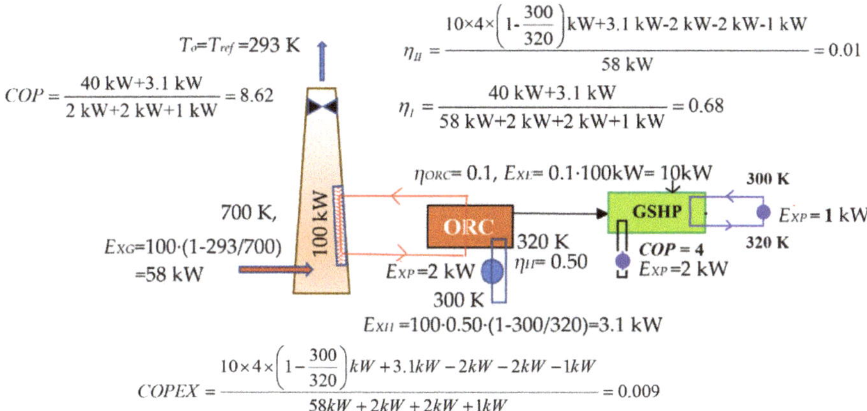

Figure 4. Exergy claim by ORC and it's waste heat where ORC power is then used in a Ground-source Heat Pump (GSHP).

2.4. Method (d): Direct use of the Thermal Exergy in ae District Energy System

Especially when a district energy system is close-by, waste heat from power plants gains additional importance. One method is to recover heat directly from the flue gas. Based on the First-Law, Xu et al. [20] had considered reclaiming heat from the ventilating air methane in a hot-air power generator. The authors had concluded that the overall system efficiency for a decarbonizing process with a hot air power generator reached 27.1%, which is better than the standalone reference systems. The arrangement of tube banks in the stack needs careful investigation. A horizontal tube or an inclined tube bank occupying the whole cross-sectional area of the stack may be effective but reduces the natural draft. Tube banks that are attached to the stack walls, which leave a large portion of the cross section free for the gas flow, may be preferable, but a careful case-by-case analysis needs to be conducted. In this aspect only, Erguvan and MacPhee [21] had performed energy and exergy analyses for several tube bank configurations in waste heat recovery applications. The method of direct heat recovery from the flue gas is depicted in Figure 5 with the relevant exergy formulations.

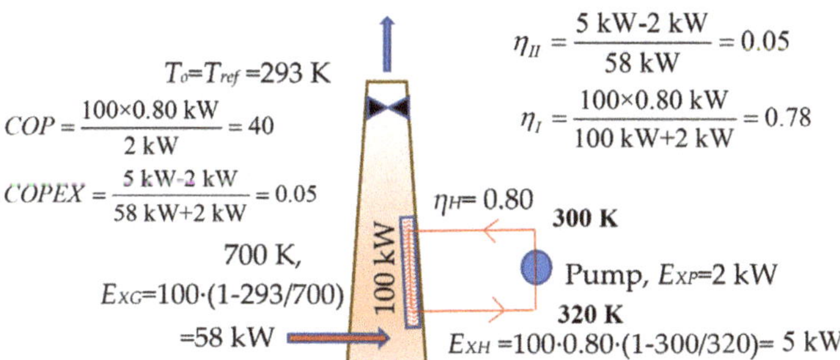

Figure 5. Thermal exergy recovery.

2.5. The Necessity for an Exergy-Based Model

There is a necessity for developing an exergy-based model for a thorough comparison of methods (a) to (d). In addition, the heat that is optimally recovered at the stack(s) of a large power plant, Q_f, may be utilized in a district energy system. In this case, the distance between the power plant and the district, L, which is subject to the condition $L < L_{max}$, where L_{max} is the maximum distance, should be exergetically rational as well as the exergy that is demanded by the pumping stations, E_{XPS}. Exergy rationality for both of the loops is defined in Equations (3) and (4).

$$Q_D \cdot \left(1 - \frac{T_{Do}}{T_{Di}}\right) > E_{XPS} \qquad \text{(Condition 1)} \qquad (3)$$

$$Q_f \cdot \left(1 - \frac{T_{fo}}{T_{fi}}\right) > (E_{XCP} + E_{XF}) \qquad \text{(Condition 2)} \qquad (4)$$

2.6. Characterization of the District Energy Model

Figure 6 shows the new model, which enables the user to analyze several design parameters and environmental conditions, along with plant performance variables that are related to the stack. Performance constants, namely (*c*) and (*a*) regarding the exergy demand, E_{XCP} of the circulation pump mainly depend upon the circuit length, pump type, and piping material and size. In the case study to be shown later in this paper, these terms are assumed to be 15 and 0.3, respectively. Figure 7 then shows the exergy flow bar for a typical heat recovery process, where the average temperature of the flue gas is 700 K. The heat recovery loop (P loop) operates between 313 K and 363 K, the average outdoor temperature in the summer T_o (also, the reference temperature, T_{ref}), which renders a critical draft condition as compared to the winter at 293 K. Subsequently, one of the metrics of the Rational Exergy Management Model (REMM) as the REMM Efficiency, ψ_R [22], as provided in Equation (5), is 0.24 [23]. This is a similar value with a solar water heater system, since the major unit exergy destruction is prior to the heat recovery, and only a low-enthalpy heat is recovered without power generation upstream, like ORC.

$$\psi_R = \frac{\varepsilon_{dem}}{\varepsilon_{sup}} = \frac{\left(1 - \frac{313}{363}\right)}{\left(1 - \frac{293}{700}\right)} = 0.24 \qquad (5)$$

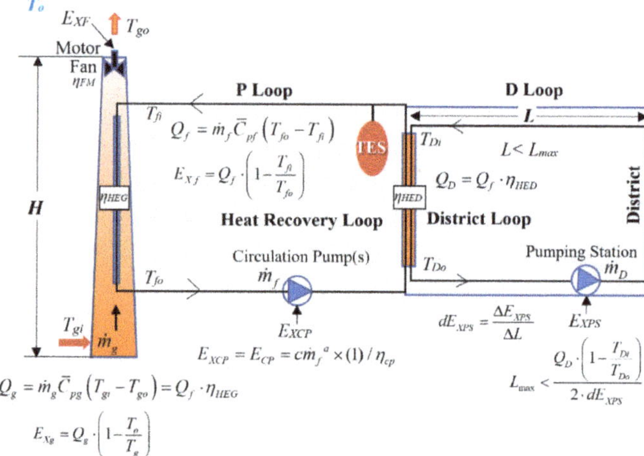

Figure 6. Flue gas heat recovery and district energy system model.

Although the waste heat recovery might seem to be economically beneficial and environmentally useful, if the exergy of the electrical energy that is used for the draught fan, along with pumping and driving other ancillary devices, is higher than the recovered lower exergy-heat, then the system will not contribute to the economy and the environment at large.

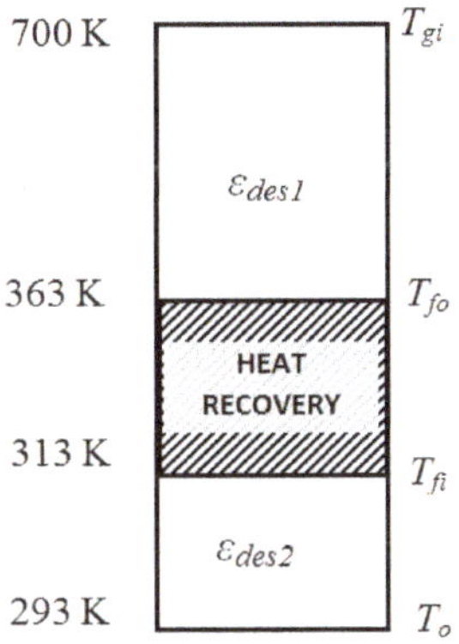

Figure 7. Exergy flow bar for the heat recovery of the flue gas in the stack.

2.6.1. Circuit P Loop: Exergy-based Optimum Heat Recovery Drive Unit for Industrial Stacks

This condition may be expressed by the objective function in Equation (6), which represents the net exergy gain in this circuit. The net exergy gain, E_{XNET} must be maximized by minimizing E_{XF} and E_{XCP} while exergy of E_{xf} is claimed from the flue gas. If the district energy system is integrated, then E_{xD} replaces E_{xf} and E_{XPS} is introduced (Equation (7)). In certain cases, the reduction of the size of the cooling tower due to the conversion of waste heat rejection to useful heat in the stack may be substantial. Subsequently, an exergy saving term that is associated with the reduction of ancillary power demand of the cooling towers may be added to the right-hand side of the other original formulations for this research work, as given below (Equations (7)–(11)). In Equation (10), the specific heat needs to be calculated at the average fluid temperature. In Equation (11), the coefficient c and the exponent a are approximated coefficients for the pump characteristics.

$$E_{XNET} = E_{xf} - E_{XF} - E_{XCP} \qquad \text{(Maximize)} \tag{6}$$

$$E_{XNET} = E_{xD} - E_{XF} - E_{XCP} - E_{XPS} \tag{7}$$

$$E_{Xf} = Q_f \cdot \varepsilon_f \tag{8}$$

$$\varepsilon_f = \left(1 - \frac{T_{fi}}{T_{fo}}\right) \tag{9}$$

$$Q_f = \dot{m}_f \cdot \overline{C}_{pf} \cdot \rho_f \cdot \left(T_{fo} - T_{fi}\right) \tag{10}$$

$$E_{XCP} = c \cdot \dot{m}_f{}^a \times (1)/\eta_{cp} = E_{CP} \tag{11}$$

2.6.2. Natural Draught Pressure and E_{XF}

Since heat is drawn from the hot flue gas in the stack, the flue gas cools down and the average flue gas temperature in the stack T_g reduces, which reduces the natural draught pressure (P_D). Equation (12) shows the relationship between T_g and P_D [24]. According to this equation, draught decreases with a decrease in the average flue gas temperature. At a given time t, the change in the draught pressure,

ΔP_D, only depends on the amount of decrease in ΔT_g if some heat is recovered from the flue gas, as given in Equation (14).

$$P_D = C \cdot P_{atm} \cdot H \cdot \left(\frac{1}{T_o} - \frac{1}{T_g} \right) = C' \cdot \left(\frac{1}{T_o} - \frac{1}{T_g} \right) \qquad \text{(kPa)} \qquad (12)$$

where,

$$T_g \sim \frac{T_{gin} + T_{go}}{2} \qquad (13)$$

$$\Delta P_D = \left(\frac{C'}{T_g{}^2} \right) \cdot \Delta T_g \qquad \text{(kPa)} \qquad (14)$$

where,

C	A constant for coal-fired plant stacks, 0.0342 K·m^{-1}
C'	$C \cdot P_{atm} \cdot H$
P_{atm}	Atmospheric pressure at the given elevation at the given time, kPa
H	Stack height, m
T_g	Average flue gas temperature in the stack, K
T_{gi}	Hot flue gas entry temperature to the stack, K
T_{go}	Flue gas exit temperature from the stack, K.

In order to resume the original natural draught, the installation of a new hot-gas type industrial forced-draught fan in the stack or oversizing, an existing one will be necessary. For new power plants with flue gas heat recovery, the stack height may be selected higher in order to avoid a fan, but in this case, the embodiment recovery of energy and exergy of the extra stack material and construction must be considered against energy and exergy savings by the associated heat recovery. The motors of these fans may operate on on-site power in a thermal power plant. Especially, if the power plant is a combined-cycle plant, the average power generation efficiency is about 0.52 in Europe [13]. For other industrial plants, like in the textile industry, electricity is mostly received from the grid. The average efficiency of grid-feeding power plants, especially in developing countries, is about 30% or even less. This efficiency, when coupled with transmission and transformation losses, makes the case even worse based on the total CO_2 emissions responsibility in a broader perspective if the fuel mix mainly consists of fossil fuels [25]. This additional CO_2 emissions responsibility in terms of the equivalent useful heat that is to be lost in a typical thermal power plant also needs to be considered in the calculations if the grid power is used in the heat recovery system. The draught fan simultaneously needs to maintain the original $\dot{m}_g$, which is given by Equation (15) [24].

$$\dot{m}_g = 0.65 \cdot \overline{A}_g \sqrt{2gH \cdot \left(1 - \left[\frac{T_o}{T_g} \right] \right)} \qquad \left(\text{m}^3/\text{s} \right) \qquad (15)$$

Equation (15) takes the molar mass of the flue gas and the outside air as equal, and the frictional resistance and heat losses in the stack walls and any heat exchanger inside the stack to be negligible. Here, $\overline{A}_g$ is the average cross-sectional area of the stack, in the unit of m^2, along with stack height (H). On the other hand, if T_g decreases, then $\dot{m}_g$ also decreases:

$$\Delta \dot{m}_g = \frac{0.65 \cdot \overline{A}_g}{2\sqrt{2gH \cdot \left(1 - \frac{T_o}{T_g} \right)}} \cdot \left(\frac{T_o}{T_g{}^2} \right) \cdot \Delta T_g \qquad \left(\text{m}^3/\text{s} \right) \qquad (16)$$

Therefore, while the draught fan needs to compensate for both the decreases in the gas flow rate and the decrease in the draught pressure, the required fan power, E_F (almost equal to fan exergy demand), at a given electric motor and fan efficiency of η_{FM} will consist of the product of ΔP_D and $\Delta \dot{m}_g$, as given in Equation (17).

$$E_F = E_{XF} = \frac{\Delta P_D \cdot \Delta \dot{m}_g}{\eta_{FM}} \qquad \text{(kW)} \qquad (17)$$

2.7. Rating Metrics

Four metrics were identified for rating the performance of waste heat recovery from the flue gas. These metrics may be related to CO_2 emissions responsibility, energy efficiency, and exergy efficiency.

2.7.1. Performance Coefficients

The metrics that are related to performance coefficients are given in Equations (18)–(20). In Equation (18), the exergy-based coefficient of performance ($COPEX$) is a product of the COP and the unit exergy gain from waste heat recovery, $\Delta \varepsilon_f$. Usually, the unit exergy gain, namely $(1 - T_{fi}/T_{fo})$, is small. Therefore, $COPEX$ will be far less than one, if a proper design recovery system is installed and the fluid flow rate is not dynamically controlled with an exergy-based control algorithm and a frequency-controlled electric motor for the circulation pump. This means that the heat recovery system is not rational. For example, if these temperatures are 300 K and 360 K, respectively, then the unit exergy of the fluid at the supply point, $\Delta \varepsilon_f$ is only 0.167 W/W. From Equation (18), the traditional COP value of the recovery system must be greater than 6 in order to satisfy the condition of $COPEX$ approaching one. This condition imposes strong design constraints on E_{XF} and E_{XCP}. This is an early indication regarding the importance and relevance of the Second-Law of Thermodynamics before attempting an economic and environmental analysis about waste heat recovery from the flue gas. A similar discussion of the author outside the context of waste heat recovery from flue gas is also valid for ground-source heat pumps [26]. In this case the heat pump only consumes electrical energy and provides thermal energy. Then:

$$COPEX = COP \cdot \Delta \varepsilon_f = COP \times \left(1 - \frac{T_{fi}}{T_{fo}}\right) \qquad (COPEX \to 1) \qquad (18)$$

$$COPEX = \frac{E_{XNET}}{E_{XF} + E_{XCP}} \qquad (19)$$

$$COP = \frac{Q}{E_{XF} + E_{XCP}} \qquad (COP > 1) \qquad (20)$$

2.7.2. Fuel Savings

The exergy-based Primary Energy Savings Ratio, PES_R, in percentage is based on 1 W of power generation. The original PES term in EU/2004/8/EC [13] was modified by the REMM model [25,27]. In Equation (21), $CHPE_\eta$ is a known value (power generation efficiency) for the given plant at design conditions [27]. The term 0.73 corresponds to $(1 - \psi_{Rref})$, which is related to the reference value of $(1 - 0.27)$ for the on-site CHP applications. The values 0.52 and 0.80 are the reference efficiencies for separate power and heat generation, respectively [13], based on 1 W of power generation.

$$PES_R = \left(1 - \frac{1}{([CHPE\eta \cdot (\frac{1}{0.52} + \frac{Q_f}{0.80})] \cdot \frac{0.73}{(1-\psi_R)})}\right) \times 100 \qquad (21)$$

2.7.3. Carbon Dioxide Emissions Replacement

The third metric is related to the CO_2 emission replacement rate. The CO_2 emissions savings rate (kg CO_2/s) that is attributable to net exergy recovery, E_{XNET}, from the flue gas for a pulverized coal power plant with coal properties of adiabatic flame temperature of 2850 K and 30,000 kJ/kg lower heating value (LHV) with no condensation in the stack and the wet coal input, as well as the CO_2 content of 3.6 kg CO_2/kg coal, was derived, which is given in Equation (22) [28,29]:

$$\dot{C}O_2 = E_{XNET} \times 1.7 \cdot 10^{-4} \tag{22}$$

2.7.4. Thermal Efficiencies

As the fourth metric, the First-Law and Second-Law efficiencies, η_I and η_{II}, may be monitored for the rating and evaluation of the system, as given in Equations (23) and (24).

$$\eta_I = \frac{Q_f}{Q_g + E_{CP}} \tag{23}$$

$$\eta_{II} = \frac{E_{Xf} - E_{XCP}}{E_{Xg}} \tag{24}$$

3. Results

Table 1 compares methods (a) to (d) with source exergy at 58 kW prior to the selection of a method. As observed from Table 1, method (d) has the highest thermal efficiency (η_I) and the second highest COP, η_{II} and $COPEX$. Other methods have greater inconsistencies. Method (d) delivers heat at 340 K in this example, which is sufficiently high for district energy systems and it may also be used in absorption/adsorption cooling equipment. Provided that the distance between the plant and the district is not too far, method (d) is selected as a feasible option for practical applications. In the case study, a coal-based thermal power plant is going to be retrofitted with a flue gas heat recovery system. The design inputs and conditions are given in Table 2. The equations that are given in the method may be expressed in terms of the fluid (water) flow rate, $\dot{m}_f$. Thus, by introducing the flow rate $\dot{m}_f$, in all terms, then taking a derivative, and equating it to zero to find the optimum flow rate. Equation (5) may be maximized. This will be a time-dependent solution in terms of all time-dependent variables that either depend on atmospheric conditions and part-load conditions, like T_g, $\dot{m}_g$, T_o, T_{fo}, T_{fi}, and variables that are related to the D Loop, as identified in previous Figure 6.

Table 1. Comparison of the four methods with source exergy at 58 kW.

Method	Claimed Exergy, kW			η_{II}	Exergy Input, kW	η_I	COP	$COPEX$ (Colum 3/Colum 5) (Including Source Exergy)
	Net Electrical Exergy, kW	Net Thermal Exergy, kW	Total Net Exergy, kW					
	1	2	3	4	5	6	7	
a	3	3.4	6.4	0.11	5 + 58	0.60	12.6	0.10
b	7	2.5	9.5	0.164	3 + 58	0.48	16.7	**0.156**
c	0	0.6	0.6	0.01	5 + 58	0.68	8.62	0.009
d	−2	5	3	0.15	2 + 58	0.78	**40**	0.05

Table 2. Design data for the case study.

Variable	Value	Unit and Comments
T_o	293	K (Outdoor temperature)
T_{gi}	600	K (Flue gas inlet temperature, constant)
T_{go}	600	K (Initial value, varies with heat recovery)
T_g	600	K (Initial value, varies with heat recovery)
$\dot{m}_g$	86.154	m^3/s (Design value, varies with heat recovery)
C_{pg}	1.151	kJ/kg·K at 750 K
C_{pf}	4.187	kJ/kg·K at 300 K
E_g	15982.64	kW

Table 2. *Cont.*

Variable	Value	Unit and Comments
E_{xg}	8177.783	kW
A_g	5	m^2 (Average cross-sectional stack area)
H	79	m (Stack height)
h	860	m (Altitude from sea level)
P_{atm}	91.41271	kPa (Corrected with altitude)
T_{fo}	363	K (Design input)
T_{fi}	345	K (Design output)
ΔT_f	18	K ($T_{fo} - T_{fi}$)
T_{fav}	354	K ($T_{fi} + \Delta T_f/2$)
η_{HE}	0.85	Assumed constant
η_{FM}	0.8	Assumed constant
ρ_f	986.1687	kg/m^3 (Corrected for temperature)
ρ_g	0.525	kg/m^3 (To be corrected for temperature and gas composition)
η_{cp}	0.75	Assumed constant

3.1. Results of the Analysis of the P Loop-Case Sdudy

In this study, a step-by-step solution is introduced, where the flow rate is changed incrementally to find the optimum point, which is subjected to the condition that the flue gas outlet temperature is not lower than the condensation temperature, T_{gc}, which is often taken at 420 K for coal combustion and low Sulfur content fuel oils [30]. Reducing the temperature of the flue gas in the stack may be associated with other adverse phenomena. It is very important that the minimum flue gas temperature is maintained and not violated in order to prevent the acid dew point from being reached. Figure 8 shows the results for the given initial and operating conditions. According to Figure 8 with a flue gas temperature of 600 K, there is an optimum fluid flow rate, which maximizes the net exergy gain from the flue gas. However, for certain cases of design and operational variables, an optimum fluid flow rate may not be found. In such a case, the solution approaches either the lowest flow rate or the highest flow rate, within an economical range. The variation of the maximum E_{NET} points for different outdoor-air temperatures when the fluid flow rate is fixed at each optimum point is shown in Figure 9. According to Figure 9, E_{NET} decreases with an increase in the outdoor-air temperature. If the inlet temperature of the flue gas is 500 K instead of 600 K, then *COPEX* decreases below the threshold value of one and the optimum solution for the fluid flow rate approaches the lower bound, which is shown in Figure 10.

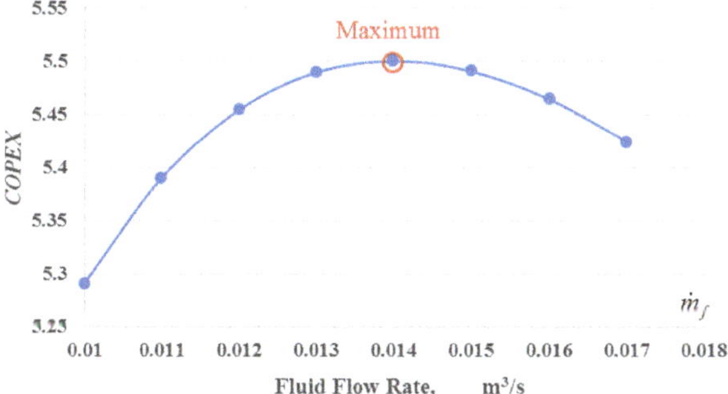

Figure 8. Variation of exergy-based coefficient of performance (*COPEX*) with a fluid flow rate with flue gas temperature 600 K. Note: Here Flue gas input exergy is ignored in calculating *COPEX*.

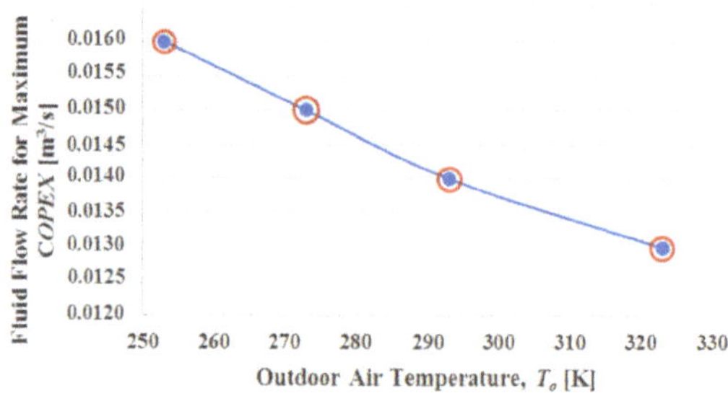

Figure 9. Variation of *COPEX* with outdoor air temperature, T_0. Note: Here flue gas input exergy is included.

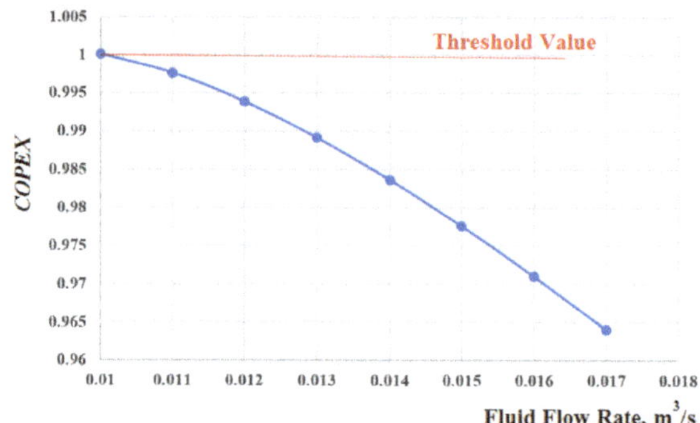

Figure 10. Variation of COPEX with a fluid flow rate with flue gas temperature 500 K.

According to Figure 10, the maximum *COPEX* that is possible is just one at the lowest flow rate, which means that the exergy that is gained by the circulating fluid is equaled to the total exergy demand of the parasitic equipment, such as the stack fan and the circulating pump. In other words, the net exergy gain is zero. These results show that the performance is highly sensitive to the flue gas inlet temperature, outdoor temperature, and the flow rate of the fluid in the P loop. Therefore, the circulation pump must be dynamically and digitally controlled to maintain *COPEX* as greater than one. There may be cases when the heat recovery system needs to be stopped. The solution starts with a minimum fluid flow rate and the temperature decrease in the flue gas in the stack is calculated first (see Equation (25)). Subsequently, the other performance variables are calculated, as provided in Table 3.

$$T_{go} = T_{gi} - \left(\frac{\dot{m}_f}{\dot{m}_g} \right) \cdot \left(\frac{C_{pf}}{C_{pg}} \right) \left(\frac{\rho_f}{\rho_g} \right) \cdot \left(T_{fo} - T_{fi} \right) \tag{25}$$

Table 3. Calculations for the case study. Note: Here Flue gas input exergy is ignored in calculating *COPEX*.

$\dot{m}_f$ (m³/s)	$\dot{m}_g$ (m³/s)	$\Delta \dot{m}_g$ (m³/s)	T_{go} (K)	ΔT_g (K)	ΔT_f (K)	E_{xf} (kW)	ΔP_D (Pa)	E_{XF} (kW)	E_{XCP} (kW)	*COPEX* (Dimensionless)
0.01	85.73	0.42	587.87	6.07	18	36.85	3.69	1.94	5.024	5.291502
0.011	85.69	0.46	586.65	6.67	18	40.54	4.06	2.35	5.169	5.390105
0.012	85.65	0.51	585.44	7.28	18	44.23	4.43	2.80	5.306	5.454215
0.013	85.60	0.55	584.22	7.89	18	47.91	4.79	3.29	5.435	5.489238
0.014	85.56	0.59	583.01	8.49	18	51.60	5.16	3.82	5.557	5.499905
0.015	85.52	0.64	581.80	9.10	18	55.28	5.53	4.40	5.674	5.490328
0.016	85.48	0.68	580.58	9.71	18	58.97	5.90	5.01	5.785	5.464054
0.017	85.43	0.72	579.37	10.31	18	62.65	6.27	5.66	5.891	5.424127

3.2. Implications of the Analysis for the D Loop

The heat recovered at the stack(s) of a large power plant, Q_f may be utilized in a district energy system. Such a D Loop, as identified in the previous Figure 6, should also consider the distance between the power plant and the district L, which is subject to the condition $L \leq L_{max}$ and the exergy that is demanded by the pumping stations, E_{XPS}. Exergy rationality for this condition is defined in Equation (26):

$$Q_D \cdot \left(1 - \frac{T_{Dout}}{T_{Din}}\right) > E_{XPS} \tag{26}$$

where,

$$Q_D = Q_f \cdot \eta_{HED} \tag{27}$$

The exergy demand of power stations in the district circuit is a function of Q_D for a given ΔT in the circuit, which is typically 20 K at the design conditions in district heating. The constant term a_o in Equation (29) is an empirical value, which is 0.6 km. In Equation (30), n depends upon the exergy of district heating supply in terms of T_{do}. The value of T_{ref} is 283.15 K (the average ground temperature in winter). 333.15 K represents a traditional district heating supply fluid temperature of 60 °C.

$$\Delta T = (T_{Din} - T_{Do}) \tag{28}$$

$$L \leq L_{max} = a_o + \left(\frac{Q_D}{1000}\right)^n \times \left(\frac{\Delta T}{20}\right)^{1.3} \qquad (Q_D > 1000 \text{ kW·h}, \ \Delta T \leq 30 \text{ °C}) \tag{29}$$

$$n = 0.6 \times \left(\frac{\left(1 - \frac{T_{ref}}{T_{Do}}\right)}{\left(1 - \frac{T_{ref}}{333.15}\right)}\right)^{(1/3)} \tag{30}$$

$$T_{Do} = T_{fo} - \Delta T_{HED} \tag{31}$$

$$T_{fo} = T_{go} - \Delta T_{HEG} \tag{32}$$

The typical temperature drop in the heat exchangers in a district energy system is targeted for 2.5 K each. Therefore:

$$T_{Do} \sim T_{go} - 5 \text{ K} \tag{33}$$

$$dE_{XPS} = \frac{\Delta E_{XPS}}{\Delta L} \tag{34}$$

$$L_{max} < \frac{Q_D \cdot \left(1 - \frac{T_{Di}}{T_{Do}}\right)}{2 \cdot dE_{XPS}} \tag{35}$$

3.3. Development of the Control Unit to Maximize the Exergy Gain

The maximization of the net exergy gain and *COPEX* is related to the flow rate of the fluid through the circulating pump in the P loop first. All of the relevant inputs are gathered every five minutes and then processed for the optimum flow rate for this given time increment. The processed information for the optimum flow rate is fed to the driver of the variable-speed pump. The optimum pump speed is then checked for condensation of the flue gas in the stack ($T_{go} > T_{gc}$) and the pump capacity. The flue gas temperature entering the stack fan is also checked against the temperature resistance of the fan. Efficiency changes in the heat exchangers are purposefully ignored. Figure 11 shows a simple flowchart of the process. If the thermal demand on the district side is lower than the optimum heat output at the same time interval, and then the surplus heat is stored in a thermal storage system (TES) before the necessary amount is sent to the district through a heat exchanger.

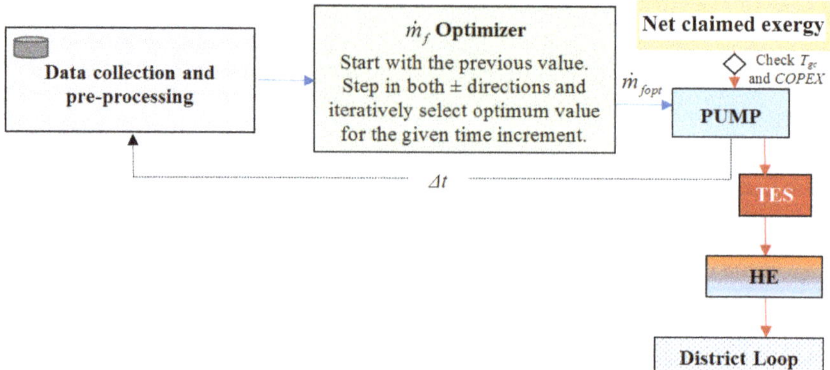

Figure 11. Data processing and control of the pump speed.

The data that is collected along with pollution variables, like the CO_2 concentration and the sulfur content, is quite crucial in monitoring the environmental impact of all industrial installations on an instantaneous basis. Thermal power plants are not an exception as a major player in global warming. Therefore, such performance and control data that is available in thermal power plants may be incorporated into a central Internet-based control center and the data re-processed for a national or even global scale to optimize not all the thermal power plants, but also all of the industrial installations. This will then complete the exergy economy loop, which is very important for decoupling sustainable growth from CO_2 emissions [31].

4. Conclusions

Waste heat recovery from the flue gas in thermal power plant stacks are important from environmental and economical points of view, especially in developing countries, where the plant efficiencies are low and mostly in relatively close vicinity of urban areas. This provides the opportunity for replacing some CO_2 emissions in the urban area if the recovered heat is distributed in the form of a district heating system. However, there are certain challenges, one of the most important of which is that the design of such recovery systems need to be based on the Second-Law of Thermodynamics due to the fact that the unit exergy of the heat that is claimed is much lower than the unit exergy of electricity that is used to drive the related ancillaries and installed draught fan to the stack. In this respect, the minimum COP of the heat recovery system, on average, has to be 6.0. This poses serious challenges for the design of the gas-to-water type of heat exchangers. This study has shown that, not only in a poor design, but also at any time during the operation, the net exergy gain may be negative (*COPEX* < 1), because the exergetic performance is very sensitive to operational and climatological conditions. The case study has further exemplified this possibility, such that COPEX may become less than one any time, even with incremental variations in the dynamic conditions. In order to avoid this risk, a complete analytical model was presented, which enables the analysis of such a system in terms of the optimum water flow rate in the closed heat reclaiming circuit for the maximum exergetic coefficient of performance at given outdoor and operating conditions of the power plant.

In plant retrofits for this purpose, however, the stack should not be replaced with a new one with a new heat exchanger inside. Instead, the original stack must be retrofitted with minimal work. In fact, this is the second challenge. In order to avoid this challenge on a large-scale plant, the heat recovery unit may be a separate unit before the stack. However, this will disturb the original design performance of the stack and a larger fan might be needed in addition to other ancillaries that are related to the separate heat recovery unit. A cassette type heat exchanger or heat exchangers may be inserted at different levels of the stack with minimum invasion and minimum disturbance to the flow [32,33]. There are other studies in the literature, which bring the heat source to condensation temperature and then remove the moisture and recover more heat [34]. However, a similar approach will increase the exergy demand of the fan system to discharge the flue gas, since there will be almost no natural draught left in the remaining voyage of the flue gas. Under all of these considerations, the model that is developed in this study addresses multiple challenges and promises optimal heat recovery controls. Case study and performance calculations regarding four methods showed that the definition of *COP* is limited to economic evaluations while the "ambient, free" resources like air, sea water, and ground are ignored. In fact, especially when waste heat in the industry is considered, the waste heat needs to be considered as an input energy and exergy, while it is not so abundant like air or sea water and they have definitely a quantifiably potential of added value in the energy sector.

Funding: This research received no external funding.

Acknowledgments: The manuscript is a revised and expanded version of an original scientific contribution that was presented at the 13th Conference on Sustainable Development of Energy, Water and Environment Systems (SDEWES) that was held during 30 September and 4 October 2018 in Palermo, Italy entitled "An Exergy-Based Optimum Control Algorithm for Rational Utilization of Waste Heat from the Flue Gas of Coal-Fired Power Plants." Four models for flue gas exergy recovery were identified and compared explicitly in terms of *COP*, *COPEX*, First and Second-Law efficiencies and the direct thermal utilization in a district energy system model was selected by using this new comparison algorithm. All text was revised and four new figures were added.

Conflicts of Interest: The author declares no conflict of interest.

Nomenclature

$\overline{A}_g$	Average cross-sectional area of the stack, m^2
a_0	Constant distance in L_{max} equation, km
a, c	Performance constants of the circulating pump
C	A constant for the coal-fired plant stack height, 0.0342 K·m^{-1}
C_p	Specific heat, kJ/kg·K
$CHPE\eta$	Partial first-law efficiency of electric power generation, dimensionless
COP	Coefficient of performance, dimensionless
$COPEX$	Exergetic coefficient of performance, dimensionless
H	Stack height, m
L	Distance between the district heating system and the power plant (one way), m
E, Q	Thermal power, kW
E_X	Exergy, kW
F	Hourly coal consumption, kg/h
$\dot{m}_g$	Gas flow rate, m^3/s
$\dot{m}_f$	Heat transfer fluid flow rate, m^3/s
n	power in L_{max} equation
P	Pressure, kPa
P_D	Natural draught pressure, kPa
PES	Primary energy savings ratio, %
PES_R	Exergy-based primary energy savings ratio, %
S	Sulphur content, %
T	Temperature, K
Greek Symbols	
ε	Unit exergy, kW/kW
$\Delta\varepsilon$	Unit exergy gain, kW/kW
η	Efficiency, dimensionless
ψ_R	REMM efficiency, dimensionless
Δ	Difference
ρ	Density, kg/m^3

Subscripts

atm	Atmospheric
CP	Circulating pump
D	District, draught
dem	Demand
des	Destroyed
Di	Return from the district loop
Do	To the district loop
I	First-law
II	Second-law
F	Stack fan
FM	Fan motor
f	Fluid (water in P Loop)
fi	Inlet fluid (water)
fo	Outlet (supply) fluid
g, G	flue gas (to/at the stack)
gc	Gas condensation
gi	Inlet gas (to the stack)
go	Outlet gas (from the stack)
H	Stack heat exchanger
max	Maximum
NET	Net
o	Outdoor
PS	Pumping station
R	Rational exergy management related
ref	Reference
sup	Supply
X	Exergy

Acronyms

CFPP	Coal-fired power plant
CHP	Combined heat and power
DE	District energy
EC	European commission
EU	European union
EC	European council
GSHP	Ground-source heat pump
HE	Heat exchanger
ORC	Organic-Rankine cycle
REMM	Rational exergy management model
TEG	Thermo-electric generator
TES	Thermal energy storage
TPG	Thermal to electric power conversion

References

1. The Economic Times. The Badarpur Plant's Effect on Air Pollution and Why it Needs to be Shut Down. 2015. Available online: https://economictimes.indiatimes.com/the-badarpur-plants-effect-on-air-pollution-and-why-it-needs-to-be-shut-down/changetheair_show/53669369.cms (accessed on 16 February 2019).
2. Irfan, U.; Vox. How Delhi Became the Most Polluted City on Earth. Available online: https://www.vox.com/energy-and-environment/2017/11/22/16666808/india-air-pollution-new-delhi (accessed on 25 November 2017).
3. Kilkis, B.I.; Gungor, C. Effective Use of Low-Enthalpy Renewable Energy Resources in District Heating and Cooling. In Proceedings of the National Clean Energy Symposium, Istanbul, Turkey, 15–17 November 2000; pp. 731–738. (In Turkish)

4. Kilkis, I.B. Ein aus verschiedenen Kreislaufen zusammengesetztes, geotermisches Energiesystem für die Stadt Denizli, Geothermie-Energie der Zukunft, Geeste. In Proceedings of the Tagusband der 4. Geothermischen Fachtagung in Konstanz, Germany, 18–20 September 1997; pp. 413–424.

5. Saghafifar, M.; Omar, A.; Mohammadi, K.; Alashkar, A.; Gadalla, M. A review of unconventional bottoming cycles for waste heat recovery: Part I—Analysis, design, and optimization. *Energy Convers. Manag.*. in press. [CrossRef]

6. Wang, D.; Bao AKunc, W.; Liss, W. Coal power plant flue gas waste heat and water recovery. *Appl. Energy* **2012**, *91*, 341–348. [CrossRef]

7. Suresh, M.V.J.J.; Reddy, K.S.; Kolar, A.K. Thermodynamic analysis of a coal fired power plant repowered with pressurized pulverized coal combustion. *J. Power Energy* **2011**. [CrossRef]

8. Lu, V.; Sun, F.; Shi, Y. Exergy Analysis of Advanced Boiler Flue Gas Heat Recovery System in Power Plant. In Proceedings of the CSEE, Cluj-Napoca, Romania, 2–5 May 2012.

9. Kaushik, S.C.; Reddy, V.S.; Tyagi, S.K. Energy and exergy analyses of thermal power plants: A review. *Renew. Sustain. Energy Rev.* **2011**, *15*, 1857–1872. [CrossRef]

10. Woudstra, N.; Woudstra, T.; Pirone, A.; van der Stelt, T. Thermodynamic evaluation of combined cycle plants. *Energy Convers. Manag.* **2010**, *51*, 1099–1110. [CrossRef]

11. Kilkis, B. Analysis of cogeneration systems and their environmental benefits. *TTMD J.* **2007**, *48/26*, 15.

12. Kilkis, Ş.; Kilkis, B. Integrated circular economy and education model to address aspects of an energy-water-food nexus in a dairy facility and local contexts. *J. Clean. Prod.* **2017**, *167*, 1084–1098. [CrossRef]

13. EU; The European Parliament; Council of the European Union. Directive 2010/31/EU of the European Parliament and of the Council of 19 May 2010 on the energy performance of buildings. *Off. J. Eur. Union* **2010**, *53*, 13–55.

14. Ando, O.H., Jr.; Calderon, N.H.; De Souza, S.S. Characterization of a Thermoelectric Generator (TEG) system for waste heat recovery. *Energies* **2018**, *11*, 1555. [CrossRef]

15. Memon, S.; Khawaja, N.T. Experimental and analytical simulation analyses on the electrical performance of thermoelectric generator modules for direct and concentrated quartz-halogen heat harvesting. *Energies* **2018**, *11*, 3315. [CrossRef]

16. Oyewunmi, O.; Markides, C.; Lazova, M.; Kaya, A.; van den Broek, M.; De Paepe, M. Case study of an Organic Rankine Cycle (ORC) for waste heat recovery from an Electric Arc Furnace (EAF). *Energies* **2017**, *10*, 649. [CrossRef]

17. Markus, P.; Dieter, B. Thermo-economic evaluation of modular organic rankine cycles for waste heat recovery over a broad range of heat source temperatures and capacities. *Energies* **2017**, *10*, 269. [CrossRef]

18. He, T.; Shi, R.; Peng, J.; Zhuge, W.; Zhang, Y. Waste heat recovery of a PEMFC system by using organic rankine cycle. *Energies* **2016**, *9*, 267. [CrossRef]

19. Ayachi, F.; Ksayer, E.B.; Neveu, P. Exergy assessment of recovery solutions from dry and moist gas available at medium temperature. *Energies* **2012**, *5*, 718–730. [CrossRef]

20. Xu, C.; Gao, Y.; Zhang, Q.; Zhang, G.; Xu, G. Thermodynamic, economic and environmental evaluation of an improved ventilation air methane-based hot air power cycle integrated with a de-carbonization oxy-coal combustion power plant. *Energies* **2018**, *11*, 1434. [CrossRef]

21. Erguvan, M.; MacPhee, D.W. Energy and exergy analyses of tube banks in waste heat recovery applications. *Energies* **2018**, *11*, 2094. [CrossRef]

22. Kılkış, Ş. A Rational Exergy Management Model to Curb CO_2 Emissions in the Exergy-Aware Built Environments of the Future. Doctoral Thesis, Division of Building Technology School of Architecture and the Built Environment KTH Royal Institute of Technology, Stockholm, Sweden, 2011.

23. Bingöl, E.; Kilkis, B.; Eralp, C. Exergy based performance analysis of high-efficiency poly-generation systems for sustainable building applications. *Energy Build.* **2011**, *43*, 3074–3081. [CrossRef]

24. PPIC. Power Plants Information Center, Flue Gas Stack. 2018. Available online: http://powerplantstechnology.blogspot.com/2010/08/flue-gas-stack.html (accessed on 16 February 2019).

25. Kilkis, B. An Exergy-Rational Model for Rating Sensible Air-to-Air Heat Recovery Systems in Sustainable Buildings. In Proceedings of the 3rd SEE SDEWES Conference, Novi Sad, Serbia, 30 June–4 July 2018.

26. Kılkış, B.; Kılkış, Ş. Hydrogen economy model for nearly net-zero cities with exergy rationale and energy-water nexus. *Energies* **2018**, *11*, 1226. [CrossRef]

27. Kilkis, B.; Kilkis, S. Exergetic Optimization of Generated Electric Power Split in a Heat Pump Coupled Poly-Generation System. In Proceedings of the Energy Sustainability Conference, Long Beach, CA, USA, 27–30 July 2007; pp. 211–218. [CrossRef]
28. Pershing, D.W.; Wendt, J.O.L. Relative Contributions of volatile nitrogen and char nitrogen to NOx emissions from pulverized coal flames. *Ind. Eng. Chem. Process Des. Dev.* **1979**, *18*, 60–66. [CrossRef]
29. Hong, B.D.; Slatic, E.R. *Energy Information Administration, Carbon Dioxide Emission Factors for Coal*; Quarterly Coal Report, DOE/EIA-0121(94/Q1); DOE/EIA: Washington, DC, USA, 1994; pp. 1–8. Available online: https://www.eia.gov/coal/production/quarterly/co2_article/co2.html.
30. Cheng, M.T.; Zeng, T.H. Calculation and Analysis of Acid Dew-Point Temperature in Coal-Fired Boiler Gas. In Proceedings of the 5th International Conference on Advanced Design and Manufacturing Engineering (ICADME), Shenzhen, China, 19–20 September 2015; pp. 615–617.
31. Kılkış, B. *Sustainability and Decarbonization Efforts of the EU: Potential Benefits of Joining Energy Quality (Exergy) and Energy Quantity (Energy) in EU Directives, A State of-the-Art Survey and Recommendations*; Exclusive EU Position Report of TTMD; TTMD: Ankara, Turkey, 2017.
32. Lu, P.C.; Fu, T.T.; Garg, S.C.; Novakowski, G. *Boiler Stack Gas Heat Recovery*; NCEL Technical Note, N-1776; Defence Technical Information Centre: Fort Belvoir, VA, USA, 1987; Available online: http://www.dtic.mil/dtic/tr/fulltext/u2/a187419.pdf.
33. Karaoglu, C.; Ozbek, A. District heating and power generation-based flue gas waste heat recovery. *Eur. Mech. Sci.* **2017**, *1*, 63–68.
34. Franco, A.; Villani, M. Optimal design of binary cycle power plants for water-dominated, medium-temperature geothermal fields. *Geothermics* **2009**, *38*, 379–391. [CrossRef]

Article

Numerical Study on Heat Transfer Performance in Packed Bed

Shicheng Wang, Chenyi Xu, Wei Liu and Zhichun Liu *

School of Energy and Power engineering, Huazhong University of Science and Technology, Wuhan 430074, China; shichengwang@hust.edu.cn (S.W.); cy_xu@hust.edu.cn (C.X.); w_liu@hust.edu.cn (W.L.)
* Correspondence: zcliu@hust.edu.cn

Received: 17 December 2018; Accepted: 22 January 2019; Published: 28 January 2019

Abstract: Packed beds are widely used in industries and it is of great significance to enhance the heat transfer between gas and solid states inside the bed. In this paper, numerical simulation method is adopted to investigate the heat transfer principle in the bed at particle scale, and to develop the direct enhanced heat transfer methods in packed beds. The gas is treated as continuous phase and solved by Computational Fluid Dynamics (CFD), while the particles are treated as discrete phase and solved by the Discrete Element Method (DEM); taking entransy dissipation to evaluate the heat transfer process. Considering the overall performance and entransy dissipation, the results show that, compared with the uniform particle size distribution, radial distribution of multiparticle size can effectively improve the heat transfer performance because it optimizes the velocity and temperature field, reduces the equivalent thermal resistance of convection heat transfer process, and the temperature of outlet gas increases significantly, which indicates the heat quality of the gas has been greatly improved. The increase in distribution thickness obviously enhances heat transfer performance without reducing the equivalent thermal resistance in the bed. The result is of great importance for guiding practical engineering applications.

Keywords: Discrete Element Model; gas–solid flow; heat transfer enhancement; entransy dissipation; numerical simulation; optimization

---

1. Introduction

Packed beds are widely used in various industry process, such as catalytic reactors, high temperature gas-cooled nuclear reactors, absorption towers, and so on [1,2]. The structure of the packed reactor is simple and the efficiency is high, so it is the most commonly used reactor in industrial production and scientific research. However, the heat transfer coefficient of the packed bed is relatively low, which is very detrimental for high temperature reactors, such as nuclear reactors. Therefore it is very import to improve the heat transfer performance, cool the bed, and raise the outlet gas temperature of the packed bed reactors.

The flow structure inside the packed bed is complex, so flow maldistribution exists, especially when the tube-to-particle d_t/d_p ratio is small. The ratio d_t/d_p affects the properties near the wall region because the porosity is large here [3–6], and the changes of porosity and velocity may cover the core area of the bed, where the heat transfer mainly occurs [7,8]. The flow maldistribution is obvious with low tube-to-particle ratio ($d_t/d_p < 15$) and will seriously affect the heat transfer or reaction in the bed [9], which determines the design of the bed. When the bed is packed with uniform size particles, there exists a large void fraction near the wall region and it can be called wall effects [10]. Several studies aim to reveal the flow and heat transfer characteristics in the packed bed. However, the correlations of heat transfer in the packed bed with small value of d_t/d_p is hardly to satisfy all the packed beds [4,11]. There are no standard empirical correlations which can be applied to all the range of tube-to-particle ratios. So, various effective parameters under steady state are proposed, including effective thermal conductivities [12], overall heat transfer coefficient, and effective transport parameters [13–15]. Recently, with the development of computer technology, the Computational Fluid Dynamics (CFD) method is adopted to save the time and economic cost. Nijemeisland [16] found that local heat transfer rates did not correlate statistically with the local flow field but related to large scale flow structures—the CFD method is adopted to study the velocity and temperature distribution. The method of composite packing can improve the heat transfer efficiency and heat flux of packed bed with low tube-to-particle ratio and restrain the wall effects [17], compared with the randomly packing. By studying the creeping, transition, and turbulent flow with tube-to-particle ratio d_t/d_p equals to 3 and 10, Reddy [11] concluded that the wall effects decrease as the ratio increases in the creeping and turbulent regimes. Many researches focus on the flow and heat transfer characteristics with low tube-to-particle ratio using CFD methods [18,19], some of them compared the effects of different computational models on the result. The CFD results show that the wall effects and flow maldistribution are obvious. However, the CFD modeling ignores the interaction among particles, which is important in dense gas–solid flow.

Theoretically, the flow in the packed bed includes particle motion, fluid flow, and the interactions between particles and fluid, between particles and particles, so particle scale study is necessary to get the information of particles and has been a research focus in the past decades [20]. The Computational Fluid Dynamics coupled with Discrete Element Method (CFD-DEM) approach has been fully developed and widely used in granular flows and fluidized beds [21,22]. The motion of discrete particles is solved by the Newton's second law, which is known as Discrete Element Method (DEM [23]), while the flow of continuum fluid is solved by the locally averaged Navier-Stokes equations, which is known as CFD. It combines the CFD for the continuum fluid and DEM for the discrete particles, and this method is able to capture the particle physics compared with CFD methods [21]. The coupling between fluids and particles is performed by Eulerian–Lagrangian framework for dense flow [24]. Some researchers apply this method to study the flow behavior of fluidized beds [24–26], and find that the simulation results agree well with experimental results, which means this method is reliable enough; the CFD-DEM methods also have some other applications in industrial [27–30]. However, in terms of packed beds, few studies apply CFD-DEM approach to research the flow and heat transfer characteristics and enhance the heat transfer inside the bed. J. Yang [17] used the discrete element model to generate the randomly packed bed, but not involved in the calculation for the flow and heat transfer. H. Wu [31] studied the thermal radiation of high temperature gas-cooled reactor using CFD-DEM approach.

In order to better understand the nature of the convective heat transfer, Guo el al. [32,33] proposed the field synergy principle, then further developed by Liu et al. [34,35], to guide the design of convective heat trnsfer process with higher heat transfer efficiency and lower flow resistance. With the development of the computation techonology, there are a tendency to design heat transfer unit or system based on the combination of the CFD and optimization algorithm [36–39]. In the current work, to better compare the performance of different parameter configurations, a new physical quantity entransy is introduced, which represents the heat transfer ability of an object [40], and the expression of entransy dissipation is derived from the entransy balance equation. It concluded that the entransy dissipation can be used to measure the irreversibility of the heating or cooling process [41]; moreover, the entransy dissipation extremum principle is proposed as a criterion to guide the optimization of heat transfer process [42]. However, this criterion is different from minimum entropy production principle. Chen et al. [43] pointed that the minimum entropy production principle should be adopted to minimize the usable energy dissipation, while the entransy dissipation extremum principle should be adopted to maximum the heat transfer ability. For heat transfer process only to heat or cool fluids, entransy dissipation is more suitable as a measure of irreversibility. Actually, entransy is an unconserved quantity during the heat transfer process, the entransy dissipation is inevitable, so based on the concept of entransy dissipation, the equivalent thermal resistance of the multidimensional problem is defined, and the goal of the heat transfer optimization is to minimize the equivalent thermal resistance [42].

From all available literatures about packed beds, most focus on the uniform size particles with CFD methods, and aimed to reveal the flow and heat transfer characteristics. However, the effects of multisizes particle mixing and its distribution are still not clear, and few studies have optimized the heat transfer performance inside the bed. J. Yang [17] pointed that the method of composite packing of different particle size can restrain the wall effects. Therefore, based on the CFD-DEM methods, a full numerical simulation is carried out for packed beds with low tube-to-particle ratios in this work. When the wall effects are obvious, the fluid will flow away from the wall region and heat transfer in the center is poor, so the optimal objective is to restrain wall effects, reduce the porosity near the wall, and strengthen the heat transfer in the core area. Considering particle motion and heat conduction between particles, the effects of radial distribution of particle size and the distribution thickness on the heat transfer and fluids flow are discussed. Entransy dissipation is used as criteria to evaluate the performance of different parameter configurations. The result is of great significance to design the packed beds reactors and reduce the volume of beds, especially for high temperature gas-cooled reactors.

2. Mathematical Model

In the CFD-DEM coupling approach, the discrete element model is based on the so called soft sphere model, and the gas phase is modeled as a continuum. The force and motion of particles are tracked at particle-scale level, and the key assumption is that the time step is small enough that the disturbances propagation distance is no more than one particle, so the velocity and acceleration of each individual particle are constant in one time step, so the interaction between particles within each time step can be ignored, and, after the end of one time step, the information of the interaction between particles will be updated and will be the start of the next step [44]; the macroscopic behavior of particles clusters is the cumulative result of the particle-scale behavior. The interaction of solid particles on gas phase is considered as the source of mass, momentum and energy equations. In current work, the mass exchange between particle and gas is neglected. The model description is given below.

2.1. Governing Equations for Solid Particles

There are two types of motion for a particle in moving bed—translation and rotation—and the motion of individual particles is determined by Newton's second law of motion, while the Hertz–Mindlin contact theory is adopted for the interaction between particles. So, the governing Equations of particle i with mass m_i and moment of inertia I_i can be written as

$$m_i \frac{d\vec{v_i}}{dt} = \sum_{j=1}^{kc} (\vec{F}_{c,ij} + \vec{F}_{d,ij}) + \vec{F}_{f,i} + m_i \vec{g} \tag{1}$$

$$I_i \frac{d\vec{\omega_i}}{dt} = \sum_{j=1}^{kc} (\vec{M}_{t,ij} + \vec{M}_{r,ij}) \tag{2}$$

where $\vec{v}_i$ and $\vec{\omega}_i$ are the translation velocities and rotation velocities of the particle i, and k_c is the number of particles interacting with particle i. $\vec{F}_{c,ij}$ and $\vec{F}_{d,ij}$ are the contact force and non-contact force respectively, $\vec{F}_{f,i}$ is the particle–fluid interaction force acting on the particle i, $m_i \vec{g}$ represents the gravitational force. $\vec{M}_{t,ij}$ and $\vec{M}_{r,ij}$ are the torques generated by tangential force and rolling friction force. In this work, the expressions of forces and torques have been listed in the literature [20].

Contacts between particles are significant for dense phase such that conductive heat transfer must be taken into account. The heat flux between the particles is defined as Equation (3), and the contact area is incorporated in the heat transfer coefficient h_c, as shown in Equation (4). The temperature change over time of each particle is updated explicitly by Equation (5).

$$Q_{ij} = h_c \Delta T_{ij} \tag{3}$$

$$h_c = \frac{4k_i k_j}{k_i + k_j} \left(\frac{3F_N r^*}{4E^*}\right)^{1/3} \tag{4}$$

$$m_i C_p^i \frac{dT_i}{dt} = \sum_{j=1}^{ki} Q_{ij} + Q_{if} + Q_{i,rad} \tag{5}$$

where ΔT_{ij} is the temperature difference between particle i and particle j, k_i and k_j are the thermal conductivity of particle i and j, respectively, and F_N is the contact normal force. r^* is the geometric mean of the particles radii according to the Hertz–Mindlin contact theory and E^* is the effective Young's modulus. The heat flux Q_{ij} is the heat conduction flux between particles i and j, and Q_{if} is the convection heat flux between fluid and solid particles. $Q_{i,rad}$ is the radiation heat flux between particle i and its surrounding environment.

2.2. Governing Equations for Gas Phase

The gas is treated as a continuum phase, and for the incompressible fluid, the fluid field can be solved by continuity, momentum, and energy conservation equations. In the current work, the standard $k - \varepsilon$ model is adopted to solve fluid flow, and the governing Equations are given as follows.

Continuity Equation:

$$\frac{\partial(\rho_f \varepsilon_f)}{\partial t} + \nabla \cdot (\rho_f \varepsilon_f \vec{u}) \tag{6}$$

Momentum Equation:

$$\frac{\partial(\rho_f \varepsilon_f \vec{u})}{\partial t} + \nabla \cdot (\rho_f \varepsilon_f \vec{u} \vec{u}) = -\varepsilon_f \nabla p + \nabla \cdot (\mu_f \varepsilon_f \nabla \vec{u}) + \vec{S}_m \tag{7}$$

Energy Equation:

$$\frac{\partial(\rho_f \varepsilon_f C_{p,f} T_f)}{\partial t} + \nabla \cdot (\rho_f \varepsilon_f \vec{u} C_{p,f} T_f) = \nabla \cdot (\lambda_f \varepsilon_f \nabla T_f) + \vec{S}_e \tag{8}$$

where ρ_f, $\vec{u}$, and p are the fluid density, velocity, and pressure. ε_f and T_f are the porosity and temperature of the fluid and $C_{p,f}$ and λ_f are the specific heat capacity and thermal conductivity of fluid. $\vec{S}_m$ is the momentum source term due to the effect of solid particles on fluid motion and $\vec{S}_e$ is the energy source term because of the heat exchange between fluids and particles, CFD and DEM are coupled by the source terms $\vec{S}_m$ and $\vec{S}_e$, the expressions of $\vec{S}_m, \vec{S}_e$ are given in detail [45].

2.3. Heat Transfer Models

The fluid and particles are obtained by solving Equations (1), (2), (6) and (7); while the heat transfer between particles and fluid is governed by Equations (5) and (8). In the present work, the temperature of particles is up to 800 K, so there are three heat transfer mechanisms considered—heat conduction, heat convection, and particle radiation—where the convection heat transfer between fluid and the wall is not considered and the wall is regarded as adiabatic. The models used to describe the different heat transfer mechanisms have been described in detail [33], and for the sake of simplicity, are no longer described here.

In order to better describe the heat transfer performance in the packed beds, several parameters are defined here. The Reynolds number and pore scale hydraulic diameter d_k are defined as follows.

$$Re_{ep} = \frac{\rho_f(\vec{u} - \vec{v}_i)d_h}{\mu_f} \tag{9}$$

$$d_h = 4\frac{\varepsilon_f}{1 - \varepsilon_f}\left(\frac{V_p}{A_p}\right) \tag{10}$$

The heat transfer coefficient and the Nusselt number are defined below.

$$h_{sf} = \frac{Q_f}{(\overline{\overline{T}}_p - \overline{T}_f)} \tag{11}$$

$$Nu_{sf} = \frac{h_{sf}d_h}{k_f} \tag{12}$$

where, Q_f is total heat flux of the heat convection in the packed beds and $\overline{\overline{T}}_p$ and $\overline{T}_f$ are the average temperature of the particles and fluids in the beds, respectively. Because of the improvement in heat transfer, at the cost of the increase of pressure drop, in order to compare the overall performance of different configurations, the expression is defined:

$$\frac{(Nu/Nu_0)}{(f/f_0)^{1/3}} \tag{13}$$

where the Nu_0 and f_0 is the Nusselt number and friction factor of the packed with uniform particle size distribution.

2.4. Entransy Dissipation

As described above, the entransy dissipation is introduced to describe the heat transfer ability. Different from the energy destruction minimization [46,47], for the simple heating or cooling process, entransy dissipation is more suitable as a measure of the irreversibility of the heat transfer process. The heat transfer process studied in this paper is to cool the high temperature particles with cold air, so it is suitable to take entransy dissipation as a measure of heat transfer performance.

According to a previously described definition [43], the entransy of flow fluid in the opening system is as follows.

$$G = \frac{1}{2}HT = \dot{m}C_pT^2 \tag{14}$$

where H is the enthalpy of the flow fluid and $\dot{m}$ is the mass flow rate.

Entransy is not conserved and will dissipate during fluid flow, the entransy dissipation rate of unit volume and unit time is

$$\Phi_h = -\dot{q}.\nabla T = k(\nabla T)^2 \tag{15}$$

Equation (15) is derived from the entransy balance equation, which is derived from the thermal energy conservation equation, and the Φ_h can be considered as the local entransy dissipation. The detailed derivation has been given by the researchers [41], and it is omitted here due to the complexity of the derivation. According to the entransy dissipation extremum principle, when the boundary temperature difference is given, maximizing the entransy dissipation leads to maximum boundary heat flux, which leads to the minimum of equivalent thermal resistance, which is the best performance for the heat transfer [48]. It can be expressed as

$$\Delta T\delta\dot{Q}_t = \delta\iiint \frac{1}{k}|q|^2 dV = 0 \tag{16}$$

The entransy dissipation is the function of heat flux and temperature gradient, as shown in Equation (11), applying the variational method [49] to the entransy balance equation, Equation (12) can be obtained. At a given temperature difference, the equivalent thermal resistance is defined as the ratio of the square of temperature different divided by the entransy dissipation, which can be written as

$$R_h = \frac{\Delta T^2}{\Phi} \text{ or } R_h = \frac{\left(\overline{\Delta T}\right)^2}{\Phi} \tag{17}$$

where, $\overline{\Delta T}$ is the average temperature difference. The smaller R_h is, the better the heat transfer performance and the stronger the heat transfer ability is. Due to entransy dissipation is the function of heat flux and the temperature gradient, the total entransy dissipation, because of the finite temperature difference in the packed bed, is defined as follows.

$$\Phi = \int (T_h - T_c) dQ \tag{18}$$

where, T_h and T_c are the temperature of high particles and temperature of cold fluid. So the T–Q graph can be used to visually represent the entransy dissipation of the heating or cooling process.

3. Simulation Conditions

The physical model of packed bed used in this work and the boundary conditions is shown in Figure 1. There are two sections of the physical model: packed section and outlet section. Considering the complexity of the gas flow in the particle section, and avoiding the influence of the outlet reversed flow, the outlet extends 90 mm downstream. The inlet is specified as a velocity inlet, the outlet is specified as a pressure outlet, and the wall is considered as adiabatic. The packed section is filled with particles with different sizes. Detailed geometric parameters of the packed bed and boundary conditions are shown in Table 1. Firstly, the particles with temperature of 800 K fill with the packed section, until the steady state of the particles is achieved, and then the gas is introduced from the bottom with velocity of 5 m/s, with temperature of 293 K, and after sufficient heat exchange between particles and gas, the gas flows out at the outlet and will be recycled for other use.

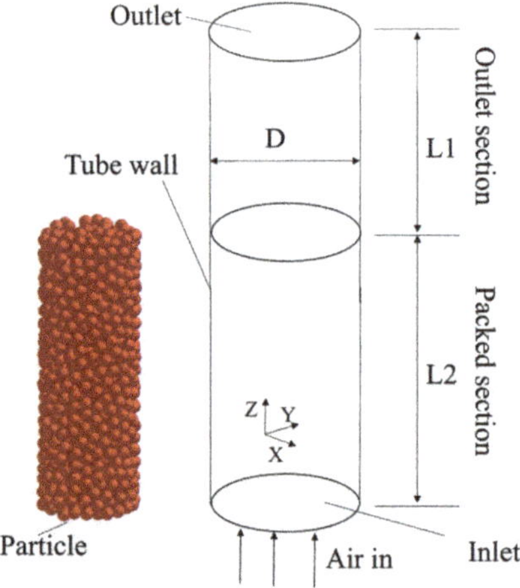

Figure 1. Computational physical model.

Table 1. Detailed parameters of the physical model.

Bed Geometry	Diameter D (mm)	40
	Particle domain L2 (mm)	180
	Extended domain L1(mm)	90
Particle	diameter d (mm)	3/4/5
	Density (kg/m^3)	4540
	thermal conductivity (W/(m.K)	24.6
	thermal capacity Cp (J/(kg.K)	630
	initial temperature (K)	800
Gas	Density (kg/m^3)	0.9944
	Thermal conductivity (W/(m.K)	0.03066
	thermal capacity Cp (J/(kg.K)	1009
	initial velocity (m/s)	5
	Initial temperature (K)	293
	dynamic viscosity (m^2/s)	2.13e-05
Contact parameter	Poisson ratio	0.3
	Young's modulus (Mpa)	64.52
	rolling friction coefficient	0.01
	static friction coefficient	0.545
	restitution coefficient	0.2

In order to study the strengthening effects of the mixing of particles with different sizes, the packed beds with different particle size distribution are shown in Figure 2. Five kinds of different particle size distributions in the radial direction are studied here, namely, random packing with uniform size of 5 mm (d = 5 mm), random mixing distribution with size of 4–5 mm (mixing 4–5 mm), radial distribution of 4–5 mm (d = 4–5 mm), random mixing distribution of 3–5 mm (mixing 3–5 mm), and radial distribution of 3–5 mm (d = 3–5 mm). When the particles are distributed along the radial direction, the central region is filled with basic particles with diameter of 5 mm, while the near wall region is filled with small particles with diameter of 4 mm or 3 mm, as shown in Figure 2c,e, the average porosity of mixing 4–5 mm and d = 4–5 mm, mixing 3–5 mm, and d = 3–5 mm are the same, respectively.

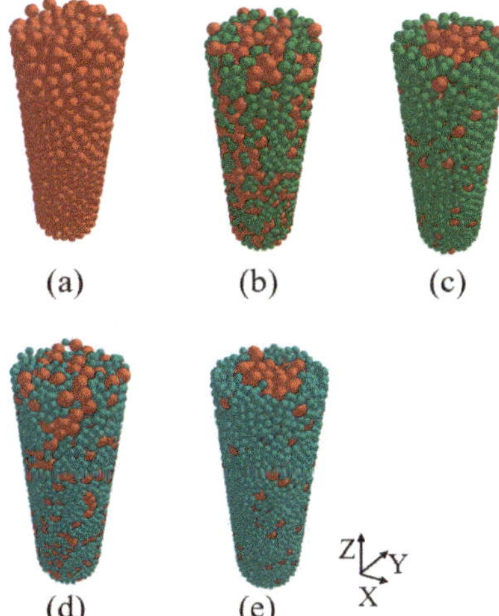

Figure 2. Radial distribution of different particle sizes: (**a**) random packing with single size particles of 5 mm; (**b**) random mixing of 4–5 mm; (**c**) radial distribution of 4–5 mm; (**d**) random mixing of 3–5 mm; and (**e**) radial distribution of 3–5 mm.

In order to further study the influence of particle size distribution on the heat transfer process and the range of the wall effects in the bed, packed beds with different distribution thickness are studied, as shown in Figure 3. There are three cases with different distribution thickness: case1 (the same as the d = 3–5 mm in Figure 2e), case2, and case3, and the near wall region of the three cases are packed with small particles of 3 mm, as shown in Figure 3b–d. Meanwhile, the packed beds filled with big particles with diameter of 5 mm and small particles with diameter of 3 mm are also studied here as comparison, as shown in Figure 3a,e. All the particles in the packed beds are generated by discrete element methods.

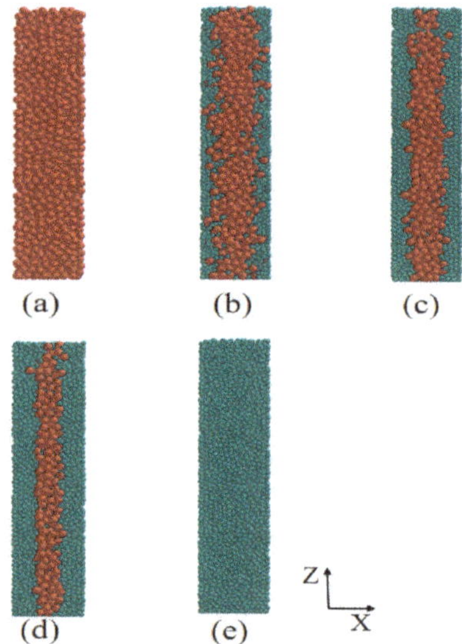

Figure 3. Radial distribution of different distribution thickness: (**a**) single particle size of 5 mm; (**b**) case1; (**c**) case2; (**d**) case3; and (**e**) single particle size of 3 mm.

The CFD-DEM method is used to study the flow and heat transfer characteristics between gas and particles, to meet the key assumptions, the time step is selected carefully. For the cases with the minimum particle diameter of 3 mm, 4 mm, and 5 mm, the time step is selected as 1e-5s, 2e-5s, and 2e-5s, respectively. In the case d = 5 mm, there are 1927 particles with diameter 5 mm; and in the case d = 4–5 mm and mixing 4–5 mm, the number of particles with diameter of 4 mm and 5 mm are 1931 and 953, respectively. In the case d = 3–5 mm and mixing 3–5 mm, the number of particles of 3mm and 5 mm are 4723 and 983, respectively. For case2, the number of particles of 3 mm and 5 mm are 6687 and 582, respectively. For case3, the number of particles with diameter of 3 mm and 5 mm are 7870 and 320, respectively. For case d = 3 mm, there are 9393 particles with diameter 3 mm.

4. Results and Discussion

4.1. Model Validation

The packed bed which is randomly packed with uniform size of 5 mm is selected for model validation. And the empirical correlations proposed by Sug Lee [50] are chosen for the validation of friction factor, and the empirical correlations proposed by Demirel [43] are chosen for the validation of heat transfer process, which can be expressed by Nusslet number. The friction factor is calculated by the following expressions [37].

$$\Delta p = 4f \frac{\rho u^2}{2} \frac{1}{d_p} \tag{19}$$

$$4f = \frac{12.5(1-\varepsilon_f)^2}{\varepsilon_f^3}(29.32\mathrm{Re}_p^{-1} + 1.56\mathrm{Re}_p^{-n} + 0.1) \tag{20}$$

where n is the factor related to porosity, and can be expressed as

$$n = 0.352 + 0.1\varepsilon_f + 0.275\varepsilon_f^2 \tag{21}$$

and the Nusslet number concluded by Demirel [43] for spheres can expressed as

$$Nu = 0.217\mathrm{Re}_p^{0.756} \tag{22}$$

The CFD-DEM results of friction factor and Nusslet number are compared with the empirical correlation results, which are shown in Figure 4, and the quantitative are given in Table 2. As can be seen from Figure 4a, the numerical simulation results agree well with the results of correlations of Sug Lee [50], and the maximum deviation is 10.25% when Re_p number is 6592.9; and, as can be found from Figure 4b, the numerical simulation results are in good agreement with the correlation results proposed by Demirel [49], and the maximum deviation is 6.65% when the R_p number is 5405.4. Therefore, it can be concluded that the CFD-DEM approach used in the present work is reliable to simulate the fluid flow and heat transfer in packed beds with low tube-to-particle ratio.

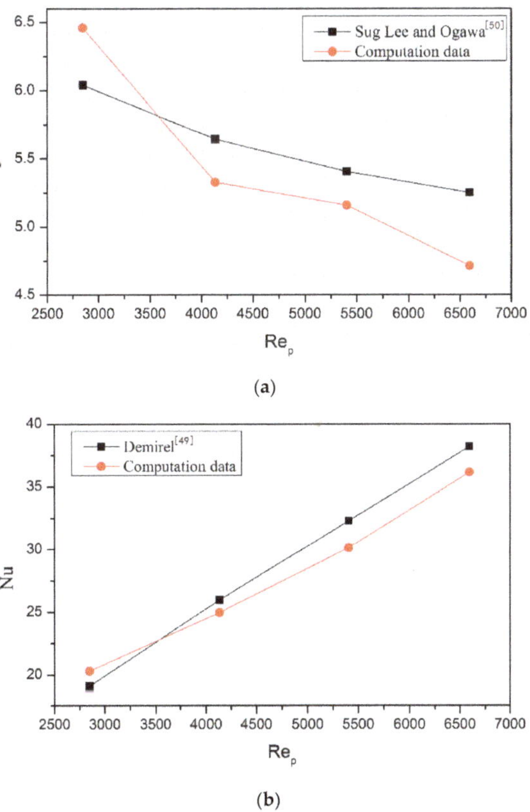

(a)

(b)

Figure 4. Validation of the computation model compared with empirical correlations: (**a**) friction factor and (**b**) Nusselt number.

Table 2. Quantitative comparison of model validation.

	Re$_p$	2849.9	4132.4	5405.3	6592.8
	Sug Lee and Ogawa [50]	19.108	25.962	32.270	38.200
Nu	Computation data	20.290	24.950	30.124	36.148
	Deviation	6.2%	3.9%	6.6%	5.4%
	Demirel [49]	6.041	5.642	5.404	5.249
f	Computation data	6.459	5.326	5.157	4.712
	Deviation (%)	6.9%	5.6%	4.5%	10.2%

4.2. The Effect of Particle Diameter Distribution

The radial distribution of particle with different sizes determines the distribution of porosity along the radial direction. To analyze the results conveniently, four typical cross-sections are selected, as shown in Figure 5, the sections of Z = 0.07922 m, Z = 0.06922 m, and Z = 0.05922 m are XY planes close to the outlet, and the plane of Y-section is ZX plane that is at the location of Y = 0 m. In the present work, the tube-to-particle ratio is relatively low ($8 < d_t/d_p < 13.3$), as discussed above, for low tube-to-particle ratio, the porosity near the wall region is larger than that in the central region when the bed is packed with particles of uniform size; and the gas flows away from the near wall region without sufficient heat exchange with the solid particles, so higher gas velocity can be found near wall region, as shown in Figure 6b, meanwhile, the temperature of the gas is lower than that in the central region, as shown in Figure 6a. Also it can be found from Figure 7 that there is large difference in temperature along the radial direction from the wall to the center of the tube, and the max temperature difference is almost 10 °C near the wall region (one diameter from the wall to the center in the radial direction) at the different cross-sections, which means that the wall effects is obvious. So it is important to decrease the porosity near wall region, improve the velocity, and optimize the flow field and temperature field.

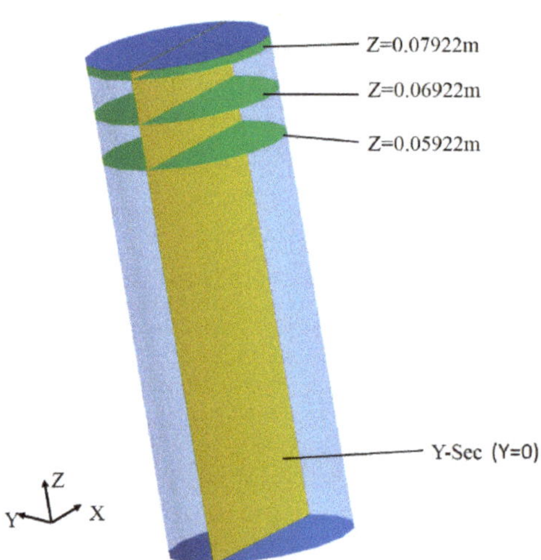

Figure 5. Typical cross-sections in the packed bed.

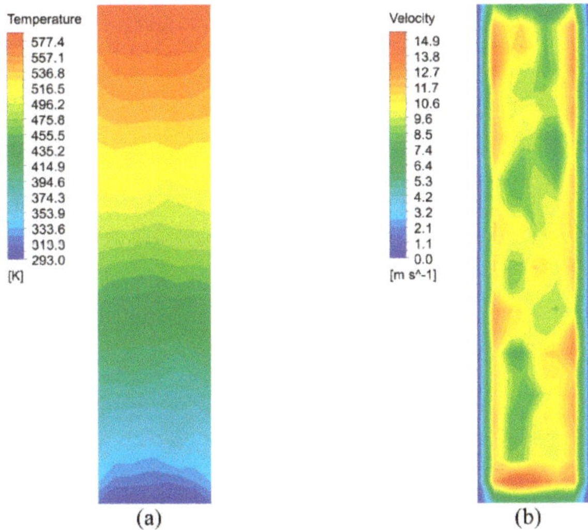

Figure 6. (**a**) Temperature distribution at Y-sec and (**b**) velocity distribution at Y-sec.

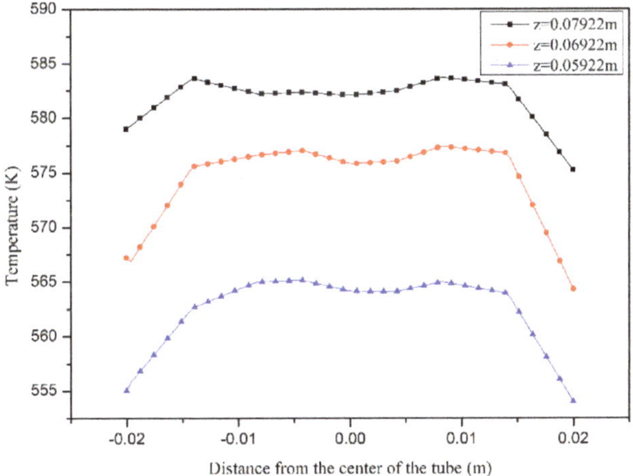

Figure 7. Radial distribution of temperature of particle size 5 mm.

Figure 8 compares the Nusselt number, friction factor at the different Re numbers, as shown in Figure 8a, and the Nu of multiparticle size distributions, which is higher than that of uniform particle size distribution at different Re, meaning that the heat transfer performance is better, and the heat transfer performance of case d = 3–5 mm is the best. However, the improvement of heat transfer performance is at the cost of the increase of pressure drop, as shown in Figure 8b, the better heat transfer performance corresponds to the higher friction factor. Figure 8c compares the overall performance of different configurations, with the performance of uniform particle size distribution as a reference. It can be seen that at low Re, the overall performance of multiparticle size distribution is better than reference, but with the increase of Re, the overall performance becomes worse than the reference, which is mainly because the pressure drop increase too fast. So at low Re, d = 3–5 mm is recommend because of the better overall performance.

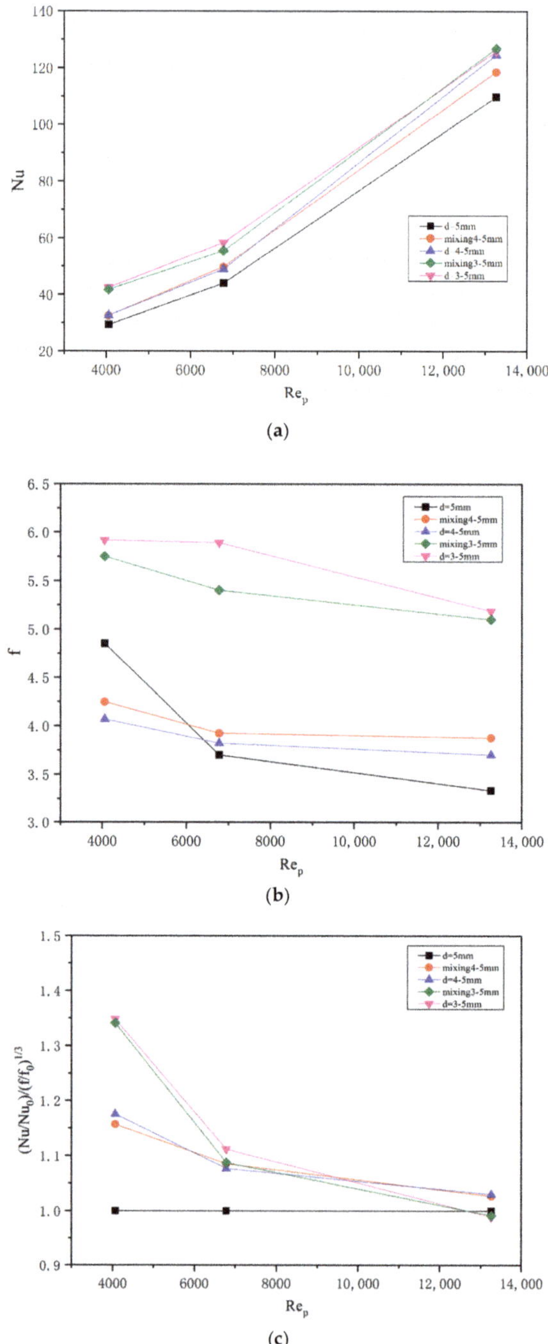

Figure 8. Performance comparisons of different particle size distributions: (**a**) Nusselt number; (**b**) friction factor; and (**c**) the overall performance.

To analyze the nature of the different heat transfer performances, the temperature and velocity distributions of different particle size distribution at section Y-Sec are shown in Figures 9 and 10 (Re = 6399). It is found that, due to the small particles are placed in the near wall region, the velocity distribution throughout the bed is more uniform compared with the uniform particle size distribution, and the temperature of the near wall region is close to that of the central region in the case d = 4–5 mm, as shown in Figures 9c and 10c. Moreover, with the decrease of particle size near wall region (d = 3–5 mm), the porosity decreases further, so the velocity throughout the bed has been increased, so the heat transfer in the core area is enhanced, as shown in Figures 9e and 10e, the temperature at the outlet is higher, as listed in Table 3. The radial temperature distributions of five different particle size distributions at the same section (Z = 0.07922 m) are shown in Figure 11. The uniform particle size distribution d = 5 mm, and the random distribution mixing 4–5 mm, mixing 3–5 mm have similar temperature distributions in the radial direction, namely, the temperature in the near wall region is lower than that in the central region, so it indicates the random mixing of big and small particles cannot restrain the wall effects. While the cases of d = 4–5 mm and d = 3–5 mm can improve the temperature in the near wall region, they also restrain the wall effects and optimize heat transfer in the beds.

Figure 9. Local temperature distributions of different particle size distributions (Y-Sec).

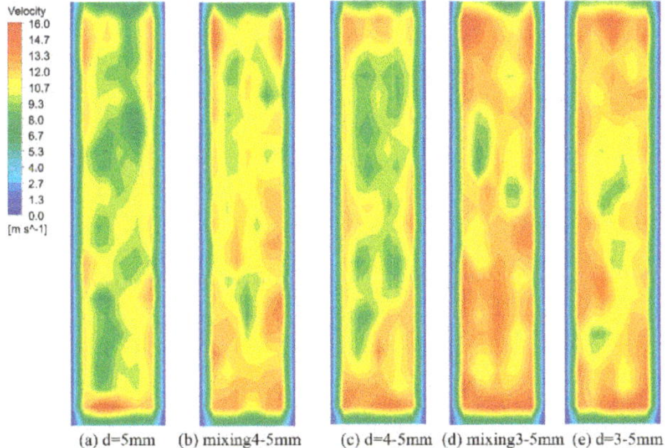

Figure 10. Local velocity distributions of different particle size distributions (Y-Sec).

Table 3. Outlet temperature and average porosity of different distributions (Re = 6399).

Particle Size Distribution	d = 5 mm	Mixing 4–5 mm	d = 4–5 mm	Mixing 3–5 mm	d = 3–5 mm (Case1)	Case2	Case3	d = 3 mm
Outlet temperature (K)	584.4	605.1	606.8	624.2	636.3	652.4	662.2	672.1
Average porosity	0.4397	0.4177	0.4177	0.4127	0.4127	0.4114	0.589	0.41

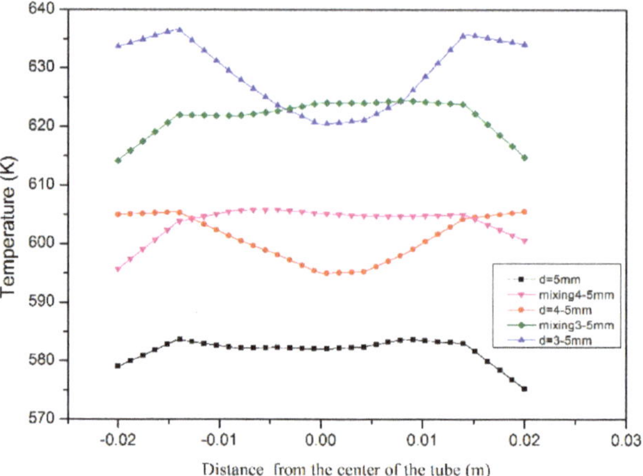

Figure 11. Radial temperature distribution of different particle size distributions (section Z = 0.07922 m).

The effects on heat transfer also can be found from the change of the isothermal lines, for the uniform particle size d = 5 mm, as shown in Figure 9a, the isothermal lines are concave, and both the temperature gradient and heat flux are toward to the center of the tube, but the gas velocity in the center is low, which indicates the heat transfer in the central area is weak; while for d = 4–5 mm and d = 3–5 mm, the isothermal lines are convex, so both the temperature gradient and heat flow are toward to the near wall region, and the velocity in the near wall region is relatively high, as shown in Figure 9c–e, as a result, the heat transfer throughout the beds are improved. Due to the fact that the velocity of the case of 3–5 mm distribution is higher than that of 4–5 mm, there is more heat flux in the d = 3–5 mm, so the outlet gas temperature is higher, as listed in Table 3. The isothermal lines in the cases of mixing 4–5 mm and mixing 3–5 mm are similar to the uniform size distributions, as shown in Figure 9b,d, the heat transfer in the bed is not optimal. So the radial distribution of particle size can obviously enhance the heat transfer in the bed.

In order to evaluate the loss of the heat transfer ability in the process, the T–Q graph and the equivalent thermal resistance is shown in Figure 12. The larger the area surround by the curve is, the greater the entransy dissipation is, and, according to the minimum thermal resistance principle [36], the heat flux is larger, which corresponds to the better heat transfer performance when the boundary temperature is constant, as shown in Figure 12a; the equivalent thermal resistance is smaller, as shown in Figure 12b. It can be seen from Figure 12 that, the area enclosed by the curve corresponding to the distribution of d = 3–5 mm is the largest, and the equivalent thermal resistance is the smallest, so for the d = 3–5 mm, the heat transfer ability between the gas and solid is utilized in the greatest extent. It should be noticed that there are small difference of the thermal difference between the case of d = 3–5 mm and mixing 3–5 mm, d = 4–5 mm and mixing 4–5 mm, but the difference among d = 5 mm, d = 4–5mm and d = 3–5 mm is obvious.

As a result, considering the overall performance and the loss and heat transfer ability, it can be concluded that the improvement of heat transfer effect is at the cost of the loss of heat transfer ability. And the radial particle size distribution of 3–5 mm is the optimal configuration.

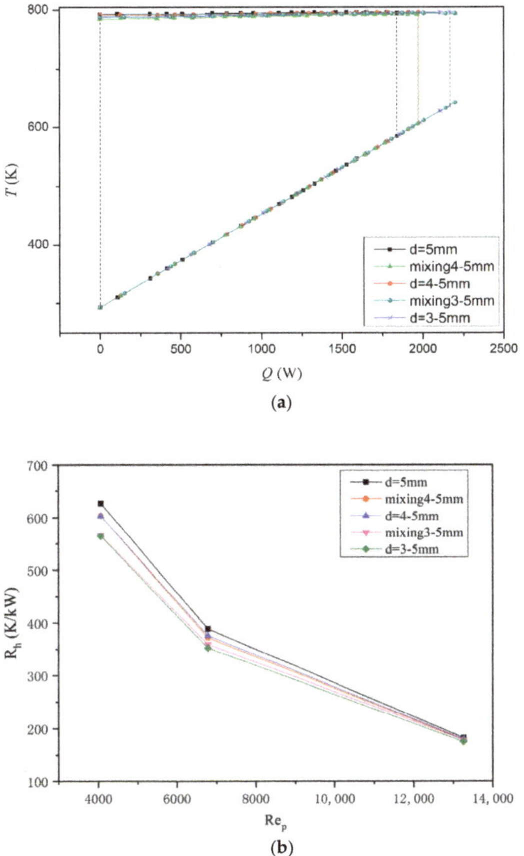

Figure 12. (**a**) T–Q graph of different particle size distributions. (**b**) Equivalent thermal resistance.

4.3. The Effect of Distribution Thickness

As discussed above, the radial distribution of particle size can obviously improve the heat transfer performance in the bed, in order to study the range of the wall effects the influence of radial distribution thickness on heat transfer is investigated, as shown in Figure 3.

The flow and heat transfer performance are greatly affected by the distribution thickness, as shown in Figure 13, with the increase in distribution thickness, the Nu increases at different Re, but the f also increases. However, with further increase in distribution thickness, the multiparticle size distribution becomes the uniform size distribution of $d = 3$ mm, and the heat transfer performance decrease obviously. The overall performance is shown in Figure 13c, and the performance of the case3 is best, the case $d = 3$ mm is worse than case $d = 5$ mm, which indicates that for the uniform size distribution, the decrease of particle size cannot improve the performance of the beds, and the increase of the outlet temperature is due to the increase of the heat transfer area. The temperature distributions along the radial directions at the same section ($Z = 0.07922$ m) are shown in Figure 14. It is clearly that the average temperature increases with the increase of distribution thickness, and there is a big jump of temperature when the radial distribution of particles starting from uniform size (from the $d = 5$ mm) to the case1. Moreover, the temperature distribution trends of case1, case2, and case3 are the same: the temperature increases first and then decreases along the wall to the center of the tube, the turning point of temperature change is one particle diameter away from the wall, so it can be inferred that the wall effects are just one diameter away from the wall

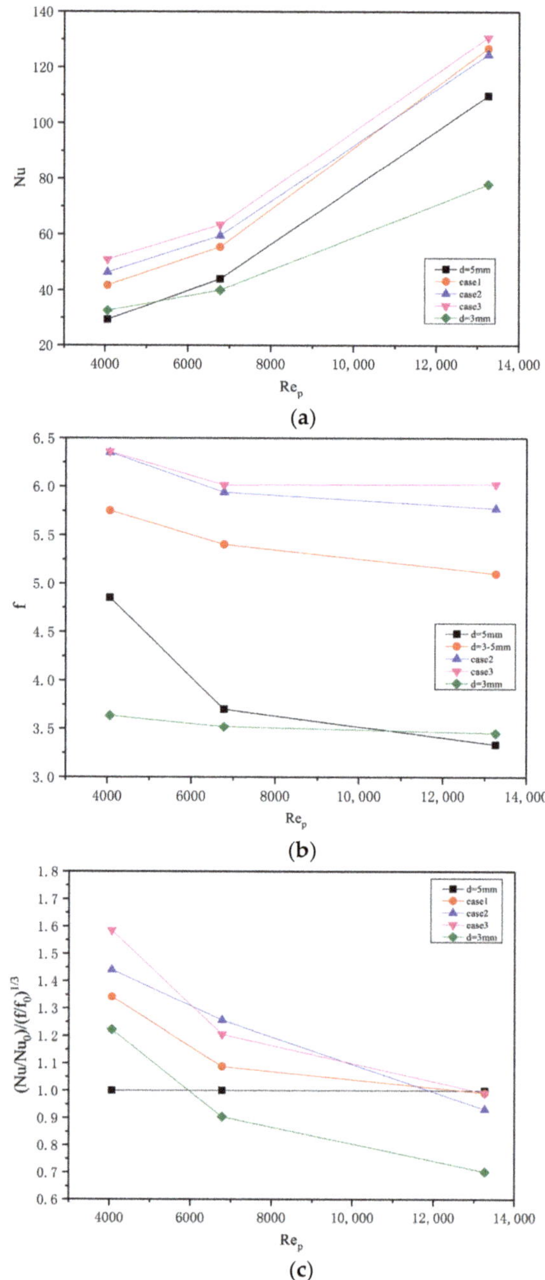

Figure 13. Performance comparisons of different radial distribution thickness: (**a**) Nusselt number; (**b**) friction factor; and (**c**) the overall performance.

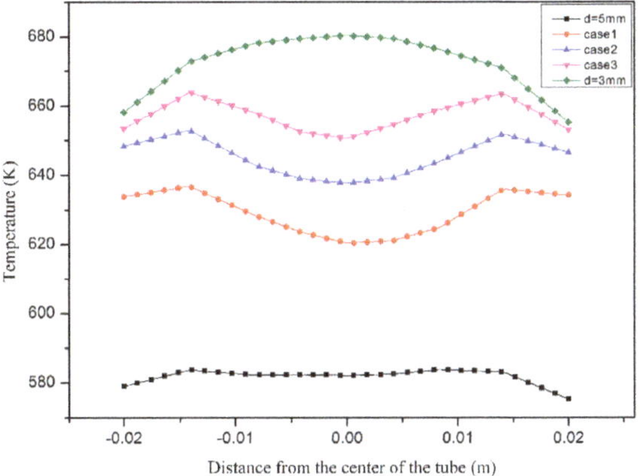

Figure 14. Radial temperature distribution of different radial distribution thickness (section Z = 0.07922 m).

The different performance can be analyzed from the flow and temperature filed, as shown in Figures 15 and 16 (Re = 6399). With the increase of the radial distribution thickness (from Figure 3a–e), the big local flow channels formed in the central region decrease, and the local gas velocity increases, which leads to the improvement of heat transfer throughout the bed, and the outlet temperature is obviously increased, as listed in Table 3. But as the distribution thickness further increases (Figure 3e), the flow and heat transfer characteristics are the same as that of the uniform particle size. For the uniform size of d = 5 mm and d = 3 mm, the isothermal lines change in the similar trend along the radial direction from the tube wall to the center, namely, the isothermal lines are concave, the wall effects are obvious, the outlet temperature of d = 3 mm is higher than that of d = 5 mm—because the average porosity of the whole bed is lower and the heat transfer area is larger–but the heat transfer performance in the bed is not improved.

Figure 15. Temperature distribution of different radial distribution thickness (Y-Sec).

The entransy dissipation and the equivalent thermal resistance of the five different distributions are shown in Figure 17, the area enclosed by the curve increases gradually with the change of the distribution thickness. So it shows that the entransy dissipation and the heat flux Q increase obviously when the particle size changes from uniform size of d = 5 mm to the case1, and the equivalent thermal resistances decrease obviously, as shown in Figure 17b. The obvious increase in entransy dissipation and heat flux from the uniform size of d = 5 mm to the case1 occurs because the heat transfer performance is obviously improved. However, the difference of thermal resistance of case1, case2, and case3 is very small, and this is because the change of the distribution thickness do not improved the temperature and flow filed so much, as shown in Figures 14 and 15, the fields of case1, case2, and case3 are similar, and the temperature gradient is similar, so there is a small difference in entransy dissipation and the equivalent thermal resistance.

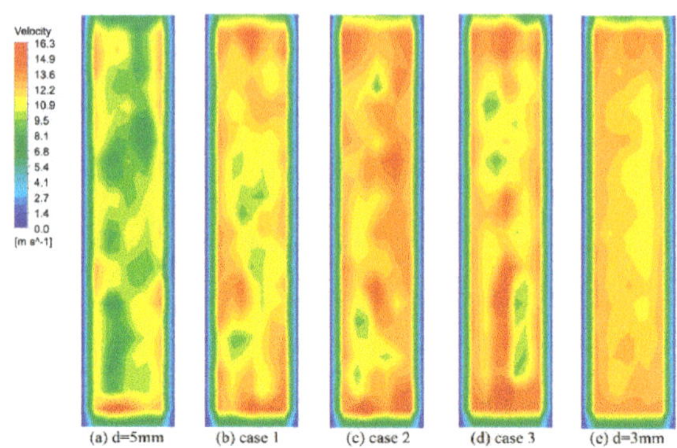

Figure 16. Velocity distribution of different radial distribution thickness (Y-Sec).

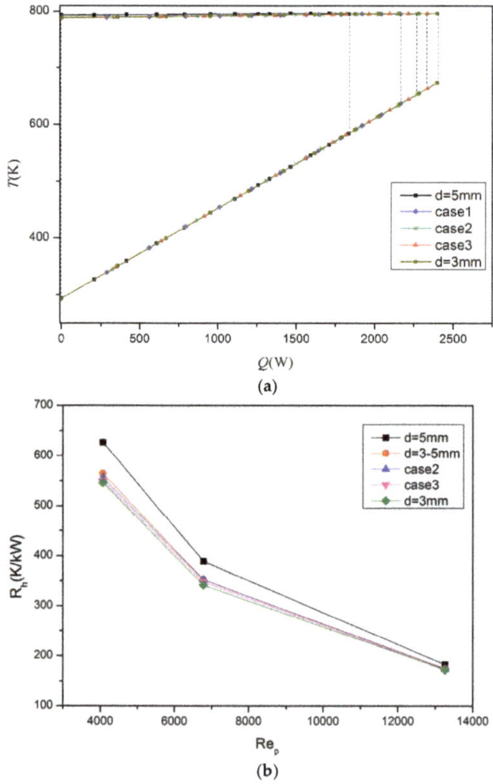

Figure 17. (**a**) T–Q graph of different radial distribution thickness and (**b**) equivalent thermal resistance.

As a result, considering the overall performance and the entransy dissipation, it can be concluded that the range of wall effects is just one particle diameter away from the wall, the heat transfer effects can be obviously improved by filling small particles in the near wall region, and the increase of distribution thickness can improve the heat transfer performance. The wall effects can be well restrained by the radial distribution of the particle size.

5. Conclusions

In the present paper, by adopting the CFD-DEM coupling method, the effects of radial distribution on particle size and distribution thickness on heat transfer are studied, the utilization of the heat transfer ability and the equivalent thermal resistance in the packed bed is analyzed. The main findings are as follows.

1. By changing the radial distribution of the particle size in the bed the velocity distribution and temperature distribution in the bed can be effectively improved, the wall effects are well restrained, and the heat transfer ability between the gas and solid is utilized in the greatest extent.
2. The range of wall effects is just one particle diameter (5 mm) away from the wall, and the heat transfer performance can be obviously improved by filling small particles in the near wall region.
3. The increase of distribution thickness can obviously improve the heat transfer effects, and the equivalent thermal resistance is reduced compared to the uniform size distribution.

The configurations in this work has improved the heat transfer in the packed bed, and increased the outlet temperature, it is important for energy saving, emission, and industrial applications.

Author Contributions: Conceptuallization: Z.L. and W.L.; methodology: S.W.; software: S.W.; validation: S.W.; resources: S.W.; formal analysis: S.W.; investigation: S.W.; datacuration: S.W.; writing: S.W.; review and editing: Z.L.; visualization: S.W.; supervision: S.W.; project administration: S.W.; funding acquisition: C.X.

Funding: The financial supporting was provided by National Key R&D Program of China (No. 2017YFB0603501).

Acknowledgments: We would like to acknowledge financial supports for this work provided by National Key R&D Program of China (No. 2017YFB0603501).

Conflicts of Interest: The authors declare no conflict of interest.

Nomenclature

A_p	Surface area of particle (m^2)
$C_{p,f}$	Specific heat capacity of gas (J/kg/K)
$C_{p,i}$	Specific heat capacity of particle i (J/kg/K)
D	The diameter of bed (m)
d_h	pore scale hydraulic diameter (m)
d_p	Particle diameter (m)
f	Friction factor
G	Entransy (W.K)
I_i	Moment of inertia of particle i (kg/m^2)
$L1$	The length of extended domain (m)
$L2$	The length of particle domain (m)
m_i	Mas of particle i (kg)
ω_i	Angular velocity of particle i (rad/s)
p	Pressure of gas (Pa)
Q	Heat flow (J/s)
T_h	equivalent thermal resistance (K/W)
T_f	Temperature of gas (T)
T_i	Temperature of particle i (T)
V_p	Volume of particle (m^3)
v_i	Velocity of particle i (m/s)
$\vec{S}_e$	Energy source

$\vec{S}_m$	Momentum source
ρ_f	Density of gas (kg/m^3)
ε_f	Porosity of bed
Φ	entransy dissipation (W.K)
ΔT	Temperature difference (K)
∇T	Temperature gradient (K/m)
k	Thermal conductivity

References

1. Bu, S.S.; Yang, J.; Zhou, M.; Li, S.Y.; Wang, Q.W.; Guo, Z.X. On contact point modifications for forced convective heat transfer analysis in a structured packed bed of spheres. *Nucl. Eng. Des.* **2014**, *270*, 21–33. [CrossRef]
2. Yang, J.; Wang, J.; Bu, S.S.; Zeng, M.; Wang, Q.W.; Nakayama, A. Experimental analysis of forced convective heat transfer in novel structured packed beds of particles. *Chem. Eng.* **2012**, *71*, 126–137. [CrossRef]
3. Dixon, A.G.; Di Constanzo, M.A.; Soucy, B.A. Fluid-phase radial transport in packed beds of low tube-to-particle diameter ratio. *Heat Mass Transf.* **1984**, *27*, 1701–1713. [CrossRef]
4. Dixon, A.G. Heat transfer in fixed beds at very low (<4) tube-to-particle diameter ratio. *Eng. Chem. Res.* **1997**, *36*, 3053–3064.
5. Borkink, J.G.H.; Westerterp, K.R. Influence of tube and particle diameter on heat transfer in packed beds. *AIChE J.* **1992**, *38*, 703–715. [CrossRef]
6. Gunn, D.J.; Ahmad, M.M.; Sabri, M.N. Radial heat transfer to fixed beds of particles. *Chem. Eng. Sci.* **1987**, *42*, 2163–2171. [CrossRef]
7. Ge, Y.; Liu, Z.C.; Liu, W. Multi-objective genetic optimization of the heat transfer for tube inserted with porous media. *Int. J. Heat Mass Transf.* **2016**, *101*, 981–987. [CrossRef]
8. Zheng, N.; Liu, P.; Shan, F.; Liu, Z.; Liu, W. Effects of rib arrangements on the flow pattern and heat transfer in an internally ribbed heat exchanger tube. *Int. J. Therm. Sci.* **2016**, *101*, 93–105. [CrossRef]
9. Reddy, R.K.; Joshi, J.B. CFD modeling of pressure drop and drag coefficient in fixed beds: Wall effects. *Particuology* **2010**, *8*, 37–43. [CrossRef]
10. Muller, G.E. Radial void fraction distribution in randomly packed fixed beds of uniformly sized spheres in cylindrical containers. *Powder Technol.* **1992**, *72*, 269–275. [CrossRef]
11. McGreavy, C.; Foumeny, E.A.; Javed, K.H. Characterization of transport properties for fixed bed in terms of local bed structure and flow distribution. *Chem. Eng. Sci.* **1986**, *41*, 787–797. [CrossRef]
12. De Beer, M.; Du Toit, C.G.; Rousseau, P.G. A methodology to investigate the contribution of conduction and radiation heat transfer to the effective thermal conductivity of packed graphite pebble beds, including the wall effect. *Nucl. Eng. Des.* **2017**, *314*, 67–81. [CrossRef]
13. Coberly, C.A.; Marshall, W.R. Temperature gradient in gas stream flowing through fixed granular beds. *Chem. Eng. Prog.* **1951**, *47*, 141–150.
14. Ferreira, L.M.; Castro, J.A.M.; Rodrigues, A.E. An analytical and experimental study of heat transfer in fixed bed. *Int. J. Heat Mass Transf.* **2002**, *45*, 951–961. [CrossRef]
15. Collier, A.P.; Hayhurst, A.N.; Richardson, J.L.; Scott, S.A. The heat transfer coefficient between a particle and a bed (packed of fluidized) of much larger particles. *Chem. Eng. Sci.* **2004**, *59*, 4613–4620. [CrossRef]
16. Nijemeisland, M.; Dixon, A.G. CFD study of fluid flow and wall heat transfer in a fixed bed of spheres. *AIChE J.* **2004**, *50*, 906–921. [CrossRef]
17. Yang, J.; Wu, J.Q.; Zhou, L.; Wang, Q.W. Computational study of fluid flow and heat transfer in composite packed beds of spheres with low tube to particle diameter ratio. *Nucl. Eng. Des.* **2016**, *300*, 85–96. [CrossRef]
18. Yang, J.; Wang, Q.W.; Zeng, M.; Nakayama, A. Computational study of forced convective heat transfer in structured packed beds with spherical or ellipsoidal particles. *Chem. Eng. Sci.* **2010**, *65*, 726–738. [CrossRef]
19. Guardo Coussirat, M.; ALarrayoz, M.; Recasens, F.; Egusquiza, E. Influence of the turbulence model in CFD modeling of wall-to-fluid heat transfer in packed beds. *Chem. Eng. Sci.* **2005**, *60*, 1733–1742. [CrossRef]
20. Zhu, H.P.; Zhou, Z.Y.; Yang, R.Y.; Yu, A.B. Discrete particle simulation of particulate systems: Theoretical developments. *Chem. Eng. Sci.* **2003**, *62*, 3378–3396. [CrossRef]

21. Li, Y.; Ji, W. Acceleratio of coupled granular flow and fluid flow simulations in pebble bed energy systems. *Nucl. Eng. Des.* **2013**, *258*, 275–283. [CrossRef]

22. Zhao, Y.; Jiang, M.; Lium, Y.; Zheng, J. Particle-scale simulation of the flow and heat transfer behaviors in fluidized bed with immersed tube. *AIChE J.* **2009**, *55*, 3109–3124. [CrossRef]

23. Cundall, P.A.; Strack, O.D.L. A discrete numerical model for granular assemblies. *Geotechnique* **1979**, *29*, 47. [CrossRef]

24. Zhong, W.; Yu, A.B.; Zhou, C.; Xie, J.; Zhang, H. CFD simulation of dense particulate reaction system: Approaches, recent advances and applications. *Chem. Eng. Sci.* **2016**, *140*, 16–43. [CrossRef]

25. Chu, K.W.; Yu, A.B. Numerical simulation of complex particle–fluid flows. *Powder Technol.* **2008**, *179*, 104–114. [CrossRef]

26. Gui, N.; Fan, J.R.; Luo, K. DEM–LES study of 3-D bubbling fluidized bed with immersed tubes. *Chem. Eng. Sci.* **2008**, *63*, 3654–3663. [CrossRef]

27. Zhou, H.; Flamant, G.; Gauthier, D.; Flitris, Y. Simulation of coal combustion in a bubbling fluidized bed by distinct element method. *Chem. Eng. Res. Des.* **2003**, *81*, 1144–1149. [CrossRef]

28. Li, J.T.; Mason, D.J. Application of the discrete element modelling in air drying of particulate solids. *Dry. Technol.* **2002**, *20*, 255–282. [CrossRef]

29. Natsui, S.; Nogami, H.; Ueda, U.; Kano, J.; Inoue, R.; Ariyama, T. Simultaneous threedimensional analysis of gas-solid flow in blast furnace by combining discrete element method and computational fluid dynamics. *ISIJ Int.* **2011**, *51*, 41–50. [CrossRef]

30. Yang, W.J.; Zhou, Z.Y.; Yu, A.B. Particle scale studies of heat transfer in a moving bed. *Powder Technol.* **2015**, *281*, 99–111. [CrossRef]

31. Wu, H.; Gui, N.; Yang, X.T.; Tu, J.Y.; Jiang, S.Y. Numerical simulation of heat transfer in packed pebble beds: CFD-DEM coupled with particle thermal radiation. *Int. J. Heat Mass Transf.* **2017**, *110*, 393–405. [CrossRef]

32. Guo, Z.Y.; Li, D.Y.; Wang, B.X. A novel concept for convective heat transfer enhancement. *Int. J. Heat Mass Transf.* **1998**, *41*, 2221–2225. [CrossRef]

33. Guo, Z.Y.; Tao, W.Q.; Shah, R. The field synergy (coordination) principle and its applications in enhancing single phase convective heat transfer. *Int. J. Heat Mass Transf.* **2005**, *48*, 1797–1807. [CrossRef]

34. Liu, W.; Liu, Z.C.; Guo, Z.Y. Physical quantity synergy in laminar flow field of convective heat transfer and analysis of heat transfer enhancement. *Chin. Sci. Bull.* **2009**, *54*, 3579–3586. [CrossRef]

35. Liu, W.; Liu, P.; Dong, Z.M.; Yang, K.; Liu, Z.C. A study on the multi-field synergy principle of convective heat and mass transfer enhancement. *Int. J. Heat Mass Transf.* **2019**, *134*, 722–734. [CrossRef]

36. Sun, H.N.; Ge, Y.; Liu, W.; Liu, Z.C. Geometric optimization of two-stage thermoelectric generator using genetic algorithm and thermodynamic analysis. *Energy* **2019**, *171*, 37–48. [CrossRef]

37. Ge, Y.; Shan, F.; Liu, Z.C.; Liu, W. Optimal structural design of a heat sink with laminar single-phase flow using computational fluid dynamics based multi-objective genetic algorithm. *J. Heat Transfer* **2018**, *9*, 13–18. [CrossRef]

38. Ge, Y.; Liu, Z.C.; Sun, H.N.; Liu, W. Optimal design of a segmented thermoelectric generator based on three-dimensional numerical simulation and multi-objective genetic algorithm. *Energy* **2018**, *147*, 1060–1069. [CrossRef]

39. Ge, Y.; Wang, S.C.; Liu, Z.C.; Liu, W. Optimal shape design of a minichannel heat sink applying multi-objective optimization algorithm and three-dimensional numerical method. *Appl. Thermal Eng.* **2019**, *148*, 120–128. [CrossRef]

40. Guo, Z.Y.; Zhu, H.Y.; Liang, X.G. Entransy-A physical quantity describing heat transfer ability. *Heat Mass Transf.* **2007**, *50*, 2545–2556. [CrossRef]

41. Chen, Q.; Zhu, H.Y.; Pan, N.; Guo, Z.Y. An alternative criterion heat transfer optimization. *Math. Phys. Eng. Sci.* **2011**, *467*, 1012–1028. [CrossRef]

42. Chen, Q.; Liang, X.G.; Guo, Z.Y. Entransy Theory for the optimization of heat transfer—A review and update. *Heat Mass Transf.* **2013**, *63*, 65–81. [CrossRef]

43. Chen, Q.; Wang, M.; Pan, N.; Guo, Z.-Y. Optimization principles for convective heat transfer. *Energy* **2009**, *34*, 1199–1206. [CrossRef]

44. Malone, F.; Xu, B.H. Particle-scale simulation of heat transfer in liquid-fluidized beds. *Powder Technol.* **2008**, *184*, 189–204. [CrossRef]

45. Zhou, Z.Y.; Yu, A.B.; Zulli, P. Particle scale study of heat transfer in packed and bubbling fluidized beds. *AIChE J.* **2009**, *55*, 868–884. [CrossRef]
46. Wang, J.; Liu, W.; Liu, Z. The application of exergy destruction minimization in convective heat transfer optimization. *Appl. Therm. Eng.* **2015**, *88*, 384–390. [CrossRef]
47. Liu, W.; Liu, Z.; Jia, H. Entransy expression of the second law of thermodynamics and its application to optimization in heat transfer process. *Int. J. Heat Transf.* **2011**, *54*, 3049–3059. [CrossRef]
48. Cheng, X.G.; Li, Z.X.; Guo, Z.Y. Variational principles in heat conduction. *J. Eng. Thermophys.* **2004**, *25*, 457–459.
49. Demirel, Y.; Sharma, R.N.; Al-Ali, H.H. On the effective heat transfer parameters in a packed bed. *Int. J. Heat Mass Transf.* **2000**, *43*, 327–332. [CrossRef]
50. Sug Lee, J.; Ogawa, K. Pressure drop through packed beds. *J. Chem. Eng. Jpn.* **1974**, *27*, 691–693.

Article

Temperature Disturbance Management in a Heat Exchanger Network for Maximum Energy Recovery Considering Economic Analysis

Ainur Munirah Hafizan [1], Jiří Jaromír Klemeš [2,*], Sharifah Rafidah Wan Alwi [1], Zainuddin Abdul Manan [1] and Mohd Kamaruddin Abd Hamid [1]

[1] Process Systems Engineering Centre (PROSPECT), Research Institute for Sustainable Environment (RISE), School of Chemical and Energy Engineering, Universiti Teknologi Malaysia (UTM), 81310 Johor Bahru, Johor, Malaysia; ainurhafizan@yahoo.com (A.M.H.); syarifah@utm.my (S.R.W.A.); dr.zain@utm.my (Z.A.M.); kamaruddinhamid@utm.my (M.K.A.H.)

[2] Sustainable Process Integration Laboratory – SPIL, NETME Centre, Faculty of Mechanical Engineering, Brno University of Technology – VUT BRNO, Technická 2896/2, 616 69 Brno, Czech Republic

* Correspondence: klemes@fme.vutbr.cz; Tel.: +420-541-144-985

Received: 31 December 2018; Accepted: 11 February 2019; Published: 13 February 2019

Abstract: The design of heat exchanger networks (HEN) in the process industry has largely focused on minimisation of operating and capital costs using techniques such as pinch analysis or mathematical modelling. Aspects of operability and flexibility, including issues of disturbances affecting downstream processes during the operation of highly integrated HEN, still need development. This work presents a methodology to manage temperature disturbances in a HEN design to achieve maximum heat recovery, considering the impact of supply temperature fluctuations on utility consumption, heat exchanger sizing, bypass placement and economic performance. Key observations have been made and new heuristics are proposed to guide heat exchanger sizing to consider disturbances and bypass placement for cases above and below the HEN pinch point. Application of the methodology on two case studies shows that the impact of supply temperature fluctuations on downstream heat exchangers can be reduced through instant propagation of the disturbances to heaters or coolers. Where possible, the disturbances have been capitalised upon for additional heat recovery using the pinch analysis plus-minus principle as a guide. Results of the case study show that the HEN with maximum HE area yields economic savings of up to 15% per year relative to the HEN with a nominal HE area.

Keywords: pinch analysis; heat exchanger network (HEN) design; plus-minus principle; supply temperature; disturbances; maximum energy recovery; bypass; economic evaluation

---

1. Introduction

Heat integration has been a well-established energy saving technique for the chemical process industry since the global energy crisis in the 1970s [1]. There has also been extensive development of pinch analysis methodologies for industrial heat exchanger network (HEN) design focusing on minimisation of operating and capital costs [2]. The developed methodologies typically assume that process parameters such as supply/target temperatures and stream flowrates are fixed [3]. In practice, these process parameters may fluctuate due to plant start-ups/shut-downs, changes in feed or product demand as well as quality, changes in environmental conditions and other operational disturbances. The impact of these parameter changes influences energy related-decision making as a step for efficient energy management in the industry [4,5]. The extent to which a HEN is able to cope with disturbances is known as flexibility [3]. Previous works have included flexibility, safety, controllability and operability in the pre-design stage.

Marselle et al. [6] pioneered the study of operability considerations for HENs. They proposed a manual combination of several resilient designs with maximum energy efficiency under different worst-case scenarios. Hafizan et al. [7] proposed the controllability of HEN under uncertainty by having a large heat exchanger area. This method is able to recover maximum heat with minimum utility. The paper was also partly presented at the 13th SDEWES conference and proceedings. Linnhoff and Kotjabasakis [8] introduced the concept of downstream paths to identify disturbance propagation paths through HEN. Supply temperature and feed flowrate disturbances occurring at the feed stream of heat exchangers can affect the target temperatures of processes located downstream of the heat exchanger path. HEN modification, therefore, may need to be performed to reject the process disturbances.

Escobar et al. [9] introduced a computational framework for the synthesis of controllable and flexible HENs over a specified range of inlet temperatures and flow rate variations using a decentralised control system. This framework comprised two stages—the HEN design and the operability analysis and adjustment of control variables during operation in the presence of uncertain parameters. Hafizan et al. [10] developed a pinch analysis-based methodology, which considered both safety and operability aspects in HEN synthesis. The concept of downstream paths suggested by Linnhoff and Kotjabasakis [8] was used for the assessment of flexibility and structural controllability. Čuček and Kravanja [11] proposed a novel three-step methodology for HEN retrofit with fixed and flexible designs for large-scale total sites (TS). This method modified the HEN by forming profitable heat exchanger matches with improved utility consumption as well as by proposing intermediate utility production.

Most recent works have developed a multi-period formulation for the synthesis of flexible HENs. Isafiade and Short [12] proposed a three-step approach for improving the degree of flexibility of a multi-period HEN with unequal periods. However, this method is unable to cater for all the possible uncertainties of process parameters. Miranda et al. [13] recently proposed three sequential steps represented by linear programming (LP), multi-linear programming (MILP) and non-linear programming (NLP) for the synthesis of multi-period HEN. The same heat transfer can be operated under different operating conditions in the multi-period [13]. Kang and Liu [14] developed a three-step method for designing a flexible multi-period HEN when the disturbances of operating parameters occurred in sub-periods. The flexibility of nominal multi-period HEN is first determined and analysed prior to identifying the bottlenecks. As a next step, the design of flexible multi-period HEN is finalised by solving the sub-period debottlenecking model.

Several authors have considered the operability issues at the early stages of the process design. The need for this s widely accepted and has motivated the integration of process design and control (IPDC). Narraway and Perkins [15] and Walsh and Perkins [16] were the among the earliest to take into account the general mathematical programming techniques for the simultaneous design and control problem using dynamic process models. Recently, Abu Bakar et al. [17] introduced a new model-based IPDC of HEN which is decomposed into four hierarchical sequential stages. The proposed methodology recommends a solution that satisfies the design, control and economic criteria.

The optimal HEN with unclassified hot/cold process streams was discussed in the work of Kong et al. [18], Quirante et al. [19] and Onishi et al. [20]. All these works depend on the process operating conditions to finalise the classification of process streams. Kong et al. [18] presented mixed-integer nonlinear programming (MINLP) for the heat integration model which accounts for unclassified process streams and variable stream temperatures and flowrates. Quirante et al. [19] later extended the disjunctive model of the pinch location method proposed by Quirante et al. [21] and work by Kong et al. [18] for the simultaneous process optimisation and heat integration. This work is also extended for the isothermal process streams, multiple utilities and area estimation of HEN. The area estimation is done based on the vertical heat transfer between the hot and cold balanced composite curves. Onishi et al. [20] proposed an optimisation model to enhance the work and HENs energy efficiency and cost-effective synthesis considering the unclassified process streams. It combined the methods of mathematical programming and pinch location while adjusting the pressure and temperature of unclassified streams.

State-of-the-art studies on HEN flexibility, show that there are a few key limitations associated with the existing methods. In these previous works, the optimal design of HEN synthesis operated under uncertain operating parameters were considered with an appropriate strategy for control and operation. The control variables were assumed to be adjusted during the operation to improve the flexibility of HEN. However, several possible HEN configurations were needed for each of the scenarios in order to increase the flexibility and some exchanger area adjustment was required during the operation. Besides that, heuristics to guide heat exchanger sizing and bypass placement, and which can be applied in all cases have not been introduced. The understanding of how the temperature fluctuation in HEN affects the amount of heat recovery is still not clearly understood.

This work presents a methodology to manage temperature disturbances in HEN design for maximum heat recovery. The impact of the supply temperature fluctuations on utility consumption, heat exchanger sizing and bypass placement are studied to ensure the target temperatures of affected streams are achieved. At the same time, reducing the impact of the fluctuation on downstream heat exchangers and the immediately propagation of the disturbances to heaters or coolers is desired. Where possible, taking advantage of the disturbances for additional heat recovery is also desirable. The plus-minus principle for process changes are used and new heuristics are introduced to guide heat exchanger sizing and bypass placement. Linnhoff and Vredeveld [22] have introduced the plus-minus principle for visualising the impact of process modifications on the minimum utility target using the composite curves (CC). Chew et al. [23] applied the plus-minus principle for process modification aimed at maximising energy savings for total site heat integration (TSHI). This methodology enabled designers to identify the potential process changes to maximise energy recovery and reduce utility consumption. Song et al. [24] further implemented the plus-minus principle for inter-plant heat integration (IPHI) for case studies involving threshold problems. The proposed methodology provides a simple technique of rapidly assessing the effect of supply temperature (T_S) fluctuations in heat recovery and utility reduction without the need for detailed process simulation.

2. Methodology

This section describes the methodology that was developed to manage supply temperature disturbances through modification of a conventional heat exchange network. By planning the right size for the heat exchangers, and by utilising the bypass streams, a HEN can be designed with the flexibility to cope with supply temperature disturbances. An illustrative case study of a HEN experiencing supply temperature disturbances on each process stream is used to demonstrate the applicability of the methodology. For the case study, disturbances are assumed to occur at all supply temperatures $(T_{SH1}, T_{SH2}, T_{SH3}, T_{SC1}, T_{SC2}$ and $T_{SC3})$ with a deviation of $\pm 5\,°C$. The manipulated variables for process control include heat exchangers, bypasses and utility flowrates of coolers and heaters. The controlled variables are all the target temperatures $(T_{tH1}, T_{tH2}, T_{tH3}, T_{tC1}, T_{tC2}$ and $T_{tC3})$.

2.1. Step 1: Stream Data Extraction with Disturbances

Table 1 shows the stream data which are used to illustrate the effect of disturbances on maximum energy recovery HEN. There are three hot streams (H1, H2 and H3) and three cold streams (C1, C2 and C3) involved in the process. The required data for the pinch study includes the supply temperature, T_s; target temperature, T_t; heat capacity flowrate, FC_p; enthalpy, ΔH and the supply temperature fluctuation temperature range. The minimum temperature difference, ΔT_{min} is set as $10\,°C$.

Table 1. Stream data for nominal operation.

Stream	Supply Temperature, T_s (°C)	Target Temperature, T_t (°C)	Heat Capacity Flowrate, FC_P (MW/°C)	Enthalpy, ΔH (MW)
Hot 1 (H1)	310 ± 5	270	3	−120
Hot 2 (H2)	235 ± 5	120	4	−460
Hot 3 (H3)	270 ± 5	60	5	−1050
Cold 1 (C1)	40 ± 5	220	3	540
Cold 2 (C2)	90 ± 5	290	5	1000
Cold 3 (C3)	240 ± 5	300	7	420

2.2. Step 2: Perform Maximum Energy Recovery Targeting for the Nominal Case

The maximum energy recovery (MER) targets are determined for the nominal case (without disturbances) by using pinch analysis targeting methods such as the problem table algorithm or composite curves by Linnhoff and Flower [25] or streams temperature vs enthalpy plot (STEP) by Wan Alwi and Manan [26]. The targeted minimum hot utility requirement Q_{Hmin} is 450 MW and the minimum cold utility requirement Q_{Cmin} is 180 MW. The hot pinch temperature, $T_{Pinch, hot}$ is at 250 °C and the cold pinch temperature, $T_{Pinch, cold}$ is at 240 °C. Figure 1 shows the composite curves of the nominal case.

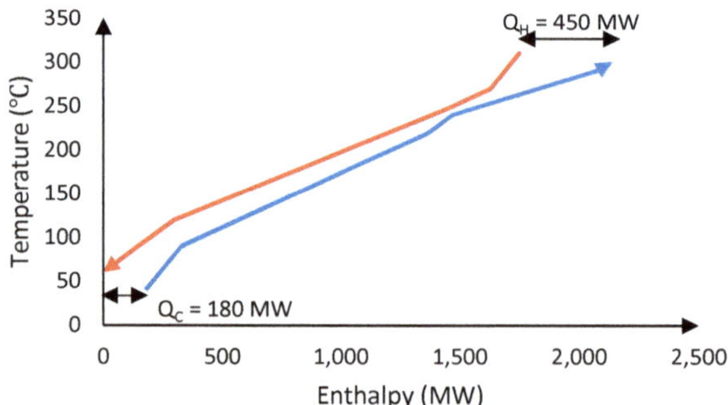

Figure 1. Composite curves for the nominal case.

2.3. Step 3: Construct the Grid Diagram (GD)

The nominal maximum energy recovery (MER) network is designed using the pinch design method by Linnhoff and Flower [25] and drawn on the grid diagram (GD) shown in Figure 2. Based on the feasibility criteria proposed by Linnhoff and Hindmarsh [27], streams matching above the pinch have to meet the criterion of $CP_{COLD} \geq CP_{HOT}$ and streams matching below the pinch have to meet the criterion of $CP_{HOT} \geq CP_{COLD}$, at the pinch location. Equations (1) and (2) are used to determine the enthalpy balances for hot and cold streams.

$$Q_{HE,hot} = FC_{p_{hot}} \cdot (T_{hot,in} - T_{hot,out}) \tag{1}$$

$$Q_{HE,cold} = FC_{p_{cold}} \cdot (T_{cold,in} - T_{cold,out}) \tag{2}$$

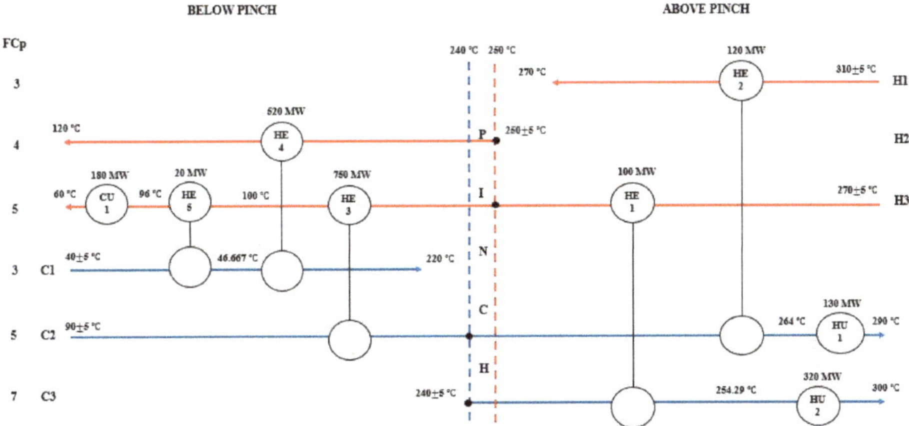

Figure 2. Maximum energy recovery (MER) heat exchanger network (HEN) design for the nominal case.

2.4. Step 4: Manage Fluctuating Ts (Ts Disturbances) in HEN to Achieve MER

The steps to manage fluctuating T_S in HEN to achieve T_t and MER can be defined in three stages. The first stage describes the effect of increasing or decreasing T_S on the hot and cold streams energy requirement. The second stage provides the fundamental theory of the plus-minus principle. The final stage explains the effect of T_S fluctuation on utilities based on the plus-minus principle. Heuristics are proposed for each disturbance scenario that necessitates bypass placement and valve opening, as well as the correct heat exchanger sizing to maximise utility savings.

2.4.1. Effect of Increasing or Decreasing T_S on Hot and Cold Streams

Figure 3 shows the effect of increasing or decreasing T_S for a hot stream. The target temperature, T_t is assumed to be maintained. An example of this situation is a fluctuating reactor exit temperature. Before the stream enters the next unit operation, the T_t should be maintained at the setpoint value. From Figure 3, it can be seen that increasing T_S for hot stream results in an increase in enthalpy, ΔH, while decreasing T_S results in a decrease in ΔH.

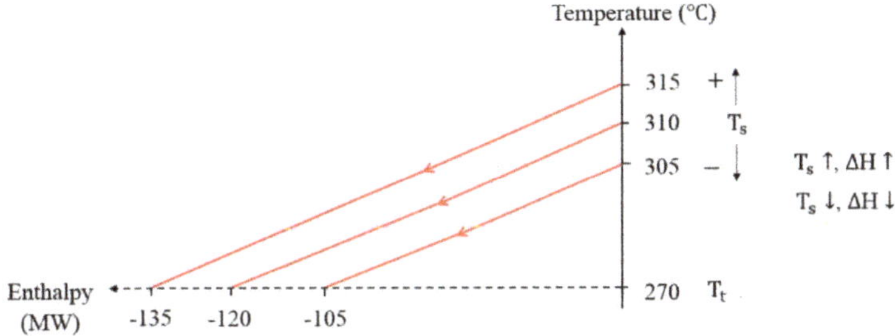

Figure 3. Effect of increasing or decreasing supply temperature (T_S) on a hot stream.

On the other hand, Figure 4 shows the effect of increasing or decreasing T_S on a cold stream. From the figure, it can be seen that an increase in T_S results in an increase in ΔH while a decrease in T_S results in a decrease in ΔH. The target temperature, T_t is also assumed to be maintained. An example of this situation is a feed stream coming from a storage tank experiencing temperature fluctuations due to changes in ambient conditions as a result of weather changes in a four-season country.

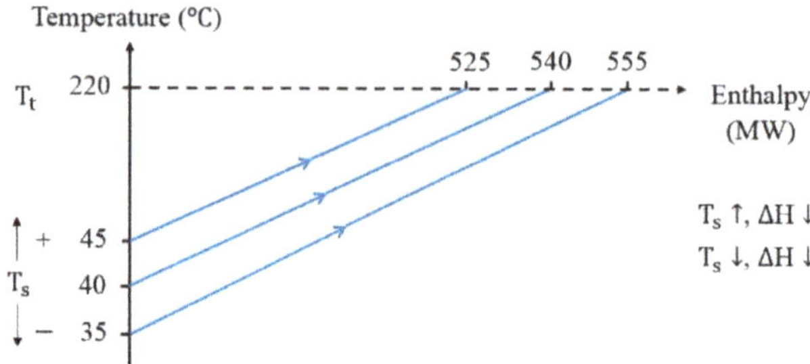

Figure 4. Effect of increasing or decreasing T_S on a cold stream.

2.4.2. The Plus-Minus Principle Concept

Figure 5 shows the effect of plus-minus principle on composite curves (CC), as explained by Linnhoff and Vredeveld [22]. Above the pinch, increasing the enthalpy of a hot stream and decreasing the enthalpy of a cold stream decreases the minimum hot utility target, $Q_{H,min}$. Doing the reverse above the pinch increases $Q_{H,min}$. Below the pinch, increasing the enthalpy of a cold stream and decreasing the enthalpy of a hot stream decreases the minimum cold utility target, $Q_{C,min}$. Doing the reverse below the pinch increases $Q_{C,min}$.

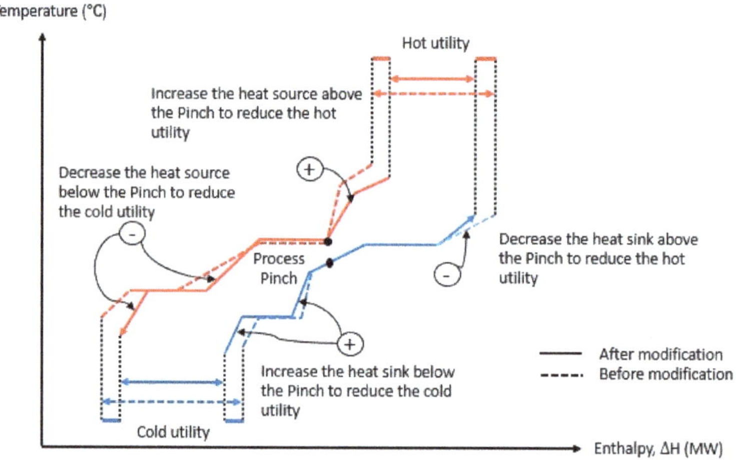

Figure 5. The plus-minus principle (amended from [22]).

2.4.3. Control Mechanism Decision by Using the Plus-Minus Principle

The plus-minus principle can be used to describe the effect of disturbances on HEN operation, and how HEN can be controlled. As shown in Step 4(i), the increase or decrease in supply temperature, T_S of the hot or cold stream results in an increase in either Q_H or Q_C; or a decrease in Q_H or Q_C. For effective control of T_t, utility units such as heaters and coolers have been widely used in process plants. Utility heaters/coolers are typically placed after a series of heat exchangers to supplement stream heating or cooling. A bypass, on the other hand, functions as a mechanism for controlling process stream parameters including for disturbance rejection.

A bypass placed on the hot stream side can be described using Equation (3) while a bypass placed on the cold stream side can be described using Equation (4) [28]. T_t^h represents the target temperature of the heat exchanger at hot the stream side, T_t^c represents the target temperature of the heat exchanger at a cold stream side; T_o^h represents the outlet temperature of the heat exchanger at hot stream side, T_o^c

represents the outlet temperature of the heat exchanger at a cold stream side; T_s^h represents the supply temperature of the heat exchanger at hot stream side, T_s^c represents the supply temperature of heat exchanger at a cold stream side; u^b represents the bypass fraction; u^h represents the stream fraction at hot stream side, u^c represents the stream fraction at cold stream side. Application of Equations (1) and (2) is explained next, in the context of heuristics development.

$$T_t^h = \left(1 - u^h\right)T_o^h + u^b T_s^h \tag{3}$$

$$T_t^c = (1 - u^c)T_o^c + u^b T_s^c \tag{4}$$

New heuristics have been introduced in this work as guides for the appropriate placement of a bypass and sizing of the heat exchanger that can reduce hot and cold utilities in cases with recurring Ts disturbances. Each proposed heuristic shall refer to the plus-minus principle of process changes. Applicability of the heuristics is explained using a case study.

Below the pinch:

Observation 1. Decreasing T_S for a hot or cold stream below the pinch results in decreasing Q_C.

The first observation states that decreasing Ts for a hot or cold stream located below the pinch results in Q_C decreasing. For example, the T_S of cold stream C1 decreases from 40 °C to 35 °C. As shown in Figure 4, decreasing T_S on cold stream C1 increases C1 enthalpy by 15 MW. Based on the plus-minus principle, since stream C1 is located below the pinch, it can be used to recover more energy. This results in a reduction of CU2 cold utility from 180 MW to 165 MW, which is desirable. In order to control the target temperature of stream C1 at 220 °C, the C1 flowrate entering HE5 is selected as a manipulated variable. The duty of HE5 is increased from 20 MW to 35 MW to allow more heat to be exchanged. To be able to do this, HE5 should be designed with a bigger area to accommodate up to 35 MW heat duty.

Observation 2. Below the pinch, the point where Q_C equals zero is the limit for hot or cold stream T_S to decrease. Further decrease in Ts leads to a penalty in Q_H.

It is observed that there exists a limit for T_S to decrease for the hot or cold stream below the pinch, i.e., at the point of zero Qc. Any additional decrease in T_S for hot or cold stream leads to a penalty of Q_H. Figure 6a,b illustrate Observation 2 involving the disturbance scenario for HE4. It can be seen for the case that when Ts of stream H2 decrease, the enthalpy decrease exceeds the MER of cold stream C1, leading to a penalty of hot utility (see Figure 6b). This situation is also illustrated using the plus-minus principle shown in Figure 7. Moreover, decreasing T_S for the hot stream at the pinch point or where the heat exchanger inlet has $\Delta T = \Delta T_{min}$ also incurs a penalty at the other side stream. However, in this case, cold stream C1 does not end at the pinch and the difference of temperature between $T_{S,H2}$ and $T_{t,C1}$ is 25 °C, which is more than T_{min}.

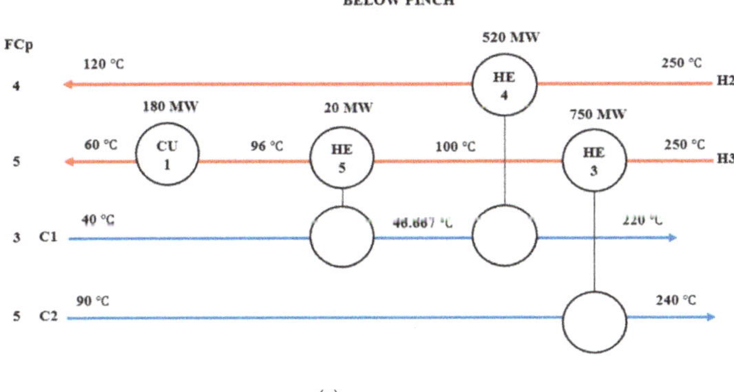

(a)

Figure 6. *Cont.*

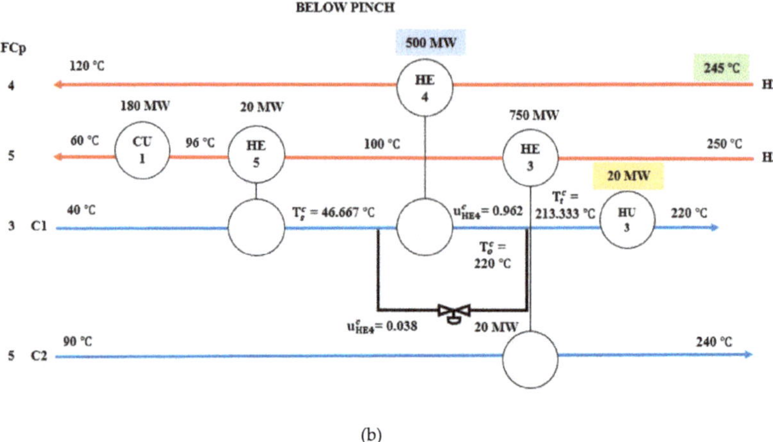

Figure 6. (**a**) HEN design with nominal T_S for hot stream H2; (**b**) HEN design with T_S decrease by 5 °C for hot stream H2.

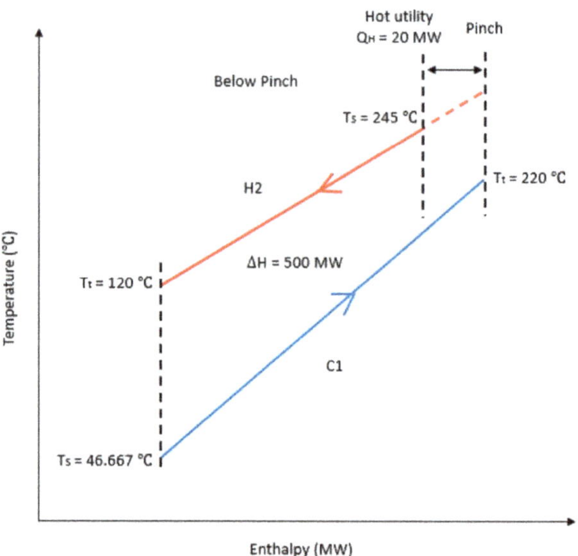

Figure 7. The plus-minus principle with decreasing T_S by 5 °C for hot stream H2 at HE4.

Observation 3. Increasing T_S for a hot or cold stream below the pinch results in increasing Q_C.

Based on this third observation, increasing T_S for a hot or cold stream located below the pinch causes Q_C to increase. To illustrate this, consider an increase in T_S for the cold stream C1 from 40 °C to 45 °C due to a disturbance. Figure 4 shows that the increase in T_S for the C1 resulted in the enthalpy for C1 decreasing by 15 MW. Since the Ts of stream C1 is located below the pinch, based on the plus-minus principle, the increase in Ts resulted in an increase in CU2 from 180 MW to 195 MW as shown in Figure 8. For this scenario, the bypass stream is selected as the manipulated variable as this deviation means the heat duty of heat exchanger HE5 is not high enough to keep the T_t at 220 °C. Equation (2) is used to calculate the bypass fraction for the bypass placed on the cold streamside. In this scenario, supply temperature, T_s^c of HE5 at the cold stream side is 45 °C. The decreasing of the heat duty of

HE 5 from 20 MW to 5MW caused the target temperature of HE5 at cold stream side, T_t^c increased to 46.667 °C. As the outlet temperature, T_o^c of HE5 with maximum duty at cold stream C1 is 56.667 °C, the bypass is placed on the cold stream side. Now the valve is opened at a bypass fraction, u_{HE5}^b of 0.857. The bypass is calculated by rearranging Equation (2). As the heat duty of heat exchanger HE5 is not high enough to achieve the final target temperature of H3, the cold utility of cooler C2 is increased to absorb the remaining heat.

$T_t^c = 46.667$ °C; $T_o^c = 56.667$ °C; $T_s^c = 45$ °C

Rearranged Equation (2) to gain the value of u^b and u^c,

$$T_t^c = \left(1 - u^b\right)T_o^c + u^b T_s^c$$
$$u^b = 0.857;\ u^c = 0.143$$

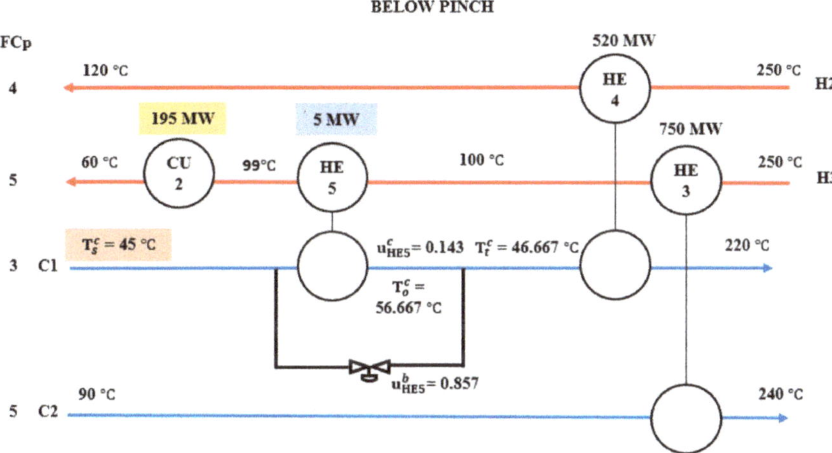

Figure 8. HEN design with T_S increased for cold stream below the pinch.

Observation 4. Below the pinch, size the heat exchanger to achieve the maximum energy recovery target when T_S decreases, and the cooler to achieve the higher utility when T_S increases.

Observation 4 states that the heat exchanger size below the pinch should be designed to achieve the maximum energy recovery when T_S decreases and the cooler size should be designed to achieve the higher utility when T_S increases. Previously, the impact of the increase or decrease of T_S on cold stream C1 was shown. In the case of decreasing T_S, more heat is allowed to be exchanged. It is preferable to design HE5 with a bigger area to accommodate up to at 35 MW heat duty instead of 20 MW heat duty for the nominal case. CU2 cold utility should also be designed with the duty of 195 MW instead of the nominal case with the duty of 180 MW in order to absorb the remaining heat when T_S increases. Figure 9a–c illustrate this situation by using the plus-minus principle. The bypass stream is used to control the duty of HE5 during the disturbances as shown in Figure 10a–c, while the CU2 cold utility is used to absorb the remaining heat.

Above the pinch:

Observation 5. Increasing T_S for a hot or cold stream above the pinch results in decreasing Q_H.

Observation 5 states that, increasing T_S for hot or cold stream located below the pinch results in Q_H decreasing. For example, T_S of hot stream H1 increases from 310 °C to 315 °C. As shown in Figure 3, the increasing T_S on hot stream H1 increases H1 enthalpy by 15 MW. Based on the plus-minus principle, since stream H1 is located above the pinch, it can be used to recover more energy. This would result in a reduction of HU1 hot utility from 130 MW to 115 MW, which is desirable (as shown in Figure 11. In order to control the target temperature of stream H1 at 270 °C, the H1 flowrate entering HE2 is selected as a manipulated variable. The duty of HE2 is increased from 120 MW to 135 MW to allow more heat to be exchanged. In order to attain it, HE2 should be designed with a bigger area to accommodate up to 135 MW heat duty.

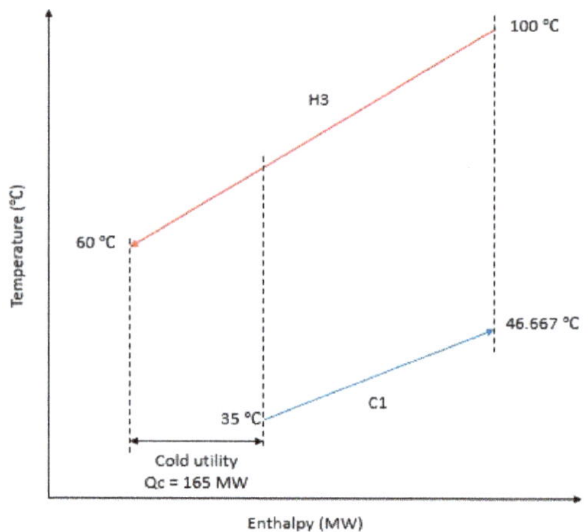

(a)

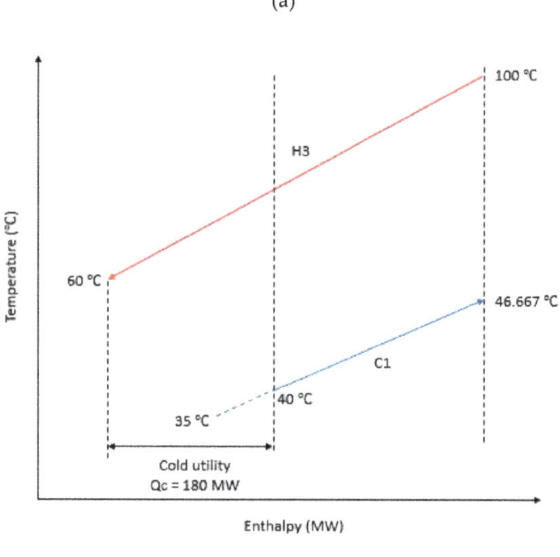

(b)

Figure 9. *Cont.*

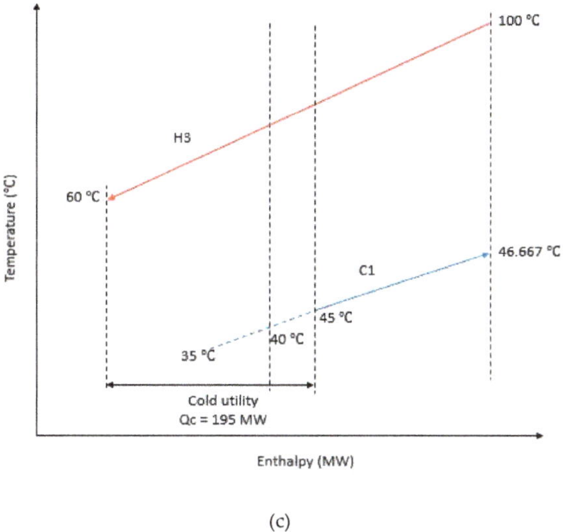

(c)

Figure 9. (**a**) The plus-minus principle with T_S decreased by 5 °C for C1; (**b**) the plus-minus principle of nominal T_S for C1; (**c**) The plus-minus principle with T_S increased by 5 °C for C1.

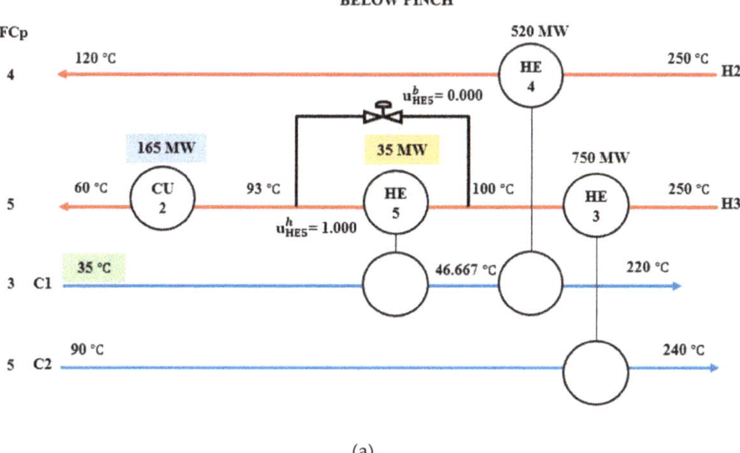

(a)

Figure 10. *Cont.*

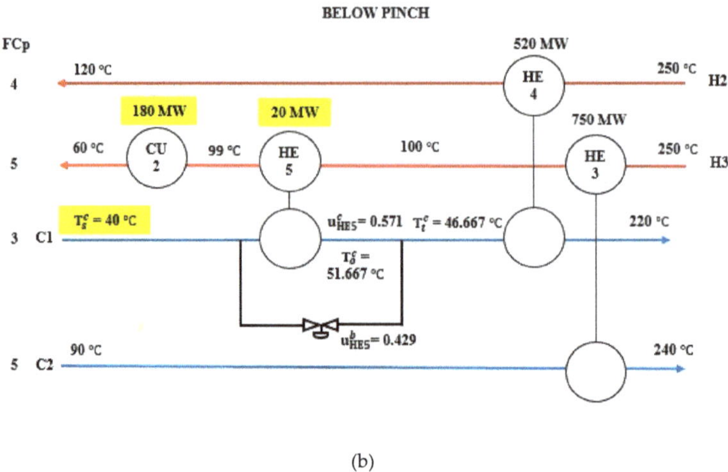

(b)

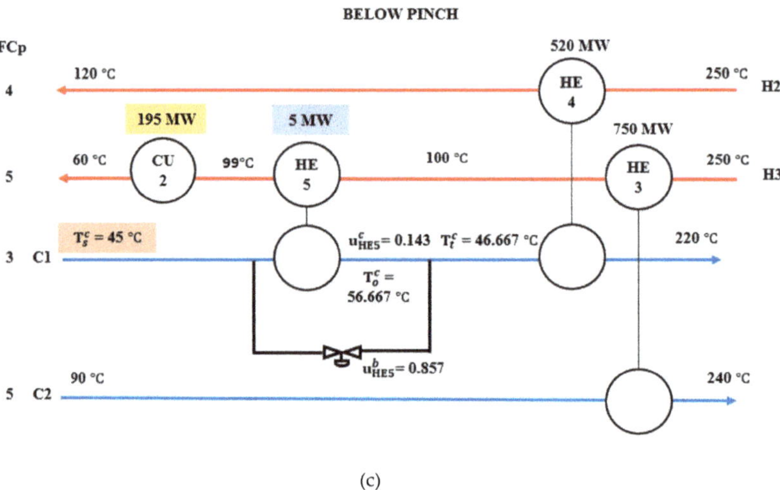

(c)

Figure 10. (**a**) HEN design with T_S decreased by 5 °C for C1. HE5 is designed with a bigger area to accommodate up to 35 MW heat duty; (**b**) HEN design with nominal T_S for C1. HE5 is designed with a bigger area and heat duty at 20 MW; (**c**) HEN design with T_S increased by 5 °C for C1. HE5 is designed with a bigger size and heat duty at 5 MW.

Observation 6. Above the pinch, the point where Q_H equals zero is the limit for hot or cold stream T_S to increase. Further increases in T_S lead to a penalty of Q_C.

As explained in Observation 2, the same situation can also occur above the pinch. There is a limit for T_S to increase for hot or cold streams above the pinch, i.e., at the point of zero Q_H. Further increases in T_S for the hot or cold stream lead to a penalty of Q_C. Besides that, increasing T_S for the hot stream at the pinch or where the heat exchanger inlet has $\Delta T = \Delta T_{min}$ also leads to a penalty at the other side stream. Figure 12 shows that T_S increases for cold stream C3 from 240 °C to 245 °C. The nominal value of T_S on cold stream C3 and the outlet temperature of HE1 on the hot stream side is at the pinch temperature. The ΔT of HE1 decrease to 5 °C, which is less than the original ΔT_{min} of 10 °C. This cause the enthalpy of cold stream C3 reduced to 385 MW. HE1 with the duty of 100 MW led the hot utility HU2 minimised to 285 MW (see Figure 12). This situation is also illustrated by the plus-minus principle shown in Figure 13a,b. Figure 13a shows the nominal temperature of stream C3 matching with stream H3 with ΔT_{min} of 10 °C. Figure 13b shows that, increasing of T_S on cold stream results in ΔT less than ΔT_{min}.

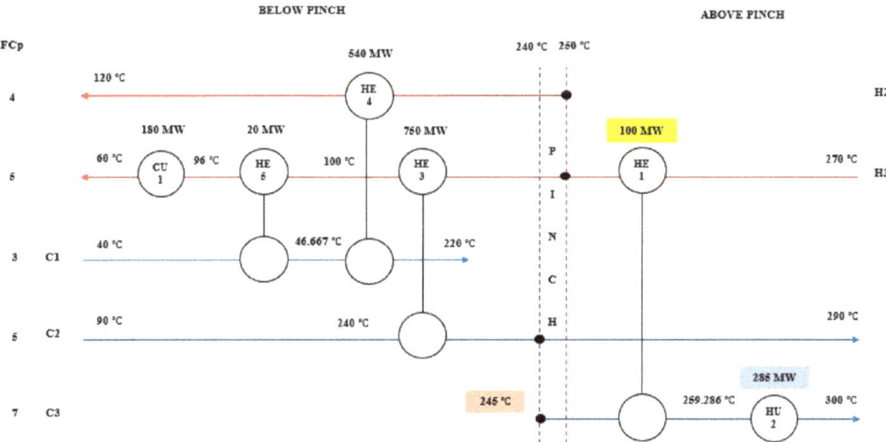

Figure 11. HEN design with increasing T_S for the hot stream H1 above the pinch.

Figure 12. HEN design with increasing T_S for cold stream at the pinch.

Observation 7. Decreasing T_S for a hot or cold stream above the pinch results in increasing Q_H.

As stated in observation 7, decreasing T_S for hot or cold stream located above the pinch causes Q_H to increase. To illustrate this, consider a decrease in T_S for the hot stream H1 from 310 °C to 305 °C. Figure 3 shows that the decrease in T_S for the H1 resulted in the enthalpy for H1 to decrease by 15 MW. Since the T_S of stream H1 is located above the pinch, based on the plus-minus principle, the decrease in T_S resulted in an increase in hot utility HU1 from 130 MW to 145 MW as shown in Figure 14. For this scenario, the bypass stream is selected as a manipulated variable as this deviation means the heat duty of heat exchanger HE2 is not high enough to keep the target temperature at 270 °C. Equation (1) is used to calculate the bypass fraction for the bypass placed on the hot streamside. In this scenario, the supply temperature, T^h_s of HE2 at the hot stream side is 305 °C. The decrease in heat duty of HE2 from 120 MW to 105 MW caused the target temperature of heat exchanger HE2 at hot stream side, T^h_t to decrease to 270 °C. As the outlet temperature, T^h_o of HE2 with maximum duty at hot stream H1 is 260 °C, the bypass is placed on the hot stream side. Then, the valve is opened at a bypass fraction,

u^b_{HE2} of 0.286. The bypass is calculated by rearranging Equation (1). As the heat duty of heat exchanger HE2 is not high enough to achieve the final target temperature of C2, the hot utility of heater HU1 is increased in order to satisfy the remaining heat.

$T^h_t = 270\ °C;\ T^h_o = 260\ °C;\ T^h_s = 305\ °C$

Rearranged Equation (1) to gain the value of u^b and u^c,

$T^h_t = \left(1 - u^b\right)T^h_o + u^b T^h_s$

$u^b = 0.222;\ u^h = 0.778$

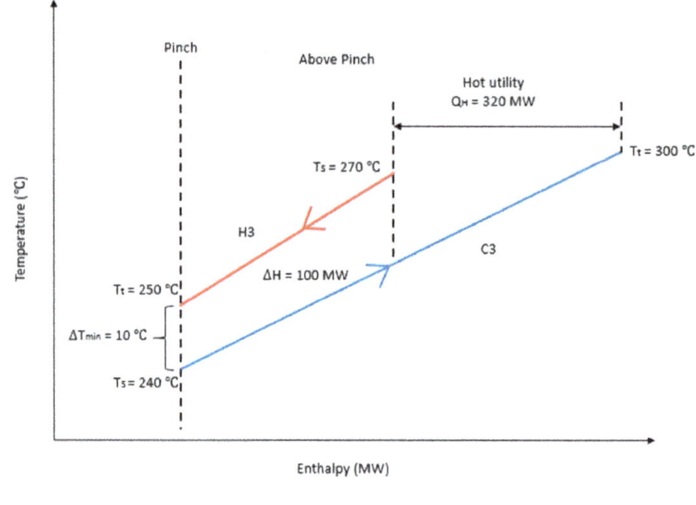

(a)

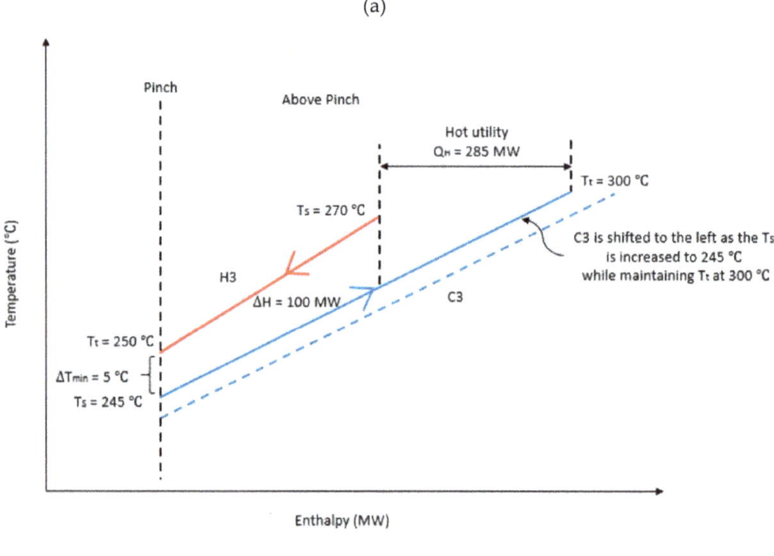

(b)

Figure 13. (**a**) The plus-minus principle of nominal T_S for cold stream at the pinch with $T_{min} = 10\ °C$; (**b**) The plus-minus principle of infeasible matching with increasing T_S for the cold stream at the pinch.

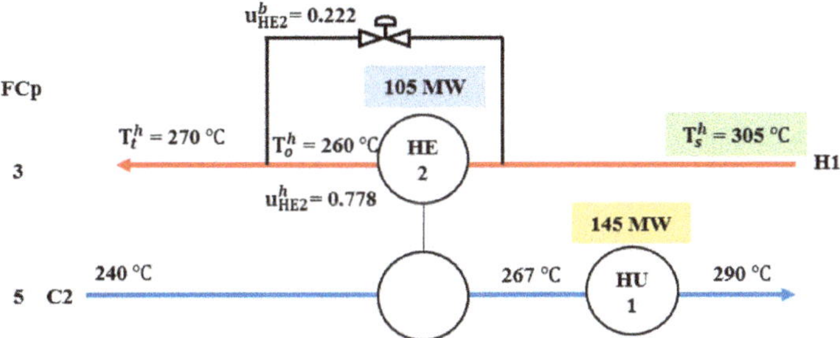

Figure 14. HEN design with T_S decreased for hot stream H1 above the pinch.

Observation 8. Above the pinch, size the heat exchanger to achieve the maximum energy recovery when T_S increase, and the heater to achieve the minimum utility when T_S decrease.

Observation 8 states that the heat exchanger size above the pinch should be designed to achieve the maximum energy recovery when T_S increases, and the heater size should be designed to achieve the higher utility when T_S decreases. Previously, the impact of increase or decrease in Ts on hot stream H1 was shown. In the case of increasing Ts, more heat is allowed to be exchanged. It is preferable to design HE2 with a bigger area to accommodate up to 135 MW heat duty instead of 120 MW for the nominal case. HU1 hot utility should also be designed with the bigger capacity of 145 MW heat duty instead of the nominal case of 130 MW, in order to cater for the remaining heat when T_S decreases. Figure 15a–c illustrate this situation by using the plus-minus principle. The bypass stream is used to control the duty of HE2 when a disturbance occurs as shown in Figure 16a–c, while HU1 hot utility is used is to satisfy the remaining heat.

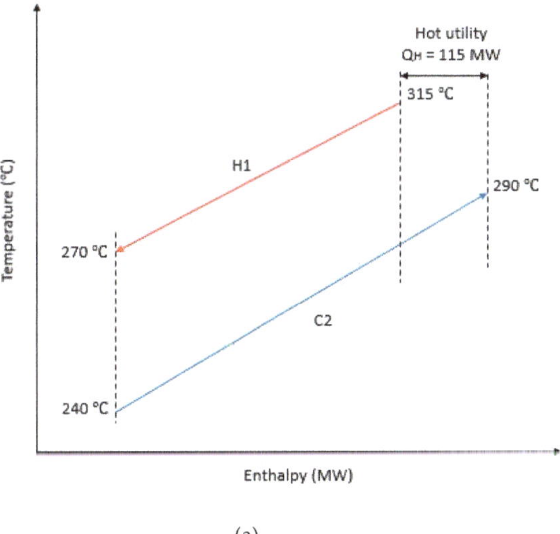

(a)

Figure 15. *Cont.*

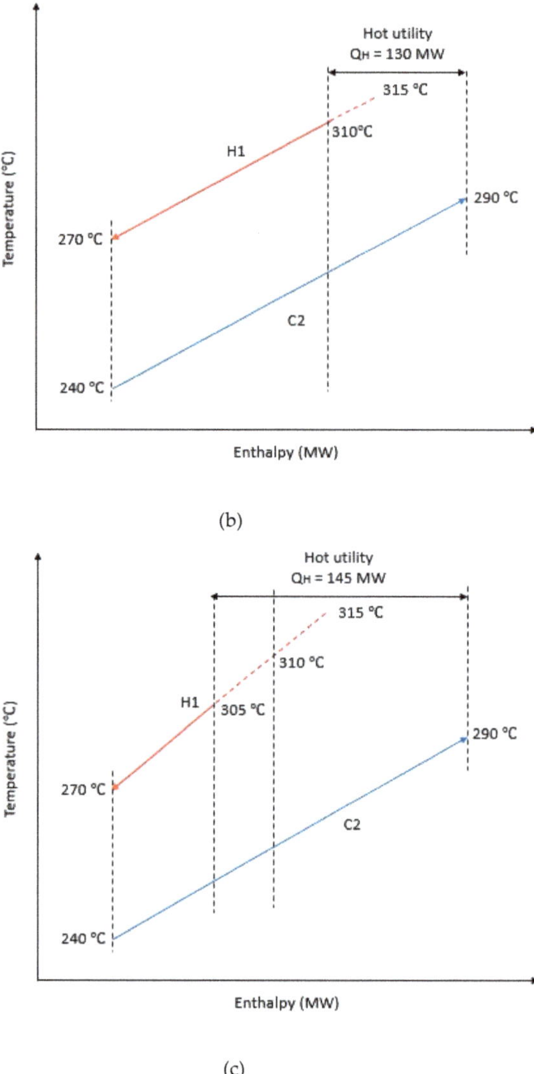

(b)

(c)

Figure 15. (**a**) The plus-minus principle with increasing T_S by 5 °C for H1; (**b**) The PLUS-MINUS principle for nominal T_S for H1 considering maximum size of HE and a bypass; (**c**) The plus-minus principle with decreasing T_S by 5 °C for H1 considering maximum size of HE and a bypass.

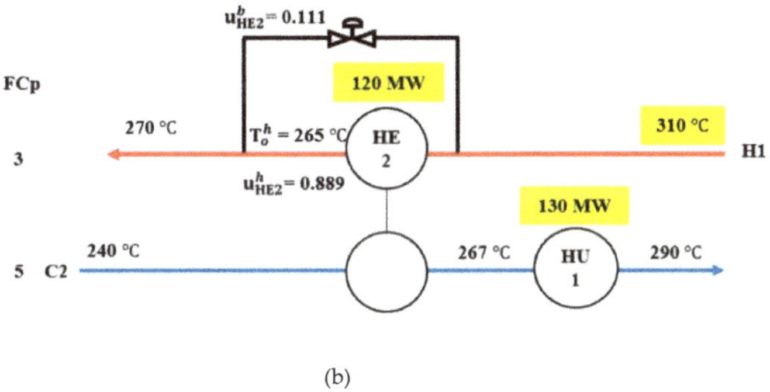

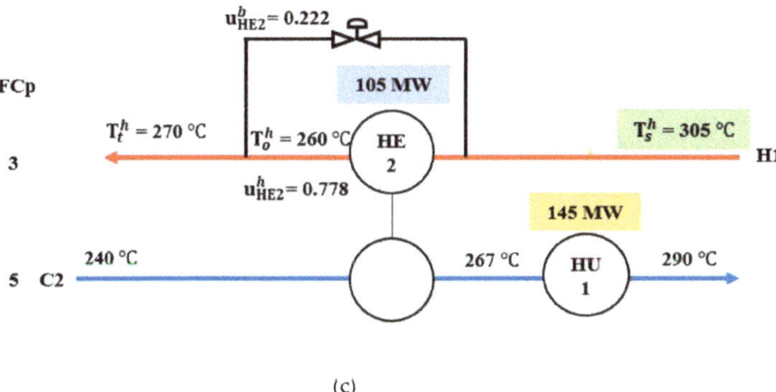

Figure 16. (**a**) HEN design with T_S increase by 5 °C for H1. HE2 is designed with a bigger area to accommodate up to 135 MW heat duty; (**b**) HEN design for nominal T_S for H1. HE2 is designed with a bigger area and 120 MW heat duty; (**c**) HEN design with T_S decrease by 5 °C for H1. HE2 is designed with bigger area and 105 MW heat duty.

Based on Observations 1 to 8, two heuristics are proposed for HEN design cases above and below the pinch to allow for flexibility and controllability in achieving maximum energy recovery:

(1)   Heuristic 1: Bypass placement to cater for disturbances in MER HEN

      a.   A bypass should be placed at the disturbed stream if the T_S value increases or decreases on either above or below the pinch and if $\Delta T \neq 0$.

      b.   A bypass should be placed on the other side of the disturbed stream if the T_S value is touching the pinch point, and if $\Delta T = 0$ or the enthalpy is less than the enthalpy of another side of the disturbed stream.

(2)   Heuristic 2: Heat exchanger sizing to cater for disturbances in MER HEN

      a.   Size a heat exchanger to cater to the highest amount of energy to be exchanged, considering all disturbances scenario.

      b.   Size a utility heat exchanger to cater to the highest amount of utility needed, considering all disturbances scenario.

After all the heuristics have been applied to the HEN design to cater for all the possible scenarios, there are some issues that must be checked:

(1) After adjusting heat exchanger duties, temperature feasibility test should be done for all the affected streams. If temperature infeasibility occurs, designers should consider redistributing the duty to the utilities.

(2) The bypass fraction should be calculated for all possible cases considering the biggest heat exchanger size.

Table 2 summarises the effects of cold and hot streams' supply temperature disturbances on hot and cold utilities as explained in the heuristics.

Table 2. Summary of cold and hot stream supply temperature disturbances on hot and cold utilities.

	$T_{S,\,hot}$	Effect on Q_C	Effect on Q_H	$T_{S,\,cold}$	Effect on Q_C	Effect on Q_H
Below Pinch	Decrease but HE ΔT still $\geq$ ΔT_{min} or $\Delta T < \Delta T_{min}$ (Observation 1)	Decrease	None unless the reduction of the hot enthalpy is more than the available cold stream for MER (Observation 2)	Decrease (Observation 1)	Decrease	None
	Decrease but HE $\Delta T = 0$ (Observation 2)	Decrease	Increase	-	-	-
	Increase (Observation 3)	Increase	None	Increase (Observation 3)	Increase	None
Above Pinch	Increase (Observation 5)	None	Decrease	Increase but HE ΔT still $\geq \Delta T_{min}$ or $\Delta T < \Delta T_{min}$ (Observation 5)	None unless the reduction of the cold enthalpy is more than the available hot stream for MER (Observation 6)	Decrease
	-	-	-	Increase but HE $\Delta T = 0$ (Observation 6)	Increase	Decrease
	Decrease (Observation 7)	None	Increase	Decrease (Observation 7)	None	Increase

3. Final Sizing and Bypass Design for MER HEN with Disturbance on T_S

The heuristics proposed are applied for designing a HEN that is flexible to disturbances. Analysis of the effect of disturbances on each stream in the HEN is summarised in Table 3. It is shown that disturbances can cause either positive or negative impacts on the Q_H and Q_C of HEN. The bypass fraction is determined based on the heuristics proposed. Figure 17 shows the final HEN with the bypass placed. Each stream applied different heuristics according to the scenario described previously.

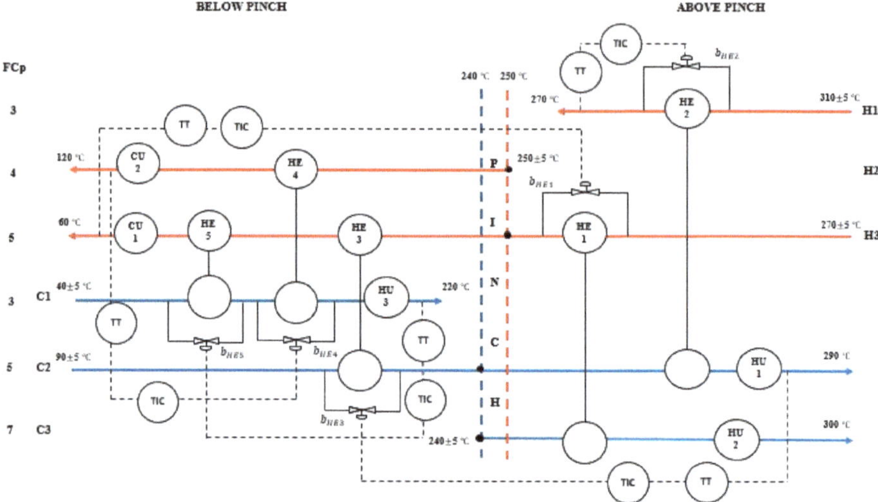

Figure 17. Overall HEN with bypass placement.

Table 3. Analysis of bypass and utility requirements of HEN design.

Stream	Stream Position	Disturbance at T_S (°C)	HE Affected	HE Max. Size (MW)	HE Duty during Disturbance (MW)	Bypass Name	Bypass Fraction (1)	Bypass Location	HU or CU Affected	The Maximum Size of HU or CU		The Duty of HU or CU during a Disturbance	
										HU (MW)	CU (MW)	HU (MW)	CU (MW)
H1	Above Pinch	+5 °C	HE2	135	135	b_{HE2}	0.000	Stream H1	HU1	145	-	115	-
		Nominal	-		120		0.111		-			130	-
		−5 °C	HE2		105		0.222		HU1			145	-
H2	Below Pinch	+5 °C	HE4	520	520	b_{HE4}	0.000	Stream C1	CU1	20	20	0	20
		Nominal	-		520		0.000		-			0	0
		−5 °C	HE4		500		0.038		HU3			20	0
H3	Across Pinch	+5 °C	HE1	125	125	b_{HE1}	0.000	Stream H3	HU2	320	180	295	180
		Nominal	-		125		0.000		HU2 & CU2			295	155
		−5 °C	HE1		100		0.200		CU2			320	155
C1	Below Pinch	+5 °C	HE5	35	5	b_{HE5}	0.857	Stream C1	CU2	-	195	-	195
		Nominal	-		20		0.429		-			-	180
		−5 °C	HE5		35		0.000		CU2			-	165
C2	Across Pinch	+5 °C	HE3	775	750	b_{HE3}	0.032	Stream C2	HU1	130	180	105	180
		Nominal	-		775		0.000		HU1 & CU2			105	155
		−5 °C	HE3		775		0.000		CU2			130	155
C3	Above Pinch	+5 °C	HE1	100	100	-	-	-	HU2	355	-	285	-
		Nominal	-		100		-		-			320	-
		−5 °C	HE1		100		-		HU2			355	-

4. Case Studies

Illustrative case studies extracted from the literature are used to verify the applicability and accuracy of the proposed methodology for optimal HEN synthesis, considering the uncertainties at the supply temperature.

Case Study 1

The first example is based on a methanol synthesis process adapted from Kijevčanin et al. [29]. The Case Study 1 has ΔT_{min} of 20 °C. The pinch temperature is at 357.2 °C. The minimum heating requirement ($Q_{H,min}$) is 1953.88 kW while the minimum cooling requirement ($Q_{C,min}$) is 3463.53 kW. Expected variations of ±10 °C in the inlet temperature of streams H1 and C3 are assumed to vary from their nominal values. The case study consists of eight hot streams and three cold streams as shown in Table 4. Four scenarios involving the variations of the inlet temperatures between streams H1 and C3 are observed (Table 5).

Table 4. Stream data of Case Study 1 [29].

Stream	Supply Temperature, T_S (°C)	Target Temperature, T_t (°C)	Heat Capacity Flowrate, FC_P (kW/°C)	Enthalpy, ΔH (kW)
Hot 1 (H1)	424.2 ± 10	120	4.6	−1399.32
Hot 2 (H2)	342.1	120	5	−1110.50
Hot 3 (H3)	342.2	120	5	−1111.00
Hot 4 (H4)	343.1	160	5.1	−933.81
Hot 5 (H5)	403.1	210	5.2	−1004.12
Hot 6 (H6)	60.6	30	2.5	−76.50
Hot 7 (H7)	98.9	30	1.1	−75.79
Hot 8 (H8)	76.4	30	0.3	−13.92
Cold 1 (C1)	349.7	450	19.2	1925.76
Cold 2 (C2)	37.7	450	5	2061.50
Cold 3 (C3)	14.5 ± 10	70	4.1	227.55

Table 5. Uncertain parameters for the considered inlet temperatures.

Scenario	$T_{S,H1}$ (°C)	$T_{S,C3}$ (°C)
A	434.2	24.5
B	414.2	4.5
C	434.2	4.5
D	414.2	24.5

All the scenarios are applied in designing a HEN with the nominal HE area (previous work) and the maximum HE area (this work). Figure 18 shows the HEN at the nominal condition with the bypass placement but with different HE area and bypass fraction.

Analysis of the effect of disturbances on streams H1 and C3 for the nominal HE area (previous work) is shown in Table 6 while the maximum HE area (this work) is shown in Table 7. It is observed that the duty of HE2 increased to 319.70 kW when the supply temperature of hot stream H1 located across or above the pinch increased. This decreased the duty of hot utility HU2 to 181.80 kW. However, if the HE2 area is maintained at the nominal size, the duty of hot utility is also maintained at the nominal value. On the other hand, the duty of HE2 for both works decreased to 227.70 kW when the supply temperature of hot stream H1 is decreased. This led to an increase of hot utility HU2 to 273.80 kW in both works. For the cold stream C3 located below the pinch, it is observed that the duty of HE5 decreased to 186.60 kW when the supply temperature is increased for both works. Both designs required large HU2 utility loads to cope with the deficit enthalpy of HE5. On the contrary, the duty of HE5 can be increased to 256.60 kW, at the maximum HE area when the supply temperature of C3 decreased. Consequently, the minimum cold utility for CU2 can be obtained.

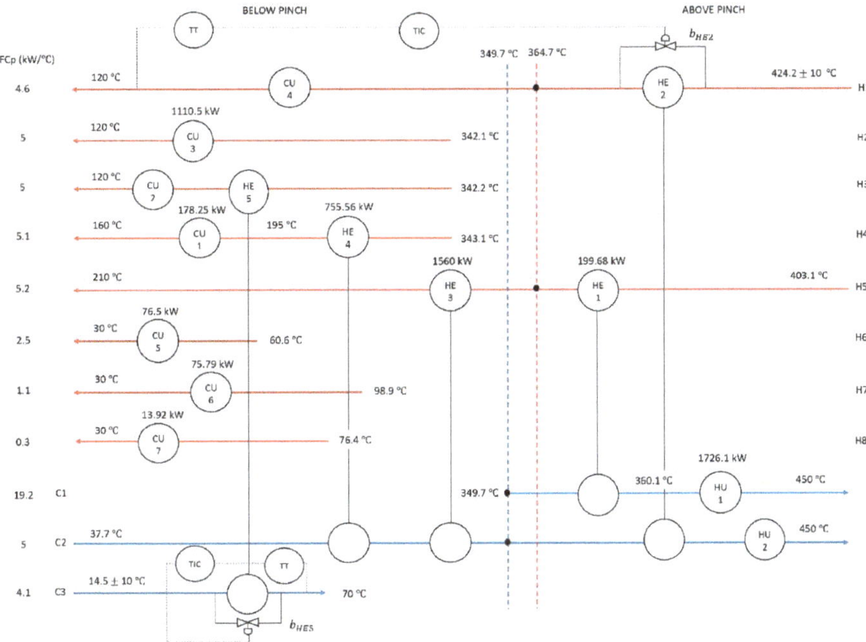

Figure 18. Overall HEN with bypass placement for Case Study 1.

The impact of the changes on the economics for the cases of nominal HE area and maximum HE area were analysed by comparing the annualised capital and utility costs for both HEN using Equation (5) [30]. The basic rule to target for cost-effective minimum utilities is to maximise the use of higher temperature cold utilities and lower temperature hot utilities. The type of utilities suggested based on temperature interval is shown in Table 8. The rates for the utilities refer to Sun et al. [31].

$$\text{Annualised capital cost} = \text{Annualised factor} \times \left(1300 + 1000\,A^{0.83}\right) \tag{5}$$

where the annualised factor is 0.298.

Table 9 compares the heat recovery and economic performance of HEN with nominal HE area and maximum HE area for all the scenarios. Results of this study show that the new heuristics can guide the user to manage temperature disturbances in HEN design for maximum heat recovery with minimum total costs. Although this work has a high annualised capital cost due to larger HE area compared to the previous work, the annualised total costs for scenarios A, B and C is still much cheaper than the previous work. For scenario D, the total annualised cost is 0.16% higher due to the large HE area, but the utility load remains the same.

Table 6. Analysis of bypass and utility requirements of HEN design with nominal HE area.

Stream	Stream Position	Disturbance at T_S (°C)	HE Affected	HE Max. Size (kW)	HE Duty during Disturbance (kW)	Bypass Name	Bypass Fraction	Bypass Location	HU or CU Affected	The Maximum Size of HU or CU		The Duty of HU or CU during a Disturbance	
										HU (kW)	CU (kW)	HU (kW)	CU (kW)
H1	Across Pinch	+10 °C	HE2	273.70	273.70	b_{HE2}	0.000	Stream H1	HU2	273.80	-	227.80	-
		Nominal	-		273.70		0.000		-			227.80	-
		−10 °C	HE2		227.70		0.168		HU2			273.80	-
C3	Below Pinch	+10 °C	HE5	227.55	186.60	b_{HE5}	0.000	Stream C3	CU2	-	924.45	-	924.45
		Nominal	-		227.55		0.000		-				-	883.45
		−10 °C	HE5		227.55		0.153		CU2				-	883.45

Table 7. Analysis of bypass and utility requirements of HEN design with maximum HE area.

Stream	Stream Position	Disturbance at T_S (°C)	HE Affected	HE Max. Size (kW)	HE Duty during Disturbance (kW)	Bypass Name	Bypass Fraction (1)	Bypass Location	HU or CU Affected	The Maximum Size of HU or CU		The Duty of HU or CU during a Disturbance	
										HU (kW)	CU (kW)	HU (kW)	CU (kW)
H1	Across Pinch	+10 °C	HE2	319.70	319.70	b_{HE2}	0.000	Stream H1	HU2	273.80	-	181.80	-
		Nominal	-		273.70		0.252		-			227.80	-
		−10 °C	HE2		227.70		0.288		HU2			273.80	-
C3	Below Pinch	+10 °C	HE5	268.60	186.60	b_{HE5}	0.305	Stream C3	CU2	-	924.45	-	924.45
		Nominal	-		227.55		0.265		-				-	883.45
		−10 °C	HE5		268.60		0.000		CU2				-	842.45

Table 8. Multiple utilities data [19].

Utilities	T_S (°C)	T_t (°C)	Rate (USD/kW·y)
Hot oil	350	320	311.280
High pressure steam (HPS)	255	254	249.024
Tempered water (TW)	70	80	31.128
Cooling water (CW)	25	30	41.504

Table 9. Comparison results of HEN total annual cost for Case Study 1.

	Previous Work (with Nominal HE Area)				This Work (with Maximum HE Area)			
	Scenario A	Scenario B	Scenario C	Scenario D	Scenario A	Scenario B	Scenario C	Scenario D
Cold utility (kW)	924.45	883.45	883.45	924.45	924.45	842.45	842.45	924.45
Hot utility (kW)	227.80	273.80	227.80	273.80	181.80	273.80	181.80	273.80
Total disturbed HE area (m^2)	77.774				82.186			
Annualised capital cost (\$/y)	5705.92				5901.70			
Annualised utility cost (\$/y)	99,685.86	112,728.50	98.409.62	114,004.74	85,366.98	111,452.25	82,814.49	114,004.74
Annualised total cost (\$/y)	105,391.78	118,434.42	104,115.54	119,710.66	91,268.68	117,353.95	88,716.19	119,906.44

5. Case Study 2

In this case study, the proposed methodology is applied to solve an illustrative example with three hot streams and three cold streams. The data used for this case study are adapted from the work of Escobar et al. [9]. The nominal data for the problem is listed in Table 10. The expected variations in the inlet temperatures are assumed ± 10 K with ΔT_{min} of 10 K. In contrast to Escobar et al. [9], the nominal configuration of HEN is maintained. The maximum heat exchanger area approach is applied to increase the flexibility of HEN due to the uncertainty of operating conditions. The uncertain parameters considered in the design are given in Table 11.

Table 10. Stream data of Case Study 2 [9].

Stream	Supply Temperature, T_S (K)	Target Temperature, T_t (K)	Heat Capacity Flowrate, FC_P (kW/K)	h (kW·m^2/K)
Hot 1 (H1)	583 ± 10	323	1.4	0.16
Hot 2 (H2)	723 ± 10	553	2.0	0.16
Cold 1 (C1)	313 ± 10	393	3.0	0.16
Cold 2 (C2)	388 ± 10	553	2.0	0.16
CU	303	323		0.16
HU	573	573		0.16

Exchanger capital cost ($/y) = 8333.3 + 641.7 Area (m^2)
Annualisation factor = 0.2/y
Cost of cooling utility = 60.576 ($ /kW·y)
Cost of heating utility = 171.428 ($/kW·y)

Table 11. Uncertain parameters for the points considered.

Iteration	$T_{S, H1}$ (K)	$T_{S, H2}$ (K)	$T_{S, C1}$ (K)	$T_{S, C2}$ (K)
1	583	723	313	388
2	573	713	303	378

Initially, the MER for the nominal case (without disturbances) is determined by using pinch analysis targeting methods. For this case study, the nominal condition corresponds to the first iteration. By using the same (nominal) HEN configuration, in order for the target temperature to be achieved, the enthalpy for the cold streams is required to be increased while the enthalpy for the hot streams is decreased. Thus, the utility consumption (cold utility) can be reduced. At the same time, the area of the heat exchangers is required to be at the maximum size for the HEN design to be feasible. Figure 19 shows the HEN at the nominal condition with the bypass placement. On the other hand, Escobar et al. [9] suggested two different HEN designs for each iteration as the nominal design is not feasible for the variations up to 10 K in the inlet temperatures. The steps where the critical point is added to the nominal conditions and the multi-period optimisation problem needs to be repeated until the flexibility is accomplished. The new HEN configuration is designed with high flexibility. The TAC is comprised of the annualised utility cost and annualised capital cost. The TAC of this work for the first iteration is much higher than the previous work as the same HEN design with maximum heat exchanger area is applied. However, the TAC for this work with the consideration of uncertainty is $25,986.93/y, comprising $3028.80/y associated with operating expenses (utility consumption) and $22,958.13/y to capital investment. This work gives 4%/y lower TAC compared to the work of Escobar et al. [9]. Although it has a higher capital cost, the utility consumption is reduced. This method is able to give positive effects even though the HEN has uncertain operating parameters, as it provides better utility usage and would be a good approach for considering a reduction in the environmental impact associated with the use of fossil-based energy sources. The results of this work and those of Escobar et al. [9] are compared in Table 12.

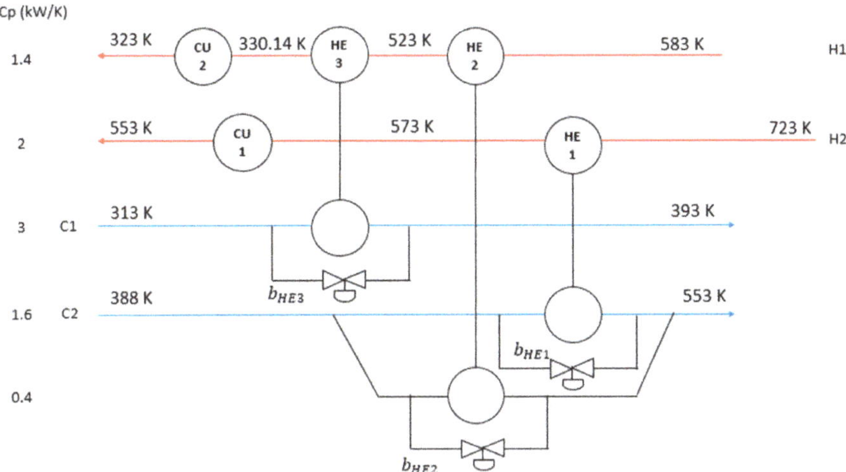

Figure 19. Overall HEN with bypass placement for Case Study 2.

Table 12. Comparison results of HEN total annual cost for Case Study 2.

Method	Escobar et al. [9]		This Work	
Iteration	1	2	1	2
Annualised utility cost ($/y)	8117.00	5573.00	8117.00	3028.80
Total affected area (m^2)	81.417	103.631	113.9538	113.9538
Annualised capital cost ($/y)	17,115.70	21,633.30	22,958.13	22,958.13
Total annualised cost ($/y)	25,232.70	27,206.30	31,075.13	25,986.93

6. Conclusions

Heat exchanger network configuration can influence disturbance propagation and process behaviour, as well as limit process controllability and operability. A systematic methodology has been developed in this work to manage temperature disturbances and make a heat exchanger network more flexible and operable toward achieving the maximum heat recovery. The method considers the impact of supply temperature fluctuations on utility consumption, heat exchanger sizing and bypass placement. The maximum energy recovery targets are determined for the nominal case (without disturbances) by using pinch analysis targeting methods. The steps to manage the fluctuating supply temperature in the heat exchanger network to achieve target temperature and maximum energy recovery are defined in three stages: (1) the effect of increasing or decreasing supply temperature on the hot and cold streams energy requirement; (2) the fundamental theory of the plus-minus principle; (3) the effect of T_S fluctuation on utilities based on the plus-minus principle. From all these steps, key observations were made and new heuristics based on the plus-minus principle of pinch analysis have been introduced for heat exchanger network design in maximising utility savings. The heuristics were applied by simulating various scenarios of disturbances occurring in the process streams of the heat exchanger network. Guidelines on the sizing of heat exchangers and bypass locations and fractions have also been proposed. The grid diagram and temperature vs enthalpy plot were used to illustrate the effect of changes in design and operating parameters. This approach involves a trade-off between the utility cost and capital cost of the affected heat exchangers. Application of the method on two case studies showed that the configuration of the heat exchanger network is maintained for all the scenarios. In addition, the exchanger areas are designed at the maximum size with a bypass to handle the most critical uncertainty of operating conditions while minimising the utility usage. It showed that this method improved the annualised utility cost by up to 89%. Previous works in the literature required the exchanger area adjustment during the operation and have several heat exchanger network configurations for each scenario. Nevertheless, in industry, the configuration and area should be fixed during the operation even with unforeseen uncertainty. The modification of heat exchanger network design during the operation will be costly. This work is beneficial as it has a reduced utility usage and a higher capital cost.

Major novelties introduced by this work include:

- The approach is able to handle the most critical uncertainty of operating conditions with the maximum heat exchanger areas while minimising the utility usage.
- The nominal configuration of HEN based on pinch analysis is maintained and controlled by the bypass.
- The effect of uncertain operating conditions on the heat exchange and utility consumption can be easily visualised through the plus-minus principle.
- The proposed new heuristics based on the plus-minus principle of pinch analysis can be applied for all problems.

However, overdesign factors could affect the annualised capital cost of heat exchanger networks. Future research should also include the probability of disturbances occurring in the heat exchanger network. This probability may influence the requirement of the maximum heat exchanger size in the design as well as the total annualised cost.

Author Contributions: Conceptualisation, S.R.W.A.; Methodology, A.M.H., S.R.W.A.; Validation, A.M.H.; Formal analysis, A.M.H., S.R.W.A.; Writing—original draft preparation, A.M.H.; writing—review and editing, S.R.W.A., Z.A.M., J.J.K.; supervision, S.R.W.A., Z.A.M., J.J.K., M.K.A.H.

Acknowledgments: The authors thank Universiti Teknologi Malaysia (UTM) for providing research funds for this project under Vote Number Q.J130000.3509.05G96, Q.J130000.2509.19H34, the Fundamental Research Grant Scheme under Ministry of Education with Vote Number R.J130000.7809.4F918 and from the EC project for Sustainable Process Integration Laboratory-SPIL (Project No. CZ.02.1.01/0.0/0.0/15_003/0000456) funded by Czech Republic Operational Program Research and Development, Education, Priority 1: Strengthening capacity for quality research in collaboration agreement with Universiti Teknologi Malaysia (UTM).

Nomenclature

FC_p	Heat capacity flowrate, kW/°C or MW/°C
$Q_{C,min}$	Minimum cold utility, kW or MW
$Q_{H,min}$	Minimum hot utility, kW or MW
T_0^h	Outlet hot temperature, °C
T_0^c	Outlet cold temperature, °C
T_{pinch}	Pinch temperature, °C
T_s	Supply temperature, °C
T_s^h	Supply hot temperature, °C
T_s^c	Supply cold temperature, °C
T_t	Target temperature, °C
T_t^h	Target hot temperature, °C
T_t^c	Target cold temperature, °C
u^h	Hot stream fraction
u^b	Bypass fraction
u^c	Cold stream fraction
ΔH	Enthalpy, kW or MW

References

1. Klemeš, J.J.; Kravanja, Z. Forty Years of Heat Integration: Pinch Analysis (PA) and Mathematical Programming (MP). *Curr. Opin. Chem. Eng.* **2013**, *2*, 461–474. [CrossRef]
2. Klemeš, J.J.; Varbanov, P.S.; Walmsley, T.G.; Jia, X. New Directions in the Implementation of Pinch Methodology (PM). *Renew. Sustain. Energy Rev.* **2018**, *98*, 439–468. [CrossRef]
3. Isafiade, A.J.; Short, M. Synthesis of Mass Exchange Networks for Single and Multiple Periods of Operations Considering Detailed Cost Functions and Column Performance. *Process Saf. Environ. Prot.* **2016**, *103*, 391–404. [CrossRef]
4. Sa, A.; Thollander, P.; Rafiee, M. Industrial Energy Management Systems and Energy-Related Decision-Making. *Energies* **2018**, *11*, 2784. [CrossRef]
5. Thollander, P.; Palm, J. Industrial Energy Management Decision Making for Improved Energy Efficiency—Strategic System Perspectives and Situated Action in Combination. *Energies* **2015**, *8*, 5694–5703. [CrossRef]
6. Marselle, D.F.; Morari, M.; Rudd, D.F. Design of Resilient Processing Plants—II Design and Control of Energy Management Systems. *Chem. Eng. Sci.* **1982**, *37*, 259–270. [CrossRef]
7. Hafizan, A.M.; Alwi, S.R.W.; Manan, Z.A.; Klemeš, J.J.; Hamid, M.K.A. Temperature Disturbance Management in Heat Exchanger Network for Maximum Energy Recovery. In Proceedings of the 13th SDWES Conference, Palermo, Italy, 30 September–4 October 2018; SDEWES2018-0484.
8. Linnhoff, B.; Kotjabasakis, E. Downstream Paths for Operable Process Design. *Chem. Eng. Prog.* **1986**, *82*, 23–28.
9. Escobar, M.; Trierweiler, J.O.; Grossmann, I.E. Simultaneous Synthesis of Heat Exchanger Networks with Operability Considerations: Flexibility and Controllability. *Comput. Chem. Eng.* **2013**, *55*, 158–180. [CrossRef]
10. Hafizan, A.M.; Alwi, S.R.W.; Manan, Z.A.; Klemeš, J.J. Optimal Heat Exchanger Network Synthesis with Operability and Safety Considerations. *Clean Technol. Environ. Policy* **2016**, *18*, 2381–2400. [CrossRef]
11. Čuček, L.; Kravanja, Z. A Procedure for the Retrofitting of Large-Scale Heat Exchanger Networks for Fixed and Flexible Designs. *Chem. Eng. Trans.* **2015**, *45*, 109–114.
12. Isafiade, A.J.; Short, M. Simultaneous Synthesis of Flexible Heat Exchanger Networks for Unequal Multi-Period Operations. *Process Saf. Environ. Prot.* **2016**, *103*, 377–390. [CrossRef]
13. Miranda, C.B.; Costa, C.B.; Caballero, J.A.; Ravagnani, M.A. Optimal Synthesis of Multiperiod Heat Exchanger Networks: A Sequential Approach. *Appl. Therm. Eng.* **2017**, *115*, 1187–1202. [CrossRef]
14. Kang, L.; Liu, Y. A Three-Step Method to Improve the Flexibility of Multiperiod Heat Exchanger Networks. *Process Integr. Optim. Sustain.* **2018**, *2*, 169–181. [CrossRef]
15. Narraway, L.T.; Perkins, J.D. Selection of Process Control Structure Based on Linear Dynamic Economics. *Ind. Eng. Chem. Res.* **1993**, *32*, 2681–2692. [CrossRef]
16. Walsh, S.; Perkins, J. Application of Integrated Process and Control System Design to Waste Water Neutralisation. *Comput. Chem. Eng.* **1994**, *18*, S183–S187. [CrossRef]
17. Bakar, S.H.A.; Hamid, M.K.A.; Alwi, S.R.W.; Manan, Z.A. Flexible and Operable Heat Exchanger Networks. *Chem. Eng.* **2013**, *32*, 1297–1302.
18. Kong, L.; Avadiappan, V.; Huang, K.; Maravelias, C.T. Simultaneous Chemical Process Synthesis and Heat Integration with Unclassified Hot/Cold Process Streams. *Comput. Chem. Eng.* **2017**, *101*, 210–225. [CrossRef]
19. Quirante, N.; Grossmann, I.E.; Caballero, J.A. Disjunctive Model for the Simultaneous Optimization and Heat Integration with Unclassified Streams and Area Estimation. *Comput. Chem. Eng.* **2018**, *108*, 217–231. [CrossRef]
20. Onishi, V.C.; Quirante, N.; Ravagnani, M.A.; Caballero, J.A. Optimal Synthesis of Work and Heat Exchangers Networks Considering Unclassified Process Streams at Sub and Above-Ambient Conditions. *Appl. Energy* **2018**, *224*, 567–581. [CrossRef]
21. Quirante, N.; Caballero, J.A.; Grossmann, I.E. A Novel Disjunctive Model for the Simultaneous Optimization and Heat Integration. *Comput. Chem. Eng.* **2017**, *96*, 149–168. [CrossRef]
22. Linnhoff, B.; Vredeveld, R. Pinch Technology has Come of Age. *Chem. Eng. Prog.* **2017**, *80*, 33–40.
23. Chew, K.H.; Klemeš, J.J.; Alwi, S.R.W.; Manan, Z.A. Process Modifications to Maximise Energy Savings in Total Site Heat Integration. *Appl. Therm. Eng.* **2015**, *78*, 731–739. [CrossRef]

24. Song, R.; Tang, Q.; Wang, Y.; Feng, X.; El-Halwagi, M.M. The Implementation of Inter-Plant Heat Integration Among Multiple Plants. Part I: A Novel Screening Algorithm. *Energy* **2017**, *140*, 1018–1029. [CrossRef]
25. Linnhoff, B.; Flower, J.R. Synthesis of Heat Exchanger Networks: I. Systematic Generation of Energy Optimal Networks. *AIChE J.* **1978**, *24*, 633–642. [CrossRef]
26. Alwi, S.R.W.; Manan, Z.A. STEP—A New Graphical Tool for Simultaneous Targeting and Design of A Heat Exchanger Network. *Chem. Eng. J.* **2010**, *162*, 106–121. [CrossRef]
27. Linnhoff, B.; Hindmarsh, E. The Pinch Design Method for Heat Exchanger Networks. *Chem. Eng. Sci.* **1983**, *38*, 745–763. [CrossRef]
28. Rathjens, M.; Bohnenstädt, T.; Fieg, G.; Engel, O. Synthesis of Heat Exchanger Networks Taking into Account Cost and Dynamic Considerations. *Procedia Eng.* **2016**, *157*, 341–348. [CrossRef]
29. Kijevčanin, M.L.; Đorđević, B.; Ocić, O.; Crnomarković, M.; Marić, M.; Šerbanović, S.P. Energy and Economy Savings in the Process of Methanol Synthesis Using Pinch Technology. *J. Serbian Chem. Soc.* **2004**, *69*, 827–837. [CrossRef]
30. Na, J.; Jung, J.; Park, C.; Han, C. Simultaneous Synthesis of a Heat Exchanger Network with Multiple Utilities Using Utility Substages. *Comput. Chem. Eng.* **2015**, *79*, 70–79. [CrossRef]
31. Sun, K.N.; Alwi, S.R.W.; Manan, Z.A. Heat Exchanger Network Cost Optimization Considering Multiple Utilities and Different Types of Heat Exchangers. *Comput. Chem. Eng.* **2013**, *49*, 194–204. [CrossRef]

Article

Techno-Economic Assessment of Turboexpander Application at Natural Gas Regulation Stations

Szymon Kuczyński *, Mariusz Łaciak, Andrzej Olijnyk, Adam Szurlej and Tomasz Włodek

AGH University of Science and Technology, Drilling, Oil and Gas Faculty, 30-059 Krakow, Poland; laciak@agh.edu.pl (M.Ł.); aoliinyk@agh.edu.pl (A.O.); szua@agh.edu.pl (A.S.); twlodek@agh.edu.pl (T.W.)
* Correspondence: szymon.kuczynski@agh.edu.pl; Tel.: +48-12-617-2247

Received: 31 December 2018; Accepted: 20 February 2019; Published: 24 February 2019

Abstract: During the natural gas pipeline transportation process, gas stream pressure is reduced at natural gas regulation stations (GRS). Natural gas pressure reduction is accompanied by energy dissipation which results in irreversible exergy losses in the gas stream. Energy loss depends on the thermodynamic parameters of the natural gas stream on inlet and outlet gas pressure regulation and metering stations. Recovered energy can be used for electricity generation when the pressure regulator is replaced with an expander to drive electric energy generation. To ensure the correct operation of the system, the natural gas stream should be heated, on inlet to expander. This temperature should be higher than the gas stream during choking in the pressure regulator. The purpose of this research was to investigate GRS operational parameters which influence the efficiency of the gas expansion process and to determine selection criteria for a cost-effective application of turboexpanders at selected GRS, instead of pressure regulators. The main novelty presented in this paper shows investigation on discounted payback period (DPP) equation which depends on the annual average natural gas flow rate through the analyzed GRS, average annual level of gas expansion, average annual natural gas purchase price, average annual produced electrical energy sale price and CAPEX.

Keywords: natural gas; natural gas regulation station; turboexpander; pressure regulator; energy recovery; energy conversion; energy system analysis

---

1. Introduction

Natural gas is usually transported over long distances through pipelines at high pressures. In order to distribute the gas locally at points along the pipeline, the pressure must be significantly reduced before it is supplied to local distribution systems [1,2]. Currently, most pressure-reducing stations use expansion valves to reduce pressure [3,4]. When there is no heat transfer to or from the environment and if no work is done, the process of choked flow of the natural gas stream, regardless of its type, is an isenthalpic process [5]. In choked flow at constant enthalpy, the gas temperature changes. This change in gas temperature, once the pressure has been reduced, is associated with the so-called Joule–Thomson effect which has a negative value for methane which is the main component of natural gas (in the range of pressures and temperatures in the gas industry) [6]. For high-methane natural gas, it can be assumed with sufficient accuracy that the temperature drop equals approximately 0.4 °C/bar of pressure reduction [7]. While the gas flows through the station, the reduced natural gas temperature might produce hydrates inside the gas pipeline, ice plugs (especially in remotes for regulators) and icing of fittings, or contribute to the unstable functioning of a regulator work, etc. [8]. If the pressure regulator is replaced by expansion devices (expansion machine), the process will be isentropic (i.e., the system will perform external work, so that a significant part of the mechanical work will be recovered). However, in this case, the temperature drop is greater [9]. The choice of the type of expansion machine (gas expansion turbine or piston expansion device) which cooperates with a power generator depends on the size of the GRS and gas flow parameters [10]. Due to this, it requires additional heating of natural gas which leads to additional costs. It is assumed that in order to generate 1 kWh of electricity by an expander, it is necessary to supply 1.15 kWh of heat [11]. The Polish natural gas transmission system has 1041 outlet points, and most of them are gas pressure regulation stations [12], but only a few dozen are suitable to use expansion machines and to produce economically viable electricity. The quantity of produced electricity depends on the gas flow rate through the gas regulation station (GRS) and the pressure reduction [13,14]. Some GRSs have large capacities, but the pressure reduction at these stations is quite low. On the other hand, there are small-capacity GRSs where the pressure can be reduced as much as 10 times relative to the inlet pressure. Such diversity in GRS technical parameters makes it difficult to select stations which could be equipped with expanders. Natural gas consumption in countries with high annual variability of the ambient temperature is highly irregular, which makes this selection even more difficult. This is because an expander's operational parameters deviate from nominal values, which has a negative impact on the efficiency of the device [12]. The economic impact which results from the application of expanders is also affected by the cost of the expander itself, as well as by the total investment cost [15]. Additionally, the sale price of produced electricity and the cost of natural gas heating has a large impact on the economic effect [16]. The purpose of this research was to develop an equation to assess the economic efficiency which results from the application of turboexpanders at individual stations.

Vortex turboexpanders usually have quite a high efficiency, in the range of 70–90%, which depends on the gas flow rate. Seasonal fluctuations in gas flow through the GRS cause an unequal generation of electricity by the turboexpander. Significant deviations in natural gas flow through the expander from the nominal numbers of flow indicated by the device manufacturer results in, among other things, a decrease in the efficiency of the device, and hence, less electricity production, even with large volumes of natural gas flow. In addition, a critically small or large natural gas flow through the expander can lead to its damage. Therefore, it is not recommended to use an expander as the only reducing device at the GRS. A typical diagram of the reduction and metering station is shown in Figure 1.

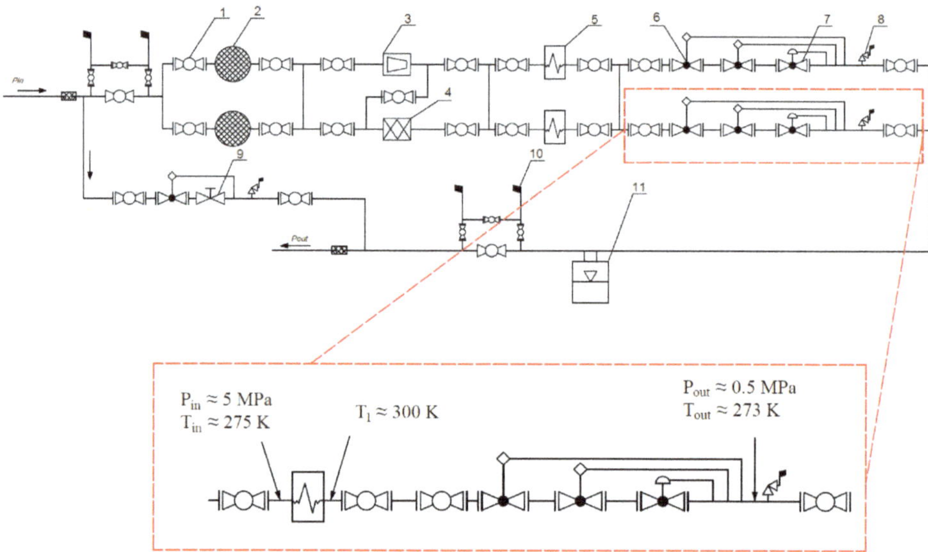

Figure 1. Scheme of the natural gas Reducing and Metering Station (without turboexpander). 1—Valve; 2—Filter; 3—Turbine gas meter; 4—Ultrasound gas meter; 5—Heat exchanger; 6—Shut down valve; 7—Pressure regulator; 8—Pressure drop valve; 9—Regulation valve; 10—Ventilation valve; 11—Gas odorizer.

A slightly more expensive, but much safer solution is to design a station which contains a traditional reduction sequence with an expander installed on it and a traditional reduction sequence which contains standard pressure reducers, as shown in Figure 2.

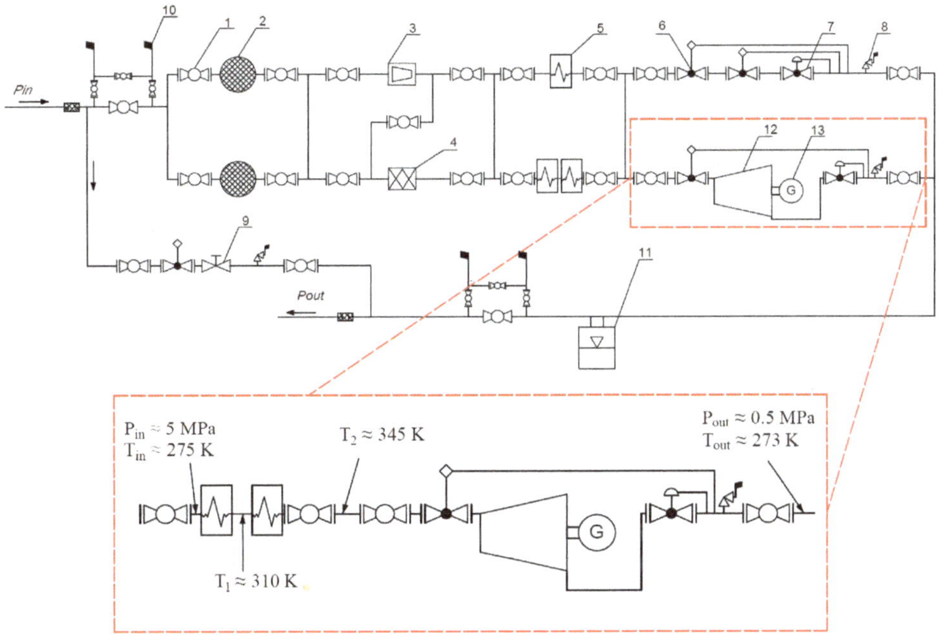

Figure 2. Schemes of Reducing and Metering Station (with turboexpander). 1—Valve; 2—Filter; 3—Turbine gas meter; 4—Ultrasound gas meter; 5—Heat exchanger; 6—Shut down valve; 7—Pressure regulator; 8—Pressure drop valve; 9—Regulation valve; 10—Ventilation valve; 11—Gas odorizer; 12—Turboexpander; 13—Generator.

This solution ensures reliability and continuity of operation of the GRS as part of the natural gas transmission and distribution system. In situations where gas flow is too high or too low, the station's automatic control system or dispatching services will connect the reduction sequence, and the reduction will only be performed with a traditional reduction sequence. When natural gas flow stabilizes (back to the flow range recommended by the expander manufacturer), it will be possible to run the expander as a basic reduction element.

Due to the larger temperature drop caused by isentropic expansion, the reduction line which contains the expander should have a larger heat exchanger, or as shown in the diagram above, two heat exchangers in series. To ensure a stable pressure level after expansion, an ordinary regulator is recommended after the expander because gas pressure at the expander outlet depends on the gas pressure at the expander inlet. Application of the reducer can provide a stable pressure at the GRS outlet.

Issues related to the application of turboexpanders at GRS were discussed in scientific papers. Arabkoohsar et al., who presented a technical and economic investigation on the gas stations to check which type of power productive gas expansion stations are suitable to be accompanied with, combined heat and power systems [16]. The integration possibility of a pressure reduction station with low-temperature heat sources was studied by Borelli et al. [17]. They stated that the novel proposed system configuration has many advantages in terms of the opportunity to exploit low-enthalpy heat sources, highly efficient primary-conversion-technology utilizations, and integration with renewable sources. Li et al. simulated the usage of a single-screw expander in the Jiyuanzhongyu natural gas letdown station in Henan province. The conclusion of their work was that the single-screw expander has great prospects in small-scale waste heat pressure energy recovery systems, as well as in new and renewable energy applications [18]. Stanek et al. presented the usefulness of exergy analysis, which in contrast to energy analysis, provides tools to indicate and quantify the origin of the electricity generated by a turboexpander [19]. Similar research was performed by Prilutskii, who compared the application of turbo- and piston expanders [20].

New technical solutions are also being developed to expand the possibilities of turboexpanders' application at small GRS. Barbarelli et al. developed a new microturbine typology which can operate with compressible fluids like steam or gases [21]. The novelty of this turbine occurs in its capability to operate with a low rotational speed with respect to the change in other operational parameters (i.e., flow rate, expansion ratio) without excessive loss of efficiency [22].

In most publications, the effectiveness of expander use is determined by economic indicators, whereas Lo Cascio et al. presented key performance indicators (KPIs) for integrated natural gas pressure reduction stations with energy recovery. Presented KPIs covered the different aspects of energy recovery in GRS. The waste energy recovery (WER) index is fundamental to assess the efficiency of the recovery process. Other indicators, such as the recovery ratio (RR) and the carbon emission recovery index (CER), provide information related to the status of the recovery process and carbon emission reduction, respectively [23]. In addition, GRS equipment optimization was discussed by Lo Cascio. Energy recovery from pressure reduction in natural gas networks was addressed through a structured retrofitting approach (SRA) aimed at optimal design. The SRA optimization technique permits the identification of the best system configuration in the GRS retrofit, particularly with regard to the turbo expander model and heat supplier technology. All the different issues of the integrated system design are addressed with the optimization technique [24].

2. Calculation Model Description

In order to develop selection criteria for gas regulation stations with turboexpanders, an equation was introduced including one factor which depends on five other independent parameters. This equation presents the impact of five independent parameters to determine the factor which indicates (or not) the application validity of the expander at the considered GRS. The discounted payback period (DPP) was chosen as a factor to present the economic efficiency of expanders in GRS. Some selected and discussed independent parameters are: annual average gas flow through the analyzed GRS, average annual level of gas expansion, average annual sale price of generated electrical energy, average purchase price of natural gas, and total investment cost. A mathematical model was developed to determine the impact of independent factors on the discounted payback period (DPP). Results of modeling were used as entry data for a statistical model (nonlinear polynomial regression). The calculation model consists of three sections: thermodynamic model, economic model, and statistical model.

2.1. Thermodynamic Model

Assuming that natural gas expansion in an expander is isentropic, the amount of electrical energy which can be produced can be calculated using the following equation [25]

$$W = \eta_o \cdot \eta_{eg} \cdot \eta_m \cdot M \cdot \Delta H \tag{1}$$

The gas enthalpy was calculated using [26,27]:

$$H = H_0 + (G - G_0) + T \cdot (S - S_0) \tag{2}$$

The entropy of the gas was determined by using:

$$S - S_0 = \frac{-\partial}{\partial T}(G - G^o)_p \tag{3}$$

Then, Gibbs energy was calculated by using:

$$G - G^o = \int_{P_0}^{P} V \cdot dP = \int_{0}^{P} V \cdot dP + \int_{P_0}^{0} V \cdot dP \tag{4}$$

where $V = \frac{R \cdot T}{P_0}$, which after substitution will get,

$$G - G^o = \int_{0}^{P} \left(V - \frac{R \cdot T}{P}\right) \cdot dP + \int_{P_0}^{0} \left(V + \frac{R \cdot T}{P}\right) \cdot dP \tag{5}$$

$$G - G^o = \int_{0}^{P} \left(V - \frac{R \cdot T}{P}\right) \cdot dP + R \cdot T \cdot \ln \frac{P}{P_0} \tag{6}$$

By presenting the Clapeyron Equation as $Z = \frac{P \cdot V}{R \cdot T} \rightarrow V = \frac{Z \cdot R \cdot T}{P}$ we can write the following formula:

$$G - G^o = \int_{0}^{P} \left(\frac{z \cdot R \cdot T}{P} - \frac{R \cdot T}{P}\right) \cdot dP + R \cdot T \cdot \ln \frac{P}{P_0} \tag{7}$$

$$G - G^o = R \cdot T \int_{0}^{P} \frac{1}{P}(Z - 1) \cdot dP + R \cdot T \cdot \ln \frac{P}{P_0} \tag{8}$$

$$G - G^o = R \cdot T \int_{0}^{P} (Z - 1) \cdot d \ln P + R \cdot T \cdot \ln \frac{P}{P_0} \tag{9}$$

Based on GERG 88 equation of state (EOS) [28],

$$Z = 1 + \frac{B}{R \cdot T} \cdot P + \frac{C - B^2}{(R \cdot T)^2} \cdot P^2 \tag{10}$$

where:

$$B = \sum_{i=1}^{n} \sum_{j=1}^{n} x_i \cdot x_j \cdot B_{ij}(T) \tag{11}$$

$$C = \sum_{i=1}^{n} \sum_{j=1}^{n} \sum_{k=1}^{n} x_i \cdot x_j \cdot x_k \cdot C_{ijk}(T) \tag{12}$$

$$B_{ij}(T) = b_{ij}^{(0)} + b_{ij}^{(1)} \cdot T + b_{ij}^{(2)} \cdot T^2 \tag{13}$$

$$C_{ijk}(T) = c_{ijk}^{(0)} + c_{ijk}^{(1)} \cdot T + c_{ijk}^{(2)} \cdot T^2 \tag{14}$$

By substituting Equation (10) into Equation (11) we obtain,

$$G - G^o = R \cdot T \int_{0}^{P} \left(\frac{B}{R \cdot T} \cdot P + \frac{C - B^2}{(R \cdot T)^2} \cdot P^2\right) \cdot d \ln P + R \cdot T \cdot \ln \frac{P}{P_0} \tag{15}$$

By solving Equation (15) we obtain the Gibbs energy:

$$G - G^o = RB \cdot P + \frac{(C - B^2) \cdot P^2}{2 \cdot R \cdot T} + R \cdot T \cdot \ln \frac{P}{P_0} \tag{16}$$

From Equation (16), it is possible to derive the entropy equation. By substituting Equation (9) to Equation (3) we obtain

$$S - S_0 = R \int_0^P \left(1 - Z - T \cdot \left(\frac{\partial Z}{\partial T} \right)_P \right) \cdot d \ln P - R \cdot \ln \frac{P}{P_0} \tag{17}$$

By substituting Equation (10) into Equation (17) we obtain:

$$S - S_0 = R \int_0^P \left(-\frac{B}{R \cdot T} \cdot P + \frac{C - B^2}{(R \cdot T)^2} \cdot P^2 - T \cdot \left(\frac{\partial Z}{\partial T} \right)_P \right) \cdot d \ln P - R \cdot \ln \frac{P}{P_0} \tag{18}$$

where:

$$\left(\frac{\partial Z}{\partial T} \right)_P = \frac{P}{R \cdot T} \cdot \left(\frac{dB}{bT} - \frac{B}{T} \right) + \frac{P^2}{(R \cdot T)^2} \cdot \left(\frac{dC}{bT} - 2 \cdot B \cdot \frac{dB}{bT} \right) + \frac{2 \cdot P^2}{R^2 \cdot T^3} \cdot \left(B^2 - C \right) \tag{19}$$

Combining Equation (18) and (19) and integrating and simplifying the newly created dependence, we get,

$$S - S_0 = \frac{-2 \cdot R \cdot T \cdot B \cdot P - 2 \cdot R \cdot T^2 \cdot P \cdot \left(\frac{dB}{bT} - \frac{B}{T} \right) - T \cdot P^2 \left(\frac{dC}{bT} - 2 \cdot B \cdot \frac{dB}{bT} \right) + P^2 \cdot (C - B^2)}{2 \cdot R \cdot T^2} - R \cdot \ln \frac{P}{P_0} \tag{20}$$

By substituting Equations (16) and (20) into Equation (2) and then simplifying, we obtain the equation of natural gas enthalpy (21)

$$H = H_0 + \frac{2 \cdot P^2 \cdot (C - B^2) - 2 \cdot R \cdot T^2 \cdot P \cdot \left(\frac{dB}{bT} - \frac{B}{T} \right) - P^2 \left(\frac{dC}{bT} - 2 \cdot B \cdot \frac{dB}{bT} \right)}{2 \cdot R \cdot T} \tag{21}$$

where:

$$H_0 = \int_0^T C_{p0} \cdot dT \tag{22}$$

The coefficient of mechanical efficiency of the expander η_m is calculated using Equation (23) [29,30]

$$\eta_m = \left(A \cdot \left(\frac{Q}{Q_n} \right)^2 + B \cdot \left(\frac{Q}{Q_n} \right) + C \right) \cdot \eta_n \tag{23}$$

Equation (23) was obtained using an empirical method based on the average operating characteristic of turboexpanders. Figure 3 shows an example of the relationship between the efficiency μ / μ_n of the expander and gas flow Q / Q_n.

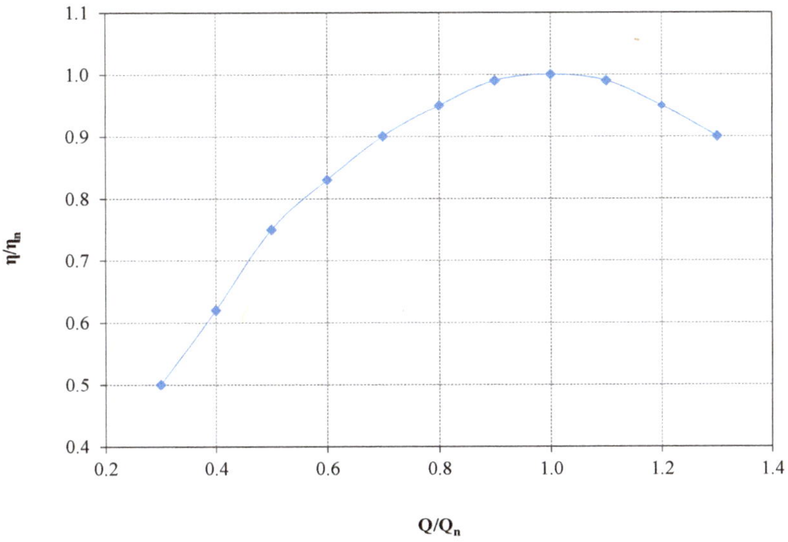

Figure 3. Relationship between the ratio of real and minimal efficiency vs. the ratio of real and nominal gas flow.

Apart from the generated electrical energy, the amount of heat used for heating up the gas before expansion is calculated using the thermodynamic model. The amount of heat needed for this purpose depends on the temperature drop due to the expansion. Gas expansion in an expander is an isotropic process, therefore, the drop in gas temperature will be much greater than in the case of normal pressure reduction and the associated Joule–Thomson effect. The gas temperature after expansion is determined using Equation (24):

$$T_2 = T_1 - \eta_o \cdot (T_1 - T_{2ad}) \tag{24}$$

where T_{2ad} is temperature of gas after expansion in the adiabatic process, calculated using Equation (25),

$$T_{2ad} = T_1 \cdot \left(\frac{P_2}{P_1}\right)^{\frac{k-1}{k}} \tag{25}$$

Due to technical and operational reasons, the gas temperature at the outlet of GRS cannot be lower than 273 K. Therefore, the gas temperature after heating is calculated using Equation (26),

$$T_1 = \frac{273}{1 - \eta_o \cdot \left(1 - \left(\frac{P_2}{P_1}\right)^{\frac{k-1}{k}}\right)} \tag{26}$$

The amount of natural gas needed to supply the necessary amount of heat (to preheat the gas before expansion) is calculated from Equation (27),

$$W_{heat} = \frac{M \cdot C_p \cdot \Delta T}{3600 \cdot \eta_{kgas\ boiler}} \tag{27}$$

where ΔT is the difference between gas temperature after heating, T_1, and gas temperature at the inlet to GRS, T_{in}.

$$\Delta T = T_1 - T_{in} \tag{28}$$

In order to carry out a reliable economic analysis of the use of the expander, only the additional amount of heat needed to heat natural gas should be taken into account. The demand for additional heat is caused by the additional temperature drop compared to the use pressure reducer.

The temperature difference between the required gas temperature after heating, in the case of a turboexpander, and the gas temperature after heating, in the case of a pressure reducer, is given by the formula:

$$\Delta T_{add} = T_1 - (4 \cdot \Delta P - T_{in}) \tag{29}$$

Hence, an additional amount of natural gas to supply additional heat, to preheat the gas is given by Equation (30),

$$W_{heat\_add} = \frac{M \cdot C_p \cdot \Delta T_{add}}{3600 \cdot \eta_{gas\ boiler}} \tag{30}$$

Based on the amount of produced electricity and amount of natural gas needed to provide the necessary amount of heat, the economic effect of using a turboexpander can be assessed.

2.2. Economic Model

The economic efficiency of the investment can be based on the discounted payback period (DPP) [31]:

$$DPP = \frac{CAPEX}{DCF_t} \tag{31}$$

where discounted cash flow (DCF) is given by Equation (32) [32],

$$DCF_t = \left[\frac{CF_1}{(1+r)^1}\right] + \left[\frac{CF_2}{(1+r)^2}\right] + \ldots + \left[\frac{CF_t}{(1+r)^t}\right] \tag{32}$$

Cash flow (CF) was calculated by using Equation (33) [33]:

$$CF = (W_{ee} \cdot EE_{price} - W_{heat_{add}} \cdot NG_{price}) - OPEX \tag{33}$$

The total capital expenditure includes the cost of design works and telemetry. Unexpected expenditures can be assessed as double the purchase cost of the device. The purchase cost of the turboexpander is determined using Equation (34) [34],

$$KZU = \alpha \cdot N_{el}^{-\beta} \tag{34}$$

Additional operational costs (OPEX) related to the operation and maintenance of the expander can be assumed to be 2% of the total investment cost. Another expenditure is the cost of natural gas used for heating up the gas before expansion. From the perspective of the operator, the difference in cash flow in the present state and after installation of the expander was also assumed. This means that the cost of heating up the gas in the existing system should be subtracted from the cost of heating it in the system equipped with an expander. These costs are paid by the operator, even though there is no income from the sales of electrical energy. In other words, only the cost of warming (above the level at which the Joule–Thomson compensation effect is reached in the regulators) was included.

The only positive element in the cash flow calculations will be the income from the sale of generated electrical energy.

2.3. Statistical Model

In order to determine the relationship between the discounted payback period and defined independent functions, a linear regression model was used, for which model coefficients were evaluated with the least squares method [35]. A function of dependent and independent parameters was described using Equation (35),

$$Y = X \cdot b + e \tag{35}$$

In the least squares method, a vector of coefficients b, for which Equation (36) has minimum significance, is the solution of Equation (35).

$$F(b) = \sum_{i=1}^{n} e_i^2 = (Y - X \cdot b)^T \cdot (Y - X \cdot b) = e^T \cdot e \tag{36}$$

Equation (37) is the necessity and sufficiency condition of minimum significance for function (36),

$$\frac{\partial F}{\partial b} = 2 \cdot X^T \cdot X \cdot b - 2 \cdot X^T \cdot Y = 0 \tag{37}$$

Hence Equation (38) is:

$$X^T \cdot X \cdot b = X^T \cdot Y \tag{38}$$

Matrix $X^T \cdot X$ has the size $m \times m$ and the following structure:

$$X^T \cdot X = \begin{vmatrix} n & \sum x_{i1} & \cdots & \sum x_{ik} \\ \sum x_{i1} & \sum x_{i1}^2 & \cdots & \sum x_{i1} \cdot x_{ik} \\ \vdots & \vdots & \vdots & \vdots \\ \sum x_{ik} & \sum x_{i1} \cdot x_{ik} & \cdots & \sum x_{ik}^2 \end{vmatrix} \tag{39}$$

Vector $X^T \cdot Y$ has m projections:

$$X^T \cdot Y = \begin{vmatrix} \sum y_i \\ \sum y_i \cdot x_{i1} \\ \vdots \\ \sum y_i \cdot x_{ik} \end{vmatrix} \tag{40}$$

The solution obtained with the least squares method is defined with matrix expression (41):

$$b = \left(X^T \cdot X \right)^{-1} \cdot \left(X^T \cdot Y \right) \tag{41}$$

The efficiency of the regression was evaluated with a determination coefficient, R^2. For a polynomial regression, the determination coefficient could be determined using Equation (42):

$$R^2 = 1 - \frac{(y - X \cdot b)^T \cdot (Y - X \cdot b)}{(Y - {}^{\backslash}y)^T \cdot (Y - {}^{\backslash}y)} \tag{42}$$

where ${}^{\backslash}y$ is a vector of magnitude n, which consists of medium significance (43):

$$^{\backslash}y = \frac{1}{n} \cdot \sum_{i=1}^{n} Y_i \tag{43}$$

3. Entry Data for Calculation Model

Real production data from fifteen operational GRS were used as entry data for the calculation model, as this should create a model which is closest to real-world conditions, and consequently, yield a higher accuracy of the searched dependence between independent factors and objective function.

Figures 4–7 present an hourly plot of gas flow intensity under normal conditions in the analyzed GRS as a function of time, as well as pressure level, at the inlet to the GRS for the selected four stations.

The plot covers a year of GRS operation. Summer minima and winter peaks of gas flow are associated with the seasonality of natural gas consumption in Poland and are clearly visible in the plot. The pressure at the GRS outlet is maintained at 0.3 MPa.

The economic analysis was based on the following assumptions:

- average sale price of produced electrical energy—38.96 USD/MWh [36]
- average gas price—17.24 USD/MWh [36]
- discount rate—5%

Figures 4–7 show how GRS operating parameters differ during the year. Reduction and Metering Stations are mainly used in medium-pressure networks, which supply natural gas to municipal customers. The share of natural gas in the energy balance of electricity production in 2017 in Poland was 3.61% [37]. In autumn and winter, due to the lowering of the ambient temperature, natural gas consumption strongly increases, which is shown in Figures 4–7. However, when the ambient temperature rises in summer, the consumption of natural gas decreases significantly. Such strong fluctuations in GRS performance parameters mean that the process of turboexpander selection for individual GRS is complicated [38,39]. This does not allow for maintenance of a stable level of electricity production because by choosing an expander with a nominal capacity close to the maximum natural gas flow in autumn and winter, it will not be possible to produce enough electricity during the summer, since gas flow will be too low and this will reduce the efficiency of the expansion device [40,41].

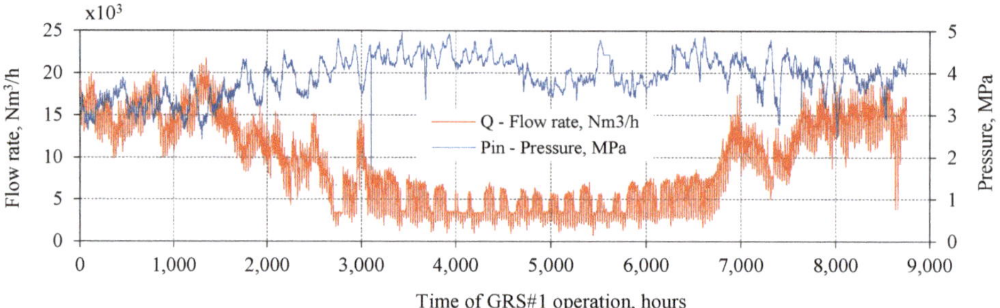

Figure 4. Natural gas flow and pressure at regulation stations GRS #1 (before expansion) over a period of one year.

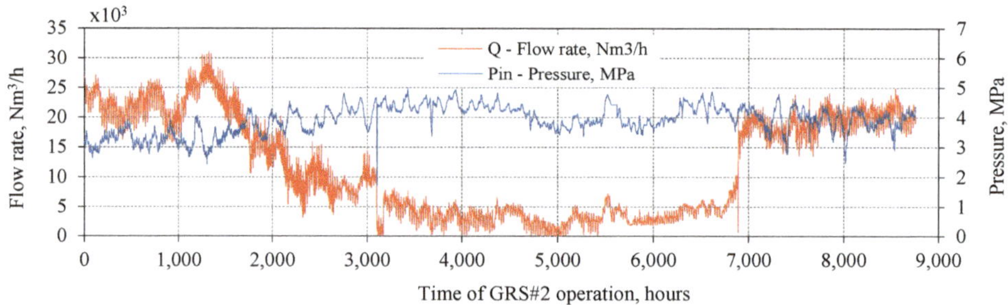

Figure 5. Natural gas flow and pressure at gas regulation stations (GRS) #2 (before expansion) over a period of one year.

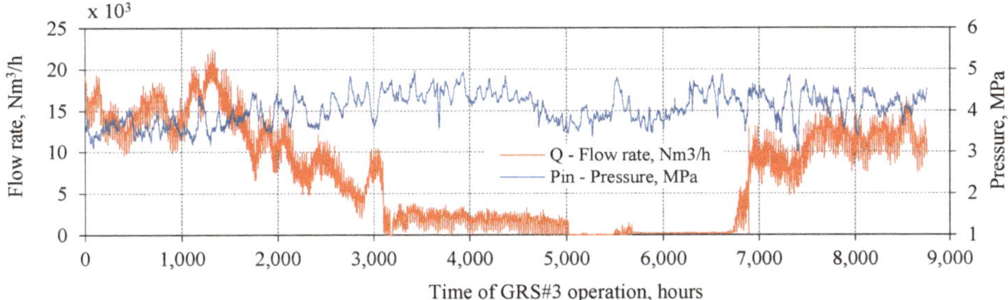

Figure 6. Natural gas flow and pressure at regulation stations GRS #3 (before expansion) over a period of one year.

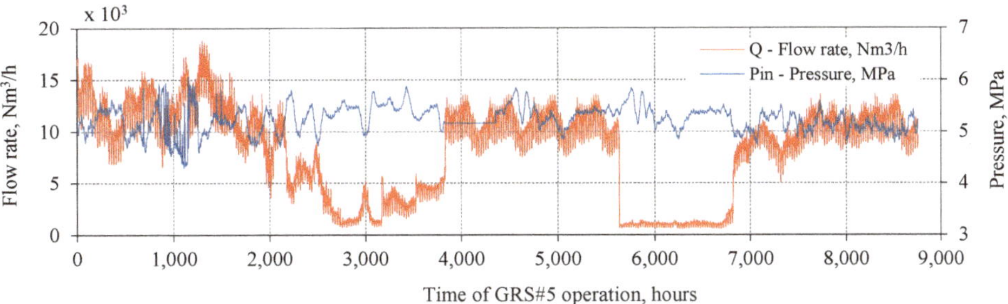

Figure 7. Natural gas flow and at pressure regulation stations GRS #5 (before expansion) over a period of one year.

The annual fluctuations in the GRS # 1 work parameters, presented in Figure 4, are the smallest among those presented. It is seen that in the summer months the daily irregularity of gas consumption during weekends decreases. The decrease in gas flow presented in Figure 4 is larger than presented in Figure 3 but not as drastic as in the case of GRS # 3 (operation profile is shown in Figure 5). In the range of 5000–6800 h, GRS # 3 almost doesn't function, and the expansion device does not produce electricity. The lack of gas flow through the station negatively affects the economic effect of the expander as the investment costs incurred for the purchase and installation of the expander device are not recovered.

The performance profile of GRS #5 presented in Figure 6, differs significantly from the GRSs discussed above. In this case, a significant increase in gas flow occurs in the range of 3800–5600 h, after which a significant reduction in gas flow is observed. The presented performance profile indicates that GRS #5 is one of the natural gas supply stations which provide gas to a ring-shaped gas distribution pipeline network in a large city. For the range of 3800–5600 h, all or almost all of the city's natural gas demand was covered by GRS #5. At the time of a dramatic drop in the flow (5600 h), the city's gas supply was most likely switched to other stations.

Another similar method to solve the problem of GRS flow decay during the summer is the ring system at the distribution network. A ring system is characteristic for large cities which are powered by several GRSs. In the winter, all of these GRSs work at maximum power, while in the summer, the gas flow to each of them is small. If it were possible to manage the gas flow in such a way that most of the city's gas demand would be covered by only one GRS, it would be an ideal place to use the turboexpander, since there would be no large seasonal fluctuations in gas flow through the station [42].

The pressure level for all presented stations within the analyzed range of time did not change drastically. Visible pressure fluctuations are caused by weekly unequal natural gas consumption. On non-working days, consumption is smaller, which means that the pressure in the network increases, whereas when consumption increases, the pressure in the network decreases. The ability to cover small peaks of gas consumption is called the cumulative capacity of the transmission network. It results in a gas pressure increase in the transmission and distribution network during the night hours and a gas pressure reduction in the morning and evening hours. The transmission network also reacts with an increase in pressure on non-working days [43].

4. Results of Calculations

Based on data as described in the previous section, a computer simulation was performed to examine technical and economic effects related to the application of vortex turboexpanders at 15 selected Reduction and Metering Stations. Figures 8–15 present generated electricity, heat demand for natural gas heating, as well as expander operational efficiency calculated on the basis of Equation (22) for the selected (as described in the previous chapter) four GRSs.

For all stations, except GRS #5 in the range of 3800–5200 h, heat demand for natural gas heating before expansion is higher than the amount of generated electricity. This may seem unreasonable because more energy is lost than produced. However, this is because natural gas, which we burn for natural gas heating (before expansion), is a primary energy carrier and costs four times less than generated electricity [26]. For this reason, the application of expanders is reasonable from an economic point of view.

Figure 8 presents electricity generation and heat demand for GRS #1. The relationship between the amount of generated electricity and heat demand is linear. The increase in electricity generation, caused by the increase of gas flow through the expander, results in an increased heat demand for natural gas heating before expansion. The only deviation from this rule occurs in the time interval 1200–1300 h. In the indicated time range, heat demand increased despite the decline in electricity generation. Data analysis of GRS #1 operation profile, see Figure 4, showed that in the described time interval, gas flow increased simultaneously and inlet pressure was reduced. In the discussed time interval, there was a significant reduction in the efficiency of the turboexpander operation, see Figure 9.

Turboexpander efficiency reduction is caused by exceeding the nominal gas flow rate for which the efficiency of the expansion device is the largest.

A combination of several negative factors such as lowering of inlet pressure and exceeding nominal gas flow caused a reduction in electricity production. On the other hand, the increase in heat demand was due to the increase in gas flow, which needed to be heated before expansion [44].

A similar phenomenon was observed for GRS #2. The profile of electricity generation and heat demand is presented in Figure 10. Turboexpander efficiency changes are shown in Figure 11. For GRS #2, the phenomenon described above is more pronounced and occurs twice.

For the purpose of this research, the authors acquired real data which covered one year of operation for 15 natural gas reduction and measurement stations (GRS). The assessment of selected GRS economic efficiency depends on the determination of the discount payback period (DPP). DPP is calculated on the basis of cumulative annual electricity production and cumulative annual heat demand (heat carrier is natural gas consumed by GRS's own needs).

On the basis of real data which covered one year of operation for 15 natural gas reduction and measurement stations (GRS), it was possible to obtain 15 dependent functions Y_j which describe the statistical model. Based on only 15 dependent functions Y_j and 15 independent parameters X_i ($i = 1, \dots, 5$), it was not possible to determine the relationship between Y_j and X_i with high accuracy.

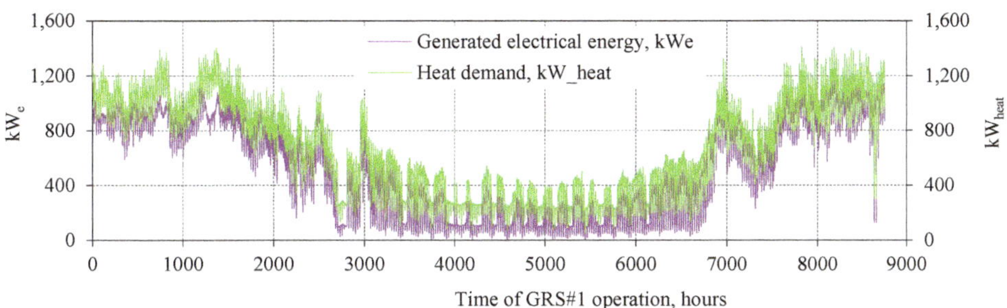

Figure 8. Generated electrical energy and heat demand at GRS #1 over a period of one year.

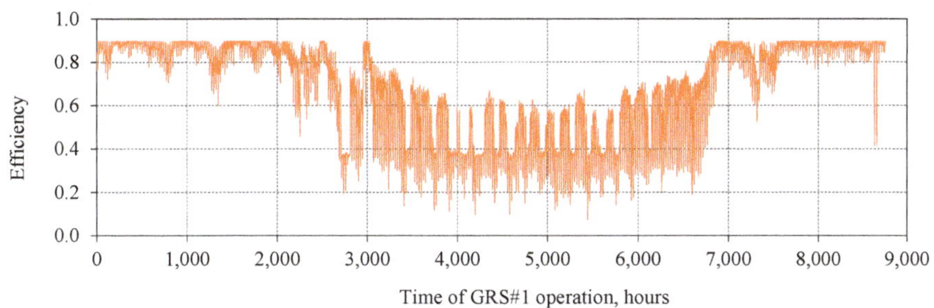

Figure 9. Expander efficiency for GRS #1 over a period of one year.

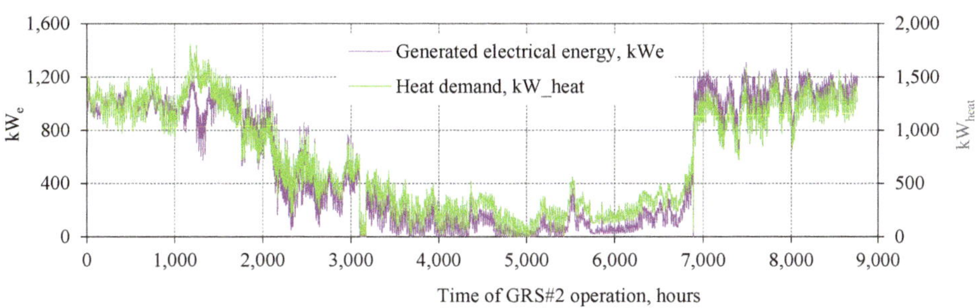

Figure 10. Generated electrical energy and heat demand at GRS #2 over a period of one year.

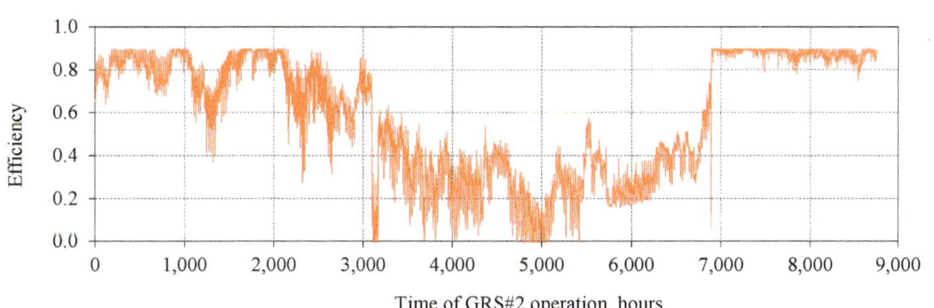

Figure 11. Expander efficiency for GRS #2 over a period of one year.

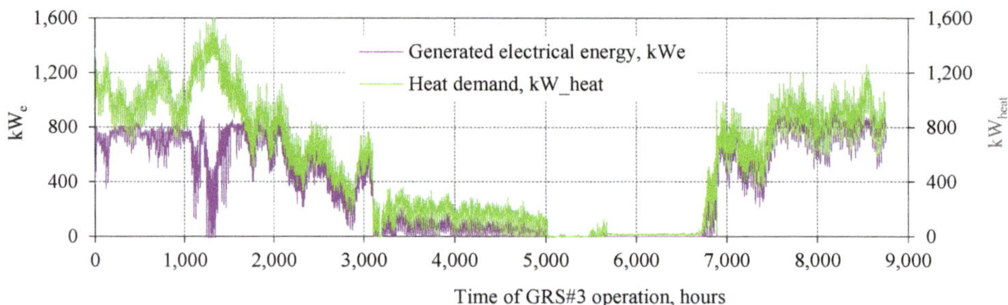

Figure 12. Generated electrical energy and heat demand at GRS #3 over a period of one year.

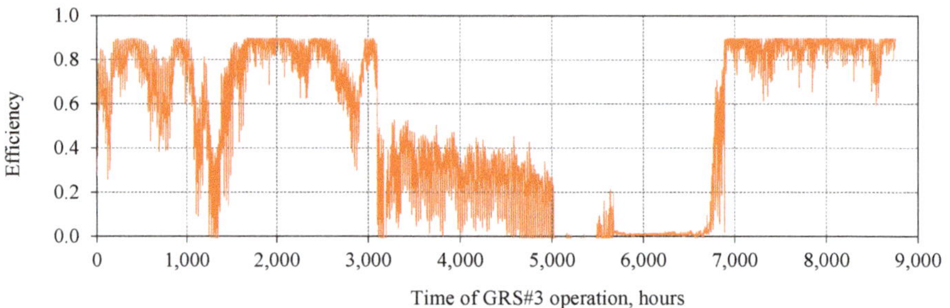

Figure 13. Expander efficiency for GRS #3 over a period of one year.

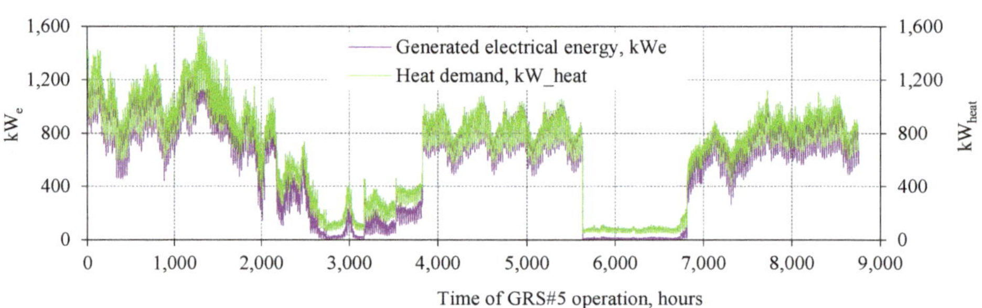

Figure 14. Generated electrical energy and heat demand at GRS #5 over a period of one year.

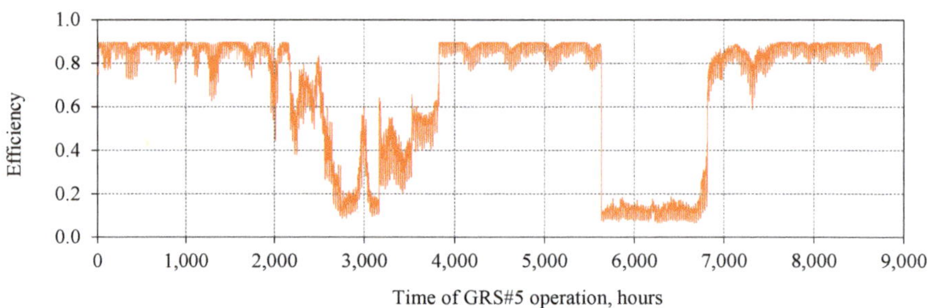

Figure 15. Expander efficiency for GRS #5 over a period of one year.

In addition, the performance characteristics of GRS vary year to year, mainly due to, e.g., climate reasons (mild or severe winter) or change (increase or decrease) in the number of natural gas consumers.

Therefore, an attempt to estimate the dependence between selected parameters X_i, based on one year of measurements cannot ensure sufficient accuracy of the developed equation, Y_j. Unfortunately, during the research, the authors did not get access to operational data of other GRSs. To increase the quantity of input data in the statistical model, data extension was applied to the operational parameters of each GRS.

Natural gas flow rate, GRS inlet pressure, and capital expenditures (CAPEX) data have been extended. Data extension of these parameters was performed with the creation of artificial values which were determined in steps of 10% in the range from -50% to $+50\%$ for each real value of a given parameter, X_i. Whereas, data of natural gas purchase price and generated electricity sell price have been extended in the range from -10% to $+10\%$ with a step of 2%.

As a result of data extension, an additional 750 samples were obtained. In total, 765 dependent functions Y_j and 765 values of X_i parameters were used for analysis which was performed in the developed model. Samples for which DPP was greater than 40 years were removed. Finally, 607 data sets were used to develop the statistical analysis.

By application of the multiple linear regression method, formula (44) was obtained:

$$Y_j = 45.227 - 10^{-4} \cdot X_1 - 1.38 \cdot X_2 + 0.0313 \cdot X_3 - 0.0243 \cdot X_4 + 3.114 \cdot 10^{-4} \cdot X_5 \tag{44}$$

Formula (44) describes the relationship between discounted payback period (DPP) as a dependent parameter and selected independent parameters such as annual average gas flow rate (X_1), average annual expansion level (X_2), natural gas purchase price (X_3), produced electrical energy sale price (X_4), and capital expenditures (CAPEX) (X_5). The coefficient of determination R^2 for the obtained equation (44) equals 0.543. The closer the coefficient of determination is to 1.00, the more accurate the regression model is. To improve the accuracy of the developed model, the influence of the individual independent parameter (X_i) on parameter Y_j (DPP) was studied. Dependencies are presented in Figure 16.

Figure 16 (top) shows the dependence of the discounted payback period (DPP) vs. average annual gas flow through GRS X_1 (highest gas flow affects a faster return on investment) and DPP vs. average annual level of gas expansion X_2. A higher level of gas expansion results in additional work which is performed by the expander, thus more electricity is generated which influences the faster reimbursement of costs incurred for the investment.

The relation presented in Figure 16 (middle) shows the dependence of DPP vs. purchase price of natural gas X_3 (the increase in the purchase price of natural gas adversely affects on DPP) and DPP vs. sale price of generated electrical energy X_4 (increase in the sale price of electricity produced positively affects the DPP).

The dependence of DPP vs. capital expenditures X_5 (CAPEX) is presented in Figure 15 (bottom). Usually low investment costs make the payback time of invested capital shorter. However, the presented results contradict this statement. Despite the fact that the price of a large expander is much higher than the price of a small one, a large turbo expander can operate at higher gas flow rates and can generate larger volumes of electricity. As a result, a properly selected and installed turboexpander on GRS can generate more electricity, which will be converted into positive cash flow.

Capital expenditures X_5 (CAPEX) depend on the installed electricity generation unit capacity. The amount of generated electricity is a function of the gas flow and expansion level. As a result, the dependence of DPP vs. capital expenditures X_5 (CAPEX) is similar to the dependency curve of DPP vs. average annual gas flow through GRS X_1 and DPP vs. average annual level of gas expansion X_2.

None of the independent individual parameters X_i have a linear influence on the dependent parameter Y_j, as shown in Equation (44).

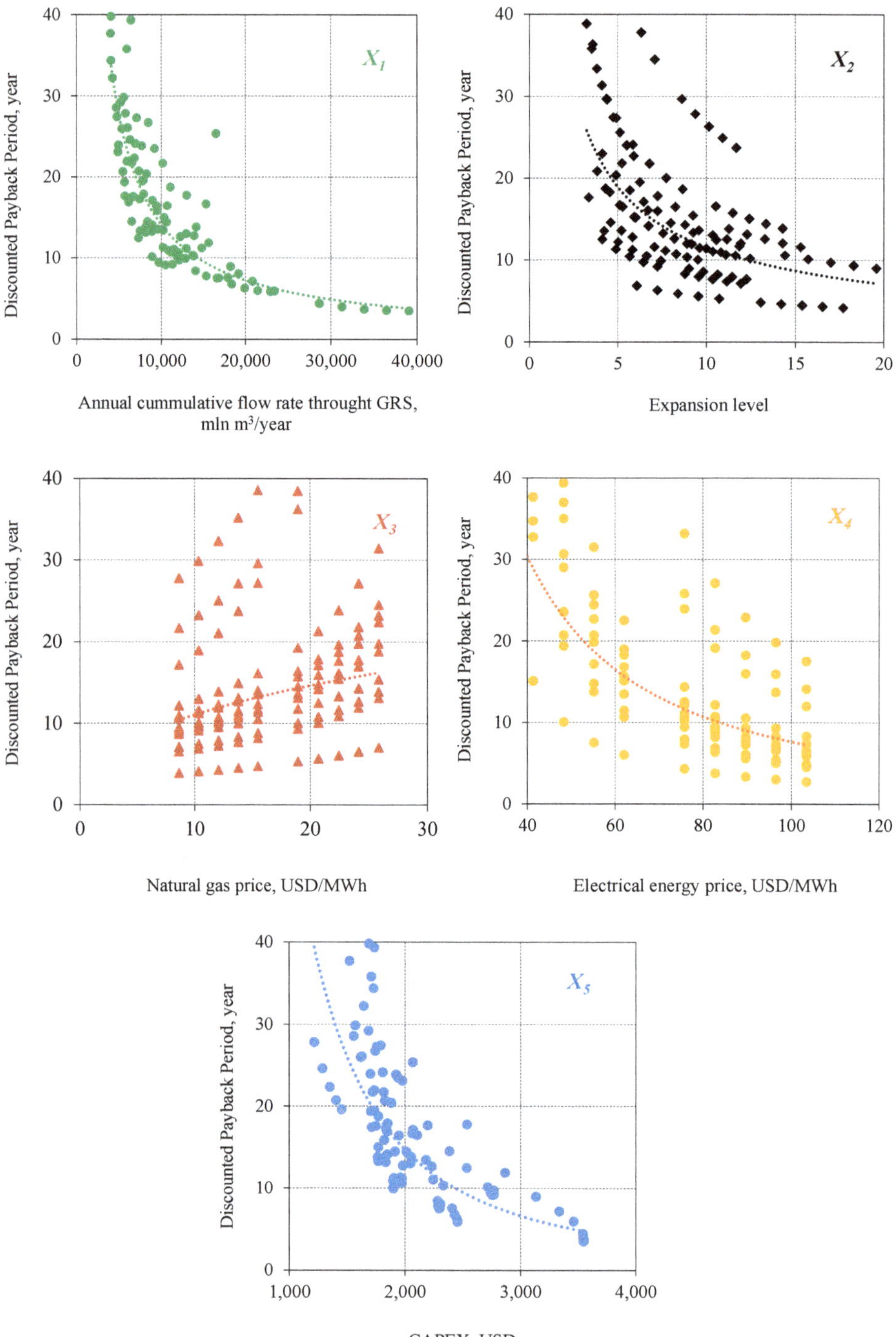

Figure 16. Discounted Payback Period (Y_j) vs. individual independent parameter (X_i).

Because the influence of most factors on DPP is nonlinear, as shown in Figure 16, the statistical model should be modified. The previous form (44):

$$Y_j = b_0 + b_1 \cdot X_1 + b_2 \cdot X_2 + \ldots + b_5 \cdot X_5 + e \tag{45}$$

was replaced with (45):

$$Y_j = b_0 + b_1 \cdot X_1^{\beta_1} + b_2 \cdot e^{X_2 \beta_2} + \ldots + b_5 \cdot X_5 + e \tag{46}$$

To restore the statistical model to the linear polynomial regression, the variable should be changed to: $Z_1 = X_1^{\beta_1}$, $Z_2 = e^{X_2 \beta_2}$ etc., after which a linear regression with new independent variables will be obtained [35]:

$$Y_j = b_0 + b_1 \cdot Z_1 + b_2 \cdot Z_2 + \ldots + b_5 \cdot Z_5 + e \tag{47}$$

By using the least-squares method, relations between the parameters X_i and Y_j were determined, which are graphically shown in Figure 16. After the statistical analysis, Equation (48) was obtained, which is dependent on the discounted payback period (DPP) and annual average natural gas flow rate through the analyzed GRS (X_1), average annual level of gas expansion (X_2), average annual natural gas purchase price (X_3), average annual produced electrical energy sale price (X_4) and CAPEX (X_5).

$$Y_j = -52.91 + 1.1 \cdot 10^5 \cdot X_1^{-0.965} + 65.34 \cdot X_2^{-0.71} + 1.39 \cdot X_3^{0.403} + 5.732 \cdot 10^5 \cdot X_4^{-1.506} - 1.906 \cdot 10^9 \cdot X_5^{-1.973} \tag{48}$$

The coefficient of determination R^2 for the obtained Equation (48) equals 0.768, which is much better than for the multiple linear model (44). The obtained Equation (48) allows for sufficient accuracy to validate the appropriateness of using turboexpanders on individual GRS. It is recommended to apply the above solution at the first stage of individual GRS selection for expansion installation. To make a final investment decision, a more detailed economic analysis, that was previously described with Equation (48), needs to be carried out for each selected station.

5. Discussion and Conclusions

Due to high efficiency and relatively low cost, expansion machines can be widely used in natural gas transmission and distribution systems. Turboexpanders can be used where gas flow is present (average annual volumes of at least a few thousand normal cubic meters per hour) and the reduction of gas pressure takes place. This means that expansion machines can be used at control nodes, reduction and metering stations at industrial facilities, and reduction and metering stations at natural gas distribution networks, which are the most common elements of a natural gas network.

The main factor which negatively affects the economy of turboexpander application is seasonal unevenness in natural gas consumption (summer minima and winter peaks). When the gas flow significantly deviates from the nominal gas flow at which the expander's operational efficiency is the highest, the production of electricity drastically decreases due to the decrease in turboexpander efficiency. This negatively affects its economic efficiency, thus extending the payback period and reducing the cash flow. A solution to the above problem is possible if the distribution network operates in a ring system. In this case, in the summer, gas flow can be regulated and directed to GRS with an installed expander. In this way, it is possible to avoid gas flow reduction to the GRS with the expander. A lack of seasonal irregularities will positively affect the amount of electricity generation, and thus, will improve the economic efficiency of turboexpander application.

The choice of which GRS an expander should be installed on, and the assessment of its economic efficiency is complicated and requires knowledge of thermodynamics, mechanics, natural gas engineering, and economics.

The formula developed and presented in this article allows a quick and easy evaluation of the applicability of an expander on the selected GRS I°. The discounted payback period (DPP) was the basis for the assessment of the investment profitability.

From a statistical point of view, the discounted payback period (DPP) is a dependent parameter, while independent parameters are the selected annual average gas flow rate (X_1), average annual expansion level (X_2), natural gas purchase price (X_3), produced electrical energy sale price (X_4), and capital expenditures (CAPEX) (X_5).

The influence of independent parameters on the dependent parameter was analyzed. It was observed that this relationship was not linear. The equation developed by the multiple linear regression method had low accuracy, as determined by R^2. The coefficient of determination R^2 for the obtained Equation (44) was 0.543.

After the implementation of non-linear multiple regression, Equation (48) was obtained, which is characterized by a much higher coefficient of determination $R^2 = 0.768$. The obtained Equation (48) allows sufficient accuracy to validate the appropriateness of using turboexpanders on a given individual GRS. It is recommended to use the above solution as the first step in the selection of the station for the installation of expansion machines. To make a final investment decision, it is necessary to perform a detailed economic analysis for each selected station which was previously examined with the developed Equation (48).

Author Contributions: Conceptualization, S.K., M.Ł. and A.O.; Formal analysis, S.K. and A.O.; Funding acquisition, M.Ł. and A.S.; Investigation, S.K. and A.O.; Methodology, A.O.; Supervision, M.Ł. and A.S.; Validation, S.K. and A.O.; Visualization, S.K., A.O. and T.W.; Writing—original draft, A.O.; Writing—review & editing, S.K. and T.W.

Funding: This work has received funding from the Statutory Research of Natural Gas Department at Drilling Oil & Gas Faculty, no. 11.11.190.555

Conflicts of Interest: The authors declare no conflict of interest.

Nomenclature

Symbols

A, B, C	coefficients depending on the expander's type and model; provided by the producer
B	second virial coefficient, $m^3/kmol$
$b_{ij}^{(0)}$	zero-order term in the expansion of B, $m^3/kmol$
$b_{ij}^{(1)}$	coefficient of first-order term in the expansion of B, $m^3/kmol \cdot K$
$b_{ij}^{(2)}$	coefficient of second-order term in the expansion of B, $m^3/kmol \cdot K^2$
b	vector of coefficient
C	third virial coefficient, $m^6/kmol^2$
$c_{ijk}^{(0)}$	zero-order term in the expansion of C, $m^6/kmol^2$
$c_{ijk}^{(1)}$	coefficient of first-order term in the expansion of C, $m^6/kmol^2 \cdot K$
$c_{ijk}^{(2)}$	coefficient of second-order term in the expansion of C, $m^6/kmol^2 \cdot K^2$
e	residual vector
G	Gibbs energy, kJ/kg,
G_0	Gibbs energy of ideal-gas, kJ/kg,
H	enthalpy kJ/kg
H_0	ideal-gas enthalpy kJ/kg,
k	exponent of gas adiabat
M	mass flow rate through the expander, kg/s
N_{el}	electric power produced in the power plant
P_1, P_2	absolute pressure before and after reduction, respectively, Pa
S	entropy kJ/kg·K
S_0	ideal-gas entropy kJ/kg·K,
Q	real flow rate through the expander, m^3/s
Q_n	nominal flow rate through the expander, m^3/s
R	gas constant, kJ/kg·K
r	discount rate
R^2	determination coefficient

t	last period for which cumulative DCF is negative
T_1	gas temperature before expansion, K
T_{in}	inlet temperature, K
ΔT_{add}	temperature difference, K
T_{2ad}	gas temperature in the adiabatic process, K
W	generated stream of electrical energy, kW
W_{heat}	thermal energy, kW_heat
V	volume, m^3.
X_i	independent function
X_1	average annual gas flow through GRS, mln m^3
X_2	average annual level of gas expansion
X_3	purchase price of natural gas, USD/MWh
X_4	sale price of generated electrical energy, USD/kWh
X_5	capital expenditures (CAPEX), mln USD
Y_j	dependent function, Discounted Payback Period (DPP), years
$'y$	vector of magnitude n
Z	compressibility factor

Greek Letters

α, β	coefficients depending on the expander's type and producer
ΔH	difference in natural gas enthalpy before and after expansion, kJ/kg
η_o	coefficient of adiabatic efficiency of the expander
η_{eg}	coefficient of efficiency of electricity generator
η_m	coefficient of mechanical efficiency of the expander
η_n	nominal efficiency of the expander

Acronyms

CER	carbon emission recovery index
CF_i	cash flow in *i*-th year
DCF	discounted cash flow
DPP	discounted payback period
EOS	equation of state
GERG 88	selected equation of state
GRS	natural gas regulation stations / natural gas reducing and metering station
KPI	key performance indicator
KZU	purchase cost of the device, USD/kW
RR	recovery ratio
SRA	structured retrofitting approach
WER	waste energy recovery index

References

1. Howard, C.; Oosthuizen, P.; Peppley, B. An investigation of the performance of a hybrid turboexpander-fuel cell system for power recovery at natural gas pressure reduction stations. *Appl. Therm. Eng.* **2011**, *31*, 2165–2170. [CrossRef]
2. Chaczykowski, M.; Osiadacz, A.J.; Uilhoorn, F.E. Exergy-based analysis of gas transmission system with application to Yamal-Europe pipeline. *Appl. Energy* **2011**, *88*, 2219–2230. [CrossRef]
3. Janusz, P.; Liszka, K.; Łaciak, M.; Oliinyk, A.; Susak, A. An analysis of the changes of the composition of natural gas transported in high-pressure gas pipelines. *AGH Drill. Oil Gas* **2015**, *32*, 713–722. [CrossRef]
4. Janusz, P.; Liszka, K.; Łaciak, M.; Smulski, R.; Oiinyk, A.; Susak, O. Applicability of equations for pressure losses in transmission gas pipelines. *AGH Drill. Oil Gas* **2015**, *32*, 525–538. [CrossRef]
5. Kuczyński, S.; Łaciak, M.; Oliinyk, A.; Szurlej, A.; Włodek, T. Determining selection criteria for reducing and metering stations for the use of turboexpanders. In Proceedings of the 1st Latin American Conference on Sustainable Development of Energy, Water and Environment Systems, Rio de Janeiro, Brazil, 28–31 January 2018; p. 219.

6. Jean-Jacques, B.; Bruno, D.; François, M. Modelling of a pressure regulator. *Int. J. Press. Vessel. Pip.* **2007**, *84*, 234–243.

7. Nasrifar, K.; Bolland, O. Prediction of thermodynamic properties of natural gas mixtures using 10 equations of state including a new cubic two-constant equation of state. *J. Pet. Sci. Eng.* **2006**, *51*, 253–266. [CrossRef]

8. Bisio, G. Thermodynamic analysis of the use of pressure exergy of natural gas. *Energy* **1995**, *20*, 161–167. [CrossRef]

9. Kostowski, W.J.; Usón, S. Thermoeconomic assessment of a natural gas expansion system integrated with a co-generation unit. *Appl. Energy* **2013**, *101*, 58–66. [CrossRef]

10. Daneshi, H.; Zadeh, H.K.; Choobari, A.L. Turboexpander as a distributed generator. In Proceedings of the Power and Energy Society General Meeting-Conversion and Delivery of Electrical Energy in the 21st Century, Pittsburgh, PA, USA, 20–24 July 2008; pp. 1–7.

11. Łaciak, M.; Smulski, R. Possible uses of natural gas expansion energy in reduction stations for production of electricity. *Nowoczesne Techniki i Technologie Bezwykopowe* **2002**, *1*, 20–22.

12. Gas Transmission Operator GAZ-SYSTEM S.A. Available online: http://en.gaz-system.pl/ (accessed on 30 November 2018).

13. Augustin, T. Gas expansion power plants of newest design- efficient power generation on the background of changing economical conditions. *Gas Waerme Int.* **2000**, *49*, 146–148.

14. Sadeghi, H.; Fouladi, H.; Kazemi, H.A. Power generation through recovery of natural gas pressure loss and waste heat energy of steam power plant in expansion turbine. In Proceedings of the ECOS 2004, Guanajuato, Mexico, 7–9 July 2004; pp. 1255–1263.

15. Comakli, K. Economic and environmental comparison of natural gas fired conventional and condensing combi boilers. *J. Energy Inst.* **2008**, *81*, 242–246. [CrossRef]

16. Arabkoohsar, A.; Gharahchomaghloo, Z.; Farzaneh-Gord, M.; Koury, R.N.; Deymi-Dashtebayaz, M. An energetic and economic analysis of power productive gas expansion stations for employing combined heat and power. *Energy* **2017**, *133*, 737–748. [CrossRef]

17. Borelli, D.; Devia, F.; Lo Cascio, E.; Schenone, C. Energy recovery from natural gas pressure reduction stations: Integration with low temperature heat sources. *Energy Convers. Manag.* **2018**, *159*, 274–283. [CrossRef]

18. Li, G.; Wu, Y.; Zhang, Y.; Zhi, R.; Wang, J.; Ma, C. Performance study on a single-screw expander for a small-scale pressure recovery system. *Energies* **2017**, *10*, 6. [CrossRef]

19. Stanek, W.; Gazda, W.; Kostowski, W.; Usón, S. Exergo-ecological assessment of multigeneration energy systems. In *Thermodynamics for Sustainable Management of Natural Resources*; Springer: Cham, Switzerland, 2017; pp. 405–442.

20. Prilutskii, A.I. Use of piston expanders in plants utilizing energy of compressed natural gas. *Chem. Pet. Eng.* **2008**, *44*, 149–156. [CrossRef]

21. Barbarelli, S.; Florio, G.; Scornaienchi, N.M. Theoretical and experimental analysis of a new compressible flow small power turbine prototype. *Int. J. Heat Technol.* **2017**, *35*, s391–s398. [CrossRef]

22. Barbarelli, S.; Florio, G.; Scornaienchi, N.M. Developing of a small power turbine recovering energy from low enthalpy steams or waste gases: Design, building and experimental measurements. *Therm. Sci. Eng. Prog.* **2018**, *6*, 346–354. [CrossRef]

23. Lo Cascio, E.; Borelli, D.; Devia, F.; Schenone, C. Key performance indicators for integrated natural gas pressure reduction stations with energy recovery. *Energy Convers. Manag.* **2018**, *164*, 219–229. [CrossRef]

24. Lo Cascio, E.; Von Friesen, M.P.; Schenone, C. Optimal retrofitting of natural gas pressure reduction stations for energy recovery. *Energy* **2018**, *153*, 387–399. [CrossRef]

25. Osiadacz, A.; Chaczykowski, M.; Kwestarz, M. An evaluation of the possibilities of using turboexpanders at pressure regulator stations. *J. Power Technol.* **2017**, *97*, 289–294.

26. Wu, S.; Rios-Mercado, R.Z.; Boyd, E.A.; Scott, L.R. Model relaxations for the fuel cost minimization of steady-state gas pipeline networks. *Math. Comput. Model.* **2000**, *31*, 197–220. [CrossRef]

27. Holm, J. Application of turboexpanders for energy conservation. *J. Turbomach* **1983**, *24*, 26–31.

28. Jaeschke, M.; Audibert, S.; van Caneghem, P.; Humphreys, A.E.; Janssen-van Rosmalen, R.; Pellei, Q.; Michels, J.P. Accurate Prediction of Compressibility Factors by the GERG Virial Equation (includes associated paper 23568). *SPE Prod. Eng.* **1991**, *6*, 343–349. [CrossRef]

29. Faddeev, I.P. Turbo expanders to utilize the pressure of natural gas delivered to Saint Petersburg and industrial centers. *Chem. Pet. Eng.* **1998**, *34*, 704–711. [CrossRef]

30. Poživil, J. Use of expansion turbines in natural gas pressure reduction stations. *Acta Montan. Slovaca* **2004**, *3*, 258–260.

31. Valero, A.; Correas, L.; Zaleta, A.; Lazzaretto, A.; Verda, V.; Reini, M.; Rangel, V. On the thermoeconomic approach to the diagnosis of energy system malfunctions: Part 2. Malfunction definitions and assessment. *Energy* **2004**, *29*, 1889–1907. [CrossRef]

32. Valero, A.; Usón, S.; Torres, C.; Valero, A. Application of thermoeconomics to industrial ecology. *Entropy* **2010**, *12*, 591–612. [CrossRef]

33. Usón, S.; Valero, A.; Agudelo, A. Thermoeconomics and industrial symbiosis. Effect of by-product integration in cost assessment. *Energy* **2012**, *45*, 43–51.

34. Skorek, J. *Ocena Efektywności Energetycznej i Ekonomicznej Gazowych Układów Kogeneracyjnych Małej Mocy*; Wydaw. Politechniki Śląskiej: Gliwice, Poland, 2002. (In Polish)

35. Воскобойников, Ю.Е. Построение Регрессионных Моделей в Пакете MathCAD: Учеб. Пособие; НГАСУ: Новосибирск, Russia, 2009. (In Russian)

36. Urząd Regulacji Energetyki. *Sprawozdanie z Działalności Prezesa Urzędu Regulacji Energetyki w 2016 r*; Energy Regulatory Office: Warsaw, Poland, 2017. (In Polish)

37. Urząd Regulacji Energetyki. *Sprawozdanie z Działalności Prezesa Urzędu Regulacji Energetyki w 2017 r*; Energy Regulatory Office: Warsaw, Poland, 2018. (In Polish)

38. Borelli, D.; Devia, F.; Brunenghi, M.M.; Schenone, M.; Spoladore, A. Waste Energy Recovery from Natural Gas Distribution Network: CELSIUS Project Demonstrator in Genoa. *Sustainability* **2015**, *7*, 16703–16719. [CrossRef]

39. Kolasiński, P.; Pomorski, M.; Błasiak, P.; Rak, J. Use of Rolling Piston Expanders for Energy Regeneration in Natural Gas Pressure Reduction Stations—Selected Thermodynamic Issues. *Appl. Sci.* **2017**, *7*, 535. [CrossRef]

40. Borelli, D.; Devia, F.; Lo Cascio, E.; Schenone, M.; Spoladore, A. Combined Production and Conversion of Energy in an Urban Integrated System. *Energies* **2016**, *9*, 817. [CrossRef]

41. Miller, I.; Gençer, E.; O'Sullivan, F.A. General Model for Estimating Emissions from Integrated Power Generation and Energy Storage. Case Study: Integration of Solar Photovoltaic Power and Wind Power with Batteries. *Processes* **2018**, *6*, 267. [CrossRef]

42. Garmsiri, S.; Rosen, M.; Smith, G. Integration of Wind Energy, Hydrogen and Natural Gas Pipeline Systems to Meet Community and Transportation Energy Needs: A Parametric Study. *Sustainability* **2014**, *6*, 2506–2526. [CrossRef]

43. Mostert, C.; Ostrander, B.; Bringezu, S.; Kneiske, T. Comparing Electrical Energy Storage Technologies Regarding Their Material and Carbon Footprint. *Energies* **2018**, *11*, 3386. [CrossRef]

44. Yousif, M.; Ai, Q.; Gao, Y.; Wattoo, W.; Jiang, Z.; Hao, R. Application of Particle Swarm Optimization to a Scheduling Strategy for Microgrids Coupled with Natural Gas Networks. *Energies* **2018**, *11*, 3499. [CrossRef]

Article

A New Method of Selecting the Airlift Pump Optimum Efficiency at Low Submergence Ratios with the Use of Image Analysis

Grzegorz Ligus [1], Daniel Zając [2], Maciej Masiukiewicz [1] and Stanisław Anweiler [1,*

[1] Faculty of Mechanical Engineering, Opole University of Technology, ul. Mikołajczyka 5, 45-271 Opole, Poland; g.ligus@po.opole.pl (G.L.); m.masiukiewicz@po.opole.pl (M.M.)
[2] Kelvion Sp. z o.o.., ul. Kobaltowa 2, 45-641 Opole, Poland; daniel.zajac@kelvion.com
* Correspondence: s.anweiler@po.opole.pl; Tel.: +48-888-272-878

Received: 2 January 2019; Accepted: 19 February 2019; Published: 22 February 2019

Abstract: This paper presents experimental studies on the optimization of two-phase fluid flow in an airlift pump. Airlift pumps, also known as mammoth pumps, are devices applied for vertical transport of liquids with the use of gas. Their operating principle involves the existence of a density gradient. This paper reports the results of experimental studies into the hydrodynamic effects of the airlift pump. The studies involved optical imaging of two-phase gas-liquid flow in a riser pipe. The visualization was performed with high-speed visualization techniques. The studies used a transparent model of airlift pump with a rectangular cross-section of the riser. The assessment of the airlift pump operation is based on the image grey-level analysis to provide the identification of two-phase flow regimes. The scope of the study also involved the determination of void fraction and pressure drops. The tests were carried out in a channel with dimensions $35 \times 20 \times 2045$ mm with the gas flux range 0.2–15.0 m^3/h. For the assessment of the two-phase flow pattern Probability Density Function (PDF) was applied. On the basis of the obtained results, a new method for selecting the optimum operating regime of airlift pump was derived. This method provides the finding of stability and efficiency of liquid transport. It can also be applied to determine the correlation between the total lifting efficiency and the required gas flux for proper operation of the airlift pump.

Keywords: two-phase flow regime; airlift pump; void fraction; image analysis; efficiency optimization

---

1. Introduction

An airlift pump in comparison to other equipment used for liquid transportation doesn't have moving parts. This represents an advantage in applications where there is a need for reliability, simplicity and compact design. The literature reports a variety of applications in which airlift pumps find uses as an optimum solution. A very extensive and detailed review of the literature in the field of airlift pumps is presented in [1], where the authors discuss the theory of operations, numerical and experimental methods possible to apply in the airlift pumping research. They also indicate the most common areas of application of airlift pumps. In many works, the authors indicate the possibility of changing the geometry of airlift pumps. In the work [2] the authors suggested different crossectional shapes of risers. Work [3] shows that the performance increment can be obtained by grouping airlift pumps into batteries. Other studies found that the system of gas distribution and the liquid reception is also affecting the operating characteristics of an airlift pump. In [4] a comparison of airstones and open-ended tubes was investigated. In [5] steady and pulsating injection modes with the use of four air distributor geometry namely radial, axial, dual and swirl were analysed. The authors in [6] propose the application of four types of the nozzle (point, star, ring, double-ring) for artificial upwelling systems. Different angles of the gas distributor installation were investigated in [7].

Any analysis concerned with the exploitation of an airlift pump requires knowledge regarding the hydrodynamics of two-phase gas-liquid flow. One of the main problems with two-phase flows is the identification of flow structures and flow parameters. Flow patterns are formed by the contact between the gas and liquid or solid phases. Three-phase flow is discussed, among others in [8,9]. The classification of flow regimes of airlift pumps is similar to the classifications for gas-liquid flow in vertical channels [10]. There are solutions for pumps with an inclined riser for which the classification is the same as in work [11]. Sometimes in the airlift pump, the following regimes are distinguished: bubble, slug, churn, wispy-annular, annular and transition patterns in the boundary region [12]. In paper [13] additional structure development in the riser pipe was presented. The flow patterns determine the heat, mass and momentum transfer [14]. Specific patterns also affect the value of the pressure drop [15]. The pressure drop is particularly important, especially when pumping liquids with higher viscosities [16].

The transient structures are difficult to describe but are very interesting from the process optimization point of view. These types of transient structures are unstable. This results in the flow disturbance and the efficiency of the liquid transport decreases. It is best to examine them using imaging and numerical modelling techniques because they are non-invasive methods. We can find many papers that prove high efficiency of image analysis methods. In the work [17] two- and three-phase flow identification in pipe bends were presented. In [18] two-phase flow pattern was recognized in the minichannel. Computer image analysis has also been successfully used for the investigation of gas-solid systems [19,20]. Another visual research method often used during the hydrodynamic investigation is Particle Image Velocimetry (PIV). This method allows determination of vector velocity fields and evaluation of flow vorticity [21] and also can be applied in the air-lift pump research [22].

Two-phase gas-liquid flows are considerably affected by the device's geometry and by the flow parameters. Research in this area is presented in [23], where the authors analyse the effects of tube diameter and submergence ratio on bubble flow patterns. The efficiency of the pump also changes with the change of temperature as demonstrated in the work [24]. The pumping performance and the effectiveness of an airlift pump have been investigated with the use of optical methods [25]. A slight difference in the flow parameters of both phases can cause significant changes in the flow regime [26]. Therefore, flow maps are used to help identify the structure and improve the efficiency of the airlift pump. Maps for two-phase flow in vertical channels are commonly known and used [27]. In contrast, for aeration pumps, maps of this type were developed by [12], and the correlation of flow structures and the efficiency of liquid transport was made [28,29]. Hydrodynamics of two-phase flows is concerned with the identification of the flow patterns and parameters experimentally and numerically [30,31]. In addition, they determine the overall efficiency of an airlift pump for the two-phase system [32], and three-phase systems [9].

The issues concerning the submergence ratio analysis in correlation to liquid phase velocity, efficiency and void fraction in the airlift pump are widely discussed problems [33]. The studies in [34] contain remarks indicating that airlift pumps operating at low submergence ratios are particularly vulnerable to the existence of such phenomena. In order to minimize the unfavorable influence of low submergence ratio, various design changes are applied. Figure 1 shows the schemes of submergence ratio ($\varepsilon = H_s/H$) in the airlift pump, which is an essential feature in the analysis of the energy efficiency and is commonly used as the ratio of the submergence depth of the pump (H_s) to its total height (H).

It seems to be difficult to establish an economically optimum condition for operational performance of the airlift pump since the theoretical model to predict the flow characteristics in full detail is still not complete [12]. The established dependence between various design parameters and two-phase flow regimes focuses on the optimization of the operating performance. The proposed method guarantees the finding of the efficiency optimization and stability of liquid transport throughout the operation of an airlift pump.

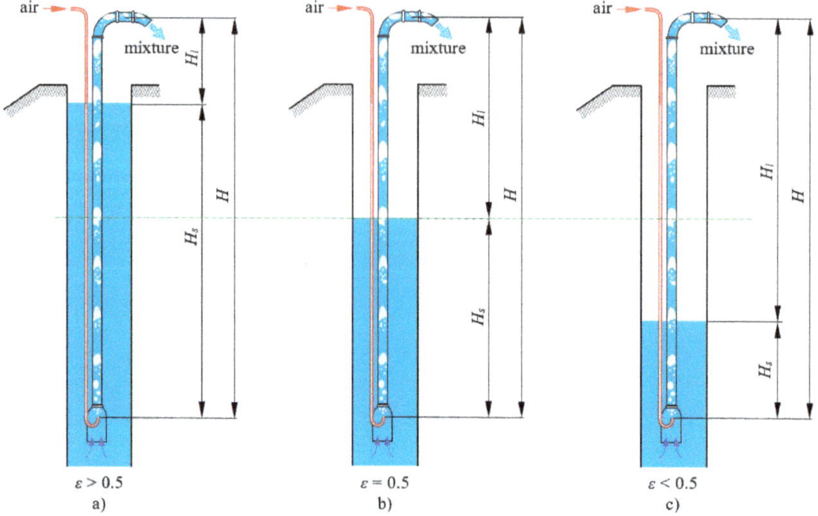

Figure 1. The scheme of submergence ratio (ε) in the airlift pumping systems. (**a**) high submergence ratio $\varepsilon > 0.5$; (**b**) medium submergence ratio $\varepsilon = 0.5$; (**c**) low submergence ratio $\varepsilon < 0.5$.

2. Measurement Setup

The conducted experiments involved the airlift pump with rectangular channel shown in Figure 2.

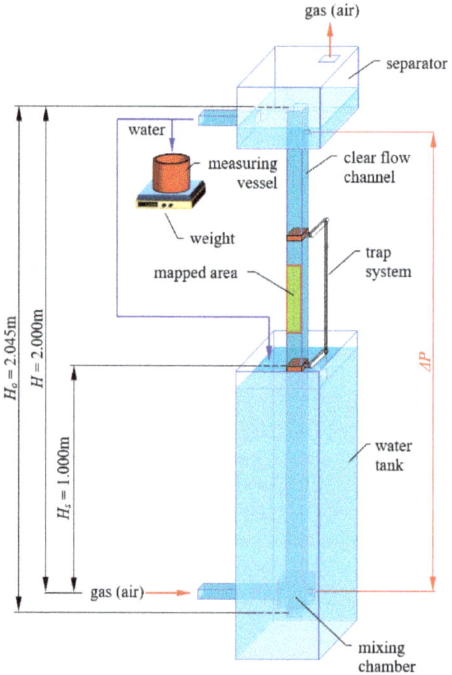

Figure 2. Measurement setup—airlift pumping system.

The measurement channel with the internal dimensions of 35 × 20 × 2045 mm was made of polymethyl methacrylate (PMMA). The riser pipe was equipped with an inlet with a gas distributor in form of the mixing chamber. The mixing chamber is equipped with the strainer. The gas distribution system enabled three gas phase distribution modes:

(1) feeding the gas phase through the perforated nozzle (eight holes with the diameter of 2 mm with a symmetrical distribution on two walls in the channel);
(2) feeding the gas phase through the slot nozzle (two slots with the dimensions of 2 × 20 mm);
(3) feeding the gas phase through the ceramic porous element (two porous bodies with the 45% porosity, particle size 300–500 μm and dimensions of 30 × 20 mm).

The selection of the total length of segments forming the riser of the airlift pump meant that the analyzed pump had a low submergence ratio (submergence ratio of ε = 0.5 see Figure 1b). The airlift pump applied liquid feed through an inlet that was submerged 1 m below the surface. After a lifting height of 1 m was achieved, the liquid was reversed in an overflow vessel to the liquid tank. In this way, the system applied a closed liquid circulation. The gas phase circulated in the open loop and was reversed into the atmosphere. The setup was also equipped with the metering of: gas phase flow rate using a rotameter, liquid phase flow rate using the weight method, pressure drop in airlift pump using a differential manometer, pressure of the gas phase at the inlet to the airlift pump using pressure transducer, gas void fraction using the trap method.

The optical path of the setup was formed by a HCC1000 series CMOS fast imaging camera (Vosskühler, Osnabrück, Germany) installed in the parallel plane of the measurement channel and a bright field as shown in Figure 3.

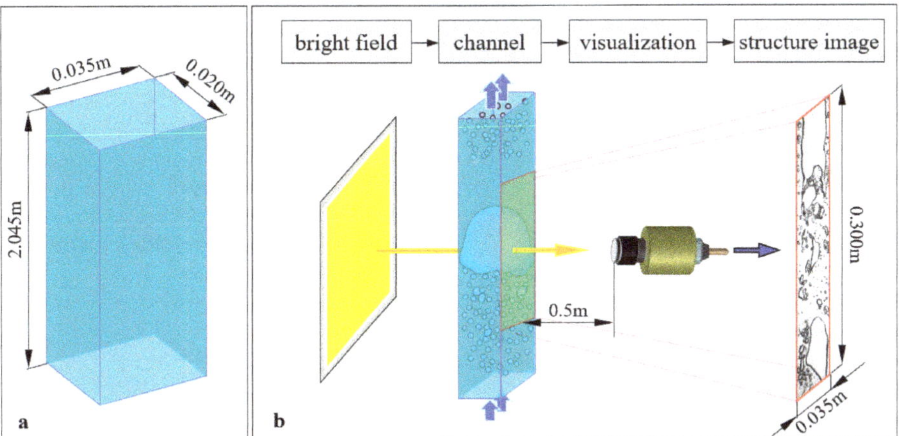

Figure 3. Measurement setup—visualization technique. (**a**) the flow channel dimensions; (**b**) the optical path.

The illumination of the measurement section was realized by using a light source of four 1000 W halogen reflectors. The halogens were connected to a four-channel Lite-Puter AX-415 controller (Lite Puter Enterprise Co., Ltd, Taipei, Taiwan). As a result, smooth and independent control of the operating parameters was provided. The registration of images applied photo captures with the resolution of 1024 × 256 pixels and two frequencies: 260 Hz for slow flow phenomena (i.e., for gas flow rates up to 4 m^3/h) and 465 Hz for fast occurring phenomena (i.e., for gas flow rates above 4 m^3/h). Table 1 provides details of the meters used in the research.

Table 1. Details of the meters used in the research.

Measurement Device	Description
Rotameter ROS-10	Measuring range: 0.11–1.1 m^3/h Standard accuracy class 2.5 according to PN-85/M-42371 Absolute error: 0.0275
Rotameter ROS-16	Measurement range: 0.5–5.0 m^3/h Standard accuracy class 2.5 according to PN-85/M-42371 Absolute error: 0.125
Rotameter RDN-25	Measurement range: 3.0–30.0 m^3/h Standard accuracy class 2.5 according to PN-85/M-42371 Absolute error: 0.75
Scale Radwag WPT5	Measuring range: 0.5–3000 g Accuracy: 0.5 g
PELTRON PXWD 0.2 Differential pressure transducer	Measuring range: 0–20 kPa Standard accuracy class 0.25 according to PN-85/M-42371 Absolute error: 50 Pa
Timer	Accuracy: 0.01 s

3. Methods

The proposed approach to the assessment of the influence of the two-phase flow regime on airlift pumping system performance applies digital image analysis of the phenomena in the riser pipe of the apparatus. The data derived from the imaging was used as the basic source of information for the further dynamic image analysis. The experiments involved three pump configurations which differed the type of gas distributor. The research data was a series of flow images recorded during variable flow rates of gas in the range from 0.2 to 15.0 m^3/h. Figure 4 presents the organizational diagram of the performed experiments.

Throughout the tests, characteristic fluctuations of the image's grey level were recorded for each registered image that is relative to the regime of the airlift pump. This occurred as a result of the spatial distribution of the gas and liquid phases. The captured relations between the grey level recorded by the imaging system and the registered flow regime led to the assessment of two-phase flow patterns. The above was achieved by application of the probability density function (PDF).

Figure 5 contains the reference PDF spectra corresponding to the variations in grey level depending on the flow regimes that were formed during the two-phase flow. Those are typical spectra for the generally accepted flow regime nomenclature, described in the literature [35,36]. Bubble and froth flows have a single maximum, with a note that the spectrum for froth flow is wider or its image is fuzzy. For the case of slug flow, the recorded images contain noticeable separated peaks corresponding to the liquid and gas phases. A similar spectrum is observed for the case of slug flow, yet in this case, the peaks are less clearly separated.

Because the flow in the two-phase flow depends on the flow resistance, the analysis of the pressure drop was connected with the two-phase regime on the stability and efficiency of the liquid transport. The knowledge of the gas void fraction R_G (also denoted in the literature as α) is necessary to calculate the pressure drop during two-phase flow for the energy efficiency assessment. The most commonly used method of determining the volume fraction α is the Zuber-Findlay method [37]. This is one of the drift flux methods, which allows the determination of the volume fraction ($\alpha = W_{GO}/W_G$) as the ratio of the apparent gas phase velocity to the actual gas phase velocity. A detailed discussion of this method has been included in the paper [38]. In the reported study, a derivative of the Zuber-Findlay method was applied for calculations of the gas void fraction, that is particularly recommended to use for airlift pumps by Reinemann et al. [39]. The void fraction of the gas was determined on the basis of the following equations:

$$R_G = \frac{Q'_G}{C_0 \cdot (Q'_L + Q'_G) + W'_{Ts}} \tag{1}$$

$$Q'_L = \frac{W_{LO}}{(g \cdot D)^{1/2}} \tag{2}$$

$$Q'_G = \frac{W_{GO}}{(g \cdot D)^{1/2}}$$

(3)

$$W'_{Ts} = 0.352 \cdot \left(1 - 318 \cdot \frac{\sigma}{\rho_L \cdot g \cdot D^2} - 14.77 \cdot \left(\frac{\sigma}{\rho_L \cdot g \cdot D^2} \right)^2 \right)$$

(4)

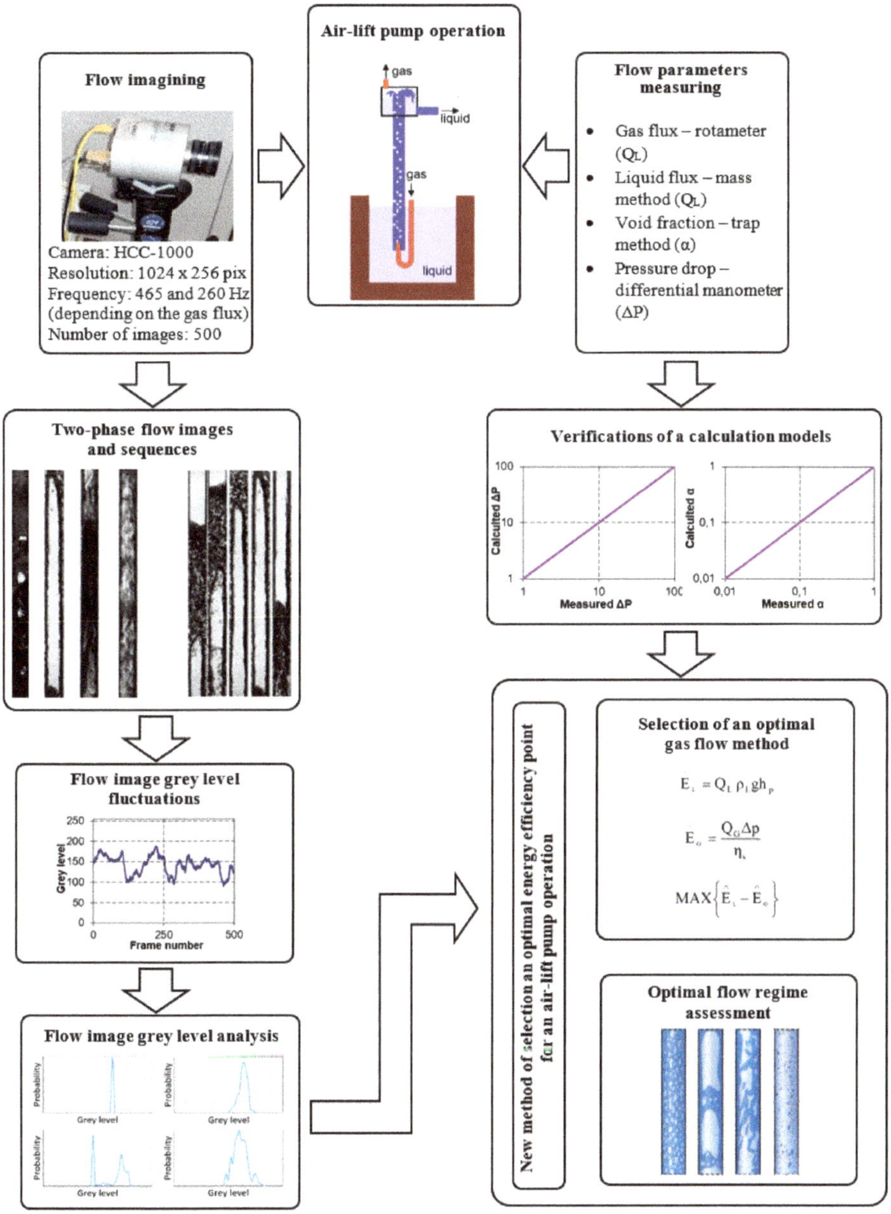

Figure 4. Methods used for selecting the airlift pump optimum efficiency at low submergence ratios.

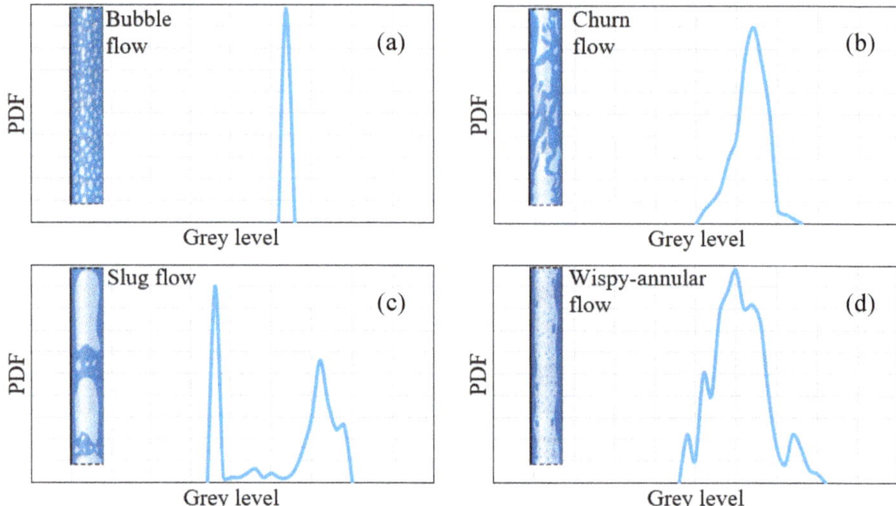

Figure 5. Reference spectra of the probability density function for two-phase flows [35]: (**a**) bubble flow, (**b**) churn flow, (**c**) slug flow, and (**d**) wispy-annular flow.

The value of the pressure drop was derived using the Lockhart-Martinelli method, as it is recommended for use in gas-liquid flows [40]. In this method, pressure drop ΔP_{2F} resulting from a flow of two-phase mixtures includes two components—hydrostatic pressure and friction pressure:

$$\Delta P_{2P} = \Delta P_{2P,H} + \Delta P_{2P,T} \tag{5}$$

In Equation (5), the pressure drops resulting from the hydrostatic pressure-induced changes are described by the following dependence:

$$\Delta P_{2P,H} = \rho_{2P}\cdot g\cdot H \tag{6}$$

where:

$$\rho_{2P} = R_G\cdot\rho_G + (1 - R_G)\cdot\rho_L \tag{7}$$

The friction pressure component was derived under the assumption that each of the phases flows separately through the entire cross-section of the channel riser:

$$\left(\frac{\partial P}{\partial z}\right)_{2P,T} = \left(\frac{\partial P}{\partial z}\right)_L \cdot \Phi_L^2 \tag{8}$$

$$\left(\frac{\partial P}{\partial z}\right)_{2P,T} = \left(\frac{\partial P}{\partial z}\right)_G \cdot \Phi_G^2 \tag{9}$$

where Φ_G and Φ_L are the equation multipliers, which were determined on the basis of the Lockhart–Martinelli parameter defined from the Equation (10) [40]:

$$X = \left(\frac{\left(\frac{\partial P}{\partial z}\right)_L}{\left(\frac{\partial P}{\partial z}\right)_G}\right)^{0.5} \tag{10}$$

In this manner, the calculation of the pressure drops involved the determination of the Lockhart-Martinelli parameter (X) and registration of the value of the multiplier Φ dedicated to the particular type of flow (laminar or turbulent) according to the chart shown in Figure 6. To carry out these calculations, it is necessary to know the additional Reynolds numbers for both phases.

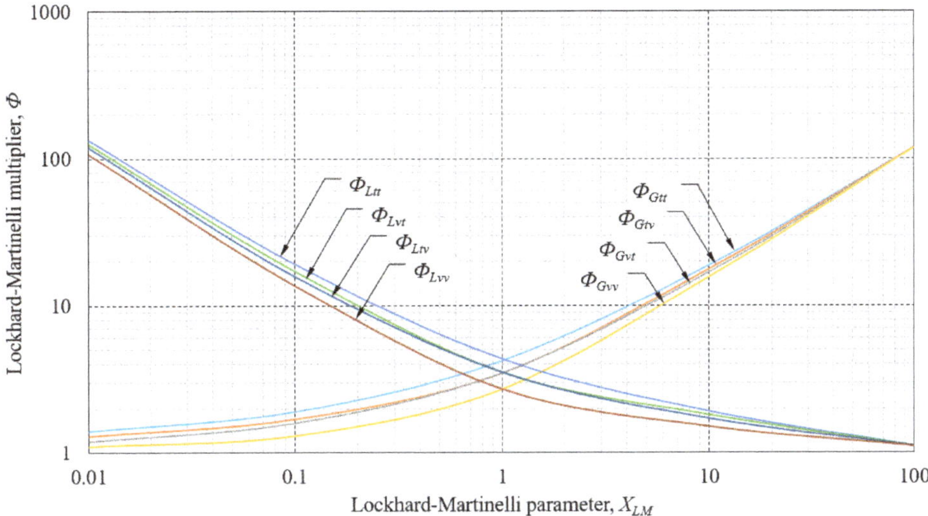

Figure 6. Chart for Lockhart-Martinelli corrections [40].

As a result, the specific pressure gradient was derived and related to the length of the riser pipe in the airlift pump H (see Figure 2) on the basis of the following relation:

$$\left(\frac{\Delta P}{H}\right)_i = \lambda_i \cdot \frac{w_{i,0}^2 \cdot \rho_i}{2 \cdot d} \tag{11}$$

For $Re_i \leq 2100$:

$$\lambda_i = \frac{64}{Re_i} \tag{12}$$

For $Re_i > 2100$:

$$\lambda_i = \frac{0.3164}{Re_i^{0.25}} \tag{13}$$

The model of assessing the operation of the airlift pump involved an assumption regarding the possibility of relating pressure drops, liquid transport efficiency and the type of two-phase flow regime. The optimization criterion that fulfills this assumption is based on the determination of the greatest difference between the energy flux needed to overcome the pressure drop and the useful energy flux of the liquid phase that is lifted to the lifting height H_L (see Figure 1).

The energy flux that results the gas phase flow is expressed by the relation:

$$\dot{F}_G = \frac{Q_G \cdot \Delta P}{\eta_s} \tag{14}$$

The energy flux resulting from the liquid phase flux is defined as follows:

$$\dot{E}_L = Q_L \cdot \rho_L \cdot g \cdot h_p \tag{15}$$

To achieve a correlation between the energy flux at the input to the system (resulting from gas phase flow) and the energy flux at the output of the system (resulting from liquid phase flow), the standardization of these parameters was performed in relation to their maximum values. Finally, the optimization criterion was defined as:

$$MAX\{\hat{E}_L - \hat{E}_G\} \tag{16}$$

The idea of the developed optimum regime selection for the studied airlift pump is illustrated in Figure 7. The idea of selecting a gas flux that guarantees optimal energy efficiency is not a new method in general, whereas in the case of optimization of the airlift pump has not been described so far in the literature. Therefore, it is considered a valuable method for optimizing this type of equipment.

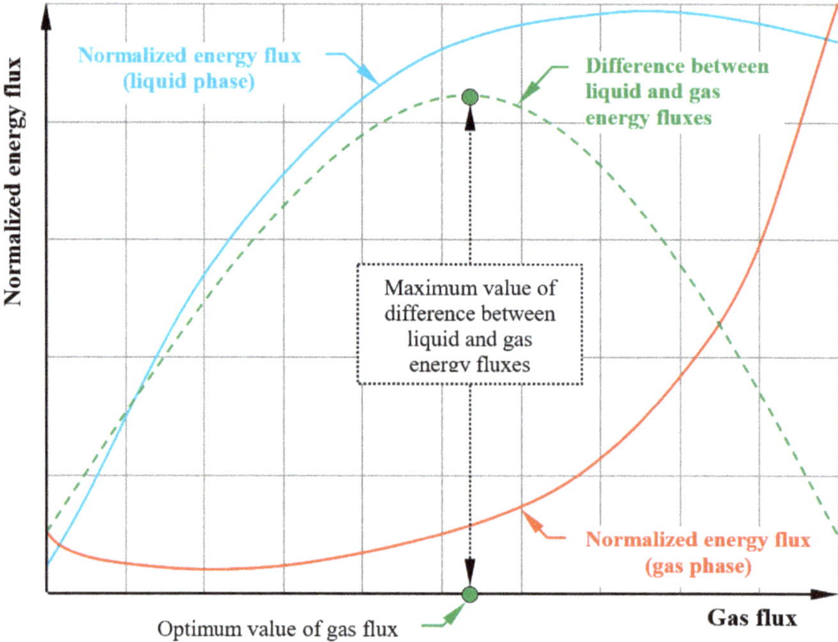

Figure 7. The idea of an optimization method for an airlift pump operation.

With the purpose of verifying the proposed method of optimizing the airlift pump, the analysis of the efficiency of the apparatus was conducted with regard to the flow rates of the phases at the inlet. The calculations used the dependence derived by Reinemann et al. [39] (Equation (17)) which depends on the basis of dimensionless volumetric gas flow:

$$\eta = \frac{Q'_L}{C_0 \cdot (Q'_L + Q'_G) + W'_{Ts} - Q'_G} \tag{17}$$

where the value of C_0 coefficient depends on the type of the flow (laminar or turbulent) and has a value of 1.2 and 2.0, respectively.

From the point of view of optimizing the energy consumption, the most important need is to get as much water as possible with the smallest possible gas flux. For the performance analysis, due to the fact that the liquid flux was measured by the weight method, an uncertainty calculation was made for it as for functional dependencies. As it turns out that the biggest uncertainty obtained does not exceed 1%.

4. Results and Discussion

The initial point in the present study involved the identification of two-phase gas-liquid flow patterns. The use of an optical method provided the visualization of phase concentrations. Table 2 contains examples of visualizations with the fluctuations of the image's grey level.

Table 2. Flow pattern and grey level fluctuation in an airlift pump.

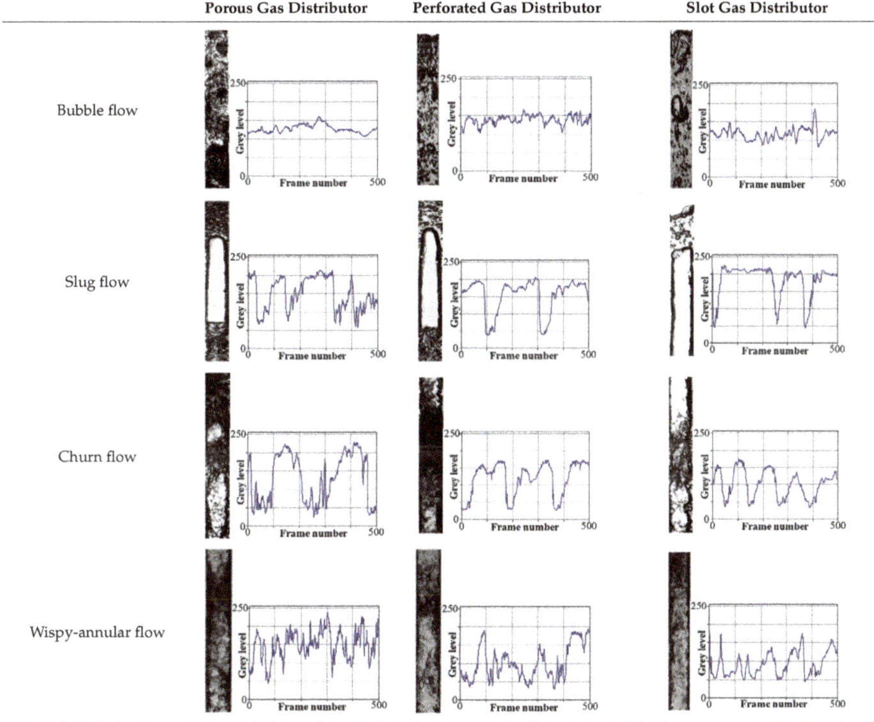

In Table 2 one can see how different distributors behave for a specific flow structure. It was demonstrated that different two-phase flow regimes can occur under identical flow conditions, yet, under the condition that other types of gas distribution systems are used. Similar results are reported in other papers [26,41,42]. This fact confirms the considerable effect of the gas phase distributor on two-phase flow regimes. For the case of airlift pumps, where the driving momentum is generated by the gas phase, the adequate selection of the gas distributor geometry plays a fundamental role. Thus, airlift pump can be considered as a device whose operation is determined by the role of the air distribution system. Which directly results from the research carried out.

Probability density functions (PDF) were calculated for the image's grey levels of each of the flow regimes with the purpose of identifying the flow patterns according to [35,36] and resulted in the courses shown in Figure 8.

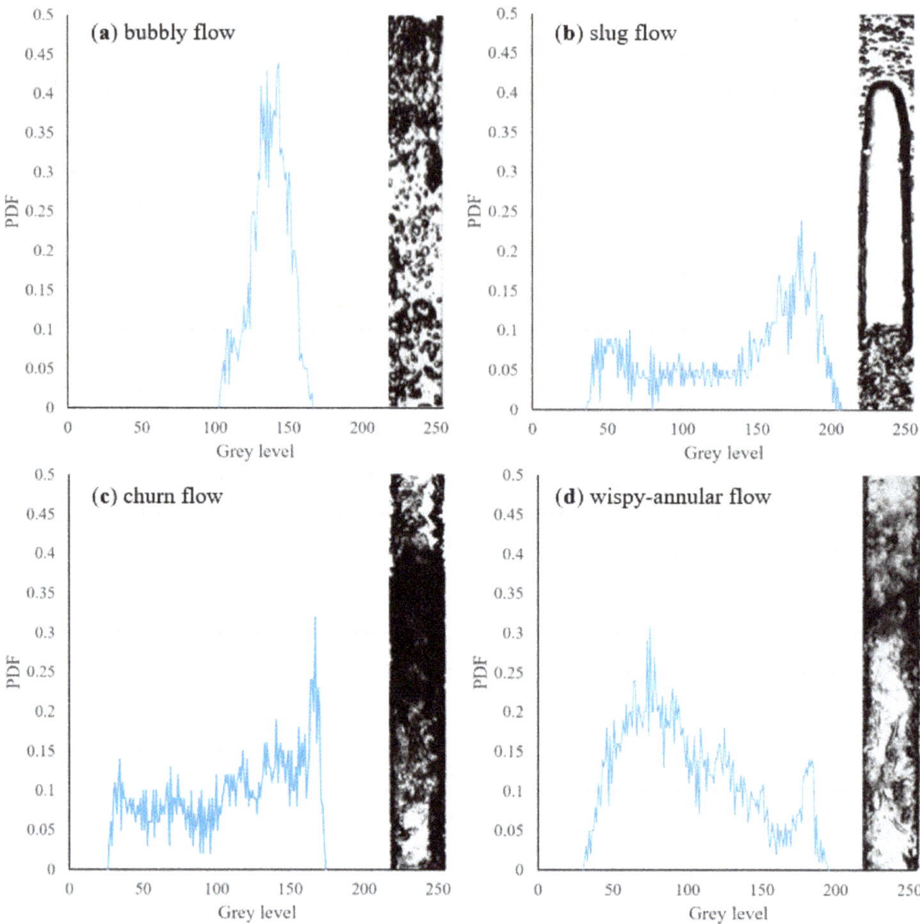

Figure 8. Probability density function spectrum for different flow regimes through an airlift pump: (**a**) bubbly flow; (**b**) slug flow; (**c**) churn flow; (**d**) wispy-annular flow.

The subsequent step involved a comparison between the resulting function spectra shown in Figure 5 with the reference (see Figure 3). The comparative analysis offers a conclusion regarding considerable discrepancies resulting from the singularities of two-phase flow through an airlift pump.

The occurrence of large volumes of small bubbles, whose presence is specific both to the froth flow and to the plug and slug regimes contributes to the deformations of the PDF spectrum. The approach that offered the possibility to identify heterogeneous two-phase flow regimes applied an analysis involving the visual verification of the registered grey. The results are shown in Figure 9. The black line represents the course of the fluctuations of the grey level in the model plug flow. Such fluctuations could provide high reliability of the application of the method using the assessment of the probability function spectrum.

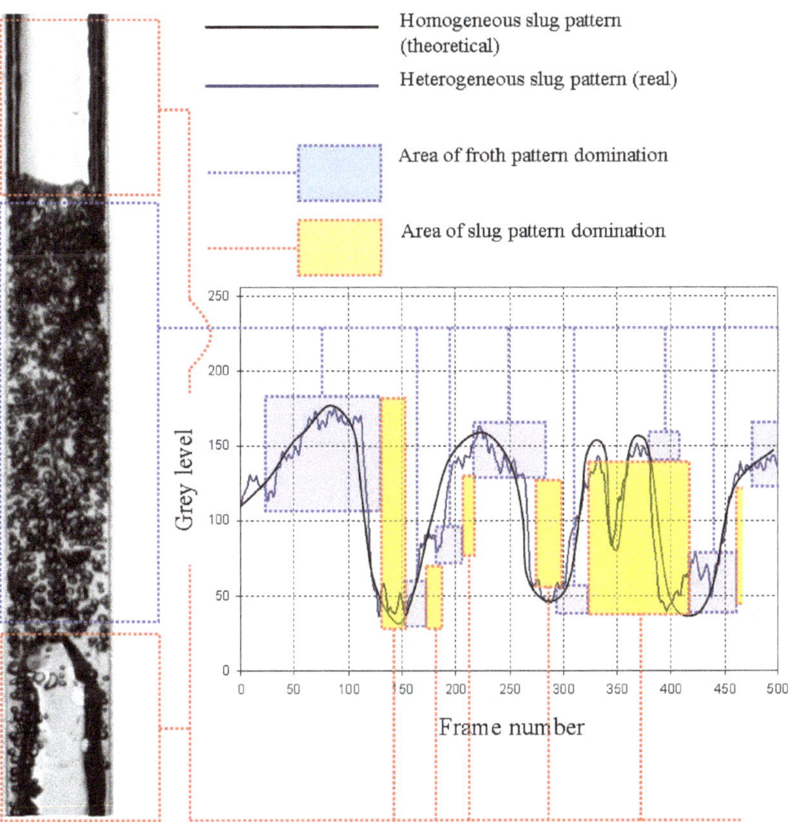

Figure 9. Influence of heterogeneous flow pattern on grey level fluctuations.

In accordance with the methods applied in this study, the void fraction was derived on the basis of Equation (1). The results of the calculations of gas phase void fraction were compared with the values measured by the trapping method and the results can be found in Figure 10. For all measurement series, 67% of the measurement points are found within the range of the relative error of 25%, which is an admissible result. This is a common error for 2P flows. The greatest level of the conformity between the results, i.e., 80% was established for the case of the porous distributor. In comparison, the values of 58% and 55% were established for the perforated and slotted distributors, respectively. Due to the dependence of the void fraction and two-phase mixture density, a similar level of conformity between the calculations and measurements was obtained for pressure drops as well (see Figure 10). The results of the void fractions offer a conclusion regarding the application of the adequate computational model for the conditions of the flow in the analyzed airlift pump (see Table 3).

In the present study, the pressure drop was one of the elements applied in the assessment of the energy flux delivered to the examined airlift pump. The determination of pressure drops applied the Lockhart–Martinelli method [40]. Such an approach offers satisfactory results with regard to two-phase gas-liquid mixture flow. An additional advantage of the applied method is a wide range of flow patterns. This plays a fundamental role in airlift pumps, where flow turbulence and disturbances appear. Figure 10 contains also the results of measured and calculated values of the pressure drops. The greatest discrepancies between the calculations and measurements are found within the initial operating of the airlift pump. This is caused by the low liquid flow rate at the outlet of the riser pipe. The characteristics of the liquid transport in an airlift pump involves liquid circulation in the riser pump until the designed transport velocity of the liquid is obtained. Within this initial measurement range, the smallest error was recorded for the slotted distributor, whereas the greatest for the porous one. As we mentioned, this correlates well with the results of measurement regarding the void fraction. We can also remark here that the slotted distributor generates the smallest local pressure drop and creates the conditions of the liquid transport development very late in the process. In contrast, the porous distributor operating for low parameters of the liquid phase provides the greatest transport efficiency.

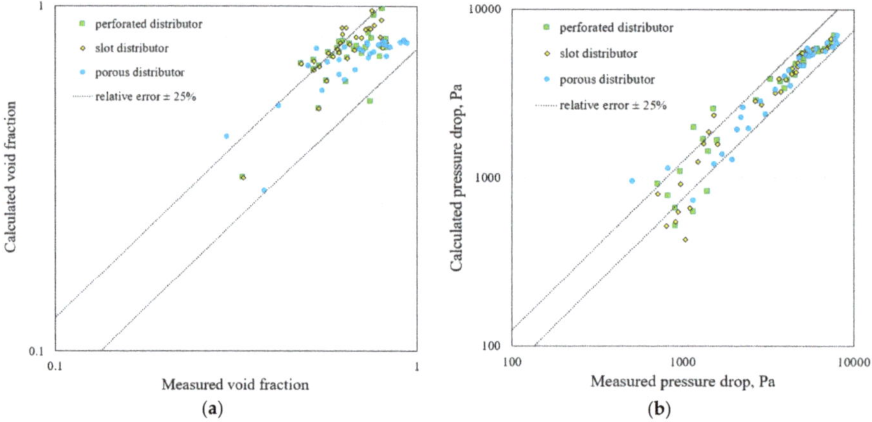

Figure 10. Comparison between the (**a**) measured with the calculated void fraction in accordance with the Reinemann drift-flux model and (**b**) comparison of measured and calculated pressure drop.

Table 3. Impact of gas distributor for an airlift pump operating.

Type of Gas Distributor	Impact on Two-Phase Flow Mixture	Impact on Pressure Drop	Impact on Void Friction
Porous	The high intensity of the heterogeneous flow High stability of liquid transport A small number of local circulations and vortices in the liquid phase	High values of pressure drops Greatest validity of a Lockhart-Martinelli model Exceeding the error limit in the small and medium gas flux range	Greatest validity of a Reinemann model Exceeding the error limit in the small and medium gas flux range
Perforated	Medium intensity of the heterogeneous flow Medium stability of liquid transport A small number of local circulations and vortices in the liquid phase for bubble structures, significant for other structures	Medium values of pressure drops Medium validity of a Lockhart-Martinelli model Exceeding the error limit in the small and medium gas flux range	Medium validity of a Reinemann model Exceeding the error limit in the small and medium gas flux range
Slot	Low intensity of the heterogeneous flow Low stability of liquid transport A significant number of local circulations and vortices in the liquid phase	Low values of pressure drops Medium validity of a Lockhart-Martinelli model Exceeding the error limit in the small and medium gas flux range	Medium validity of a Reinemann model Exceeding the error limit in the high gas flux range

The internal circulation of the liquid phase in the riser was minimized as a consequence of using a porous distributor. The perforated distributor was characterized by intermediate characteristics in comparison to the two other designs discussed above. The presented discrepancies cease to exist when the effective liquid lifting starts. Except for the initial operating range of the airlift pump, where hydrodynamic singularities are observed associated with the special characteristics of two-phase flow, the conformity of the measured and calculated values of the pressure drop is fully acceptable (See Figure 10). This remark is relevant in the context of the simplicity and the general characteristics of the Lockhart–Martinelli calculation method. The presented method fulfills the condition that makes it applicable to the calculations aimed at optimization of the airlift pumps.

The identification of an adequate operating regime of an airlift pump represents the key parameter which determines the exploitation cost of the apparatus. In the conducted study, the operating cost goal was achieved by the using energy fluxes of the phase mixture to determine the energy efficiency of the vertical liquid transport. It was observed that as a result of applying a common coordinate system to visualize the coefficients of the energy parameter, it is possible to determine a region in which the goal of liquid transport maximization can be accompanied by energy minimization.

The objective of the presented method was to identify the point that offers a compromise between the maximization of liquid lifting and minimization of energy input. This objective was realized as a result of testing the variability of the function representing the difference between the normalized energy fluxes of the two phases. The point in which a maximum difference occurred was assigned with the energy flux of the gas phase corresponding to it so as to guarantee the optimum operating regime of the airlift pump.

The selection of optimum gas phase flow rates is presented in Figure 8. The optimum operating range of the perforated and slotted distributors is found in the range of the plug flow regime. For the porous distributor, this range shifts in the direction of the greater flow rates and are located in the slug flow regime. However, the slug regime is optimum due to the greater normalized flow rate of the transported liquid phase ($\Delta E = 0.1$) despite the greater amount of energy needed to pump the liquid. The effect was reflected in the increased stability of the liquid transport over time. For the perforated and slotted distributors, plug flow was characterized by intensive pulsations and existence of plugs in the flow accompanied by considerable cyclic changes in the efficiency of the liquid transport. The fluctuations in the charts representing liquid flow rates were smoother. This fact is also reflected in the analysis of the fluctuations of the grey level of the two slug and churn patterns described in Table 2.

The results of the finding of an optimum operating regime selection of the airlift pump are presented in Figure 11. This graph shows flow characteristics with the optimum energy points' localization. The overall performance of the airlift pump was determined on the basis of Equation (17) and the optimum efficiency from Equation (16). In Figure 12, the optimum efficiency (marked with ●) is located between the points representing maximum efficiency (marked with ■) and maximum pump capacity (marked with ▲).

The efficiency analysis depends on the gas flow rate and the distributor type. The maximum efficiency of the aeration pump is not identical to the evenness of liquid transport. Most often, at the highest efficiency, the liquid flow is pulsating. And that is why we propose using the optimum efficiency, which in this case is better than the maximum performance.

The construction of a gas distributor should ensure the maximum amount of bubbles in the widest range of the gas stream, resulting in the largest liquid transport. Despite the fact that in our case the porous distributor was characterized by the highest pressure drop, it generated the largest liquid transport for the largest range of gas flux. In addition, the point corresponding to the best performance of the apparatus does not correspond to the highest efficiency of liquid transport. For gas distributors characterized by an increased pressure drop (such as porous distributors), the optimum operating regime shifts towards the maximum efficiency. For this reason, this type of distributor can be considered as optimum for airlift pumps operating at low submergence ratios.

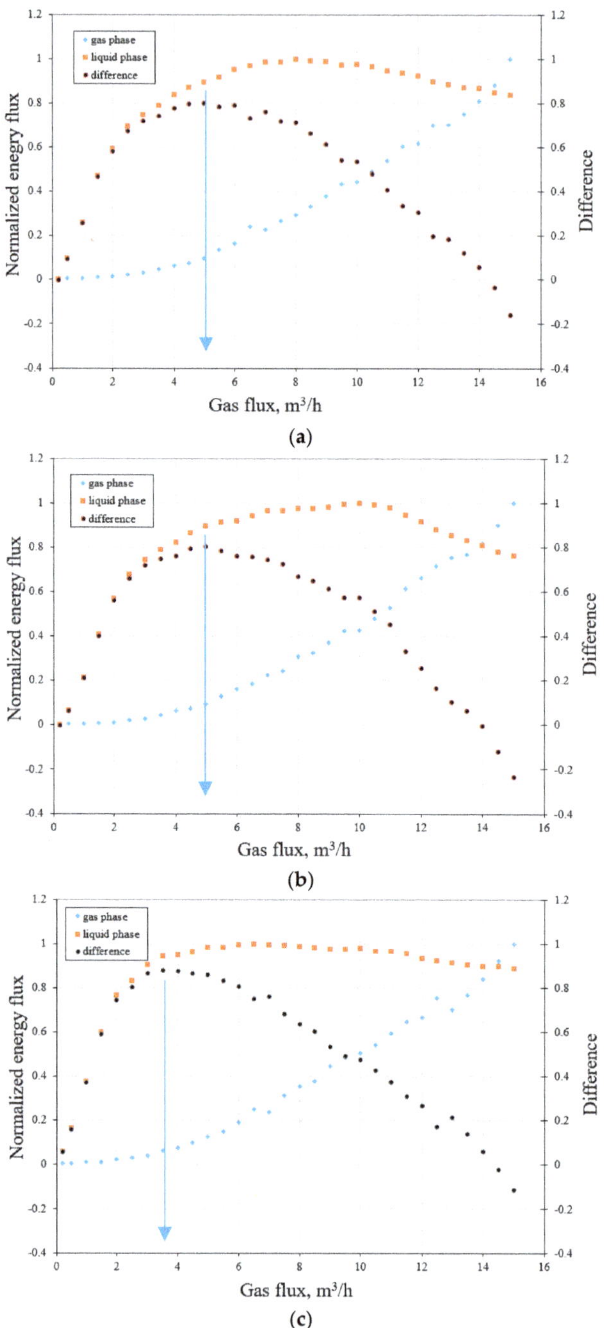

Figure 11. Optimum flow parameters for an airlift pump for $H_l = 1.0$ m and different gas distributors: (**a**) perforated distributor; (**b**) slotted distributor; (**c**) porous distributor.

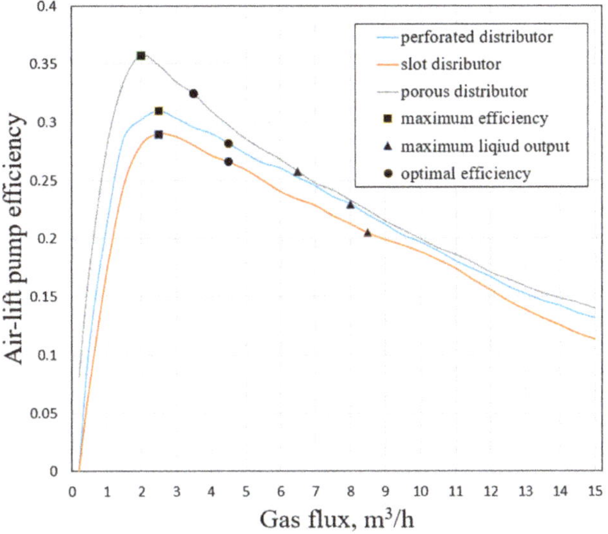

Figure 12. Findings of the proposed method for selecting an optimum operating regime of the airlift pump.

5. Conclusions

The study was concerned with the description of hydrodynamics and energy efficiency of airlift pumps. The analysis of the operation of the existing apparatus the assessment of the operating parameters of the pump with the use of an optimization algorithm was applied.

The following methods were used to develop the method of analysis: analysis of the grey level of the image, trap method for measuring gas volume, measurement of pressure drop by means of a differential manometer, measurement of liquid phase strips by gravimetric method. In addition, calculation methods such as the Lockhart-Martinelli method and the drift-flux method were used. The energy flow fed into the system with the potential energy stream at the given height was correlated. The experiments led to the following conclusions:

Fluctuations of the grey level of the registered images corresponding to the two-phase mixture flow in the riser of an airlift pump are reflected in the variable pump performance and continuity of the liquid transport. The determination of the normalized fluxes of the liquid flow demonstrates the important role of the system of the gas phase distribution. The type of gas distributor needs to be accounted at the design stage, in particular in the apparatus operating under low submergence ratios.

The proposed method for the selection of the airlift pump operating parameters offers the determination of optimum energy efficiency (see ● in Figure 12) at low submergence ratios.

The operating regime of the apparatus within the discussed flow parameters offers the best compromise between the liquid lifting efficiency and exploitation cost of the system (i.e., the cost of generating the air feeding the system).

Author Contributions: Data curation, M.M.; Investigation, G.L.; Methodology, G.L.; Resources, D.Z.; Software, D.Z.; Supervision, S.A.; Visualization, M.M.; Writing—Original Draft, G.L. and D.Z.; Writing—Review & Editing, M.M. and S.A.

Funding: This research received no external funding.

Conflicts of Interest: The authors declare no conflict of interest.

Nomenclature

λ	coefficient of resistance dependent on the Reynolds number (Re) and the relative roughness of the pipe, -
ε	submergence ratio, -
η	airlift pump overall efficiency, -
η_s	compressor efficiency, -
ρ_L	liquid density, kg/m^3
ρ_{2P}	two phase mixture density, kg/m^3
ρ_G	gas density, kg/m^3
σ	liquid phase surface tension, N/m
C_0	liquid slug velocity profile coefficient ($C_0 = 1.2$ for high Re, $C_0 = 2.0$ for low Re)
D	riser pipe diameter, m
$\dot{E}_L$	energy flux resulting from the liquid phase, W
$\dot{E}_G$	energy flux resulting from the gas phase, W
$\hat{E}_L$	normalized energy flux (liquid phase), -
$\hat{E}_G$	normalized energy flux (gas phase), -
g	gravitational acceleration, m/s
Φ_G	Lockhart-Martinelli equation multiplier for gas phase, -
Φ_{Gtt}	L-M multiplier for the gas phase, where the liquid phase is turbulent and the gas phase is also turbulent, -
Φ_{Gtv}	L-M multiplier for the gas phase where the liquid phase is turbulent but the gas phase is laminar (viscous), -
Φ_{Gvt}	L-M multiplier for the gas phase where the liquid phase is laminar (viscous) but the gas phase is turbulent, -
Φ_{Gvv}	L-M multiplier for the gas phase where the liquid phase is laminar (viscous) and the gas phase also laminar (viscous), -
Φ_L	L-M multiplier for liquid phase, -
Φ_{Ltt}	L-M multiplier for the liquid phase where the liquid phase is turbulent and the gas phase is also turbulent, -
Φ_{Ltv}	L-M multiplier for the liquid phase where the liquid phase is turbulent but the gas phase is laminar (viscous), -
Φ_{Lvt}	L-M multiplier for the liquid phase where the liquid phase is laminar (viscous) but the gas phase flowing alone in the channel is turbulent, -
Φ_{Lvv}	L-M multiplier for the liquid phase where the liquid phase is laminar (viscous) and the gas phase is also laminar (viscous), -
H	total lift, m
H_l	pump lift height, m
Ho	overall height of the pump, m
H_s	immersion depth, m
i	respective phase (i.e., gas or liquid) as subscript
R_G	Void friction, -
ΔP	total pressure drop related to gas flux considering two-phase flow,
$\Delta P_{2F,H}$	pressure drop caused by a change in hydrostatic energy,
$\Delta P_{2F,T}$	pressure drop caused by friction,
Q_G	volume of gas, m^3/h
Q'_L	dimensionless liquid flow coefficient, -
Q_L	volume of liquid, m^3/h
Q'_L	dimensionless gas flow coefficient, -
W'_{Ts}	dimensionless coefficient of plug movement in stationary liquid, -
W_G	gas velocity, m/s
W_{GO}	superficial velocity of gas phase, m/s
W_{LO}	superficial velocity of liquid phase, m/s

References

1. Hanafizadeh, P.; Ghorbani, B. Review study on airlift pumping systems. *Multiph. Sci. Technol.* **2012**, *24*, 323–362. [CrossRef]
2. Kumar, E.A.; Kumar, K.R.V.; Ramayya, A.V. Augmentation of airlift pump performance with tapered upriser pipe. An experimental study. *J. Mech. Eng. Div.* **2003**, *84*, 114–119.
3. Yoo, K.H.; Masser, M.P.; Hawcroft, B.A. An in-pond raceway system incorporating removal of fish waste. *Aquac. Eng.* **1994**, *14*, 175–187. [CrossRef]
4. Loyless, J.C.; Malone, R.C. Evaluation of air-lift pump capabilities for water delivery, aeration, and degasification for application to recirculating aquaculture systems. *Aquac. Eng.* **1988**, *18*, 117–133. [CrossRef]
5. Ahmed, W.H.; Aman, A.M.; Badr, H.M.; Al-Qutub, A.M. Air injection methods: The key to better performance of airlift pump. *Exp. Therm. Fluids Sci.* **2016**, *70*, 354–365. [CrossRef]
6. Qiang, Y.; Fan, W.; Xiao, C.; Pan, Y.; Chen, Y. Effects of operating parameters and injection method on the performance of an artificial upwelling by using airlift pump. *Appl. Ocean Res.* **2018**, *78*, 212–222. [CrossRef]
7. Deendarlianto; Supraba, I.; Majid, A.I.; Pradecta, M.R.; Indarto; Widyaparaga, A. Experimental investigation on the flow behavior during the solid particles lifting in a micro-bubble generator type airlift pump system. *Case Stud. Therm. Eng.* **2019**, *13*, 100386. [CrossRef]
8. Kassab, S.Z.; Kandil, H.A.; Warda, H.A.; Ahmed, W.H. Experimental and analytical investigations of airlift pumps operating in three-phase flow. *Chem. Eng. J.* **2007**, *131*, 273–281. [CrossRef]
9. Fujimoto, H.; Nagadani, T.; Takuda, H. Performance characteristics of a gas-liquid-solid airlift pump. *Int. J. Multiph. Flow* **2005**, *31*, 1116–1133. [CrossRef]
10. Masiukiewicz, M.; Anweiler, S. Two-phase flow phenomena assessment in minichannels for compact heat exchangers using image analysis methods. *Energy Convers. Manag.* **2015**, *104*, 44–54. [CrossRef]
11. Qi, D.; Zou, H.; Ding, Y.; Luo, W.; Yang, J. Engineering Simulation Tests on Multiphase Flow in Middle-and High-Yield Slanted Well Bores. *Energies* **2018**, *11*, 2591. [CrossRef]
12. Samaras, V.C.; Margaris, D.P. Two-phase flow regime maps for air-lift pump vertical upward gas-liquid flow. *Int. J. Multiph. Flow* **2005**, *31*, 757–766. [CrossRef]
13. Apazidis, N. Influence of buble expansion and relative velocity on the performance and stability of an airlift pump. *Int. J. Multiph. Flow* **1985**, *11*, 459–475. [CrossRef]
14. Wasilewski, M. Analysis of the effects of temperature and the share of solid and gas phases on the process of separation in a cyclone suspension preheater. *Sep. Purif. Technol.* **2016**, *168*, 114–123. [CrossRef]
15. Pietrzak, M.; Placzek, M.; Witczak, S. Comparison of pressure drop correlations for two-phase flow in small diameter channels. *Przem. Chem.* **2017**, *96*, 592–597.
16. Ganat, T.A.; Hrairi, M. Gas–Liquid Two-Phase Upward Flow through a Vertical Pipe: Influence of Pressure Drop on the Measurement of Fluid Flow Rate. *Energies* **2018**, *11*, 2937. [CrossRef]
17. Pietrzak, M.; Witczak, S. Flow patterns and void fractions of phases during gas–liquid two-phase and gas–liquid–liquid three-phase flow in U-bends. *Int. J. Heat Fluid Flow* **2013**, *44*, 700–710. [CrossRef]
18. Sikora, M. Flow structures during refrigerants condensation. *J. Mech. Energy Eng.* **2017**, *1*, 101–106.
19. Anweiler, S. Development of videogrammetry as a tool for gas-particle fluidization research. *J. Environ. Manag.* **2017**, *203*, 942–949. [CrossRef]
20. Anweiler, S.; Ulbrich, R. Flow pattern for different fluidization apparatuses. *Inzynieria Chemiczna i Procesowa* **2004**, *25*, 577–582.
21. Ligus, G.; Wasilewski, M. Impact of stirrer rotational speed on liquid circulation in a rectangular vessel—A study applying DPIV. *E3S Web Conf.* **2018**, *44*, 00096. [CrossRef]
22. Wang, Z.; Kang, Y.; Wang, X.; Wu, S.; Li, X. Investigation of the hydrodynamics of slug flow in airlift pumps. *Chin. J. Chem. Eng.* **2018**, *26*, 2391–2402. [CrossRef]
23. Kim, S.H.; Sohn, C.H.; Hwang, J.Y. Effect of tube diameter and submergence ratio on bubble pattern and performance of airlift pump. *Int. J. Multiph. Flow* **2014**, *58*, 195–204. [CrossRef]
24. Oueslati, A.; Megriche, A. The effect of liquid temperature on the performance of an airlift pump. *Enrgy Procedia* **2017**, *119*, 693–701. [CrossRef]
25. Abed, K.A. Operational criteria of performance of airlift pump. *J. Inst. Eng.* **2003**, *84*, 92–94.
26. Deckwer, W.D. *Bubble Column Reactors*; John Wiley and Sons: New York, NY, USA, 1991.

27. Mishima, K.; Hibiki, T. Some characteristics of air-water two-phase flow in small diameter vertical tubes. *Int. J. Multiph. Flow* **1996**, *22*, 703–712. [CrossRef]
28. Kassab, S.; Kandil, H.A.; Warda, H.A.; Ahmed, W.H. Air-lift pump characteristics under two-phase flow conditions. *Int. J. Heat Fluid Flow* **2009**, *30*, 88–98. [CrossRef]
29. Hanafizadeh, P.; Ghanbarzadeh, S.; Saidi, M.H. Visual technique for detection of gas-liquid two-phase flow regime in the airlift pump. *J. Pet. Sci. Eng.* **2011**, *75*, 327–335. [CrossRef]
30. Borsuk, G.; Pochwala, S.; Wydrych, J. Numerical methods in processes of design and operation in pneumatic conveying systems. In *Engineering Mechanics Book Series: Engineering Mechanics*; Zolotarev, I., Radolf, V., Eds.; Institute of Thermomechanics, Academy of Sciences of the Czech Republic: Prague, Czech Republic, 2016; pp. 83–86.
31. Wasilewski, M.; Duda, J. Multicriteria optimisation of first-stage cyclones in the clinker burning system by means of numerical modelling and experimental research. *Powder Technol.* **2016**, *289*, 143–158. [CrossRef]
32. De Cachard, F.; Delhaje, J.M. A slug-churn flow model for small-diameter airlift pumps. *Int. J. Multiph. Flow* **1996**, *22*, 627–649. [CrossRef]
33. Tighzert, H.; Brahimi, M.; Kechround, N.; Benabbas, F. Effect of submergence ratio on the liquid phase velocity, efficiency and void fraction in air-lift pump. *J. Pet. Sci. Eng.* **2013**, *110*, 155–161. [CrossRef]
34. Tramba, A.; Topalidou, A.; Kastrinakis, E.G.; Nychas, S.G. Visual study of an airlift pump operating at low submergance ration. *Can. J. Chem. Eng.* **1995**, *73*, 755–764. [CrossRef]
35. Lubbesmeyer, D.; Leoni, B. Fluid-velocity measurements and flow pattern identification by noise-analysis of light-beam signals. *Int. J. Multiph. Flow* **1983**, *9*, 665–679. [CrossRef]
36. Lubbesmeyer, D. Possibilities of flow-pattern identification by noise techniques. *Prog. Nucl. Energy* **1982**, *9*, 13–21. [CrossRef]
37. Zuber, N.; Findlay, J.A. Average volumetric concentration in two-phase flow systems. *J. Heat Transf.* **1965**, *87*, 453–468. [CrossRef]
38. Masiukiewicz, M.; Anweiler, S. Two-phase flow structure assessment based on dynamic image analysis. *Flow Meas. Instrum.* **2019**, *65*, 195–202. [CrossRef]
39. Reinemann, D.J.; Parlange, J.Y.; Timmons, M.B. Theory of small-diameter airlift pump. *Int. J. Multiph. Flow* **1990**, *16*, 113–122. [CrossRef]
40. Lockhart, R.W.; Martinelli, R.C. Proposed Correlation of data for isothermal two-phase flow. *Chem. Eng. Prog.* **1949**, *45*, 39–48.
41. Masiukiewicz, M.; Zając, D.; Ulbrich, R. Image analysis applied to pattern recognition of two-phase flow in two-dimensional bed. *Inzynieria Chemiczna i Procesowa* **2001**, *22*, 917–922.
42. Ulbrich, R.; Krótkiewicz, M.; Szmolke, N.; Anweiler, S.; Masiukiewicz, M.; Zajac, D. Recognition of two-phase flow patterns with the use of dynamic image analysis. *Proc. Inst. Mech. Eng. Part E J. Process Mech. Eng.* **2002**, *216*, 227–233. [CrossRef]

Article

A New Method Based on Thermal Response Tests for Determining Effective Thermal Conductivity and Borehole Resistivity for Borehole Heat Exchangers

Aneta Sapińska-Sliwa [1], Marc A. Rosen [2], Andrzej Gonet [1], Joanna Kowalczyk [3] and Tomasz Sliwa [1,*]

[1] Faculty of Drilling, Oil and Gas, Department of Drilling and Geoengineering, Laboratory of Geoenergetics, AGH University of Science and Technology (AGH UST), al. Mickiewicza 30, 30-059 Krakow, Poland; ans@agh.edu.pl (A.S.-S.); gonet@agh.edu.pl (A.G.)

[2] Faculty of Engineering and Applied Science, University of Ontario Institute of Technology, 2000 Simcoe Street North, Oshawa, ON L1H 7K4, Canada; marc.rosen@uoit.ca

[3] Faculty of Mathematics and Natural Sciences, University of Rzeszow, Al. Rejtana 16c, 35-959 Rzeszów, Poland; jkowalcz@univ.rzeszow.pl

* Correspondence: sliwa@agh.edu.pl; Tel.: +48-12-617-22-17

Received: 10 January 2019; Accepted: 17 March 2019; Published: 20 March 2019

Abstract: Research on borehole heat exchangers is described on the development of a method for the determination, based on thermal response tests, of the effective thermal conductivity and the thermal resistivity for borehole heat exchangers. This advance is important, because underground thermal energy storage increasingly consists of systems with a large number of borehole heat exchangers, and their effective thermal conductivities and thermal resistivities are significant parameters in the performance of the system (whether it contains a single borehole or a field of boreholes). Borehole thermal energy storages provide a particularly beneficial method for using ground energy as a clean thermal energy supply. This benefit is especially relevant in cities with significant smog in winter. Here, the authors describe, in detail, the development of a formula that is a basis for the thermal response test that is derived from Fourier's Law, utilizing a new way of describing the basic parameters of the thermal response test, i.e., the effective thermal conductivity and the thermal resistivity. The new method is based on the resistivity equation, for which a solution giving a linear regression with zero directional coefficient is found. Experimental tests were performed and analyzed in support of the theory, with an emphasis on the interpretation differences that stem from the scope of the test.

Keywords: geoenergetics; ground source heat pumps; borehole heat exchangers; thermal response test; borehole thermal energy storage

---

1. Introduction

A significant increase of new heating and heating/cooling installations that is based on heat pumps and borehole heat exchangers (BHE) has been recently observed in many countries, including Switzerland [1], Germany [2], Sweden [3], Canada [4], and the United States [5]. Borehole thermal energy storage (BTES) permits the extraction of heat from the ground for heating in winter and the extraction of cool (i.e., the input of heat) for air conditioning in summer [6,7]. A BTES is a type of geoenergetic system, which also includes energy systems that are based on geothermal waters. Geothermal energy utilisation is usually more problematic when it is connected with geothermal water rather than the ground.

The energy efficiency of a BHE mostly depends on the thermal conductivity of the underground rock mass. Other construction parameters also influence the energy efficiency. There are various types of BHEs, with the most typical being:

- single U-tube;
- multi U-tube;
- coaxial [8,9];
- helical [10,11]; and,
- BHE in piles [12].

BHEs can be vertically drilled. Alternatively, the BHE construction technology of Geothermal Radial Drilling (GRD) allows for directional (oblique) wells [13]. GRD provides the possibility of using the ground under buildings for thermal energy storage by the placing of boreholes. The effectiveness for various BHEs of different constructions has been compared. For instance, the effective thermal conductivity and performance in closed-loop vertical ground heat exchangers have been compared [14] and evaluated with TRTs [15], while the performances of the U-tube, concentric tube, and standing column well ground heat exchangers have been compared while using simulation [16]. Additionally, the thermal performances have been assessed for various types of underground heat exchangers [17] and for borehole heat exchangers specifically using TRTs [18]. The latter study included an analysis and comparison of interpretation methods.

BTES efficiencies in the literature consider such factors as borehole array geometry [19], heat transfer from the surroundings [20,21], grout parameters [22], freezing of underground water [23], and underground water flow. Studies that account for underground water flow include examinations of the influence of groundwater on: pile geothermal heat exchanger with cast-in spiral coils [24], closed-loop ground-source heat pump systems [25], the heat transfer in ground heat exchangers [26], the optimization of large-scale ground-coupled heat pump systems [27] and vertical closed-loop geothermal systems [28], the simulation of borehole heat exchangers [29], and the performance of geothermal heat exchangers [30]. Reference [31] describes many factors concerning coaxial BHEs.

The thermal response test (TRT) is an accurate and conventional method for the determination of the thermal properties of different cases of borehole heat exchangers [32]. A TRT is typically performed in large installations (over 100 kW) for an exploration BHE. According to [33], the TRT is an economic test for a lower capacity limit of about 30 kW. When the thermal parameters of the analyzed borehole heat exchanger are known, it is possible to establish the number of boreholes that are needed to satisfy heating and/or cooling demands.

Much work is being carried out to improve TRTs. For instance, a distributed TRT approach is described in [34]. The measuring process during a disturbed TRT (i.e., one with thermal sensors inside the BHE) affects the results. Sensors inside a BHE cause local turbulences and pressure losses, rendering the measuring unsettled/distorted. New methods for the interpretation of TRT results using statistics are described in [35,36]. The factors that are considered in a TRT are described in [37], for example, outdoor air temperature. Error analysis for a TRT is described in [38].

A TRT can be performed using various methods, e.g., the method that is described by Eskilson [39] and the computer code Earth Energy Designer (EED), which was developed following that method. Additionally, the extent of an underground thermal energy storage, i.e., the number and distribution of borehole heat exchangers, can be determined with numerical modeling, e.g., [40] or with commercial simulators [41].

Thermal response tests and mathematical modeling can also be used for determining the energy efficiency for thermal purposes of oil and gas wells, which have the potential for conversion to borehole heat exchangers [42]. This application has attracted increasing interest in recent years, especially for deep borehole heat exchangers, and it has correspondingly been subject to analysis by energy specialists in academe and industry.

BHE modeling also needs to account for thermal stresses. Doing so is essential for large installations that are connected with thermal waters and heat recovery from enhanced geothermal systems [43,44] and large BHE fields. The methods of thermal stress calculation that are used in the oil and gas industry can be applied for geothermal wells (and BHEs) [13].

In this article, we propose and verify a new method of establishing effective thermal conductivity of BHEs and assessing the usefulness of this method for utilization with thermal response tests. Existing methods provide the BHE thermal resistivity (R_b) as a function of the time of the TRT, whereas it should be constant in time. The method that is proposed here rectifies this shortcoming. That is, it is often possible to observe a change in R_b with time during a TRT [45]. Sometimes the change is an increase and at other times a decrease. Since the formula includes the effective thermal conductivity of the BHE (λ_{eff}), the value of λ_{eff} can be found, which yields a constant function R_b with time.

The novelty of this article lies in it presenting a new enhancement to a methodology. The methodology is described in the paper and verified while using the analyses of two TRTs. The main contribution and scientific significance is that it assists efforts to interpret TRTs better, so as to avoid tests providing erroneous or problematic results. Problematic TRTs occur quite often in practice.

The proposed new approach is based on the equation for the thermal resistivity of a BHE R_b in the function of the thermal conductivity of a rock mass λ or effective thermal conductivity λ_{eff}, as determined on the basis of a TRT. The new methodology is based on the assumption that R_b does not depend on the duration of the TRT, an assumption that corresponds to reality.

The basic assumption of the new approach can be expressed by the formula $R_b = f(t) = $ constant. The equation $R_b = f(t)$ can be developed into the dependence of $R_b = kt + b$. According to the model assumption $kt + b = $ constant, the equation is spilled if and only if $k = 0$. The determination of thermal conductivity λ_x and BHE thermal resistivity R_b is reduced in this methodology to determining such a value of λ_x, to obtain $k = 0$. Subsequently, we get $R_b = b$.

2. Thermal Response Test Mathematical Background

According to the well-established Kelvin infinite line source theory thermal response test was developed [46]. The TRT methodology is based on the partial differential equation form of the Fourier thermal conductivity equation, which describes the dynamic dependence of temperature T on the distance from heat exchanger r and duration of the test t, i.e., determines $T = T(r, t)$. The equation has the following form:

$$\frac{\partial^2 T}{\partial r^2} + \frac{1}{r}\frac{\partial T}{\partial r} = \frac{\rho c_p}{\lambda}\frac{\partial T}{\partial t} \tag{1}$$

One method of solving such a partial differential equation involves substitution. This transforms the partial differential Equation (1) to an ordinary differential equation. Perina [47] has used this approach to describe the Theis equation in hydrogeology, which gives the pressure distribution $p = p(r,t)$. To use this approach, we let

$$u = \frac{r^2 \rho c_p}{4t\lambda} \tag{2}$$

and

$$\rho c_p = \frac{\lambda}{\alpha} \tag{3}$$

Subsequently, we can show that

$$u = \frac{r^2}{4\alpha t} \tag{4}$$

and Equation (1) assumes the following form:

$$\frac{\partial^2 T}{\partial r^2} + \frac{1}{r}\frac{\partial T}{\partial r} = \frac{1}{\alpha}\frac{\partial T}{\partial t} \tag{5}$$

Returning to the Substitution (4), we finally obtain

$$T(r,t) = T_0 + \frac{q}{k\pi\lambda} \int\limits_{\frac{r^2}{4\alpha\lambda}}^{\infty} \frac{e^{-x}}{x} dx \tag{6}$$

In view of the substitution in Equation (4), and by substituting the integral in (6) with an approximate expression, we obtain

$$T(r,t) = T_0 + \frac{q}{4\pi\lambda} \left[\ln\left(\frac{4\alpha t}{r^2}\right) - \gamma \right] \tag{7}$$

Regarding initial and boundary conditions, it is noted that the solution obtained is not numerical. Rather, it is analytical and the idea of the mathematical model of the TRT is based on an infinite linear heat source. Accordingly, we do not solve the differential equation in a finite region. The (linear) source has a length that corresponds to the borehole depth. The time of the TRT is limited (max. 100 h). The initial temperature corresponds to the natural temperature distribution, as seen in Figure 1, but it is normally approximated with one initial temperature: T_o.

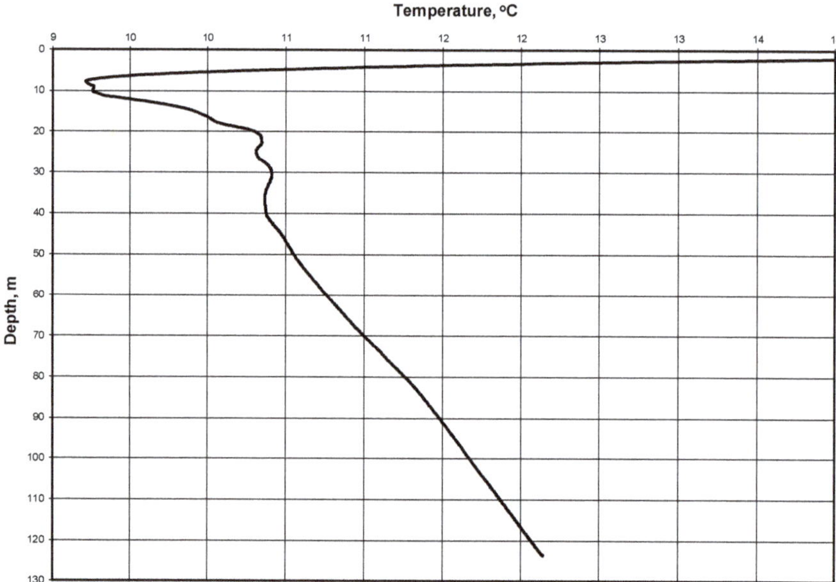

Figure 1. Natural (undisturbed) temperature profile of the BHE in Żarów.

3. Two Thermal Response Tests

The Thermal Response Test (TRT), which is sometimes called the Geothermal Response Test (GRT), is a suitable method in determining the effective thermal conductivity of the ground and the borehole thermal resistance (or the thermal conductivity of the borehole fill). A temperature curve is obtained, which can be evaluated by several methods. The resulting thermal conductivity is based on the total heat transport in the ground. Other effects, like convective heat transport (in permeable layers with groundwater), and further disturbances are automatically included, so it may be more correct to speak of an "effective" thermal conductivity λ_{eff}. The test equipment can be made in such a way that it can be easily transported to the site, e.g., on a light trailer (Figure 2) [48]. In short, a TRT relies on forcing the closed circulation of a heat carrier that is heated with a constant heating power. The temperature change with time is analyzed.

The TRT involves introducing and collecting energy from a borehole heat exchanger (Figure 2). During the tests, the heat carrier is most frequently heated at a constant heating power P, which is measured, and the temperatures T_1 and T_2 are then recorded. The heating power is maintained by switching the heaters on and off. The automation system takes into account the variability with the temperature of the heat carrier density ρ and the specific heat at constant pressure c_p.

This section describes two actual thermal response tests. The Laboratory of Geoenergetics, Faculty of Drilling, Oil, and Gas, AGH University of Science and Technology in Krakow, Poland performed the tests and analyses [49,50]. The tests were carried out to illustrate and compare the results from the old and new methods of TRT interpretation.

For the sake of interpretation, it is noted that the TRT was performed for a borehole heat exchanger, the geological profile of which is presented in Table 1. The first test was performed in Żarów (Dolnośląskie Region, Poland) in 2011. The average heating power during the test was $P = 5920$ W and the volumetric flow rate of the carrier was 16 dm^3·min^{-1} (Figure 3a). The control of the volume value of the heat carrier volumetric flow rate was carried out with a rotary (windmill) flowmeter with an accuracy of 0.25 dm^3 per one impulse. The flow measurements have the task of only visual control of the correctness of the TRT execution, where the relationship Q = const should be satisfied. Similarly, for the value of heating power P, which is calculated on the basis of temperature measurements from the dependence $P = Q\rho c\Delta T$, the visual relationship (Figure 3a) should be observed during the entire test to ensure $P = $ const.

The borehole heat exchanger is $H = 120$ m deep, so the heat exchange per unit of depth is $q = 49.34$ W·m^{-1}. In the linear heat source model, the unit heating power is assumed to be uniform. A constant temperature is also assumed in the model for the heat carrier and the ground. Table 2 presents the design of the borehole heat exchanger.

A second TRT was performed in the BHE at the Laboratory of Geoenergetics in the university, where the lithological profile is as described in Table 3.

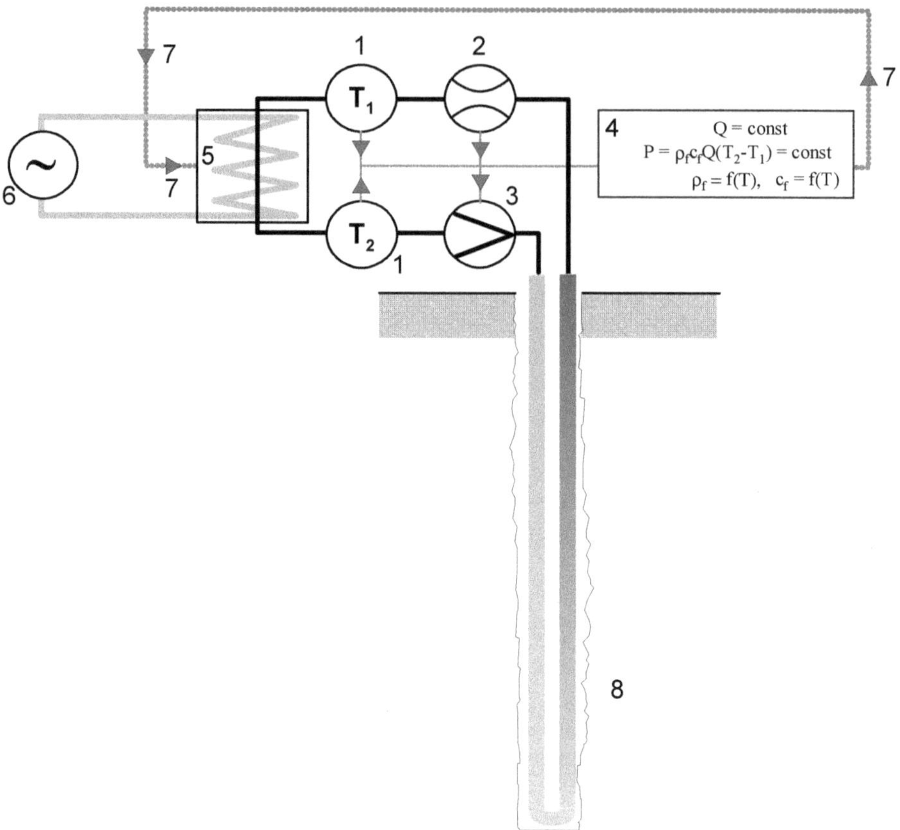

Figure 2. Schematic of thermal response test devices and operation. Legend: 1–thermometer, with absolute error 0.1 °C, 2–flowmeter, 3–pump, 4–control computer (stabilisation of thermal power and record the data), 5–set of heaters, 6–current source, 7–heater control signal, 8–borehole heat exchanger, Q–flow rate of heat carrier, P–heat flow rate (power), T_1–temperature of heat carrier (outflow from borehole heat exchangers (BHE)), T_1–temperature of heat carrier (inflow to BHE), $\rho_f = f(T)$–density of heat carrier as a function of temperature, and $c_f = f(T)$–specific heat of heat carrier as a function of temperature.

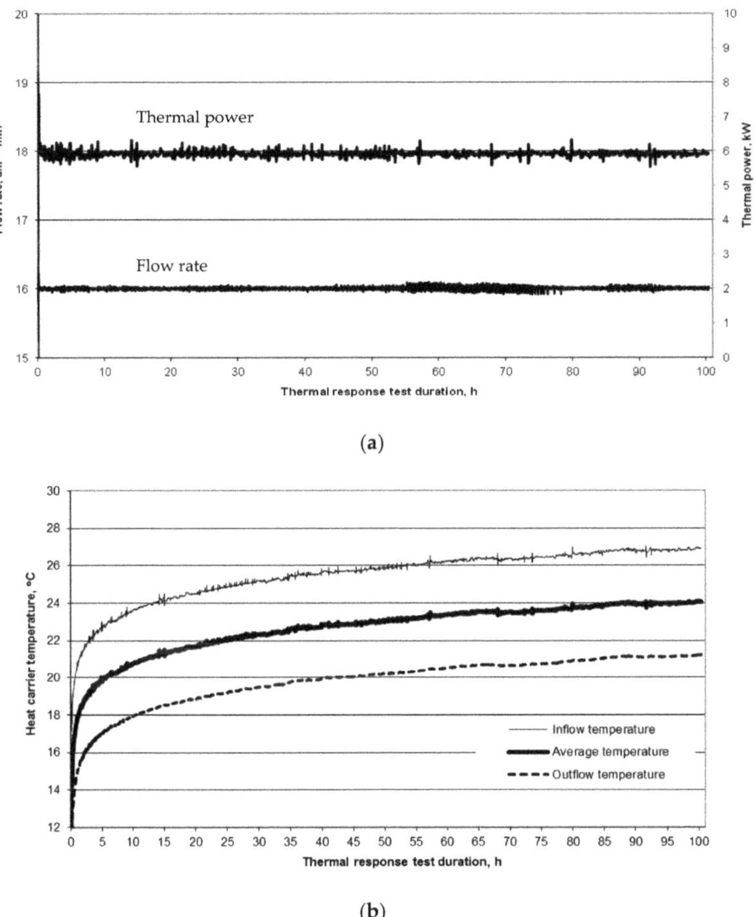

Figure 3. (**a**) Variation during Thermal Response Test (TRT) of heating power and carrier volumetric flow rate with time; (**b**) results of thermal response test.

Table 1. Lithological profile of borehole in Żarów, Poland.

Lithology	Top, m Below Surface	Bottom, m Below Surface	Thickness, m	Thermal Conductivity, $W \cdot m^{-1} \cdot K^{-1}$	Volumetric Thermal Capacity, $MJ \cdot m^{-3} \cdot K^{-1}$	Thermal Diffusivity, $10^{-6} \cdot m^2 \cdot s^{-1}$
Soil	0.0	0.3	0.3	0.80	2.00	0.40
Clayey sandy gravel	0.3	6.0	5.7	1.50	1.90	0.79
Sandy gravel	6.0	14.0	8.0	1.50	2.10	0.71
Gravel	14.0	16.0	2.0	1.80	2.10	0.86
Clay	16.0	18.0	2.0	1.50	2.00	0.75
Silt	18.0	20.0	2.0	1.20	1.90	0.63
Brown coal	20.0	22.0	2.0	0.30	1.80	0.17
Medium sand	22.0	36.0	14.0	2.10	2.00	1.05
Coarse sand and gravel	36.0	87.0	51.0	2.00	2.10	0.95
Granite detritus	87.0	90.0	3.0	3.00	2.10	1.43
Metamorphic rocks–amphibiolites	90.0	120.0	30.0	2.90	2.60	1.12
Weighted mean				2.15	2.17	0.97

Table 2. Design of borehole heat exchanger in Żarów and at Laboratory of Geoenergetics, Faculty of Drilling, Oil, and Gas, AGH University of Science and Technology in Kraków [51].

No.	Parameter	Value for BHE in Żarów	Value for BHE of Laboratory	Schematic
1	Design of borehole heat exchanger	single U-tube	single U-tube	
2	Borehole diameter D_o (diameter of drilling bit)	Interval 0–90 m with diameter 193 mm, interval 90–120 m with diameter 125 mm (on average 176 mm)	143 mm	

Table 2. *Cont.*

No.	Parameter	Value for BHE in Żarów	Value for BHE of Laboratory	Schematic
3	Depth of borehole, H_b	121 m	78 m	-
4	Depth of deposition of borehole tubes, H	120 m	78 m	-
5	Distance between axes of borehole tubes k	50 mm	80 mm	
6	Type of material used for sealing BHE tubes	Gravel (2–8 mm, with $\lambda = 1.8\ \mathrm{Wm^{-1}K^{-1}}$ assumed)	Gravel (with $\lambda_g = 1.8\ \mathrm{Wm^{-1}K^{-1}}$)	
7	Outer diameter of BHE tubes, d_z	40 mm	40 mm	

Table 2. *Cont.*

No.	Parameter	Value for BHE in Żarów	Value for BHE of Laboratory	Schematic
8	Thickness of BHE tube wall, b	3.7 mm	2.4 mm	
9	Material of borehole tubes	Polyethylene PE 100 (λ = 0.42 Wm^{-1}K^{-1})	Polyethylene PE 100 (λ = 0.42 Wm^{-1}K^{-1})	

Table 3. Lithological profile of borehole at AGH-UST in Krakow, Poland [51].

Lithology	Top, m Below Surface	Bottom, m Below Surface	Thickness, m	Thermal Conductivity, W·m^{-1}·K^{-1}	Volumetric Thermal Capacity, MJ·m^{-3}·K^{-1}	Thermal Diffusivity, 10^{-6}·m^2·s^{-1}
Anthropogenic (dark grey embankment with debris)	1.8	2.2	0.4	1.60	2.00	0.80
Alluvion (grey soil)	2.2	2.6	0.4	1.60	2.20	0.73
Fine sand, dusty and slightly clayey	2.6	4.0	1.4	1.00	2.00	0.50
Fine sand	4.0	6.0	2.0	1.20	2.50	0.48
Sandy gravel and gravel	6.0	15.0	9.0	1.80	2.40	0.75
Grey silt	15.0	30.0	15.0	2.20	2.30	0.96
Grey silt shale	30.0	78.0	48.0	2.10	2.30	0.91
Weighted mean				2.04	2.31	0.88

The values of D_o, d_z, H_b, b, and H (in Equation (8)) in Table 2 were provided by a company making a BHE without any data on measurement accuracy. The value of λ_g comes from the literature [51].

Figure 3b presents the curves illustrating the data during the TRT test, where the dependence on the duration of the test is observed. The average thermal conductivity that is assumed for rocks of 2.15 W·m^{-1}·K^{-1} is based on data in the literature [51], as is the average volumetric specific heat of rocks in the profile of 2.17 MJ·m^{-3}·K^{-1} [52,53]. A 35% propylene glycol solution is used as the heat carrier. At 20 °C, the specific heat of the carrier is 3810 J·kg^{-1}·K^{-1} and its density is 1028 kg·m^{-3}.

The average temperature of the rock mass is determined on the basis of the heat carrier circulation, without heating (i.e., before the TRT heating phase begins). The return flow temperature is 11.1 °C and the mean measured air temperature is 16.1 °C. The mean natural temperature in the borehole can also be determined on the basis of temperature logging [54]. The mean temperature of the rock mass, based on the temperature profile (Figure 1), is 11.00 °C. The NIMO-T (Non-wired Immersible Measuring Object for Temperature) was used for temperature profiling in BHE. The relative error of the temperature measurement was 0.0015 °C and the absolute error was 0.1 °C [54].

The thermal diffusivity $\alpha = 0.97 \cdot 10^{-6}$ m^2·s^{-1} is calculated using data from the literature [51] and Equation (3).

The mean temperature of the heat carrier flowing into the BHE during the test is 25.48 °C and the mean return flow temperature is 19.82 °C. Thus, the mean temperature difference is 5.67 °C and the mean temperature of the heat carrier is 22.65 °C.

Figure 4 shows the characteristic times that are used for the interpretation of TRT results. There, the points are denoted, as follows: t_0 the beginning of heating phase of the test (heaters on), t_1 the slope point of the curve (time of the first complete circulation loop of the heat carrier), t_2 the time corresponding to t = 5 $r_o^2\alpha^{-1}$, t_3 the time corresponding to t = 20 $r_o^2\alpha^{-1}$, t_4 the half-time of the heating phase of the test, and t_5 the end of the heating phase of the test (heaters off).

The following values were obtained during the test: $t_0 = 0$, $t_1 = 480$ s (0.13 h), $t_2 = 46348$ s (12.87 h), $t_3 = 185391$ s (51.50 h), $t_4 = 180660$ s (50.18 h), and $t_5 = 361320$ s (100.37 h). Linear regression analysis is used for determining coefficients of line slope in the semi-logarithmic system (log t) for the following time intervals:

- from t_0 to t_5
- from t_1 to t_5
- from t_2 to t_5
- from t_0 to t_2
- from t_0 to t_3
- from t_2 to t_3
- from t_1 to t_4
- from t_3 to t_5.

The values of the slope coefficient k and the effective thermal conductivity λ_{eff} are calculated on the basis of Equation (7), as follows:

$$\lambda_{eff} = \frac{P}{4\pi Hk} = \frac{q}{4\pi k} \tag{8}$$

Table 4 lists the results, including the average values of the BHE thermal resistivity R_b and values for various data intervals. The value of R_b is calculated, as follows:

$$R_b = \frac{1}{q}[T_{av}(t) - T_0] - \frac{1}{4\pi\lambda}\left[\ln\frac{4\alpha t}{r_o^2} - \gamma\right] \tag{9}$$

Here, values for λ can be taken from the literature ($\lambda = 2.15$ W·m^{-1}·K^{-1} from Table 1) or calculated with Equation (7).

In Table 5, the TRT results are shown for a BHE belonging to the Laboratory of Geoenergetics, Faculty of Drilling, Oil, and Gas AGH–University of Science and Technology for a heat carrier volumetric flow rate of 12 dm^3·min^{-1} and heating power P = 4000 W. The lithological profile of the borehole was described earlier (Table 5), as was the design of the borehole heat exchanger (Table 2).

The correlation coefficient for two TRTs was calculated. It concerned the temperatures dependence of the heat carrier and the duration of TRT. In both cases, the correlation coefficient had a higher value than 0.925.

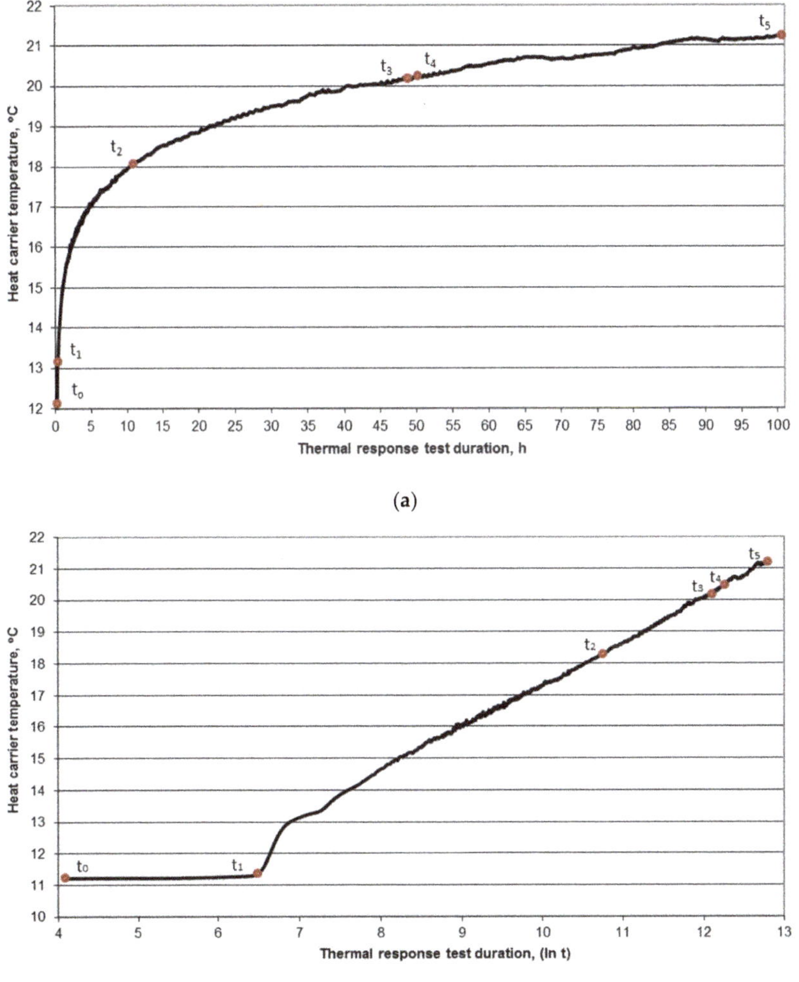

Figure 4. Variation of temperature of heat carrier flowing out of BHE with (**a**) time and (**b**) logarithm of time, showing key characteristic times.

Table 4. Thermal resistivity R_b and effective thermal conductivity λ_{eff} of BHE, as determined for various data ranges (TRT in Żarów).

Data	Full Range of Data	From Slope to End of TRT	From $t = 5\,r_o^2\alpha^{-1}$ to End of TRT	From Beginning to $t = 5\,r_o^2\alpha^{-1}$	From Beginning to $t = 20\,r_o^2\alpha^{-1}$	From $t = 5\,r_o^2\alpha^{-1}$ to $t = 20\,r_o^2\alpha^{-1}$	From Slope to Half Time of Full Range of Data	From $t = 20\,r_o^2\alpha^{-1}$ to End of TRT
	t_0 to t_5	t_1 to t_5	t_2 to t_5	t_0 to t_2	t_0 to t_3	t_2 to t_3	t_1 to t_4	t_3 to t_5
R_b for λ_{eff} according to the literature (Equation (9)), $\mathrm{m\cdot K\cdot W^{-1}}$	0.1094	0.1093	0.1069	0.1265	0.1144	0.1103	0.1144	0.1041
Relative change of R_b with respect to full measurement range, %	0.00	0.09	2.29	__−15.63__	−4.57	−0.82	−4.57	4.84
R_b for λ_{eff} according to TRT (Equation (9)), $\mathrm{m\cdot K\cdot W^{-1}}$	0.1303	0.1303	0.1303	0.1308	0.1299	0.1296	0.1299	0.1307
Relative change of R_b with respect to full measurement range, %	0.00	0.00	0.00	−0.38	0.31	__0.54__	0.31	−0.31
Effective thermal conductivity λ_{eff} in BHE (Equation (8)) for temperature of inflow heat carrier, $\mathrm{W\cdot m^{-1}\cdot K^{-1}}$	2.8132	2.7993	2.6959	2.9879	2.8821	2.718	2.8633	2.7749
Effective thermal conductivity λ_{eff} in BHE (Equation (8)) for temperature of outflow heat carrier, $\mathrm{W\cdot m^{-1}\cdot K^{-1}}$	2.8022	2.787	2.697	2.943	2.8619	2.7156	2.8389	2.7472
Effective thermal conductivity λ_{eff} in BHE (Equation (8)) for mean temperature of heat carrier, $\mathrm{W\cdot m^{-1}\cdot K^{-1}}$	2.8077	2.7931	2.6965	2.9653	2.872	2.7168	2.8511	2.761
Relative change of λ_{eff} with respect to full measurement range, %	0.00	0.52	3.96	__−5.61__	−2.29	3.24	−1.55	1.66

Table 5. Thermal resistivity R_b and effective thermal conductivity λ_{eff} of BHE, as determined for various data intervals (TRT at Laboratory of Geoenergetics).

Data	Full Range of Data	From Slope to End of TRT	From $t = 5\, r_o^2 \alpha^{-1}$ to End of TRT	From Beginning to $t = 5\, r_o^2 \alpha^{-1}$	From Beginning to $t = 20\, r_o^2 \alpha^{-1}$	From $t = 5\, r_o^2 \alpha^{-1}$ to $t = 20\, r_o^2 \alpha^{-1}$	From Slope to Half Time of Full Range Of Data	From $t = 20\, r_o^2 \alpha^{-1}$ to End of TRT
	t_0 to t_5	t_1 to t_5	t_2 to t_5	t_0 to t_2	t_0 to t_3	t_2 to t_3	t_1 to t_4	t_3 to t_5
R_b for λ_{eff} according to the literature (Equation (9)), m·K·W^{-1}	0.1383	0.1382	0.1389	0.1313	0.1364	0.1382	0.1371	0.1391
Relative change of R_b with respect to full measurement range, %	0.00	0.07	−0.43	**5.06**	1.37	0.07	0.87	−0.58
R_b for λ_{eff} according to TRT (Equation (9)), m·K·W^{-1}	0.1276	0.1274	0.1274	0.1299	0.1299	0.1299	0.1289	0.1264
Relative change of R_b with respect to full measurement range, %	0.00	0.16	0.16	**−1.80**	**−1.80**	**−1.80**	−1.02	0.94
Effective thermal conductivity λ_{eff} in BHE (Equation (8)) for temperature of inflow heat carrier, W·m^{-1}·K^{-1}	1.9659	1.9296	2.0082	2.2368	1.9496	1.8620	1.8779	2.1439
Effective thermal conductivity λ_{eff} in BHE (Equation (8)) for temperature of outflow heat carrier, W·m^{-1}·K^{-1}	1.9750	1.9275	2.0079	2.3576	1.9754	1.8613	1.8745	2.1442
Effective thermal conductivity λ_{eff} in BHE (Equation (8)) for mean temperature of heat carrier, W·m^{-1}·K^{-1}	1.9705	1.9286	2.0080	2.2956	1.9624	1.8616	1.8762	2.1440
Relative change of λ_{eff} with respect to full measurement range, %	0.00	2.13	−1.90	**−16.50**	0.41	5.53	4.79	−8.80

4. A New Way of Determining Parameters from TRTs

Although the test (Figure 3b) was carried out almost ideally, there are discrepancies in the BHE thermal conductivity and thermal resistivity values. These discrepancies stem from the different ranges of data that are assumed for the analyses. Therefore, a new way of determining the parameters λ_{eff} and R_b is proposed.

Figure 5 presents a graph showing the dependence of BHE thermal resistivity on test duration. The curves for R_{b1} and R_{b4} are more 'linear' in Figure 5 than the curves for R_{b2}, R_{b3}, and R_{b5}. Both of the curves only differ in the value of λ from Equation (8).

The proposed approach involves determining a value of λ for which linear regression that is based on $R_b = f(t)$ assumes the form of a function $R_b = kt + b$ with a slope coefficient k of zero. The task of determining λ and R_b values reduces to finding a λ value, for which $k = 0$, after which we have $R_b = b$. For the TRT that was performed in Żarów, we determined the following pairs of values meeting this requirement: $\lambda_{eff} = 2.77$ W·m^{-1}·K^{-1} and $R_b = 0.129$ m·K·W^{-1}. This contrasts with the values obtained with the traditional method of $\lambda = 2.81$ W·m^{-1}·K^{-1} and $R_b = 0.130$ m·K·W^{-1}.

For the TRT performed at the BHE of the Laboratory of Geoenergetics, we find $\lambda = 1.98$ W·m^{-1}·K^{-1} and $R_b = 0.134$ m·K·W^{-1} (whereas with the traditional method $\lambda = 1.97$ W·m^{-1}·K^{-1} and $R_b = 0.128$ m·K·W^{-1}).

Figure 5 shows the relation of R_b with time from the TRT for the test in Żarów. The curves of $R_b = f(t)$ only vary due to the value of λ in Equation (9). The waveforms represent the following:

- R_{b1}–graph of BHE thermal resistance vs. time for the conductivity λ_{eff} calculated for data from a time equal $5r_0^2 \cdot \alpha^{-1}$ to the end of the heating phase of TRT ($\lambda_{eff} = 2.70$ W·m^{-1}·K^{-1})
- R_{b2}–graph of BHE thermal resistance vs. time for the conductivity $\lambda = 125\%$ λ_{eff} ($\lambda = 3.37$ W·m^{-1}·K^{-1});
- R_{b3}–graph of BHE thermal resistance vs. time for the conductivity $\lambda = 75\%$ λ_{eff} ($\lambda = 2.03$ W·m^{-1}·K^{-1});
- R_{b4}–graph of BHE thermal resistance vs. time for the conductivity when linear regression yields a constant function (for which $k = 0$ in the function $R_b = kt + b$), according to the new method described above ($\lambda = 2.80$ W·m^{-1}·K^{-1}); and,
- R_{b5}–graph of BHE thermal resistance vs. time for the conductivity based on data in the literature, as in Table 1 ($\lambda = 2.15$ W·m^{-1}·K^{-1}).

The values of R_b and λ_{eff} for various time intervals calculated based on the new methodology are listed in the Table 6 for the TRT at Żarów and in Table 7 for the TRT at the Laboratory of Geoenergetics.

Table 6. BHE thermal resistivity R_b and effective thermal conductivity λ_{eff}, determined using the new method for various data intervals (for TRT in Żarów).

Data	Full Range of Data	From Slope to End of TRT	From $t = 5\,r_o^2\alpha^{-1}$ to End of TRT	From Beginning to $t = 5\,r_o^2\alpha^{-1}$	From Beginning to $t = 20\,r_o^2\alpha^{-1}$	From $t = 5\,r_o^2\alpha^{-1}$ to $t = 20\,r_o^2\alpha^{-1}$	From Slope to Half Time of Full Range of Data	From $t = 20\,r_o^2\alpha^{-1}$ to End of TRT
	t_0 to t_5	t_1 to t_5	t_2 to t_5	t_0 to t_2	t_0 to t_3	t_2 to t_3	t_1 to t_4	t_3 to t_5
R_b for λ_{eff} according to new method, m·K·W^{-1}	0.1290	0.1290	0.1270	0.1320	0.1310	0.1270	0.1300	0.1290
Relative change of R_a with respect to full measurement range, %	0.00	0.00	1.55	$-$**2.33**	$-$1.55	1.55	$-$0.78	0.00
Effective thermal conductivity λ_{eff} in BHE for mean temperature of heat carrier according to new method, W·m^{-1}·K^{-1}	2.7666	2.7620	2.6970	2.9745	2.8337	2.7173	2.8262	2.7598
Relative change of λ_{eff} with respect to full measurement range, %	0.00	0.17	2.52	$-$**7.51**	$-$2.43	1.78	$-$2.15	0.25

Table 7. BHE thermal resistivity R_b and effective thermal conductivity λ_{eff}, determined using the new method for various data intervals (for TRT at Laboratory of Geoenergetics).

Data	Full Range of Data	From Slope to End of TRT	From $t = 5\,r_o^2\alpha^{-1}$ to End of TRT	From Beginning to $t = 5\,r_o^2\alpha^{-1}$	From Beginning to $t = 20\,r_o^2\alpha^{-1}$	From $t = 5\,r_o^2\alpha^{-1}$ to $t = 20\,r_o^2\alpha^{-1}$	From Slope to Half Time of Full Range of Data	From $t = 20\,r_o^2\alpha^{-1}$ to End of TRT
	t_0 to t_5	t_1 to t_5	t_2 to t_5	t_0 to t_2	t_0 to t_3	t_2 to t_3	t_1 to t_4	t_3 to t_5
R_b for λ_{eff} according to new method, m·K·W^{-1}	0.1340	0.1340	0.1380	0.1310	0.1300	0.1290	0.1300	0.1470
Relative change of R_b with respect to full measurement range, %	0.00	0.00	−2.99	2.24	2.99	3.73	2.99	−9.70
Effective thermal conductivity λ_{eff} in BHE for mean temperature of heat carrier according to new method, W·m^{-1}·K^{-1}	1.9829	1.9738	2.0397	1.9872	1.8714	1.8588	1.8968	2.1751
Relative change of λ_{eff} with respect to full measurement range, %	0.00	0.46	−2.86	−0.22	5.62	6.26	4.34	−9.69

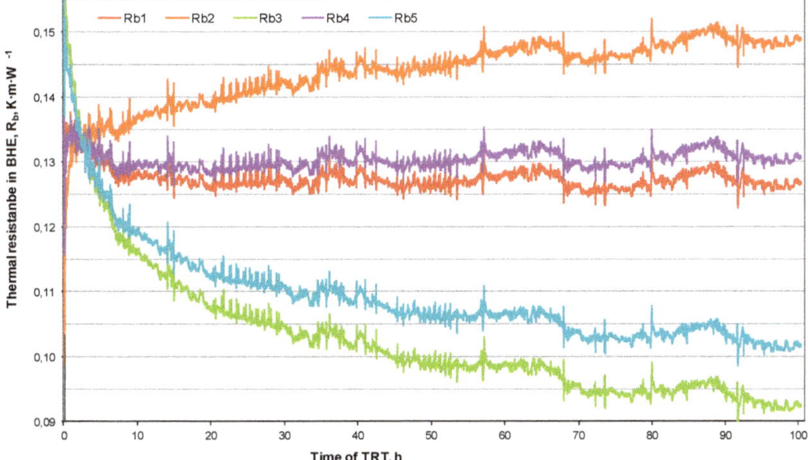

Figure 5. Thermal resistivity of BHE vs. TRT time, where R_{b1}, R_{b2}, R_{b3}, R_{b4}, and R_{b5} are described in the text.

5. Results and Discussion

The thermal response test is the most favored way of determining the basic BHE parameters. The effective thermal conductivity λ_{eff} is the most important value for characterizing a BHE, and it is used for determining its energy efficiency. The effective thermal conductivity is mainly dependent on the thermal conductivity of rocks λ, especially when the BHE has been correctly performed. In reality, λ_{eff} is also dependent on the heat transfer resistivity between the heating agent circulating in the BHE and the rock mass. This heat transfer resistivity accounts for:

- transfer of heat from the heating agent to the material (most frequently U-tubes), which depends on, among other factors, its viscosity;
- heat conduction through the material (U-tube), which is affected by its thermal conductivity, e.g., for the case of a polyethylene tube $\lambda = 0.42\ \mathrm{W \cdot m^{-1} \cdot K^{-1}}$;
- heat flow between the material of U-tube and the BHE filling/sealing material, where discontinuities may occur; and,
- Heat flow between the BHE filling/sealing material and rock mass, where some discontinuities may be encountered.

An analysis of the experimental and analytical results reveals that, despite a correctly performed TRT, the values of λ_{eff} are not constant with the time of test. This is caused by the assumed duration of the test. Such differences with respect to the value that was obtained for the full time of the test (100 h) may as great as 16.50%, relative to the traditional method (Table 5), and 9.69% for the new method (Table 7). The greatest percentage difference between the values of λ_{eff} and R_b are underlined in bold in Tables 4–7. The relative change of R_b or λ_{eff} with respect to the full measurement range is calculated in %, relative to the values in the first time interval (t_0 to t_5). Accordingly, the percentage difference for this interval (i.e., the full range of data, t_0 to t_5) is always zero.

The new method is observed to be more accurate and stable in time for calculating the effective thermal conductivity λ_{eff} in BHEs. However, when analyzing the BHE thermal resistivity R_b, larger discrepancies can be observed for relative deviations from the basic value (for the full test duration). A maximum deviation of 1.80% for the traditional method (Table 5) and 9.70% for the new method is observed (Table 7). The greatest differences are observed for the TRT performed at the Laboratory of Geoenergetics, Faculty of Drilling, Oil, and Gas, AGH University of Science and Technology. For TRTs performed in Żarów, the corresponding deviations are much smaller. Therefore, more analyses of TRT data are needed while using the new methodology to assess these discrepancies, and that is the topic of ongoing research by the authors.

Many more analyses of TRT results are needed, along with the corresponding statistical analyses, to choose a better method for the interpretation of TRT results. In practice, there is no ideal TRT. The functional variation of temperature with time has many distortions. Simultaneously with improving TRT interpretations, the TRT measuring procedure also needs improvement. The inflow of material at the outside temperature should be reduced/eliminated, and a reliable automatic system is needed for maintaining a constant heating power when the variable voltages are present in the electrical network. Both of these requirements are being addressed at the Laboratory of Geoenergetics.

The accuracy of the calculation of the effective thermal conductivity coefficient and the thermal resistance has not been extensively examined in this article. That is because the target of this article is to describe the new methodology. Research by the present authors is ongoing to assess the precision of the results that were obtained with the new method, and it is expected to be reported soon.

6. Conclusions

The thermal response test is the most accurate way of determining parameter values of borehole heat exchangers. The effective thermal conductivity λ_{eff} and thermal resistivity of borehole R_b can be used in the design of an appropriate number of borehole exchangers for a given heating power demand and for a given time duration.

However, when interpreting the thermal response test, there are sometimes problems with the resulting values. That is, the values of thermal conductivity λ_{eff} and thermal resistivity R_b can differ depending on the assumed range of data, especially the time data. Various values of basic parameters are seen to be obtained, even for correctly performed tests, when analyzing various TRT time intervals.

The proposed method of determining basic TRT parameters is based on the BHE thermal resistivity R_b equation. This dependence (Equation (9)) is also observed to be a function of effective thermal conductivity λ_{eff} of the borehole heat exchanger. It is suggested that, a pair of the test results, i.e., effective thermal conductivity λ_{eff} and BHE thermal resistivity R_b, can predict the dependence of resistivity as a function of time, such that the slope coefficient of the regression line that is based on this approach is zero.

It is concluded from the analyses that the proposed new method of determining the values of the basic parameters of a BHE is more accurate and independent of thermal response test duration. The differences that were obtained for various TRT times with the proposed method for λ_{eff} are lower than with the traditional method. However, larger differences are obtained for R_b. Further work to assess the usefulness of this method in the interpretation of TRT data appears to be merited.

Author Contributions: Conceptualization, A.S.-S., A.G. and M.A.R.; methodology, A.S.-S.; software, T.S.; validation, A.S.-S. and T.S.; formal analysis, J.K.; investigation, T.S.; resources, A.S.-S. and A.G.; data curation, T.S.; writing, reviewing and editing of manuscript, A.S.-S., T.S. and M.A.R.; visualization, T.S.; supervision, A.S.-S.; project administration, A.S.-S.; funding acquisition, A.S.-S. and T.S.

Funding: This research was funded by statutory research programme at the Faculty of Drilling, Oil and Gas, AGH University of Science and Technology in Krakow, Poland, grant number 11.11.190.555.

Conflicts of Interest: The authors declare no conflict of interest.

Nomenclature/Glossary

b	thickness of pipe of U-tube (mm)
c_p	specific heat at constant pressure of ground (J kg^{-1} K^{-1})
c_f	specific heat of heat carrier (J kg^{-1} K^{-1})
D_o	borehole diameter (m)
d_z	outer diameter of pipe of U-tube (m)
H	depth of BHE (m)

H_b depth of borehole (m)
k theoretical directional factor of TRT
P thermal power of TRT (W)
q unit thermal power (W m^{-1})
Q flow rate of heat carrier (m^3·s^{-1})
r radial distance from vertical axis of borehole heat exchanger (m)
r_o borehole heat exchanger radius (m)
R_b BHE thermal resistivity (m K W^{-1})
T temperature of ground (K)
T_o initial temperature (K)
T_{av} mean temperature of heat carrier in BHE during TRT (K)
T_1 temperature of heat carrier (outflow from BHE) (K)
T_2 temperature of heat carrier (inflow to BHE) (K)
t time (s)
u variable
α thermal diffusivity of ground (m^2 s^{-1})
λ thermal conductivity of ground (W m^{-1} K^{-1})
λ_{eff} effective thermal conductivity in BHE (W m^{-1} K^{-1})
λ_x thermal conductivity obtained by new method (W m^{-1} K^{-1})
λ_g thermal conductivity of grout (gravel) (W m^{-1} K^{-1})
ρ density of ground (kg m^{-3})
ρ_f density of heat carrier (kg·m^{-1})
γ Euler constant ($\gamma = 0.5772156$)
ΔT mean temperature difference T_2–T_1 (K)

References

1. Rybach, L.; Eugster, W.J. Sustainability aspects of geothermal heat pump operation, with experience from Switzerland. *Geothermics* **2010**, *39*, 365–369. [CrossRef]
2. Weber, J.; Ganz, B.; Schellschmidt, R.; Sanner, B.; Schulz, R. Geothermal energy use in Germany. In Proceedings of the World Geothermal Congress, Melbourne, Australia, 19–25 April 2015.
3. Gehlin, S.; Andersson, O.; Bjelm, L.; Alm, P.G.; Rosberg, J.E. Country update for Sweden. In Proceedings of the World Geothermal Congress, Melbourne, Australia, 19–25 April 2015.
4. Raymond, J.; Malo, M.; Tanguay, D.; Grasby, S.; Bakhteyar, F. Direct utilization of geothermal energy from coast to coast: A review of current applications and research in Canada. In Proceedings of the World Geothermal Congress, Melbourne, Australia, 19–25 April 2015.
5. Boyd, T.L.; Sifford, A.; Lund, J.W. The United States of America country update 2015. In Proceedings of the World Geothermal Congress, Melbourne, Australia, 19–25 April 2015.
6. Dincer, I.; Rosen, M.A. *Thermal Energy Storage: Systems and Applications*, 2nd ed.; Wiley: London, UK, 2011.
7. Kizilkan, O.; Dincer, I. Borehole thermal energy storage system for heating applications: Thermodynamic performance assessment. *Energy Convers. Manag.* **2015**, *90*, 53–61. [CrossRef]
8. Sliwa, T.; Rosen, M.A. Natural and artificial methods for regeneration of heat resources for borehole heat exchangers to enhance the sustainability of underground thermal storages: A review. *Sustainability* **2015**, *7*, 13104–13125. [CrossRef]
9. Beier, R.A.; Acuña, J.; Mogensen, P.; Palm, B. Borehole resistance and vertical temperature profiles in coaxial borehole heat exchangers. *Appl. Energy* **2013**, *102*, 665–675. [CrossRef]
10. Zarrella, A., De Carli, M. Heat transfer analysis of short helical borehole heat exchangers. *Appl. Energy* **2013**, *102*, 1477–1491. [CrossRef]
11. Zarrella, A.; Capozza, A.; De Carli, M. Analysis of short helical and double U-tube borehole heat exchangers: A simulation-based comparison. *Appl. Energy* **2013**, *112*, 358–370. [CrossRef]
12. Li, M.; Lai, A.C.K. Heat-source solutions to heat conduction in anisotropic media with application to pile and borehole ground heat exchangers. *Appl. Energy* **2012**, *96*, 451–458. [CrossRef]
13. Knez, D. Stress State Analysis in Aspect of Wellbore Drilling Direction. *Arch. Min. Sci.* **2014**, *59*, 69–74. [CrossRef]

14. Lee, C.; Moonseo, P.; Sunhong, M.; Shin-Hyung, K.; Byonghu, S.; Hangseok, C. Comparison of effective thermal conductivity in closed-loop vertical ground heat exchangers. *Appl. Therm. Eng.* **2011**, *31*, 3669–3676. [CrossRef]

15. Lee, C.; Moonseo, P.; The-Bao, N.; Byonghu, S.; Jong, M.C.; Hangseok, C. Performance evaluation of closed-loop vertical ground heat exchangers by conducting in-situ thermal response tests. *Renew. Energy* **2012**, *42*, 77–83. [CrossRef]

16. Yavuzturk, C.; Chiasson, A.D. Performance analysis of U-tube, concentric tube, and standing column well ground heat exchangers using a system simulation approach. *ASHRAE Trans.* **2002**, *108*, 925.

17. Li, X.; Yan, C.; Zhihao, C.; Jun, Z. Thermal performances of different types of underground heat exchangers. *Energy Build.* **2006**, *38*, 543–547. [CrossRef]

18. Zarrella, A.; Emmi, G.; Graci, S.; De Carli, M.; Cultrera, M.; Santa, G.D.; Galgaro, A.; Bertermann, D.; Müller, J.; Pockelé, L.; et al. Thermal Response Testing Results of Different Types of Borehole Heat Exchangers: An Analysis and Comparison of Interpretation Methods. *Energies* **2017**, *10*, 801. [CrossRef]

19. Kurevija, T.; Vulin, D.; Krapec, V. Effect of borehole array geometry and thermal interferences on geothermal heat pump system. *Energy Convers. Manag.* **2012**, *60*, 134–142. [CrossRef]

20. Jaszczur, M.; Polepszyc, I.; Sapinska-Sliwa, A. Numerical analysis of the boundary conditions model impact on the estimation of heat resources in the ground. *Pol. J. Environ. Stud.* **2015**, *24*, 60–66.

21. Jaszczur, M.; Polepszyc, I.; Sapińska-Śliwa, A.; Gonet, A. An analysis of the numerical model influence on the ground temperature profile determination. *J. Therm. Sci.* **2017**, *26*, 82–88. [CrossRef]

22. Sliwa, T.; Rosen, M.A. Efficiensy analysis of borehole heat exchangers as grout varies via thermal response test simulation. *Geothermics* **2017**, *69*, 132–138. [CrossRef]

23. Eslami-nejad, P.; Bernier, M. Freezing of geothermal borehole surroundings: A numerical and experimental assessment with applications. *Appl. Energy* **2012**, *98*, 333–345. [CrossRef]

24. Deqi Wang, D.; Lu, L.; Zhang, W.; Cui, P. Numerical and analytical analysis of groundwater influence on the pile geothermal heat exchanger with cast-in spiral coils. *Appl. Energy* **2015**, *160*, 705–714. [CrossRef]

25. Chiasson, A.C.; Rees, S.J.; Spitler, J.D. A preliminary assessment of the effects of ground-water flow on closed-loop ground-source heat pump systems. *ASHRAE Trans.* **2000**, *106*, 380–393.

26. Diao, N.; Li, Q.; Fang, Z. Heat transfer in ground heat exchangers with groundwater advection. *Int. J. Therm. Sci.* **2004**, *43*, 1203–1211. [CrossRef]

27. Fujii, H.; Itoi, R.; Fujii, J.; Uchida, Y. Optimizing the design of large-scale ground-coupled heat pump systems using groundwater and heat transport modelling. *Geothermics* **2005**, *34*, 347–364. [CrossRef]

28. Hecht-Méndez, J.; de Paly, M.; Beck, M.; Bayer, P. Optimization of energy extraction for vertical closed-loop geothermal systems considering groundwater flow. *Energy Convers. Manag.* **2013**, *66*, 1–10. [CrossRef]

29. Molina-Giraldo, N.; Blum, P.; Zhu, K.; Bayer, P.; Fang, Z. A moving finite line source model to simulate borehole heat exchangers with groundwater advection. *Int. J. Therm. Sci.* **2011**, *50*, 2506–2513. [CrossRef]

30. Fan, R.; Jiang, Y.Q.; Yao, Y.; Shiming, D.; Ma, Z.L. A study on the performance of a geothermal heat exchanger under coupled heat conduction and groundwater advection. *Energy* **2007**, *32*, 2199–2209. [CrossRef]

31. Sliwa, T.; Nowosiad, T.; Vytyaz, O.; Sapinska-Sliwa, A. Study on efficiency of deep borehole heat exchangers. *SOCAR Proc.* **2016**, *2*, 29–42. [CrossRef]

32. Luo, J.; Rohn, J.; Bayer, M.; Priess, A. Thermal Efficiency Comparison of Borehole Heat Exchangers with Different Drillhole Diameters. *Energies* **2013**, *6*, 4187–4206. [CrossRef]

33. Sanner, B.; Mands, E.; Sauer, M.; Grundmann, E. Economic aspects of thermal response test–advantages, technical improvements, commercial application. In Proceedings of the Effstock 2009: International Conference on Thermal Energy Storage, Stockholm, Sweden, 14–17 June 2009; pp. 1–9.

34. Acuña, J.; Palm, B. Distributed thermal response tests on pipe-in-pipe borehole heat exchangers. *Appl. Energy* **2013**, *109*, 312–320. [CrossRef]

35. Zhang, L.; Zhang, Q.; Huang, G.; Du, Y. A p(t)-linear average method to estimate the thermal parameters of the borehole heat exchangers for in situ thermal response test. *Appl. Energy* **2014**, *131*, 211–221. [CrossRef]

36. Choi, W.; Ooka, R. Interpretation of disturbed data in thermal response tests using the infinite line source model and numerical parameter estimation method. *Appl. Energy* **2015**, *148*, 476–488. [CrossRef]

37. Li, Y.; Mao, J.; Geng, S.; Han, X.; Zhang, H. Evaluation of thermal short-circuiting and influence on thermal response test for borehole heat exchanger. *Geothermics* **2014**, *50*, 136–147. [CrossRef]

38. Witte, H.L.J. Error analysis of thermal response tests. *Appl. Energy* **2013**, *109*, 302–311. [CrossRef]

39. Eskilson, P. Thermal Analyses of Heat Extraction Boreholes. Ph.D. Thesis, Department of Mathematical Physics, Lund Institute of Technology, Lund, Sweden, 1987.

40. Sliwa, T.; Gonet, A. Theoretical model of borehole heat exchanger. *J. Energy Resour. Technol.* **2005**, *127*, 142–148. [CrossRef]

41. Sliwa, T.; Gołaś, A.; Wołoszyn, J.; Gonet, A. Numerical model of borehole heat exchanger in ANSYS CFX software. *Arch. Min. Sci.* **2012**, *57*, 375–390.

42. Sliwa, T.; Kotyza, J. Application of existing wells as ground heat source for heat pumps in Poland. *Appl. Energy* **2003**, *74*, 3–8. [CrossRef]

43. Baria, R.; Michelet, S.; Baumgaertner, J.; Dyer, B.; Gerard, A.; Nicholls, J.; Hettkamp, T.; Teza, D.; Soma, N.; Asanuma, H.; et al. Microseismic monitoring of the world's largest potential HDR reservoir. In Proceedings of the 29th Workshop on Geothermal Reservoir Engineering, Stanford University, Stanford, CA, USA, 26–28 January 2004.

44. Valley, B.; Evans, K.F. *Stress Orientation at the Basel Geothermal Site from Wellbore Failure Analysis in BS1, Report to Geopower Basel AG for Swiss Deep Heat Mining Project Basel. ETH Report Nr.: ETH 3465/56*; Ingenieurgeologie ETH: Zürich, Switzerland, 2006.

45. Lamarche, L.; Raymond, J.; Koubikana Pambou, C.H. Evaluation of the Internal and Borehole Resistances during Thermal Response Tests and Impact on Ground Heat Exchanger Design. *Energies* **2018**, *11*, 38. [CrossRef]

46. Badenes, B.; Mateo Pla, M.Á.; Lemus-Zúñiga, L.G.; Sáiz Mauleón, B.; Urchueguía, J.F. On the Influence of Operational and Control Parameters in Thermal Response Testing of Borehole Heat Exchangers. *Energies* **2017**, *10*, 1328. [CrossRef]

47. Perina, T. Derivation of the Theis (1935) Equation by Substitution. *Ground Water* **2010**, *48*, 6–7. [CrossRef]

48. Sanner, B.; Hellström, G.; Spitler, J.; Gehlin, S. Thermal Response Test–Current Status and World-Wide Application. In Proceedings of the World Geothermal Congress 2005, Antalya, Turkey, 24–29 April 2005.

49. Sliwa, T.; Gonet, A. Otworowe wymienniki ciepła jako źródło ciepła lub chłodu na przykładzie Laboratorium Geoenergetyki WWNiG AGH. *Wiertnictwo Nafta Gaz* **2011**, *28*, 419–430. (In Polish)

50. Sliwa, T.; Rosen, M.A.; Jezuit, Z. Use of oil boreholes in the Carpathians in geoenergetics systems historical and conceptual review. *Res. J. Environ. Sci.* **2014**, *8*, 231–242. [CrossRef]

51. Gonet, A.; Sliwa, T.; Stryczek, S.; Sapinska-Sliwa, A.; Pająk, L.; Jaszczur, M.; Złotkowski, A. *Metodyka Identyfikacji Potencjału Cieplnego Górotworu wraz z Technologią Wykonywania i Eksploatacji Otworowych Wymienników Ciepła*; Gonet, A., Ed.; Wydawnictwa Uczelniane AGH: Krakow, Poland, 2011. (In Polish)

52. Plewa, S. *Rozkład Parametrów Geotermalnych na Obszarze Polski*; Wydawnictwo CPPGSMiE PAN: Krakow, Poland, 1994. (In Polish)

53. Somerton, W.H. *Thermal Properties and Temperature-Related Behavior of Rock/Fluid Systems*; Elsevier: Amsterdam, The Netherlands, 1992.

54. Rohner, E.; Rybach, L.; Schärli, U. A New, Small, Wireless Instrument to Determine Ground Thermal Conductivity In-Situ for Borehole Heat Exchanger Design. In Proceedings of the World Geothermal Congress, Antalya, Turkey, 24–29 April 2005; pp. 1–4.

 *energies*

Article

Hydraulic Experiments on a Small-Scale Wave Energy Converter with an Unconventional Dummy Pto

Luca Martinelli *, Matteo Volpato, Chiara Favaretto and Piero Ruol

Department of Civil, Architectural and Environmental Engineering, Università di Padova, 35122 Padova, Italy; matteo.volpato@unipd.it (M.V.); chiara.favaretto@dicea.unipd.it (C.F.); piero.ruol@unipd.it (P.R.)
* Correspondence: luca.martinelli@unipd.it

Received: 28 December 2018; Accepted: 25 March 2019; Published: 29 March 2019

Abstract: This paper investigates on a Wave Energy Converter (WEC) named Energy & Protection, 4th generation (EP4). The WEC couples the energy harvesting function with the purpose of protecting the coast from erosion. It is formed by a flap rolling with a single degree of freedom around a lower hinge. Small-scale tests were carried out in the wave flume of the maritime group of Padua University, aiming at the evaluation of the device efficiency. The test peculiarity is represented by the system used to simulate the Power Take Off (PTO). Such dummy PTO permits a free rotation of two degrees before engaging the shaft, allowing the flap to gain some inertia, and then applying a constant resistive moment. The EP4 was observed to reach a 35% efficiency, under short regular waves. The effects, in terms of coastal protection, are small but not negligible, at least for the shortest waves.

Keywords: wave energy converters; Power Take Off; EP4; latching; wave flume; floating; moorings; renewable electricity generation systems; SDEWES 2018

---

1. Introduction

The availability of renewable energy in the oceans [1] has long inspired many inventors of wave energy converters. In Europe, since the 1990s, several devices were developed by north-west European companies [2]. Conversely, due to scarce wave energy resources in the Italian seas [3,4], for many years the Italian contribution to the development of wave energy converters has been limited to the pioneering work of Boccotti, through the REWEC 3 J-type WEC, patented in 1998 [5].

The first WECs developed in the North European Countries reached a rather high technological readiness level [6,7] but suffered from the harsh oceanic environment at demonstration phase. Many developers eventually closed their companies or suspended their activities, losing part of their competitive advantage in favor of more recent concepts, so that now it is not too late to start with new patents.

In view of such problematic R&D processes of WECs, some authors even suggested that a mild wave climate, like the Mediterranean one [8,9], is suited to the development and, for some cleverly conceived concepts, to the commercial phase. In recent years, ENEA (the Italian National Agency for New Technologies, Energy and Sustainable Economic Development) started to gain interest in the marine energy sector, fostering new initiatives.

As a result, Italian inventors started proposing many new concepts. In the scientific literature we can find, beside the REWEC 3, the ISWEC [10], OBREC [11], SeaBreath [4], ShoWED [12], and DEIM [13], Tecnomac (EDS, [14]). Other devices have not published the tests in the scientific literature, but were presented to business events, e.g., 40South Energy (through several devices, the most recent being the interesting H24 module, http://www.40southenergy.com/2018/09/2294/), Swaths, Generma, Onda, WaveAbsorber, WEM/WOM, EP4.

The last of these devices is the one studied in this paper. The full name is Energy and Protection, 4th generation (EP4), patented by Dario Bernardi. In order to overcome the disadvantages of the

---

low energetic content of the Italian sea, the EP4 is designed to achieve a secondary objective, i.e., the protection of the coast from the wave action, thus providing a defense against erosion (similarly to other concepts, see [15]).

Following a well-established R&D roadmap (see e.g., [16]), the device was tested in a physical model hydraulic laboratory under regular waves, on a small scale. The inventor built a 1:10 model, with no generator, and tested it briefly at sea, to achieve a proof of concept, i.e., the ability to rotate a shaft. The same model (integrated to fit the flume frames) has been tested in the wave flume of Padova University.

The aim of the paper is to assess the device effectiveness in terms of power production and ability to protect the rear beach.

For a real WEC, the hydraulic, electrical or mechanical Power Take Off (PTO) is the system that allows the generation of electricity. On a small scale model, the real physical PTO cannot be modeled in detail, and no electricity is produced. To maintain the same overall dynamic behavior, a system is used that just mimics the damping and resistance effects, restraining the WEC movements in similitude to the physical PTO. Such system is termed dummy PTO.

It is well known [17,18] that an appropriate control technology has the capability to significantly affect the amount of energy taken from WECs, and the choice of the PTO significantly affects the final result. Therefore, the dummy PTO was built in agreement with the inventor's requirements.

Finally, the restraining force and movements were measured, in order to obtain the amount of energy dissipated by the system, to be interpreted as the "converted energy".

The paper layout is as follows. After this introduction, a classification of the different types of dummy PTOs used in the hydraulic tests is presented. Then, the experiments are described, including the facility, the scale model of the EP4, the selected dummy PTO, the test programme, the instrumentation and the analysis methods. Results are then described in terms of free oscillations, energy harvesting capacity and wave attenuation. Finally, the conclusions are drawn, with some comments on future developments.

2. Classification of Dummy PTOs

All WEC devices, under the wave action, are able to produce some movements or to spin a wheel. The ability to produce a movement (or rotation, e.g., of a turbine wheel) is considered a "proof of concept", i.e., a proof that wave energy has been converted into kinetic energy in some instant.

In order to evaluate the amount of harvested energy it is possible to restrain the movement with a force (or torque, in case of a turbine), that mimics the effect of a real PTO. Such restraining force (torque) R is sometimes called the PTO "load". The device that allows to apply such load is sometimes called "dummy" PTO.

The product of R by the velocity (or angular velocity) v is the power dissipated by the dummy PTO. The average of the dissipated power is assumed to be the "converted energy" E_c:

$$E_c = <R(t)\, v(t)> \tag{1}$$

Obviously, the velocity v is directly affected by the load. If the load is too large the device is totally restrained, with $v = 0$, and from Equation 1 we see that $E_c = 0$. If $R = 0$, v achieves the maximum value but, again, $E_c = 0$. The optimal value of the load is the one that gives the largest value of E_c.

It should be pointed out that the optimal $R(t)$ is not constant, and may depend on the present and future state (position and velocity) of the WEC.

At prototype scale, the load is likely to be controlled (through the inverter) in real time, with substantial advantages in the achieved efficiency. During physical model tests, a similar system can be achieved with motor controlled by a PLC (available at the University of Padova) enabling the "active" PTO to exert any reaction force, based on the measured information of the system state (see e.g., by [19]). This investigation, however, increases significantly the costs of the experiments.

This level of accuracy is not suited for devices at an initial stage of development, such as the case of the EP4. More frequently, passive PTOs are designed. On the one hand, the type of PTO significantly changes the overall performance at the lab scale and it must be selected with care. On the other hand, the inventor is usually not financially supported and the cost of the experiment must be kept extremely low.

A simple classification of the existing passive PTOs is given below, and it is based on the achieved value of R:

(1) R is constant. It is achieved by applying a friction force [16]. As a mere example, we may imagine that the floating WEC induces the vertical movement of a bar. Then, the bar is forced to slither between jaws. The jaws apply a different friction force, based on the setting.

(2) $R(t)$ is proportional to the velocity $v(t)$. It is achieved by oil filled dashpots or pneumatic dampers [20]. In the example above, the bar may be attached to a recycled pneumatic cylinder. By varying the hole in the air cylinder, the piston proportionality constant is modified.

(3) R is affected by the position. This is the case selected for the experiments, as described in Sub-Section 3.2. Another example is the pre-stressed piston applied to the Weptos in [21], who described the possible benefits of a negative stiffness mechanism in terms of combined wave–PTO interaction (later confirmed by [22]).

The passive PTO is usually realized with adjustable (and repeatable) "settings". The optimal load is found by selecting the "setting" that gives the larger value of E_c, usually under regular waves.

The settings are assumed to depend on the wave period but not on wave height. There is no proof that the ideal load found under regular wave conditions (of given period) is optimal also under irregular waves (of same peak period), although this has been verified for some experiments (e.g., [23]).

A final remark should address the PTO design load. The cost of the real PTO depends on the maximum value of power that may be converted. When the input power exceeds a threshold, the device may be either limited or disconnected, entering a "safe" mode. In [24], a quick way to select the design value on the basis of the hydraulic model tests is suggested. Whenever the power exceeds the design value, and this is a frequent case for irregular extreme waves, it is not realistic to assume that energy is harvested. Therefore Equation 1 should be modified on the basis of the PTO expected behavior.

3. Experimental Investigation on the WEC

3.1. The Facility

Physical model tests on the EP4 were carried out in the 36 m $\times$ 1.0 m $\times$ 1.4 m wave flume of Padova University (Figure 1). The wavemaker is an oleodynamic roto-translational paddle equipped with a hardware wave absorption system. To perform the tests, a fixed bottom was used. For average tide conditions, water depth was 0.5 m at the paddle, and 0.4 m at the structure. An array composed of four wave gauges (WG) was located 9.2 m in front of the model to measure the incident and reflected waves. Another gauge was placed 2.0 m behind the device. The instrumentation used in the tests also comprised of a load cell and a video-camera used to monitor the displacements. A MATLAB-based code extracts the flap rotation in time $\theta(t)$. After a lowpass filter, the signal is derived in time to obtain the rotational velocity $v(t)$.

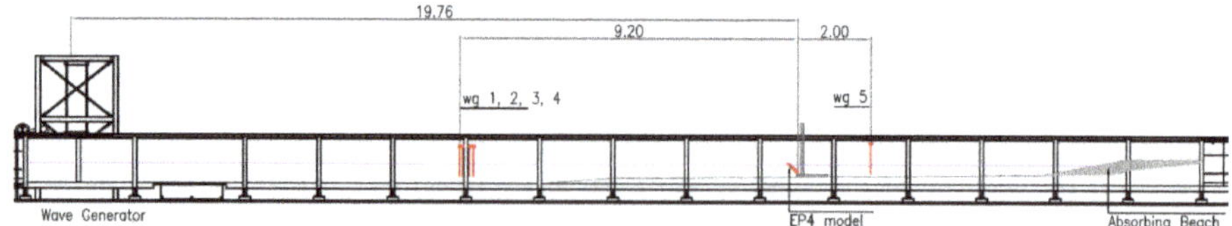

Figure 1. Wave flume and test setup with wave gauge and structure positions 3.2. The EP4 device.

The EP4 is a floater hinged at the base, free to oscillate, and connected by a chain to an upper shaft (Figure 2). The floater is 84 cm long, 37 cm wide, with total volume 0.013 m^3. In the initial position it bent toward the incident waves (the wave generator is at the left, in the figure). A stabilizing bar (black in the figure), placed at the side of the center of floatation, assures that only one position is stable (unless the water level was extremely low, in which case there are two stable positions). The model was built in scale 1:10.

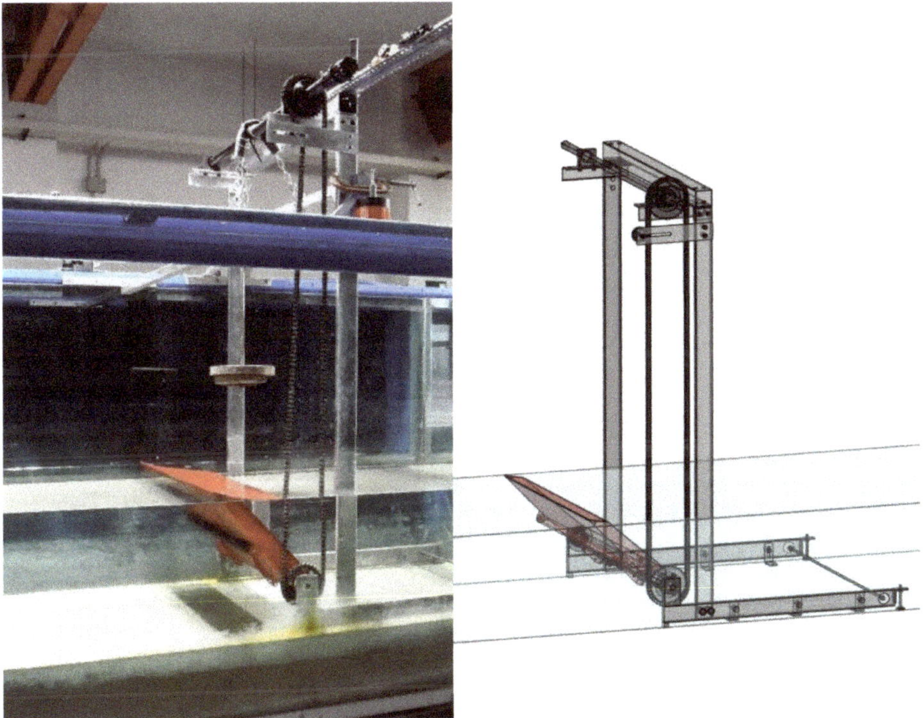

Figure 2. EP4 in the flume, with hanging load that applies a friction restraining the bar rotations.

The chain in Figure 2 was attached to an upper disk. The most simple PTO is the application of a constant resistive friction force to the disk (i.e., case no. 1 in the previous Subsection), and the actual force time history $R(t)$ is represented in Figure 3, upper panel. Obviously the resistive force changed sign whenever the oscillation (the sinusoidal continuous line) reversed its direction. The dashed line represents the device velocity. A sinusoidal behavior is plotted for simplicity, being very similar to the actual measured signal $v(t)$.

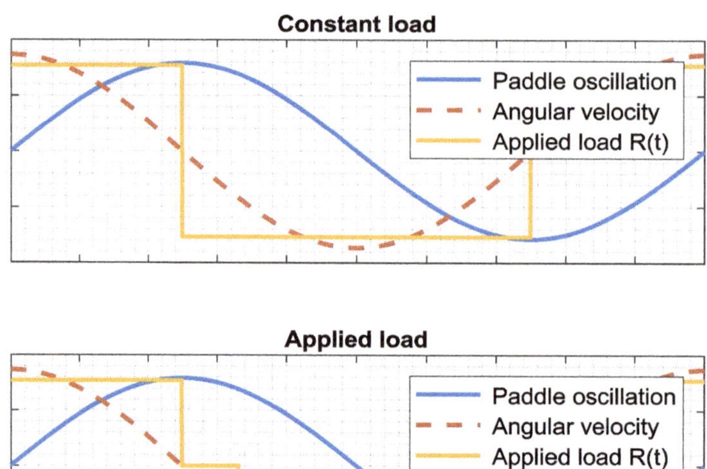

Figure 3. Scheme of the time history of the applied PTO load.

When the oscillation changed direction, i.e., when the paddle was at the extreme right or extreme left, the velocity was null. In these conditions, the friction force restrained the device movements, with no benefit, since the velocity was low and therefore the total contribution to the produced energy was minimal (Equation (1)).

By a simple procedure, i.e., enlarging the hole that connects the crown gear to the shaft (Figure 4), it was possible to delay the moment when the upper shaft was engaged.

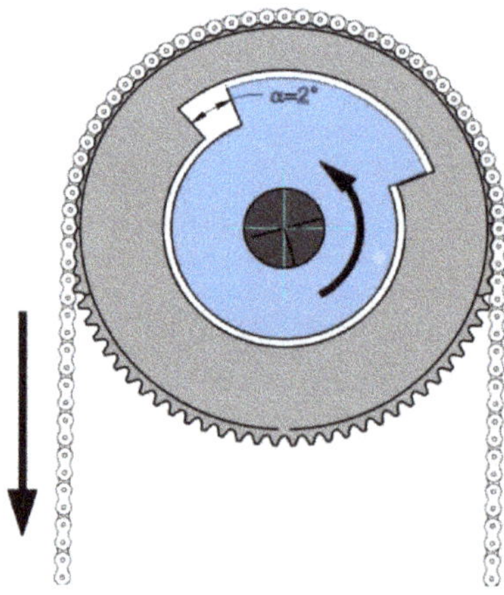

Figure 4. Details of the dummy PTO.

Therefore, the force was applied with some delay, after the inversion of the motion. Figure 3 lower panel shows this case. The rotating flap can therefore accelerated, arriving to a larger value of the velocity (with no useless restraining force).

A similar delay would occur in a possible improved design where the shaft is coupled to a "double gear and clutch system" used to spin a flywheel in the same direction. In order to engage with the flywheel, the velocity must exceed a certain threshold, or else the clutch does not engage. Therefore, the selected dummy PTO is coherent with the future developments.

A few settings of the PTO loads were preliminarily analyzed, and two were found more significant. In both cases, the restraining forces engaged after a rotation of approximately 1 degree. The two settings of the PTO were measured in static conditions to be 5.5 and 7.7 Nm. The friction force under dynamic conditions, used for the evaluation of E_c (Equation (1)), was assessed to be much lower and approximately half of this value, and equal to $L_1 = 2.9$ Nm, $L_2 = 4.1$ Nm when the shaft was engaged, 0 otherwise (see Figure 3, lower panel).

3.2. Test Programme

The test programme includes an initial analysis of the free oscillation, with evaluation of the natural period of oscillation T_N, that was found to be 6 s. This value appears to be too large (19 s at prototype scale) to favor resonance, and this is a first noteworthy conclusion.

Then, 3 water levels were considered, i.e., target depth, low tide and high tide. The scaled values are 0.40 m, 0.35 m and 0.45 m respectively. Free oscillations were measured for all water depth, whereas the power production was only evaluated for d = 0.40 m (see Table 1). Wave height and period vary in the range 2–8 cm and 1–4 s, according to Table 2 (free oscillations) and Table 3 (with the applied load). A number of additional tests were also carried out with different loads, following in a confused pattern, that need not be presented here as they do not contribute to the conclusions.

Table 1. Summary test programme.

	Free	Optimal Load (R = 4.1 Nm)	Other Loads
d = 35 cm	X		
d = 40 cm	X	X	X
d = 45 cm	X		

Table 2. Free oscillation tests.

	T = 1 s	T = 2 s	T = 4 s
H = 2 cm	X	X	X
H = 4 cm	X	X	X
H = 6 cm	X	X	X
H = 8 cm	X	X	X

Table 3. Tests with PTO load.

	T = 1 s	T = 2 s	T = 4 s
H = 2 cm			
H = 4 cm	X	X	X
H = 6 cm			
H = 8 cm	X	X	X

4. Results

4.1. RAO

The first result presents the Response Amplitude Operator (RAO) measured in the absence of applied load (free oscillation). The RAO is defined as the ratio between the total extension of the oscillation and the incident wave height. Note that although the generated waves are regular, the signal is not "exactly periodic" in time. The incident wave component is separated from the reflected one by an array of wave gauges, through the Zelt–Skielbreya [25] method. The incident wave height used to evaluate the RAO is assumed to be the H_{rms} characteristic value of the incident component. In analogy, the oscillation amplitude used for the RAO is the rms value (among the possible averages, the "rms" is chosen since this characteristic value preserves the total energy). The generated period is found to be equal to the target.

Figures 5–7 present the RAO for the different water depths, function of the period T and the target wave height H.

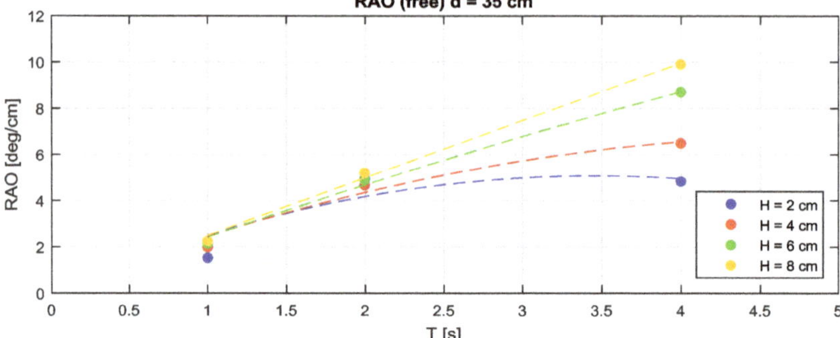

Figure 5. Measured RAO. Free oscillations, water depth = 0.35 cm.

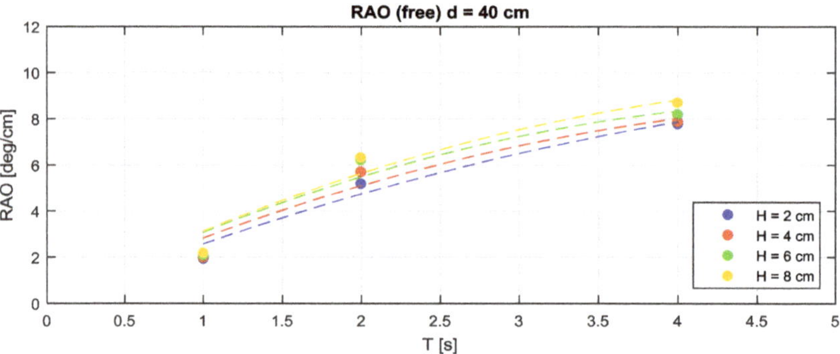

Figure 6. Measured RAO. Free oscillations, water depth = 0.40 cm.

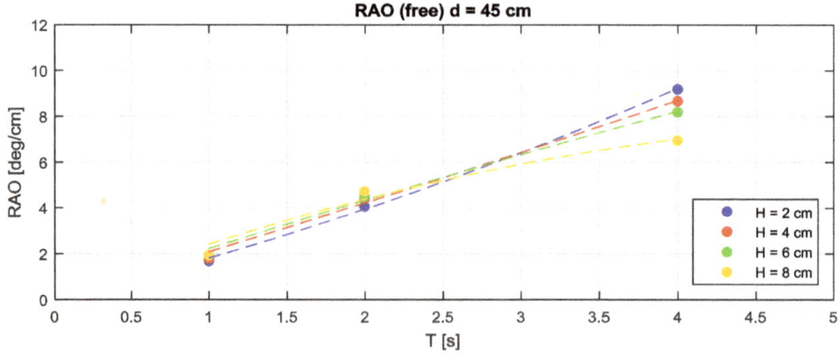

Figure 7. Measured RAO. Free oscillations, water depth = 0.45 cm.

In all cases, the RAO grows almost proportionally with the tested wave periods (as a consequence of the horizontal stretching of the wave orbits with T ranging from 1–4 s).

For T = 1 and 2 s, the RAO behavior and, consequently, the oscillation angle, is similar for the three water depths (d = 0.35, 0.40, 0.45 m). In particular, for T = 1 s (i.e., 3 s at prototype scale) the device oscillations is approximately 2°/cm, independent from H and d. For instance, for H = 8 cm, the oscillations are 2°/cm × 8 cm = 16°, and there is one oscillation every second. For T= 2 s, the RAO is approximately 6°/cm.

For T = 4 s, the RAO depends significantly both on water depth and on wave height: (a) for H = 2 and 4 cm, the oscillation increases with increasing water depths; (b) for H = 6 cm the oscillation is constant for the three tested water depths; (c) for H = 8 cm, the oscillation decreases with water depth.

To explain this behavior, a number of factors should be considered. The torque applied by the wave depends on the wave frequency, the wave orbital shape, the oscillation range (i.e., the obliquities spanned by the device during its motion), the nonlinear shape of the wave (higher crests and longer troughs) and, the abrupt non-linear effect due to possible submergence of the device during its movement, so that high and small waves induce qualitatively different responses. The device oscillation range depends on water depth since the initial equilibrium position of the floater changes for the three cases. For d = 0.35 m, 0.40 m and 0.45 m, the initial rest angle (with respect to the horizontal) is approximately 35°, 50° and 60°.

When the device is restrained by the PTO, the device movements are significantly reduced and the RAO is obviously much smaller. Figure 8 shows the RAO for the case with L2 = 4.1 Nm, that was found to give a slightly larger power for all wave periods. Figure 8 shows that the RAO measured in correspondence to H = 8 cm was consistently larger than the case H = 4 cm.

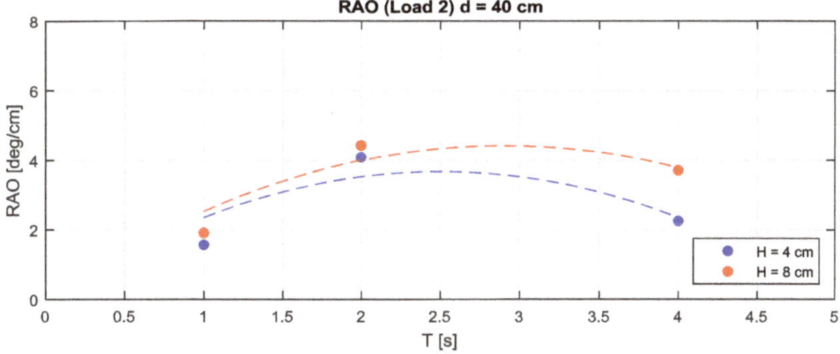

Figure 8. Measured RAO. Optimal load applied, water depth = 0.40 cm.

4.2. Measured Efficiency

The efficiency is obtained as the ratio between the converted energy E_c and the incident wave energy, $E_i = \rho g H^2_i/8$.

The incident wave energy flux is proportional to the group celerity and therefore, quantitatively, the highest power production is achieved for longer waves. However, in relative terms, the device efficiency decreases with the wave period, having an opposite behavior of the RAO achieved without load. In fact, larger RAOs do not correspond to a more effective transfer mechanism from wave to mechanical energy, and in fact the largest oscillations are achieved in stationary conditions, when the wave energy transfer mechanism is low. The RAO is rather affected by the wave kinematic.

Figure 9 shows the efficiency measured in the wave flume. The higher efficiency is approximately 35%, measured for T = 1 s. The case H = 8 cm is more relevant, being associated to a larger power output.

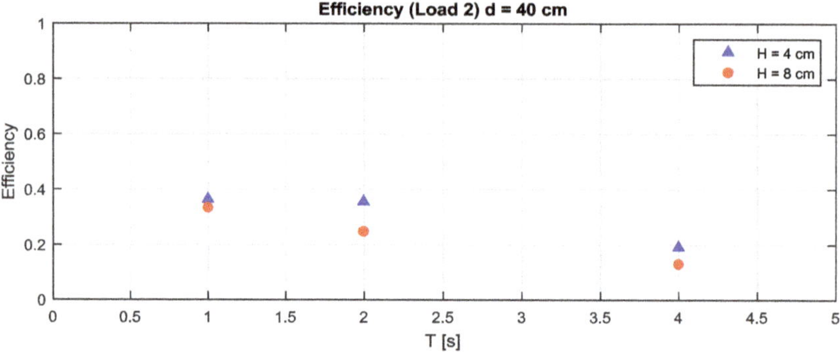

Figure 9. Measured Efficiency. Optimal load applied, water depth = 40 cm.

4.3. Wave Attenuation

The wave reflection and transmission are certainly affected by the device movements and hence, as seen in the previous chapters, by the PTO load. Figures 10 and 11 show, for the case with optimal PTO load, the reflection coefficients defined as the ratio between reflected and incident wave height, and the transmission coefficient defined as the ratio between transmitted and incident wave height. As expected, the transmission coefficient increases for long periods. It is in fact easy to understand that the device cannot limit the transmission of slow oscillations (long period).

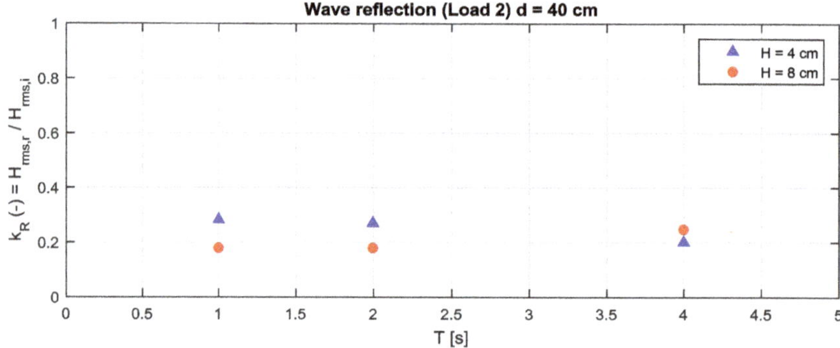

Figure 10. Reflection coefficient. Optimal load applied, water depth = 40 cm.

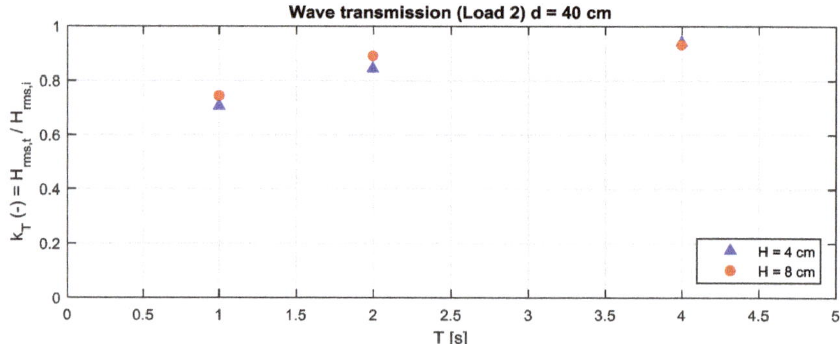

Figure 11. Transmission coefficient. Optimal load applied, water depth = 40 cm.

4.4. Energy Balance

The obtained results (k_R, k_T and η) are checked by means of an energy balance condition. Energy flux conservation requires that, once the stationary conditions are reached, the incident energy flux is given by the sum of reflected, transmitted, dissipated and converted ones. For horizontal bed, the energy flows with constant group celerity, and since the channel width is also constant, the equation can be written as balance of energies:

$$E_i = E_r + E_t + E_c + E_d \tag{2}$$

The wave energy is proportional to the square of the wave height. If such proportionality factor is $\rho g/8$, Equation (2) becomes:

$$\rho g/8\, H^2_i = \rho g/8\, H^2_r + \rho g/8\, H^2_t + E_c + E_d \tag{3}$$

Let's define the efficiency η as the ratio between converted and incident wave energy, and ε as the sum of the measurement errors and the ratio between the energy dissipated by the device movements and the incident wave energy. Dividing all terms in Equation (3) by the incident wave energy we find:

$$1 = k^2_r + k^2_t + \eta + \varepsilon \tag{4}$$

Figure 12 shows this balance, accounting for the first 3 terms in the RHS of Equation (4). It shows how the incident energy is distributed into reflected, transmitted and harvested energy, among the different tested wave periods.

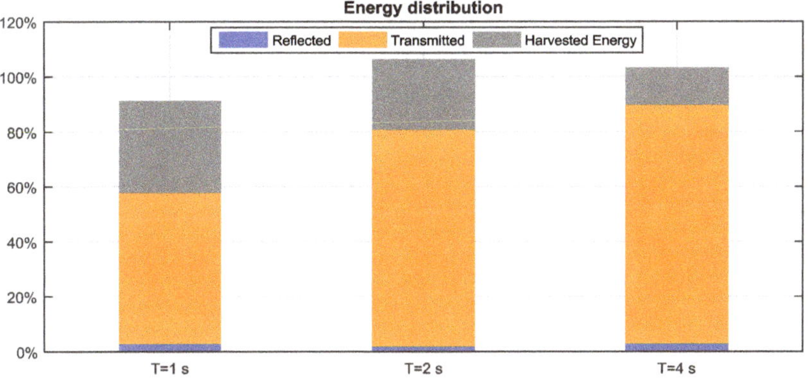

Figure 12. Distribution of incident wave energy. Optimal load applied, H = 8 cm, d = 40 cm.

The sum of the three terms is expected to be always lower than 1 (i.e., 100%), since the dissipations in water are usually not small (in absence of moment errors, $\varepsilon > 0$). However, even assuming $\varepsilon = 0$, for T > 2 s, the sum is larger than 100%, and this can only be ascribed to measurement errors. However, the error is not significant (slightly above 5%) and the measured quantities may be considered sufficiently accurate.

5. Conclusions

The dynamics and the efficiency of a new WEC named EP4 were tested in the wave flume of Padova University. Tests were performed in scale 1:10 with respect to a possible application in front of a low-lying sandy coast, at 4 m water depths, where the installation could benefit by its double purpose of harvesting the wave energy and protecting the shore.

The EP4 efficiency was obtained with a novel control system characterized by a PTO load delayed with respect to the device movements.

The device appears to resonate at a very high period (19 s at prototype scale), and it is suggested that the design is modified to reduce the added mass associated to its movements, so that the natural periods are close to the main wave periods, to utilize the motion resonance effect and therefore to achieve even better performance under short waves.

The peculiar PTO used to restrain the device motion resulted very efficient and easy to build. The resistive force, simulating the generator, was designed to be zero when the flap changes direction, in order to allow the WEC to gain some kinetic energy. Rotations were accurately measured by a HD video camera, and automatically post-processed to obtain the rotational amplitude in time. Incident and reflected waves were measured by an array of 4 wave gauges, and transmission by a 5th gauge.

The device efficiency was found to be 35% for periods of 1 s (3 s at full scale) and a little less for longer periods, characterized by a higher energy content. Since only a single value of PTO delay was tested, an optimized control strategy may lead to larger efficiency also for long waves.

Irregular wave conditions, naturally associated with a lower performance, will probably be tested through additional experiments. In this case, the PTO will be equipped with a flywheel to smooth the irregularities of the wave field. To force the rotation of the shaft along the same direction, a gear and clutch system can be used: during the motion inversion, before the flywheel is engaged, the application of the resistive load would be delayed in similitude to the tested conditions.

The performance in terms of coastal protection followed the expectations: The measured wave transmission coefficient was respectively 75% and 90% for waves with period of 1 and 2 s (3 to 6 s at full scale).

Author Contributions: Conceptualization, L.M., M.V., P.R., C.F.; Methodology, L.M., M.V., P.R., C.F.; Investigation, L.M., M.V., P.R., C.F.; Writing and Editing, L.M., M.V., P.R., C.F.; Supervision, L.M., M.V., P.R., C.F.

Funding: This research received no external funding.

Acknowledgments: We thank Dario Bernardi, inventor of the EP4, for sharing the information on his device. This paper is a deep revision of the note presented at SDEWES 2018 Conference, code SDEWES2018.00422, with title "EXPERIMENTAL INVESTIGATION ON THE DYNAMICS OF A FLOATING WEC WITH PTO PHASE CONTR" and selected for resubmission at Energies.

Conflicts of Interest: The authors declare no conflict of interest.

References

1. Barstow, S.; Mørk, G.; Lønseth, L.; Mathisen, J.P. WorldWaves wave energy resource assessments from the deep ocean to the coast. *J. Energy Power Eng.* **2011**, *5*, 730–742.
2. Magagna, D.; Monfardini, R.; Uihlein, A. *JRC Ocean Energy Status Report 2016 Edition*; Technology, Market and Economic Aspects of Ocean Energy in Europe; Publications Office of the European Union: Luxembourg, 2016.
3. Vicinanza, D.; Cappietti, L.; Ferrante, V.; Contestabile, P. Estimation of the wave energy in the Italian offshore. *J. Coast. Res.* **2011**, *64*, 613–617.
4. Martinelli, L.; Pezzutto, P.; Ruol, P. Experimentally based model to size the geometry of a new OWC device, with reference to the Mediterranean Sea wave environment. *Energies* **2013**, *6*, 4696–4720. [CrossRef]
5. Boccotti, P. Comparison between a U-OWC and a conventional OWC. *Ocean Eng.* **2007**, *34*, 799–805. [CrossRef]
6. Parmeggiani, S.; Kofoed, J.P.; Friis-Madsen, E. Experimental study related to the mooring design for the 1.5 MW Wave Dragon WEC demonstrator at DanWEC. *Energies* **2013**, *6*, 1863–1886. [CrossRef]
7. Castro-Santos, L.; Silva, D.; Bento, A.; Salvação, N.; Guedes Soares, C. Economic Feasibility of Wave Energy Farms in Portugal. *Energies* **2018**, *11*, 3149. [CrossRef]
8. Liberti, L.; Carillo, A.; Sannino, G. Wave energy resource assessment in the Mediterranean, the Italian perspective. *Renew. Energy* **2013**, *50*, 938–949. [CrossRef]
9. Arena, F.; Barbaro, G. The Natural Ocean Engineering Laboratory, NOEL, in Reggio Calabria, Italy: A Commentary and Announcement. *J. Coast. Res.* **2013**, *29*, vii–x. [CrossRef]

10. Bracco, G.; Cagninei, A.; Giorcelli, E.; Mattiazzo, G.; Poggi, D.; Raffero, M. Experimental validation of the ISWEC wave to PTO model. *Ocean Eng.* **2016**, *120*, 40–51. [CrossRef]

11. Contestabile, P.; Di Lauro, E.; Buccino, M.; Vicinanza, D. Economic assessment of Overtopping BReakwater for Energy Conversion (OBREC): A case study in Western Australia. *Sustainability* **2016**, *9*, 51. [CrossRef]

12. Martinelli, L.; Ruol, P.; Favaretto, C. Hybrid structure combining a wave energy converter and a floating breakwater. In Proceedings of the 26th International Ocean and Polar Engineering Conference, Rhodes, Greece, 26 June–2 July 2016; International Society of Offshore and Polar Engineers: Houston, TX, USA, 2016.

13. Franzitta, V.; Catrini, P.; Curto, D. Wave energy assessment along Sicilian coastline, based on DEIM point absorber. *Energies* **2017**, *10*, 376. [CrossRef]

14. Negri, M.; Malavasi, S. Wave Energy Harnessing in Shallow Water through Oscillating Bodies. *Energies* **2018**, *11*, 2730. [CrossRef]

15. Mendoza, E.; Silva, R.; Zanuttigh, B.; Angelelli, E.; Andersen, T.L.; Martinelli, L.; Nørgaard, J.; Ruol, P. Beach response to wave energy converter farms acting as coastal defence. *Coast. Eng.* **2014**, *87*, 97–111. [CrossRef]

16. Pecher, A.; Kofoed, J.; Espedal, J.; Hagberg, S. Results of an experimental study of the Langlee wave energy converter. In Proceedings of the Twentieth International Offshore and Polar Engineering Conference, Beijing, China, 20–25 June 2010; Volume 1, pp. 877–885.

17. Falnes, J. *Principles for Capture of Energy from Ocean Waves. Phase Control and Optimum Oscillation*; Department of Physics, NTNU: Trondheim, Norway, 1997.

18. Falnes, J. Optimum control of oscillation of wave-energy converters. *Int. J. Offshore Polar Eng.* **2002**, *12*, 147–155.

19. Beatty, S.; Ferri, F.; Bocking, B.; Kofoed, J.P.; Buckham, B. Power take-off simulation for scale model testing of wave energy converters. *Energies* **2017**, *10*, 973. [CrossRef]

20. Flocard, F.; Finnigan, T.D. Experimental investigation of power capture from pitching point absorbers. In Proceedings of the Eight European Wave and Tidal Energy Conference, Upsala, Sweden, 7–10 September 2009.

21. Peretta, S.; Ruol, P.; Martinelli, L.; Tetu, A.; Kofoed, J.P. Effect of a negative stiffness mechanism on the performance of the Weptos rotors. In Proceedings of the VI International Conference on Computational Methods in Marine Engineering (MARINE 2015), Rome, Italy, 15–17 June 2015; pp. 58–72.

22. Têtu, A.; Ferri, F.; Kramer, M.; Todalshaug, J. Physical and Mathematical Modeling of a Wave Energy Converter Equipped with a Negative Spring Mechanism for Phase Control. *Energies* **2018**, *11*, 2362. [CrossRef]

23. Martinelli, L.; Zanuttigh, B.; Kofoed, J.P. Selection of design power of wave energy converters based on wave basin experiments. *Renew. Energy* **2011**, *36*, 3124–3132. [CrossRef]

24. Martinelli, L.; Zanuttigh, B. Effects of Mooring Compliancy on the Mooring Forces, Power Production, and Dynamics of a Floating Wave Activated Body Energy Converter. *Energies* **2018**, *11*, 3535. [CrossRef]

25. Zelt, J.A.; Skjelbreia, J. Estimating incident and reflected wave fields using an arbitrary number of wave gauges. In Proceedings of the 23rd International Conference on Coastal Engineering (ICCE), Venice, Italy, 4–9 October 1992; ASCE: Reston, VA, USA, 1992; Volume 1, pp. 777–789.

Article

Torrefaction as a Valorization Method Used Prior to the Gasification of Sewage Sludge

Halina Pawlak-Kruczek [1,*], Mateusz Wnukowski [1], Lukasz Niedzwiecki [1], Michał Czerep [1], Mateusz Kowal [1], Krystian Krochmalny [1], Jacek Zgóra [1], Michał Ostrycharczyk [1], Marcin Baranowski [1], Wilhelm Jan Tic [2,3] and Joanna Guziałowska-Tic [2]

[1] Department of Boilers, Burners and Energy Systems; Faculty of Mechanical and Power Engineering; Wroclaw University of Science and Technology, 50-370 Wrocław, Poland; mateusz.wnukowski@pwr.edu.pl (M.W.); lukasz.niedzwiecki@pwr.edu.pl (L.N.); michal.czerep@pwr.edu.pl (M.C.); mateusz.kowal@pwr.edu.pl (M.K.); krystian.krochmalny@pwr.edu.pl (K.K.); jacek.zgora@pwr.edu.pl (J.Z.); michal.ostrycharczyk@pwr.edu.pl (M.O.); marcin.baranowski@pwr.wroc.pl (M.B.)
[2] Department of Environmental Engineering, Opole University of Technology, 45-758 Opole, Poland; w.tic@po.opole.pl (W.J.T.); j.guzialowska@po.opole.pl (J.G.-T.)
[3] West Technology &Trading Polska Sp. z o.o., 45-641 Opole, Poland
[*] Correspondence: halina.pawlak@pwr.edu.pl; Tel.: +48-71-320-3942

Received: 30 November 2018; Accepted: 2 January 2019; Published: 6 January 2019

Abstract: The gasification and torrefaction of sewage sludge have the potential to make the thermal utilization of sewage sludge fully sustainable, thus limiting the use of expensive fossil fuels in the process. This includes sustainability in terms of electricity consumption. Although a great deal of work has been performed so far regarding the gasification of sewage sludge and some investigations have been performed in the area of its torrefaction, there is still a gap in terms of the influence of the torrefaction of the sewage sludge on its subsequent gasification. This study presents the results from the torrefaction tests, performed on a pilot scale reactor, as well as two consecutive steam gasification tests, performed in an allothermal fixed bed gasifier, in order to determine if torrefaction can be deemed as a primary method of the reduction of tar content for the producer gas, from the aforementioned gasification process. A comparative analysis is performed based on the results obtained during both tests, with special emphasis on the concentrations of condensable compounds (tars). The obtained results show that the torrefaction of sewage sludge, performed prior to gasification, can indeed have a positive influence on the gas quality. This is beneficial especially in terms of the content of heavy tars with melting points above 40 °C.

Keywords: sewage sludge; torrefaction; steam gasification; tars

---

1. Introduction

1.1. Introduction

Sewage sludge is a residue of wastewater processing, is biologically active and consists of water, organic matter, including dead and alive pathogens, as well as organic and inorganic contaminants such as polycyclic aromatic hydrocarbons (PAHs) and heavy metals [1–3]. Utilization methods leading to stabilization and safe recycling are gradually replacing storage, landfilling and land-spreading. In the EU countries, novel methods are becoming increasingly popular [4], due to both environmental and economic reasons, as landfilling is deemed to be the most costly way to dispose of the sewage sludge [5].

Land-spreading is typically the most economical way to dispose of sewage sludge [5]. However, the cost of this can be subject to significant changes, depending on the distance between the sewage treatment plant and the location of the land where spreading takes place [5]. Moreover, any odor-related regulations, as well as the EU Nitrate Directive, might make this practice increasingly difficult. Incineration and co-incineration are also feasible options. However, relatively high moisture and ash content have negative influences on the combustion itself, as well as on the fuel logistics. This is the main obstacle in using the effect of scale in large power plants and combined heat and power (CHP) plants as well as in the cement industry, due to transportation costs. Novel thermal processes are currently a subject of active investigation, due to increasingly common restrictions on landfilling [6].

For example, in Poland (Figure 1), thermal treatment is an increasingly popular route for the utilization of the sewage sludge [7]. It is not difficult to notice from Figure 1 that the use of sewage sludge for land-spreading is fairly stable, and landfilling shows a constant decrease, whereas thermal utilization is increasingly important. In Poland, there are currently at least 45 installations for the drying of sewage sludge, mostly drum and tape dryers, as well as 12 installations using solar energy [8]. Incineration may be performed in existing incineration units (at least 11) that are based on fluidized bed (mostly) and grate furnaces [8–10]. Moreover, incineration is possible in 13 facilities of cement producers in Poland [11], as well as in municipal waste incineration facilities which are completed or in an advanced stage of construction (close to commissioning) in 17 different cities [12,13]. In all of these cases, logistics is critical for the economic feasibility of the solution; therefore, the problem of sewage sludge is the most severe in the case of small and medium size towns, without their own thermal utilization facilities, with limited possibilities of local land-spreading. It should not be overlooked that state-of-the-art thermal utilization leaves the problem of ashes unresolved. However, there are emerging technologies which allow ash to be used for the production of fertilizers [14].

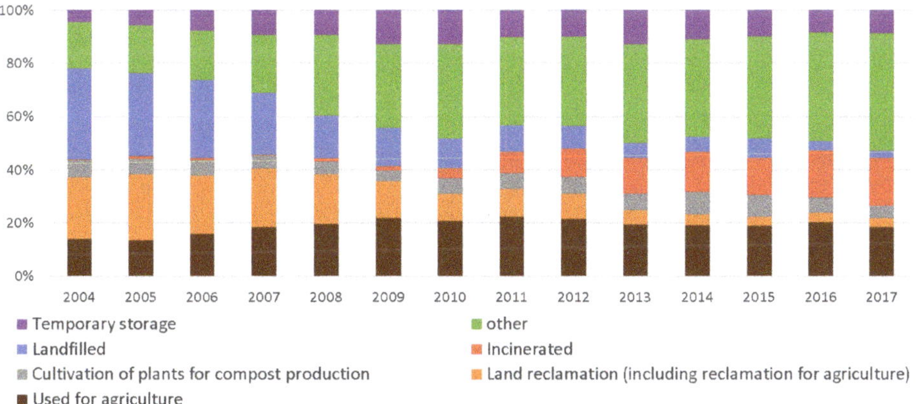

Figure 1. Changing trends for sewage sludge utilization in Poland according to the Main Statistical Office of Poland (GUS) [7] ("Used for agriculture" means the cultivation of all marketed crops, including crops designed to produce fodder; "Landfilled" is used exclusively for deposition in locations which have an official status as landfill areas).

1.2. Torrefaction of Sewage Sludge

Torrefaction is a thermal treatment, performed under a pressure close to ambient and elevated temperature (typically 250 °C to 300 °C) [15]. It can be performed in anaerobic conditions or with the presence of oxygen. Typically, oxygen is present when the flue gases are used as a heat source for direct torrefaction systems [15]. However, in the case of indirectly heated reactors, torgas can also contain some oxygen, due to leakages. Torgas is a by-product of the process that can consist mainly of condensable hydrocarbons, water, carbon dioxide. Some small amounts of carbon monoxide, as well as negligible amount of hydrogen, methane and other permanent gases (hydrocarbons), are also present [15–17].

Some fundamental work laying the foundations for the thermal treatment of sewage sludge has been performed, using TGA (thermo-gravimetric analysis) and DTG (differential thermo-gravimetric) techniques [18–20]. In general, very little has been published strictly on the torrefaction of sewage sludge, whereas some works have reported the results of experiments performed using materials other than sewage sludge, such as different types of industrial sludges.

Pulka et al. determined that torrefaction causes an increase in the higher heating value (HHV) of the pretreated material on a dry ash-free (daf) basis [21]. The significance of this increase was not substantial, due to the increased ash content of the torrefied samples [21]. An increase in ash content and HHV was also observed in another work of Poudel et al. for the torrefaction of sewage sludge [22], as well as for the torrefaction of sewage sludge blends with waste wood [23]. The successful torrefaction of sewage sludge, both using a fluidized bed reactor [24] and auger reactor [25], was performed by Atienza-Martinez et al. Residence times, used during preformed investions, were relevant for the practical operation of the torrefaction installations (13 to 35 min for the auger reactor and 3.6 up to 10.2 min for fluidized bed). A decrease in the energy density was observed in both cases (dry basis).

Huang et al. used a laboratory scale batch reactor to perform the torrefaction of waste from pulp industries [26]. The obtained energy densification ratios varied between 1.26 and 1.5, depending on the process parameters [26]. Huang et al. investigated the microwave co-torrefaction of sewage sludge with Leucaena and noticed a synergetic effect of the use of combined feedstocks for torrefaction [27,28]. An increase in the HHV on a dry ash-free basis was reported, and an HHV of 48 MJ/kg was obtained [27]. It was possible to achieve ratios of O/C and H/C similar to anthracite [27]. Huang et al. investigated the kinetics of the torrefaction of sewage sludge, using the simplified distributed activation energy model [29]. T.X. Do et al. used a series of heat and energy balance calculations to assess the performance of a hypothetical plant, using fry-drying and torrefaction as unit operations, along with a steam boiler using a part of the product [30]. The capability of achieving self-sufficiency, with an additional output of 33% of the dry solid mass, originally fed to the dryer, was reported [30]. The feasibility of using additives for the torrefaction of sewage sludge was also investigated by Pawlak-Kruczek et al. [31]. The study showed that the addition of lignite resulted in an improved heating value of the produced torgas, whereas the addition of CaO resulted in a relatively smaller amount of complex hydrocarbons present in torgas [31].

Peckyte and Baltrenaite studied the properties of carbonized products obtained from residues of various types of the sludges from paper and leather industries [32]. The performed research indicated that the form of biochar restrained the leaching of heavy metals [32], despite the concentrations being considerable when compared with the restrictions set by the regulations [33]. Wang et al. performed an assessment of the environmental effects of the carbonization of sewage sludge and concluded that it may have an overall positive environmental impact in comparison with landfilling and incineration [34].

Various studies reported results from the torrefaction of various types of feedstocks, performed under atmospheres containing oxygen, ranging from values close to 0% up to 15% [35–41]. However, for the case of the torrefaction of sewage sludge under the presence of oxygen, there is little information to be found in the literature. Pawlak-Kruczek, et al. presented results of such a study, performed in a laboratory-scale batch reactor and compared torrefaction in the presence of oxygen with torrefaction in vapothermal conditions [42]. The study concluded that similar results, in terms of mass and energy yields, can be achieved under comparably lower temperatures for vapothermal torrefaction [42].

1.3. Gasification of Raw and Torrefied Sewage Sludge

Gasification is a process of the conversion of a solid fuel into a mixture of gases such as hydrogen, carbon monoxide, carbon dioxide, methane and other hydrocarbons, often called producer gases [43]. The mixture produced for the purpose of chemical synthesis is called a syngas [43]. The optimization of this process typically aims to obtain a gas with a reasonably high calorific value and a minimum amount of impurities. Among these, tars are especially problematic from the practical point of view as they can condense or even become solid at ambient temperature, thus causing deposition problems [43]. According to this definition, tars are all organic contaminants with a molecular mass larger than that of benzene [44] (which amounts to 78.11 Da [45]). The limit of tar content for IC engines is typically from 50 up to 100 mg/m^3 [46–48], although some of the published works state lower values [49,50]. The typical tar content for a producer gas is higher and can reach the orders of magnitude of 1 to 10 g/m^3 [50]. However, the staged gasification of biomass can result in a tar content as low as 50 mg/m^3 [50].

A significant amount of work has been performed to date on the gasification of sewage sludge. Werle reported a decreased temperature and increased concentration of combustible components of producer gases with an increase in the oxygen content of the sludge [51]. Schweizer et al. observed a hydrogen content exceeding 40% during the steam gasification of sewage sludge in a laboratory-scale fluidized bed gasifier [52]. In another work, Werle determined that the laminar flame speed increased with the increasing hydrogen content of the producer gas [53]. Werle and Dudziak assessed that it is possible to use producer gas from sewage sludge in spark-ignition engines [54]. However, Szwaja et al. determined that producer gas from sewage sludge requires a 40% addition of methane to obtain a satisfactory performance of a spark-ignition engine [55]. In another study, Werle confirmed that increased air temperature, at the inlet of a fixed bed gasifier, resulted in an increased yield of combustible compounds during the gasification of sewage sludge [56]. Calvo et al. reported a hydrogen content varying between 21.0% and 20.7% and tar content between 0.846 and 0.585 g/m^3 of the gas from gasification of sewage sludge in a simple atmospheric fluidized bed gasifier [57]. Werle and Dudziak found that tars from the gasification of sewage sludge consisted mostly of phenols and their derivatives [1]. Akkache et al. observed a hydrogen content during the steam gasification of sewage sludge in a laboratory-scale fixed bed rig exceeding 30% [58]. Reed et al. investigated trace element distribution in the gasification of sewage sludge and determined that condensed phase may contain various species containing Ca, ammonium chloride (NH_4Cl), as well as various species containing barium, mercury and zinc [59]. The presence of the latter in the gases could not be explained by existing thermodynamic models [59]. Judex et al. published results from existing sewage sludge gasification plants in Balingen and Manheim (Germany), with respective processing capacities of 1950 t/a and 5000 t/a of dry sewage sludge [60]. Producer gas from the fluidized bed gasifiers on average had lower heating values (LHVs) of 3.2 MJ/m^3 and 4.7 MJ/m^3, respectively [60]. The Balingen gasifier worked with an average gasification temperature of 820 °C, with an average excess air ratio (λ) of 0.33, whereas the gasifier in Manheim worked with an average gasification temperature of 870 °C, with an average excess air ratio (λ) of 0.28 [60]. Sewage sludge in Manheim had a comparably higher carbon content (30.0 %$_{dry}$) and lower ash content (39.5 %$_{dry}$) in comparison with sewage sludge from Balingen, having 16.9 %$_{dry}$ of carbon and 57.0 %$_{dry}$ of ash, respectively [60]. Hydrogen content was not significantly different: on average, 13.1% in Balingen and 13.3% in Manheim [60]. On average, a higher CO content (13.8% comparing to 8.1%) was measured in Manheim, whereas higher CO_2 content was measured in Balingen (16.7% compared to 13.0%) [60]. The average methane content measured in Manheim (4.2%) was roughly double that measured in Balingen (2.1%) [60]. Kokalj et al. proposed using the plasma gasification of sewage sludge as an alternative mean of energy accumulation combined with sewage sludge utilization, as the work proposed the storage of the producer gas, which would be used during peak load and produced in off-peak time [61]. Huang et al. postulated the co-gasification of sewage sludge with torrefied biomass and performed calculations, using the thermodynamic equilibrium model, based upon the Gibbs free energy minimization [62]. The calculation showed that the optimum mixing ratio of wet sewage sludge was between 30% and 55%, depending on the gasification temperature [62]. In general, torrefaction offers enhancements for the subsequent gasification process from a general thermodynamic point of view [63], as well as in terms of a potentially decreased tar content [64]. Striūgas et al. performed research on the use of a plasma reactor for cleaning the gas from a downdraft gasification of the sewage sludge [48]. The study revealed that the use of plasma treatment for the gas produced in a downdraft gasifier can reduce the tar content down to 90 mg/mN^3 [48]. However, there is hardly any information on the gasification of torrefied sewage sludge, let alone any experimental investigation trying to assess the effect of torrefaction on the quality of the gas produced, using this fuel.

2. Novelty, Relevance, Goals and Scope of Work

Although a lot of work has been performed to date on the gasification of sewage sludge, and some investigations have been performed in the area of its torrefaction, there is still a gap in terms of the influence of the torrefaction of the sewage sludge on its subsequent gasification. The attempt to address this gap is an important novelty aspect of this paper. Moreover, studies on the torrefaction of the sewage sludge published to date have focused on the torrefaction performed on bench-scale rigs. On the other hand, the torrefaction experiment performed within the scope of this study was conducted on a pilot-scale device. This study, in essence, is a continuation of the previously published paper [65] containing the concept of a novel, fully sustainable installation for the thermal utilization of sewage sludge.

In order to achieve a completely sustainable installation, it is necessary for it to not only match its own heat requirements, but also its requirements for electricity supply. The easiest way to perform this is to utilize existing infrastructure and use the gas from gasification in an existing CHP unit that utilizes biogas from the anaerobic digestion of the sludge. This, however, introduces the requirement of a reasonably clean producer gas.

Of course, cleaning might be necessary anyway. Nonetheless, its extent can potentially be limited by using torrefaction as a primary method of the reduction of the tar content of the producer gas. Moreover, while the ability to achieve high concentrations of hydrogen in the producer gas can be valuable, it might also be detrimental in terms of the subsequent use of the gas in an engine, due to potential issues with knocking.

This study presents the results from the torrefaction tests performed on a pilot-scale reactor, as well as two consecutive steam gasification tests, performed in an allothermal fixed bed gasifier. A relatively flat bed of the feedstock was used in order to simulate the steam gasification of the traveling grate, in order to validate the feasibility of this concept.

The most important goal of this study was the determination of whether torrefaction can be deemed as a primary method of the reduction of tar content for the gas, produced in the aforementioned gasification process. Moreover, this study introduces a novel way to present the results of tar analysis, named the tar deposition profile diagram. In this diagram, concentrations of individual tarry compounds are arranged according to their respective melting points, which is deemed to be critical in terms of their deposition. This diagram can be used as a simple tool to qualitatively assess the deposition potential of the producer gas, for example in gas coolers, by looking at the difference in the gas temperature between the inlet and the outlet of such a device and checking it against the diagram.

3. Materials and Methods

Samples of the sewage sludge for the suite of torrefaction experiments were obtained after the process of their drying in a commercial-scale paddle dryer, already described in a different paper [65]. Samples of wet sewage sludge used for the drying tests were originally obtained after the fermentation and mechanical dewatering stages of the sewage treatment, performed at the sewage treatment plant in Brzeg Dolny. Standard proximate analysis and the ultimate analysis of both raw sewage sludge and torrefied product were performed according to European standards. References of all the relevant standard procedures are presented in Table 1 (required accuracies are stated in the respective standards). The fusibility of the ashes was assessed using the standardized characteristic temperatures method, as specified in a Polish Standard PN-ISO 540:2001.

Approximately 150 kg of sewage sludge, dried in an industrial-scale paddle dryer, was torrefied using a multistage tape reactor (Figure 2), belonging to the research group of Boilers, Combustion and

Thermal Processes at Wroclaw University of Science and Technology. Drying tests were described in detail in another paper [65].

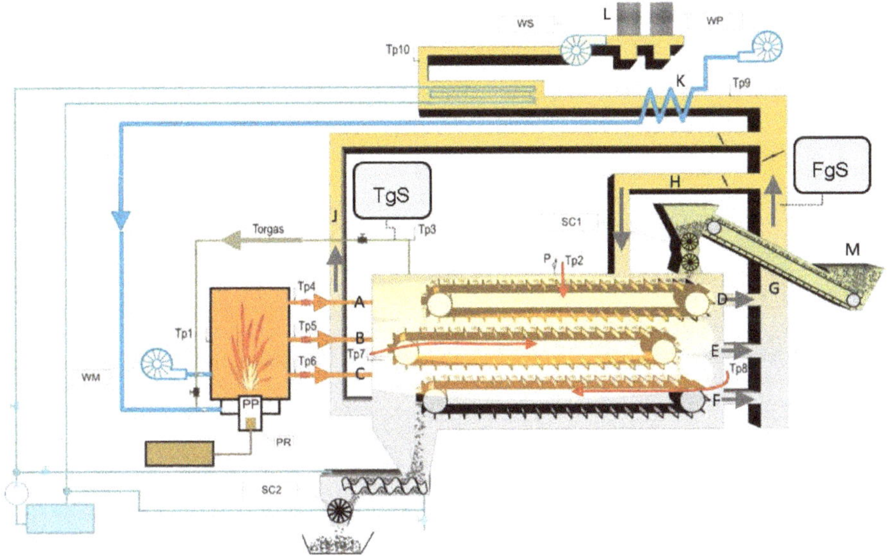

Figure 2. Multistage tape dryer/torrefier developed by Wroclaw University of Science and Technology (Tp—thermocouple; TgS—torgas sampling port; FgS—flue gas sampling port; WP—preheated secondary air blower; WM—primary air blower; WS—flue gas extraction fan; SC1—airlock at the inlet; SC2—airlock at the outlet; PR—pressure regulator; PP—oil burner; P—pressure gauge; A,B,C—ducts delivering hot flue gases to the inside of the shelves; D,E,F—ducts for evacuation of the flue gases out of the shelves; G—main flue gas extraction duct; H—flue gas recirculation to the freeboard of the reactor; J—duct for extraction of the gases out of the freeboard of the reactor; K—combustion air pre-heater; L—bag filters).

The patented technology presented in Figure 2 uses the external heating of the processed material, through the surface of three metal plates. The material is moved through the reactor by a chain conveying system and falls down from one plate into another, until it reaches the output auger, with a water jacket. The raw material is fed from the hopper to the reactor by another chain conveyor. To prevent the leakage of torgas to the surrounding area, airlocks are installed on both ends of the reactor.

The combustion chamber, with a ceramic refractory located beside the torrefier, is used for the burning of the torgas. An oil burner is used as a source of startup heat and as a pilot flame source during normal operation. The torrefier can also be used as a dryer, and in this type of operation, the oil burner is the main source of the heat. When working in torrefaction mode, the installation is operating reasonably close to the autothermal point, whereas the supply of auxiliary heat is determined by the moisture content of the feedstock.

The main advantage of this design is the possibility to significantly minimize the risk of the agglomeration of the particles, due to sticky tars condensing on the surfaces, as processed material only touches the hot surfaces of the reactor. In this way, the operational cost could be reduced by reducing the risk of emergency shutdowns due to clogging.

During the performed experiment, dried sewage sludge was gradually fed from the hopper to the reactor and subsequently torrefied. The average temperature, measured under the top plates of the shelves, was determined to be 391.9 °C. This was an arithmetic average and the calculation was based on the values measured by three thermocouples installed inside each of the shelves, close to the middle of the respective top plates (Tp2, Tp7 and Tp8—see Figure 2). The velocity of the chains along the length of the reactor, with the scrapers attached to them, was set in a way to allow the average total residence time of 20 min, under the assumption that there was no distribution of the particle residence time in the reactor.

In order to assess the severity of the torrefaction process, parameters of mass yield (Y_m) and energy yield (Y_e) were used as performance indicators typically used for that purpose [15,31,66,67]. Mass yield was assessed, using the volatile matter content of both feedstock and product, as proposed by Weber et al. [67]:

$$Y_m = \frac{1 - VM_{feedstock}}{1 - VM_{product}}$$

where

Y_m—mass yield, -;
VM—respective volatile matter content of feedstock and product, $\%_{dry}$.

The well-established formula was used for the calculation of the energy yield [15,21,23]:

$$Y_e = Y_m \cdot \frac{HHV_{feedstock}}{HHV_{product}}$$

where

Y_e—energy yield, -;
HHV—respective higher heating value of feedstock and product, MJ/kg.

Gasification tests were performed using a laboratory scale allothermal batch gasifier (Figure 3), heated by a mantle made of 3 band heaters, installed on the side walls of the reactor. The temperature of the reactor was controlled by a PLC (Programmable Logic Controller), with a type K thermocouple installed inside of the ceramic refractory of one of the band heaters. The temperature of the mantle was set to 900 °C for both tests. This resulted in an average temperature at the edge of the feedstock layer of approximately 675 °C, with temperatures in the freeboard area varying between 760 °C and 350 °C, depending on the height.

Gasification was performed for both raw and torrefied sewage sludge samples, using steam as a gasifying agent. Steam was produced using an electrically heated steam generator, which had been calibrated prior to the performed tests. The calibration involved the measurement of the temperature of the steam at the outlet of the steam generator, as well as the mass flow rate of the steam, measured gravimetrically by condensing in the tank. The calibration allowed the setting of the control knob of the steam generator to achieve the desired temperature and mass flow rate of the live steam. For both tests, the steam generator was set in a way allowing a constant generation of 1000 g of live steam per hour, with an outlet temperature of 96 °C. Both samples were sieved through a set of calibrated sieves (ISO 3310-2 compliant), using a sieve shaker. A stack of the sieves with apertures of 1 mm, 2 mm, 3.15 mm, 4 mm, 5 mm and 6.3 mm was used.

A sample basket with a diameter of 30 cm and height of 45 cm, made of a stainless-steel mesh with an aperture of 0.5 mm, was used for the holding of a layer of 5-cm thick sample material in both of the cases. The basket was placed inside of the hot reactor, which was subsequently closed. It was anticipated that the layer of the material will resemble a layer on a grate in the case of a subsequent scaling up of the gasifier. This was expected to give some indication of the feasibility of the use of a traveling grate in the conceptual gasifier. Steam was fed directly under the bottom mesh (grate) of the sample basket.

A sample of the produced gas was taken from the top of the reactor and put through a series of impinger bottles filled with isopropanol. First, the impinger bottle was installed, using a laboratory grip, in the vicinity of the gas outlet, in order to minimize the length of the teflon hose, connecting the gas outlet of the gasifier with the aforementioned impinger bottle (Figure 3). In this way, the loss of the condensable by-products of gasification was minimized. A series of three impinger bottles, connected to the outlet of the first impinger, was immersed in a PLC-controlled cooling bath SD 07R-20. The bath was filled with ethylene glycol and the temperature was set to be -15 °C. After leaving the series of impinger bottles, dry, cold gas went through the conditioner, with an in-built pump, which helped to overcome the pressure drop introduced by the series of impingers. This allowed the sampling of the gas with a sufficiently high volumetric flow rate (at least 1.0 L/min as required by the analyzer).

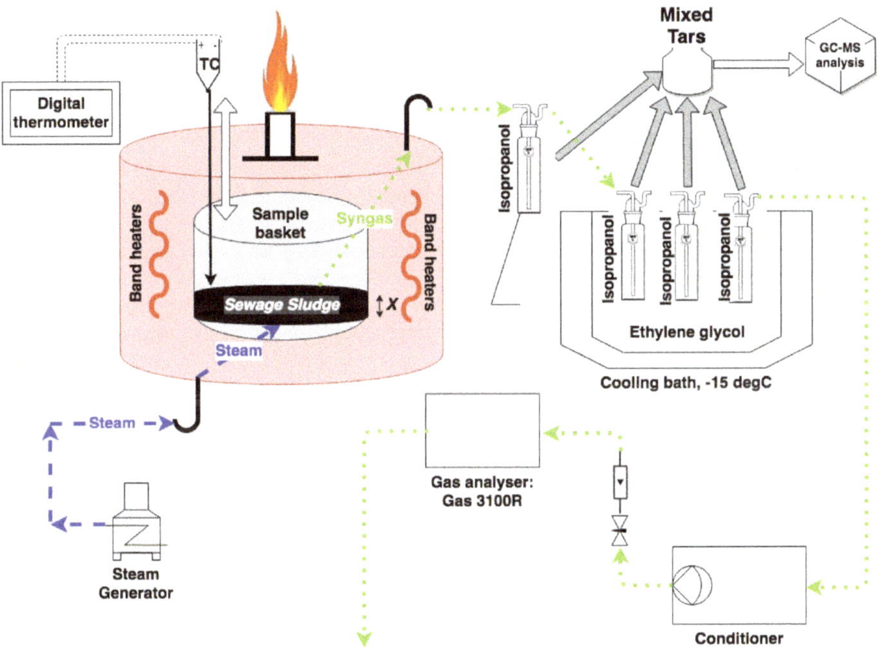

Figure 3. Allothermal gasifier: diagram of the test rig (TC—type K thermocouple; X—thickness of the sample layer).

The composition of permanent gases in both cold, dry producer gas and torgas were determined on-line using the Gas 3100R analyzer (manufactured by G.E.I.T Europe bvba, Bunsbeek, Belgium and supplied by Atut Sp. Z O.O. Lublin, Poland). This analyzer uses NDIR (non-dispersive infra-red) sensors for measurements of CO_2, CO, CH_4 and C_xH_y (light hydrocarbons, given as an equivalent of methane). A TCD (thermal conductivity detector) sensor is used to measure the H_2 content, whereas an electrochemical sensor is used for the determination of the O_2 content. The analyzer was calibrated using nitrogen with a purity of 5.0 before each measurement. The precision of Gas 3100R is 1% of the measuring range for CO_2, CO, CH_4, C_xH_y and 2% of the measuring range in the case of the TCD sensor for H_2 and electrochemical sensor for O_2. The measuring ranges were as follows: CO_2, 20%; CO, 40%; CH_4, 10%; C_xH_y, 5%; H_2, 55%; and 25% in the case of O_2. Gas 3100R has a linearity drift of 1% of measuring range per week, both for zero and for span. The excess of the gas was burned using a flare installed on the top cover.

Samples of the solutions from impinger bottles were mixed together in an Erlenmeyer flask of sufficient volume after the experiment. The mixed sample was subsequently analyzed using GC-MS, which consisted of an Agilent 7820-A chromatograph (manufactured by Agilent Technologies, Palo Alto, CA, USA) and an Agilent 5977B MSD spectrometer (Agilent Technologies, Palo Alto,

CA, USA). In the chromatograph, a Stabilwax-DA column (Restek, Benner Circle, Bellefonte, PA, USA) was used. Helium was used as a carrier gas (1.5 mL/min). The heating-up program was set to achieve 50 °C in 5 min and subsequently heat the column with a ramp of 10 °C/min, until the temperature of 200 °C was reached, and held for another 20 min afterwards. The data obtained with GC/MS was analyzed using the base peak chromatograms (BPC). Mixed samples were analyzed three times, and average values are presented in this study. A few major compounds, detected in the tar mixtures, were the subject of quantitative analysis in addition to the qualitative analysis. The quantitative analysis was performed using the calibration curves done for each of the compounds. Calibration curves were made using four points which corresponded to four known concentrations of the particular compound in the solvent (isopropanol). Five repetitions were done for each of the determined points. Reference substances and isopropanol of chromatographic grade were used for the calibration.

Temperature distribution was measured along the height of the sample basket, using a first-class K type thermocouple and a digital thermometer. Measurement was performed in an empty gasifier. The thermocouple was inserted from the top of the gasifier through an opening located half way between the central axis of the basket and its circumference. The thermocouple was gradually moved down every 5 cm, starting from the top edge of the sampling basket, until the tip reached the depth of 40 cm. Then, the tip was moved down by 2.5 cm.

4. Results

The results of the proximate and ultimate analysis of both raw and torrefied sewage sludge are presented in Table 1 below. It is clear, when comparing it with the original material, that sewage sludge changed due to torrefaction. However, contrary to the expectations, a higher heating value of the torrefied sewage sludge decreased in comparison to the raw material.

Table 1. Proximate and ultimate analysis of the torrefied sewage sludge.

Test	Symbol	Value		Unit	Standard Procedure
		Raw	Torrefied		
Moisture content [1]	MC	26.2	2.30	%	EN ISO 18134-2:2015
Volatile matter content	VM ^{d}	58.1	44.40	%	EN 15148:2009
Ash content	A ^{d}	32.5	45.67	%	EN ISO 1822:2015
Higher heating value	HHV	15,700	10,300	kJ/kg	EN 14918:2009
Lower heating value [2]	LHV	10,939	10,006	kJ/kg	EN 14918:2009
Carbon content	C ^{d}	27.89	15.83	%	EN ISO 16948:2015
Hydrogen content	H ^{d}	6.67	2.92	%	EN ISO 16948:2015
Nitrogen content	N ^{d}	4.36	4.18	%	EN ISO 16948:2015
Sulfur content	S ^{d}	0.29	0.27	%	EN ISO 16994:2016
Oxygen content	O ^{d}	28.80	31.13	%	EN ISO 16993:2015

[1] Wet basis; [2] Calculated using the formula from the standard; ^{d} Dry basis.

LHV also decreased, but only slightly (Table 1). The volatile matter content decreased significantly, as expected. Ash content was subject to a significant increase (Table 1). Particle size distribution (Figure 4) was similar both for raw and torrefied sewage sludge, even with a slightly higher number of fine particles ($d > 1$ mm) for the case of raw biomass. The calculated d_{50} diameter was 1292 μm for raw and 1442 μm for torrefied sewage sludge. The fusibility of the ashes was very similar in the case of both raw and torrefied sewage sludge (Table 2).

Table 2. Fusibility of the ashes for raw and torrefied sewage sludge under the reducing conditions (± 20 °C for all presented results).

Characteristic Temperature	Raw Sewage Sludge	Torrefied Sewage Sludge	Unit
Deformation temperature (DT)	1000	1010	°C
Sphere temperature (ST)	1020	1050	°C
Hemisphere temperature (HT)	1110	1110	°C
Flow temperature (FT)	1210	1210	°C

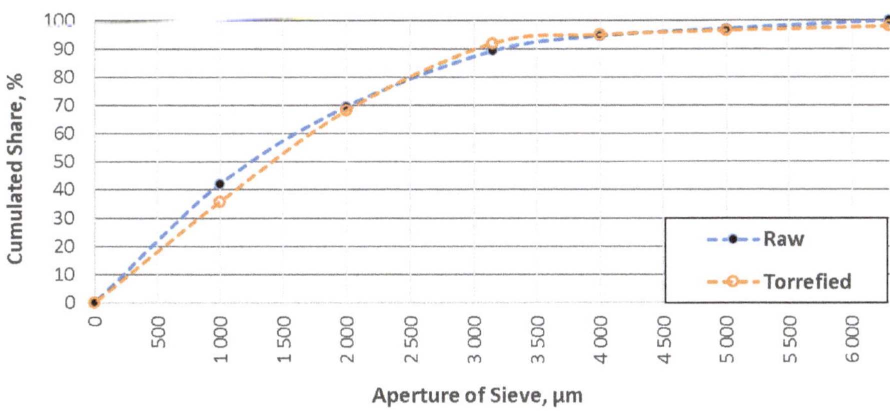

Figure 4. Particle size distribution of raw (pre-dried) and torrefied sewage sludge.

Torrefaction was performed in an oxidizing atmosphere, as can be clearly determined from the composition of torgas presented in Figure 5. Among the permanent gaseous products of torrefaction, CO_2 was prevalent. Nonetheless, CO content was substantial. Moreover, the content of hydrogen, methane and other light hydrocarbons (C_nH_m) was significant when torrefaction processes were considered. The average LHV of the permanent gases produced from the torgas was relatively high (15.10 MJ/m^3). The calculated mass yield was 0.754, whereas the calculated energy yield was 0.494.

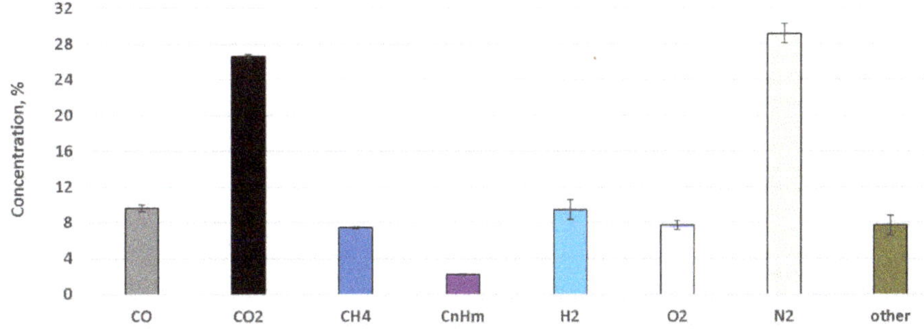

Figure 5. Composition of permanent gases present in torgas, measured during the torrefaction test of the pilot-scale installation.

There were significant differences in the average composition of the gas obtained during both experiments (see Figure 6). The most notable was the difference in the hydrogen content, which was much higher in the case of the gasification of raw sewage sludge in comparison to the gasification of torrefied sewage sludge. On the other hand, the concentration of methane, as well as the concentration of other light hydrocarbons (C_nH_m), was substantially higher during the gasification of the torrefied sewage sludge. The contents of carbon dioxide and carbon monoxide were also higher for the experiment performed with torrefied feedstock. As a consequence of their respective compositions, dry producer gases from the gasification of raw sewage sludge had an average lower heating value of 13.51 MJ/m^3, whereas the gas produced from torrefied sewage sludge reached 17.51 MJ/m^3 on average.

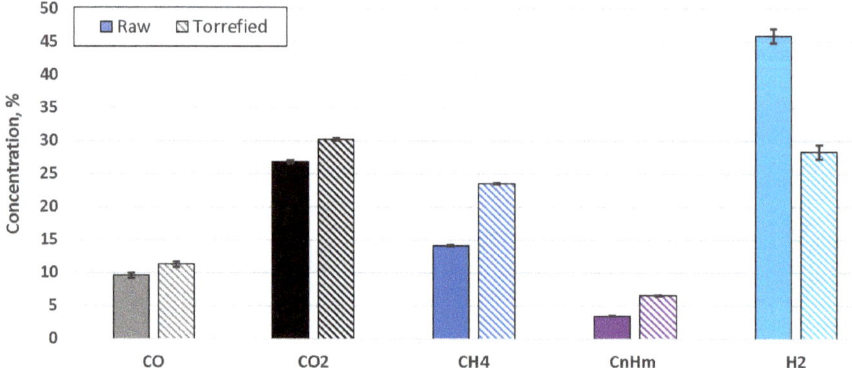

Figure 6. Composition of the dry producer gases obtained during both gasification experiments.

The condensable compounds detected in the samples of solvent (isopropanol) solution are reported in Tables 3 and 4. The results presented in Table 3 are qualitative, whereas the results presented in Table 4 are quantitative. The concentrations of respective compounds are given as per kg of the feedstock. The names of all the compounds are the official IUPAC names. Melting points, boiling points and average masses were taken from the on-line database of the Royal Society of Chemistry [45]. Average values were used. The deposition profiles of tars collected during both experiments are presented in the diagram below (Figure 7). It was assumed that the likelihood of the tar deposition corresponds with the respective melting points of the compounds.

Table 3. Condensable compounds, identified using GC-MS analysis—qualitative analysis.

Compound	Form.	Boil. Point	Melt. Point	Avg. Mass	Relative Area of the Peak [2]			
					Raw		Torrefied	
					Value	SD [1]	Value	SD [1]
		°C	°C	Da	%	%	%	%
Toluene	C_7H_8	111	−95	92.14	21.41	0.64	32.60	0.39
Propiononitrile	C_3H_5N	97	−92	55.08 [3]	0.14	0.03	1.11	0.02
2-Methylpyridine	C_6H_7N	128	−70	93.12	1.40	0.09	1.64	0.02
o-Xylene	C_8H_{10}	140	−48	106.16	0.10	0.03	1.40	0.02
3-Methyl-1H-pyrrole	C_5H_7N	144	−48	81.12	0.12	0.01	1.40	0.03
Pyridine	C_5H_5N	115	−42	79.10	6.23	0.21	3.12	0.03
Thiophene	C_4H_4S	84	−38	84.14	2.01	0.11	0.26	0.01
2-Methyl-1H-pyrrole	C_5H_7N	147	−36	81.12	0.51	0.01	0.69	0.02
Styrene	C_8H_8	145	−31	104.15	4.54	0.26	7.26	0.05
3-Methylbenzonitrile	C_8H_7N	210	−25	117.15	0.10	0.01	- [5]	-
Benzeneacetonitrile	C_8H_7N	234	−24	117.15	0.23	0.01	0.30	0.01
1H-Pyrrole	C_4H_5N	130	−23	67.09 [3]	2.71	0.11	3.84	0.02
2-Methylnaphthalene	$C_{11}H_{10}$	242	−22	142.20	2.24	0.09	1.93	0.03
1-Methylnaphthalene	$C_{11}H_{10}$	242	−22	142.20	1.41	0.07	1.63	0.04
1-Benzofuran	C_8H_6O	174	−18	118.13	0.40	0.02	0.34	0.01
Quinoline	C_9H_7N	237	−15	129.16	3.32	0.07	0.91	0.01
Benzonitrile	C_7H_5N	191	−13	103.12	5.32	0.15	2.86	0.03
2-Methylbenzonitrile	C_8H_7N	205	−13	117.14	0.31	0.02	0.11	0.01
1H-Indene	C_9H_8	181	−2	116.16	4.83	0.21	2.68	0.02
2-Methylquinoline	$C_{10}H_9N$	247	−2	143.18	0.21	0.01	0.13	0.01
m-Cresol	C_7H_8O	203	11	108.12	-	-	0.40	0.07
Acetic acid	$C_2H_4O_2$	118	17	60.05 [3]	0.23	0.05	3.51	0.01
3-Methylpyridine	C_6H_7N	144	18	93.13	0.51	0.03	0.61	0.01
Isoquinoline	C_9H_7N	242	26	129.16	0.80	0.01	0.12	0.01

Table 3. *Cont.*

Compound	Form.	Boil. Point	Melt. Point	Avg. Mass	Relative Area of the Peak [2]			
					Raw		Torrefied	
					Value	SD [1]	Value	SD [1]
		°C	°C	Da	%	%	%	%
4-Methylbenzonitrile	C_8H_7N	218	28	117.15	0.22	0.01	0.13	0.01
Phenol	C_6H_6O	182	41	94.11	1.81	0.07	5.56	0.12
p-Cresol	C_7H_8O	202	41	108.14	0.43	0.02	3.88	0.07
1H-Indole	C_8H_7N	254	53	117.15	4.03	0.15	2.68	0.04
3-Pyridinamine	$C_5H_6N_2$	250	62	94.11	0.21	0.03	1.16	0.01
2-Naphthonitrile	$C_{11}H_7N$	157	67	153.18	0.72	0.01	-	-
Naphthalene	$C_{10}H_8$	218	81	128.17	22.00	0.60	8.19	0.12
Acenaphthylene	$C_{12}H_8$	280	93	152.19	4.21	0.09	1.05	0.01
Phenanthrene	$C_{14}H_{10}$	338	100	178.23	3.13	0.01	0.45	0.03
9H-Fluorene	$C_{13}H_{10}$	295	115	166.22	0.70	0.02	-	-
5,5-Dimethyl-2,4-imidazolidinedione	$C_5H_8N_2$	n.a. [4]	175	128.13	0.51	0.05	4.65	0.13
3,3′-Sulfanediyldipropanenitrile	$C_6H_8N_2$	n.a.	n.a.	140.21	2.41	0.16	0.62	0.04
2-Benzothiophene	C_8H_6S	n.a.	n.a.	134.20	1.14	0.04	0.20	0.01

[1] Standard deviation; [2] relative to the total area of all identified peaks; [3] molecular mass lower than that of benzene (78.11 Da); [4] information not available; [5] not detected.

Table 4. Quantitative analysis of selected condensable compounds, identified using GC-MS analysis.

Compound	Form.	Boil. Point	Melt. Point	Avg. Mass	Concentration in the Producer Gas			
					Raw		Torrefied	
					Value	SD	Value	SD
		°C	°C	Da	mg/m³	mg/m³	mg/m³	mg/m³
Toluene	C_7H_8	111	−95	92.14	2229.1	91.4	3211.0	61.0
Pyridine	C_5H_5N	115	−42	79.10	1170.0	37.4	453.8	3.7
1-Methylnaphthalene	$C_{11}H_{10}$	242	−22	142.20	225.6	7.7	149.4	3.5
Benzonitrile	C_7H_5N	191	−13	103.12	688.2	18.2	243.0	3.5
1H-Indene	C_9H_8	181	−2	116.16	574.7	25.3	243.6	0.9
Acetic acid	$C_2H_4O_2$	118	17	60.05 [1]	752.3	32.0	1680.3	5.2
Phenol	C_6H_6O	182	41	94.11	521.7	10.7	531.0	11.8
p-Cresol	C_7H_8O	202	41	108.14	198.4	3.7	444.9	7.9
Naphthalene	$C_{10}H_8$	218	81	128.17	1367.6	47.7	384.7	8.5
Acenaphthylene	$C_{12}H_8$	280	93	152.19	263.9	6.4	62.0	1.5

[1] Average molecular mass lower than that of benzene (78.11 Da).

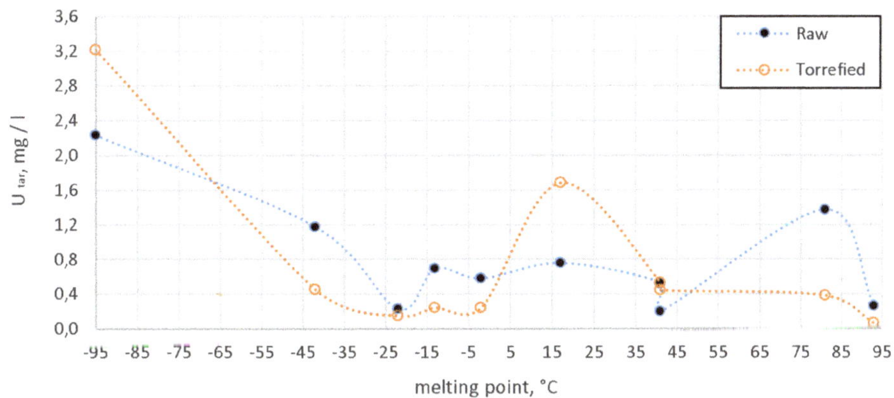

Figure 7. Tar deposition profile diagram: content of major tarry compounds in the produced gas arranged according to their respective melting points.

5. Discussion

The average temperature of the torrefaction process was high (391.9 °C) in comparison to a typical range of the torrefaction temperatures for biomass [15]. However, some additional factors should be taken into account. Firstly, sewage sludge is not a typical lignocellulosic biomass. It contains proteins which are cracked into more simple compounds, such as amines, as well as nitrile and nitrogen-containing heterocyclic compounds, at temperatures between 300 °C and 500 °C [25]. Moreover, temperatures below 400 °C are considered to be low in terms of the thermal decomposition of the unsaturated fatty acids [25]. Additionally, practical difficulties of the measurements performed at a pilot-scale installation should be taken into account. The design of the reactor, with scrapers moving the material on top of the shelves, makes it impossible to measure the actual process temperature inside the layer of the torrefied material. Therefore, an average from three thermocouples installed close to the top plate of the shelves seems to be the best way to indicate the process temperature. It should be taken into account that the heat transfer between the flue gases inside of the shelves (convection), heat conduction of the shelves themselves, as well as heat transfer between the top side of the shelves and the torrefied material, and heat transfer within the bed itself, can influence the actual temperature. This temperature would undoubtfully be lower than the temperature used in the manuscript for the characterization of the process.

In terms of the torrefaction, some counterintuitive results have been obtained. Typically, torrefaction leads to the increase in the heating value of the processed biomass [15–17,68–70]. In the case of this study, both HHV and LHV decreased as a consequence of the performed treatment (Table 1). Moreover, the carbon content decreased after the torrefaction. In general, it seems plausible to justify this by a couple of factors, such as the severity of the process (temperature), torrefaction in the oxidizing atmosphere (see Figure 5), as well as the autocatalytic effect of the inorganic fraction of the material. It is clear from Table 1 that ash content increased significantly during torrefaction. Moreover, the initial ash content of the sewage sludge, prior to torrefaction, was high.

The severity of the torrefaction resulted in a relatively low mass yield of 0.754, which is similar to the value determined by Poudel et al. (approx. 0.63 for 400 °C) [22]. However, the aforementioned sharp decrease in HHV, caused by the significant increase in the ash content, resulted in a very low energy yield of 0.494, significantly lower than in the case of the work of Poudel et al. (approx. 0.61 for 400 °C) [22]. This difference between the literature results and results reported by this study could probably be attributed to the differences between the samples of the sewage sludge, as that used in the work of Poudel et al. had an initial ash content of approximately 15%, which increased up to approximately 26% after torrefaction at 400 °C. In the case of torrefaction in a pilot-scale reactor, as presented in this study, slightly less than half of the energy contained in the original material remained in the product. Such a disproportion between mass and energy yields indicates a significant part of the chemical energy going into torgas, which finds confirmation in the heating value of dry torgas (15.10 MJ/m^3).

This would indicate a good combustibility of the obtained torgas. In the context of the previous paper with a novel installation concept [65], such a high share of the energy going to torgas would not be disadvantageous, as the introduction of torgas into the gasifier was proposed in that particular solution. Moreover, relatively high concentrations of combustible compounds in torgas, such as methane, light hydrocarbons and carbon monoxide, seem beneficial from that point of view as well. The oxygen content detected in the torgas can be explained by the fact that it was not completely air-tight. Overall, for the pilot-scale installation, indirect methods seem better for the determination of the mass balance, such as that developed by Weber et al. [67]. The evidence could be found in respective d_{50} diameters and particle size distributions of both feedstock and product; i.e., the d_{50} was higher for the torrefied material, whereas the content of fine particles was higher for the raw sewage sludge. Common sense would dictate that in such a reactor, a part of the material will be always broken down into smaller particles, due to the mechanical attrition caused by the scrapers. The most plausible explanation of the obtained results seems to be the loss of fine particles within the cavities of the reactor. This would introduce an additional bias if a direct method was to be used for the purpose of ensuring the mass balance of an experiment in such a reactor.

Gasification testing, performed using raw sewage sludge, confirmed the results presented in other literature sources, in terms of a very high hydrogen content which can be achieved during the steam gasification of sewage sludge. In general, the performed tests confirmed the hypothesis that the torrefaction of the sewage sludge was beneficial with respect to its subsequent gasification. Producer gas from the gasification of torrefied sewage sludge was more calorific, with an average LVH of 17.15 MJ/m^3 in comparison to the average LHV of 13.51 MJ/m^3 for the gas produced using raw sewage sludge. This was, most likely, caused by significantly higher concentration of methane and other light hydrocarbons (C_nH_m) as well as a slightly higher concentration of carbon monoxide (Figure 6), for the case of the gasification of torrefied material. The hydrogen concentration was lower during the test with torrefied sewage sludge. It seems plausible to attribute this difference to the higher extent of the gas–char reactions. The torrefied sewage sludge had already been partially devolatilized, prior to the gasification. It is typical for the torrefaction process to increase the reactivity of the material by influencing its surface, and such results have already been reported for other types of biomass [71]. The influence of the torrefaction on the surface of the torrefied sewage sludge and its reactivity should be a subject of further investigation. The differences in the obtained average gas compositions between both experiments might also be attributed to some extent to the autocatalytic effects due to the fact that the ash content of the torrefied sewage sludge was significantly higher, in comparison to the raw sewage sludge (see Table 1).

The composition of the producer gas obtained during the gasification of torrefied sewage sludge can be considered slightly advantageous if the producer gas is supposed to be used in an engine, as hydrogen typically causes knocking issues.

The results presented in this paper indicate a positive influence of torrefaction with respect to the subsequent gasification of sewage sludge, especially in the context of the composition of tars. Tar deposition profiles (Figure 7) based on the quantitative analysis of tars for both samples demonstrate clearly that the torrefaction of the sewage sludge resulted in a decreased content of heavy tars, with their respective melting points between 40 °C and 95 °C, in the gas. The use of the producer gas in the engine requires its cooling to approximately 30 °C prior to the intake. This typically introduces a requirement to install a gas cooler. The aforementioned tars exist either as solids or as highly viscous liquids within the range of the temperatures mentioned above. In the longer term, this could lead to the deposition of heavy tars, leading to maintenance problems and frequent shutdowns, caused by the blockages due to the buildup of the clogs of tars and particulates sticking to the cool surfaces. Indeed, torrefaction clearly caused concentrations of naphthalene and acenaphthylene to be lower in comparison to the gas produced using raw sewage sludge. It is possible that the slightly lower moisture content of the torrefied sludge likely caused different temperature gradients across the bed of material, probably resulting in higher temperatures on average.

Overall, the decrease in the concentration of the condensable compounds was not tremendous and decreased by 7.4% as the total tar content decreased from 7.99 g/m^3 to 7.40 g/m^3. However, this could be attributed mostly to a substantial increase in the content of toluene, which increased from 2.23 g/m^3 to 3.21 g/m^3. The melting point of toluene is -95 °C and its viscosity in ambient conditions is not much different to water. The decrease in the concentration of the compounds with melting points higher than 40 °C was substantial (39.5%). The content of selected tar species deemed to be the most problematic decreased from 2.35 g/m^3 to 1.42 g/m^3, which is still substantial. Nonetheless, it should not be overlooked that the temperature of the gasification process was relatively low (675 °C). Moreover, it seems reasonable to expect that such a decrease would help in terms of a subsequent cleaning of the gas. The authors believe that the optimization of both steps (torrefaction and gasification) could lead to a further decrease of the tar content. The idea of gasification on the traveling grate needs further refinement, as the study showed practical difficultie, that could occur in terms of the ability of the gasifier to achieve higher temperatures, as the gasifier was allothermal and used electric heating. Some other types of heat source would be needed for the gasifier with steam as the only gasification agent, as there is no chance for this solution to maintain the autothermal character of the process. Plasma gasification, with steam as the gasification agent, should be taken into consideration. On a larger scale, this would, however, require a source of plasma which would be easy to maintain.

6. Conclusions

A suite of performed experiments demonstrated clearly the potential of torrefaction as a viable primary method for decreasing the tar content of the producer gas. Torrefaction led to the improvement of the calorific value of the producer gas and its quality. The severity of the torrefaction was high, which resulted in a good quality of the torgas and a relatively high amount of the chemical energy going to torgas. Even though it could potentially lead to the thermal runaway of the installation, this solution seems to be beneficial, as torgas could potentially be mixed with the producer gas in the gasifier. More research is needed in two main research areas:

- Since the test was performed in an allothermal gasifier, investigation is needed to confirm if the gasification process can be authothermal;
- Tests performed on a pilot-scale gasifier should be performed in order to confirm if such an improvement in tar composition is indeed sufficient to significantly improve the maintenance of such a gasifier and decrease the frequency of the shutdowns.

Further optimization of both the torrefaction and gasification of sewage sludge is needed to reap all the benefits of the combination of these two technologies. More studies are needed on both the gasification and torrefaction of the sewage sludge, using many samples of different origins, as the differences between the treatment facilities could lead to significant differences in terms of the obtained results. Moreover, it would be interesting to perform some more detailed studies on the influence of the composition of the inorganic fraction of the sewage sludge on the torrefaction and gasification processes. In light of the performed experiments, the feasibility of the concept of the gasification on traveling grates still cannot be fully confirmed. Further refinement of the concept is needed.

Author Contributions: H.P.-K., W.J.T., J.G.-T., conceived the concept of the installation as well as the concept of the paper. They analyzed the results obtained from the measurements and took part in writing the paper. Moreover, H.P.-K. played an advisory role, checked and approved the study in its present form. M.W. designed the sampling system for the measurement of the concentration of tars in the producer gas and performed GC-MS analysis. L.N. has written a significant part of the paper, made the tar deposition profile diagram and took part in the making of the sampling diagram for gasification tests. K.K. designed the sampling system for the measuring of gaseous compounds of torgas and producer gas, and performed the ultimate analysis of both materials. M.K. performed gasification experiments, took part in torrefaction experiments and performed proximate analysis. J.Z., M.C., M.O. and M.B. took part in the torrefaction tests, the making of the diagram of the pilot scale installation and the analysis of the results. Moreover, M.C. prepared the overview of the utilization of sewage sludge in Poland.

Funding: This work was supported by Regional Operational Program for the Opole Voivodship 2014–2020, Priority Axis 01 Innovation in the Economy, Measure 1.1 Innovation in the Economy. Key Project No. RPOP.01.01.00-16-44/2016.

Conflicts of Interest: The authors declare no conflict of interest.

References

1. Werle, S.; Dudziak, M. Analysis of organic and inorganic contaminants in dried sewage sludge and by-products of dried sewage sludge gasification. *Energies* **2014**, *7*, 462–476. [CrossRef]
2. Kacprzak, M.; Neczaj, E.; Fijałkowski, K.; Grobelak, A.; Grosser, A.; Worwag, M.; Rorat, A.; Brattebo, H.; Almås, Å.; Singh, B.R. Sewage sludge disposal strategies for sustainable development. *Environ. Res.* **2017**, *156*, 39–46. [CrossRef]
3. Lee, L.H.; Wu, T.Y.; Shak, K.P.Y.; Lim, S.L.; Ng, K.Y.; Nguyen, M.N.; Teoh, W.H. Sustainable approach to biotransform industrial sludge into organic fertilizer via vermicomposting: A mini-review. *J. Chem. Technol. Biotechnol.* **2018**, *93*, 925–935. [CrossRef]
4. Cieślik, B.M.; Namieśnik, J.; Konieczka, P. Review of sewage sludge management: Standards, regulations and analytical methods. *J. Clean. Prod.* **2015**, *90*, 1–15. [CrossRef]

5. Andersen, A. *Disposal and Recycling Routes for Sewage Sludge Part 4: Economic Report*; Office for Official Publications of the European Communities: Luxembourg, 2002.

6. Werle, S.; Wilk, R.K. A review of methods for the thermal utilization of sewage sludge: The Polish perspective. *Renew. Energy* **2010**, *35*, 1914–1919. [CrossRef]

7. GUS, Departament Badań Regionalnych i Środowiska. *Ochrona Środowiska 2017*; GUS: Warszawa, Poland, 2017; p. 551.

8. Werle, S. Sewage Sludge-to-Energy Management in Eastern Europe: A Polish Perspective. *Ecol. Chem. Eng. S* **2015**, *22*, 459–469. [CrossRef]

9. Pajak, T. Thermal Treatment as sustainable sewage sludge management. *Environ. Prot. Eng.* **2013**, *39*, 41–53. [CrossRef]

10. Bień, J. Zagospodarowanie komunalnych osadów ściekowych metodami termicznymi. *Inż. Ochr. Śr.* **2012**, *15*, 439–450.

11. Bień, J.; Bień, B. Utilisation of Municipal Sewage Sludge by Thermal Methods in the Face of Storage Disallowing. *Inż. Ekol.* **2015**, *45*, 36–43. [CrossRef]

12. Jurczyk, M.P.; Pająk, T. Initial Operating Experience with the New Polish Waste-to-Energy Plants. In *Waste Management*; Thomé-Kozmiensky, K.J., Thiel, S., Eds.; TK Verlag: Neuruppin, Germany, 2016.

13. Cyranka, M.; Jurczyk, M.; Pająk, T. Municipal Waste-to-Energy plants in Poland—Current projects. *E3S Web Conf.* **2016**, *10*, 00070. [CrossRef]

14. Gorazda, K.; Tarko, B.; Wzorek, Z.; Kominko, H.; Nowak, A.K.; Kulczycka, J.; Henclik, A.; Smol, M. Fertilisers production from ashes after sewage sludge combustion—A strategy towards sustainable development. *Environ. Res.* **2017**, *154*, 171–180. [CrossRef] [PubMed]

15. Moscicki, K.J.; Niedzwiecki, L.; Owczarek, P.; Wnukowski, M. Commoditization of biomass: Dry torrefaction and pelletization-a review. *J. Power Technol.* **2014**, *94*, 233–249.

16. Van der Stelt, M.J.C.; Gerhauser, H.; Kiel, J.H.A.; Ptasinski, K.J. Biomass upgrading by torrefaction for the production of biofuels: A review. *Biomass Bioenergy* **2011**, *35*, 3748–3762. [CrossRef]

17. Tumuluru, J.S.; Sokhansanj, S.; Hess, J.R.; Wright, C.T.; Boardman, R.D. A review on biomass torrefcation process and product properties for energy applications. *Ind. Biotechnol.* **2011**, *7*, 384–401. [CrossRef]

18. Nyakuma, B.B.; Magdziarz, A.; Werle, S. Physicochemical, Thermal and Kinetic Analysis of Sewage Sludge. *Proc. ECOpole* **2016**, *10*, 473–480. [CrossRef]

19. Magdziarz, A.; Wilk, M.; Kosturkiewicz, B. Investigation of sewage sludge preparation for combustion process. *Chem. Process Eng. Inz. Chem. Proces.* **2011**, *32*, 299–309. [CrossRef]

20. Magdziarz, A.; Werle, S. Analysis of the combustion and pyrolysis of dried sewage sludge by TGA and MS. *Waste Manag.* **2014**, *34*, 174–179. [CrossRef]

21. Pulka, J.; Wiśniewski, D.; Gołaszewski, J.; Białowiec, A. Is the biochar produced from sewage sludge a good quality solid fuel? *Arch. Environ. Prot.* **2016**, *42*, 125–134. [CrossRef]

22. Poudel, J.; Ohm, T.I.; Lee, S.H.; Oh, S.C. A study on torrefaction of sewage sludge to enhance solid fuel qualities. *Waste Manag.* **2015**, *40*, 112–118. [CrossRef]

23. Poudel, J.; Karki, S.; Gu, J.H.; Lim, Y.; Oh, S.C. Effect of Co-Torrefaction on the Properties of Sewage Sludge and Waste Wood to Enhance Solid Fuel Qualities. *J. Residuals Sci. Technol.* **2017**, *14*, 23–36. [CrossRef]

24. Atienza-Martínez, M.; Fonts, I.; ábrego, J.; Ceamanos, J.; Gea, G. Sewage sludge torrefaction in a fluidized bed reactor. *Chem. Eng. J.* **2013**, *222*, 534–545. [CrossRef]

25. Atienza-Martínez, M.; Mastral, J.F.; Ábrego, J.; Ceamanos, J.; Gea, G. Sewage sludge torrefaction in an auger reactor. *Energy Fuels* **2015**, *29*, 160–170. [CrossRef]

26. Huang, M.; Chang, C.C.; Yuan, M.H.; Chang, C.Y.; Wu, C.H.; Shie, J.L.; Chen, Y.H.; Chen, Y.H.; Ho, C.; Chang, W.R.; et al. Production of torrefied solid bio-fuel from pulp industry waste. *Energies* **2017**, *10*, 910. [CrossRef]

27. Huang, Y.-F.; Sung, H.-T.; Chiueh, P.-T.; Lo, S.-L. Microwave torrefaction of sewage sludge and leucaena. *J. Taiwan Inst. Chem. Eng.* **2017**, *70*, 236–243. [CrossRef]

28. Huang, Y.-F.; Sung, H.-T.; Chiueh, P.-T.; Lo, S.-L. Co-torrefaction of sewage sludge and leucaena by using microwave heating. *Energy* **2016**, *116*, 1–7. [CrossRef]

29. Huang, Y.W.; Chen, M.Q.; Luo, H.F. Nonisothermal torrefaction kinetics of sewage sludge using the simplified distributed activation energy model. *Chem. Eng. J.* **2016**, *298*, 154–161. [CrossRef]

30. Do, T.X.; Lim, Y.; Cho, H.; Shim, J.; Yoo, J.; Rho, K.; Choi, S.-G.; Park, B.-Y. Process modeling and energy consumption of fry-drying and torrefaction of organic solid waste. *Dry Technol.* **2017**, *35*, 754–765. [CrossRef]
31. Pawlak-Kruczek, H.; Krochmalny, K.K.; Wnukowski, M.; Niedzwiecki, L. Slow pyrolysis of the sewage sludge with additives: Calcium oxide and lignite. *J. Energy Resour. Technol.* **2018**, *140*. [CrossRef]
32. Pečkytė, J.; Baltrėnaitė, E. Assessment of heavy metals leaching from (bio) char obtained from industrial sewage sludge. *Environ. Prot. Eng.* **2015**, *7*, 399–406. [CrossRef]
33. Council Decision. Council of the European Union 2003/33/EC—Council Decision establishing criteria and procedures for the acceptance of waste at landfills pursuant to Article 16 of and Annex II to Directive 1999/31/EC. *Off. J. Eur. Communities* **2003**, *L 11/27*, 27–49.
34. Wang, N.Y.; Shih, C.H.; Chiueh, P.T.; Huang, Y.F. Environmental effects of sewage sludge carbonization and other treatment alternatives. *Energies* **2013**, *6*, 871–883. [CrossRef]
35. Lasek, J.A.; Kopczyński, M.; Janusz, M.; Iluk, A.; Zuwała, J. Combustion properties of torrefied biomass obtained from flue gas-enhanced reactor. *Energy* **2017**, *119*, 362–368. [CrossRef]
36. Joshi, Y.; Di Marcello, M.; Krishnamurthy, E.; De Jong, W. Packed-Bed Torrefaction of Bagasse under Inert and Oxygenated Atmospheres. *Energy Fuels* **2015**, *29*, 5078–5087. [CrossRef]
37. Uemura, Y.; Saadon, S.; Osman, N.; Mansor, N.; Tanoue, K.I. Torrefaction of oil palm kernel shell in the presence of oxygen and carbon dioxide. *Fuel* **2015**, *144*, 171–179. [CrossRef]
38. Uemura, Y.; Omar, W.; Othman, N.A.; Yusup, S.; Tsutsui, T. Torrefaction of oil palm EFB in the presence of oxygen. *Fuel* **2013**, *103*, 156–160. [CrossRef]
39. Chen, W.H.; Zhuang, Y.Q.; Liu, S.H.; Juang, T.T.; Tsai, C.M. Product characteristics from the torrefaction of oil palm fiber pellets in inert and oxidative atmospheres. *Bioresour. Technol.* **2016**, *199*, 367–374. [CrossRef] [PubMed]
40. Chen, W.H.; Lu, K.M.; Liu, S.H.; Tsai, C.M.; Lee, W.J.; Lin, T.C. Biomass torrefaction characteristics in inert and oxidative atmospheres at various superficial velocities. *Bioresour. Technol.* **2013**, *146*, 152–160. [CrossRef]
41. Lu, K.M.; Lee, W.J.; Chen, W.H.; Liu, S.H.; Lin, T.C. Torrefaction and low temperature carbonization of oil palm fiber and eucalyptus in nitrogen and air atmospheres. *Bioresour. Technol.* **2012**, *123*, 98–105. [CrossRef]
42. Pawlak-Kruczek, H.; Wnukowski, M.; Krochmalny, K.; Kowal, M.; Baranowski, M.; Zgóra, J.; Czerep, M.; Ostrycharczyk, M.; Niedzwiecki, L. The staged thermal conversion of sewage sludge in presence of oxygen. In *The Clearwater Clean Coal Conference: Proceedings of the 43rd International Technical Conference on Clean Energy*; Sakkestad, B.A., Ed.; Published by Coal Technologies Associates Post Office Box 1130 Louisa, VA 23093 USA: Clearwater, FL, USA, 2018.
43. Reed, T.B.; Das, A. *Handbook of Biomass Downdraft Gasifier Engine Systems*; Biomass Energy Foundation: Golden, CO, USA, 1988.
44. Devi, L.; Ptasinski, K.J.; Janssen, F.J.J.G. A review of the primary measures for tar elimination in biomass gasification processes. *Biomass Bioenergy* **2002**, *24*, 125–140. [CrossRef]
45. Royal Society of Chemistry ChemSpider | Search and Share Chemistry. Available online: http://www.chemspider.com/ (accessed on 23 November 2018).
46. Asadullah, M. Biomass gasification gas cleaning for downstream applications: A comparative critical review. *Renew. Sustain. Energy Rev.* **2014**, *40*, 118–132. [CrossRef]
47. Hasler, P.; Nussbaumer, T. Gas cleaning for IC engine applications from fixed bed biomass gasification. *Biomass Bioenergy* **1999**, *16*, 385–395. [CrossRef]
48. Striūgas, N.; Valinčius, V.; Pedišius, N.; Poškas, R.; Zakarauskas, K. Investigation of sewage sludge treatment using air plasma assisted gasification. *Waste Manag.* **2017**. [CrossRef] [PubMed]
49. Brown, M.D.; Baker, E.G.; Mudge, L.K. Environmental design considerations for thermochemical biomass energy. *Biomass* **1986**, *11*, 255–270. [CrossRef]
50. Bui, T.; Loof, R.; Bhattacharya, S.C. Multi-stage reactor for thermal gasification of wood. *Energy* **1994**, *19*, 397–404. [CrossRef]
51. Werle, S. Sewage sludge gasification: Theoretical and experimental investigation. *Environ. Prot. Eng.* **2013**, *39*, 25–32. [CrossRef]
52. Schweitzer, D.; Gredinger, A.; Schmid, M.; Waizmann, G.; Beirow, M.; Spörl, R.; Scheffknecht, G. Steam gasification of wood pellets, sewage sludge and manure: Gasification performance and concentration of impurities. *Biomass Bioenergy* **2018**, *111*, 308–319. [CrossRef]

53. Werle, S. Numerical analysis of the combustible properties of sewage sludge gasification gas. *Chem. Eng. Trans.* **2015**, 1021–1026. [CrossRef]

54. Werle, S.; Dudziak, M. Evaluation of the possibility of the sewage sludge gasification gas use as a fuel. *Ecol. Chem. Eng. S* **2016**, *23*, 229–236. [CrossRef]

55. Szwaja, S.; Kovacs, V.B.; Bereczky, A.; Penninger, A. Sewage sludge producer gas enriched with methane as a fuel to a spark ignited engine. *Fuel Process. Technol.* **2013**, *110*, 160–166. [CrossRef]

56. Werle, S. Impact of feedstock properties and operating conditions on sewage sludge gasification in a fixed bed gasifier. *Waste Manag. Res.* **2014**, *32*, 954–960. [CrossRef]

57. Calvo, L.F.; García, A.I.; Otero, M. An experimental investigation of sewage sludge gasification in a fluidised bed reactor. *Sci. World J.* **2013**. [CrossRef] [PubMed]

58. Akkache, S.; Hernández, A.B.; Teixeira, G.; Gelix, F.; Roche, N.; Ferrasse, J.H. Co-gasification of wastewater sludge and different feedstock: Feasibility study. *Biomass Bioenergy* **2016**, *89*, 201–209. [CrossRef]

59. Reed, G.P.; Paterson, N.P.; Zhuo, Y.; Dugwell, D.R.; Kandiyoti, R. Trace element distribution in sewage sludge gasification: Source and temperature effects. *Energy Fuels* **2005**, *19*, 298–304. [CrossRef]

60. Judex, J.W.; Gaiffi, M.; Burgbacher, H.C. Gasification of dried sewage sludge: Status of the demonstration and the pilot plant. *Waste Manag.* **2012**, *32*, 719–723. [CrossRef] [PubMed]

61. Kokalj, F.; Arbiter, B.; Samec, N. Sewage sludge gasification as an alternative energy storage model. *Energy Convers. Manag.* **2017**, *149*, 738–747. [CrossRef]

62. Huang, Y.W.; Chen, M.Q.; Li, Q.H.; Xing, W. Hydrogen-rich syngas produced from co-gasification of wet sewage sludge and torrefied biomass in self-generated steam agent. *Energy* **2018**, *161*, 202–213. [CrossRef]

63. Prins, M.J.; Ptasinski, K.J.; Janssen, F.J.J.G. More efficient biomass gasification via torrefaction. *Energy* **2006**, *31*, 3458–3470. [CrossRef]

64. Dudyński, M.; Van Dyk, J.C.; Kwiatkowski, K.; Sosnowska, M. Biomass gasification: Influence of torrefaction on syngas production and tar formation. *Fuel Process. Technol.* **2015**, *131*, 203–212. [CrossRef]

65. Tic, W.; Guziałowska-Tic, J.; Pawlak-Kruczek, H.; Woźnikowski, E.; Zadorożny, A.; Niedźwiecki, Ł.; Wnukowski, M.; Krochmalny, K.; Czerep, M.; Ostrycharczyk, M.; et al. Novel Concept of an Installation for Sustainable Thermal Utilization of Sewage Sludge. *Energies* **2018**, *11*, 748. [CrossRef]

66. Pawlak-Kruczek, H.; Krochmalny, K.; Mościcki, K.; Zgóra, J.; Czerep, M.; Ostrycharczyk, M.; Niedźwiecki, Ł. Torrefaction of Various Types of Biomass in Laboratory Scale, Batch-Wise Isothermal Rotary Reactor and Pilot Scale, Continuous Multi-Stage Tape Reactor. *Eng. Prot. Environ.* **2017**, *20*, 457–472. [CrossRef]

67. Weber, K.; Heuer, S.; Quicker, P.; Li, T.; Løvås, T.; Scherer, V. An Alternative Approach for the Estimation of Biochar Yields. *Energy Fuels* **2018**. [CrossRef]

68. Nhuchhen, D.; Basu, P.; Acharya, B. A Comprehensive Review on Biomass Torrefaction. *Int. J. Renew. Energy Biofuels* **2014**, *2014*, 506376. [CrossRef]

69. Esseyin, A.E.; Steele, P.H.; Pittman, C.U., Jr. Current trends in the production and applications Torrefied Wood/Biomass—A review. *Bioresources* **2015**, *10*, 8812–8858. [CrossRef]

70. Chew, J.J.; Doshi, V. Recent advances in biomass pretreatment—Torrefaction fundamentals and technology. *Renew. Sustain. Energy Rev.* **2011**, *15*, 4212–4222. [CrossRef]

71. Saeed, M.A.; Andrews, G.E.; Phylaktou, H.N.; Gibbs, B.M.; Niedzwiecki, L.; Walton, R. Explosion and Flame Propagation Properties of Coarse Wood: Raw and Torrefied. In *Proceedings of the Eighth International Seminar on Fire & Explosion Hazards (ISFEH8)*; Choa, J., Molkov, V., Sunderland, P., Tamanini, F., Torero, J., Eds.; USTC Press: Hefei, China, 2016; pp. 579–588.

 *energies*

Article

Modelling of the Biomass mCHP Unit for Power Peak Shaving in the Local Electrical Grid

Michał Gliński [1],*, Carsten Bojesen [2], Witold Rybiński [1] and Sebastian Bykuć [1]

[1] Distributed Energy Department, The Szewalski Institute of Fluid-Flow Machinery Polish Academy of Sciences, 80-231 Gdańsk, Poland; witold.rybinski@imp.gda.pl (W.R.); sbykuc@imp.gda.pl (S.B.)

[2] The Faculty of Engineering and Science Department of Energy Technology Thermal Energy Systems, Aalborg University, 9220 Aalborg, Denmark; cbo@et.aau.dk

* Correspondence: mglinski@imp.gda.pl

Received: 21 December 2018; Accepted: 29 January 2019; Published: 31 January 2019

Abstract: In the article, the method and algorithm for a control strategy of the operation of a micro combined heat and power (mCHP) unit and for reducing the power consumption peaks (peak shaving) are proposed and analyzed. Two scenarios of the mCHP's operation, namely with and without the control strategy, are discussed. For calculation purposes, a boiler fired with wood pellets coupled with a Stirling engine, manufactured by ÖkoFEN, was used. These results were used to analyze two scenarios of the control strategy. In this study, the operation of mCHP was simulated using the energyPRO software. The application of this control strategy to dispersed mCHP systems allows for a very effective "peak shaving" in the local power grid. The results of calculation using the new algorithm show that the electricity generated by the mCHP system covers the total demand for power during the morning peak and reduces the evening peak by up to 71%. The application of this method also allows for a better reduction of the load of conventional grids, substations, and other equipment.

Keywords: control strategy; peak-shaving; mCHP; Stirling engine; renewable energy; energy consumption profile

---

1. Introduction

Nowadays in Poland, ≈80% of the electricity is generated in the central coal-fired power plants and transmitted through the old electric grids to distant customers [1]. In addition, most households have heating systems based on high emission coal-fired boilers, which have a low efficiency. One of the possibilities to solve these problems is a wide application of modern highly efficient a micro combined heat and power (mCHP) unit powered by renewable energy sources (RES). It could decrease the load of electric grids and reduce air pollution from fossil fuels.

In 2004, the European Directive 2004/8/EC (together with the amendments in 2012/27/EU [2]) was introduced in EU countries to increase energy efficiency and strengthen the power supply security by developing highly efficient combined heat and power production in cogeneration, i.e., CHP plants. An additional aspect is a reduction of the negative impact from the combustion of fossil fuels on the environment and on global warming. Over many years, the European Directives have been limiting the emission of greenhouses gases into the atmosphere. Therefore, in recent years in many countries, one can observe an increase in the share of energy produced using renewable energy sources (RES) and decentralized CHP plants. Nowadays, the main aim is to optimize the generation of heat and electricity as well as reduce the losses due to energy transmission.

The conversion of the production of heat and electricity using fossil fuels to the production using renewable energy sources (RES) has already started and is implemented globally to some extent. In some countries, the conversion is well underway and in others it has just started. A large part of the renewable energy sources is characterized as non-controllable and fluctuating. One of the consequences of this is that the energy supply structure will change from the demand-controlled supply to the availability-controlled consumption. Furthermore, conversion to RES should be accompanied by a detailed study of how the discrepancy between availability and consumption can be balanced and how to achieve built-in flexibility.

A sector where there is a potential for flexibility and balancing the energy consumption covers households. This sector also accounts for a major part of the total electricity and heat consumption.

The consumption of electricity is increasing but the local power transmission grid has not been changed and updated properly. Old wires and transmission equipment and their integration with RES power systems have a negative impact on the operational parameters of the power grid such as voltage reduction, frequency fluctuation, etc. Higher consumption of electricity, specifically during daytime, generated huge peaks of power consumption in the grid. These peaks have a huge impact on fluctuations in the grid and can lead to an overload or blackout [3]. That is why in many countries, the demand side management introduces the peak-time tariffs for the reduction of energy consumption in households [4–6]. Globally, morning and evening power peaks in the power grid were distinguished. Therefore, many studies focus on strategies for peak shaving in the electrical grid for better power quality and lower energy transmission losses. In many publications, battery energy storage systems (BESS) and electric vehicles (EVs) applied to peak smoothing were reviewed and analyzed [7,8]. However, these technologies are still too expensive for wide residential application.

One type of the considered devices for cogeneration in residential areas are micro combined heat and power (mCHP) systems with electric power production of 1–5 kWe. Such systems have very high overall efficiencies of total energy conversion, which can reach 90% [9,10]. Recently, mCHP units are used mainly for heat production, where electricity is generated as the support of the power grid. The interest in mCHP systems based on renewable energy sources (RES) and their development are growing. Some examples are:

mCHP system with an internal combustion engine (ICE) using methanol as fuel.
mCHP system with an external combustion engine like a Stirling engine (SE) or an organic Rankine cycle (ORC) system using biomass as fuel.
Fuel cell supplied by hydrogen.

For the purpose of this work, the ÖkoFEN mCHP unit [11] fed with biomass in the form of pellets was chosen for further analyses. It was initially assumed, and then verified, that the heating power of the device together with the appropriate heat storage meets the requirements for the heat supply of the example house described in Section 3 of this work. A higher capacity heat storage extends the possibilities for application of this mCHP unit for higher heat demand cases. There are also mCHP devices of higher power on the market; however, most of them are fueled with natural gas. Regardless of their power, they can be used for electricity peak shaving in a similar way as described in this paper.

Most publications concerning mCHP units, energy storages, and load-shifting control strategies are focused on one or two of these topics, but rarely on a combination of all three [12–14]. This article presents mCHP devices as smart power generation systems where heat and electricity are generated only during periods of high demand for electricity in a household.

This description is intended as an introduction to two articles: about peak shaving and the further about impact on the environment using biomass in distributed mCHP's instead of coal in central power plant during the peaks.

The main purpose of this article is to develop a method and algorithms that can be used for the control strategy of the operation of an mCHP unit in order to reduce its power consumption peaks (peak shaving). Two scenarios of the mCHP's operation, namely with and without the control strategy, are presented. The profiles of power and heat demand were created and used to calculate the power and heat production of a biomass-fired mCHP device that operates in an exemplary house.

For the calculation purposes, data of a boiler fired with wood pellets coupled with a Stirling engine, manufactured by ÖkoFEN, are used. The data are as follows: a nominal thermal power of 9 kWth and an electrical power of 0.6 kWe at the total efficiency of 90% (with peaks of 13 kWth and 1 kWe respectively) [11]. These results were used to analyze two scenarios of the control strategy. In the first scenario, the standard control of the mCHP unit, which depended on the water temperature in the heat storage tank, was assumed. In the second scenario, the operation time of the mCHP depends on the duration of morning and evening peaks of energy demand. In this study, the operation of the mCHP unit was simulated using the energyPRO 4.0 software. The application of this control strategy to dispersed mCHP devices allows for a very effective "peak shaving" in the local power grid. The surplus of the heat produced by the domestic mCHP device is stored in a tank and can be used for heating.

2. Review of the Literature

Different studies highlighted the advantages of an mCHP system in a typical residential building [15–17]. Martinez et al. [17] provides a broad review of available micro combined heat and power systems. The first part of this review is focused on the existing energy conversion devices with internal combustion engines (ICEs) and other propulsion systems such as Stirling engines, organic Rankine cycle (ORC) systems, and fuel cells. The second part deals with the mCHP systems supplied by renewable energy sources, mainly the solar energy-based technologies (e.g., PVT combined with an ORC system). Based on this thorough analysis, the cogeneration plant can achieve a total efficiency of 75–95%. They concluded that mCHP systems with ICEs are characterized by a higher efficiency than the other devices, but they also indicated that an mCHP system with a Stirling engine is an attractive technology thanks to external combustion where biomass or waste can be used as fuel.

Rosato et al. [18] analyzed the operational performance of mCHP units coupled with thermal storage tanks situated in several apartment buildings. They both agreed that mCHP systems are both profitable and reliable and can be more energy-efficient than traditional power systems. The mCHP technologies with microturbines and Stirling engines can provide a proper fit to the demand of each household with regard to a positive ratio between thermal and electrical energy than other solutions like ICEs or fuel cells [19]. They indicated that mCHP systems can play an important role in the household production of electricity, even exceeding the amount produced by a domestic photovoltaic (PV) plant.

Ummenhofer et al. [20] carried out experimental tests of an mCHP unit with an ICE coupled with a thermal energy storage (TES) and an electric energy storage (EES). They focused on the performance of charging and discharging the EES to improve peak time coverage of the power grid. The smart control system of the mCHP device coupled with the EES was applied. They indicated that an appropriate selection of the specific system's components with correct management of the working time of the tested mCHP system gave additional benefits associated with higher coverage during periods of peak energy demands and a shorter run time. In general, the number of run-time cycles of the mCHP unit has a greater impact on the level of self-sufficiency achieved than their respective run-time duration. Both the operation time and start-up time of the mCHP unit with the EES should be adapted to a residential setting and different peak energy demand periods.

In Reference [21], dynamic modelling was conducted to investigate the significant impact of the demand side management (DSM) on the operational performance of air supply heat pumps (ASHPs) with mCHP units for heating a residential building. Cooper et al. [21] show that the DSM can reduce the peak electricity demand of a heat pump in the local power grid. However, in the investigated scenarios, it is unlikely that the peak demands can be reduced sufficiently such that they will not exceed the capacity of the local distribution transformer if ASHPs are used in all dwellings. By using a combination of mCHP units with ASHPs, it is possible to supply heating to all dwellings without exceeding this capacity. The authors concluded that heat pumps with mCHP units do not achieve the maximum performance due to a limitation in the operation regimes of the DSM, which are optimized for the power grid.

Angrisani et al. [22] carried out simulations to assess the thermo-economic performance of an mCHP system used in households with a low energy demand. They focus on the estimation of the economic profitability of a cogeneration system since it is strongly influenced by the amount of electricity generated and consumed by its user. The results of this study indicate that the installation of mCHP systems in buildings with a low energy demand allows for increasing the percentage of electricity generated and consumed by users. It reduces the bidirectional electricity flow between the users and the power grid, and its impact on the grid due to the large diffusion of distributed generation systems. Moreover, this study shows that the load sharing approach among users with different load profiles leads to better energy and economic results compared to conventional systems. Weather conditions have a strong influence on the operational hours of the mCHP unit, which obviously affects the thermo-economic performance of the system.

3. Energy Consumption Profile

For analysis purposes, a typical 250-square-meter house, localized in Legionowo near Warsaw in Poland, was assumed. The house was built using outdated residential building methods, and its energy performance coefficient was equal 110 kWh/m^2 year, so the building had a higher heat loss to the atmosphere than new low-energy buildings. The house was inhabited by four people, two adults and two children. For working days in winter, the power consumption daily profile was assumed based on the internal reports of the Institute of Fluid-Flow Machinery, as shown on the plot in Figure 1. For the remaining seasons, the power consumption profiles were determined based on the standard customer's profile from tariff G11 from the Polish energy group ENEA. The following are examples of the power profiles for this house in central Poland.

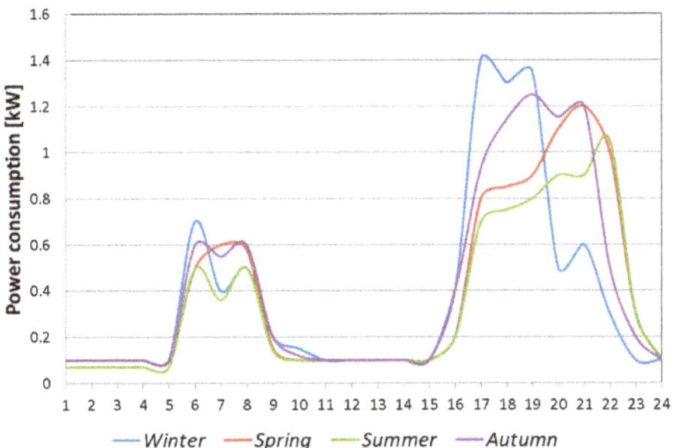

Figure 1. Average power consumption profile of one day.

In Figure 1, two peaks of power consumption are observed. These power peaks are called morning and evening peaks, and occur when the inhabitants are the most active in the house. Yearly demand for electricity in that house was calculated and was ≈3.6 MWh. The morning peaks occurred at the almost same time; however, the winter one was the highest and the summer one was the lowest. The autumn evening peak was shifted by 2 h in comparison to the winter one at 5 pm, the spring one by 4 h, and the summer one by 5 h. The winter evening peak was also the highest while the summer one was the lowest; however, the differences were higher than in the morning.

The heat demand for space heating was calculated as well. The heat loss of the building was calculated based on the heat transfer coefficient, temperature differences, and heat transfer area of the main construction elements. The heat loss due to the ventilation system was also taken into account. However, the radiation effect of the sun was not taken into account. The outdoor temperature measurements in 2015 were obtained (through the energyPRO software) from the meteorological station located in Legionowo near Warsaw. These data were used in an MS Excel 2013 program to calculate the heat demand for heating each hour and exported as a time series to the energyPRO software. The heating season started on 1 October 2015, and ended on 31 May 2015. In Figure 2, the profiles of the outdoor temperature on a representative day of four seasons are presented.

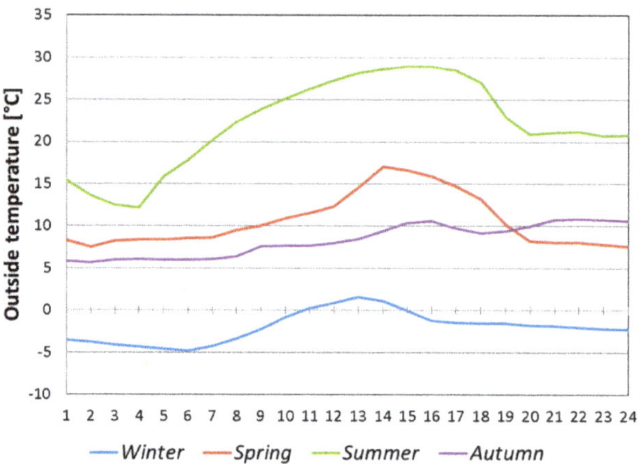

Figure 2. Outside temperature within one day in different seasons.

The temperature inside the building during the heating season was assumed to be 20 °C when the house was occupied and 16 °C otherwise. The house was occupied by four persons from 12 a.m. to 8 a.m. and from 4 p.m. to 12 a.m., and not occupied from 8 a.m. to 4 p.m.

Based on the above assumptions, the heat demand was calculated. In Figure 3, the heating profile of space heating is shown. One can notice that the heat demand strongly depended on the outside temperature and on the number of people in the house (the morning and evening peak demands). During the working day with no people inside, the heat consumption was 30–80% lower than during morning and evening activities.

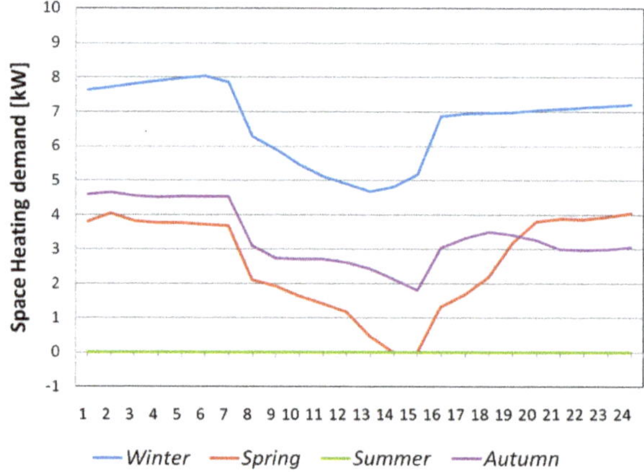

Figure 3. Heat demand for space heating within one day in different seasons.

Based on the experimental data [23], the heat demand for heating of the domestic hot water (DHW) was assumed. Figure 4 shows the profiles of the heat demand for DHW for each representative day. As can be seen on the plot, morning and evening peak heat demand can also be distinguished.

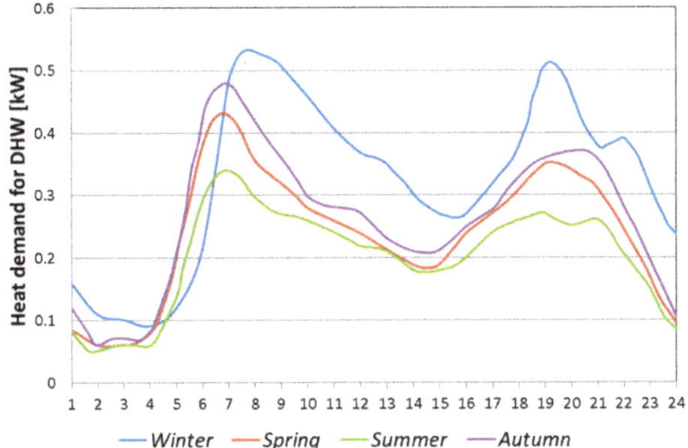

Figure 4. Heat demand for domestic hot water (DHW) within one day in different seasons.

Figure 5 shows the total heat demand (the sum of space heating and DHW heating) for a representative day for each season. The yearly demand for heat was equal 27,654 kWh and for electricity 3600 kWh as calculated using the energyPRO software. In the summer, heat demand was only for heating DHW. The greatest demand for heat took place during winter when the heat was consumed for space heating and DHW. In the transition period (spring or autumn), the heat demand was 2 times lower than during the winter season. Figure 6 presents the comparison of heat and power consumption. Each respective morning and evening peak consumption of the heat and electricity took place at the same time across seasons.

To reduce the amount of electrical energy drawn from the grid during morning and evening peak demands, both heat and electricity should be produced using an mCHP cogeneration system.

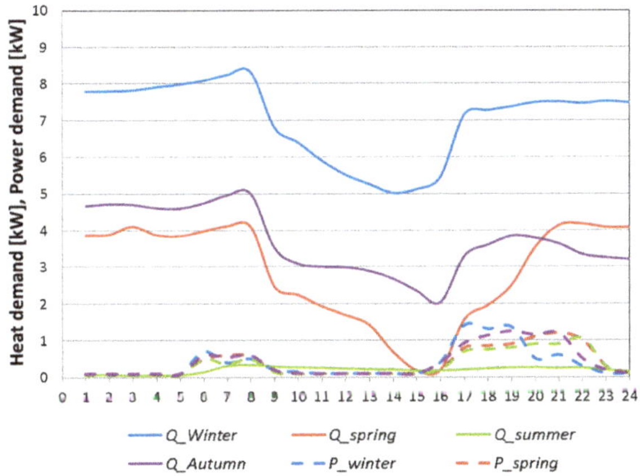

Figure 5. Total heat demand (solid line) for space heating and DHW within one day in different seasons and the assumed power demand profile (dashed line).

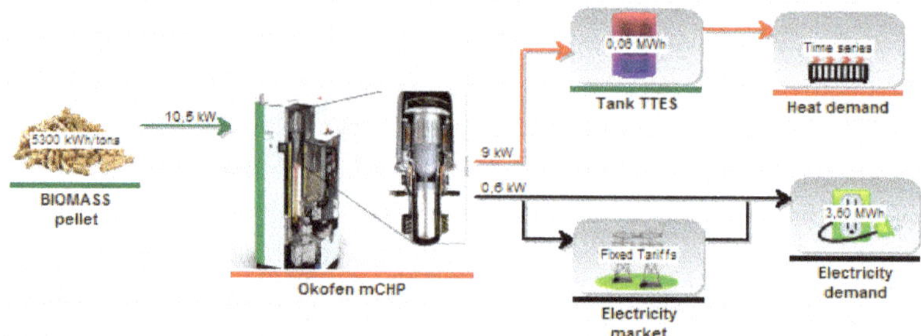

Figure 6. Scheme of the mCHP system, graphical user interface from the energyPRO 4.0 software.

4. Modelling of a Biomass-Fired mCHP Unit for Peak Shaving

The simple modelling of the operation of an mCHP system was carried out using the energyPRO 4.0 software. Figure 6 shows the main part of the developed model. For heat and electricity production, the biomass-fired boiler ÖkoFEN Pellematic Condens_e (ÖkoFEN, Niederkappel, Austria) with a Stirling engine was used. The boiler was supplied with wood pellets, which were burnt in the combustion chamber where the heat was produced. The exhaust fumes heated up the Stirling engine connected to the generator producing electricity. The hot water flowed from the mCHP system to a thermal energy storage (TES) tank and supplied the heating system of the building. The generated electricity satisfied the actual power demand of the building and it was possible to export the energy surplus to the national power grid via a direct connection. When the Stirling engine did not operate, the building was supplied from the power grid.

The results of the heat and power demand calculation for one year were used as the input time series to the control algorithm developed using the energyPRO software. In the standard control algorithm, the mCHP device operates depending on the temperature of hot water in the TES tank (i.e., on the amount of heat stored in the tank). If the temperature in the tank is too low, the mCHP device is switched on and operates until the storage is fully charged.

The standard algorithm uses the mCHP device primarily for the production of heat and its supply to the thermal energy storage tank. The amount of heat produced depends only on the storage temperature:

$$\left(\dot{Q}_{th}, P_{el} \right)_{prod} = f_{st} \left(T_{storage} \right) \tag{1}$$

The additionally produced electricity is used up in the house and/or exported to the power grid.

The authors developed the method and algorithm for the shaving of the peak electricity demands. In this algorithm, the mCHP unit with a Stirling engine was primarily used for electricity generation mainly during these high electricity demands. The amount of the electricity and heat produced depended on the electricity profile and heat demand profile, time of the occurrence of peak demands, and the storage temperature:

$$\left(\dot{Q}_{th}, P_{el} \right)_{prod} = f_{new} \left(\dot{Q}_{th,demand}, P_{el,demand}, \tau_{peak}, T_{storage} \right) \tag{2}$$

The mCHP unit was also used for heat production during peak-free periods and the produced electricity was used up in the house and/or exported to the power grid.

Description of the new algorithm:

Based on the given power demand profile the time of appearance and duration of the morning and evening power peaks were determined.

The heat demand profiles for different seasons were determined.

Based on the given power profile and determined heat demand profile, the profile of the mCHP unit's operation was adaptively determined; this prepared profile was further used to control the operation of the mCHP unit during its exploitation.

The primary goal was peak shaving.

The secondary goal was the supply of heat to the heating system and domestic hot water.

Heat was mainly produced in combination with the electricity generation during the peak energy demands. However, the mCHP unit could produce heat for longer than during the peak energy demands in order to store the highly enough amount of heat to prevent the mCHP unit from turning on during the peak-free periods.

If the amount of heat in the thermal storage was less than 10%, the mCHP unit had to also produce heat during peak-free periods.

The heat losses of the storage were calculated using the Fourier law with the default values of insulation conductivity and thickness using the energyPRO software (λ = 0.037 W/mK, δ = 100 mm).

Due to the thermal inertia of the mCHP system, a signal for the start-up of the mCHP system was sent 1 h before the peak energy demand occurred.

The amount of the electricity generated in the mCHP system that exceeded the electricity demand was exported to the power grid.

The buying cost of pellets and the varying price of electricity during peak and peak-free periods was taken into account.

The mCHP worked during the morning and evening peak energy demands in order to reduce power peaks (peak shaving).

This task-oriented description of the algorithm was performed by the low-level, internal optimization algorithms included in energyPRO. The option "Island Operation" was on, and covering the internal electricity demand took precedence over the optimization of the electricity production sent to the electricity market. The option "User Defined Operation Strategy" was set to the default profile for one electricity production unit. The electricity and heat demand profiles were given, as well as the size, insulation conductivity and specific heat of the thermal storage, its lower heat energy limit, and the ambient temperature. The minimum, nominal, and maximum electric and thermal power of the mCHP unit was also given. Based on this input data and option settings, the internal energyPRO algorithms optimized the operation profile of the electricity/heat production in the mCHP, which was the algorithm output. Each month was optimized separately; however, the state at the month's end was the initial state for the next month.

Figure 7 shows a comparison of the profiles of energy demand, its production and heat stored for the standard control, and the new algorithm during the winter. In the standard algorithm, the mCHP device operated when the heat demand was high to fill the 3-cubic-meter storage tank up to 63 kWh. In the new algorithm, the mCHP unit operated only during morning and evening times and supplied the tank with heat up to 33 kWh and 28 kWh, respectively. The charging time of the TES tank and its charge level were different for both analyzed algorithms. In the standard control algorithm, the tank was charged up to 100% capacity, but in the new algorithm, the charge level was 35–66% of the tank capacity.

Figure 8 shows the power production in the mCHP unit, the power demand, and the difference between them. When this difference is positive, it is exported to the power grid and drawn from the grid when it is negative. Looking at Figure 8, it can be seen that more power was drawn from the power grid when the standard algorithm was used.

When the mCHP unit operated using the new algorithm, it satisfied the total demand for power during the morning peak and up to 71% of the total demand during the evening peak. In the standard algorithm, the mCHP unit worked periodically and its operating time depended only on the temperature level of the hot water in the tank, irrespective of whether any peak demands occurred. As the results show, the low coverage of the electricity consumption and production is visible. The mCHP unit that used the standard control algorithm often worked during peak-free time, but unfortunately at the same time when the electricity drawn from the power grid was the cheapest. Analyzing both algorithms, it can be concluded that for the standard algorithm, over 2.6 times more electricity was drawn from the power grid during the peak energy demand. When the new control algorithm was applied, 3.7 times less electricity was exported to the power grid within the year. Thanks to the new algorithm, less electricity needed to be drawn from the power grid and energy needs were met to a greater extent.

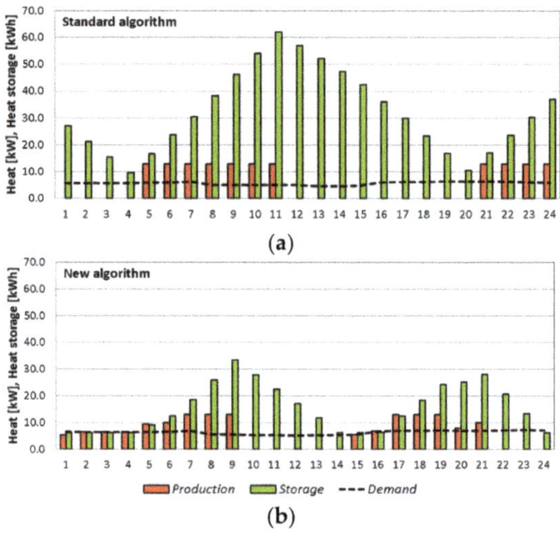

Figure 7. Profiles of the heat produced by the mCHP system (kW) and heat stored in the tank (kWh) with a capacity of $3\,\mathrm{m}^3$ during the winter season, depending on the control algorithm used: (**a**) standard algorithm, and (**b**) new algorithm.

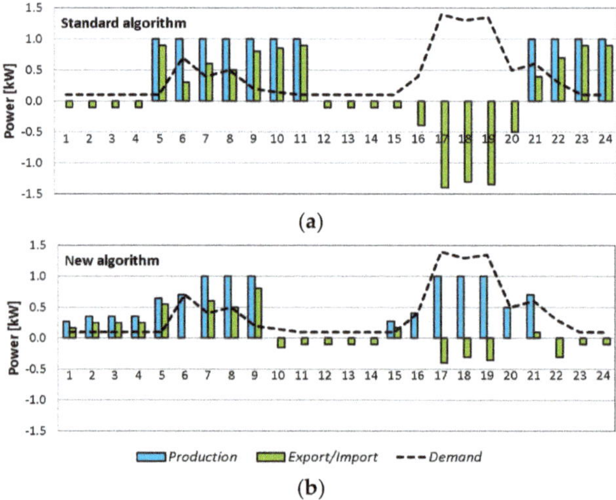

Figure 8. Profiles of the electricity demand and production in the mCHP system during winter season depending on the control algorithm used: (**a**) standard algorithm, and (**b**) new algorithm.

Figure 9 shows the different profiles of demand for electricity and its production for all four seasons in one year using the new algorithm. Having analyzed the graphs, it can be seen that the production of electricity strongly depended on the demand for heating. In the winter the morning peak was completely covered by the electricity from mCHP, while the evening peak was up to 71%. In the spring, almost all power demand was covered by the mCHP production because the power demand was lower than in winter. The morning peak was completely covered and the evening one was covered up to 83%. In the second transition heating periods (autumn), the morning peak was also completely covered; however, the evening one was less covered (up to 80%) than in spring. In the summer, the power demand was the lowest; however, both the morning and evening peaks were almost not covered by the power produced in the mCHP. This was caused by impossibility of utilization of the heat produced by the mCHP since the heat demand was the lowest in the summer.

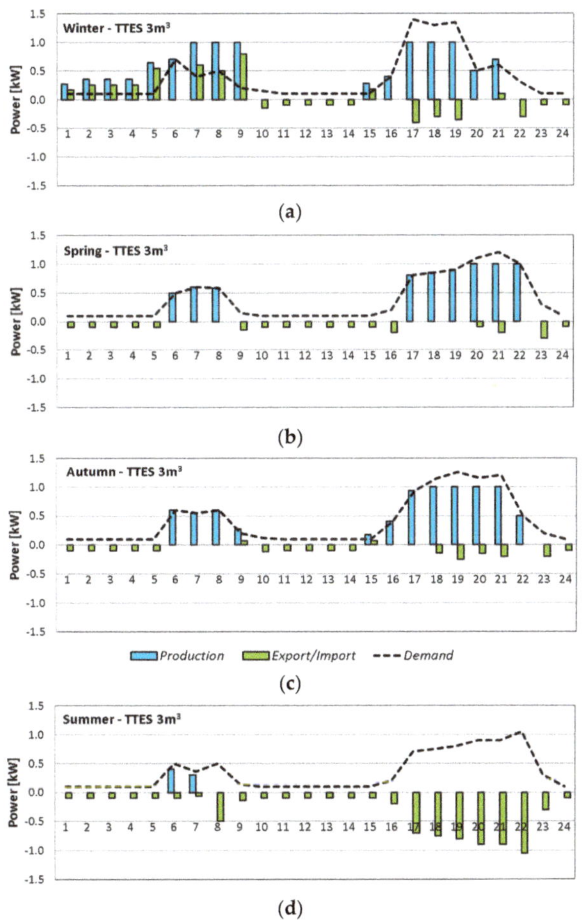

Figure 9. Comparison of the electricity demand and production in the mCHP system that used the new algorithm for each representative day of (**a**) winter, (**b**) spring, (**c**) autumn, and (**d**) summer.

Figure 10 shows a comparison of the heat demand of the building, the heat produced by the mCHP system, and the amount of heat stored in the 3-cubic-meter tank for four seasons in one year using the new algorithm. In the winter, the mCHP also operated before the morning peak to additionally produce heat due to the highest heat demand in this season. The heat was stored in the tank TES and utilized during the break between the morning and evening peaks. The mCHP was turned on before the evening peak for the same reasons as before the morning one. In the transition heating periods (spring and autumn), the mCHP operated during the morning and evening peaks only. However, the amount of heat produced during the peaks was sufficient to cover the heat demand. In the summer, the mCHP almost did not operate because the thermal storage reached its upper limit (63 kWh) and was almost full all the time due to the lowest heat demand in this season. Actually, the mCHP operated 2 h every second day only.

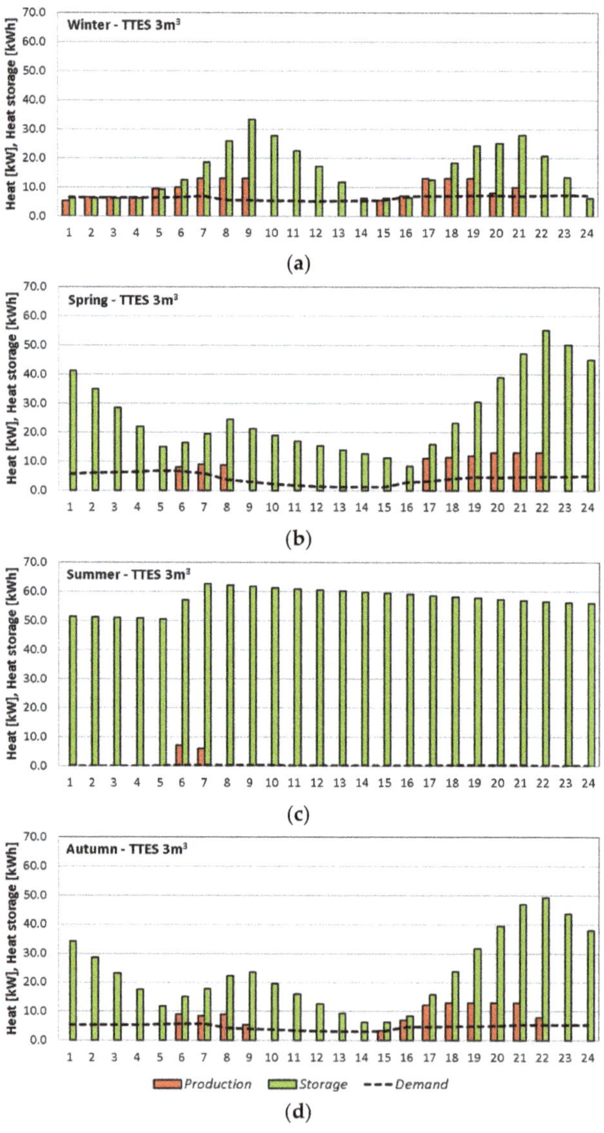

Figure 10. Comparison of the heat demand, production and storage in the mCHP system which uses the new algorithm, for each representative day of (**a**) winter, (**b**) spring, (**c**) summer, and (**d**) autumn.

In the next step of modelling the operation of the mCHP system, the efficiency of heat and electricity production for different sizes of the TES tank were compared. In Table 1, the calculation results obtained using the energyPRO software for the operation of the mCHP unit within one year with three different tank sizes are presented. The total yearly efficiency is defined as the sum of heat and electricity production divided by fuel consumption within the year. The mCHP system with the largest tank (i.e., 5-cubic-meter tank) was able to produce the highest amount of electricity. The largest tank had the highest charge level as well. This also means that the largest tank had about 43% more heat loss than the 3-cubic-meter TES tank. The mCHP unit with the 3-cubic-meter tank (63 kWh) had the highest total efficiency of cogeneration. However, this comparison of the efficiencies did not take into account the investment costs, which may be decisive when the choice of the tank is made.

Table 1. The calculation results of the mCHP unit with three different tank sizes, which operated within one year.

Volume of the TES Tank (m^3)	Total Efficiency (%)	Heat Production (kWh)	Electricity Production (kWh)	Fuel Consumption (kWh)
1	83.0	27,597	1856	35,492
3	83.5	28,850	1991	36,929
5	83.3	29,231	2021	37,507

Figure 11 shows a comparison between the application of three different tank sizes. The tank with a capacity of 1 m^3 was a little bit too small. During the evening peak, this tank was charged up to 100% capacity (31 kWh), which temporarily stopped the operation of mCHP and disturbed the process of peak shaving at 9 pm. For the 3 and 5-cubic-meter tanks, the charging and discharging processes were very similar. However, the 5-cubic-meter tank (104 kWh) was too large for this application due to higher heat loss and unused capacity.

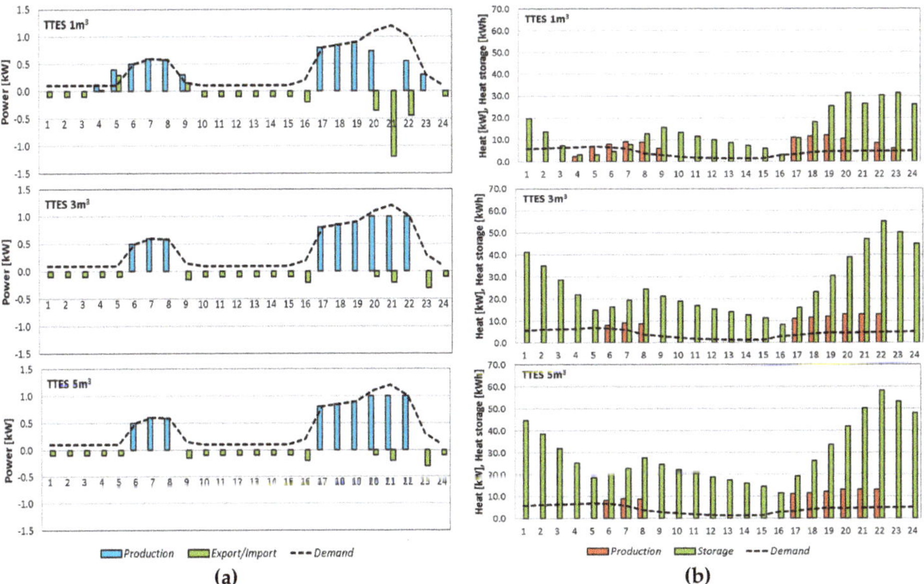

Figure 11. Comparison of the electricity demand and production (**a**), and heat demand, production, and storage (**b**) in the mCHP system, which used the new algorithm for three different tank sizes during the spring season.

5. Conclusions

This paper discussed the modelling of an mCHP device using the energyPRO 4.0 software. The new control algorithm of the mCHP device's operation, which allows for reducing peak energy demands, was analyzed. The primary goal was peak reduction, the secondary goal was heat production out of the peaks (whereas in the standard algorithm the primary goal was heat production). This algorithm used the actual power and heat demand, as well as their daily profiles in different seasons, to control the heat and power production in the mCHP system. The standard control algorithm used only the temperature in the heat storage tank. Due to employing the new algorithm, the mCHP system could satisfy the total demand for energy during the morning peak and up to 71% of the total demand for energy during the evening peak. In the new algorithm, a reduced amount of energy was drawn from the power grid and the own energy demand was met to a greater extent. The effectiveness of the reduction of peak energy demands strongly depended on the demand for heating. In the transition heating periods (spring and autumn), the energy demand was satisfied to a greater extent than during the winter period. During the transition periods, the mCHP unit was able to satisfy up to 100% of the total morning peak energy demand and up to 83% of the total evening peak. In the summer, the power demand was almost completely unsatisfied by the mCHP because of the impossibility of utilization of the produced heat.

It should be noted that the size of the mCHP unit's heat storage tank affected the efficiency of the reduction of the peak energy demands. The optimal size of the heat storage tank was chosen to be $3\,\text{m}^3$.

The very important advantage of the analyzed algorithm is the possibility of reduction of the local (within the household) peak energy demands. This individual reduction does not require the coordination between the other households in the local power grid.

Looking into these actions from a broader perspective, installing mCHP devices in many buildings in the local power grid can decrease the grid's load and provide more stable electric parameters during peak demands for energy.

Author Contributions: Conceptualization and Calculation, M.G.; Writing—Original Draft Preparation, M.G. and C.B.; Writing—Review & Editing, W.R.; Supervision, C.B. and S.B.; Project Administration and Funding Acquisition, S.B.

Funding: This research was funded by the European Union's Horizon 2020 grant number 692197.

Acknowledgments: The collaboration between the authors was made possible and was supported by the SUPREME project. The project has received funding from the European Union's Horizon 2020 research and innovation program under Grant Agreement No. 692197.

References

1. Report of Polish Transmission System Operator (PSE) for 2017. Available online: www.pse.pl (accessed on 19 April 2018).
2. EU. Directive 2012/27/EU of the European Parliament and of the Council of 25 October 2012 on energy efficiency, amending Directives 2009/125/EC and 2010/30/EU and repealing Directives 2004/8/EC and 2006/32/EC. *Off. J. Eur. Union* **2012**, *55*, 1–97.
3. Muratori, M.; Schuelke-Leech, B.A.; Rizzoni, G. Role of residential demand response in modern electricity markets. *Renew. Sustain. Energy Rev.* **2014**, *33*, 546–553. [CrossRef]
4. Lund, H.; Andersen, A.N. Optimal designs of small CHP plants in a market with fluctuating electricity prices. *Energy Convers. Manag.* **2005**, *46*, 893–904. [CrossRef]
5. Torriti, J. Price-based demand side management: Assessing the impacts of time-of-use tariffs on residential electricity demand and peak shifting in Northern Italy. *Energy* **2012**, *44*, 576–583. [CrossRef]
6. Venizelou, V.; Philippou, N.; Hadjipanayi, M.; Makrides, G.; Efthymiou, V.; Georghiou, G.E. Development of a novel time-of-use tariff algorithm for residential prosumer price-based demand side management. *Energy* **2018**, *142*, 633–646. [CrossRef]

7. Baeten, B.; Rogiers, F.; Helsen, L. Reduction of heat pump induced peak electricity use and required generation capacity through thermal energy storage and demand response. *Appl. Energy* **2017**, *195*, 184–195. [CrossRef]

8. Uddin, M.; Romlie, M.F.; Abdullah, M.F.; Abd Halim, S.; Abu Bakar, A.H.; Chia Kwang, T. A review on peak load shaving strategies. *Renew. Sustain. Energy Rev.* **2018**, *82*, 3323–3332. [CrossRef]

9. Rosato, A.; Sibilio, S.; Ciampi, G. Energy, environmental and economic dynamic performance assessment of different micro-cogeneration systems in a residential application. *Appl. Therm. Eng.* **2013**, *59*, 599–617. [CrossRef]

10. Sartor, K.; Quoilin, S.; Dewallef, P. Simulation and optimization of a CHP biomass plant and district heating network. *Appl. Energy* **2014**, *130*, 474–483. [CrossRef]

11. ÖkoFEN Forschungs-und Entwicklungs Ges.m.b.H.—mCHP Manufacturer's Website. Available online: http://www.okofen-e.com/en/pellematic_condens_e/ (accessed on 11 June 2018).

12. Allison, J.; Cowie, A.; Galloway, S.; Hand, J.; Kelly, N.J.; Stephen, B. Simulation, implementation and monitoring of heat pump load shifting using a predictive controller. *Energy Convers. Manag.* **2017**, *150*, 890–903. [CrossRef]

13. Angrisani, G.; Canelli, M.; Roselli, C.; Sasso, M. Integration between electric vehicle charging and micro-cogeneration system. *Energy Convers. Manag.* **2015**, *98*, 115–126. [CrossRef]

14. López, M.A.; De La Torre, S.; Martín, S.; Aguado, J.A. Demand-side management in smart grid operation considering electric vehicles load shifting and vehicle-to-grid support. *Int. J. Electr. Power Energy Syst.* **2015**, *64*, 689–698. [CrossRef]

15. Maghanki, M.M.; Ghobadian, B.; Najafi, G.; Galogah, R.J. Micro combined heat and power (MCHP) technologies and applications. *Renew. Sustain. Energy Rev.* **2013**, *28*, 510–524. [CrossRef]

16. Murugan, S.; Horák, B. A review of micro combined heat and power systems for residential applications. *Renew. Sustain. Energy Rev.* **2016**, *64*, 144–162. [CrossRef]

17. Martinez, S.; Michaux, G.; Salagnac, P.; Bouvier, J.-L. Micro-combined heat and power systems (micro-CHP) based on renewable energy sources. *Energy Convers. Manag.* **2017**, *154*, 262–285. [CrossRef]

18. Rosato, A.; Sibilio, S. Performance assessment of a micro-cogeneration system under realistic operating conditions. *Energy Convers. Manag.* **2013**, *70*, 149–162. [CrossRef]

19. Comodi, G.; Cioccolanti, L.; Renzi, M. Modelling the Italian household sector at the municipal scale: Micro-CHP, renewables and energy efficiency. *Energy* **2014**, *68*, 92–103. [CrossRef]

20. Ummenhofer, C.D.; Olsen, J.; Page, J.; Roediger, T. How to improve peak time coverage through a smart-controlled MCHP unit combined with thermal and electric storage systems. *Energy Build.* **2017**, *139*, 78–90. [CrossRef]

21. Cooper, S.J.G.; Hammond, G.P.; McManus, M.C.; Rogers, J.G. Impact on energy requirements and emissions of heat pumps and micro-cogenerators participating in demand side management. *Appl. Therm. Eng.* **2014**, *71*, 872–881. [CrossRef]

22. Angrisani, G.; Canelli, M.; Roselli, C.; Sasso, M. Microcogeneration in buildings with low energy demand in load sharing application. *Energy Convers. Manag.* **2015**, *100*, 78–89. [CrossRef]

23. Parker Danny, S. Research Highlights from a Large Scale Residential Monitoring Study in a Hot Climate. In Proceedings of the International Symposium on Highly Efficient Use of Energy and Reduction of its Environmental Impact, Osaka, Japan, 22–23 January 2002; pp. 108–116.

 *energies*

Article

Impact of Nanoadditives on the Performance and Combustion Characteristics of Neat Jatropha Biodiesel

Abul Kalam Hossain * and Abdul Hussain

Sustainable Environment Research Group, School of Engineering and Applied Science, Aston University, Birmingham B4 7ET, UK; abdul_369@hotmail.com
* Correspondence: a.k.hossain@aston.ac.uk; Tel.: +44-(0)-121-204-3041

Received: 27 December 2018; Accepted: 5 March 2019; Published: 10 March 2019

Abstract: Jatropha biodiesel was produced from neat jatropha oil using both esterification and transesterification processes. The free fatty acid value content of neat jatropha oil was reduced to approximately 2% from 12% through esterification. Aluminium oxide (Al_2O_3) and cerium oxide (CeO_2) nanoparticles were added separately to jatropha biodiesel in doses of 100 ppm and 50 ppm. The heating value, acid number, density, flash point temperature and kinematic viscosity of the nanoadditive fuel samples were measured and compared with the corresponding properties of neat fossil diesel and neat jatropha biodiesel. Jatropha biodiesel with 100 ppm Al_2O_3 nanoparticle (J100A100) was selected for engine testing due to its higher heating value and successful amalgamation of the Al_2O_3 nanoparticles used. The brake thermal efficiency of J100A100 fuel was about 3% higher than for neat fossil diesel, and was quite similar to that of neat jatropha biodiesel. At full load, the brake specific energy consumption of J100A100 fuel was found to be 4% higher and 6% lower than the corresponding values obtained for neat jatropha biodiesel and neat fossil diesel fuels respectively. The NOx emission was found to be 4% lower with J100A100 fuel when compared to jatropha biodiesel. The unburnt hydrocarbon and smoke emissions were decreased significantly when J100A100 fuel was used instead of neat jatropha biodiesel or neat fossil diesel fuels. Combustion characteristics showed that in almost all loads, J100A100 fuel had a higher total heat release than the reference fuels. At full load, the J100A100 fuel produced similar peak in-cylinder pressures when compared to neat fossil diesel and neat jatropha biodiesel fuels. The study concluded that J100A100 fuel produced better combustion and emission characteristics than neat jatropha biodiesel.

Keywords: biofuel; CI engine; combustion; emission; greenhouse gas; jatropha biodiesel; nanoparticle; performance

---

1. Introduction

The amount of CO_2 in the atmosphere has increased significantly since the start of the industrial era in the 18th century [1]. Fossil fuels used in the transportation and electricity (and heat) production sectors are responsible for about 40% of the total global greenhouse gas (GHG) emissions [2]. In the UK, road transports are responsible for 22% of the total UK CO_2 emissions [3]. Use of renewable biofuels instead of fossil based fuels could reduce the GHG emissions significantly [4–6]. Biodiesels, produced through transesterification of seed oils (or wastes), have diesel like physico-chemical fuel properties and may substitute fossil based diesel fuel. They are biodegradable, has higher oxygen content and cetane number [7]. Engine performance and combustion characteristics were assessed by researchers using various biodiesels and their blends with fossil diesel [8,9]. The type of feedstock used for biodiesels production affect life cycle energy and GHG emission of the transesterification process. Hence, it is important what type of crops are used for biodiesel production such as edibles and non-edibles. This has led to controversy surrounding farmland and whether it should be used for food or fuel [6]. Mofijur et al. [10] studied the engine performance characteristics operated separately with biofuels obtained from edible and non-edible feedstocks. Two biodiesels, produced from palm (edible) and jatropha (non-edible) oils were used. They found that considering the overall emission reduction potential, jatropha biodiesel was better than palm biodiesel [10,11].

The effects of various oxygenated additives on biofuels were investigated by the researchers to further improve the combustion and emission characteristics of the biofuels powered internal combustion (IC) engines. For example, nanoparticles were added to fuel mixtures to improve the engine performance and combustion characteristics; typically, metallic oxides nanoparticles were used to increase the heat release rate and thermal efficiency [12–14]. Metal-oxide nanoparticles have the ability to donate oxygen atoms to the fuel mixture and can create high surface to volume ratio; hence, they act as high reactive medium for combustion. Other advantages of adding nanoparticle additives are: increased thermal conductivity, flash point and fire point temperatures; and reduced kinematic viscosity [12,13]. The nanoparticle additives essentially behave like a catalyst. Due to high surface to volume ratio they are able to react more effectively, thus increasing the rate of fuel combusted [14,15]. Cerium oxide, aluminium oxide, cobalt oxide and zinc oxide are amongst the most popular nanoadditives due to their unique composition that aids in a more effective way of burning the fuel inside the engine cylinder [16–18]. Effective mixing of nanoparticles in the fuel mixture is important, literature reported that use of surfactant and ultrasonic machine helped to produce single phase nanoparticle fuel blend [19,20].

Furthermore, studies demonstrated that addition of cerium oxide in the fuel has the ability to reduce in-cylinder pressure, this in turn causes a decrease in the NOx emissions; in addition, due to the catalytic soot combustion characteristics, cerium oxide has the added capacity to remove soot from the particulate filter [11,21]. Razek et al. [22] investigated the effects of nanoparticle additives on jatropha biodiesel (JBD)-diesel blends. They reported that blend containing 20% JBD and 80% diesel with Al_2O_3 nanoadditives gave 12% increase in brake thermal efficiency (BTE) and 12.5% reduction in brake specific fuel consumption (BSFC). The authors reported that NOx emission was decreased by 13%; emissions of unburnt hydrocarbon (UHC) and CO gases were reduced by 10% and 29% respectively [22]. Effects of nanoadditives on waste-derived biodiesels were also investigated. A significant reduction in CO, UHC and NOx emissions were reported when poultry litter biodiesel-diesel-nanoparticle blend was used in the IC engine instead of the fuel blend without nanoparticles [23]. Compared to the fossil diesel fuel, up to 2% increase in engine power and 7.08% decrease in BSFC were observed when multi wall carbon nanotubes and nanosilver nanoparticles were added to the waste cooking oil biodiesel-diesel blends [24]. Nanoadditives enhanced the combustion characteristics of the pure fossil diesel powered engine. As a result of better combustion, the CO_2 emissions increased by up to 17.03% and CO emissions decreased by 25.17% when compared to pure fossil diesel fuel operation [24]. Up to 8% reduction in BSFC was achieved when ferrofluid nanoparticles were added to pongamia biodiesel-diesel (B20) blends [25]. The authors reported that due to improved burning, the emissions of CO and UHC gases were also decreased when compared to non-additive fuel blends [25].

The thermal efficiency was improved by about 2.2% and emissions of HC, CO and smoke were considerably decreased when copper oxide nanoparticles-mahua biodiesel-fossil diesel blends were used instead of B20 blend without nanoparticles [26]. Basha et al. [27] studied the combined effects of carbon nanotubes and diethyl ether additives on biodiesel emulsion fuels. They reported that additives gave better engine performance than pure biodiesel and pure fossil diesel [27]. In another study, carbon nanotube and ethanol was added to B2 fuel (B2E4C60) and observed a 15.52% increase in engine power and 11.73% decrease in BSFC as compared to when the engine was operated with pure fossil diesel fuel [28]. The authors also found that due to the additives, the CO and UHC emissions were decreased by 5.47% and 31.72% respectively, but the NOx emissions increased by about 12.22% [28]. Approximately 7–20% and 15–28% reductions in CO and UHC gases were observed when graphene oxide nanoparticles were added to *Ailanthus altissima* biodiesel-diesel blends (B0, B10, and B20) [29].

Furthermore, the effects of nanoadditives on a thermal barrier coated engine were also investigated. A simulation study on a coated piston showed increased temperature distribution and reduced heat flux when compared to uncoated piston [30]. Due to the reduced heat flux, an improvement in thermal efficiency by 1.75% was observed on coated engine using *Cymbopogon flexuosus* biofuel-fossil diesel blends with 20 ppm cerium oxide nanoadditive when compared to an uncoated engine using the same fuel [30]. In a separate study, *Cymbopogon flexuosus* biofuel (20%)-fossil diesel (80%) blends with various proportions of cerium oxide nanoadditives achieved up to 4.76% higher thermal efficiency and 6.6% decrease in smoke opacity as compared to biofuel-diesel blends without nanoadditives [31]. Addition of nanoparticles gave increased heat release rate and peak in-cylinder pressure; emissions of UHC, CO and NOx gases were reduced by 7%, 12.5% and 3%, respectively, at full engine load. [31]. The literature reports that in a thermal barrier coated engine, the nitrogen oxides gases were increased and emissions of UHC, CO and smoke opacity were reduced. Carbon-coated aluminium additives were added in biodiesel-diesel-ethanol blends and tested in a diesel engine to assess the engine performance and emission characteristics; the study found that B10 blend with 4% ethanol and 30 ppm nanoparticles reduced both BSFC and NOx emissions by about 6% when compared to B10 fuel (without ethanol and nanoparticles) [32]. However, the authors reported that compared to B10 fuel, the particles number (PN) emissions were increased by 2.2 times for B10-ethanol-nanoparticles fuel; on the contrary, this was decreased by about 11.8% for B10-ethanol fuel [32].

Up to 12% improvement in BTE, 30% reduction in NO emission, 60% reduction in CO emission, 44% reduction in UHC emission and 38% reduction in smoke emission were observed when both cerium oxide and alumina nanoadditives was added to B20 jatropha biodiesel blend as compared to B100 fuel [33]. Ignition delay was affected when nanoadditives were used in the fuel. Jatropha biodiesel emulsion fuel (83% jatropha biodiesel, 15% water, and 2% surfactants (Span80 and Tween80)) mixed with aluminium nanoparticles gave lower ignition delay, better engine performance and reduced emissions compared to pure jatropha biodiesel or jatropha biodiesel emulsion [19]. Another study reported that the ignition delay was deceased by about 9% when carbon nanotube and Ag nanoparticles were added to jojoba biodiesel-diesel blends [34].

Jatropha oil (JCO) is derived from the *Jatropha curcas* plant, they can be grown in unfarmable lands and can endure adverse weather conditions. Non-edible oils are the most appropriate feedstock for biodiesel production as they do not put a strain on global food demand [35]. However, the concern with non-edible feed stocks is that some crops have a high Free Fatty Acid value (FFA). The FFA value determines whether or not the oil needs to undergo an additional process (ie. esterification) before transesterification. The esterification process or 'pre-treatment' makes biodiesel production a two-step process capable of producing a high yield of fuel in a relatively short amount of time [36]. *Jatropha carcus* trees are grown in many parts of India and in Africa. Use of 100% biodiesel (B100) would provide much more emission reduction benefits than using biodiesel-diesel blends. Most studies found in the literature reported effects of nanoparticles on jatropha biodiesel-diesel blends. The aim of the current study is to investigate the performance, combustion and emission characteristics of a multi-cylinder diesel engine operated with nanoparticles—100% jatropha biodiesel fuel mixture. Initial findings of the study have been presented at the 13th SDEWES conference [37]. Two nanoparticles cerium oxide and aluminium oxides will be used in this study. Jatropha biodiesel will be produced in the lab using two stages, i.e., esterification and transesterification. Nanoadditives-J100 fuel blends will be tested in a multi-cylinder engine. The specific objectives of this study are:

- Pre-treatment of jatropha oil and production of jatropha biodiesel
- Amalgamation of nanoparticle additives into jatropha biodiesel
- Measurement of physico-chemical properties of various fuel blends
- Engine testing using nanoparticle-J100 blend and assessment of combustion, emission and performance characteristics
- Comparison of results with and without nanoparticles and recommendations

2. Materials and Methods

2.1. FFA Determination and Pre-Treatment (Acid Esterification) of JCO

Typically, biodiesel feedstocks that have a FFA value below 2% can be converted into biodiesel through a single step process known as transesterification. However, if the FFA content is > 2% than the additional process (i.e., esterification) is required. Jatropha oil was produced in Ghana and collected from a UK supplier. Isopropanol, H_2SO_4, methanol, phenolphthalein 1% and potassium hydroxide (KOH) were purchased from Sigma-Aldrich (Dorset, UK). An acid base titration method (using KOH) was used to determine the FFA content present in JCO. Alcohol to oil ratio, temperature and reaction time are important parameters for esterification process [38,39]. The formulae used by Heroor and Bharadwaj [38] was used to calculate the FFA content (in %) of the jatropha oil and biodiesel. The most favorable esterification process ascertained for the jatropha oil were: using a methanol to oil (molar) ratio of 6:1, with a H_2SO_4 catalyst concentration of 0.5%, reaction time of 45 min whilst maintaining the temperature of $40 \pm 5\,^{\circ}C$ [40]. Esterification of jatropha oil was conducted using 5% H_2SO_4 and 20% methanol at a steady temperature of $65\,^{\circ}C$ [41]. Tiwari et al. [42] reported that for esterification of jatropha oil, a methanol to oil ratio of 0.28:1 (v/v) should be used with a catalyst concentration of 1.43%, along with a reaction time of 88 min at $60\,^{\circ}C$ temperature. Based on the above literature and laboratory trials, following methods were applied for esterification of JCO:

- JCO was poured into a flask, placed on to a hot plate and heated up to $60\,^{\circ}C$
- Methanol with a ratio of 60% (w:w) of methanol to oil, was taken into another beaker. After that 1% H_2SO_4 was added to the methanol beaker. The methanol and H_2SO_4 mixture was stirred for approximately 5 min before adding the solution to the $60\,^{\circ}C$ heated JCO. Once mixed, the solution was left on the hotplate stirrer for about 90 min at a temperature of $55 \pm 5\,^{\circ}C$
- The esterified solution was then poured into a separation funnel and left for 2 hours. The top layer was waste water and methanol, the bottom layer was esterified JCO

2.2. Transesterification

The FFA content in pre-treated JCO (i.e., esterified JCO) was measured again by the titration procedure in order to make sure that the FFA value is below 2%. Literature reported effective transesterification of JCO by using a methanol to oil (molar) ratio of 9:1, and with a KOH catalyst of 0.5% [40]. In another study, the recommended amount of catalyst suggested is between 0.1% to 1% (w/w) of oils [43]. Furthermore, another literature reported that the molar ratio of methanol to oil at 6:1, NaOH concentration of 0.7% (w/w), and reaction temperature of $65\,^{\circ}C$ was found to be effective for the transesterification of JCO [41]. The optimum parameters for the transesterification of JCO were ascertained by implementing: (i) a methanol to oil ratio of 0.16 (v/v), and (ii) a constant temperature of $60\,^{\circ}C$. The KOH was used as a catalyst.

2.3. Addition of Nanoparticles to Neat Jatropha Biodiesel

Jatropha biodiesel was produced in the lab (Sections 2.1 and 2.2); nanoparticles Al_2O_3 and CeO_2, and surfactant Triton X-100 were purchased from Sigma-Aldrich. A GT Sonic ultrasonicator (Shenzhen, China) was used for mixing the nanoparticles in biodiesel-surfactant mixture. Quantities of 50 ppm and 100 ppm of nanoparticles were added to neat jatropha biodiesel (J100). The following method was adapted for nanoparticles addition into neat biodiesel:

- 1000 ppm of Triton-X100 was added to J100
- Either 50 ppm or 100 ppm of CeO_2/Al_2O_3 nanoparticles were added to J100 mixture to produce four samples: J100C50 and J100C100 with CeO_2 additives; J100A50 and J100A100 with Al_2O_3 additives (Figure 1)

- Biodiesel-nanoparticles mixtures were placed in an ultrasonicator (frequency at 40 kHz and water at 45 °C) for a duration of 45 min. After that the samples were left for 72 h at room temperature to see the stability of the mixture

Figure 1. Fuel samples (from left to right): Diesel, J100, J100A100, J100A50, J100C100 and J100C50.

2.4. Characterisation of Fuel Samples

Crucial properties that are vital to the fuel's performance in engine and exhaust emissions were measured in the lab. A Parr 6100 bomb calorimeter (Parr Instrument Company, Moline, IL, USA) was used to measure the higher heating value (HHV) in accordance to ASTM-D240 standard. Canon Fenski u-tube viscometers (CANON Instrument Company, State College, PA, USA) and thermostatic water bath ($\pm$0.1 °C) was used to measure the kinematic viscosities according to ISO 3104 having an accuracy of $\pm$0.22%. Viscosity was measured at temperatures of 40 °C and 22 °C. The densities were measured using a hydrometer in accordance to ISO 3675 standard. Flash point temperatures were measured using a Setaflash series 3 plus closed cup flash point tester (model 33000-0, STAN-HOPE SETA, UK). The test methods used were in compliance with DIN EN 22719, a part of the EN14214 standard. The acid value was measured using the same technique used for FFA measurement [30].

2.5. Engine Testing

A model LPWSBio3 three cylinder engine manufactured by Lister Petter (Teignmouth, UK) was used in the investigation (Table 1). An eddy current Froude AG80HS dynamometer (Froude Ltd., Worcester, UK) was used to measure and adjust the engine load and speed (Figure 2). The torque and speed accuracies of the dynamometer are $\pm$0.4 Nm and $\pm$1 rpm respectively. A five-gas emission analyser BEA 850 (Robert Bosch Ltd., Middlesex, UK) and smoke opacity meter (Bosch RTM 430) was used to analyse the exhaust gas components (CO, CO_2, NOx, O_2 and UHC) and to measure the smoke intensity respectively. The resolution for CO, CO_2, NOx, O_2 and UHC measurements are 0.001% vol., 0.01% vol., 1 ppm vol., 0.01% vol. and 1 ppm vol. respectively. The absorption coefficient resolution for the smoke meter is 0.01 m^{-1}. Ratio of air to fuel was also measured using the same emission analyser. A LabVIEW data acquisition system was used to log the temperatures at different locations of the engine. Combustion characteristics were evaluated using a Kistler combustion analyser. A pressure sensor (6125C11, Kistler Instruments Ltd., London, UK) and charge amplifier (Kistler 5064B11) was used to measure pressure inside the cylinder. Another pressure sensor (Kistler 4065A500A0) and amplifier (Kistler 4618A0) was used to measure the fuel line injection pressure. An optical encoder (Kistler 2614A) was used for detection of the crank angle position. The amplifiers and the encoder electronics were connected to the 'KiBox' (Kistler, model 2893AK8) for data logging. The KiBoxCockpit software (Kistler Instruments Ltd., London, UK) was used to measure and analyse various combustion parameters such as in-cylinder pressure, P-V diagram, heat release rate, combustion duration etc.

Table 1. Specifications of the engine used in the experiment.

Model/Type	LPWS Bio3 Water Cooled
No. of cylinders	3
Rated speed	1500 rpm
Continuous power at rated speed	9.9 kW
Type of fuel injection	Indirect injection with individual fuel injection pumps
Fuel pump injection timing	20_ BTDC
Continuous power fuel consumption at 1500 rpm	3.19 L/h (fossil diesel)
Exhaust gas flow	41.4 L/s at full loads at 1500 rpm

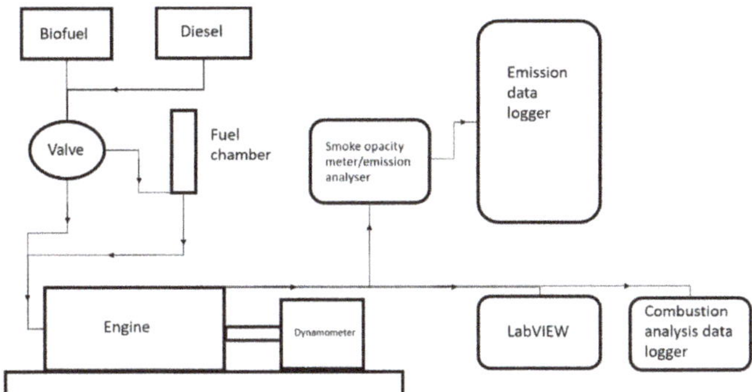

Figure 2. Schematic diagram of the engine test rig.

Fossil diesel, neat jatropha biodiesel (J100) and neat jatropha biodiesel with aluminium nanoparticles were tested in the engine. The engine was first operated with fossil diesel, then switched to neat jatropha biodiesel, and then finally operated with jatropha biodiesel-nanoparticle blend. The engine was operated at constant speed of 1500 rpm. For each test fuel, the engine was tested at six (6) different loads starting from low to full engine load. Once all data were measured and recorded at one load, the engine was ramped up to the next load using the dynamometer. At the end of test, the engine was switched back to fossil diesel and operated for about 15 min before stopping the engine.

3. Results and Discussion

3.1. Nanoparticles Addition and Fuel Characteristics

Through visual observation it was found that the Al_2O_3 nanoparticles had fully dissolved into their biodiesels and there was no sedimentation present. On the other hand, CeO_2 nanoparticles did not dissolved completely and some sedimentation was seen at the bottom of the container for both 50 ppm and 100 ppm doses. Hence, blend containing Al_2O_3 nanoparticles was chosen for engine testing due to having better diffusion characteristics. The failure of the CeO_2 not mixing fully was perhaps due to the type of surfactant used. Other surfactant such as Span 80 and Tween 80 might help to blend CeO_2 nanoparticles with J100 fuel [44].

The results given in Table 2 demonstrate that the properties of J100 with or without nanoparticles mostly comply with the EN14214 standard. The properties of the J100 biodiesel was similar to the properties reported in the literature [41]. Density of J100 and jatropha-nanoadditive blends were higher than that of fossil diesel. Greater fuel density would allow for more fuel to be pumped via the fuel line, and greater mass of the fuel can be stored in a tank [40]. The EN14214 standard states that all biodiesel fuels must have acid value lower than 0.50 mg KOH/g, it had been observed that all fuel samples except JCO were able to achieve this value. Due to the addition of both surfactant and nanoparticles, the heating values of the nanoadditive fuel blends samples were slightly lower than that of neat jatropha biodiesel.

Table 2. Measured properties of the test fuels.

Property	Units	Neat Diesel	Jatropha Curcas Oil (JCO)	Jatropha Biodiesel (J100)	J100C50	J00C100	J00A50	J00A100	EN 14214 Standards (Biodiesel)
Acid Value	mg KOH/g	0.34	13.59	0.45	0.28	0.20	0.20	0.20	<0.50
Flash point	°C	63.6	181.8	171.2	174.8	177	173.6	175.6	>101
Density	Kg/m^3	832.3	922.6	881.6	880	879	877	878	860–900
Viscosity at 22 °C	cSt	3	75.57	4.73	5.92	6.23	6.03	5.95	N/A
Viscosity at 40 °C	cSt	2.13	37.47	3.37	3.99	4.08	4.11	4.07	>3.5–5.0
HHV	MJ/kg	45.64	39.39	37.54	37.39	37.29	37.27	37.34	N/A

3.2. Engine Performance and Emission Characteristics

Pure fossil diesel, J100 and J100A100 fuels were tested in the engine. In general, the bsfc for fossil diesel operation was found to be lower than those obtained for J100 and J100A100 fuels (Figure 3). The reason why J100 and J100A100 had higher BSFC values was owed to the fact they both had higher density values and lower calorific values when compared to corresponding values for fossil diesel.

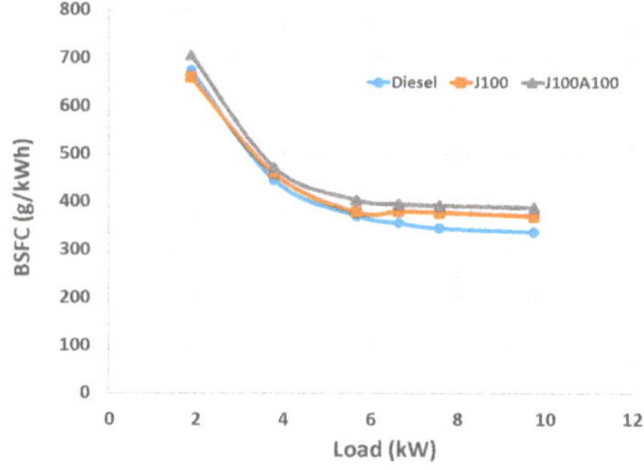

Figure 3. BSFC vs. engine load.

At about 3.8 kW engine load, the BSFC of the fossil diesel was 6% lower than that of J100A100 fuel. However, on average, a difference of 3% in BSFC was observed between J100 and J100A100 fuels. At 9.75 kW (100%) load, the BSFC of J100A100 was about 13% higher than that of fossil diesel; on the other hand, the BSFC of J100 was 4.5% lower when compared to J100A100 fuel (Figure 3). On the contrary, it was observed that the brake specific energy consumption (BSEC) value of the J100A100 fuel was lower than that of fossil diesel throughout all load range (Figure 4). At full load, BSEC of the J100A100 fuel was found to be decreased by approximately 6%. This explains that when the engine was operated with J100A100 fuel, comparatively less energy was required to produce the same power output as compared to fossil diesel operation.

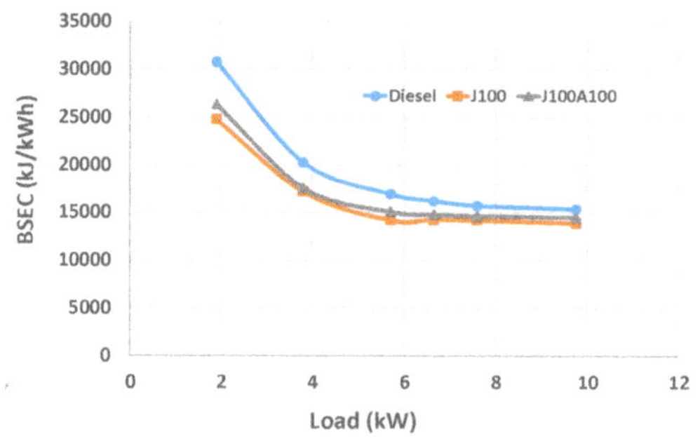

Figure 4. BSEC vs. engine load.

The brake thermal efficiency (BTE) of J100 and J100A100 fuels were higher than the corresponding values observed for fossil diesel (Figure 5). At 20% engine load (1.9 kW), the BTE values for fossil diesel, J100 and J100A100 fuels were respectively 11%, 14.52% and 13.65%. A higher BTE value for the biodiesel-nanoparticle blend might be attributed to a higher oxygen content present in the biodiesel and higher reactivity of the fuel mixture due to the nanoadditives [14,18]. On average, the BTE values of J100A100 fuel was about 3% higher than that of fossil diesel. At higher loads, the thermal efficiency of J100 was observed to be slightly higher than those obtained for J100A100 fuel. Higher viscosity of J100A100 fuel might have caused this characteristic. At higher loads, the volume of carbon monoxide (CO) produced by the biodiesel-nanoparticle blend was found to be higher than that of fossil diesel (Figure 6); on the contrary, opposite characteristic was observed at lower loads. It was believed that higher BSFC value and higher oxygen content in the J100A100 fuel caused higher CO emissions at higher loads. Similarly, at higher loads, the CO_2 emission of J100A100 fuel was found to be slightly higher than fossil diesel due to the higher BSFC value and higher oxygen content (Figure 7). The amount of UHC produced by the biodiesel blend was found to be lower than that of regular diesel (Figure 8). This was due to the fact that in the case of biodiesel-nanoparticles blend, more complete combustion took place inside the cylinder. It was observed that better combustion characteristics of the nanoadditive blend led to higher BTE (Figure 5). The J100A100 blend had the aid of an increased catalytic effect which helped in improving the overall combustion [14].

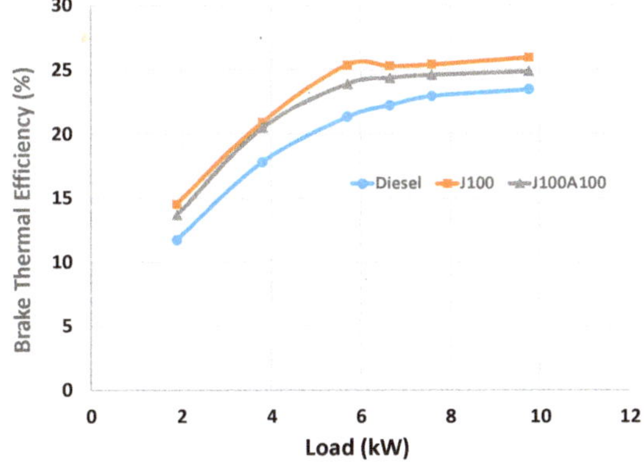

Figure 5. Brake thermal efficiency (BTE) vs. engine load.

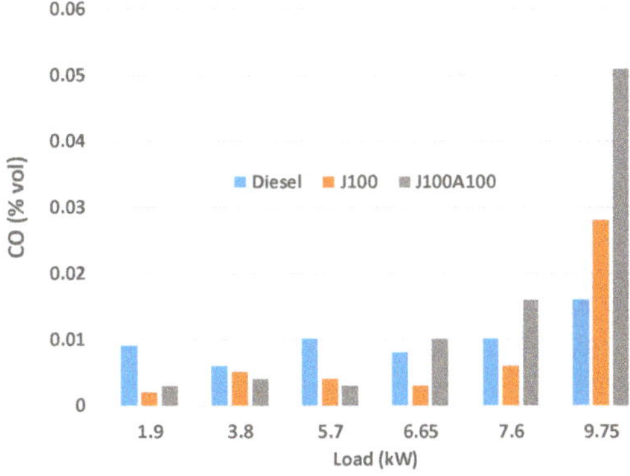

Figure 6. CO emission vs. engine load.

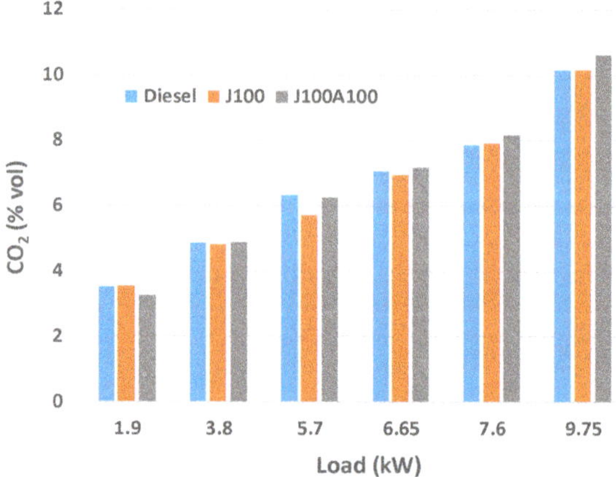

Figure 7. CO_2 emission vs. engine load.

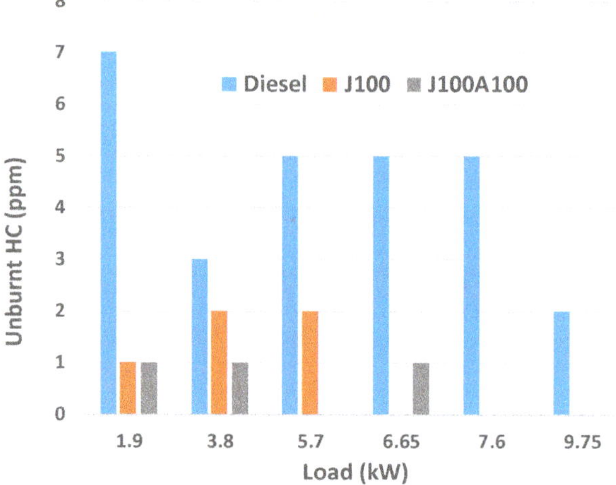

Figure 8. UHC emission vs. engine load.

It was observed that the J100 and J100-nanoparticle blend produced a greater amount of nitrogen oxide (NOx) emissions when compared to fossil diesel (Figure 9). Biodiesels intrinsically containing a greater amount of double bond molecules caused a higher adiabatic flame temperature which in turn leads to a greater concentration of NOx emissions. In the case of nanoparticle blends, an increase in combustion temperature caused due to a greater rate of reaction and conversion of the oxygen present in the nanoparticle blend led to an increased rate of NOx emissions [33]. Under most engine loads, the NOx emissions of J100 were higher than those of J100A1000 fuel. The reason for this was due to the catalytic nature of the nanoparticles present in the J100A1000 fuel, the nanoadditives broken down hydrocarbon compounds before they were able to become fully formed products [23]. The smoke opacity values for both pure biodiesel and biodiesel-nanoparticle blend were found to be much lower than those observed for fossil diesel (Figure 10). Oxygen content of the test fuel plays a significant role in the formation of smoke [45]. Better combustion due to higher oxygen content caused lower smoke levels in the case of nanoparticle-biodiesel blend [46,47]. When comparing between J100 and J100A100 fuels, it was observed that J100A100 had a lower smoke opacity values because of the greater amount of oxygen present (aided by the Al_2O_3 nanoadditive) in the J100A100 fuel.

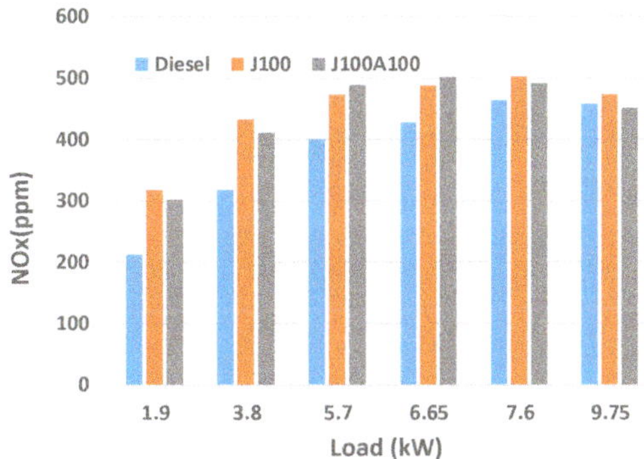

Figure 9. NOx emission vs. engine load.

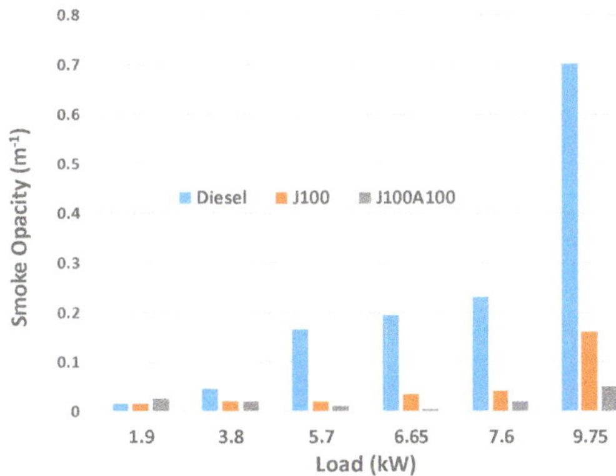

Figure 10. Smoke opacity vs. engine load.

3.3. Combustion Characteristics

Figures 11 and 12 shows in-cylinder pressures for all test fuels at 60% and 100% loads respectively. In general, increase in the in-cylinder pressures were observed with the increase in engine loads (Figures 11 and 12). At 60% engine load (Figure 11), the charge temperature was low, lower charge temperature lengthened the ignition delay period [48]. At 100% load (Figure 12), more fuel was injected into the chamber which caused the gas and wall temperatures to increase, this in turn reduced the ignition delay period. The peak in-cylinder pressures were occurred almost at the same crank angle location for all fuels. The fossil diesel gave highest peak in-cylinder pressure at 60% load; however, at 100% load, the peak in-cylinder pressures were almost equal for all fuels.

The heat release rates for both 60% and 100% engine loads are shown in Figures 13 and 14. At 60% load (Figure 13), the peak of heat release rate for fossil diesel was much higher than other fuels. The J100A100 fuel produced lowest peak heat release rate; this was caused due to the longer ignition delay period of J100A100 fuel when compared to fossil diesel and J100 fuels. The lower viscosity and better volatility traits of pure diesel fuel enabled to produce highest peak heat release rate at 60% engine load [17]. However, at 100% load (Figure 14), peaks for both pure diesel and J100A100 fuels were almost the same; this combustion characteristic suggested that at higher temperatures, the nanoadditive fuel blend were not as volatile, and not enough fuel mixture is formed in the premixed burning phase. The total heat release values at 60% and 100% loads are demonstrated in Figures 15 and 16. At 60% load (Figure 15), amongst all fuels, J100A100 gave highest amount of total heat release; which suggested that once burning started, nanoadditive fuel burnt quite quickly relative to other fuels. The oxygen donated by the nanoparticles aided accelerated burning of the J100A100 fuel [49].

It was observed that the total heat release values for all test fuels were almost same at 100% load (Figure 16). This might be attributed to the absorption of heat by high heat capacity gases such as CO_2, whose concentration increases with the increase of engine load. These gases absorb a fragment of the amount of total heat release which caused the J100 and its blend to follow a same trend [50], as demonstrated in Figure 16.

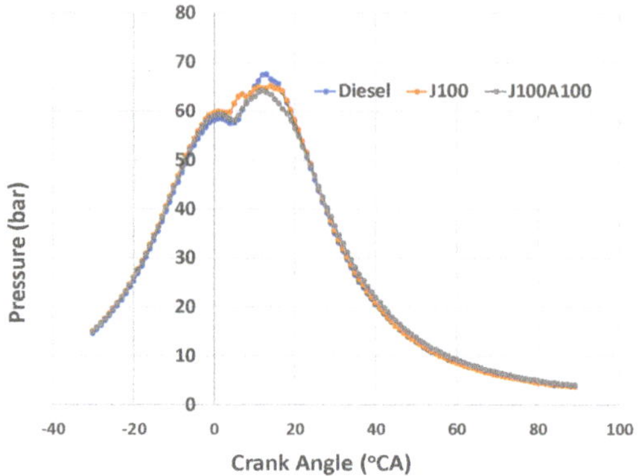

Figure 11. Cylinder pressure at 60% load.

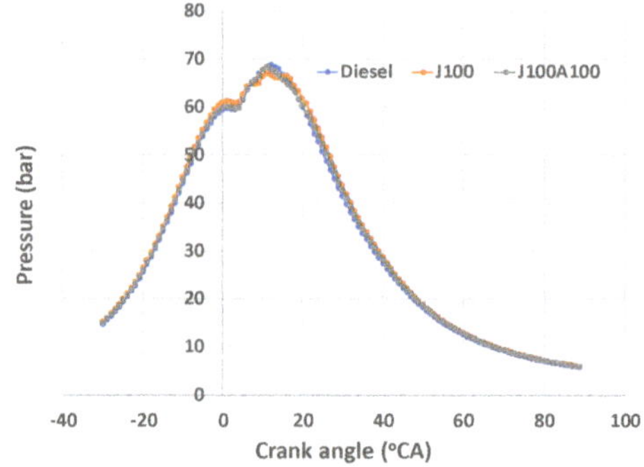

Figure 12. Cylinder pressure at 100% Load.

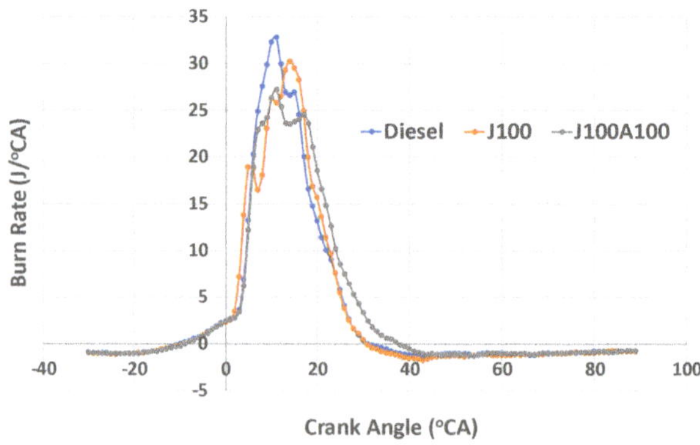

Figure 13. Heat release rate at 60% load.

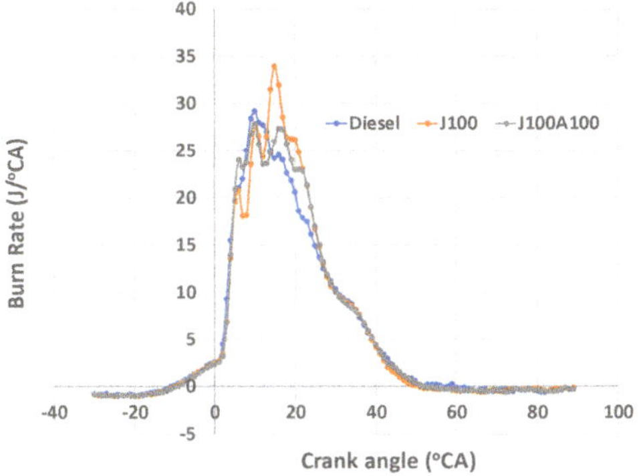

Figure 14. Heat release rate at 100% load.

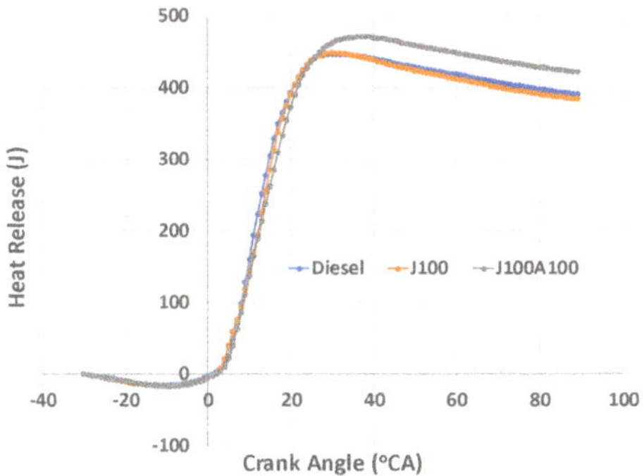

Figure 15. Total heat release at 60% load.

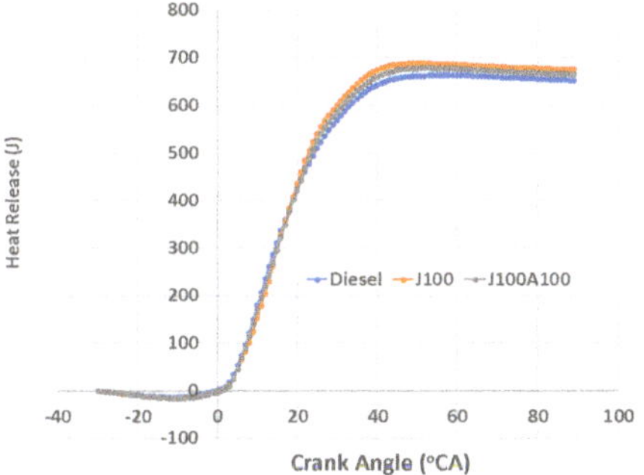

Figure 16. Total heat release at 100% load.

4. Conclusions and Recommendations

Jatropha biodiesel was produced using both esterification and transesterification processes. The addition of Al_2O_3 and CeO_2 nanoparticles to pure jatropha biodiesel was carried out by using an ultrasonicator device and surfactant Triton X-100. It was observed that the surfactant decreased the calorific values of the fuel blends. The CeO_2 nanoparticles failed to fully amalgamate with the jatropah biodiesel. For successful amalgamation of the CeO_2 nanoparticles into J100 fuel, other surfactant such as Span 80 and Tween 80 might be effective instead of Triton X-100. The J1000A100 blend was observed as having promising characteristics that would aid in the fuel's performance when tested in the IC engine. Neat jatropha biodiesel and neat fossil diesel fuels were used as a control. The findings on engine performance, combustion and emission characteristics results are summarised below:

(1) At full load, the BSEC values of J100A100 blend was found to be 4% higher and 6% lower than the corresponding values obtained for J100 and neat fossil diesel fuels respectively. On the other hand, the BSFC value of J100A100 blend was found to be higher than fossil diesel; however, on average, an improvement of 3% in BTE was observed for J100A100 fuel when compared to fossil diesel.

(2) At low loads, J100A100 fuel gave lower amount of CO emissions. On the other hand, in almost all loads, J100A100 produced smallest amount of UHC emission due to the rich oxygen content in the nanoadditive fuel blend. At higher loads, J100A100 fuel gave improved NOx emission characteristics when compared to J100 fuel.

(3) Better combustion due to nanoadditives led to least smoke opacity values when the engine was operated with J100A100 fuel. The J100A100's combustion characteristics helped display and measure the fuel's performance in the combustion chamber. The peak in-cylinder pressure for J100A100 fuel at 60% engine load was seen as being lowest when compared to other fuels; however, this trend was changed when the engine load was increased. At 100% load, the peak in-cylinder pressures for all three fuels were almost same.

(4) At 60% and 100% loads, the J100A100 blend was observed as having a constant heat release rate, this was attributed to the additives ability to provide a constant burn and not to succumb to volatility. The total heat release was found to be higher at 60% load for the J100A100 fuel; however, at 100% load, this value was equal to those obtained for other fuels.

To conclude, overall J100-Al$_2$O$_3$ nanoadditive fuel performed better when compared to fossil diesel and J100 fuels. A reduction in NOx and UHC emission as well as smoke opacity, and an overall increase in BTE were observed when compared to J100 fuel. Effect of nanoparticles in the environment is yet to be investigated and recommended as a future work. Measurement of cetane number and oxygen content in the biodiesel-nanoparticle blends are other important items for further studies.

Author Contributions: Conceptualization, A.K.H. and A.H.; methodology, A.K.H. and A.H.; validation, A.K.H. and A.H.; formal analysis, A.K.H. and A.H.; investigation, A.K.H. and A.H.; resources, A.K.H. and A.H.; data curation, A.K.H. and A.H.; writing—original draft preparation, A.K.H. and A.H.; writing—review and editing, A.K.H.; supervision, A.K.H.; project administration, A.K.H.

Funding: This research received no external funding.

Nomenclature

BSEC	Brake Specific Energy Consumption
BSFC	Brake Specific Fuel Consumption
BTE	Brake Thermal Efficiency
FFA	Free Fatty Acid
GHG	Greenhouse Gas
HHV	Higher Heating Value
J100	Neat jatropha biodiesel (100%)
J100A100	Jatropha biodiesel (100%) with 100 ppm Al$_2$O$_3$
J100A50	Jatropha biodiesel (100%) with 50 ppm Al$_2$O$_3$
J100C100	Jatropha biodiesel (100%) with 100 ppm CeO$_2$
J100C50	Jatropha biodiesel (100%) with 50 ppm CeO$_2$
JBD	Jatropha Biodiesel
JCO	Jatropha Curcas Oil
IC	Internal Combustion
UHC	Unburnt Hydrocarbons

References

1. Khoo, H.H.; Tan, R.B.H. Environmental impact evaluation of conventional fossil fuel production (oil and natural gas) and enhanced resource recovery with potential CO_2 sequestration. *Energy Fuels* **2006**, *20*, 1914–1924. [CrossRef]
2. Global Emissions by Economic Sector. Available online: https://www.epa.gov/ghgemissions/global-greenhouse-gas-emissions-data#Trends (accessed on 17 February 2019).
3. Environmental Protection UK. Car Pollution. Environmental Protection UK, 2015. Available online: https://www.environmental-protection.org.uk/policy-areas/air-quality/air-pollution-and-transport/car-pollution/ (accessed on 20 November 2018).
4. DAWN Biofuels: A Substitute for Petroleum. Available online: https://www.dawn.com/news/266470 (accessed on 15 November 2018).
5. Conserve Energy Future. Advantages and Disadvantages of Biofuels. Conserve Energy Future. Available online: https://www.conserve-energy-future.com/advantages-and-disadvantages-of-biofuels.php (accessed on 22 December 2018).
6. Hoover, F.-A.; Abraham, J. Biofuels can help solve climate change, especially with a carbon tax. *Int. J. Sustain. Energy* **2009**, *28*, 171–182. [CrossRef]
7. Giakoumis, E.G.; Sarakatsanis, C.K. A Comparative Assessment of Biodiesel Cetane Number Predictive Correlations Based on Fatty Acid Composition. *Energies* **2019**, *12*, 422. [CrossRef]
8. Kaya, T.; Kutlar, O.A.; Taskiran, O.O. Evaluation of the Effects of Biodiesel on Emissions and Performance by Comparing the Results of the New European Drive Cycle and Worldwide Harmonized Light Vehicles Test Cycle. *Energies* **2018**, *11*, 2814. [CrossRef]
9. Wan Ghazali, W.N.M.; Mamat, R.; Masjuki, H.H.; Najafi, G. Effects of biodiesel from different feedstocks on engine performance and emissions: A review. *Renew. Sustain. Energy Rev.* **2015**, *51*, 585–602. [CrossRef]
10. Mofijur, M.; Hazrat, M.A.; Rasul, M.G.; Mahmudul, H.M. Comparative Evaluation of Edible and Non-edible Oil Methyl Ester Performance in a Vehicular Engine. *Energy Procedia* **2015**, *75*, 37–43. [CrossRef]
11. Mofijur, M.; Atabani, A.E.; Masjuki, H.H.; Kalam, M.A.; Masum, B.M. A study on the effects of promising edible and non-edible biodiesel feedstocks on engine performance and emissions production: A comparative evaluation. *Renew. Sustain. Energy Rev.* **2013**, *23*, 391–401. [CrossRef]
12. Shaafi, T.; Sairam, K.; Gopinath, A.; Kumaresan, G.; Velraj, R. Effect of dispersion of various nanoadditives on the performance and emission characteristics of a CI engine fuelled with diesel, biodiesel and blends—A review. *Renew. Sustain. Energy Rev.* **2015**, *49*, 563–573. [CrossRef]
13. Saxena, V.; Kumar, N.; Saxena, V.K. A comprehensive review on combustion and stability aspects of metal nanoparticles and its additive effect on diesel and biodiesel fuelled C.I. engine. *Renew. Sustain. Energy Rev.* **2017**, *70*, 563–588. [CrossRef]
14. Aalam, C.S.; Saravanan, C.G. Effects of nano metal oxide blended Mahua biodiesel on CRDI diesel engine. *Ain Shams Eng. J.* **2017**, *8*, 689–696. [CrossRef]
15. Nanoparticles Increase Biofuel Performance. Available online: https://www.sciencedaily.com/releases/2011/04/110408075042.htm (accessed on 25 August 2018).
16. Soutter, B.W. Nanoparticles as Fuel Additives. 2012, pp. 1–3. Available online: https://www.azonano.com/article.aspx?ArticleID=3085 (accessed on 17 February 2019).
17. Arul Mozhi Selvan, V.; Anand, R.B.; Udayakumar, M. Effects of Cerium Oxide Nanoparitcle Addition in Diesel and Diesel-Biodiesel-Ethanol Blends on the performance and emission characteristics of a C1 Engine. *ARPN J. Eng. Appl. Sci.* **2009**, *4*, 1–6.
18. Gan, Y.; Qiao, L. Combustion characteristics of fuel droplets with addition of nano and micron-sized aluminum particles. *Combust. Flame* **2011**, *158*, 354–368. [CrossRef]
19. Sadhik Basha, J.; Anand, R.B. Role of nanoadditive blended biodiesel emulsion fuel on the working characteristics of a diesel engine. *J. Renew. Sustain. Energy* **2011**, *3*. [CrossRef]
20. Saraee, H.S.; Jafarmadar, S.; Taghavifar, H.; Ashraf, S.J. Reduction of emissions and fuel consumption in a compression ignition engine using nanoparticles. *Int. J. Environ. Sci. Technol.* **2015**, *12*, 2245–2252. [CrossRef]
21. Valencia, M.; López, E.; Andrade, S.; Iris, M.L.; Hurtado, N.G.; Pérez, V.R.; García, A.G.; de Lecea, C.S.M.; Bueno López, A. Evidences of the Cerium Oxide-Catalysed DPF Regeneration in a Real Diesel Engine Exhaust. Available online: https://core.ac.uk/download/pdf/32320231.pdf (accessed on 11 February 2019).

22. Razek, S.M.A.; Gad, M.S.; Thabet, O.M. Effect of Aluminum Oxide Nano-Particle in Jatropha Biodiesel on Performance, Emissions and Combustion Characteristics of D I Diesel Engine. *Ijraset* **2017**, *5*, 358–372. [CrossRef]

23. Ramesh, D.K.; Dhananjaya Kumar, J.L.; Hemanth Kumar, S.G.; Namith, V.; Basappa Jambagi, P.; Sharath, S. Study on effects of Alumina nanoparticles as additive with Poultry litter biodiesel on Performance, Combustion and Emission characteristic of Diesel engine. *Mater. Today Proc.* **2018**, *5*, 1114–1120. [CrossRef]

24. Ghanbari, M.; Najafi, G.; Ghobadian, B.; Yusaf, T.; Carlucci, A.P.; Kiani Deh Kiani, M. Performance and emission characteristics of a CI engine using nano particles additives in biodiesel-diesel blends and modeling with GP approach. *Fuel* **2017**, *202*, 699–716. [CrossRef]

25. Kumar, S.; Dinesha, P.; Bran, I. Influence of nanoparticles on the performance and emission characteristics of a biodiesel fuelled engine: An experimental analysis. *Energy* **2017**, *140*, 98–105. [CrossRef]

26. Chandrasekaran, V.; Arthanarisamy, M.; Nachiappan, P.; Dhanakotti, S.; Moorthy, B. The role of nano additives for biodiesel and diesel blended transportation fuels. *Transp. Res. Part D Transp. Environ.* **2016**, *46*, 145–156. [CrossRef]

27. Sadhik Basha, J. Impact of Carbon Nanotubes and Di-Ethyl Ether as additives with biodiesel emulsion fuels in a diesel engine—An experimental investigation. *J. Energy Inst.* **2018**, *91*, 289–303. [CrossRef]

28. Heydari-Maleney, K.; Taghizadeh-Alisaraei, A.; Ghobadian, B.; Abbaszadeh-Mayvan, A. Analyzing and evaluation of carbon nanotubes additives to diesohol-B2 fuels on performance and emission of diesel engines. *Fuel* **2017**, *196*, 110–123. [CrossRef]

29. Hoseini, S.S.; Najafi, G.; Ghobadian, B.; Mamat, R.; Ebadi, M.T.; Yusaf, T. Novel environmentally friendly fuel: The effects of nanographene oxide additives on the performance and emission characteristics of diesel engines fuelled with Ailanthus altissima biodiesel. *Renew. Energy* **2018**, *125*, 283–294. [CrossRef]

30. Dhinesh, B.; Maria Ambrose Raj, Y.; Kalaiselvan, C.; KrishnaMoorthy, R. A numerical and experimental assessment of a coated diesel engine powered by high-performance nano biofuel. *Energy Convers. Manag.* **2018**, *171*, 815–824. [CrossRef]

31. Dhinesh, B.; Niruban Bharathi, R.; Isaac JoshuaRamesh Lalvani, J.; Parthasarathy, M.; Annamalai, K. An experimental analysis on the influence of fuel borne additives on the single cylinder diesel engine powered by Cymbopogon flexuosus biofuel. *J. Energy Inst.* **2017**, *90*, 634–645. [CrossRef]

32. Wu, Q.; Xie, X.; Wang, Y.; Roskilly, T. Effect of carbon coated aluminum nanoparticles as additive to biodiesel-diesel blends on performance and emission characteristics of diesel engine. *Appl. Energy* **2018**, *221*, 597–604. [CrossRef]

33. Prabu, A. Nanoparticles as additive in biodiesel on the working characteristics of a DI diesel engine. *Ain Shams Eng. J.* **2017**, *9*, 2343–2349. [CrossRef]

34. Najafi, G. Diesel engine combustion characteristics using nano-particles in biodieseldiesel blends. *Fuel* **2018**, *212*, 668–678. [CrossRef]

35. Chhetri, A.B.; Tango, M.S.; Budge, S.M.; Watts, K.C.; Islam, M.R. Non-edible plant oils as new sources for biodiesel production. *Int. J. Mol. Sci.* **2008**, *9*, 169–180. [CrossRef] [PubMed]

36. Banković-Ilić, I.B.; Stamenković, O.S.; Veljković, V.B.; Stamenkovi, O.S.; Veljkovi, V.B.; Bankovi, I.B. Biodiesel production from non-edible plant oils. *Renew. Sustain. Energy Rev.* **2012**, *16*, 3621–3647. [CrossRef]

37. Hossain, A.K.; Hussain, A. Effect of Nanoparticle on Combustion and Emission Characteristics of neat Jatropha Biodiesel (Paper No 0552-1). In Proceedings of the 13th SDEWES Conference, Palermo, Italy, 30 September–4 October 2018.

38. Heroor, S.H.; Bharadwaj, S.D.R. Production of Bio-fuel from Crude Neem Oil and its Performance. *Int. J. Environ. Eng. Manag.* **2013**, *4*, 425–432.

39. Chongkhong, S.; Kanjaikaew, U.; Tongurai, C.; Yai, H. A Review of FFA Esterification for Biodiesel Production. *10th Int. PSU Eng. Conf.* **2012**, *1*, 1–5.

40. Patil, P.D.; Deng, S. Optimization of biodiesel production: Transesterification of edible and non-edible vegetable oils. *Fuel* **2009**, *88*, 1302–1306. [CrossRef]

41. Singh, R.; Padhi, S. Characterization of jatropha oil for tha preparation of biodiesel. *Nat. Prod. Radiance* **2009**, *8*, 127–132.

42. Kumar Tiwari, A.; Kumar, A.; Raheman, H. Biodiesel production from jatropha oil (Jatropha curcas) with high free fatty acids: An optimized process. *Biomass Bioenergy* **2007**, *31*, 569–575. [CrossRef]

43. Ma, F.; Hanna, M.A. Biodiesel production: A review. *Bioresour. Technol.* **1999**, *70*, 1–15. [CrossRef]

44. Kumar, A. Effects of Cerium Oxide Nanoparticle on Compression Ignintion Engine Performance and Emission Charchteristic when Using Water Diesel Emuslion. Master's Thesis, Thapar University, Patiala, India, 2014.
45. Enweremadu, C.C.; Rutto, H.L. Combustion, emission and engine performance characteristics of used cooking oil biodiesel—A review. *Renew. Sustain. Energy Rev.* **2010**, *14*, 2863–2873. [CrossRef]
46. Dhar, A.; Agarwal, A.K. Performance, emissions and combustion characteristics of Karanja biodiesel in a transportation engine. *Fuel* **2014**, *119*, 70–80. [CrossRef]
47. Bakeas, E.; Karavalakis, G.; Stournas, S. Biodiesel emissions profile in modern diesel vehicles. Part 1: Effect of biodiesel origin on the criteria emissions. *Sci. Total Environ.* **2011**, *409*, 1670–1676. [CrossRef] [PubMed]
48. Qi, D.H.; Chen, H.; Geng, L.M.; Bian, Y.Z.; Ren, X.C. Performance and combustion characteristics of biodiesel-diesel-methanol blend fuelled engine. *Appl. Energy* **2010**, *87*, 1679–1686. [CrossRef]
49. Kadarohman, A.; Hernani; Rohman, I.; Kusrini, R.; Astuti, R.M. Combustion characteristics of diesel fuel on one cylinder diesel engine using clove oil, eugenol, and eugenyl acetate as fuel bio-additives. *Fuel* **2012**, *98*, 73–79. [CrossRef]
50. Singh, A.P.; Agarwal, A.K. Combustion characteristics of diesel HCCI engine: An experimental investigation using external mixture formation technique. *Appl. Energy* **2012**, *99*, 116–125. [CrossRef]

Article

Techno-Economic Assessment of Bio-Energy with Carbon Capture and Storage Systems in a Typical Sugarcane Mill in Brazil †

Sara Restrepo-Valencia * and Arnaldo Walter

Department of Energy, School of Mechanical Engineering, University of Campinas—UNICAMP, Campinas 13083860, Brazil; awalter@fem.unicamp.br

* Correspondence: sara.valencia@fem.unicamp.br; Tel.: +55-19-352-13283

† The present work is an extension of the paper "A. Techno-Economic Assessment of BECCS Systems in the Brazilian Sugarcane Sector" presented at the 13th Conference on Sustainable Development of Energy, Water and Environment Systems—SDEWES Conference, 30 September–4 October, Palermo, Italy.

Received: 28 December 2018; Accepted: 4 March 2019; Published: 22 March 2019

Abstract: For significantly reducing greenhouse gas emissions, those from electricity generation should be negative by the end of the century. In this sense, bio-energy with carbon capture and storage (BECCS) technology in sugarcane mills could be crucial. This paper presents a technical and economic assessment of BECCS systems in a typical Brazilian sugarcane mill, considering the adoption of advanced—although commercial—steam cogeneration systems. The technical results are based on computational simulations, considering CO_2 capture both from fermentation (released during ethanol production) and due to biomass combustion. The post combustion capture technology based on amine was considered integrated to the mill and to the cogeneration system. A range of energy requirements and costs were taken from the literature, and different milling capacities and capturing rates were considered. Results show that CO_2 capture from both flows is technically feasible. Capturing CO_2 from fermentation is the alternative that should be prioritized as energy requirements for capturing from combustion are meaningful, with high impacts on surplus electricity. In the reference case, the cost of avoided CO_2 emissions was estimated at 62 €/t CO_2, and this can be reduced to 59 €/t CO_2 in case of more efficient technologies, or even to 48 €/t CO_2 in case of larger plants.

Keywords: bioelectricity; carbon capture; negative emissions; sugarcane; biomass; climate change

1. Introduction

In order to maintain 2 °C as the maximum increase in the global average temperature, the levels of atmospheric concentrations must be kept below 450 ppm of CO_{2eq} during the 21st century [1]. Therefore, worldwide emissions of CO_2 have to be drastically reduced in the coming decades, inducing deep changes in the energy systems [2]. This scenario requires that emissions from electricity generation should be negative by the end of the century, with fast progress in energy efficiency and promotion of low-carbon technologies. In this context, carbon capture and storage (CCS) is crucial because it represents a process by which large amounts of carbon dioxide can be captured and stored for the long term [1].

The CCS technology involves four main steps: conditioning processes to separate CO_2 into a pure stream, carbon capture itself, its compression and, finally, storage for long term periods [1]. In the case of CCS applied to power units, significant losses in efficiency are expected; for instance, the Intergovernmental Panel on Climate Change (IPCC) indicates a 9% net reduction in efficiency for coal-fired power plants (pulverized) and 7% for combined cycle gas-fired power plants [1].

Post-combustion technology consists in the removal of CO_2 from the exhaust gases. Capture by absorption is recognized as the reference technology [3,4] and considered mature for power plants [5]. The removal from flue gases uses a solvent, generally amines, to absorb CO_2 molecules, being CO_2 then released by heating or drastic pressure reductions [6]. Flue gases need to be cooled before getting in contact with the solvent: the temperatures must be between 40 and 60 °C at the entrance of absorption columns [7]. Costs and energy requirements—also called energy penalties—from a CCS unit using absorption capture depends mainly on the solvent properties. It is estimated that the heating for solvent regeneration is responsible for over 25% of the energy penalty when compression is included [7].

Combining bio-energy with carbon capture and storage (BECCS) offers the prospect of energy supply with net negative emissions and is clearly an important approach to reach the target of 2 °C. BECCS combines production of fuels and electricity from renewable biomass with carbon capture and storage of the CO_2 emitted when biomass is converted [2]. As the CO_2 is removed from the atmosphere during the growth of the raw material, life-cycle absolute emissions of BECSS could be negative [8]. In this sense, BECCS technology applied in sugarcane mills would be fundamental, contributing with very low greenhouse gas (GHG) emissions in both the transport sector (with avoided emissions due to the displacement of fossil gasoline) and in electricity generation [9].

The production of ethanol (via the fermentation of sugars) releases a pure stream of CO_2, which means there is no penalty for its separation in the CCS process. This is the most obvious option to capture CO_2 in sugarcane mills, and it is estimated that, considering the ethanol production figures of 28.5 million m^3 in Brazil, it would be possible to reduce CO_2 emissions by 27.7 million tonnes per year [9]. Carbon capture in sugarcane mills could at least double—or triple—with the adoption of CCS technologies in cogeneration systems in which residual biomass is burned—usually bagasse, and more recently, bagasse combined with straw.

This work focuses on assessing the technical and economic impacts of BECCS in a typical Brazilian sugarcane mill. First, a hypothetical sugarcane mill was selected to perform the evaluation that considers advanced steam cogeneration systems. A literature review was conducted to select the CCS technology and obtain representative data to model the integration of the CCS unit with the cogeneration plant. The feasibility analysis is based on typical costs (i.e., investments, operation and maintenance costs) and efficiencies, and the final assessment is based on the costs of CO_2 avoided emissions.

2. Materials and Methods

2.1. Cogeneration Plant

A typical Brazilian sugarcane mill, but rather representative among the more than three hundred existing mills, was considered for this study, with a 4 Mt/y (million tonnes of sugarcane crushed per year) milling capacity. The cogeneration unit would be fully integrated to the mill to supply electric power and steam to the industrial process, and also maximizing surplus electricity. The steam demand for both sugar and ethanol production was assumed equivalent to 340 kg of steam (at 2.5 bar and 137 °C) per tonne of sugarcane; this is the minimum consumption of steam currently considered economically viable [10]. The power plant would operate along the whole year (with 90% capacity factor), being as a cogeneration unit during the harvest season and as a single power plant during off-season. Biomass would be stored to assure the operation during off-season and this is already a common practice for mills that generate electricity throughout the year; in general, mills have area available for this, and the costs are not prohibitive. The cogeneration technology is the one known as condensing-extraction steam-turbine (CEST), with live steam at the highest possible pressure and temperature (120 bar/535 °C is the state-of-art in Brazilian sugarcane mills, according to [9]).

The CEST technology is very common in modern sugarcane mills. Bagasse used to be the only fuel but, recently, a blend of bagasse and straw has been used [11] due to the growing straw availability at the mill site as a consequence of mechanized harvesting. Table 1 presents a summary of the main characteristics of the reference mill operating with CEST system and burning biomass—bagasse and straw—as fuel. Bagasse availability is defined by the fiber content of the sugarcane plant (14%), i.e., 280 kg of bagasse with 50% moisture per tonne of cane. As for straw its availability at the mill was considered 50% in relation to the total amount available at the field, resulting 161 kg per tonne of cane (with 13% moisture).

Table 1. Characteristics of the reference mill and the power plant.

Parameter	Value
Power plant annual capacity factor	90%
Milling capacity (t/h)	772
Annual harvest season (h)	5184
Mill capacity factor during harvest season	90%
Total annual milling capacity (Mt/y)	4.0
Bagasse availability per tonne of sugarcane [a] (kg)	280 (50% moisture content)
Straw availability per tonne of sugarcane [b] (kg)	161 (13% moisture content)
Energy demand	
Steam process requirement per tonne of sugarcane (kg)	340
Electricity consumption per tonne of sugarcane [c] (kWh)	30
Cogeneration system—CEST	
Boiler efficiency (base LHV)	85%
Live steam parameters	120 bar/535 °C
Isentropic efficiency of the steam turbine (per body)	79%

Sources: [a] [11] for bagasse's LHV 7.52 MJ/kg; [b] [11] for straw's LHV 12.96 MJ/kg; [c] [9].

The CEST system was modelled in a non-commercial software able to simulate its integration with sugarcane mills and to estimate electricity generation [12]. The current experience with straw use as fuel has shown that problems like slagging, fouling and surface corrosion are common when the straw share is above 15–20% in the fuel blend (mass basis). These problems are due to biomass and their ash compositions, which have much more chlorine, CaO and K_2O in the case of straw compared to bagasse [13]. As it is predicted that in the considered case straw would represent about one third of the fuel input, the hypothesis is that the problems mentioned would be solved in the future. Biomass consumption—bagasse and straw—would be distributed along the year to assure fuel supply according to system's requirement.

The emissions of CO_2 from combustion were estimated considering full combustion with 30% excess air [14], and carbon content 48.6% in the dry fuel for both bagasse and straw [15]. For estimating CO_2 from fermentation, it was assumed a sugar mill with a medium to large annexed distillery (i.e., 50% of the sugarcane would be used to ethanol production). A typical Brazilian mill with such a capacity (4 Mt/y) produces both ethanol and sugar with some flexibility (in general, each output varies between 40% and 60%; basis is the sugarcane input) [16]. Therefore, CO_2 from fermentation was calculated for an ethanol production of 86.3 L per tonne of sugarcane [17] and a CO_2 production of 0.96 kg per kg of ethanol [9], resulting in an emission index 0.78 kg of CO_2 per liter of ethanol (calculated for an ethanol density of 0.809 kg/L).

2.2. CCS Unit

Carbon dioxide both from fermentation and from biomass combustion were considered to be processed in a CCS unit. CO_2 from the combustion passes through the complete CCS process (referred to as absorption, regeneration and compression), while CO_2 from fermentation only passes through the compression steps. These two streams are combined in the transportation and storage stages. It was assumed a post-combustion technology based on capture with monoethanolamine (MEA).

2.2.1. MEA Technology

Capture process by absorption is the reference technology and MEA is the incumbent solvent used. Absorption characteristics of the solvent determine energy penalties and impact the economic feasibility of capturing. In this study, parameters of the capture process based on MEA technology were taken from [7]. Solvent regeneration was considered using steam extracted at the medium-pressure stage of the steam turbine, that coincidentally is the same pressure required by the industrial process (i.e., 2.5 bar, 137 °C). The three levels of heat requirement for solvent regeneration are related with the different stages of technology development: 4.4 GJ/t CO_2 (1998 kg of steam per tonne of CO_2); 2.6 GJ/t CO_2 (1180 kg of steam per tonne of CO_2); and 1.6 GJ/t CO_2 (726 kg of steam per tonne of CO_2). It was assumed 90% as the maximum possible capture rate in the CCS unit. Absorption and regeneration take place in the unit hereafter referred to as CCS. Power requirement for flow gases treatment (at the CCS unit) was estimated at 25.84 kW per unit of exhaust gas flow, in kg/s [18]. This figure includes electricity requirement for pumps and blowers, capture pre-treatment pumping, cooling water pumping and blower duties and, finally, solvent pumping duties.

2.2.2. Compression Unit

After CO_2 separation from the exhaust gases, it goes to the compression unit, being its power requirement estimated from [19]. CO_2 is compressed from 1 bar to 150 bar in order to be transported through a pipeline. Compression is divided into two steps: first compression from 1 bar to the CO_2 critical pressure (73.9 bar), and then, in the liquid phase, a pump can be used to boost final pressure. First step was assumed as an ideal gas compression in 5-stages with intermediate cooling, and 85% isentropic efficiency per stage. Pumping requirement was calculated with isentropic efficiency of 85%. Power requirements were considered for each CO_2 stream.

2.3. Economic Performance Assessment

Cost estimations were done based on a literature review for each technology: CEST system, CCS based on MEA technology, CO_2 compression and CO_2 transport and storage at a nearby saline aquifer. All costs are presented in €$_{2014}$ in order to be coherent with the references used for CCS systems. For all equipment the useful life is 25 years, and the discount rate considered in the base case is 10% per year. This discount rate was chosen because it is a compromise taking into account the investments on generating electricity with biomass—annual rates of less than 10% would make investments more difficult—and the investments on carbon capturing and storage—in this case, due to the technology stage, the feasibility is related to a discount rate as low as possible. In any case, the results for the discount rate of 8% per year are also presented (see Section 3.2) in order to make comparisons possible with what has been published. For electricity generation, the feasibility was evaluated based on the minimum selling price (MSP). In the case of CO_2 capture, the minimum credit price (i.e., the income obtained by selling the credits of capturing CO_2) was estimated to cover all costs, including electricity that is not sold due to energy sanctions imposed by CCS.

2.3.1. Power Plant Capital Costs

The cogeneration plant at the sugarcane mill aims at self-sufficiency and the sale of surplus electricity to the grid. Capital costs were estimated ($/kW) from an updated function adapted from [20], which estimates turn-key investments in Brazilian currency (R$), including storage of biomass and connecting costs to the grid, according to Equation (1). Values in R$$_{2014}$ were converted into Euro using the exchange rate by the end of 2014 (3.23 R$/€):

$$C_{CEST} = 3578 \cdot (capacity)^{-0.334} \tag{1}$$

where C represents the specific capital costs, in €/kW installed for the CEST technology, and capacity is the total installed *capacity* in MW.

2.3.2. CCS Unit and Compression Unit Capital Costs

For the CCS unit and the compression unit, scaling—according to Equation (2)—was used to estimate capital costs (scale factor 0.6); scaling is function of CO_2 capturing capacity (CO_2 flow going to CCS and to the compression unit). Values from [4] were taken to estimate the parameters of units based on MEA technology. The costs of the three different technology levels were considered:

$$C = C_{ref} \cdot \left(\frac{Q}{Q_{ref}}\right)^{\alpha} \tag{2}$$

where C represents the capital cost, Q the capacity, α is the scaling factor and ref indicates the reference case.

2.3.3. Transport and Storage Capital Costs

In this study, CO_2 storage was considered to be at the geological formation Rio Bonito, located in the south and southeast regions of Brazil [21]. CO_2 injection shall be at least 1200 m below surface [9]. It is assumed that geological conditions are adequate to keep CO_2 stored for centuries, as required to make CCS a real alternative to mitigate GHG emissions.

Capital costs for transport and storage were estimated from [9]. It was assumed a pipeline with 10 km length, hypothesis that is coherent with the assumption that sugarcane mills are located nearby existing saline aquifers. Storage capital costs include a preliminary assessment of three wells drilled at 1200 m deep, which is a practical assumption to find a reservoir with appropriate conditions to long term storage.

2.3.4. Fuel Costs

As it was mentioned in Section 2.1, bagasse and straw are the biomasses burned in the boiler. No cost was attributed to the bagasse, as it is already available at the mills. This assumption was taken to simplify the economic analysis: in specific cases mills have the opportunity to sell some surplus bagasse to other consumers, depending on the location and the amount available. For the use of straw, the combined cost of collecting and transport to the mill was attributed, summing-up 17.76 € per tonne of straw [22].

2.3.5. Operation and Maintenance Costs

Operation and maintenance costs (O&M) of CO_2 capture (CCS and compression units) are based on [4] and were estimated, as annual values, as function of the total investment. Annual O&M costs for the cogeneration plant were assumed according to the current practices in Brazil. For CO_2 transport and storage, as a simplification, annual O&M costs were assumed at 2% of the total investment. Table 2 presents assumptions for O&M in this study.

Table 2. Assumptions for operation and maintenance costs.

Parameter	Annual Value as Function of the Total Investment
Cogeneration system—CEST	2%
CCS unit	5.8%
Compression unit	4.6%
Transport and storage	2%

2.4. Scaling Effects

The previous sections presented the hypothesis for assessing the feasibility of BECCS in sugarcane mills. Scaling effects on milling capacity were explored as a significant impact on capital costs is expected. For this reason, it was also considered a smaller milling capacity (2 Mt/y) and a larger mill (8 Mt/y). Annual milling capacity of 2 Mt of sugarcane crushed could be considered as an average mill in Brazil; [16] reports that 39% of sugarcane mills are close to this capacity. Bigger capacities, as 4 Mt/y and 8 Mt/y, are less usual—4 Mt/y is more common and few units are close to 8 Mt/y—but larger mills is the general tendency in the future.

2.5. GHG Emissions Due to the Supply of Biomass

It was assumed, by simplification, that both bagasse and straw are carbon neutral, i.e., there would be no GHG emissions due to the biomass used as fuel in the cogeneration unit. Many life cycle assessments of ethanol from sugarcane and electricity generation from bagasse assume that the bagasse is carbon neutral (it would be a residue) being all the environmental burdens imposed on ethanol and sugar, as the main final products [10,17,23]. In the case of straw, as currently there is no other use other than as fuel on the site of the mill, and because the straw is derived from a mechanized harvest that is a new practice, it is common to impose to the straw a share of the emissions of the sugarcane harvest and its transport to the mill. In this sense, the hypothesis assumed in this document is optimistic regarding the benefits of carbon capture related to cogeneration systems. However, it is important to bear in mind that approximately one third of the amount of straw that is supposed to be available as fuel at the factory site is anyway transported to the plant as impurities and, in addition, the straw represents approximately one third of the total energy input. Therefore, the simplification carried out does not imply a great distortion with respect to the benefits of carbon capture, and was considered reasonable for a preliminary evaluation of BECCS in a sugarcane mill.

3. Results and Discussion

This section is devoted to present and discuss the results and is divided into three parts. The first part presents the technical performance of the integrated BECCS systems to a hypothetical sugarcane mill. As previously mentioned, three levels of heat demand for solvent regeneration—related to the different stages of the technology—were evaluated. The second part focuses on the feasibility of carbon capture in a sugarcane mill. Finally, in the third part the effects of scale are analyzed.

3.1. Technical Performance

The simulated cogeneration system has a steam turbine with one controlled extraction at 2.5 bar and condensation of the remaining flow. Five cases were assessed: the reference case, i.e., the cogeneration plant without CCS; case 1—cogeneration plant with CCS only from CO_2 of fermentation; case 2—cogeneration plant with CCS from both fermentation and combustion, and solvent regeneration requiring 4.4 GJ/t CO_2; case 3—cogeneration plant with CCS (fermentation and combustion CO_2) and 2.6 GJ/t CO_2 as heat requirement; and case 4—cogeneration plant with CCS (fermentation and combustion CO_2) and 1.6 GJ/t CO_2 as heat requirement. For all cases, simulation includes harvest and off-harvest seasons.

Figure 1 shows the process flow diagram that represents the operation in the harvest season (i.e., with steam extraction for industrial process), with CCS. In software's basic configuration the steam turbine has three bodies. The steam flow to the deaerator (stream 1) corresponds to 2% of the steam raised and is extracted from the turbine body *b*. The stream (2) feeds the industrial process (stream 5) and the heat exchanger for regenerating the solvent (stream 4), being its thermodynamic state adjusted (in a desuperheater) to the required temperature (137 °C). Streams (6), (7) and (8) refer to condensates, being assumed 90% recovery of streams (6) and (8), but both at 90 °C; pumps for setting the pressure of condensing flows before the deaerator are omitted in Figure 1.

Performance results are presented in Table 3. In the reference case—cogeneration without CO_2 capture, there is no steam extraction going to the CCS plant and, therefore, power output is maximum. In this case, net power output was estimated at 77 MW during harvest season and 64 MW in the off-season, result that corresponds to a surplus output of 144 kWh per tonne of sugarcane (174 kWh/t generated). Electricity generation, or sold, per tonne of sugarcane crushed is an indicator commonly used to express the efficiency of electricity production in a sugarcane mill, and the result above can be compared to the predicted current best figures in sugarcane sector (130–170 kWh/t of cane) [9,23].

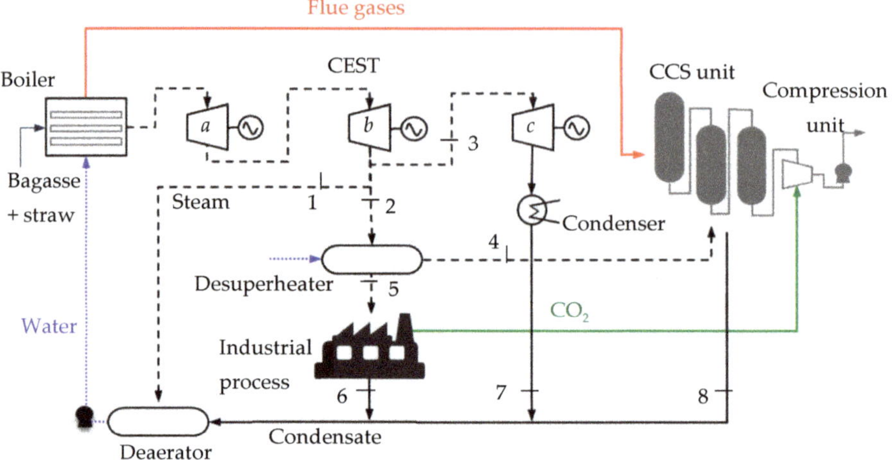

Figure 1. BECCS process flow diagram (harvest season).

Table 3. Performance results for the BECCS systems.

Parameter	Reference Case	Case 1	Case 2	Case 3	Case 4
Energy (as steam) for regeneration (GJ/t CO$_2$)	-	-	4.4	2.6	1.6
CO$_2$ emission (Mt CO$_2$/y)	1.38	1.25	0.58	0.29	0.12
Total CO$_2$ captured (Mt CO$_2$/y)	-	0.13 (10%)	0.79 (58%)	1.09 (79%)	1.26 (91%)
Harvest season					
CO$_2$ captured (combustion) (Mt CO$_2$/y)	-	-	0.43 (43%)	0.72 (73%)	0.88 (90%)
CO$_2$ captured (fermentation) (Mt CO$_2$/y)	-	0.13 (100%)	0.13 (100%)	0.13 (100%)	0.13 (100%)
Net power output (MW)	100.6	100.6	85.3	85.3	88.8
Mill demand (MW)	23.2	23.2	23.2	23.2	23.2
Power requirement for CCS unit (MW)	-	-	11.4	19.3	23.6
Compression power(combustion) (MW)	-	-	7.6	12.9	15.8
Compression power (fermentation) (MW)	-	2.4	2.4	2.4	2.4
Net power output (MW)	77.4	75	40.7	27.5	23.8
Off-season					
CO$_2$ captured (combustion) (Mt CO$_2$/y)	-	-	0.24 (90%)	0.24 (90%)	0.24 (90%)
Power output (MW)	64.1	64.1	48.8	55.0	58.6
Power requirement for CCS unit (MW)	-	-	12.3	12.3	12.3
Compression power (MW)	-	-	8.2	8.2	8.2
Net power output (MW)	64.1	64.1	28.3	34.5	38.1
General results					
Total electricity output (GWh/y)	695	682	408	356	346
Surplus electricity output (GWh/y)	575	562	288	236	226
Surplus electricity per tonne (kWh/t)	144	141	72	59	57
Energy penalty due CCS	-	2%	43%	50%	52%

Case 1 presents a special case in which capture of only CO$_2$ from fermentation was considered. As the CO$_2$ from fermentation is naturally a pure stream, and separation is not necessary, CO$_2$ goes directly to the compression unit. Power requirement for compression was estimated at 2.4 MW (2% of the net power output) and capture corresponds to 0.13 Mt CO$_2$/y, which means 10% of the total mill's emissions.

Capturing CO$_2$ from both fermentation and combustion imposes meaningful energy penalties. There is a constraint in Case 2—with 1998 kg of steam required per tonne of CO$_2$—due to steam availability: during harvest season the system is unable to supply all required steam for both the industrial process and solvent regeneration, thus forcing the reduction of the capture rate. In all cases a minimum steam flow of 3 kg/s (stream (3)) was assumed to be expanded at c. Results of Case 2 indicate that it is possible to capture only 43% of the CO_2 emitted by the combustion process. However, during the off-season, as no steam is required for the industrial process, 90% of the CO_2 from combustion gases is captured (the maximum assumed). Thus, power required for compression during harvest season is due to all CO_2 from fermentation and to the amount captured from flue gases, while during off-season all possible CO_2 captured from flue gases (90%) is compressed. In summary, annual capture rate was estimated at 58%, capturing 0.79 Mt CO_2 (0.13 from fermentation), which is a significant value for a BECCS system, taking into account what has been considered. However, in this case it is predicted a significant reduction in net power output: 48% of the total power during harvest season and 59% otherwise. The balance corresponds to the surplus of 40.7 MW and 28.3 MW, in the harvest and the off-season, respectively. Even though, results still correspond to meaningful surplus electricity regarding the current practices: 72 kWh/t of cane.

In Case 3 the heat requirement for solvent regeneration is 2.6 GJ/t CO_2, and this demand is related to a technology that is expected to be feasible by 2020 [7]. Current technologies are yet further from this parameter (3.2–3.6 GJ/t CO_2) [4,24]. With lower steam requirement per tonne of CO_2 captured (1180 kg), capture from combustion would be higher, but the maximum cannot yet be reached: 73% during harvest season. As consequence, total annual capture would be 1.09 Mt CO_2 (capture rate of 79%). The impacts on power would correspond to 68% of the total generation during the harvest season and to 37% during the off-season. Due to the higher power consumption during harvest, final surplus electricity decreases to 59 kWh/t of cane.

Finally, in Case 4, with heat requirement equivalent to 726 kg of steam per tonne of CO_2, it would be possible to supply all steam required both for the industrial process and the CCS unit, resulting in maximum capture efficiency. In this case the annual capture would be 1.26 Mt CO_2, corresponding to a capture rate of 91% (due to fermentation). The net production of energy during the harvest was further reduced (73% of the power output), but the impact during off-season was reduced to 35% of power output. The indicator of surplus electricity per tonne of sugarcane was estimated at 57 kWh.

3.2. Economic Performance

In order to compare the results presented in this paper with some presented in the literature, the economic assessment was done with all costs estimated for 2014, in Euro. In all cases the investments were supposed to correspond to a single flow in year 0 of the cash flow.

In the reference case, in which the aim is to maximize surplus electricity, the MSP was estimated taking into account all taxes and charges usually incident to this type of enterprise in Brazil. Table 4 presents the main costs and the calculated MSP for the reference case.

The MSP resulted at 48 €/MWh, a value that could be compared with the reference price set in auctions for new enterprises in 2014 (New Energy Auctions), 62 €/MWh, for biomass power units [25]. The difference is explained by the discount rate usually assumed by investors in bioelectricity (higher than the one assumed here) and by the competition in the electricity sector, which vary depending on the investments in other energy sources and the expectation of investing in new hydro power plants.

Results for the Cases 1–4 are also presented in Table 4. The estimated CO_2 credit price, based on the amount of CO_2 captured, would cover all costs (the minimum rate of attractiveness would be 10% per year) and also the loss of revenue due to less electricity sold. In this sense, the results presented in Table 4 correspond to the minimum selling price of capturing CO_2.

In Case 1, a small amount of CO_2 is captured with no meaningful energy penalty. The CO_2 minimum credit price was estimated at 21 €/t CO_2, a relative low cost for CCS, being the best opportunity in case of sugarcane mills. The estimated CO_2 price for Case 1 is basically the same presented in [9] (27.2 US$/t CO_2 in 2014, when the average exchange rate was 1.33 Euro/US$).

Table 4. Costs and economic performance indicators for the BECCS systems.

Parameter	Reference Case	Case 1	Case 2	Case 3	Case 4
Total plant costs					
Power plant (M€)	77	77	77	77	77
CO_2 capture unit (M€)	-	-	171.6	224.5	253.8
CO_2 compression unit (M€)	-	11.1	26	33.6	37
CO_2 transport and storage (M€)	-	1.3	3.0	4.0	4.5
Fuel costs (M€/y)	11.4	11.4	11.4	11.4	11.4
O&M costs					
Power plant (M€/y)	1.5	1.5	1.5	1.5	1.5
CO_2 capture unit (M€/y)	-	-	10.0	13.1	14.8
CO_2 compression unit (M€/y)	-	0.5	1.2	1.5	1.7
CO_2 transport and storage (k€/y)	-	26	62	80	89
Performance indicators					
Electricity price (MSP) (€/MWh)	48	48	48	48	48
CO_2 credit (minimum price) (€/t CO_2)	-	21	66	62	59

It can be seen from Table 4 that, in Cases 2 to 4, the capital costs due to the CCS units represent from 72% to 79% of the total investment, and from 88% to 92% of the total annual O&M costs. It is clear from these figures that carbon capture would be the main driver of investments in Cases 2 to 4, far exceeding the costs of surplus electricity production.

A comparison with the results presented by [4,18] is shown in Table 5. References consider CCS plants based on MEA technology, and both publications were used as reference for estimating costs and performance parameters. However, in both cases the estimates were done for CCS natural gas combined cycle (NGCC) power plants. MEA technology considered in [4] had a heat specific requirement equal to 3.66 GJ/t CO_2, that would be intermediate between Cases 2 and 3 in this paper, while [18] considers heat requirement similar to Case 4. As in both references the discount rate is relatively low (7–7.5%), new results related with this study—for a discount rate of 8% per year—were included in Table 5. The minimum price to be paid per tonne of CO_2 captured is relatively similar comparing the results of this study (for lower discount rate) to those presented by [18], but the cases are very different for a straight comparison.

Table 5. Comparison among similar cases of CCS in thermal power plants.

Parameter	This Study		[4]	[18]
Electricity production technology	CEST		NGCC	NGCC
Power plant capacity (MW)	100		830	557
Specific heat requirement for MEA (GJ/t CO_2)	2.6		3.66	4.4
Total CO_2 captured (Mt CO_2/y)	1.09		1.9	1.31
Discount rate	10%	8%	7.5%	7%
Base year (for costs)	2014		2014	2011
MSP of electricity (€/MWh)	48.0	44.0	90.7	49.4 [a]
CO_2 credit (€/t CO_2)	62.0	55.0	80.7	51.1 [a]

[a] Original values in US dollar were converted to euro using the average exchange rate in 2011 (0.719).

The more expensive electricity generation is, the higher the credit for capturing CO_2 would be. The results of this study are relative close to those presented by [18] because here the case is related to a cogeneration unit that mostly uses residual biomass as fuel, despite the fact that the benefits of scaling effects on electricity generation do not exist. The results presented by [5] are impacted by a higher cost of fuel. Another important aspect is that for the case reported in this paper, the minimum price to be paid for capturing CO_2 is impacted by the stream of CO_2 from fermentation (that varies from 16% of the total in Case 2, to 10% in Case 4) and that has a relatively small cost. Following the same procedure described in this paper, but not considering the capture of CO_2 from fermentation, the credit price would grow from 59 €/t CO_2 to 70 €/t CO_2 (for discount rate 10%). It is also worth mentioning that the amount of CO_2 captured per year is not much smaller (57% to 83%) than in power units that would burn natural gas, despite the much smaller installed electricity capacity (12% to 18%). Comparing biomass and natural gas, the higher carbon content per unit of energy and the much lower efficiency of electricity generation explain the huge penalties of CO_2 capture on electricity generation. Another useful comparison is with the carbon price needed for making a technology competitive. In the case of switching from NGCC to a coal power plant with CCS, considering 8% as the annual capital costs for the investments, [26] presents 85 €/t CO_2 as the break-even price. Thus, in a general sense it can be concluded that the CO_2 capture costs presented in this paper are in line with the estimates presented in the literature, but a main difference is that the BECCS system considered here is able to contribute with negative GHG emissions.

Cases 2 to 4 correspond to different stages of development of MEA based technology. Case 2 corresponds to the current commercial stage, while Case 3 represents the technology that could be available in short term. Moving from current to future technologies would impact carbon capture, with an increase of 38% on annual output as long as Case 3 is compared to Case 2; as previously mentioned, capturing in Case 2 is negatively impacted by the higher steam demand for recovering amine. On the other hand, moving from Case 2 to Case 3 significantly impact surplus electricity, with a decrease of almost 13% in the total electricity that could be sold along the year. The impact on the minimum price to be paid per tonne of CO_2 captured is less pronounced, with a reduction of 6% comparing Cases 3 and 2. The change from Case 3 to Case 4 is less pronounced (an increase of 16% in annual capture, a reduction of 2.8% in surplus electricity, and a 4.8% reduction in carbon costs).

The fact that the minimum price to be paid per tonne of CO_2 captured is almost equal in all three cases indicates that, from an economic point of view, it is not necessary to wait for advanced MEA technologies in order to go for pilot BECCS projects. Therefore, it is necessary to consider technologies that would impact less on surplus electricity. However, a very important find is that CO_2 capture from fermentation has a lower cost and a small impact on the energy balance, and should be prioritized for pilot BECCS units in Brazil.

3.3. Scaling Effects

For the considered BECCS system scaling effects are analyzed in this section. Case 3, with heat requirement for solvent regeneration equivalent to 2.6 GJ/t CO_2, was chosen to be scaled into a smaller industry (2 Mt/y) and a larger mill (8 Mt/y). The same performance parameters previously presented were considered, and costs were estimated according to the assumptions mentioned before (for 10% discount rate). Table 6 presents total plant costs and the main economic results. Scale effects are clear both on the MSP of surplus electricity and on the minimum price to be paid for capturing CO_2.

Taking as reference the price presented by [4] in the case of capture in a large combined cycle power plant (80.7 €/t CO_2), and assuming that people would be able to pay this value in the future, full carbon capture (both from fermentation and combustion) would be feasible in Brazilian sugarcane mills, but it is clear that the feasibility would be enhanced with the mill capacity. As regard with the results presented by [18], comparatively the feasibility would exist for larger mills. Considering that mills with capacity equivalent to 2 Mt/y are currently the average in Brazil, in many existing mills it would be feasible to capture CO_2. On the other hand, considering that new mills tend to be larger, in the future it would be reasonable to consider mill's location also taking into account the aim of storing CO_2 at lower costs.

Allocating all capital costs to the annual surplus electricity, the indicator varies from 1874 to 1099 €/MWh, depending on the mill size (see Table 6). This figure for the reference case [4] is only 134 €/MWh. Alternatively, allocating total capital costs to the annual amount of CO_2 captured, this indicator varies from 237 to 405 €/t CO_2 captured per year (Table 6), while this figure is 483 €/t in the case presented by [4]. It seems clear that investors should have a completely different rationale in each case: while in case of a natural gas combined power plant CO_2 capture would be a complement that should be fairly paid, in the case of carbon capture in a sugarcane mill surplus electricity produced in a cogeneration unit should no longer be the priority. In this case, the priority should be capturing CO_2, with the advantage that this enterprise would contribute with negative emissions. Indeed, whether the benefits of negative emissions would be recognized, a larger payment per tonne of CO_2 would be possible, and surplus electricity could be more competitive with other generation options.

Table 6. Total investment costs and economic performance results for different milling capacities.

Parameter	Milling Capacity (Mt/y)		
	2	**4**	**8**
Performance results			
Power plant capacity (MW)	50	100	200
CO_2 captured (Mt CO_2/y)	0.55	1.09	2.19
Electricity price (MSP) (€/MWh)	54	48	43
CO_2 credit (minimum price) (€/t CO_2)	80	62	48
Total plant costs			
Power plant (M€)	48.5	77.0	122.3
Capture unit (M€)	148.1	224.5	340.3
Compression unit (M€)	22.2	33.6	50.9
Transport and storage (M€)	2.7	4.0	6.1
Economic indicators (as function of annual outputs)			
Investment cost per tonne of CO_2 captured (€/t)	405	310	237
Investment cost per surplus electricity unit (€/MWh)	1874	1435	1099

4. Conclusions

This work aimed to explore the technical and economic feasibility of BECCS systems in the Brazilian sugarcane sector. Post-combustion technology based on MEA was considered for capturing CO_2 from biomass combustion, and three technology levels—related to heat requirements for solvent regeneration—were assessed. Results show that CO_2 capture, both from fermentation and combustion, is technically possible but energy penalties are meaningful in case of combustion, with considerable impacts on surplus electricity. Energy penalties due CCS imply deep reduction in electricity generation, varying from 43% to 52% regarding the reference case. In this sense, it is important to evaluate other technologies for capturing CO_2. The more expensive the electricity sold, the higher the price to be paid per tonne of CO_2 captured.

Comparatively, capturing CO_2 from the fermentation is the best opportunity, because of the low impact on the mill, the relatively low cost, and the benefits on the ethanol carbon footprint. Clearly, is the alternative that should be prioritized.

The high impact on electricity production in the case of biomass-based cogeneration units, compared with well-known estimates for natural gas plants, is due to the comparatively low efficiency of electricity generation. Investments and costs are far greater for capturing CO_2 than for generating surplus electricity. In this sense, the rationale for investments should be different: the priority would be capturing CO_2—resulting in net negative emissions, and selling electricity would be a second priority.

The CO_2 credits presented in this document and also in the literature (approximately € 45–80/t CO_2) are much higher than the current price of carbon in different markets (e.g., considering the CO_2 European Emission Allowances, the carbon price was less than € 10/t CO_2 in the first half of 2018 and around € 20/t CO_2 by the end of 2018 [27]), but it is important to take into account some important aspects. First, carbon markets are currently depressed due to low demand. Second, and most importantly, carbon capture and storage would be costly among the mitigation options and the large-scale implementation of CCS systems would be feasible only in the medium to long term.

Supposing that carbon capture through CCS would be a target in the future, investing in a BECCS system in a sugarcane mill would be much more effective than investing in a power plant that burns natural gas, for instance. The price to be paid would be lower, and the result would be negative emissions.

Author Contributions: Conceptualization, S.R.V. and A.W; Methodology, S.R.V. and A.W; Software, S.R.V; Validation, S.R.V. and A.W; Investigation, S.R.V. and A.W; Writing-Original S.R.V; Draft Preparation, S.R.V; Writing-Review & Editing, S.R.V. and A.W; Visualization, S.R.V; Supervision, A.W.

Funding: This research received no external funding.

Acknowledgments: The authors would like to acknowledge the support of CAPES (Coordenação de Aperfeiçoamento de Pessoal de Nível Superior, Brazil).

Conflicts of Interest: The authors declare no conflict of interest.

References

1. Intergovernmental Panel on Climate Change. *IPCC Climate Change 2014: Mitigation of Climate Change*; Edenhofer, O., Pichs-Madruga, R., Sokona, Y., Minx, J.C., Farahani, E., Kadner, S., Seyboth, K., Adler, A., Baum, I., Brunner, S., et al., Eds.; Cambridge University Press: Cambridge, UK; New York, NY, USA, 2014; ISBN 9781107654815.
2. International Energy Agency. *World Energy Outlook 2016*; OECD/IEA: Paris, France, 2016.
3. International Energy Agency. *20 Years of Carbon Capture and Storage*; OECD/IEA: Paris, France, 2016.
4. Van der Spek, M.; Ramírez, A.; Faaij, A. Challenges and uncertainties of ex ante techno-economic analysis of low TRL CO_2 capture technology: Lessons from a case study of an NGCC with Exhaust Gas Recycle and Electric Swing Adsorption. *Appl. Energy* **2017**, *208*, 920–934. [CrossRef]
5. Van der Spek, M.; Ramirez, A.; Faaij, A. Improving uncertainty evaluation of process models by using pedigree analysis. A case study on CO_2 capture with monoethanolamine. *Comput. Chem. Eng.* **2016**, *85*, 1–15. [CrossRef]
6. Luis, P.; Van der Bruggen, B. The role of membranes in post- combustion CO_2 capture. *Greenh. Gases Sci. Technol.* **2013**, *3*, 318–337. [CrossRef]
7. Peeters, A.N.M.; Faaij, A.P.C.; Turkenburg, W.C. Techno-economic analysis of natural gas combined cycles with post-combustion CO_2 absorption, including a detailed evaluation of the development potential. *Int. J. Greenh. Gas Control* **2007**, *1*, 396–417. [CrossRef]
8. Kemper, J. Biomass and carbon dioxide capture and storage: A review. *Int. J. Greenh. Gas Control* **2015**, *40*, 401–430. [CrossRef]
9. Moreira, J.R.; Romeiro, V.; Fuss, S.; Kraxner, F.; Pacca, S.A. BECCS potential in Brazil: Achieving negative emissions in ethanol and electricity production based on sugar cane bagasse and other residues. *Appl. Energy* **2016**, *179*, 55–63. [CrossRef]
10. Seabra, J.E.A.; Tao, L.; Chum, H.L.; Macedo, I.C. A techno-economic evaluation of the effects of centralized cellulosic ethanol and co-products refinery options with sugarcane mill clustering. *Biomass and Bioenergy* **2010**, *34*, 1065–1078. [CrossRef]
11. Pedroso, D.T.; Machin, E.B.; Proenza Pérez, N.; Braga, L.B.; Silveira, J.L. Technical assessment of the Biomass Integrated Gasification/Gas Turbine Combined Cycle (BIG/GTCC) incorporation in the sugarcane industry. *Renew. Energy* **2017**, *114*, 464–479. [CrossRef]
12. Walter, A.; Llagostera, J.; Ensinas, A.V.; de Maio, D.S.; Reis, M.; Leme, R.M. *Levantamento do Pontencial Nacional de Produção de Eletricidade nos Segmentos Sucro-Alcooleiro, Madeireiro e em Usinas de Beneficiamento de Arroz*; Report Núcleo Interdisciplinar de Planejamento Energético—UNICAMP: Campinas, Brazil, 2005.
13. Bizzo, W.A.; Lenço, P.C.; Carvalho, D.J.; Veiga Soto, P.J. The generation of residual biomass during the production of bio-ethanol from sugarcane, its characterization and its use in energy production. *Renew. Sustain. Energy Rev.* **2014**, *29*, 589–603. [CrossRef]
14. Li, C.; Gillum, C.; Toupin, K.; Donaldson, B. Biomass boiler energy conversion system analysis with the aid of exergy-based methods. *Energy Convers. Manag.* **2015**, *103*, 665–673. [CrossRef]
15. Walter, A.; Ensinas, A.V. Combined production of second-generation biofuels and electricity from sugarcane residues. *Energy* **2010**, *35*, 874–879. [CrossRef]

16. Moreira, M.M.R. Strategies for Expansion of the Sugar- Energy Sector and its Contributions for the Brazilian NDC. PhD Thesis, Univesity of Campinas, Campinas, Brasil, 2016.

17. Macedo, I.C.; Seabra, J.E.A.; Silva, J.E.A.R. Green house gases emissions in the production and use of ethanol from sugarcane in Brazil: The 2005/2006 averages and a prediction for 2020. *Biomass and Bioenergy* **2008**, *32*, 582–595. [CrossRef]

18. Khorshidi, Z.; Florin, N.H.; Ho, M.T.; Wiley, D.E. Techno-economic evaluation of co-firing biomass gas with natural gas in existing NGCC plants with and without CO_2 capture. *Int. J. Greenh. Gas Control* **2016**, *49*, 343–363. [CrossRef]

19. Mccollum, D.L.; Ogden, J.M. *Techno-Economic Models for Carbon Dioxide Compression, Transport, and Storage & Correlations for Estimating Carbon Dioxide Density and Viscosity*; Institute of Transportation Studies: Berkeley, California, USA, 2006.

20. Gouvello, C. *Brazil Low-carbon: Country Case Study*; The World Bank Group: Washington, DC, USA, 2010.

21. Ketzer, J.; Machado, C.; Camboim, G.; Iglesias, R. *Atlas Brasileiro de Captura e Armazenamento Geológico de CO_2*; Editora Universitároa da PUCRS, Ed.; Porto Alegre, RS, Brazil, 2016; ISBN 9788539707652.

22. Cardoso, T.D.F. Avaliação socioeconômica e ambiental de sistemas de recolhimento e uso da palha de cana-de-açúcar. PhD Thesis, Univesity of Campinas, Campinas, Brasil, 2014.

23. Seabra, J.E.A.; Macedo, I.C. Comparative analysis for power generation and ethanol production from sugarcane residual biomass in Brazil. *Energy Policy* **2011**, *39*, 421–428. [CrossRef]

24. Hetland, J.; Yowargana, P.; Leduc, S.; Kraxner, F. Carbon-negative emissions: Systemic impacts of biomass conversion. A case study on CO2 capture and storage options. *Int. J. Greenh. Gas Control* **2016**, *49*, 330–342. [CrossRef]

25. CCEE Resultado consolidado dos leilões—August 2018. Available online: https://www.ccee.org.br/portal/faces/pages_publico/inicio (accessed on 28 August 2018).

26. *The World Bank Report of the High-Level Commission on Carbon Prices*; The World Bank Group: Washington, DC, USA, 2017.

27. CO_2 European Emission Allowances Market Insider. Available online: https://markets.businessinsider.com/commodities/co2-emissionsrechte (accessed on 5 February 2019).

Article

Assessment of the Energy Potential of Chicken Manure in Poland

Mariusz Tańczuk [1,*], **Robert Junga** [1], **Alicja Kolasa-Więcek** [2] **and Patrycja Niemiec** [1]

[1]   Faculty of Mechanical Engineering, Opole University of Technology, 45-271 Opole, Poland;
      r.junga@po.opole.pl (R.J.); pat.niemiec@doktorant.po.edu.pl (P.N.)
[2]   Faculty of Natural Sciences and Technology, Opole University, 45-047 Opole, Poland; akolasa@uni.opole.pl
*   Correspondence: m.tanczuk@po.opole.pl; Tel.: +48-664-475-355

Received: 2 January 2019; Accepted: 25 March 2019; Published: 1 April 2019

Abstract: Animal waste, including chicken manure, is a category of biomass considered for application in the energy industry. Poland is leading poultry producer in Europe, with a chicken population assessed at over 176 million animals. This paper aims to determine the theoretical and technical energy potential of chicken manure in Poland. The volume of chicken manure was assessed as 4.49 million tons per year considering three particular poultry rearing systems. The physicochemical properties of examined manure specimens indicate considerable conformity with the data reported in the literature. The results of proximate and ultimate analyses confirm a considerable effect of the rearing system on the energy parameters of the manure. The heating value of the chicken manure was calculated for the high moisture material in the condition as received from the farms. The value of annual theoretical energy potential in Poland was found to be equal to around 40.38 PJ. Annual technical potential of chicken biomass determined for four different energy conversion paths occurred significantly smaller then theoretical and has the value from 9.01 PJ to 27.3 PJ. The bigger energy degradation was found for heat and electricity production via anaerobic digestion path, while fluidized bed combustion occurred the most efficient scenario.

Keywords: biomass; chicken manure; proximate and ultimate analysis; energy potential

---

1. Introduction

In recent years, changes in the energy policies of all member states of the European Union (EU) have been aimed at increasing the proportion of renewable energy sources (RES) in the energy supply. As a consequence of these changes, legal regulations are being introduced to increase the share of RES in the total energy balance of the European community. The efforts made by the specific governments have, accordingly, focused on the application of the common and accessible RES such as solar, wind, water and biomass. In particular, biomass, is a very promising type of RES, which demonstrates considerable diversity in terms of physicochemical properties leading to wider potential applications in the energy industry [1]. The determination of the energy potential of various types of biomass, including bio-wastes, is indispensable for the development of technologies of biomass to energy conversion as well as for effective implementation of existing technological solutions.

Poultry manure (excreta) is a problematic biomass waste. Due to the increasing concentration of poultry in holdings, the issue of removal and utilization of the manure plays an important role. Soil conditioning—the manner in which this type of manure has been predominantly utilized up to this time—has a number of drawbacks [2], yet it has been the most popular one. This technique is applied all over the world but causes hazards to both crops and human health as a consequence of the negative effect of decomposing manure on various components of the natural environment. The effect of excessive soil conditioning can take the following forms: eutrophication, phytotoxic effects on succeeding crops, propagation of pathogens, pollution of atmospheric air and emission of greenhouse gases [3]. In Poland, using chicken manure as a raw material in crop fertilization as well as in mushroom cultivation are two main ways of its application and utilization. According to [4] total volume of annually generated chicken manure in Poland is enough to fertilize over 20% of crops in environmentally balanced way in terms of nitrogen concertation. It should be underlined again that massive use of poultry manure as an organic fertilizer creates a strong risk of environmental degradation due to over-fertilization. Moreover, ammonia (NH_3) atmospheric emissions from chicken manure land spreading can be several dozen times greater than emissions from combustion [5]. For this reasons it is necessary to define and apply new routes of poultry manure treatment in a sustainable way for the environment not only in Poland but in every location where these wastes are utilized mainly through soil conditioning.

One of the solutions to prevent negative environmental impact of chicken manure treatment is using it for energy generation. This can be of great interest to industry, unfortunately it is still rarely used in energy sector due to its unique chemical composition. There are several ways in which chicken manure can be used for generation of energy or for deriving improved fuel. It can be treated in such processes as fermentation (digestion), esterification, pyrolysis, gasification, co-gasification, combustion and co-combustion [3,5–7]. For more efficient energy generation it is necessary to pre-process or pre-treat chicken manure by physical and/or thermo-chemical processes such as drying, torrefaction, pyrolysis, steam treatment, and ozone treatment [6].

According to [8] chicken manure can be a promising substrate for slow pyrolysis, but the product of pyrolytic conversion should by bio-coal rather than syngas. The gasification of poultry-derived waste has been limited to laboratory and small-scale applications, whereas full and pilot scale installations operating in industrial conditions are incidentally mentioned in the literature. Experiments on gasification and co-gasification systems using both fixed and fluidized bed combustors show that to obtain a high quality fuel gas, the addition of another type of biomass or coal is desirable [9]. Combustion and co-combustion of poultry derived waste is the best known process of its thermal utilization: industrial scale installations are operated worldwide, including UK, USA, Netherlands, Ireland, Australia, Belgium, France, Germany, Italy, Spain [7], confirming the economic feasibility of the application. Despite the wide range of high-temperature thermal routes of chicken manure utilizations, in Poland only digestion processing is used so far. A raw chicken manure feedstock of 690 Mg/h supplies a biogas installation of 25–30 kW located directly on the farm in Pszczyna city. Incidental larger biogas units with capacity up to 1 MW are operated in the co-fermentation mode with mixing of waste from agricultural production and processing. The way of utilizing chicken manure as a kind of alternative fuel can be of interest in particular in the countries where hard coal still forms the basic source of energy and the number of poultry producers is relatively large. Local boilers situated in the vicinity of farms appear to be an area with great potential for application of this type of biomass. The heat produced in this manner can be used for heating local buildings [10]. Advanced technologies can also be applied for distributed electricity generation on chicken farms [5].

Poland meets two basic criteria that indicate the feasibility of utilizing chicken manure for energy. The first one is associated with the considerable chicken population on farms in Poland which is a leading poultry producer in Europe. According to European Egg Processors Association (EEPA) [11], Poland was listed in fourth place among all EU states in terms of laying hens population in 2016 (11.3 % of the EU-28), next to Germany (13.7%), France (12.6%), Spain (11.4%), United Kingdom (11.0%), Italy (10.8%) and Netherlands (8.9%). According to Food and Agriculture Organization of the United Nations (FAO) [12], in 2016 Poland was listed in second place among all EU states in terms live chickens (12.0% of the EU-28), next to Germany (13.0%), France (11.9%), United Kingdom (11.1%), Italy (10.1%), Spain (10.1%) and Netherlands (7.5%). The second one is associated with the dominant share of coal-fired sources in the production of heat: coal and coal products account for 86% of total heat production in Poland [13], similar to some other European countries where coal continues to dominate in the structure of the heat production system [14,15]. Additionally, Poland has the dominant share of coal-fired sources in the production of heat and electricity supplying chicken farms.

The latter part of this work aims to determine the quantitative and energy potential of chicken manure in Poland on the basis of the proposed methodology. Energy potential was assessed regarding two potential levels: theoretical and technical. As mentioned before, there are many technical ways for conversion chicken manure into useful forms of energy such as heat and electricity. The most common and widely used and technically advanced solutions were chosen as possible and representative conversion scenarios for the case of technical potential assessment analysis:

- biogas production through anaerobic digestion process combined with cogeneration of heat and power based on internal combustion (IC) engine,
- raw manure pre-drying preceding syngas generation in updraft gasification process combined with cogeneration of heat and power based on an organic Rankine cycle (ORC) module,
- raw manure pre-drying preceding combustion in fluidized-bed combustor combined with cogeneration of heat and power based on an organic Rankine cycle (ORC) module,
- raw manure pre-drying preceding combustion in fluidized-bed combustor for direct heat generation.

2. Material, Methods and Input Data

The total energy potential of the manure can be accessed through the weight of the manure produced by the population of birds in a given area. The features of chicken manure and considering it as a potential source of energy requires an individual approach to the assessment of their resources in the specific area (e.g., in the country). It results from the fact that the territorial distribution of chicken farms is random regarding rearing systems and their locations takes form of local territorial clusters.

The assessment of the energy potential of biomass (including chicken biomass) can be made in a number of ways [16–18]. The method of assessing this potential depends on the quantity, details and characteristics of the data available to the researcher.

From the perspective of practical application of the results, the energy potential of biomass can be classified according to the so-called potential level [19–21]. In the most common classification of the energy potential, it can be distinguish between:

- theoretical potential: volume of energy that can be useful based on the availability of equipment with 100% efficiency (losses in the process are not accounted for), and under the assumption that the total available potential is harnessed for energy production, including specific collection factor and alternative use factor;
- technical potential: this part of the theoretical potential, which can be practically utilised, yet reduced due to technical restrictions (efficiency of available equipment, internal losses in the process, geographical location, energy storage losses). It is usually derived on the basis of detailed technical analysis;
- economic (market) potential: relative to fuel prices, tax rates, economic parameters and levels of subsidies. It is the part of the technical potential that can be applied after accounting for the criteria of economic tools (detailed economic analysis of profitability);
- applicable potential: energy from biomass, which can be ultimately used in energy production (usually smaller than the economic one).

2.1. Quantitative Potential of Chicken Manure

The total quantitative potential of chicken manure in Poland was determined as the sum of the potential of manure produced by laying hens and broilers. These two types are responsible for the majority of poultry production in Poland and were taken into account in a further analysis. Table 1 presents data on chicken population reared on Polish farms, according to the type of the rearing system.

Table 1. Population of chicken in Poland and its provinces according to the rearing system, as of December 2017 on the basis of [22,23], in thousand.

Province	Laying Hens			Broilers	Overall Chickens
	Cage	Litter	Free-Range	Litter	
Dolnośląskie (I)	1949.7	414.2	67.9	3463.0	5894.8
Kujawsko-Pomorskie (II)	1686.6	226.8	57.9	9842.3	11,813.6
Lubelskie (III)	1632.6	352.2	100.7	4469.5	6555.0
Lubuskie (IV)	1196.3	47.7	10.8	3463.0	4717.8
ódzkie (V)	2057.9	255.0	129.2	9422.7	11,864.8
Małopolskie (VI)	2107.0	433.9	49.4	2473.6	5063.9
Mazowieckie (VII)	9449.1	842.6	335.6	27,333.0	37,960.3
Opolskie (VIII)	337.8	475.1	37.9	3215.6	4066.4
Podkarpackie (IX)	1699.0	367.5	32.6	3339.3	5438.3
Podlaskie (X)	616.7	690.2	17.2	11,254.8	12,578.8
Pomorskie (XI)	1481.6	181.6	55.6	4250.8	5969.6
Śląskie (XII)	1595.1	685.1	19.7	6170.9	8470.8
Świętokrzyskie (XIII)	1088.5	86.7	203.1	4205.1	5583.4
Warmińsko-Mazurskie (XIV)	913.9	20.8	29.1	3620.0	4583.8
Wielkopolskie (XV)	16,773.9	554.5	290.0	17,201.2	34,819.6
Zachodniopomorskie (XVI)	1016.0	269.4	96.0	9947.9	11,329.3
Poland–overall	45,601.7	5903.3	1,532.7	123,672.7	176,710.4

The annual amount of chickens (laying hens and broilers) reared in Poland was the basis for determining the amount of manure available for energy use. For this purpose, the data on the daily production of feces (droppings) by different types of chickens was analyzed. Table 2 presents the daily production of droppings by selected chicken type. The presented data indicates that adult birds produce from 150 to 160 g of droppings, while young ones (pullets) produce from 65 to 110 g. The lowest daily volume of droppings concerns broilers.

Table 2. Daily production of droppings of various types of chickens and its density [24].

Chicken Type		Daily Volume Per Bird (g)	Density of Droppings (kg·m^{-3})
	Ready to lay pullets	100	605
Pullets	Meat pullets	110	680
	Broilers	65	622
	Laying hens	150	650
Adult hens	Parent laying hens	155	670
	Meat chickens	160	680

Due to the lack of precise data, resulting from the variability of the course of hen growth (which could provide the percentage of adult hens and broilers kept on specific farms), the further calculations here have adopted the mean value of mass of manure produced by a laying hen (110 g) and a broiler (65 g).

The amount of manure generated by chicken farms depends not only on the mass of chicken droppings but also on the amount and type of litter (for the litter rearing system). Polish farms use straw as a litter. For the purpose of further analysis and on the basis of data collected from different Polish chicken farms, it is assumed that straw used as an additive increases the amount of litter manure by 15%.

In case of broilers, the calculations were performed under the assumption that the production cycle of broilers lasts 6 weeks, followed by a shut down for maintenance of up to 3 weeks. Hence, within a year, it is possible to conduct a maximum of six production cycles taking the total time of 252 days, during which manure is produced. In the case of laying hens it was assumed that the production cycle is continuous and lasts 365 days per year.

2.2. Qualitative Potential of Chicken Manure

Qualitative potential depends on physicochemical properties of the chicken manure: moisture content, ash content, volatile matter and elemental composition. Such compositional data of the manure can be derived on the basis and proximate and ultimate analysis. Performing these analyses is essential for determining the energy potential of the chicken wastes. They are also of a great importance in context of thermal destruction of poultry manure used in energy applications.

For the needs of research the samples of the manure were collected from two representative farms situated in Opole province. The first one was a large, modern holding, which specializes in rearing laying hens in the cage system and broilers on litter. The farm has a capacity of around 160,000 birds. The other one is a small farm rearing chickens in backyard system and not exceeding 1000 birds. The averaged samples from the cage system were clean manure with a high moisture level. In contrast, the samples collected from the litter system contained some addition of feathers and straw litter. The material collected from the backyard system contained manure and more feathers, straw and tree leaves. Samples from both sites were collected in November and for the control group in January.

2.2.1. Ultimate, Proximate and Heating Value Analyses

Collected samples were weighed and the moisture loss during pre-drying was measured at 20 °C temperature after 24 h of storage. Subsequently, averaged samples were prepared and dried 105 °C until solid mass (dry matter) was obtained. Total moisture content (M_t) (calculated on a wet basis) was determined in accordance with the PN-EN 14774-1 standard. After drying and grinding, the samples were kept in Falcone tubes in a nitrogen atmosphere, secured by sealing with Parafilm and placed in an extractor.

Prepared samples were subjected to ultimate analysis by means of an Elementar Vario Macro Cube apparatus (Elementar Analysensysteme, Langenselbold, Germany). The concentrations of carbon (C), nitrogen (N), sulphur (S) and hydrogen (H) were measured, whereas the concentration of oxygen (O) was calculated by difference. Proximate analysis manure of samples was undertaken by means of a STA 449 F3 Jupiter apparatus (Netzsch, Selb, Germany) by application of DTA/TG method [25]. The sample was heated at the rate of 50 °C/min in the range of the temperatures from 80 to 1150 °C. The level of ash content (A) was measured after combustion of the samples performed in the air atmosphere, while for volatile matter content (VM), the analysis was performed in the nitrogen atmosphere. The concentration of the fixed carbon (FC) was derived by difference.

Higher heating value (HHV) was derived in an AC500 bomb calorimeter (LECO, Saint Joseph, MI, USA) in accordance with the EN-14918:2010 standard in order to determine heating value of solid biofuels. HHV (in kJ·kg^{-1}) was transformed into lower heating value in the dry state LHV_d (in kJ·kg^{-1}) in accordance with the common formula, in which the last term of the right-hand-side of the equation represents latent heat of water vapour formed due to hydrogen in the dry matter (H_d), according to Equation (1):

$$LHV_d = HHV_d - 2441.8 \, (9H_d/100) \tag{1}$$

In turn, lower heating value in the wet matter was derived by converting LHV_d in accordance with the Equation (2):

$$LHV_{ar} = (LHV_d - 24.41 \, M_t)\frac{100 - M_t}{100} \tag{2}$$

Table 3 presents the average results of ultimate and proximate analysis of the samples of chicken manure.

A synthetic overview of physicochemical parameters of the manure as a fuel for thermal conversion systems is presented in the paper [2], which provides an overview of the variability ranges of fixed carbon, volatile matter and ash content as well as higher heating value of dried chicken waste. These parameters vary widely, but they are characteristic of the analyzed type of biomass and depend on various factors such as nutrition, housing systems or environmental conditions including microclimate as well as litter type. For the purpose of this energy analysis, a distinction was made between the properties of the chicken waste from three most popular rearing systems in Poland, comparing the results obtained to the same rearing systems used in the world. The available research studies describe quite precisely the analytical composition of the manure. Unfortunately, the comparison is still difficult, because the type of rearing system, composition of bedding material or the information on whether the material has been previously processed e.g., through pre-draying, are usually not specified.

Table 3. Results of physicochemical analysis of manure for various types of poultry systems.

System	Proximate Analysis (wt. %)				Ultimate Analysis (wt.%, d)					Heating Values (MJ·kg^{-1})	
	M_{tw}	VM_d	FC_d *	A_d	C	H	N	S	O *	LHV_d	LHV_{ar}
Cage	70.9	67.5	16.9	15.6	39.67	4.72	5.49	0.40	34.12	12.744	3.201
Litter	18.1	69.9	18.1	12.0	42.86	5.57	5.50	0.68	33.39	16.546	13.189
Free-range	54.9	50.4	5.2	44.4	21.85	2.50	1.73	0.28	29.24	8.577	3.262

* By difference.

Parameters of the raw manure obtained from the cage system presented in Table 4 can be directly related to analyses of a battery-cage system located in Spain [26]. In both cases the moisture content M_{tw} was found high, even above 70%. Values of C, H and N content are also similar to those presented in the paper [26], that amount to 36.2%, 4.6% and 5.9%, respectively. A significant difference concerns the element S. In general, its value for Spanish farms manure was about 0.1% on average. The S content was 0.357% only on one of the Spanish farm, which is comparable to the results obtained in the study. The content of volatile matter VM_d and fixed carbon FC_d (roughly giving the organic matter) presented in Table 4, was 15 p.p. higher than mentioned in [26], which results from different analytical standards applied in both studies. The same conclusion applies to the ash content A_d, which in analyzed Polish farm is about 15 p.p. lower than for samples tested from Spanish farms. Despite these differences, the calorific value LHV_d of the manure from cage systems is within the range reported by [26], i.e., 2660 MJ·kg^{-1} on average.

In the case of manure samples from litter systems, the results of the performed analyses presented in Table 4 can be referred to more bibliographic sources. The moisture content M_{tw} obtained within the research, was similar to the results reported for litter systems used in different states in the USA [10], Australia [7] and Poland [27]. However, according to [28] the moisture content of poultry litter can widely range from 9.25 to 51.8%. As reported in [29] the average moisture content in the manure from farms with appropriate husbandry practices and controlled ventilation should range from 18 to 36%. Bedding with a moisture content exceeding 36% increases the risk of e.g., poultry contracting cococydiosis, internal parasites or insects [29]. The content of basic elements C, H, N and S in the sampled manure is similar to the values recorded on Irish [28] and Polish [30] farms. Based on studies presented by Lynch et al. [28] and Toptas et al. [31] the changes of these parameters in chicken manure can be relatively wide. Lynch et al. [28] suggests that these differences may be due to the type of material bedding, litter storage conditions as well as farming practices [28]. Nevertheless, taking into account the dispersion of the elements contents, the results given in Table 4 are still in the middle of these ranges. Available publications indicate that content of volatile matter VM_d in broiler litter varies from 53 to 75%, while the fixed carbon FC_d from 6 to 23% [7,10,30,32]. It may be seen that values of VM_d and FC_d presented in Table 4 are comparable to those from litter raring system located in the USA [10], Turkey [32], Ireland [28] and Poland [30]. It should be noted that the manure analyzed in the study [10] and [28] contained wood shaving while the material from Poland [30] and Turkey [32] contained straw in its composition. The same levels of ash content A_d as in presented study (approximately 12% on dry basis) were reported from other farms worldwide with wood and straw bedding materials [10,28,31,32]. According to Lynch et al. [28] the ash content can vary from 10.61 to

19.58%, depending on the bedding materials. However, there are also high ash contents reported, even up to 50% [33], explained by the admixture of additional solid materials [33]. The LHV_d of the samples from analyzed litter system (Table 3) are close to those from Turkish [32], Irish [28] and Polish [30] farms: from 15.1 to 16.7 MJ·kg^{-1}. In general, the values obtained from the tests are within the heating value ranges given by Lynch et al. [28].

The literature does not provide results regarding the physicochemical analysis of the manure produced in the free-range systems. In Poland, free-range systems are gaining popularity slowly, however such farms are usually more similar to the semi-intensive systems. It can explain the results of the analyses (for instance high ash content) presented in Table 3.

Figures 1–3 illustrate the results of measurements of lower heating value of manure from the farms examined as a function of their moisture content, with a comparison of the results reported in earlier works. The presented ranges of LHV variation can be considered as typical of chicken manure and used to determine the energy potential of manure regardless of the place of its generation.

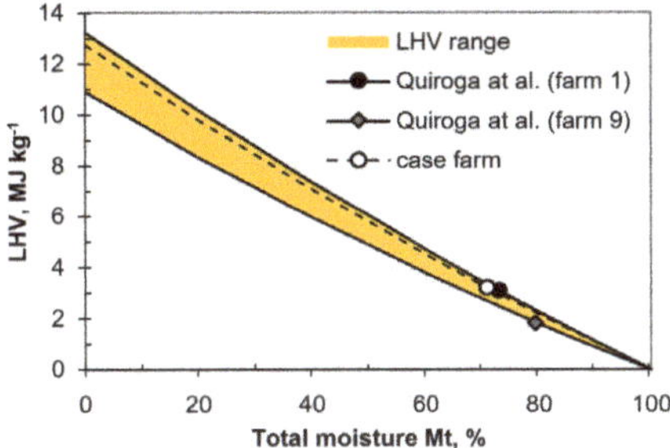

Figure 1. Lower heating value of manure from the cage system.

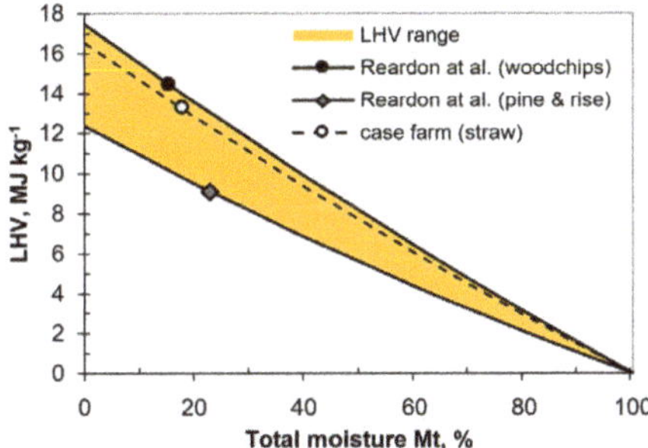

Figure 2. Lower heating value of manure from the litter system.

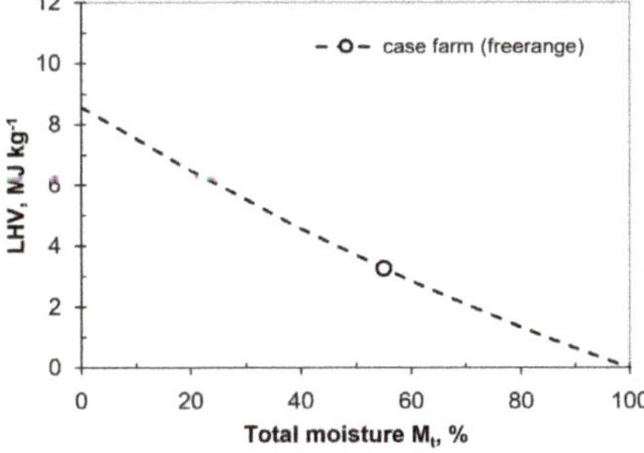

Figure 3. Lower heating value of manure from the free-range system.

2.2.2. Energy Potential Assessment

The lower heating value derived in this study was subsequently applied in the procedure of assessing energy potential of chicken manure in Poland. On the basis of the determined quantitative potential of chicken manure and results of the study on lower heating value of manure, it was possible to derive the value of the energy potential for the particular provinces in Poland. The potential derived in further part of the paper is limited to two levels:

- theoretical level resulting from the overall maximum amount of chicken manure which can be considered theoretically available for energy generation within fundamental bio-physical limits,
- technical level which takes into account energy losses due to collection, transport, storage, pre-treatment (for example drying) and energy conversion from the fuel to the useful form of energy (heat or/and electricity).

The theoretical potential was calculated on the basis of Equation (3) for three poultry systems (cage, litter and free-range):

$$E_{th,i} = B_i \cdot LHV_{ar,i} \tag{3}$$

The overall theoretical potential was derived as a sum of the potentials derived for the particular poultry systems, in accordance with Equation (4):

$$E_{th} = \sum_i E_{th,i} \tag{4}$$

The technical potential determined in the next part of the study was subject to limitations resulting from the availability of chicken manure. It was assumed that not all the manure generated in chicken farms was available for energy processing, taking into account both other possible treatment routs as well as energy losses due to collection, transport and storage. The second assumption limiting the amount of available manure is exclusion the free-range system manure volume from further calculations due to its negligible small amount and very large degree of terrain dispersion of free-range chicken farm.

Quantity of manure available for further energetic use was calculated as:

$$B_a = \sum_i B_{a,i} = \sum_i B_i \cdot a_f \tag{5}$$

The availability factor a_f was assumed to have a value of 0.8 due to the fact that part of the available manure may be utilized in other then energy generation routs: as a soil fertiliser in crop production in large-scale agriculture, in commercial mushroom cultivation and as a granulate fertiliser on sale for commercial and non-commercial gardening. The volume of manure derived from Equation (5) was the basis for further technical potential assessment. The efficiency of the conversion from chemical energy of a raw manure (representing technical potential) into useful form of energy depends on the particular transition path. The energy generation paths adopted for analysis have been shown in Table 4.

For each of the scenarios adopted, the technical potential was then determined on the basis of assumptions concerning the efficiency of conversion of stages in a given path.

Technical potential in the particular j-th path of energy conversion was calculated as a sum of values derived for manure from cage and litter breeding systems, according to Equation (6)

$$E_{tech}^{PATHj} = E_{tech,cage}^{PATH\,j} + E_{tech,litter}^{PATH\,j} \tag{6}$$

Table 4. Analyzed chicken manure energy generation routs (scenarios).

No.	Generation Path (Scenario)	Process Stages
1	PATH 1-Anaerobic digestion and cogeneration	I. Anaerobic (co)digestion II. Heat and electricty generation in IC engine
2	PATH 2-Gasification and cogeneration	I. Drying II. Gasification III. Heat and electricity generation in ORC unit
3	PATH 3-Combustion and cogeneration	I. Drying II. Fluidized-bed boiler combustion FBC III. Heat and electricity generation in ORC unit
4	PATH 4-Combustion	I. Drying II. Fluidized-bed boiler combustion FBC

Energy potential for the conversion route based on poultry manure anaerobic digestion (PATH 1) was calculated for the particular i-th breeding systems (cage or litter) as:

$$E_{tech,i}^{\text{PATH 1}} = B_{a,i} \cdot VS_i \cdot b_i \cdot LHV_{BIO,i} \cdot \eta_{th,IC} + B_{a,i} \cdot VS_i \cdot b_i \cdot LHV_{BIO,i} \cdot \eta_{el,IC} - E_{d,i} \tag{7}$$

Equation (8) was the basis for deriving energy potential available from the gasification and ORC cogeneration technology path (PATH 2):

$$E_{tech,i}^{\text{PATH 2}} = B_{dried,i} \cdot s_i \cdot LHV_{s,i} \cdot \eta_{th,ORC} + B_{dried,i} \cdot s_i \cdot LHV_{s,i} \cdot \eta_{el,ORC} - E_{d,i} \tag{8}$$

The technical potential of chicken manure in case of combustion in fluidized-bed boiler combined with ORC energy generation carried out according to PATH 3 was calculated in the following manner:

$$E_{tech,i}^{\text{PATH 3}} = B_{dried,i} \cdot LHV_{dried,i} \cdot \eta_{th,ORC} + B_{dried,i} \cdot LHV_{dried,i} \cdot \eta_{el,ORC} - E_{d,i} \tag{9}$$

The last conversion rout taken into account in the analysis is direct heat production in FCB boiler (PATH 4):

$$E_{tech,i}^{\text{PATH 4}} = B_{dried,i} \cdot LHV_{dried,i} \cdot \eta_{th} - E_{d,i} \tag{10}$$

In case of energy conversion paths that require drying the manure before further processing (PATH 2, PATH 3 and PATH 4), the amount of energy E_d used for water evaporation from manure depends on dryer efficiency resulting from specific heat and electricity demands of drying process:

$$E_{d,i} = B_{w,i} \cdot q + B_{w,i} \cdot e \tag{11}$$

q: specific heat demand for drying, kWh·kg^{-1} of evaporated water
e: specific electricity demand for drying, kWh^{-1} of evaporated water

The amount of water B_w evaporated from available raw manure B_a depends on moisture content in manure before drying M_{tw} and required moisture content after drying process $M_{tw,ad}$, according to Equation (12):

$$B_{w,i} = M_{tw} \cdot B_{a,i} - M_{tw,ad} \cdot B_{dried,i} \tag{12}$$

where the volume of the material after drying B_{dried} was determined from the mass balance of the drying process.

Set of the input data used for further assessment of technical potential can be found in Table 5.

Table 5. Input data for technical potential calculation according to assumed conversion paths.

Parameter	Unit	Value	
		Cage Manure	**Litter Manure**
	PATH 1		
VS	% by weight	84.4	88.0
b	$m^3 \cdot Mg^{-1}$ of VS	300	270
LHV_{BIO}	$MJ \cdot m^{-3}$	20	20
$\eta_{th,IC}$	-	0.50	0.50
$\eta_{el,IC}$	-	0.38	0.38
	PATH 2		
M_{tw}	% by weight	70.9	18.1
$M_{tw,ad}$	% by weight	15.0	15.0
q	$kWh \cdot kg^{-1}$ of evaporated water	0.90	0.90
e	$kWh \cdot kg^{-1}$ of evaporated water	0.08	0.08
s	$m^3 \cdot kg^{-1}$ of dried mass input	2.5	2.5
LHV_s	$MJ \cdot m^{-3}$	3.5	4.5
$\eta_{th,ORC}$	-	0.7	0.7
$\eta_{el,ORC}$	-	0.18	0.18
	PATH 3		
M_{tw}	% by weight	70.9	18.1
$M_{tw,ad}$	% by weight	35	18.1
q	$kWh \cdot kg^{-1}$ of evaporated water	0.90	-
e	$kWh \cdot kg^{-1}$ of evaporated water	0.08	-
LHV_{dried} *	$MJ \cdot Mg^{-1}$	3201	7728
$\eta_{th,ORC}$	-	0.7	0.7
$\eta_{el,ORC}$	-	0.18	0.18
	PATH 4		
M_{tw}	% by weight	70.9	18.1
$M_{tw,ad}$	% by weight	15.0	18.1
q	$kWh \cdot kg^{-1}$ of evaporated water	0.90	-
e	$kWh \cdot kg^{-1}$ of evaporated water	0.08	-
LHV_{dried} *	$MJ \cdot Mg^{-1}$	3201	7728
η_{th}	-	0.85	0.87

* derived on the basis of Equation (2).

3. Results and Discussion

3.1. Quantitative Potential

In accordance with the adopted assumptions, the annual production of chicken manure from different rearing system was determined and presented in Table 6.

The annual laying hens manure production equals to 2165 thousand Mg while volume of broiler manure is around 2330 thousand Mg. This corresponds to the total annual chicken manure production of 4495 thousand Mg. Considering the weight of chicken manure depending on the rearing system, the amount of manure from the cage system is smaller than the one from litter, and is around 1831 thousand Mg compared to 2602 thousand Mg from litter (laying hens and broilers).

The results of quantitative chicken manure potential assessment analysis have been linked with geographic data with use of a GIS based method implemented in QGIS software [34]. The map generated with this free and open source Geographic Information System is presented in Figure 4.

Table 6. Annual chicken (laying hens and broilers) manure production, in Mg.

Province	Laying Hens			Broilers	Overall Chickens
	Cage	Litter	Free-Range	Litter	
Dolnośląskie (I)	78,280	19,125	2726	65,233	165,364
Kujawsko-Pomorskie (II)	67,717	10,472	2325	185,399	265,913
Lubelskie (III)	65,549	16,262	4043	84,192	170,046
Lubuskie (IV)	48,031	2202	434	65,233	115,900
ódzkie (V)	82,625	11,774	5187	177,495	277,081
Małopolskie (VI)	84,596	20,034	1983	46,595	153,209
Mazowieckie (VII)	379,381	38,905	13,474	514,872	946,632
Opolskie (VIII)	13,563	21,937	1522	60,572	97,593
Podkarpackie (IX)	68,215	16,968	1309	62,902	149,395
Podlaskie (X)	24,761	31,868	691	212,007	269,326
Pomorskie (XI)	59,486	8385	2232	80,072	150,176
Śląskie (XII)	64,043	31,633	791	116,241	212,708
Świętokrzyskie (XIII)	43,703	4003	8154	79,211	135,072
Warmińsko-Mazurskie (XIV)	36,693	960	1168	68,190	107,012
Wielkopolskie (XV)	673,472	25,603	11,644	324,019	1,034,737
Zachodniopomorskie (XVI)	40,792	12,439	3854	187,389	244,474
Poland–overall	1,830,908	272,570	61,538	2,329,623	4,494,639

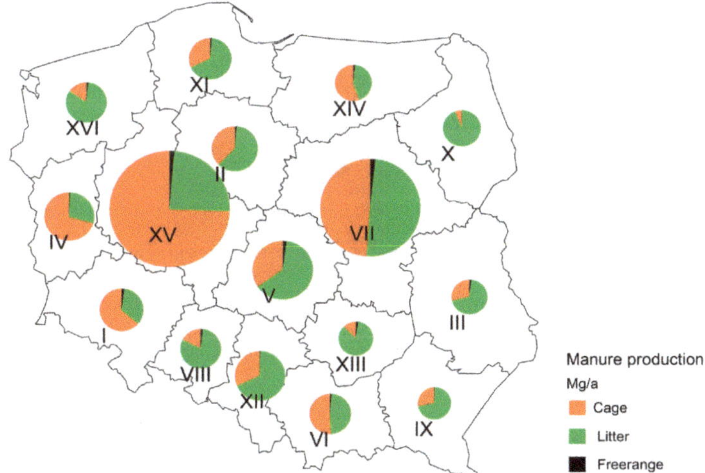

Figure 4. Annual chicken manure potential from various rearing systems in provinces of Poland.

3.2. Theoretical Qualitive Potential

The value of the theoretical energy potential for the particular provinces in Poland was derived with use of presented methodology and on the basis of input data of the quantitative potential of chicken manure in Poland (Section 3.1) and results of proximate, ultimate and heating value analyses (Section 2.2.1). The results of the calculations are summarized in Table 7. In addition, the technical potential has been presented in Figure 5 with use of QGIS software (3.4, Free Software Foundation, Boston, MA, USA).

Table 7. Annual theoretical energy potential of chicken manure in Poland, E_{th} in TJ.

Province	Cage	Litter	Free-Range	Total
Dolnośląskie (I)	250.6	1112.6	8.9	1372.1
Kujawsko-Pomorskie (II)	216.8	2583.3	7.6	2807.7
Lubelskie (III)	209.8	1324.9	13.2	1547.9
Lubuskie (IV)	153.7	889.4	1.4	1044.6
ódzkie (V)	264.5	2496.3	16.9	2777.7
Małopolskie (VI)	270.8	878.8	6.5	1156.0
Mazowieckie (VII)	1214.4	7303.8	44.0	8562.1
Opolskie (VIII)	43.4	1088.2	5.0	1136.6
Podkarpackie (IX)	218.4	1053.4	4.3	1276.0
Podlaskie (X)	79.3	3216.5	2.3	3298.0
Pomorskie (XI)	190.4	1166.7	7.3	1364.4
Śląskie (XII)	205.0	1950.3	2.6	2157.9
Świętokrzyskie (XIII)	139.9	1097.5	26.6	1264.0
Warmińsko-Mazurskie (XIV)	117.5	912.0	3.8	1033.3
Wielkopolskie (XV)	2155.8	4611.2	38.0	6804.9
Zachodniopomorskie (XVI)	130.6	2635.5	12.6	2778.7
Poland–overall	5860.7	34,320.3	200.7	40,381.8

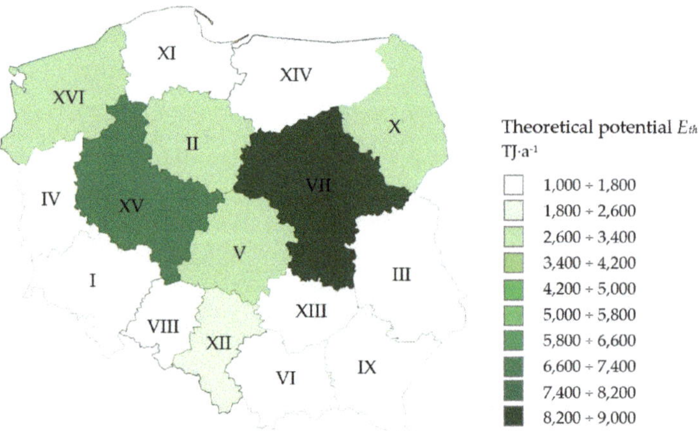

Figure 5. Energy potential of chicken manure in provinces in Poland derived on the theoretical level.

Total energy potential of chicken manure assessed for Poland equals 40,382 TJ·a^{-1}, including three analyzed rearing systems. Manure from the cage system has biggest potential: more then 34,200 TJ·a^{-1} which is close to 85% of overall chicken manure potential. Cage system provides wastes for less than 15% of total potential, whereas free-range rearing system generates manure has negligible share: hardly 0.5%. The highest chicken manure energy resources have been determined in Mazowieckie (VII) and Wielkopolskie (XV) provinces. They constitute almost 40% of the country's potential.

3.3. Thechnical Qualitive Potential

According to common classification of the energy potential, the next step in assessing the potential should be connected with the technical task of determining the efficiency of converting chemical energy of the fuel to useful forms of energy [35]. Within the research, the technical energy potential of two types of chicken manure (from cage and litter systems) has been determined according to four chosen conversion scenarios. It has been assumed that the energy generated in each scenario is in form of heat or heat and electricity in combined manner. Results of the calculation are shown in Table 8.

Table 8. Annual technical energy potential of chicken manure in Poland overall, E_{tech} in TJ.

| Generation | Cage System | | Litter System | | Total | | Overall |
Scenario	Heat	Electricity	Heat	Electricity	Heat	Electricity	Potential
PATH 1	1079	820	4051	3899	5130	3899	9029
PATH 2	1632	651	15,682	4051	17,315	4702	22,017
PATH 3	2339	796	19,220	4942	21,559	5738	27,297
PATH 4	3099	0	23,888	0	26,987	0	26,987

On the basis of the results presented in Table 8 it can be claimed that the biggest technical energy potential of chicken manure was achieved via the PATH 3 and PATH 4 conversion routes, where after drying, the manure was utilized in FBC combustor. Thereafter output heat is supplying ORC unit for electricity and useful heat generation (PATH 3) or is being directly transformed into useful heat flux (PATH 4). Relatively low potential was found for anaerobic digestion scenario (PATH 1) where overall energy production was from around 33% of the potential gained in PATH 4 to 41% of the PATH 2 potential. Significantly lower production of energy from poultry manure by anaerobic digestion compared to gasification process was noticed by Chang [36] who reported 44.5% higher energy production in syngas than that in biogas.

An additional summary of the results achieved in the research is presented in Figure 6. It can be observed that technical potential of the chicken manure is much smaller than technical one, regardless of the analyzed conversion path. It is mainly due to availability of manure assumed for the calculation (availability factor $a_f = 0.8$) as well as efficiency of energy transformation through the particular stages in the assumed scenarios.

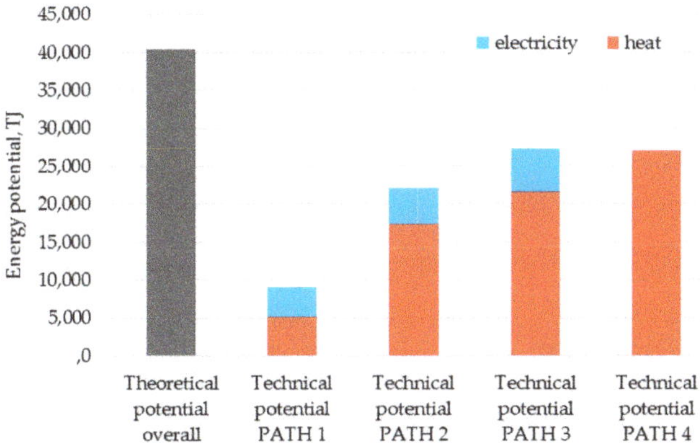

Figure 6. Energy potential of chicken manure determined for Poland according to different technical scenarios.

As shown in Figure 7, for theoretical and technical level of determined energy potential, the share of manure from the litter system is significantly high (between 79% and 89%) in comparison to the quantitative potential of raw manure (near 59%). Despite the fact that the amount of manure from the cage systems is not much smaller than the amount of manure gained from litter systems (41.3%:58.7%), the litter manure has a much greater potential for generation of useful energy. When comparing the individual paths, the highest cage share was recorded for anaerobic digestion route (PATH 1). Both the gasification pathway (PATH 2) and the combustion pathways (PATH 3 and PATH 4) are quite similar for both analyzed types of manure.

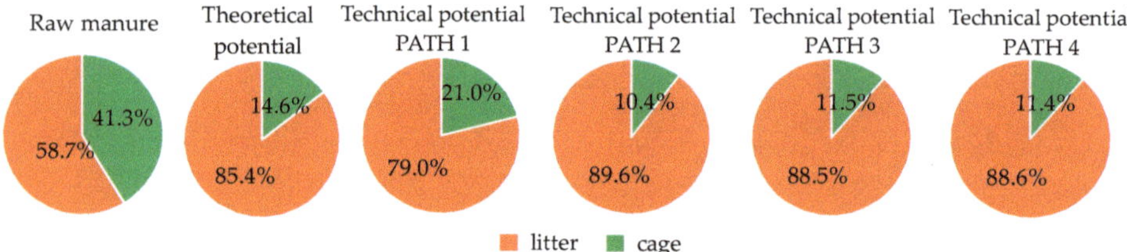

Figure 7. The contribution of particular types of manure in the determined energy potential.

3.4. Technical and Environmental Issues of Proposed Energy Generation Routes

The use of technologies adopted for estimating the technical potential of chicken manure in particular energy conversion paths requires additional discussion on the technical and environmental aspects of their application.

The chicken manure is a special type of substrate for biogas plants, due to the high content of organic substances, which through the anaerobic digestion generates biogas and biofertiliser as a by-product. The use of poultry manure as a substrate in agricultural biogas plants causes operational problems, mainly due to high concentrations of ammonium nitrogen and unfavourable ratio of organic carbon to nitrogen [37]. Therefore, proper methane fermentation of poultry manure requires balancing of C/N ratio, e.g., by introducing an appropriate amount of C-rich co-substrate. Dilution the chicken manure with water is a basic option to eliminate the effect of NH_3-N on bio-digestion inhibition. Unfortunately, the dilution increases volume of waste and declines feasibility of the method [6]. Another manners to overcome the ammonia inhibition are addition of phosphorite ore to the raw manure or implementation of air stripping technology [37]. The utility of the anaerobic digestion is also limited by the presence of hydrogen sulfide (H_2S) in biogas. H_2S has highly corrosive properties, therefore effective desulphurization of biogas before further thermal processing is required [36].

Thermochemical conversion of the poultry waste through the gasification generates syngas with calorific values significantly lower compared to the biogas, however the energy yields from gasification is much higher than from digestion. The quality of syngas and residues is mainly influenced by the type of gasification agent (air, steam or carbon dioxide), reactor configuration and operating parameters, mainly temperature and air equivalence ratio [38]. Among commonly used systems, usually fixed and fluidized bed reactors are used for chicken manure gasification purposes. The updraft fixed bed systems are preferable for fuels such as chicken manure due to its tolerance on high moisture and ash content as well as low sensitivity to fuel size and quality [38]. Nevertheless, this technology has several disadvantages, which should be considered prior to the possible application. These disadvantages include content of undesirable by-products such as tar, dust, alkalis, heavy metals, H_2S and HCl, which should be reformed, in particular, when the syngas is considered as a fuel to supply the gas turbines and IC engines. The essential difficulty in the manure gasification is the high alkali content in the ash leading to its agglomeration in the bed [39]. Attempts to prevent the phenomenon of agglomeration, by lowering the temperature in the reactor (below the ash melting point), decrease conversion efficiency and lead to the formation of excessive amounts of char.

In case of combustion pathway, the greatest drawback is associated with the large moisture and ash content in the manure, which may cause operational problems and decrease overall efficiency. Hence, drying raw manure is necessary to reduce the mass of manure and make it more suitable for further thermal processing such as combustion or gasification. The moisture content can be reduced through the use of various evaporative dryer technologies [40], integrated within the ventilation and manure transport system on the chicken farm (e.g., manure draying tunnels MDT) or being a part of the combustion system. Drying the manure on site by utilizing the exhaust air from livestock facilities significantly cuts the costs of energy required for drying process. The ash in the manure contains high amount of alkaline elements, particularly K and Na oxides, which leads to formation of low-temperature melting compounds and eutectics causing the probability of fouling and slagging [41]. High-ash poultry manure combustion may have a negative impact on the operation of boilers by its agglomeration in the bed, formation of deposits in the freeboard zone and fouling of the boiler hot surfaces and ancillary equipment [5]. Moreover, the presence of the fuel bound-Cl element increases the additional risk of high-temperature corrosion evaluation on boiler surfaces. The corrosive damages are caused by molecular chlorine-product of oxidation of hydrogen chloride contained in hot gases, as well as alkaline chlorides, formed as a result of binding with alkaline elements (K and Na oxides). Intensive research is currently being carried out to counteract these negative phenomena [5]. Another issue worth considering in case of chicken manure combustion is NO_x formation as the major air pollutant. Application of combustion technologies based on the fluidized bed combustion (FBC) can be solution to solve this disadvantage [31]. NO_x reduction in this technology is related to lowering the combustion temperature, obtaining uniform temperature distribution in the chamber and appropriate fuel to air ratio (excess air ratio). Furthermore, the high sulphur and volatile matter content in the manure forces a suitable composition of the fuel-air mixture in the boiler to ensure both high efficiency and low emissions of harmful pollutants. The composition of manure fuel requires not only the use of primary reduction methods, but also applications of systems for the intensive gases cleaning, i.e., the selective catalytic reduction (SCR) for denitrification, sorption for the removal of acid gases (SO_2, HCl) and electrostatic technique for dedusting. These techniques are usually used in large plants, but can also be applied in small combustion/cogeneration units to meet emission standards [42]. Under Polish conditions, application of any optimized manure combustion technology is an interesting alternative to coal-based heat and electricity generation, in particular on poultry farms where heat is generated with use of fossil fuels. Compared to combustion of coal, it can reduce significantly the emissions of flue gas components into the atmosphere, as shown by [5].

4. Conclusions

The research reported in the paper is dedicated to assessing the theoretical and technical energy potential of chicken manure in the particular provinces of Poland. The conducted analyses resulted in deriving the value of this potential on the basis of collected data of annual chicken manure generation as well as its physicochemical properties. The volume of the theoretical potential is promising in the context of its utilization as a source of chemical energy of the fuel useful for producing heat and electricity directly on sites where chickens are reared. Conversion of chicken waste feedstock into useful form of energy has a great potential for improving the environmental sustainability of poultry production. Taking into account the fact that the energy supplying chicken farm in Poland comes mainly from coal, the use of the manure on site for heat or/and electricity generation is all the more valuable.

The volume of chicken manure in Poland is estimated at 4.49 million tons, of which the cage system accounts for 40.7%, while the litter one accounts for 57.9%. Chicken manure from the free-range system is responsible for only 1.4% of poultry production and can be neglected regarding its energy potential. The majority of chicken manure is generated in Wielkopolskie and Mazowieckie provinces, which account for 44% of the total production of this material in Poland. The volume of chicken manure does not translate directly into the potential in terms of various physicochemical properties of manure in the analysis poultry systems. The total volume of theoretical energy potential is equal to 40.38 PJ per year.

It should be emphasized that energy potential of chicken manure derived at theoretical level disregards the loss of energy conversion in facilities used for manure conditioning (drying) as well as energy losses during production of useful energy (e.g., heat). The relevant analysis of determination of technical energy potential of chicken manure in Poland was conducted for four different energy conversion paths. Determined technical potential occurred significantly smaller then theoretical. Its value ranged from 9.01 PJ per year to 27.3 PJ per year. The bigger energy degradation was found for

Energies **2019**, *12*, 1244

heat and electricity production via anaerobic digestion path, while fluidized bed combustion occurred the most efficient scenario.

In addition, on the basis of the analysis carried out within the presented research, it should be concluded that the choice of an appropriate pathway for energetic conversion of manure use should be adapted to its availability and physicochemical parameters.

Author Contributions: Conceptualization, M.T. and R.J.; methodology, M.T. and R.J.; software, M.T. and A.K.-W.; validation, A.K.-W.; formal analysis, R.J.; investigation, M.T. and R.J.; resources, M.T., R.J. and P.N.; data curation, P.N.; writing—original draft preparation, M.T. and R.J.; writing—review and editing, M.T.; visualization, M.T.; supervision, M.T. and R.J.

Funding: This research received no external funding.

Acknowledgments: The authors are grateful for the support of the company Ferma Drobiu Trzy Koguty Sp. z o.o. which provided chicken manure for the research.

Conflicts of Interest: The authors declare no conflict of interest.

Abbreviations

a_f	availability factor
A	ash content, % by weight
b	biogas yield, $m^3 \cdot Mg^{-1}$ of VS
B	weight of chicken manure, Mg
C	carbon content, % by weight
e	specific electricity demand for drying, $kWh \cdot kg^{-1}$ of evaporated water
E	energy potential, TJ
HHV	higher heating value, $MJ \cdot Mg^{-1}$ ($MJ \cdot m^{-3}$)
H	hydrogen content, % by weight
LHV	lower heating value, $MJ \cdot Mg^{-1}$ ($MJ \cdot m^{-3}$)
M	moisture content, % by weight
N	nitrogen content, % by weight
O	oxygen content, % by weight
q	specific heat demand for drying, $kWh \cdot kg^{-1}$ of evaporated water
s	syngas yield, $m^3 \cdot kg^{-1}$ of dried mass input
S	sulphur content, % by weight
VM	volatile matter, % by weight
VS	organic matter, % by weight
Subscripts:	
a	available
ar	as received basis
bio	biogas
d	dry basis
daf	dry and ash free basis
el	electrical
i	type of rearing system
j	number of the energy conversion path (scenario)
t	total
$tech$	technical
th	theoretical, thermal
w	wet basis
η	efficiency
IC	internal combustion
FBC	fluidized bed combustion
ORC	organic Rankine cycle

References

1. Saidura, R.; Abdelaziza, E.A.; Demirbasb, A.; Hossaina, M.S.; Mekhilefc, S. A review on biomass as a fuel for boilers. *Renew. Sustain. Energy Rev.* **2011**, *15*, 2262–2289. [CrossRef]
2. Qian, X.; Lee, S.; Soto, A.-M.; Chen, G. Regression Model to Predict the Higher Heating Value of Poultry Waste from Proximate Analysis. *Resources* **2018**, *7*, 39. [CrossRef]
3. Dalólio, F.S.; Nogueira da Silva, J.; Carneiro de Oliveira, A.C.; Ferreira Tinôco, I.F.; Barbosa, R.C.; Resende, M.O.; Teixeira Albino, L.F.; Teixeira Coelho, S. Poultry Litter as Biomass Energy: A Review and Future Perspectives. *Renew. Sustain. Energy Rev.* **2017**, *76*, 941–949. [CrossRef]
4. Wieremiej, W. Usefulness of Poultry Wastes in Fertilization of Maize (*Zea mays* L.) and Their Influence on Selected Soil Properties (In Polish). Ph.D. Thesis, Siedlce University of Natural Sciences and Humanities, Siedlce, Poland, 30 March 2017.
5. Billen, P.; Costa, J.; Van der Aa, L.; Van Caneghem, J.; Vandecasteele, C. Electricity from poultry manure: A cleaner alternative to direct land application. *J. Clean. Prod.* **2015**, *96*, 467–475. [CrossRef]
6. Kelleher, B.P.; Leahy, J.J.; Henihan, A.M.; O'Dwyer, T.F.; Sutton, D.; Leahy, M.J. Advances in poultry litter disposal technology: A review. *Bioresour. Technol.* **2002**, *83*, 27–36. [CrossRef]
7. Florin, N.H.; Maddocks, A.R.; Wood, S.; Harris, A.T. High-temperature thermal destruction of poultry derived wastes for energy recovery in Australia. *Waste Manag.* **2019**, *24*, 1399–1408. [CrossRef] [PubMed]
8. Cantrell, K.B.; Hunt, P.G.; Uchimiya, M.; Novak, J.M.; Ro, K.S. Impact of pyrolysis temperature and manure source on physicochemical characteristics of biochar. *Bioresour. Technol.* **2012**, *107*, 419–428. [CrossRef]
9. Tańczuk, M.; Junga, R.; Werle, S.; Chabiński, M.; Ziółkowski, . Experimental analysis of the fixed bed gasification process of the mixtures of the chicken manure with biomass. *Renew. Energy Rev.* **2019**, *136*, 1055–1063. [CrossRef]
10. Reardon, J.P.; Lilley, A.; Brown, K.; Beard, K.; Wimberly, J.; Avens, J. *Demonstration of a Small Modular Biopower System Using Poultry Litter*; DOE SBIR Phase-I Final Report; Community Power Corporation: Englewood, CO, USA, 2001. Available online: https://www.osti.gov/servlets/purl/794292/ (accessed on 12 December 2018).
11. European Egg Processors Association. EU Statistics. Available online: http://www.eepa.info/Statistics.aspx (accessed on 12 November 2018).
12. Food and Agriculture Organization of the United Nations. FAOSTAT. Available online: http://www.fao.org/faostat/en/#data/QA (accessed on 12 November 2018).
13. International Energy Agency. *Energy Policies of IEA Countries, Poland 2016 Review*; IEA Publications: Paris, France, 2017; ISBN 978-92-64-27230-9.
14. Tańczuk, M.; Radziewicz, W.; Olszewski, E.; Skorek, J. Projected configuration of a coal-fired district heating source on the basis of comparative technical-economical optimization analysis. In Proceedings of the International Conference on Energy, Environment and Material Systems (EEMS), E3S Web of Conferences, Polanica Zdrój, Poland, 13–15 September 2017; Volume 19. [CrossRef]
15. Tańczuk, M.; Masiukiewicz, M.; Anweiler, S.; Junga, R. Technical Aspects and Energy Effects of Waste Heat Recovery from District Heating Boiler Slag. *Energies* **2018**, *11*, 796. [CrossRef]
16. Batidzirai, B.; Smeets, E.M.W.; Faaij, A.P.C. Harmonising bioenergy resource potentials: Methodological lessons from review of state of the art bioenergy potential assessments. *Renew. Sustain. Energy Rev.* **2012**, *16*, 6598–6630. [CrossRef]
17. Gonzalez-Salazara, M.A.; Morini, M.; Pinelli, M.; Spina, P.R.; Venturini, M.; Finkenrath, M.; Poganietz, W.R. Methodology for estimating biomass energy potential and its application to Colombia. *Appl. Energy* **2014**, *136*, 781–796. [CrossRef]
18. Lourinho, G.; Brito, P. Assessment of biomass energy potential in a region of Portugal (Alto Alentejo). *Energy* **2015**, *81*, 189–201. [CrossRef]
19. Nadel, S.; Shipley, A.M.; Elliott, R.N. *The Technical, Economic and Achievable Potential for Energy Efficiency in the United States: A Meta-Analysis of Recent Studies*; American Council for Energy-Efficient Economy (ACEEE): Washington, DC, USA, 2004.
20. Karaj, S.; Rehl, T.; Leis, H.; Müller, J. Analysis of biomass residues potential for electrical energy generation in Albania. *Renew. Sustain. Energy Rev.* **2010**, *14*, 493–499. [CrossRef]
21. Tańczuk, M.; Ulbrich, R. Assessment of energetic potential of biomass. *Proc. ECOpole* **2009**, *3*, 23–26.

22. Main Veterinary Inspectorate. Registers and Records Kept in Main Veterinary Inspectorate (In Polish). Available online: https://www.wetgiw.gov.pl/handel-eksport-import/rejestr-podmiotow-prowadzacych-dzialalnosc-nadzorowana (accessed on 12 November 2018).

23. Statistics Poland. Farm Animals in 2017. Available online: http://stat.gov.pl/en (accessed on 12 November 2018).

24. Dobrzański, Z. The Relationship between Modern Poultry Production Systems and the Protection of Natural and Productive Environment (In Polish). First Agricultural Portal. 2002. Available online: http://www.ppr.pl/artykul-ppr-2924.php?_resourcePK=2924 (accessed on 7 November 2018).

25. Mayoral, M.C.; Izquierdo, M.T.; Andrés, J.M.; Rubio, B. Different approaches to proximate analysis by thermogravimetry analysis. *Thermochim. Acta* **2001**, *370*, 91–97. [CrossRef]

26. Quiroga, G.; Castrillón, L.; Fernández-Nava, Y.; Marañón, E. Physico-chemical analysis and calorific values of poultry manure. *Waste Manag.* **2010**, *30*, 880–884. [CrossRef]

27. Staroń, P.; Kowalski, Z.; Staroń, A.; Banach, M. Thermal conversion of granules from feathers, meat and bone meal and poultry litter to ash with fertilizing properties. *Agric. Food Sci.* **2017**, *26*, 173–180. [CrossRef]

28. Lynch, D.; Henihan, A.M.; Bowen, B.; Lynch, D.; McDonnell, K.; Kwapinski, W.; Leahy, J.J. Utilisation of poultry litter as an energy feedstock. *Biomass BioEnergy* **2013**, *49*, 197–204. [CrossRef]

29. Trziszka, T. *Jajczarstwo: Nauka, Technologia, Praktyka*; Wydawnictwo Akademii Rolniczej we Wrocławiu: Wrocław, Poland, 2000.

30. Polesek-Karczewska, S.; Turzyński, T.; Kardaś, D.; Heda, . Front velocity in the combustion of blends of poultry litter with straw. *Fuel Process. Technol.* **2018**, *176*, 307–315. [CrossRef]

31. Toptas, A.; Yildrim, Y.; Duman, G.; Yanik, Y. Combustion behavior of different kinds of torrefied biomass and their blends with lignite. *Bioresour. Technol.* **2015**, *177*, 328–336. [CrossRef]

32. Yurdakul, S. Determination of co-combustion properties and thermal kinetics of poultry litter/coal blends using thermogravimetry. *Renew. Energy* **2016**, *89*, 215–223. [CrossRef]

33. Whitely, N.; Ozao, R.; Artiaga, R.; Cao, Y.; Pan, W.P. Multi-utilization of chicken litter as biomass source. Part I. Combustion. *Energy Fuel* **2006**, *20*, 2660–2665. [CrossRef]

34. QGIS—A Free and Open Source Geographic Information System. Available online: https://www.qgis.org/ (accessed on 23 January 2019).

35. Oliveira, M.O.; Somariva, R.; Ando Junior, O.H.; Neto, J.M.; Bretas, A.S.; Perrone, O.E.; Reversat, J.H. Biomass Electricity Generation Using Industry Poultry Waste. In Proceedings of the International Conference on Renewable Energies and Power Quality (ICREPQ'12), Santiago de Compostela, Spain, 28–30 March 2012; pp. 1650–1654.

36. Chang, F.H. Energy and sustainability comparisons of anaerobic digestion and thermal technologies for processing animal waste. In Proceedings of the 2004 ASAE Annual Meeting, Ottawa, ON, Canada, 1–4 August 2004.

37. Singh, K.; Lee, K.; Worley, J.; Risse, L.M.; Das, K.C. Anaerobic digestion of poultry litter: A review. *Appl. Eng. Agric.* **2010**, *26*, 677–688. [CrossRef]

38. Watson, J.; Zhang, Y.; Si, B.; Chen, W.-T.; de Souza, R. Gasification of biowaste: A critical review and outlooks. *Renew. Sustain. Energy Rev.* **2018**, *83*, 1–17. [CrossRef]

39. Taupe, N.; Lynch, D.; Wnetrzak, R.; Kwapinska, M.; Kwapinski, W.; Leahy, J. Updraft gasification of poultry litter at farm-scale—A case study. *Waste Manag.* **2016**, *50*, 324–333. [CrossRef]

40. Brammer, J.G.; Bridgwater, A.V. Drying technologies for an integrated gasification bio-energy plant. *Renew. Sustain. Energy Rev.* **1999**, *3*, 243–289. [CrossRef]

41. Lynch, D.; Henihant, A.M.; Kwapinski, W.; Zhang, L.; Leahy, J. Ash agglomeration and deposition during combustion of poultry litter in a bubbling fluidized-bed combustor. *Energy Fuels* **2013**, *27*, 4684–4694. [CrossRef]

42. Commission Regulation (EU) No 592/2014 of 3 June 2014 Amending Regulation (EU) No 142/2011 as Regards the Use of Animal by-Products and Derived Products as a Fuel in Combustion Plants. Available online: https:eur-lex.europa.eu/ (accessed on 12 January 2019).

Article

Techno-Economic Assessment of Solar Hydrogen Production by Means of Thermo-Chemical Cycles

Massimo Moser [1,*], Matteo Pecchi [1,2] and Thomas Fend [3]

[1]   Department of Energy System Analysis, Institute of Engineering Thermodynamics, German Aerospace Center (DLR), Pfaffenwaldring 38, 70569 Stuttgart, Germany; matteo.pecchi@natec.unibz.it

[2]   Faculty of Science and Technology, Free University of Bozen-Bolzano Universitätsplatz—Piazza Università, 5, 39100 Bozen-Bolzano, Italy

[3]   Department of Solar Chemical Engineering, Institute of Solar Research, German Aerospace Center (DLR), Linder Höhe, 51147 Köln-Porz, Germany; thomas.fend@dlr.de

*   Correspondence: massimo.moser@dlr.de; Tel.: +49-(0)711-6862-779

Received: 20 December 2018; Accepted: 22 January 2019; Published: 23 January 2019

Abstract: This paper presents the system analysis and the techno-economic assessment of selected solar hydrogen production paths based on thermochemical cycles. The analyzed solar technology is Concentrated Solar Power (CSP). Solar energy is used in order to run a two-step thermochemical cycle based on two different red-ox materials, namely nickel-ferrite and cerium dioxide (ceria). Firstly, a flexible mathematical model has been implemented to design and to operate the system. The tool is able to perform annual yield calculations based on hourly meteorological data. Secondly, a sensitivity analysis over key-design and operational techno-economic parameters has been carried out. The main outcomes are presented and critically discussed. The technical comparison of nickel-ferrite and ceria cycles showed that the integration of a large number of reactors can be optimized by considering a suitable time displacement among the activation of the single reactors working in parallel. In addition the comparison demonstrated that ceria achieves higher efficiency than nickel-ferrite (13.4% instead 6.4%), mainly because of the different kinetics. This difference leads to a lower LCOH for ceria (13.06 €/kg and 6.68 €/kg in the base case and in the best case scenario, respectively).

Keywords: solar hydrogen; concentrated solar power (CSP); two-step thermochemical cycles; simulation; techno-economic assessment

---

1. Introduction

Within the framework of the 2015 United Nations Climate Change Conference (COP 21) in Paris, it has been agreed that global warming should be limited to maximum 2 °C above pre-industrial levels. In the IEA 450 Policy Scenario, which can be considered as a consistent pathway for the fulfilment of the climate change target, the set limitation for CO_2 concentration is 450 ppm [1]. In order to reach this goal, greenhouse gas emissions in 2035 need to be reduced by 30%, with respect to 2015. Considering the continuous increase of worldwide energy consumption [2], such goal can only be reached by an acceleration in the introduction of renewable energies. However, in such an energy supply system, which is characterized by a high degree of intermittency, some form of storage is required. Among others, conversion of solar energy into chemical fuels such as hydrogen may turn out as an effective form of storage [3], since chemical storage has the key advantage of being free of self-discharge, with respect to thermal and electrical storage systems. Hydrogen can be converted into electricity by means of turbines, generators, and fuel cells, or it can be upgraded to synthetic liquid fuels [4].

1.1. Technology and Literature Review

Several processes have been considered for the production of solar hydrogen. A comprehensive overview is given in Bozoglan et al. [5]. Five different solar hydrogen techniques can be generally distinguished. One option is to use PV [6], whereas the generated power can be used to run an electrolysis plant [7]. Another option is photoelectrolysis, which consists in the splitting of water in a photoelectrochemical cell powered by solar energy. Despite substantial research effort has been carried out on this topic in the last four decades, substantial problems hinder a wide diffusion of this technology. Solar hydrogen may also be produced by microorganisms such as algae and bacteria (photobiolysis) using solar irradiation as energy source, but this technology is at an early stage of development.

Solar thermolysis has been widely assessed in the past [8]. The feasibility of such a single-step concept is hindered by the high required temperatures (above 2,500 K) as well as by the challenging separation of the produced H_2 and O_2 gases [9]. Due to these reasons, the techno-economic feasibility of such concept will be low also in the future, according to Kogan [10]. An alternative, as described by Tyner [11], is to use Concentrating Solar Power (CSP) to first generate electrical power, which can be later utilized to run an electrolysis plant. Also a combination of concentrating PV and solar power tower has been recently proposed [12]. Two-step thermochemical cycles offer two key advantages in comparison to direct solar thermolysis [13,14]: first, the gas separation is realized by the absorption of the oxygen into the reactor material, while pure hydrogen is released. Second, the required temperature is significantly lower than for thermolysis. In addition, thermodynamic analysis shows that some of those cycles can achieve—in principle—efficiencies up to 70%, if energy recovery is applied [15].

Several materials have been analyzed both theoretically and experimentally [16]. Among them are zinc, iron, tin, terbium oxides, mixed ferrite, ceria, and perovskite [17]. Cycles based on nonvolatile metals have clear advantages in comparison to sulfur- and bromine-based cycles, in terms of toxicity and corrosivity. In addition, differently from volatile metal oxide-based processes, these materials allow the construction of fixed-bed structural reactors (monoliths) with important structural strength. Among this group mixed ferrites offer the advantage of relatively low reaction temperatures and therefore they have been widely investigated. nickel-ferrite cycles belong to the more general mixed-metal ferrite cycles, whose general reactions are shown in Figure 1. [18].

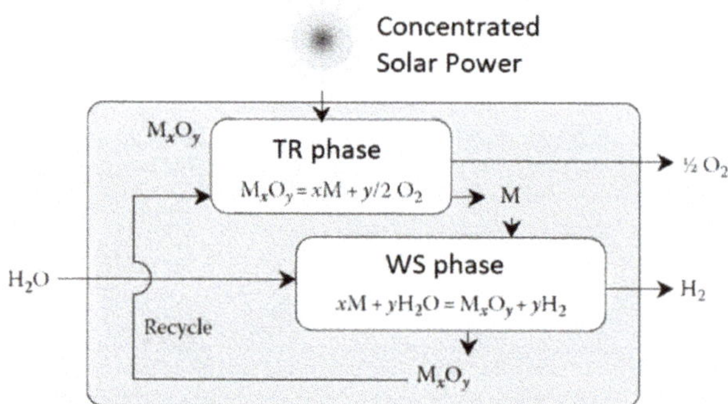

Figure 1. Basic scheme of a 2-step metal-oxide thermochemical process [18].

The first reaction is the endothermic thermal reduction step (TR), in which the metal oxide is reduced by removing part of the oxygen contained in it. The parameter δ is the non-stoichiometric factor that indicates to which extent the reaction occurs. Such non-stoichiometric factor is a function of process temperature and oxygen partial pressure [19]. In the second step—the exothermic water splitting (WS)—the MO is oxidized with water while hydrogen is released. Recently, an industrial-scale plant (750 kWh$_{th}$) which uses concentrated solar radiation to split water into hydrogen and oxygen using nickel ferrite has been realized at the Plataforma Solar de Almeria (Almeria, Spain) and is expected to produce 3 kg of hydrogen per week [20]. Within those systems, two or more reactors are used in parallel to perform the batch-processes achieving a quasi-continuous hydrogen production. The main drawbacks of this system are material sintering and a relatively low efficiency.

Within the last years, ceria has become the benchmark for non-volatile metal oxide systems. The use of ceria as an interesting material for solar hydrogen applications has risen due to its quick kinetics and its structural stability over a wide range of temperatures [21]. However, higher temperatures with respect to nickel-ferrite, are required in order to achieve satisfying values of the non-stoichiometric factor. In addition, ceria-based cycles simultaneously produce a mixture of hydrogen and carbon monoxide, with the option to further process the syngas in a Fischer-Tropsch reactor to produce liquid solar fuels [22,23]. For those reasons this work focuses on the evaluation of nickel-ferrite and ceria and compares the resulting figures with hydrogen generated by water electrolysis.

With regard to the economic evaluation of hydrogen production by means of solar thermochemical cycles, few studies exist. Graf et al. [24] presented a simplified techno-economic characterization of hybrid-sulfur cycles and a metal oxide based cycles. The calculated price for thermochemical cycles ranged from 3.5 €/kg to 12.8 €/kg in an optimistic and conservative scenario, respectively. Recently, Nicodemus compared again solar thermochemical cycles and solar PV electrolysis [25]. The focus was on the learning curves for each of the main plant components. In that paper, hydrogen production costs of two main routes were presented, i.e., via PV-electrolysis and via solar thermochemical cycles (Zn/ZnO). Depending on the assumptions regarding different policy support mechanisms, the expected LCOH for the solar thermochemical cycle were in the range between 4 €/kg and 4.5 €/kg. The paper further highlighted that the cost projections for PV and electrolysis may lead to a LCOH of 2 €/kg and 3 €/kg, calling for incentive-based policy also for thermochemical cycles in order to improve their long-term economic competitiveness.

1.2. Intention

The aim of the presented work has been the development of a flexible model for the preliminary assessment and the comparison of different two-step thermochemical cycles for solar hydrogen production, taking into account at the same time technical and economic aspects. We believe that the development of such a simplified model is relevant in order to identify key techno-economic performance parameters of such processes, and can improve the understanding over differences and similarities of different concepts and plant configurations. In addition, the produced results may be useful in order to highlight some promising paths for future research work. Indeed, the main novelty of this work consists in bridging the gap between in-depth technical studies and economic analyses. First ones typically focus on very specific technical aspects such as new materials or component design and do not include economic analysis [20–22]. In the second ones, which to the best of our knowledge assume fixed or reference plant configurations, the inter-dependence between economic input parameters and optimal plant design are neglected [24,25].

The model developed within the framework of this study allows designing an optimal number of modules per reactor for large-scale applications as a function of several technical constraints. In fact, if only one reactor would be used, a large amount of the available solar energy would be wasted. This happens since the solar field power output must be sized to fulfil the peak power demand of the reactor, but a single reactor absorbs its peak power value only for a small fraction of the cycle. A constant power requirement makes a more efficient use of the available solar power. This ideal behavior can be at least approached by running several reactors together in a module, each reactor starting its cycle with a given time displacement. The optimal number of reactors per module allows approaching a constant value for the module power consumption curve.

Moreover, the objective is the analysis of long-term cost perspectives and the comparison of solar thermochemical cycles based on CSP with alternative renewable hydrogen production routes.

In order to achieve this goal, a simplified model has been developed based on available literature and in cooperation with the DLR Institute of Solar Research.

The paper is structured as follows: after a literature review, the structure and the details of the technical model are described and discussed. A dedicated technical model has been realized for each of the plant components, i.e., solar field, single reactor, reactor groups. A distinction has been done between the materials used in the reactors. In particular, physical properties and kinetics of both nickel-ferrite and ceria have been considered based on available literature. A simplified economic model has also been implemented. Finally, a case study has been carried out. Technical and economic performances are presented and critically discussed. The technical comparison takes into account the two mentioned reactor materials. The economic analysis includes the consideration of different scenarios and takes into account learning curves for CSP. The hydrogen production cost obtained via solar thermochemical cycles are compared with those for competing systems such as electrolysis powered by photovoltaics (PV).

2. Materials and Methods

The model consists of following submodules:

- Sun power evaluation model
- Reactor model
- Kinetic models for Nickel-Ferrite and Ceria
- Plant model
- Economic model

The paper first presents the structure of the developed model. In the subsequent analysis the results of the parametric studies are shown and critically discussed. Finally, key findings and suggestions for future research activities are summarized in the conclusions and in the outlook, respectively.

2.1. Technical Model

As mentioned above, the technical model consists of four sub-models implemented in the Python programming language, which are described in the following sections. The description focuses on the overall structure of the models. Details about single components and validation of the numerical results relative to them can be found in the available literature, which has been taken as a basis for the model development. However, at the plant scale, no validation is available.

2.1.1. Sun Power Evaluation Model

The first step of the calculation evaluates the available solar power. In a CSP plant for hydrogen production (Figure 2), the incoming irradiation is reflected by the heliostat field and concentrates onto the top of a tower, where a receiver is located. Since both the current available direct normal irradiance (DNI) on the solar field and the position of the sun continuously change over the day and over the seasons, an accurate evaluation of the net usable solar power is of foremost importance.

The calculation procedure is performed according to these steps:

- Available DNI data (hourly time step for a typical meteorological year) are read from a file
- The position of the sun at each time step is calculated according to [26]. The results from this step are the solar elevation angle and the solar azimuth angle
- Given the sun position, the geographical location of the plant, the receiver capacity and the tower height, the current efficiency of the solar field η_{sf}, namely the ratio of available DNI that is directed towards the receiver, can be calculated from an efficiency matrix [27]. The interpolation among matrix points is linear. The details of the interpolation procedure of the 5-D matrix can be found in [28]. The matrix takes into account different loss mechanisms, due to the reflection of the

solar radiation towards the receiver (i.e., geometrical losses such as cosine losses, shading and blocking), and to the non-ideal behavior of the heliostats (i.e., astigmatism).

- Finally, the net available heat at the receiver is assessed as [20]:

$$\dot{Q}_{sf} = DNI \cdot \eta_{sf} \cdot A_{sf} \cdot (1 - f_{spill})$$

where Q_{sf} is further reduced by a non-dimensional spillage factor f_{spill}, as not all irradiation reflected by the heliostats hits the receiver [29].

Figure 2. CSP heliostat field and solar tower at the Plataforma Solar de Almería, Spain (DLR).

2.1.2. Single Reactor Model

The thermodynamic model for a single reactor is based on [20], which is a simplified 0-D model with energy balances on absorber and window surfaces. The main heat fluxes in and from the receiver are shown in Figure 3.

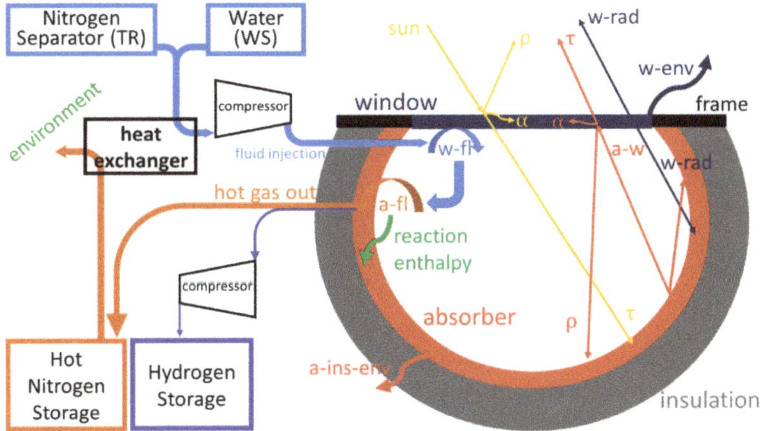

Figure 3. Thermodynamic solar reactor model, based on [20] with own adaptations.

The reactor is composed of the absorber, i.e., the metal oxide active material, and of a quartz window. In reality, the absorber is bell-shaped, however, in the model this form is approximated as a half-sphere in order to simplify the calculations. In addition, the absorber is assumed to behave as a black-body.

The available solar power (straight yellow arrow) hits the quartz window and is either reflected, absorbed or transmitted toward the absorber. The absorber reflects towards the window (straight red arrow), and also in this case the radiation is either absorbed, transmitted or reflected by the window. The absorber also loses a share of power through the insulation (curly red arrow). Also the quartz window itself loses heat by radiation and convection to the environment (straight and curly dark blue arrows). During the thermal reduction phase, nitrogen is pre-heated and injected while the absorber temperature is kept constant. The nitrogen flows through the absorber pores with an initial temperature lower than the one of the absorber, and reaches thermal equilibrium with it. The heat absorbed from the absorber to heat up the nitrogen is simply calculated knowing the gas specific heat. As nitrogen flows through the active material, this latter releases oxygen. During the water splitting phase steam is injected instead of nitrogen, while hydrogen is released, but the modelling approach is the same. The details of the calculation of each single term in the heat fluxes can be found in [20].

Nitrogen is obtained by air separation in a nitrogen generator consuming electricity, while the electricity consumption for its pumping in the system is calculated based on a fixed pressure ratio.

A heat exchanger between nitrogen and steam inlet and outlet flows fulfils the overall heat requirements (boiling and overheating). When hot nitrogen exits from the reactor, it is stored in a container with a certain thermal efficiency. This operation is not performed with water and hydrogen exiting the reactor. Whenever heat is required to heat up nitrogen or to vaporize and over heat steam, the required amount is subtracted from the value available in the storage, also in this case considering a certain efficiency to simulate the losses.

The heat balance on the absorber and on the quartz window are performed at each time step by applying a forward Euler method. The energy balance on the absorber is:

$$\frac{dT_a}{dt} = \frac{\dot{Q}_{sun,\tau} + 0.5 \cdot \dot{Q}_{w,rad} - \dot{Q}_{a-w,rad} - \dot{Q}_{a-ins,cond} - \dot{Q}_{a-fl,conv} - \dot{Q}_{h,reac}}{m_{abs} \cdot c_{p,abs}}$$

The energy balance on the window is:

$$\frac{dT_w}{dt} = \frac{\dot{Q}_{sun,\alpha} - \dot{Q}_{w,rad} + \dot{Q}_{a-w,\alpha} - \dot{Q}_{w-fl,conv} - \dot{Q}_{w-env,conv}}{m_w \cdot c_{p,w}}$$

2.1.3. Kinetic Models for Nickel-Ferrite and Ceria

The duration of each reaction phase (TR and WS) is determined by the kinetics. For nickel-ferrite a unimolecular decay law has been used as proposed by [19], based on [30]. The non-stoichiometric factor δ can be calculated linearly interpolating the results of [19] for the reaction temperature (1400 °C for the TR phase) and the oxygen partial pressure (10^{-5} bar).

From δ, the maximum value of the oxygen moles that can be extracted at equilibrium $\psi_{O_2,max}$ can be assessed, considering the mass and the molar mass of the absorber:

$$\psi_{O_2,\,max} = \frac{\delta \cdot m_a}{2 \cdot MM_a}$$

After that, the oxygen release at each time step can be assessed by integration of the kinetic differential equation:

$$\frac{d\psi_{O_2}}{dt} = -k_{reg} \cdot \psi_{O_2}$$

So it possible to compute the moles of oxygen released up to a certain instant $n_{O_2}(t)$ by integration, and to obtain the molar production rate $\dot{n}_{O_2}(t)$ as the difference of the value of $n_{O_2}(t)$ in two consecutive time steps:

$$\dot{n}_{O_2}(t) = \frac{n_{O_2}(t) - n_{O_2}(t-1)}{\delta t} = \frac{\psi_{O_2,\,max}}{\delta t}\cdot\left(e^{-k_{reg}\cdot(t-1)} - e^{-k_{reg}\cdot(t)}\right)$$

For the WS phase, a similar procedure applies, but also the steam concentration is accounted for in the differential equation:

$$\frac{d\psi_{O_2}}{dt} = -k_{ws}\cdot c_{H_2O}\cdot(\psi_{O_2,\,max} - \psi_{O_2})$$

The integration follows the same procedure as for the TR phase. The enthalpy of reaction associated to the oxygen release and to the oxygen absorption are considered fixed values and have been taken from [31], while the specific heat capacity of the ferrite have been taken from [32].

For ceria the kinetic calculation is similar, with the only difference that in this case oxygen can be released also during the transition phase between WS and TR. The value of δ is given in [33] as a function of the temperature and of the oxygen partial pressure, while the constants k_{reg} and k_{ws} has been interpolated from [34]. The enthalpy of reaction for the regeneration phase is not constant, as suggested by [33], while the enthalpy of oxygen absorption has been considered equal to the nickel-ferrite case, since no data was available in the literature.

2.1.4. Plant Model

Due to the fact that the temperature requirements for the two phases of the cycle are different—i.e., for nickel-ferrite 1400 °C and 900 °C for TR and WS, respectively—the power needed by each single reactor strongly varies over time, as shown in Figure 4 (left). The power requirement is the highest during the transition from WS to TR phase, when a heat-up takes place. During the TR phase the temperature of the absorber remains constant, while the power requirement falls to an intermediate value. The power requirement of the TR phase is not constant due to the impact of the reaction kinetics on the thermal balance. After that a cool down takes place. During this phase the power requirement falls to zero. The last phase is the WS, when the power requirement slightly increases.

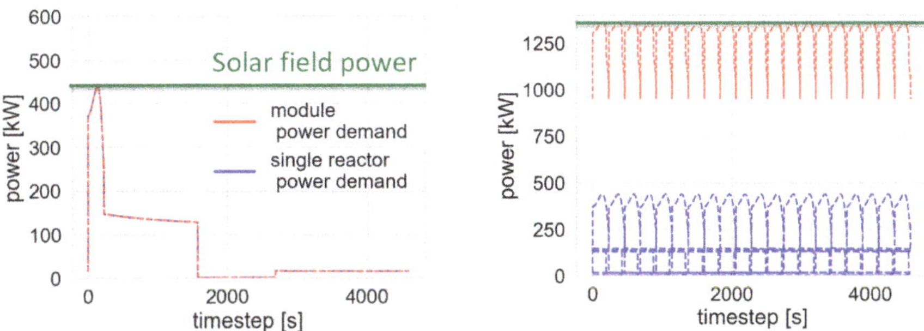

Figure 4. Heat requirement of a single nickel-ferrite reactor (**left**), and of a module (**right**) during a complete cycle.

If only one reactor is used, a large amount of the available solar energy (i.e., the difference between the peak power and the current value of the required power) would be wasted, since the solar field would need to be sized on the peak of the reactor power curve. A constant power requirement makes a more efficient use of the available solar power, (red line in Figure 4). This ideal behavior can be at least approached by running several reactors together in a module, each reactor starting is cycle with a given time displacement (Figure 4, right plot). The solar power consumption of a module is then the sum of the power consumption of its single reactors, considering the time displacement of each reactor power curve. The optimal number of reactors per module allows approaching a constant value for the module consumption curve. It is a function of the duration of the heat-up phase, which in turn is a function of the chosen absorber material. Finally, a complete plant consists of a series of modules, and in each hour of the year, the number of activated modules in the plant depends on the currently available sun power. The main results of the technical model are—for each time step—temperature value of absorber and quartz window, heat flows and solar heat requirements for a single reactor as well as for an entire module, oxygen and hydrogen production rates. In the end, the size of the main plant components is evaluated.

2.2. Economic Model

The results of the technical model are fed into the economic model in order to calculate the levelized cost of hydrogen (LCOH). The LCOH is defined as the ratio between the total annual cost and the amount of produced hydrogen, considering the energy cost for hydrogen pumping in the long-term storage:

$$LCOH = \frac{total\ annual\ cost}{m_{H_2,\ annual\ production}}$$

The total annual cost is calculated as a sum of a series of items. The most important one is the annual instalment [Mio. €/y], which is assumed to be completely covered by the loan, according to:

$$inst_{ann} = inv \cdot \frac{i \cdot (1+i)^{(t_{debt} - t_{construction})}}{(1+i)^{(t_{debt} - t_{construction}) - 1}}$$

Another cost item is the metal oxide substitution, as it is assumed that the active material is prone to deterioration and only can withstand a certain number of cycles before being substituted. Annual cost of water, technical nitrogen, nitrogen compression and a general factor for operation and maintenance and insurance are taken into account. The main assumptions used for the calculation of such cost are listed in Table 1.

Table 1. Assumption of main assumptions for the economic model (base case).

Parameter	Unit	Value
Construction time	y	2
Debt period	y	20
Plant lifetime	y	30
Interest rate	%/y	0.08
Discount rate	%/y	0.05
Absorber life	y	1
Absorber specific cost	€/kg	10
Water use for mirror cleaning	m^3/m^2	0.02
Water cost	$€/m^3$	3
N_2 pumping consumption	kWh/mol	0.001356
Pressure ratio	-	5
Adiabatic compressor efficiency	-	0.7
Cost of electricity	€/kWh	0.20
O&M cost	%inv/y	2
Insurance factor	%inv/y	0.5

The total investment *inv* is the sum of the investment for all plant components, i.e., heliostat field, thermal storage for nitrogen, hydrogen underground storage, reactor and solar tower. Some important data and the main cost assumptions for the base case are reported in Table 2. Also the hydrogen efficiency storage is reported, which considers a loss of 10% of the energy content in the gas for its pumping and extraction in the storage [35,36].

Table 2. Investment assumptions (base case).

Parameter	Unit	Value
Solar field specific investment	$€/m^2$	150
Thermal storage specific investment	$€/kWh_{th}$	30
Hydrogen storage specific investment	$€/kWh_{th}$	30
Hydrogen storage efficiency	%	90
Reactor shell cost	€/piece	5000
Specific cost of insulation	€/m	10,000
Specific cost of absorber	€/kg	10
Absorber mass	Kg	150–250
Reactor total mass	kg	700

The tower specifications (e.g., tower height as a function of receiver capacity) required for the cost calculation is based on [27]. The total reactor mass is based on internal consideration at DLR, and is supposed to be fixed for simplicity reasons. The volume available for the absorber is also fixed, since it depends on the reactor configuration based on [20]. The absorber mass depends on the density and porosity of the material, however, due to the small fraction of reactor mass in the absorber material, the total mass of the reactor is not a function of it, but it is fixed. The absorber specific cost refers to an eventual industrial scale production, and is the same for the two materials.

3. Results and Discussion

Case Study

The developed model is finally used to assess the LCOH in a case study. The selected site for the analysis is the Plataforma Solar de Almería (Almeria, Spain). Figure 5 shows a DNI distribution map of Spain. One can observe that South-East Spain is characterized by highest annual DNI values (around 2100 kWh/m^2/y). The plant is assumed to have a heliostat field with 3000 heliostats with each 40.96 m^2 surface. The nominal receiver thermal power is of 90 MW.

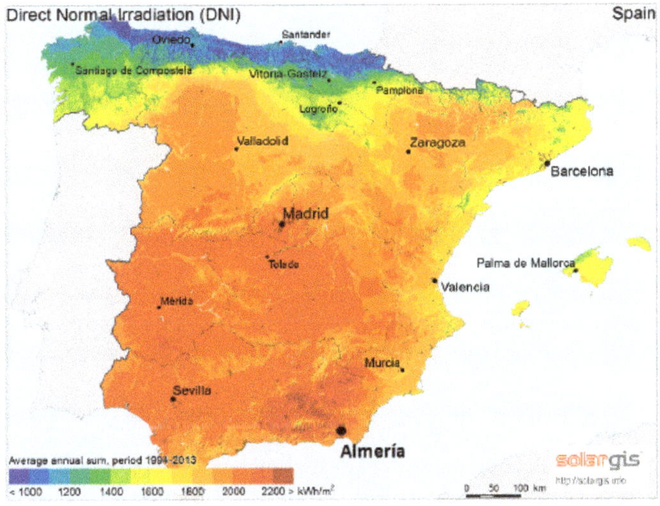

Figure 5. DNI distribution map of Spain [37].

First, key technical plant characteristics such as optimal reactor number for a module, overall energy balances and typical production profiles are shown. A comparison between Nickel-Ferrite and

Ceria is carried out. Second, the economic results for the base case are presented and a sensitivity analysis on selected parameters is carried out. Finally, a best case scenario is evaluated and its results are reported.

A view on the energy balance for a single reactor helps understanding the main loss mechanisms of such systems. A Sankey plot for a full reactor cycle is shown in Figure 6, which represents the time integral of the power fluxes during the whole cycle. The enthalpy term is related to the efficiency. The resulting reactor efficiency in the base case is 6.4% for nickel-ferrite and 13.4% for ceria. Such a difference is mainly due to the faster kinetics of ceria, which leads to a shorter cycle duration time. In both cases, the re-radiation losses through the window are very high and accounts for 59.4%–67.0% of the incoming solar energy. The geometry of the absorber and of the window has been taken as a given input and has not been optimized within the framework of this work.

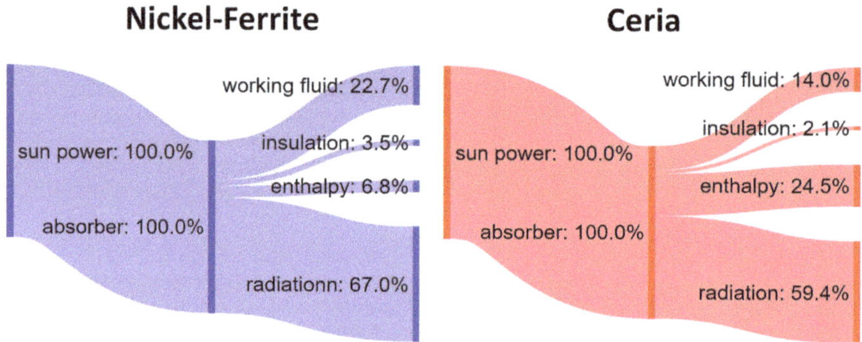

Figure 6. Sankey diagrams for nickel-ferrite (**left**) and ceria (**right**) for a full cycle.

The heat losses through the insulation have a minor impact on the system performance. A larger energy amount (from 14.0% up to 22.7%) is required for the heat-up of the working fluid entering the reactor.

As discussed above, a smart choice of the number of reactor in each module allows to dramatically reduce the non-used share of available solar power. Figure 7 shows the results of the optimization procedure. In both cases, the single reactors within one module are activated with such a time displacement that the peak power requirement of different reactors do not overlap. Then, the optimal number of modules per reactor is mainly a function of the duration of the heat-up phase. In principle, additional gains could be reached by the optimization of the heat-up profile. However, this would have been out of the scope of this preliminary work. In ceria both the temperature level of the TR phase as well as the spread between TR and WS temperature is higher than for nickel-ferrite. This leads to longer heat-up duration in ceria cycles. As a consequence, in ceria a lower number of reactors per module can be obtained than for nickel-ferrite. In the figure one can also observe that the cycle duration for nickel-ferrite is almost twice the cycle time for ceria. This is due to the different kinetics and temperature levels. In addition, the module peak power for ceria is roughly 40% higher than for nickel-ferrite.

The previous results show the behavior of single reactors and single modules under the assumption of constant available solar power. In reality, available sun power varies during the day and more than one module can be run at the same time, as shown in Figure 8.

For a given number of modules in a plant, only a certain fraction can be operated at the same time, depending of the currently available solar power. When the Sun power is equal or higher than the sum of the nominal power of a certain number of modules, such modules are heated up. Within each single module, the reactors are successively activated taking into account the time displacement between reactors. This module activation trend is visible in Figure 8, where during the morning hours the number of productive cycles is lower than the number of activated modules, since the heat-up procedure needs a certain time. Due to the fact that the cycle time in ceria is lower than in nickel-ferrite, ceria is more reactive than nickel-ferrite to variations in the available solar power.

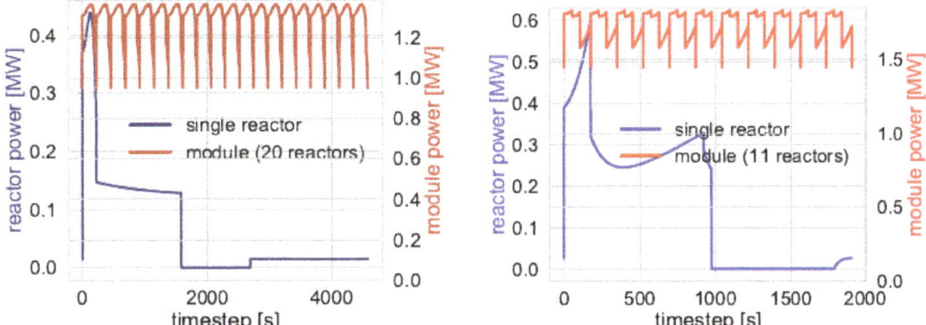

Figure 7. Solar power requirement for a single reactor and for a module (**left**: nickel-ferrite, **right**: ceria).

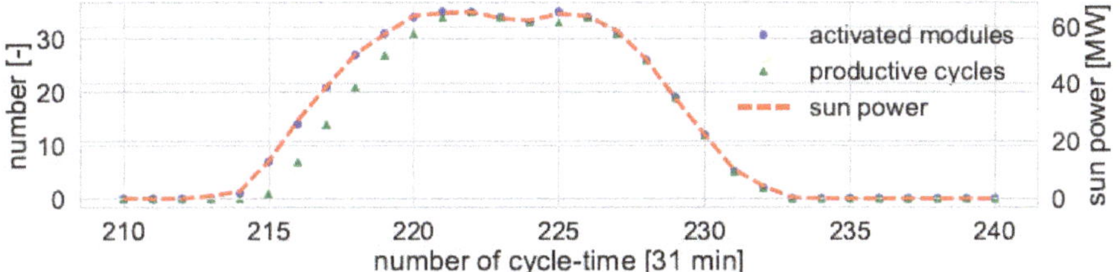

Figure 8. Plant daily performance (ceria, typical day, base case).

The total investment for the case is 35.5 Mio. € for nickel-ferrite and 32.5 Mio. € for ceria. The higher investment for nickel-ferrite is due to the higher number of reactors needed, which in turn is a result of the slower kinetics. However, since solar field investment (which is the same in both cases) accounts for ca. two thirds of the total investment, the difference is relatively small. The resulting LCOH are 38.83 €/kg for nickel-ferrite and 13.06 €/kg for ceria. Figure 9 shows the LCOH structure. In both cases, the instalment accounts for more than 60% of the LCOH, while the substitution of the active material represents ca. another 20% of the specific production cost. The operation and maintenance cost amount to ca. 15% of LCOH.

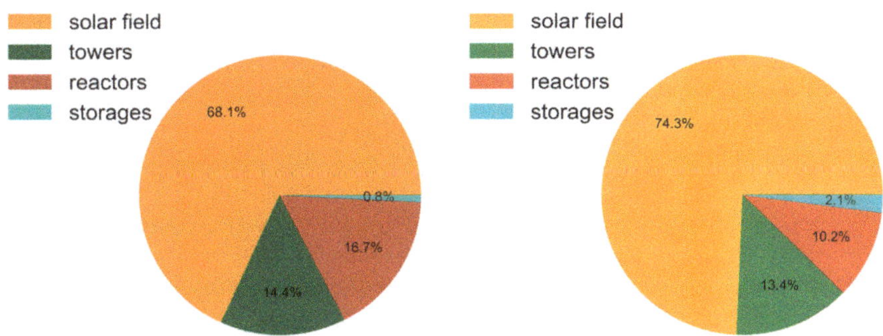

Figure 9. LCOH shares for the base case (**left**: nickel-ferrite, **right**: ceria).

The conclusion is that solar field and tower cost are—less surprisingly—the most important cost factors. Also, the reduction of the substitution of metal oxide can significantly enhance the economic figures. Such considerations are confirmed by the sensitivity analysis for the case with ceria on economic parameters in Figure 10 (the case with nickel-ferrite is similar and it is reported in the Appendix A). The reduction of the cost of the solar field turns out to be by far the most important cost driver. In addition, absorber cost and life time play an important role. On the contrary, the impact of nitrogen cost or electricity cost for compressors on LCOH is almost negligible.

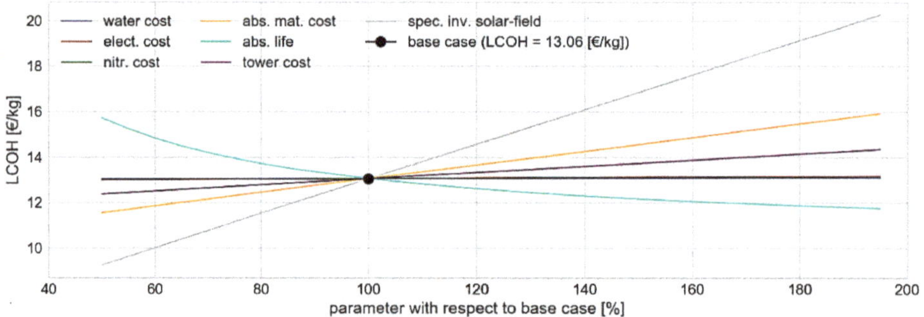

Figure 10. Sensitivity analysis of LCOH (ceria, base case).

Finally, a best case scenario for ceria has been evaluated. Nickel-ferrite has been neglected due to the higher LCOH in the base case. This scenario takes into account a reduction of the key economic parameters. In particular, heliostat field as well as tower cost are assumed to be reduced by 40%. The active material cost is halved and the absorber life time has been extended from 1 year to 20 years. These assumptions, in particular those regarding the absorber life time, are very optimistic. However, this assumption is taken as this work aims at the evaluation of the long term potentials of this technology. The LCOH of ceria in the best case accounts to 6.68 €/kg. The results compare well with the figures recently presented by Nicodemus [25].

4. Conclusions

Within this paper a techno-economic model for the evaluation and comparison of different materials for solar thermochemical hydrogen production has been developed. The novelty of the model consists in the simplified but still flexible approach, which considers at the same time technical and economic aspects. In this way, the model is able to take into account the inter-dependence between economic input parameters and optimal plant design. In addition, the model makes possible evaluating key design and operation parameters of such plants, and comparing different materials (i.e., nickel-ferrite and ceria). Furthermore, sensitivity analyses over a wide range of techno-economic parameters can be easily realized.

The technical model has been developed based on available literature and in close collaboration with the DLR Institute of Solar Research. It consists of four sections including sun power evaluation, single reactor model, kinetics for both the considered materials, and complete plant model including underground hydrogen storage facilities. The model has been used to optimize the design of a large-scale solar hydrogen production plant (90 MWth solar reactor power at design). In particular, the design optimization procedure consists on the minimization of the solar power dumping which typically occurs due to the different heat requirements of the thermal reduction and water splitting phases, respectively. For both nickel-ferrite and ceria an optimal number of reactors per module has been found. Each of the reactors within one module is then operated with a suitable time displacement relative to the other reactors of the same module. The comparison demonstrated that ceria achieves higher efficiency than nickel-ferrite (13.4% instead 6.4%), which is mainly due to the different kinetics.

This difference leads to a lower LCOH for ceria (13.06 €/kg instead of 38.83 €/kg for nickel-ferrite, both in the base case). The analysis of the cost structure highlighted the importance of reducing investment cost for solar field in order to improve the plant economics. A best case scenario with optimistic assumptions regarding investment cost and absorber life-time has been considered. LCOH in this case is 6.68 €/kg for ceria. Finally, the integral evaluation of the thermal balances over a full cycles showed that around two third of the incoming solar radiation are lost due to re-radiation through the window. The main competitor of solar thermochemical cycles for hydrogen production remains photovoltaics-powered electrolysis, which is expected to reach specific production cost between 2 €/kg and 3 €/kg in the long term perspective.

Author Contributions: Conceptualization, M.M.; Formal analysis, M.M. and M.P.; Methodology, M.M., M.P. and T.F.; Software, M.P.; Supervision, M.M.; Writing—original draft, M.M.; Writing—review & editing, M.M. and M.P.

Funding: This research was funded by the Helmholtz Association within the framework of the Program-Oriented Funding (DLR basic funding: "Future Fuels" Project).

Acknowledgments: A special thank goes to Matteo Pecchi for his extraordinary engagement during his stay at DLR.

Conflicts of Interest: The authors declare no conflict of interest. The funders had no role in the design of the study; in the collection, analyses, or interpretation of data; in the writing of the manuscript, or in the decision to publish the results.

Abbreviations

Acronyms

COP	Conference of the Parties
CSP	Concentrating Solar Power
DLR	Deutsches Zentrum für Luft-und Raumfahrt (German Aerospace Center)
IEA	International Energy Agency
MO	Metal Oxide
O&M	Operation and maintenance
PV	Photovoltaics
TR	Thermal Reduction Step
WS	Water Splitting Step

Symbols

$\dot{Q}_{h,reac}$	[MWth]	Thermal power due to the chemical reactions
$\dot{Q}_{a-fl,conv}$	[MWth]	Convective power between absorber and working fluid
$\dot{Q}_{a-ins,cond}$	[MWth]	Conductive power through the insulation
$\dot{Q}_{a-w,rad}$	[MWth]	Radiation power emitted by the absorber toward the window
$\dot{Q}_{a-w,\alpha}$	[MWth]	Radiation power from absorber, absorbed by window
$\dot{Q}_{sf}$	[MWth]	Solar field thermal power
$\dot{Q}_{sun,\alpha}$	[MWth]	Solar power absorbed
$\dot{Q}_{sun,\tau}$	[MWth]	Solar power transmitted by the absorber window
$\dot{Q}_{w,rad}$	[MWth]	Radiation power emitted by the window
$\dot{Q}_{w-env,conv}$	[MWth]	Convective power between window and environment
$\dot{Q}_{w-fl,conv}$	[MWth]	Convective power between window and working fluid
A_{sf}	[m^2]	Heliostat surface
T_a	[K]	Temperature of the absorber
T_w	[K]	Temperature of the window
LCOH	[€/kg]	Levelized cost of hydrogen
$c_{p,abs}$	[kJ/kg/K]	Specific heat of the absorber
$c_{p,w}$	[kJ/kg/K]	Specific heat capacity of the window
f_{spill}	[-]	Non-dimensional spillage factor
k_{reg}	[s^{-1}]	Rate constant for regeneration step

m_{abs}	[kg]	Absorber mass
m_w	[kg]	Windows mass
$t_{construction}$	[y]	Construction time
t_{debt}	[y]	Debt period
η_{sf}	[-]	Solar field efficiency
ψ_{O_2}	[mol]	Removable O2 at equilibrium
DNI	[W/m^2]	Direct Normal Irradiance
dt	[s]	Time step
i	[%/y]	Interest rate
inv	[Mio. €]	Total investment

Appendix A

Table A1. Demonstrations of solar thermochemical hydrogen production (adapted from [17]).

Parameter	Non-volatile Metal Oxide	Volatile Metal	Sulfur-Based
Project	HYDROSOL-Plant [20]	Solar production of zinc and hydrogen [38]	SOL2HY2 [39]
Years	2014–2017	2008–2015	2013–2016
Location	PSA, Spain	Odeillo, France	Jülich, Germany
Cycle Type	Doped ferrites	Zinc oxide	Modified hybrid sulfur
Reactor Type	Stationary monolithic honeycomb with CPC and quartz glass window	Windowed cavity with oxide particles	Stationary catalyst coated monolithic absorber with CPC and quartz glass window
Scale	750 kW$_{th}$	100 kW$_{th}$	100 kW$_{th}$
Max. Temp.	1400 °C	1750 °C	1200 °C
Challenges	Sintering, low efficiency	Low heat-up	Windows scale-up

Table A2. Structure of the heliostat field efficiency matrix, as a function of solar elevation angle (vertical axis) and solar azimuth angle (horizontal axis)–example for intercept power 11.5 MWth and latitude 20 °N.

Azimuth → Sun height ↓	0°	15°	30°	45°	60°	75°	90°	105°	120°	135°	150°	165°	180°
0°	0	0	0	0	0	0	0	0	0	0	0	0	0
1.2°	0.0913	0.0941	0.0909	0.0966	0.0999	0.1027	0.1112	0.123	0.1303	0.1389	0.1413	0.1459	0.143
5.2°	0.1417	0.1458	0.1446	0.1549	0.1621	0.1715	0.1815	0.2005	0.2106	0.2233	0.2269	0.2356	0.2307
15°	0.2883	0.2951	0.2988	0.3185	0.3318	0.3558	0.3687	0.4036	0.423	0.4479	0.4577	0.4743	0.4623
24.9°	0.4072	0.4121	0.4166	0.4389	0.464	0.4864	0.4972	0.5414	0.5711	0.5919	0.6059	0.6255	0.6241
34.8°	0.4877	0.4904	0.4943	0.5166	0.5441	0.5621	0.5823	0.6152	0.6471	0.6636	0.6762	0.6949	0.7003
44.6°	0.5511	0.5473	0.5499	0.5706	0.5957	0.6112	0.6284	0.6587	0.688	0.7021	0.7123	0.7297	0.7405
59.4°	0.6348	0.6367	0.637	0.6538	0.6676	0.6841	0.6956	0.7202	0.7374	0.7529	0.759	0.7733	0.7765
75.2°	0.7138	0.7149	0.7184	0.7239	0.7312	0.7399	0.7494	0.7591	0.7684	0.7766	0.7829	0.787	0.7885
90°	0.7695	0.7695	0.7695	0.7695	0.7695	0.7695	0.7695	0.7695	0.7695	0.7695	0.7695	0.7695	0.7695

Table A3. List of ranges used for the economic parametric study.

Parameter	Unit	Base Case	Min	Max
Water cost	€/m^3	3	1.5	6
Electricity cost	€/kWh	0.20	0.10	0.40
Absorber life time	y	1	0.5	2
Absorber specific cost	€/kg	10	5	20
N$_2$ electricity consumption	kWh/mol	0.001356	0.000678	0.002712
Tower cost factor	%	100	50	200
Solar field specific investment	€/m^2	150	75	300

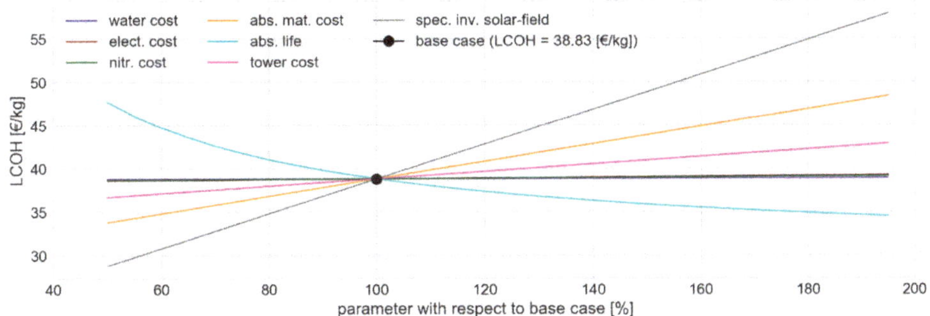

Figure A1. Sensitivity analysis of LHC (nickel-Ferrite, base case).

References

1. International Energy Agency (IEA). World Energy Outlook 2015. Available online: www.iea.org/publications/freepublications/publication/WEO2015.pdf (accessed on 11 November 2017).
2. British Petroleum (BP). BP Global. BP Energy Outlook 2017 Edition. Available online: https://www.bp.com/en/global/corporate/energy-economics/energy-outlook/energy-outlook-downloads.html (accessed on 22 January 2019).
3. Wang, F.-C.; Hsiao, Y.-S.; Yang, Y.-Z. The Optimization of Hybrid Power Systems with Renewable Energy and Hydrogen Generation. *Energies* **2018**, *11*, 1948. [CrossRef]
4. SolarPACES—Solar Power and Chemical Energy Systems Implementing Agreement of the International Energy Agency. *Solar Fuels from Concentrated Sunlight*. 2010 Report. Available online: stage-ste.eu (accessed on 22 January 2019).
5. Bozoglan, E.; Midilli, A.; Hepbasli, A. Sustainable assessment of solar hydrogen production techniques. *Energy* **2012**, *46*, 85–93. [CrossRef]
6. Kalogirou, S.A. Chapter 9. In *Photovoltaic Systems, Solar Energy Engineering*, 2nd ed.; Processes and Systems; Academic Press: Cambridge, MA, USA, 2014; pp. 481–540. [CrossRef]
7. Kovač, A.; Marciuš, D.; Budin, L. Solar hydrogen production via alkaline water electrolysis. *Int. J. Hydrogen Energy* **2018**. [CrossRef]
8. Baykara, S.Z. Experimental solar water thermolysis. *Int. J. Hydrogen Energy* **2004**, *29*, 1459–1469. [CrossRef]
9. Ihara, S. *Direct Thermal Decomposition of Water*; Ohta, T., Ed.; Solar energy systems, Pergamon Press: New York, NY, USA, 1989; pp. 59–79.
10. Kogan, A. Direct solar thermal splitting of water and on site separation of the products—Theoretical evaluation of the hydrogen yield. *Int. J. Hydrogen Energy* **1997**, *22*, 481–486. [CrossRef]
11. Tyner, C.E.; Kolb, G.J.; Meinecke, W.; Trieb, F. Concentrating Solar Power in 1999—An IEA/SolarPACES Summary of Status and Future Prospects. *SolarPACES* **1999**, *9*. [CrossRef]
12. Kaleibari, S.; Yanping, Z.; Abanades, S. Solar-driven high temperature hydrogen production via integrated spectrally split concentrated photovoltaics (SSCPV) and solar power tower. *Int. J. Hydrogen Energy* **2018**. [CrossRef]
13. Agrafiotis, C.; Roeb, M.; Sattler, C. A review on solar thermal syngas production via redox pair-based water/carbon dioxide splitting thermochemical cycles. *Renew. Sustain. Energy Rev.* **2015**, *42*, 254–285. [CrossRef]
14. Villafán-Vidales, H.I.; Arancibia-Bulnes, C.A.; Riveros-Rosas, D., Romero-Paredes, H.; Estrada, C.A. An overview of the solar thermochemical processes for hydrogen and syngas production: Reactors, and facilities. *Renew. Sustain. Energy Rev.* **2017**, *75*, 894–908. [CrossRef]
15. Miller, J.E.; Allendorf, M.D.; Diver, R.B.; Evans, L.R.; Siegel, N.P.; Stuecker, J.N. Metal oxide composites and structures for ultra-high temperature solar thermochemical cycles. *J. Mater. Sci.* **2008**, *43*, 4714–4728. [CrossRef]

16. Bhosale, R.; Kumar, A.; AlMomani, F.; Ghosh, U.; Saad Anis, M.; Kakosimos, K.; Shende, R.; Rosen, M.A. Solar Hydrogen Production via a Samarium Oxide-Based Thermochemical Water Splitting Cycle. *Energies* **2016**, *9*, 316. [CrossRef]

17. Hinkley, J.; Agrafiotis, C. *Solar Thermal Energy and its Conversion to Solar Fuels via Thermochemical Processes, Polygeneration with Polystorage for Chemical and Energy Hubs*; Academic Press: Cambridge, MA, USA, 2019; pp. 247–286. [CrossRef]

18. Steinfeld, A. Solar thermochemical production of hydrogen—A review. *Sol. Energy* **2005**, *78*, 603–615. [CrossRef]

19. Lange, M. Efficiency Analysis of Solar Driven Two-Step Thermochemical Water Splitting Processes Based on Metal Oxide Redox Pairs. Ph.D. Thesis, RWTH Aachen University, Aachen, Germany, 2014; p. 153. Available online: http://publications.rwth-aachen.de/record/444960 (accessed on 22 January 2019).

20. Säck, J.P.; Breuer, S.; Cotelli, P.; Houaijia, A.; Lange, M.; Wullenkord, M.; Spenke, C.; Roeb, M.; Sattler, C. High temperature hydrogen production: Design of a 750KW demonstration plant for a two-step thermochemical cycle. *Sol. Energy* **2016**, *135*, 232–241. [CrossRef]

21. Furler, P.; Scheffe, J.; Steinfeld, A. Syngas production by simultaneous splitting of water and carbon dioxide via ceria redox reactions in a high temperature solar reactor. *Energy Environ. Sci.* **2012**, *5*, 6098–6103. [CrossRef]

22. Marxer, D.; Furler, P.; Scheffe, J.; Geerlings, H.; Falter, C.; Batteiger, V.; Sizmann, A.; Steinfeld, A. Demonstration of the entire production chain to renewable kerosene via solar thermochemical splitting of H_2O and CO_2. *Energy Fuel* **2015**, *29*, 3241–3250. [CrossRef]

23. Marxer, D.; Furler, P.; Takacs, M.; Steinfeld, A. Solar thermochemical splitting of CO_2 into separate streams of CO and O_2 with high selectivity, stability, conversion, and efficiency. *Energy Environ. Sci.* **2017**, *10*, 1142–1149. [CrossRef]

24. Graf, D.; Monnerie, N.; Roeb, M.; Schmitz, M.; Sattler, C. Economic comparison of solar hydrogen generation by means of thermochemical cycles and electrolysis. *Int. J. Hydrogen Energy* **2008**, *33*, 4511–4519. [CrossRef]

25. Nicodemus, J.H. Technological learning and the future of solar H_2: A component learning comparison of solar thermochemical cycles and electrolysis with solar PV. *Energy Policy* **2018**, *120*, 100–109. [CrossRef]

26. Schenk, H. *ENERMENA—Yield Analysis for Parabolic trough Solar Thermal Power Plants*; Internal Report; Deutsches Zentrum für Luft und Raumfahrt (DLR): Köln, Germany, 2011.

27. Dersch, J.; Schwarzbözl, P.; Richert, T. Annual Yield Analysis of Solar Tower Power Plants with GREENIUS. *J. Sol. Energy Eng.* **2011**, *133*. [CrossRef]

28. Pecchi, M. System Analysis of Solar Hydrogen Production—Mathematical Modelling and Techno-Economic Assessment. Master's Thesis, Germany and University of Trento, Trento, Italy, 2017.

29. Garcia, L.; Burisch, M.; Sanchez, M. Spillage estimation in a heliostats field for solar field optimization. *Energy Procedia* **2015**, *69*, 1269–1276. [CrossRef]

30. Agrafiotis, C.; Zygogianni, A.; Pagkoura, C.; Kostoglou, M.; Konstandopoulos, A.G. Hydrogen production via solar-aided water splitting thermochemical cycles with nickel ferrite: Experiments and modelling. *AIChE J.* **2013**, *59*, 1213–1225. [CrossRef]

31. Diver, R.B.; Miller, J.E.; Allendorf, M.D.; Siegel, N.P.; Hogan, R.E. Solar Thermochemical Water-Splitting Ferrite-Cycle Heat Engines. *J. Sol. Energy Eng.* **2008**, *130*, 041001. [CrossRef]

32. Nelson, A.T.; White, J.T.; Andersson, D.A.; Aguiar, J.A.; McClellan, K.J.; Byler, D.D.; Short, M.P.; Stanek, C.R. Thermal Expansion, Heat Capacity, and Thermal Conductivity of Nickel Ferrite ($NiFe_2O_4$). *J. Am. Ceram. Soc.* **2014**, *97*, 1559–1565. [CrossRef]

33. Furler, P. Solar Thermochemical CO_2 and H_2O Splitting via Ceria Redox Reactions. Ph.D. Thesis, ETH, Zurich, Switzerland, 2014.

34. Chueh, W.C.; Haile, S.M. A thermochemical study of ceria: Exploiting an old material for new modes of energy conversion and CO_2 mitigation. *Philos. Trans. R. Soc. Lond. A Math. Phys. Eng. Sci.* **2010**, *368*, 3269–3294. [CrossRef]

35. Parks, G.; Boyd, R.; Cornish, J.; Remick, R. *Hydrogen Station Compression, Storage, and Dispensing Technical Status and Costs*; NREL Technical Report NREL/BK-6A10-58564; NREL: Golden, CO, USA, 2014. Available online: www.nrel.gov (accessed on 22 January 2019).

36. Nitsch, J.; Pregger, T.; Naegler, T.; Heide, D.; de Tena, D.L.; Trieb, F.; Scholz, Y.; Nienhaus, K.; Gerhardt, N.; Sterner, M.; et al. *Langfristszenarien und Strategien für den Ausbau der Erneuerbaren Energien in Deutschland bei Berücksichtigung der Entwicklung in Europa und Global*; Report for the German Ministry of Environment; DLR, IWES, IFNE: Stuttgart, Germany, 2011. (In German)

37. Solargis—Weather Data and Software for Solar Power Investments. Available online: https://solargis.com/ (accessed on 19 December 2018).

38. Villasmil, W.; Brkic, M.; Wuillemin, D.; Meier, A.; Steinfeld, A. Pilot scale demonstration of a 100-kWth solar thermochemical plant for the thermal dissociation of ZnO. *J. Sol. Energy Eng.* **2013**, *136*, 011016. [CrossRef]

39. Odorizzi, S. ENGINSOFT S. SOL2HY2—Solar to Hydrogen Hybrid Cycles Project Publishable Summary. 2016. Available online: http://cordis.europa.eu/docs/results/325/325320/final1-sol2hy2-publishable-summary-final-report-24-2.pdf (accessed on 22 January 2019).

Article

Thermodynamic and Technical Issues of Hydrogen and Methane-Hydrogen Mixtures Pipeline Transmission

Szymon Kuczyński, Mariusz Łaciak, Andrzej Olijnyk, Adam Szurlej and Tomasz Włodek *

AGH University of Science and Technology, Drilling, Oil and Gas Faculty, Krakow PL30059, Poland; skuczyns@agh.edu.pl (S.K.); laciak@agh.edu.pl (M.Ł.); aoliinyk@agh.edu.pl (A.O.); szua@agh.edu.pl (A.S.)
* Correspondence: twlodek@agh.edu.pl; Tel.: +48-12-617-3668

Received: 27 December 2018; Accepted: 4 February 2019; Published: 12 February 2019

Abstract: The use of hydrogen as a non-emission energy carrier is important for the innovative development of the power-generation industry. Transmission pipelines are the most efficient and economic method of transporting large quantities of hydrogen in a number of variants. A comprehensive hydraulic analysis of hydrogen transmission at a mass flow rate of 0.3 to 3.0 kg/s (volume flow rates from 12,000 Nm^3/h to 120,000 Nm^3/h) was performed. The methodology was based on flow simulation in a pipeline for assumed boundary conditions as well as modeling of fluid thermodynamic parameters for pure hydrogen and its mixtures with methane. The assumed outlet pressure was 24 bar (g). The pipeline diameter and required inlet pressure were calculated for these parameters. The change in temperature was analyzed as a function of the pipeline length for a given real heat transfer model; the assumed temperatures were 5 and 25 °C. The impact of hydrogen on natural gas transmission is another important issue. The performed analysis revealed that the maximum participation of hydrogen in natural gas should not exceed 15%–20%, or it has a negative impact on natural gas quality. In the case of a mixture of 85% methane and 15% hydrogen, the required outlet pressure is 10% lower than for pure methane. The obtained results present various possibilities of pipeline transmission of hydrogen at large distances. Moreover, the changes in basic thermodynamic parameters have been presented as a function of pipeline length for the adopted assumptions.

Keywords: hydrogen; hydrogen pipelines; hydrogen transmission; pipeline transmission; pressure drop; energy storage

---

1. Introduction

Recent trends in modern economies are focused on greenhouse gas emissions reduction and mitigation of climate change effects. Countries around the world have begun to shift their energy production to renewable energy sources (RES). Energy from renewable sources may help mitigate emissions from traditional fossil fuel energy generation [1]. Implementation of EU energy policy requires investment in power technologies based on RES. The dynamics of RES development and application can be traced to the basis of its installed capacity. According to data from 2016 [2], the highest increase of installed power was observed for wind farms (i.e., 12,490 MW, or 51% of all new installed capacity in EU) and solar energy power plants (i.e., 6700 MW, or 27.4% of all new installed capacity in EU). By 2040, RES-based EU technologies will constitute 80% of new installed power, while after 2030, wind energy is predicted to become the leading electrical energy source [3]. Wind energy, with significant growth in the RES share it represents, will cause problems associated with an uneven generation of electrical energy, resulting from variable atmospheric conditions [4]. Frequently, high electricity generation is possible during periods of low demand for electrical energy (e.g., days off), whereas during periods of higher demand (evening peak), production is much lower. Moreover, the use of renewable energy from solar and wind farms is connected with power transmission system problems because of the irregularity and instability of energy supply [5,6].

Accordingly, a significant development in energy storage technology is required to increase application of RES in electrical energy generation sector. Power-to-Gas is an example of such technology. By using this technology, electrical energy can be converted to gaseous fuel (hydrogen). Hydrogen as an energy carrier can store the largest quantities of energy and has high energy content per mass unit [4]. This makes hydrogen technology very advantageous from a technological point of view [7]. Hydrogen has great importance as a promising green energy carrier, but practically does not occur in nature freely, hence it is not the primary source of energy. Hydrogen is usually generated as a secondary energy carrier from primary sources such as natural gas or wind energy. Hydrogen will play an important role in the world energy mix in the future [8].

Requirements regarding the proportion of renewable energy sources in national electricity systems have been described in Directive 2009/28/EC. The general objective of achieving a 20% content of RES usage by 2020 in the European Union was included in this statement [7,9]. Most of the power generated in the field of renewable energy sources was comprised of onshore and offshore wind farms. Therefore, for the further development of renewable energy usage in the power generation sector, it will be necessary to make progress in the use of energy storage technologies. The use of hydrogen as an energy storage technology allows the largest amount of energy storage and is distinguished by having the highest power output. Thus, this technology is very beneficial in technical terms [10–12]. In the near future, the natural gas system will be used for transmission of the growing volume of alternative fuels (e.g., hydrogen and biomethane), which will be added to traditional natural gas mixtures.

Recently, projects and concepts for the construction of energy storage sites in salt caverns have been the main focus related to the development of renewable energy sources. Hydrogen obtained during the withdrawal of salt caverns has to be transported to the place of its utilization. Pipeline transmission of hydrogen and the possibility of hydrogen addition to the natural gas transmission system are still new solutions requiring further research, and have been confirmed by real applications in a small number of cases. Specific thermodynamic analysis of methane–hydrogen mixtures is a main novelty of this research, in particular the use of hydrogen to improve natural gas flow parameters in the pipeline while maintaining quality requirements.

2. Basics of Hydrogen Transmission

Hydrogen is presented as an effective energy carrier which can be used efficiently with minimum greenhouse gas emissions in the processes in which it is involved [13]. The use of hydrogen in the long term for the purpose of balancing energy production with the demand of the electrical energy market is a real solution to the problem of excessive amounts and shortages of energy on the market. Hydrogen is used for storing surplus electrical energy. Salt caverns, as an efficient source of hydrogen storage, provide good conditions for injection and production [14–17]. Salt caverns are located in sites which have specific geological conditions; therefore, in many cases hydrogen must be transported across long distances from the storage site. The idea of energy storage in salt caverns with the use of hydrogen is presented in Figure 1. Over recent years the pipeline transport of hydrogen has been considered mainly as an integral and significant element of the renewable energy system [18]. On the other hand, the industrial use of hydrogen is widespread, where the production of hydrogen is closely associated with specific technological processes. Hydrogen is transmitted through pipelines of various diameters and lengths, depending on the way in which hydrogen will be used. Pipeline transmission is one of the cheapest methods for transporting large quantities of hydrogen [19,20]. Pipelines for the transmission of hydrogen at long distances exist in a few places worldwide, with a total length of 4500 km.

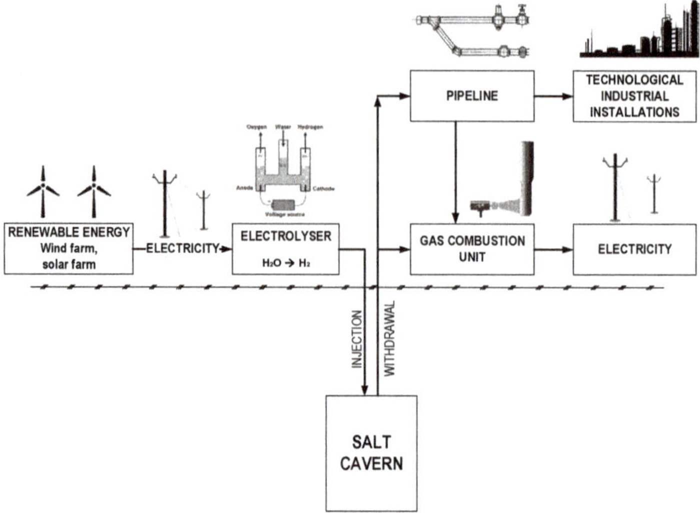

Figure 1. Scheme of hydrogen energy storage processes.

Hydrogen is commonly considered a gas similar to methane (i.e., the main component of natural gas). Therefore, most technological requirements for hydrogen transmission pipelines are identical to those of natural gas pipelines, with certain modifications regarding safety, infrastructure, and materials. These conditions must be met before the transmission of hydrogen is initiated through a pipeline network. Hydrogen has its specific set of physical and chemical properties which makes the pipeline transmission of this gas significantly different from that of natural gas. Due to physicochemical properties, hydrogen pipeline transmission is much more difficult than natural gas. Even compressed hydrogen can provide only about a third of the energy when compared to methane per unit of volume [21].

Natural gas transport and distribution networks are very well developed; therefore, the hydrogen pipeline system is directly compared to the natural gas transmission system [22]. The natural gas system consists of gas compression stations, pipelines, gas stations and gas storage facilities, including caverns [23]. The gas compression station provides the energy needed to generate gas flow at a given rate and pressure. In the case of natural gas, gas networks are divided into high pressure transmission pipelines and medium or low-pressure distribution pipelines. In the case of hydrogen transmission, the distribution to smaller receiving centers worldwide is quite rare, excepting some locations, such as the Leuna industrial district in Germany. In most cases, hydrogen transportation is considered as a transmission from point A to point B, for a technologically justified purpose. Selected issues related to hydrogen pipeline transmission have been presented recently in a number of papers [24–29].

3. Model Development

3.1. Basic Assumptions

The hydraulic analysis of hydrogen transmission through the pipelines was based on a number of technological assumptions (e.g., mass flow rate from 0.3 to 3.0 kg/s, which corresponds to volume flow rate from 12,000 Nm^3/h to 120,000 Nm^3/h). The recommended outlet pressure for an exemplary technological installation was assumed to be 24 bar (g). In the presented exemplary case, the medium inlet temperature was set at 5 °C and the ambient temperature was 15 °C. For heat flow analysis, the pipeline was located one meter deep. The length of the exemplary pipeline chosen for analysis was 100 km.

3.2. Hydraulic Friction Factor

The zone of turbulent flow in rough pipes consists of a transient flow zone and developed roughness zone. In the transient flow zone, the linear friction coefficient (λ) depends on the Reynolds number (Re) and relative roughness (ε): $\lambda = f(Re, \varepsilon)$. Colebrook and White presented the following formula for the linear friction coefficient λ in this zone [30]:

$$\frac{1}{\sqrt{\lambda}} = -2\lg\left(\frac{2.51}{Re \cdot \sqrt{\lambda}} + \frac{\varepsilon}{3.71}\right) \tag{1}$$

The Colebrook-White Equation (1) was systematically analyzed from a theoretical and experimental point of view and, as a result, considered to be the most accurate of all relationships determining λ coefficient in the transient flow zone. For the zone with a full roughness impact $\lambda = f(\varepsilon)$, the linear friction coefficient is analyzed with the Prandtl-Nikuradse Equation (2) [31]:

$$\frac{1}{\sqrt{\lambda}} = -2\lg\frac{\varepsilon}{3.71} \tag{2}$$

Equation (2) was established by Prandtl, based on experiments conducted by Nikuradse on pipes with artificial sand roughness. The Prandtl–Nikuradse equation could be used on the following assumed criterion:

$$Re \cdot \varepsilon \cdot \sqrt{\lambda} \geq 200 \tag{3}$$

3.3. Pressure Drop Analysis

The pressure drop in gas pipelines can be determined with one of the hydraulic equations commonly applied for high pressure gas pipelines (e.g., Renouard equation, Panhandle equation). In this paper, authors applied the General Flow Equation (4), directly derived from the Bernoulli equation [32]:

$$p_1^2 - p_2^2 = 7.569 \cdot 10^5 \cdot ZT_{in}Ld\lambda \cdot \left(\frac{p_b}{T_b}\right)^2 \cdot \frac{Q_n^2}{D^5} \tag{4}$$

where p_1, p_2—pressure at respectively inlet and outlet of pipeline, Z—compressibility factor, T_{in}—temperature in pipeline, L—length of pipeline, d—relative density, Q_n—volume flow rate, D—inner diameter.

The difference of levels on a given pipeline section has a significant impact on the flow hydraulics. In pressure drop equations, the influence of terrain elevation change is accounted for with the use of an equivalent length of the pipeline L_e, which is the pipeline length corrected for the impact of terrain elevation with respect to the level of initial point of the pipeline or pipeline section [32].

$$L_e = \frac{L(\exp(s) - 1)}{s} \tag{5}$$

where s—dimensionless parameter which determines the influence of terrain elevation, which depends on temperature (T), compressibility factor (Z), relative density (d) and elevation level difference (Δh).

Dimensionless parameter which describes the terrain elevation impact is defined as:

$$s = 0.0684d\left(\frac{\Delta h}{T \cdot Z}\right) \tag{6}$$

If a pipeline of length L is divided into a number of sections (i.e., L_1, L_2, L_3, etc.) in which the terrain level significantly changes, then the parameter j should be introduced in such a way that the impact of the terrain elevation could be determined for each segment of the pipeline.

$$j = \frac{\exp(s) - 1}{s} \tag{7}$$

In this case, the pipeline equivalent length L_e accounts for the effect of the terrain elevation for each pipeline section with the dependence:

$$L_e = j_1 L_1 + j_2 L_2 \exp(s_1) + j_3 L_3 \exp(s_2) + \dots \tag{8}$$

3.4. Temperature Changes along the Pipeline

The temperature change basic model as a function of pipeline length was determined with the dependence combining the total heat transfer coefficient (U), mass flow rate ($\dot{M}$) and isobaric heat capacity (Cp) [32]:

$$\int_{T_{in}}^{T(x)} \frac{dT_{in}}{T_{in} - T_{out}} = \int_{0}^{L_r} \frac{U \cdot \pi \cdot D \cdot dx}{\dot{M} \cdot Cp} \tag{9}$$

For a pipeline with length L and diameter D, with the Joule–Thomson effect (μ_{JT}) change of temperature of the transmitted medium from T_1 to T_2 can be described with the equation [33]:

$$T_2 = T_{out} + (T_1 - T_{out}) \exp\left(\frac{-L \cdot U \cdot \pi \cdot D}{\dot{M} \cdot Cp}\right) + \frac{M}{\pi \cdot \dot{U} \cdot D}\left(\mu_{JT} \cdot Cp \cdot \frac{dp}{dL}\right)\left(1 - \exp\left(\frac{-L \cdot U \cdot \pi \cdot D}{\dot{M} \cdot Cp}\right)\right) \tag{10}$$

3.5. Heat Transfer Analysis

The basic principles of heat transfer analysis were formulated with heat flow laws based on conduction (Fourier law), convection (Newton law), and radiation (Stefan–Boltzman law) in association with the first law of thermodynamics. Conduction and convection are most the most important parameters for the heat flow transferred to the pipeline. Conduction, described with the Fourier law (according to cylindrical coordinates) for a pipeline in dynamic conditions, has the following form [34]:

$$\frac{1}{r}\frac{\partial}{\partial r}\left(k \cdot r \frac{\partial T}{\partial r}\right) = \rho \cdot Cp \frac{\partial T}{\partial \tau} \tag{11}$$

where r—radius, k—thermal conductivity, ρ—density, τ—time, Cp—isobaric heat capacity.

Under static conditions, the right side of Equation (11) equals zero; therefore, the total heat flow per unit of length of the pipeline between the medium within the pipeline and the environment is

$$Q = -2\pi r \cdot L \cdot k \frac{\partial T}{\partial r} \tag{12}$$

After integration of Equation (12), we have

$$Q = \frac{2\pi r \cdot L \cdot k \cdot (T_n - T_{n+1})}{\ln\left(\frac{r_{n+1}}{r_n}\right)} \tag{13}$$

The most important parameter which determines the ability of a particular cylindrical obstacle to heat transfer is the heat transfer coefficient (U). Taking into account convection and conductivity effects for a pipeline with complex parameters, the heat transfer coefficient equals [34,35]

$$U = \frac{1}{\frac{1}{\alpha_{in}} + \frac{r_{in}}{k_p}\ln\left(\frac{r_{out}}{r_{in}}\right) + \frac{r_{in}}{k_{iso}}\ln\left(\frac{r_{iso}}{r_{out}}\right) + \frac{r_{in}}{k_{ground}}\ln\left(\frac{2z_x}{r_{iso}}\right) + \frac{r_{in}}{z_x \alpha_{out}}} \tag{14}$$

In order to determine the heat transfer in the presented case (Figure 2), the thermal conductivity coefficient was analyzed for the pipeline wall (k_p), thermal insulation (k_{iso}), and ground in which the pipeline was buried (k_{ground}). R_{in} and R_{out} represent the inner and outer radius of the pipeline, respectively. r_{iso} is the radius including thermal insulation, and z_x is the depth of pipeline deposition in the ground. Convection effects are described by α_{in}—inner coefficient of heat penetration (assumed or determined with, for example, the Dittus–Boelter formula) and α_{out}—outer coefficient of heat penetration defined with Equation (15) for a pipeline seated in the ground at a depth z_x [35]:

$$\alpha_{out} = \frac{2 \cdot k_{ground}}{D_{tot} \cdot \ln\left[\frac{2z_x + \sqrt{4z_x^2 - D_{tot}^2}}{D_{tot}}\right]} \tag{15}$$

where D_{tot} is the total diameter of pipeline with thermal insulation.

Calculations of the overall heat transfer coefficient for pure hydrogen and pure methane as a function of the pipeline diameter are presented in Figure 3. Other parameters, for example, thermal insulation or burial depth of the gas pipeline, have remained unchanged. Calculations were made for the same volume flow rates in three variants specified in the presented case (12,000 Nm^3/h, 40,000 Nm^3/h, 120,000 Nm^3/h). Obtained results are similar for methane and hydrogen for assumed volume flow rates.

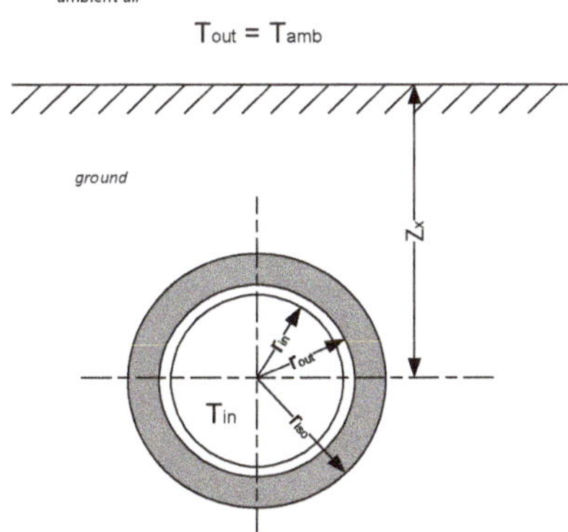

Figure 2. Cross section of an exemplary pipeline.

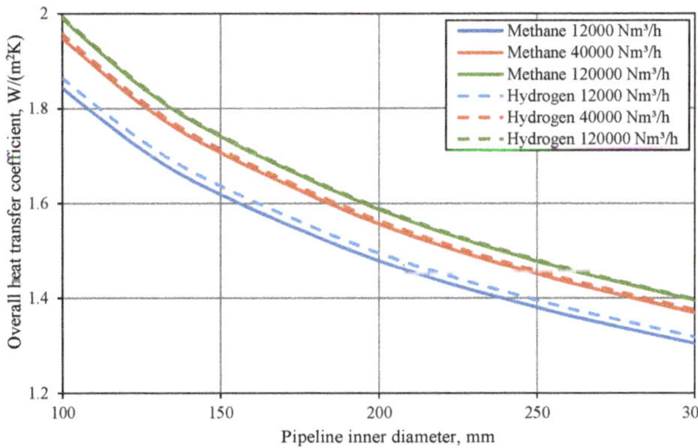

Figure 3. Overall heat transfer coefficient as a function of pipeline inner diameter for different flow rates of pure methane and pure hydrogen.

The final equation for total heat flux has the following form:

$$Q = \frac{2\pi r_{in} L(T_{in} - T_{out})}{\frac{1}{\alpha_{in}} + \frac{r_{in}}{k_p} \ln\left(\frac{r_{out}}{r_{in}}\right) + \frac{r_{in}}{k_{iso}} \ln\left(\frac{r_{iso}}{r_{out}}\right) + \frac{r_{in}}{k_{ground}} \ln\left(\frac{2z_x}{r_{iso}}\right) + \frac{r_{in}}{z_x \alpha_{out}}} \tag{16}$$

3.6. Real Gas Behavior and Thermodynamic Parameters Description

Thermodynamic parameters of transmitted hydrogen and methane/hydrogen mixtures as real gas are calculated with the Peng–Robinson equation of state commonly applied in the oil and gas industry [36]:

$$p = \frac{RT}{v - b_m} - \frac{a_m}{v(v + b_m) + b(v - b_m)} \tag{17}$$

The parameters of the equation of state a_m and b_m are based on classic mixing rules. They depend on the critical parameters of the analyzed gas. Compressibility factor Z is determined on the basis of a polynomial form of the Peng–Robinson equation of state:

$$Z^3 + (B - 1)Z^2 + (A - 3B^2 - 2B)Z + (B^3 + B^2 - AB) = 0 \tag{18}$$

where A and B are dimensionless parameters of equation of state, depending on the present temperature and pressure conditions:

$$A = \frac{a_m p}{R^2 T^2} \tag{19a}$$

$$B = \frac{b_m p}{RT} \tag{19b}$$

The compressibility factor is a key parameter while determining the pressure drop in the pipeline, and changes in most of the thermodynamic parameters of the analyzed gas as well as changes in temperature is a function of the pipeline length. Another important parameter is the density of the transmitted gas, which is determined with the general form of the equation of state:

$$\rho = \frac{p}{ZRT} \tag{20}$$

Equation (10) also makes use of specific heat capacity at a constant pressure (Cp) and the Joule–Thomson coefficient (μ_{JT}), which have a significant impact on temperature changes of the transmitted hydrogen [37,38]:

$$\begin{aligned}
Cp = Cp_{id} + Cp_r = Cp_{id} + R \cdot \left(T\left(\frac{\partial Z}{\partial T}\right)_p + Z - 1\right) + \frac{T\frac{da_m}{dT} - a_m}{2\sqrt{2}b_m} \cdot \\
\left[\frac{\left(\frac{\partial Z}{\partial T}\right)_p + (1+\sqrt{2})\left(\frac{\partial B}{\partial T}\right)_p}{Z + (1+\sqrt{2})B} - \frac{\left(\frac{\partial Z}{\partial T}\right)_p - (\sqrt{2}-1)\left(\frac{\partial B}{\partial T}\right)_p}{Z - (\sqrt{2}-1)B}\right] + \frac{T\frac{d^2 a_m}{dT^2}}{2\sqrt{2}b_m} \cdot \ln\left(\frac{Z+(1+\sqrt{2})B}{Z+(1-\sqrt{2})B}\right)
\end{aligned} \tag{21}$$

where Cp_{id}—isobaric heat capacity for ideal gas, Cp_r—residual part of isobaric heat capacity.

The Joule–Thomson coefficient can be expressed with specific heat capacity at constant pressure:

$$\mu_{JT} = \frac{1}{Cp}\left[T\left(\frac{\partial v}{\partial T}\right)_p - v\right] \tag{22}$$

Using the real gas Law (18), Equation (22) can be written as

$$\mu_{JT} = \frac{1}{Cp}\left[\frac{T}{Z \cdot \rho}\left(\frac{\partial Z}{\partial T}\right)_p\right] \qquad (23)$$

where

$$\left(\frac{\partial Z}{\partial T}\right)_p = \frac{\left(\frac{\partial A}{\partial T}\right)_p(B - Z) + \left(\frac{\partial B}{\partial T}\right)_p\left(6BZ + 2Z - 3B^2 - 2B + A - Z^2\right)}{3Z^2 + 2(B - 1)^2 + (A - 2B - 3B^2)} \qquad (24)$$

In the case of hydrogen, the description of the Joule–Thomson effect has a special meaning. Unlike for natural gas, the Joule–Thomson coefficient for hydrogen is negative, which means that hydrogen temperature increases with isenthalpic expansion.

3.7. Pipeline Diameter Selection

Prior to the hydraulic analysis of the pipeline, the optimum diameter of the pipeline should be determined for parameters such as assumed working pressure, length of the pipeline, roughness, etc. The selection of the diameter is also important for determination of the inlet pressure to the pipeline (includes associated costs) and possibility to compress the medium. In the analyzed case, the diameter was determined with a function of inlet pressure at the beginning of the pipeline for the assumed outlet pressure at the end of the pipeline of 24 bar (g). Calculations were based on the General Flow Equation (4), which directly stems from the Bernoulli law. The presented equation also contains an element (Δh) referring to the change of the elevation of the pipeline with respect to the assumed reference level [32,39]:

$$D = \sqrt[5]{\frac{16\lambda \cdot Z^2 \cdot R^2 \cdot T^2 \cdot L \cdot \dot{M}^2}{\pi^2 \cdot (Z \cdot R \cdot T \cdot (p_1^2 - p_2^2) - 2 \cdot g \cdot P_{av}^2 \cdot \Delta h)}} \qquad (25)$$

Calculations were performed for pure hydrogen (Figure 4) and methane/hydrogen mixtures with a maximum hydrogen content of 15% mol (Figure 5).

It should be noted that from the perspective of mass flow rate, the recommended diameters for the pure hydrogen are much smaller than for methane/hydrogen mixtures, which results from the low mass of hydrogen (Figures 4 and 5).

The recommended diameters for the methane/hydrogen mixture are much larger, as the density of pure hydrogen under normal conditions equals 0.0898 kg/m^3, while the density of a mixture of methane and 15% hydrogen is 0.6223 kg/m^3.

Recommended pipeline diameters for the assumed flow rates of pure hydrogen and methane/(15%)hydrogen mixtures are presented in Table 1. Evidently, transmission of the same volume of methane in a mixture with 15% hydrogen content required much larger diameters.

Table 1. Recommended pipeline diameters for pure hydrogen and methane/hydrogen mixture transmission.

Volume Flow Rate, Nm^3/h	Pure Hydrogen	Methane (85%) Hydrogen (15%)
	Diameter, mm	Diameter, mm
12,000	100–150	125–200
10,000	150 250	250 300
80,000	200–300	300–400
120,000	250–400	350–500

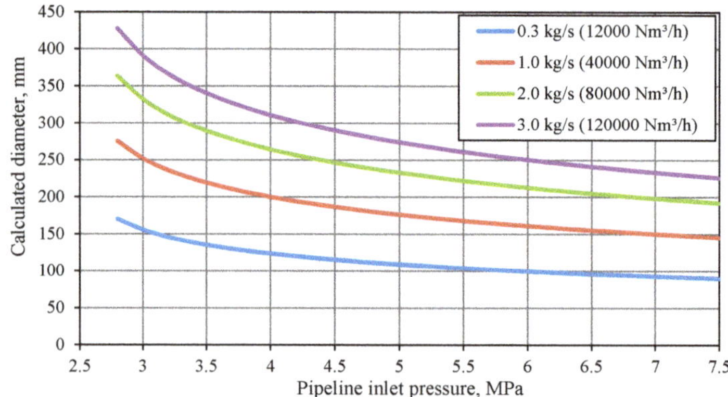

Figure 4. Diameter calculation for pure hydrogen pipeline as a function of pipeline inlet pressure.

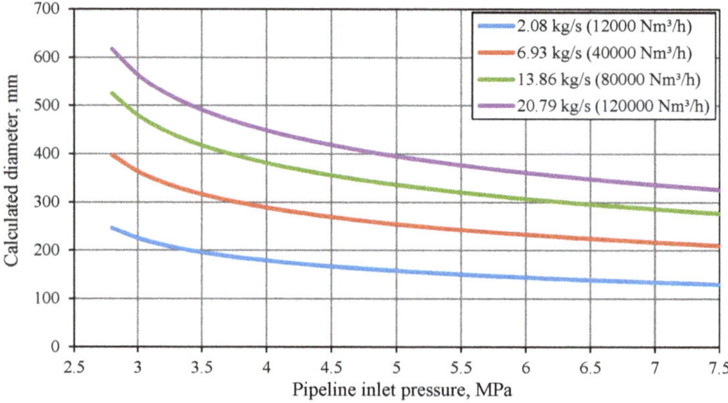

Figure 5. Diameter calculation for methane–hydrogen (15%) mixture pipeline as a function of pipeline inlet pressure.

4. Calculation Results

Flow modeling in a hydrogen transmission pipeline covers the analysis of changes in pressure and temperature as a function of length of the pipeline. The changes of compressibility factor, density, specific heat, and the Joule–Thomson effect as parameters especially important for hydrogen pipeline transmission were presented for one of the variants. A comparative analysis was also performed for the methane/hydrogen mixture and mass flow rate of 40,000 Nm^3/h.

4.1. Pressure Drop

The profiles of pressure and temperature changes for recommended pipeline diameters, assumed mass flow rates of 0.3, 1.0, 2.0 and 3.0 kg/s, (volume flow rates: 12,000 Nm^3/h, 40,000 Nm^3/h, 80,000 Nm^3/h and 120,000 Nm^3/h) and hydrogen inlet temperature of 5 °C are presented in Figures 6–9. Results of modelling the pressure drop profiles confirmed the preliminary calculations of the pipeline diameter. For the smallest recommended diameters for a selected flow rate, the inlet pressure in the pipeline was from 6.01 to 6.76 MPa (Figures 6–9), while for the largest recommended diameters the inlet pressure ranged from 2.91 to 3.2 MPa (Figures 6–9). The analysis of temperature changes revealed that the transmitted hydrogen warms up more slowly for smaller diameters and higher flow rates.

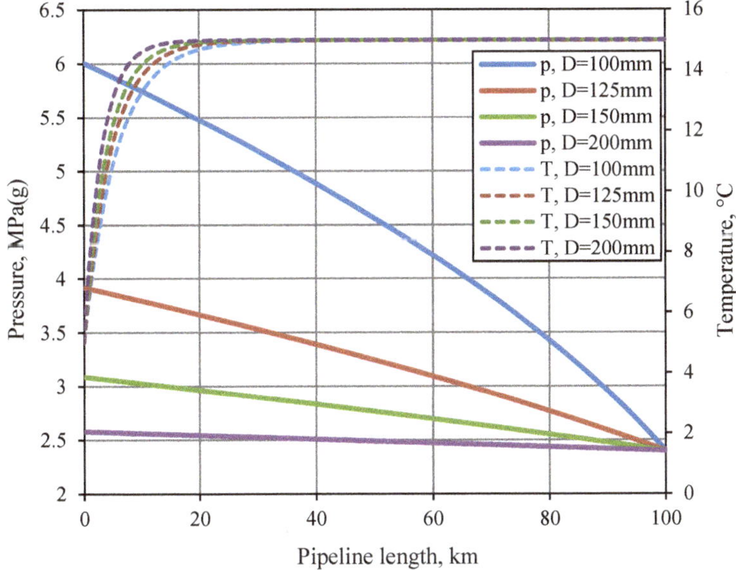

Figure 6. Pressure (*p*) (continuous lines) and temperature (*T*) (dashed lines) changes for hydrogen mass flow rate of 0.3 kg/s (12,000 Nm^3/h) and different diameters.

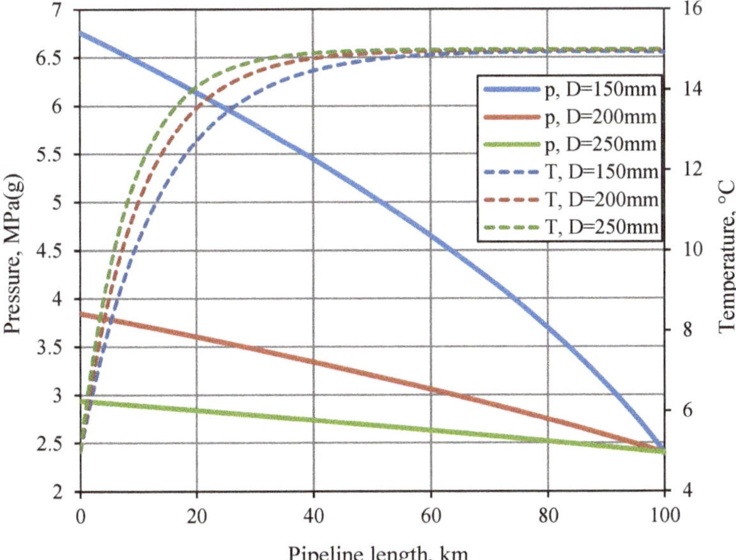

Figure 7. Pressure (*p*) (continuous lines) and temperature (*T*) (dashed lines) changes for hydrogen mass flow rate of 1.0 kg/s (40,000 Nm^3/h) for different diameters.

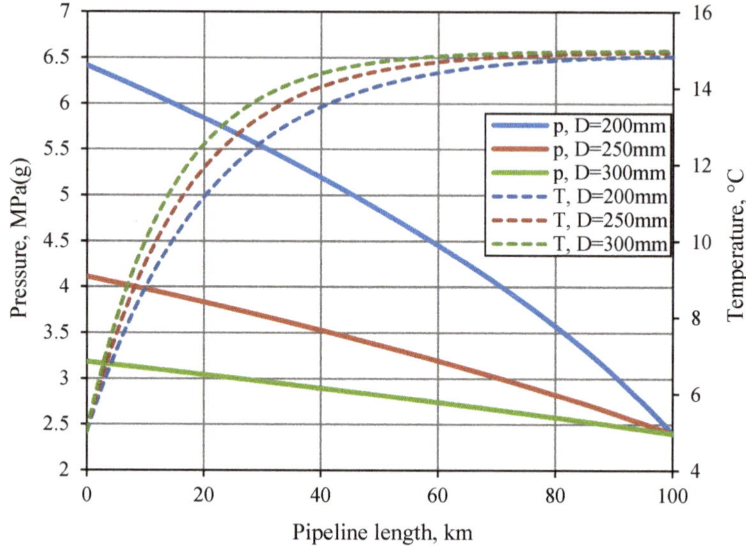

Figure 8. Pressure (p) (continuous lines) and temperature (T) (dashed lines) changes for hydrogen mass flow rate of 2.0 kg/s (80,000 Nm^3/h) for different diameters.

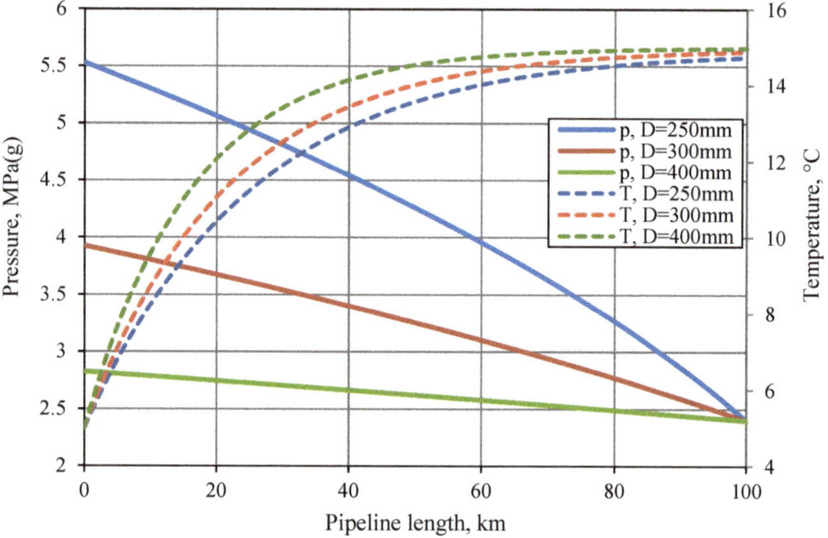

Figure 9. Pressure (p) (continuous lines) and temperature (T) (dashed lines) changes for hydrogen mass flow rate of 3.0 kg/s (120,000 Nm^3/h) for different diameters.

4.2. Thermodynamic Parameters Analysis

The analysis of selected thermodynamic parameters, such as density, flow rate, compressibility factor, isobaric heat capacity and Joule–Thomson coefficient were presented for hydrogen transmission at a flow rate of 40,000 Nm^3/h (1.0 kg/s). Figure 10 illustrates the changes in hydrogen density and its flow rate as a function of pipeline length for selected pipeline diameters.

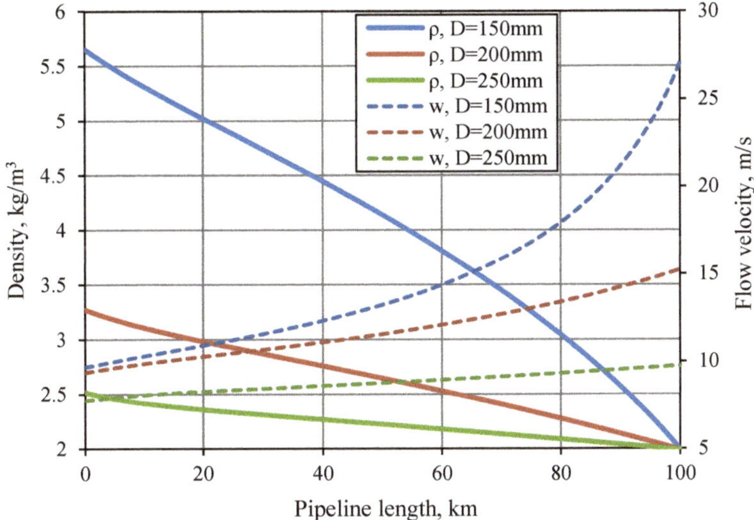

Figure 10. Density (ρ) (continuous lines) and flow velocity (w) (dashed lines) changes for hydrogen mass flow rate 1.0 kg/s (40,000 Nm^3/h).

The density change is directly connected to compressibility factor Z change, as presented in Figure 11. Compressibility factor Z for hydrogen is higher than unity even for relatively low pressure values. This is one of the most characteristic properties of hydrogen, which distinguishes it from most of the real gases. For instance, compressibility factor $Z > 1$ for natural gas is usually only at pressures equal or greater than 40 MPa.

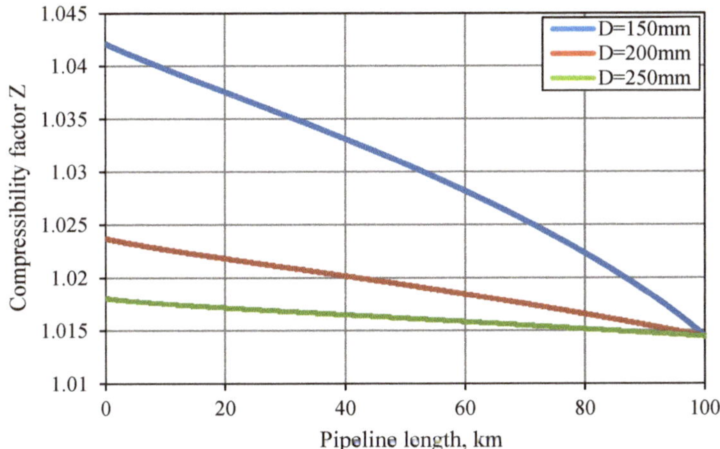

Figure 11. Compressibility factor changes for hydrogen mass flow rate 1.0 kg/s (40,000 Nm^3/h).

Another characteristic parameter of hydrogen is the Joule–Thomson coefficient. The Joule–Thomson effect occurs in temperature change of the gas during isenthalpic pressure drop. For most real gases, the Joule–Thomson effect is positive (i.e., gas temperature decreases with pressure reduction). In the case of hydrogen (Figure 12), the opposite effect occurs (i.e., the Joule–Thomson coefficient is negative): with a rapid change in pressure, the hydrogen temperature increases. This is theoretically possible for the hydrogen transmission pipelines, with considerable pressure drops per unit of pipeline length (e.g., with pipeline diameters which are too small or very high flow rates).

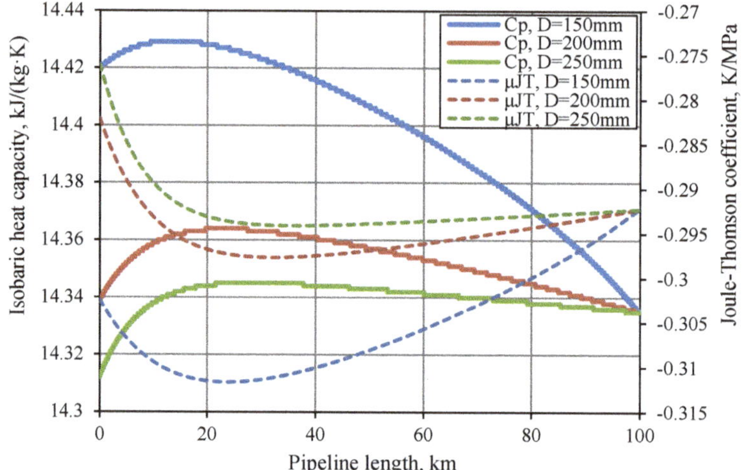

Figure 12. Isobaric heat capacity (*Cp*) (continuous lines) and Joule–Thomson coefficient (μ_{JT}) (dashed lines) changes for hydrogen mass flow rate 1.0 kg/s (40,000 Nm^3/h).

4.3. Methane-Hydrogen Mixtures

Hydrogen can also be transmitted through natural gas pipelines as an additive to natural gas. This is one of the alternative methods of hydrogen transportation. Numerous analyses and studies devoted to this issue have been performed recently [21,29,40]. Hydrogen has different thermodynamic parameters compared to methane, which is the main component of natural gas. This causes significant changes in the flow conditions of natural gas that contains hydrogen. Industrial practice and scientific publications indicate that the maximum admissible molar fraction of hydrogen in a mixture with natural gas should not exceed 15%. The maximum hydrogen content in the natural gas transmission system suggested in the United States should be in the range of 5%–15%. Several European countries have introduced limits on the content of hydrogen in natural gas pipeline systems from 0.1% to 12% by volume. The maximum hydrogen content usually depends on the technical conditions for a given pipeline [41]. Hydrogen significantly influences natural gas transmission conditions. The basic advantage is lowering the pressure drop of natural gas transmitted with a hydrogen admixture, and possibility to transmit natural gas across longer distances without additional gas compression stations. Unfortunately, the hydrogen content in natural gas significantly deteriorates the energy parameters and calorific value of natural gas. The changes in higher heating value (HHV) and lower heating value (LHV), calorific value of the mixture, and Wobbe index as a function of hydrogen content in the methane mixture are presented in Figure 13.

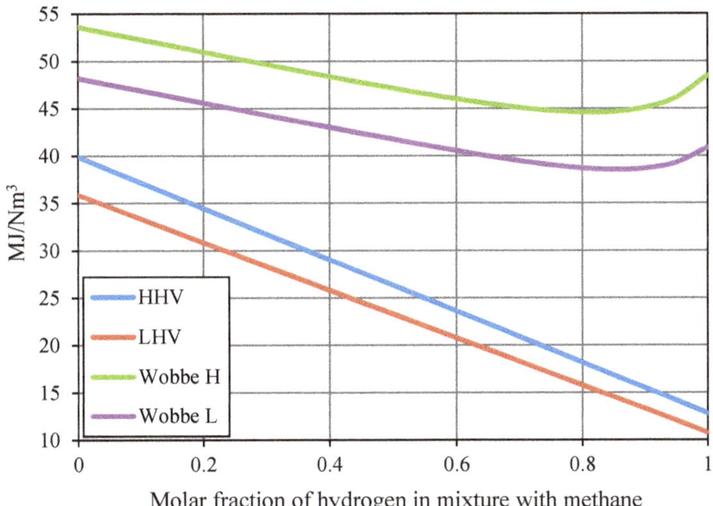

Figure 13. Changes of higher heating value (HHV), lower heating value (LHV) and Wobbe indexes as a function of hydrogen molar fraction in mixture with methane.

The pressure and temperature changes as a function of pipeline length for a mixture of methane and 15% hydrogen transmitted at a flow rate of 40,000 Nm^3/h are presented in Figure 14 for three recommended pipeline diameters. The change in thermodynamic parameters was also analyzed as a function of molar fraction of hydrogen in the mixture for a 250 mm diameter.

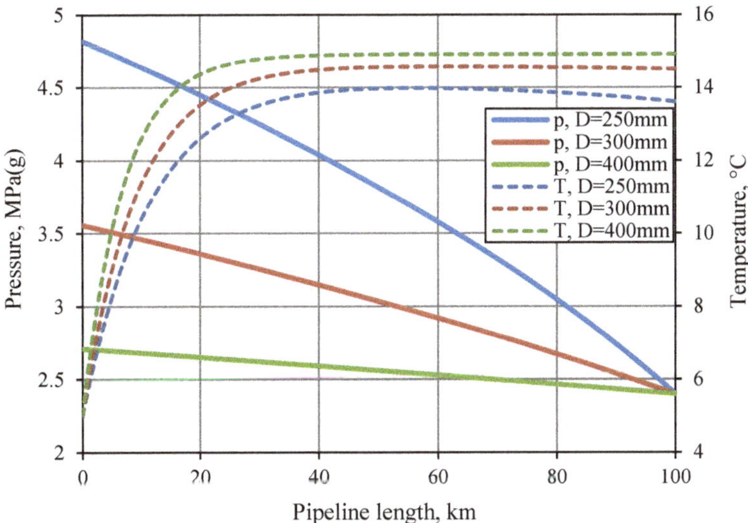

Figure 14. Pressure (*p*) (continuous lines) and temperature (*T*) (dashed lines) changes for methane–15% hydrogen mixture mass flow rate of 6.93 kg/s (40,000 Nm^3/h) for different diameters.

This analysis of the hydrogen molar fraction in the hydrogen/methane mixture confirms the effect of pressure drop in the analyzed pipeline. The required inlet pipeline pressure, for a 15% hydrogen content in the gas mixture, is approximately 10% lower when compared to pure methane (Figure 15). An increase in hydrogen molar content caused the temperature of the analyzed gas mixture to more rapidly approach the ambient temperature. On the other hand, the Joule–Thomson effect for methane caused slight cooling of the analyzed mixtures, and for pure hydrogen this effect did not take place.

The variability of analyzed thermodynamic parameters of a methane/hydrogen mixture as a function of hydrogen molar fraction is shown in Figures 16–18. The analyzed mixtures contained a maximum 15% of hydrogen; studies and practice have shown that this amount of hydrogen in the methane or natural gas mixture does not significantly affect the transmission parameters of the pipeline. However, it should be noted that a change in density is important: it can be altered by up to 20% for a 15% content of hydrogen in the mixture when compared to pure methane. Apart from a considerable lowering of density, the Joule–Thomson effect is also lowered with an increase in hydrogen content. With increased hydrogen content, its specific heat also increases for a unit of mass due to the high calorific value of hydrogen per unit of mass. The compressibility factor has been also observed to grow significantly. The addition of hydrogen improves to some extent the natural gas transmission conditions by pressure drop reduction in the pipeline. However, hydrogen content above 15%–20% in the gas mixture significantly influences the calorific value of natural gas. The change in thermodynamic conditions with an increased hydrogen content may also affect the natural gas transmission system (i.e., gas compression stations or gas reduction stations). Another important issue is selection of the material for pipeline construction in light of the hydrogen corrosion case [42], which affects the cost of its construction [20,43].

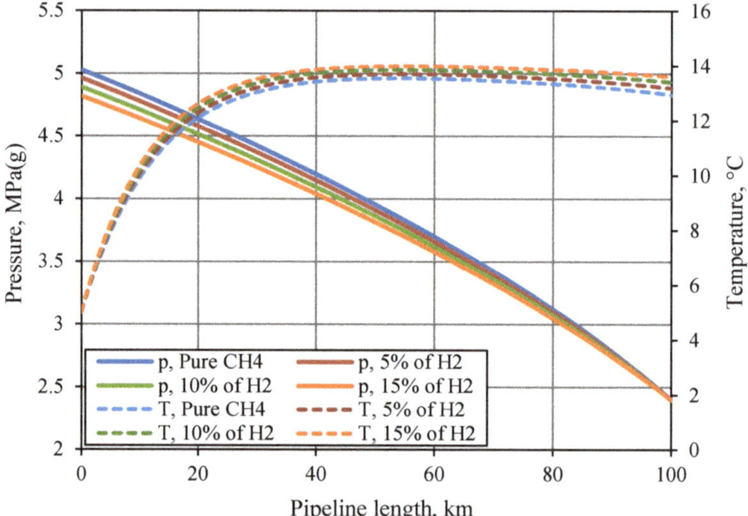

Figure 15. Pressure (*p*) (continuous line) and temperature (*T*) (dashed line) changes as a function of hydrogen content in mixture with methane and pipeline length.

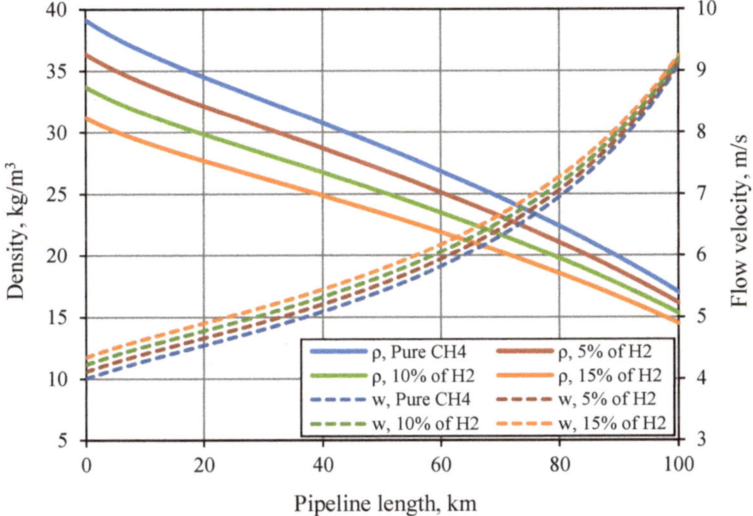

Figure 16. Density (ρ) (continuous lines) and flow velocity (w) (dashed lines) changes as a function of hydrogen content and pipeline length.

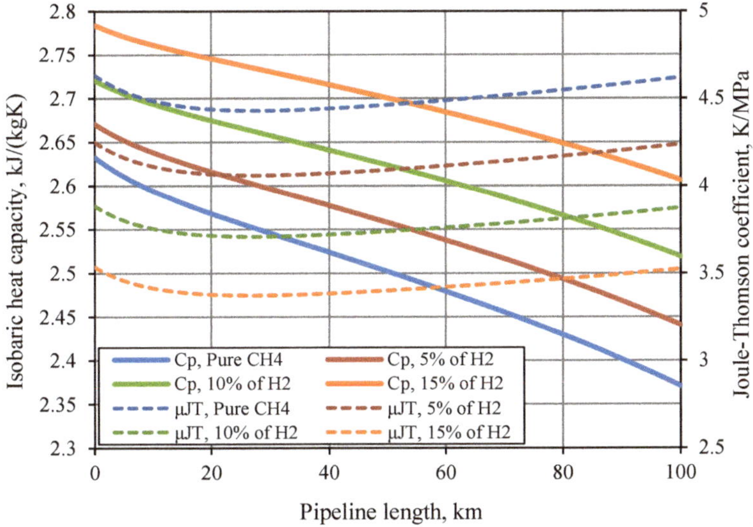

Figure 17. Isobaric heat capacity (Cp) (continuous lines) and Joule–Thomson coefficient (μ_{JT}) (dashed lines) changes as a function of hydrogen content and pipeline length.

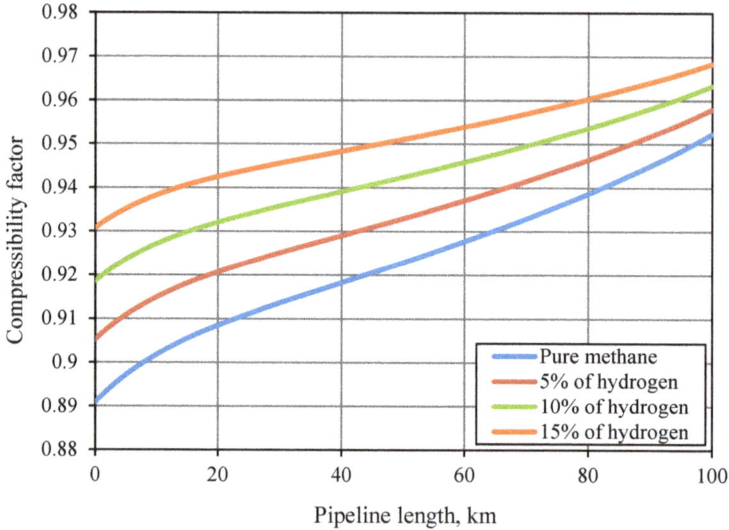

Figure 18. Compressibility factor changes as a function of hydrogen content and pipeline length.

5. Discussion

Hydrogen pipeline transmission analysis discussed an example of 100 km long pipeline and the required outlet pressure of 24 bar (g). Pressure drop and temperature change profiles were determined for selected pipeline diameters and mass flow rates from 0.3 to 3.0 kg/s, which corresponds to volume flow rates from 12,000 to 120,000 Nm^3/h of hydrogen. An analogous analysis has been performed for a mixture of methane and hydrogen with a maximum hydrogen content of 15%. Due to the fact that hydrogen has a significantly lower mass than methane, the pipeline transmission of methane and hydrogen mixtures requires larger pipeline diameters for similar volume flow rates. On the basis of the determined profiles of pressure and temperature changes in the pipeline, a detailed analysis of thermodynamic parameters was performed. This analysis was important from a hydrogen and methane–hydrogen mixture pipeline transmission perspective. In particular, the analysis included density, flow velocity, isobaric heat capacity, the compressibility factor, and Joule-Thomson coefficient. The last two parameters, which significantly affect the conditions of transport, are specific to hydrogen. Hydrogen has negative values of Joule-Thomson coefficient and the compressibility factor value exceeds 1.0 in the range of relatively low pressures. Due to different thermodynamic parameters, the hydrogen content in a mixture with methane causes significant changes in natural gas pipeline transport conditions. The most important change is the reduction of pressure drop, which allows an increase in the distance at which it is possible to transport the assumed volume of natural gas. Significant differences also occur in the temperature change profiles in the pipeline, as the hydrogen content reduces the positive Joule-Thomson effect for natural gas. It should be noted that in the simulation it was assumed that full heat transfer between the transported gas and the ambient environment may occur. The molar fraction of hydrogen in a mixture with natural gas may have a beneficial influence on the conditions of its transmission. However, hydrogen content should not exceed 15%–20% in the mixture. With a hydrogen content above 20% in the mixture with methane, the higher heating value (HHV) drops below 35 MJ/Nm^3; thus, natural gas loses its calorific value (below values described by norms and standards), even though the upper Wobbe index remains within accepted norms.

6. Conclusions

The main objective of this paper was to analyze the possibilities of pipeline transmission of hydrogen and methane/hydrogen mixtures. This analysis has been performed to determine the impact of hydrogen content on the conditions of natural gas transmission, the main component of which is methane. It should be emphasized that hydrogen will play an increasingly important role as an energy carrier in the global economy, particularly for energy storage. The economic environment for the use of hydrogen should be favorable in the coming years. A steady increase of renewable energy content in the total energy balance in all regions of the world, and the growing irregularities in power generation and usage, will have an important role in the field of hydrogen utilization. Hydrogen pipeline transmission is the most effective method for transporting significant amounts of hydrogen over long distances, in particular for its storage, when the appropriate storage site for hydrogen is located at a considerable distance from the source of its generation due to geological conditions (suitable locations for salt cavern). In addition, technological and technical issues related to pipeline transmission of natural gas with an increased hydrogen content should be considered, where a significant change in thermodynamic parameters may also affect the operational conditions of installations associated with the transmission system. An additional problem is the impact of the increased molar fraction of hydrogen on the pipe material.

Author Contributions: Conceptualization, T.W. and M.Ł.; formal analysis, T.W.; funding acquisition, M.Ł. and A.S.; investigation, T.W. and S.K.; methodology, T.W.; supervision, M.Ł. and A.S.; validation, T.W. and A.O.; visualization, T.W.; writing—original draft, T.W.; writing—review and editing, S.K. and T.W.

Funding: This work received funding from the Statutory Research of Natural Gas Department at Drilling Oil & Gas Faculty, no. 11.11.190.555.

Conflicts of Interest: The authors declare no conflict of interest.

Nomenclature

a_m	Peng Robinson EOS parameter, $N \cdot m^4/mol^2$
b_m	Peng Robinson EOS parameter (co-volume), m^3/mol
A, B	Dimensionless Peng Robinson EOS parameters
Cp	Isobaric heat capacity, $J/(kg \cdot K)$ or $J/(mol \cdot K)$
Cp_{id}	Ideal gas isobaric heat capacity, $J/(kg \cdot K)$ or $J/(mol \cdot K)$
Cp_r	Real gas (residual) isobaric heat capacity, $J/(kg \cdot K)$ or $J/(mol \cdot K)$
d	Relative density of gas
D	Pipeline inner diameter, m
g	Gravity constant, m/s^2
Δh	Elevation level difference, m
k	Thermal conductivity, $W/(m \cdot K)$
L	Pipeline segment length, m
L_e	Equivalent pipeline segment length, m
$\dot{M}$	Mass flow rate, kg/s
P	Pressure, Pa
p_1	Pipeline inlet pressure, Pa
p_2	Pipeline outlet pressure, Pa
p_{av}	Pipeline average pressure, Pa
p_b	Base pressure, Pa
Q_n	Volume flow rate under normal conditions, Nm^3/s
R	Pipeline inner radius, m
R	Gas constant, $J/(mol \cdot K)$
Re	Reynolds number

T	Fluid temperature, K
T_b	Base temperature, K
T_{in}	Temperature in pipeline, K
T_{out}	Ambient temperature, K
U	Overall heat transfer coefficient, W/(m²·K)
V	Molar volume, m³/mol
Z	Compressibility factor,
z_x	Depth of pipeline burial, m
A	Convective heat transfer coefficient, W/(m²·K)
E	Relative pipeline roughness
Λ	Linear friction factor
μ_{JT}	Joule–Thomson coefficient, K/Pa
P	Fluid density, kg/m³

Abbreviations

RES	Renewable energy sources
HHV	Higher heating value
LHV	Lower heating value
J–T	Joule–Thomson

References

1. Kharel, S.; Shabani, B. Hydrogen as a long-term large-scale energy storage solution to support renewables. *Energies* **2018**, *11*, 2825. [CrossRef]
2. Wind EUROPE 2017, Wind in Power. 2016 European statistics. Available online: http://windeurope.org (accessed on 20 December 2017).
3. International Energy Agency. *World Energy Outlook 2017*; International Energy Agency: Paris, France, 2017.
4. Kuczyński, S.; Łaciak, M.; Oliinyk, A.; Szurlej, A.; Włodek, T. Thermodynamic and technical issues of hydrogen and methane-hydrogen mixtures high-pressure pipeline transmission. In Proceedings of the 1st Latin American Conference on Sustainable Development of Energy Water and Environmental Systems-LA SDEWES, Rio de Janeiro, Brazil, 28–31 January 2018.
5. Fu, Z.; Lu, K.; Zhu, Y. Thermal system analysis and optimization of large-scale compressed air energy storage (CAES). *Energies* **2015**, *8*, 8873–8886. [CrossRef]
6. Abbaspour, M.; Satkin, M.; Mohammadi-Ivatloo, B.; Lotfi, F.H.; Noorollahi, Y. Optimal operation scheduling of wind power integrated with compressed air energy storage (CAES). *Renew. Energy* **2013**, *51*, 53–59. [CrossRef]
7. Blew, W.; Foltynowicz, M.; Kowalczyk, A. Projekt HESTOR. Innowacyjne metody magazynowania i wykorzystania energii z OZE. *Pol. Chem.* **2016**, *2*, 10–14.
8. Hosseini, S.F.; Wahid, M.A. Hydrogen production from renewable and sustainable energy resources: Promising green energy carrier for clean development. *Renew. Sustain. Energy Rev.* **2016**, *57*, 850–866. [CrossRef]
9. Union, E. Directive 2009/28/EC of The European Parliament and of The Council of 23 April 2009 on the promotion of the use of energy from renewable sources. *Off. J. Eur. Union* **2009**, *OJL 140*, 16–62.
10. Blacharski, T.; Kogut, K.; Szurlej, A. The perspectives for the use of hydrogen for electricity storage considering the foreign experience. *E3S Web of Conf.* **2017**, *14*, 1–10. [CrossRef]
11. Ball, M.; Basile, A.; Veziroglu, T.N. *Compendium of Hydrogen Energy: Hydrogen Use, Safety and the Hydrogen Economy*; Woodhead Publishing: Cambridge, UK, 2015.
12. Włodek, T.; Kuczyński, S.; Łaciak, M.; Szurlej, A. Possibilities and Selected Aspects of Hydrogen Energy Storage. 2nd World Congress on Petroleum and Refinery. *J. Environ. Biotechnol.* **2017**, *8*, 48.
13. Simon, J.; Ferriz, A.M.; Correas, L.C. HyUnder–hydrogen underground storage at large scale: Case study Spain. *Energy Procedia* **2015**, *73*, 136–144. [CrossRef]
14. Tarkowski, R. Perspectives of using the geological subsurface for hydrogen storage in Poland. *Int. J. Hydrog. Energy* **2017**, *42*, 347–355. [CrossRef]

15. Ślizowski, J.; Smulski, R.; Nagy, S.; Burliga, S.; Polański, K. Tightness of hydrogen storage caverns in salt deposits. *AGH Drill. Oil Gas* **2017**, *34*, 397–409. [CrossRef]

16. Lankof, L.; Polański, K.; Ślizowski, J.; Tomaszewska, B. Possibility of energy storage in salt caverns. *AGH Drill. Oil Gas* **2016**, *33*, 405–415. [CrossRef]

17. Ślizowski, J.; Urbańczyk, K.; Łaciak, M.; Lankof, L.; Serbin, K. Efektywność magazynowania gazu ziemnego i wodoru w kawernach solnych. *Przemysł Chemiczny Chem. Ind.* **2017**, *96*, 994–998.

18. Witkowski, A.; Rusin, A.; Majkut, M.; Stolecka, K. Comprehensive analysis of hydrogen compression and pipeline transportation from thermodynamics and safety aspects. *Energy* **2017**, *141*, 2508–2518. [CrossRef]

19. Lins, P.; Almeida, A. Multidimensional risk analysis of hydrogen pipelines. *Int. J. Hydrog. Energy* **2012**, *37*, 13545–13554. [CrossRef]

20. Gupta, R.; Basile, A.; Veziroglu, T.N. *Compendium of Hydrogen Energy: Hydrogen Storage, Distribution and Infrastructure*; Woodhead Publishing: Cambridge, UK, 2016.

21. Haeseldonckx, D.; D'haeseleer, W. The use of the natural-gas pipeline infrastructure for hydrogen transport in a changing market structure. *Int. J. Hydrog. Energy* **2007**, *32*, 1381–1386. [CrossRef]

22. Blacharski, T.; Janusz, P.; Kaliski, M.; Zabrzeski, Ł. The effect of hydrogen transported through gas pipelines on the performance of natural gas grid *AGH Drill. Oil Gas* **2016**, *33*, 515–529.

23. Serbin, K.; Ślizowski, J.; Urbańczyk, K.; Nagy, S. The influence of thermodynamic effects on gas storage cavern convergence. *Int. J. Rock Mech. Min. Sci.* **2015**, *79*, 166–171. [CrossRef]

24. De Vries, H.; Florisson, O.; Thiekstra, G.C. Safe operation of natural gas appliances fueled with hydrogen/natural gas mixtures (progress obtained in the NaturalHy-project). In Proceedings of the International Conference on Hydrogen Safety, San Sebastian, Spain, 11–13 September 2007.

25. Wurm, J.; Pasteris, R.F. The transmission of gaseous hydrogen. In Proceedings of the Fall Meeting of the Society of Petroleum Engineers of AIME, Las Vegas, NV, USA, 30 September–3 October 1973; p. 4526.

26. Peet, Y.; Saguut, P.; Charron, Y. Pressure loss reduction in hydrogen pipeline by surface restructuring. *Int. J. Hydrog. Energy* **2009**, *34*, 8964–8973. [CrossRef]

27. Takahashi, K. Transportation of Hydrogen by pipeline. In *Energy Carriers and Conversion Systems*, 2nd ed.; Ohta, T., Ed.; EOLSS: Paris, France, 2004.

28. Włodek, T.; Łaciak, M.; Kurowska, K.; Węgrzyn, Ł. Thermodynamic analysis of hydrogen pipeline transportation—Selected aspects. *AGH Drill. Oil Gas* **2016**, *33*, 379–396. [CrossRef]

29. Dodds, P.E.; Demoullin, S. Conversion of the UK gas system to transport hydrogen. *Int. J. Hydrog. Energy* **2013**, *38*, 7189–7200. [CrossRef]

30. Colebrook, C.F.; White, C.M. Experiments with fluid friction in roughened pipes. *Proc. R. Soc.* **1937**, *161*, 367–381.

31. Nagy, S. *Vademecum Gazownika—Vademecum of the Gasifier*; SITPNIG: Krakow, Poland, 2014; Volume 1.

32. Shashi Menon, E. *Gas Pipeline Hydraulics*; CRC PRESS: Boca Raton, FL, USA, 2005.

33. Duan, J.; Wang, W.; Zhang, Y.; Liu, H.; Lin, B.; Gong, J. Calculation on inner wall temperature in oil-gas pipe flow. *J. Cent. South Univ.* **2012**, *19*, 1932–1937. [CrossRef]

34. Incropera, F.P.; DeWitt, D.P. *Introduction to Heat Transfer*, 3rd ed.; John Wiley & Sons Inc.: New York, NY, USA, 1966.

35. Bai, Y.; Bai, Q. *Subsea Pipelines and Risers*; Elsevier: Cambridge, UK, 2005.

36. Peng, D.Y.; Robinson, D.B. A new two-constant equation of state. *Ind. Eng. Chem. Fundam.* **1976**, *15*, 59–64. [CrossRef]

37. Pratt, M.R. Thermodynamic properties involving derivatives: Using the Peng-Robinson equation of state. *Chem. Eng. Educ.* **2001**, *35*, 112–115.

38. Tarom, N.; Hossain, M.M. A practical method for the evaluation of the Joule Thomson effects to predict flowing temperature profile in gas producing wells. *J. Nat. Gas Sci. Eng.* **2015**, *26*, 1080–1090. [CrossRef]

39. Nagy, S.; Łaciak, M.; Włodek, T. Analiza możliwości bezpiecznego rurociągowego transportu wodoru. Report from Work Package 7, Project GEKON-HESTOR. Unpublished work. 2016.

40. Bedel, L.; Junker, M. Natural gas pipelines for hydrogen transportation. In Proceedings of the WHEC Conference Session, Lyon, France, 13–16 June 2006.

41. Odgen, J.; Myers Ja, A.; Scheitrum, D.; McDonald, Z.; Miller, M. Natural gas as a bridge to hydrogen transportation fuel: Insights from the literature. *Energy Policy* **2018**, *115*, 317–329.

42. Meng, B.; Gu, C.; Zhang, L.; Zhou, C.; Li, X.; Zhao, Y.; Zheng, J.; Chen, X.; Han, Y. Hydrogen effects on X80 pipeline steel in high-pressure natural gas/hydrogen mixtures. *Int. J. Hydrog. Energy* **2017**, *42*, 7404–7412. [CrossRef]
43. Fekete, J.R.; Sowards, J.W.; Amaro, R.L. Economic impact of applying high strength steels in hydrogen gas pipelines. *Int. J. Hydrog. Energy* **2015**, *40*, 10547–10558. [CrossRef]

MDPI

St. Alban-Anlage 66

4052 Basel

Switzerland

Tel. +41 61 683 77 34

Fax +41 61 302 89 18

www.mdpi.com

Energies Editorial Office

E-mail: energies@mdpi.com

www.mdpi.com/journal/energies

www.ingramcontent.com/pod-product-compliance
Lightning Source LLC
LaVergne TN
LVHW071502180726
843512LV00014B/998